VOLUME 7
1950 – PRESENT

Science and Its Times

Understanding the Social Significance of Scientific Discovery

VOLUME 7
1950 – PRESENT

Science and Its Times

Understanding the Social Significance of Scientific Discovery

Neil Schlager, Editor
Josh Lauer, Associate Editor

Produced by Schlager Information Group

Detroit
New York
San Francisco
London
Boston
Woodbridge, CT

Science and Its Times

VOLUME 7

1950-present

NEIL SCHLAGER, *Editor*
JOSH LAUER, *Associate Editor*

GALE GROUP STAFF

Amy Loerch Strumolo, *Project Coordinator*
Christine B. Jeryan, *Contributing Editor*

Mary K. Fyke, *Editorial Technical Specialist*

Maria Franklin, *Permissions Manager*
Margaret A. Chamberlain, *Permissions Specialist*
Shalice Shah-Caldwell, *Permissions Associate*

Mary Beth Trimper, *Production Director*
Evi Seoud, *Assistant Production Manager*
Wendy Blurton, *Senior Buyer*

Cynthia D. Baldwin, *Product Design Manager*
Tracey Rowens, *Senior Art Director*
Barbara Yarrow, *Graphic Services Manager*
Randy Bassett, *Image Database Supervisor*
Mike Logusz, *Imaging Specialist*
Pamela A. Reed, *Photography Coordinator*
Leitha Etheridge-Sims *Junior Image Cataloger*

ISBN: 0-7876-3939-7

Printed in the United States of America
10 9 8 7 6 5 4 3 2 1

Contents

Contents

1950-present

Mathematics

Medicine

Physical Sciences

Technology and Invention

Preface

The interaction of science and society is increasingly a focal point of high school studies, and with good reason: by exploring the achievements of science within their historical context, students can better understand a given event, era, or culture. This cross-disciplinary approach to science is at the heart of *Science and Its Times.*

Readers of *Science and Its Times* will find a comprehensive treatment of the history of science, including specific events, issues, and trends through history as well as the scientists who set in motion—or who were influenced by—those events. From the ancient world's invention of the plowshare and development of seafaring vessels; to the Renaissance-era conflict between the Catholic Church and scientists advocating a sun-centered solar system; to the development of modern surgery in the nineteenth century; and to the mass migration of European scientists to the United States as a result of Adolf Hitler's Nazi regime in Germany during the 1930s and 1940s, science's involvement in human progress—and sometimes brutality—is indisputable.

While science has had an enormous impact on society, that impact has often worked in the opposite direction, with social norms greatly influencing the course of scientific achievement through the ages. In the same way, just as history can not be viewed as an unbroken line of ever-expanding progress, neither can science be seen as a string of ever-more amazing triumphs. *Science and Its Times* aims to present the history of science within its historical context—a context marked not only by genius and stunning invention but also by war, disease, bigotry, and persecution.

Format of the Series

Science and Its Times is divided into seven volumes, each covering a distinct time period:

Volume 1: 2000 B.C.-699 A.D.

Volume 2: 700-1449

Volume 3: 1450-1699

Volume 4: 1700-1799

Volume 5: 1800-1899

Volume 6: 1900-1949

Volume 7: 1950-present

Dividing the history of science according to such strict chronological subsets has its own drawbacks. Many scientific events—and scientists themselves—overlap two different time periods. Also, throughout history it has been common for the impact of a certain scientific advancement to fall much later than the advancement itself. Readers looking for information about a topic should begin their search by checking the index at the back of each volume. Readers perusing more than one volume may find the same scientist featured in two different volumes.

Readers should also be aware that many scientists worked in more than one discipline during their lives. In such cases, scientists may be featured in two different chapters in the same volume. To facilitate searches for a specific person or subject, main entries on a given person or subject are indicated by bold-faced page numbers in the index.

Within each volume, material is divided into chapters according to subject area. For volumes 5, 6, and 7, these areas are: Exploration and Discovery, Life Sciences, Mathematics, Medicine, Physical Sciences, and Technology and Invention. For volumes 1, 2, 3, and 4, readers will find that the Life Sciences and Medicine chapters have been combined into a single section, reflecting the historical union of these disciplines before 1800.

Arrangement of Volume 7: 1950-present

Volume 7 begins with two notable sections in the frontmatter: a general introduction to science and society during the period, and a general chronology that presents key scientific events during the period alongside key world historical events.

The volume is then organized into six chapters, corresponding to the six subject areas listed above in "Format of the Series." Within each chapter, readers will find the following entry types:

Chronology of Key Events: Notable events in the subject area during the nineteenth century are featured in this section.

Overview: This essay provides an overview of important trends, issues, and scientists in the subject area during the nineteenth century.

Topical Essays: Ranging between 1,500 and 2,000 words, these essays discuss notable events, issues, and trends in a given subject area. Each essay includes a Further Reading section that points users to additional sources of information on the topic, including books, articles, and web sites.

Biographical Sketches: Key scientists during the era are featured in entries ranging between 500 and 1,000 words in length.

Biographical Mentions: Additional brief biographical entries on notable scientists during the era.

Bibliography of Primary Source Documents: These annotated bibliographic listings feature key books and articles pertaining to the subject area.

Following the final chapter are two additional sections: a general bibliography of sources related to the history of science, and a general subject index. Readers are urged to make heavy use of the index, because many scientists and topics are discussed in several different entries.

A note should be made about the arrangement of individual entries within each chapter: while the long and short biographical sketches are arranged alphabetically according to the scientist's surname, the topical essays lend themselves to no such easy arrangement. Again, readers looking for a specific topic should consult the index. Readers wanting to browse the list of essays in a given subject area can refer to the table of contents in the book's frontmatter.

Additional Features

Throughout each volume readers will find sidebars whose purpose is to feature interesting events or issues that otherwise might be overlooked. These sidebars add an engaging element to the more straightforward presentation of science and its times in the rest of the entries. In addition, the volume contains photographs, illustrations, and maps scattered throughout the chapters.

Comments and Suggestions

Your comments on this series and suggestions for future editions are welcome. Please write: The Editor, *Science and Its Times,* Gale Group, 27500 Drake Road, Farmington Hills, MI 48331.

Advisory Board

Contributors

Mark H. Allenbaugh
Lecturer
George Washington University

Peter J. Andrews
Freelance Writer

Janet Bale
Freelance Writer and Editor

Bob Batchelor
Writer
Arter & Hadden LLP

Katherine Batchelor
Independent Scholar and Writer

Sherri Chasin Calvo
Freelance Writer

Geri Clark
Science Writer

Brooke E. Coates
Freelance Writer
Professor of English

David A. DeWitt
Assistant Professor of Biology
Liberty University

Philip Downey
Freelance Writer

Thomas Drucker
Graduate Student, Department of Philosophy
University of Wisconsin

H. J. Eisenman
Professor of History
University of Missouri-Rolla

Ellen Elghobashi
Freelance Writer

Nathan L. Ensmenger
History & Sociology of Science
University of Pennsylvania

Randolph Fillmore
Freelance Science Writer

Richard Fitzgerald
Freelance Writer

Maura C. Flannery
Professor of Biology
St. John's University, New York

Donald R. Franceschetti
Distinguished Service Professor of Physics and Chemistry
The University of Memphis

Jean-François Gauvin
Historian of Science
Musée Stewart au Fort de l'Ile Sainte-Hélène, Montréal

Phillip H. Gochenour
Freelance Editor and Writer

Brook Ellen Hall
Professor of Biology
California State University at Sacramento

Diane K. Hawkins
Head, Reference Services—Health Sciences Library
SUNY Upstate Medical University

Robert Hendrick
Professor of History
St. John's University, New York

Jessica Bryn Henig
History of Science Student
Smith College

James J. Hoffmann
Diablo Valley College

Mary Hrovat
Freelance Writer

Leslie Hutchinson
Freelance Writer

P. Andrew Karam
Environmental Medicine Department
University of Rochester

Evelyn B. Kelly
Professor of Education
Saint Leo University, Florida

Rebecca Brookfield Kinraide
Freelance Writer

Israel Kleiner
Professor of Mathematics
York University

Judson Knight
Freelance Writer

Lyndall Landauer
Professor of History
Lake Tahoe Community College

Josh Lauer
Freelance Editor
Lauer InfoText Inc.

Adrienne Wilmoth Lerner
Division of History, Politics, and International Studies
Oglethorpe University

Brenda Wilmoth Lerner
Science Correspondent

K. Lee Lerner
Prof. Fellow (r), Science Research & Policy Institute
Advanced Physics, Chemistry and Mathematics, Shaw School

Eric v. d. Luft
Curator of Historical Collections
SUNY Upstate Medical University

Elaine McClarnand MacKinnon
Assistant Professor of History
State University of West Georgia

Lois N. Magner
Professor Emerita
Purdue University

Ann T. Marsden
Writer

Kyla Maslaniec
Freelance Writer

William McPeak
Independent Scholar
Institute for Historical Study (San Francisco)

Duncan J. Melville
Associate Professor of Mathematics
St. Lawrence University

Leslie Mertz
Biologist and Freelance Science Writer

Kelli Miller
Freelance Writer

J. William Moncrief
Professor of Chemistry
Lyon College

Heather Moncrief-Mullane
Masters of Education
Wake Forest University

Stacey R. Murray
Freelance Writer

Ashok Muthukrishnan
Freelance Writer

Lisa Nocks
Historian of Technology and Culture

Stephen D. Norton
Committee on the History & Philosophy of Science
University of Maryland, College Park

Brian Regal
Historian
Mary Baker Eddy Library

Sue Rabbitt Roff
Cookson Senior Research Fellow
Centre for Medical Education
Dundee University Medical School

Michelle Rose
Freelance Science Writer

Steve Ruskin
Freelance Writer

Elizabeth D. Schafer
Independent Scholar

Neil Schlager
Freelance Editor
Schlager Information Group

Keir B. Sterling
Historian, U.S. Army Combined Arms Support Command
Fort Lee, Virginia

Gary S. Stoudt
Professor of Mathematics
Indiana University of Pennsylvania

George Suarez

Rebecca J. Timmons
Instructor
Westark College

Todd Timmons
Mathematics Department
Westark College

David Tulloch
Graduate Student
Victoria University of Wellington, New Zealand

A. Bowdoin Van Riper
Adjunct Professor of History
Southern Polytechnic State University

Stephanie Watson
Freelance Writer

Richard Weikart
Associate Professor of History
California State University, Stanislaus

Giselle Weiss
Freelance Writer

A.J. Wright
Librarian
Department of Anesthesiology
School of Medicine
University of Alabama at Birmingham

Michael T. Yancey
Freelance Writer

Introduction: 1950–present

Overview

The second half of the twentieth century saw the most rapid increase in scientific knowledge of any time in history. Particularly amazing progress was made in genetics and biotechnology, computer technology, astronomy, and medical science. By the turn of the century, a greater fraction of the population used, developed, and relied on science and technology than ever before, and people increasingly looked to science and technology for answers to their most pressing questions. There were problems, however. Larger and more powerful nuclear weapons, devastating industrial accidents, and environmental degradation showed that no blessing is unmixed.

Despite its many advantages, technology's dark side and its ominous achievements made many people fear that it threatened the future of both humanity and the Earth. This is one of the fundamental dichotomies of the period. The other is that although increasingly driven by scientific and technological advances, the era was marked by a renewed interest in religion and a groundswell of anti-technology sentiment. This is perhaps an unavoidable consequence of the fact that science and technology, by themselves, are neither good nor evil, but can be used for both.

Emerging technologies also strained social relations. Many worried that an increasingly hi-tech world would cease to value people as individuals. The wealth and opportunity enjoyed by people of the First World were often resented by their less-advanced and less-developed counterparts in the Third World. Finally, many felt that science and religion were mutually exclusive, and that embracing one meant rejecting the other. These tensions—technology versus conservation, science versus religion, progress versus individuality, rich nations versus poor undeveloped countries— shaped the world in which we live, and will continue to influence the future.

In the developed world technology and science have become almost indistinguishably woven into our daily lives and our society. Their increasing presence is reflected in newspaper and television reports. Science fiction is routinely popular, as evidenced by *Andromeda Strain, Coma,* and *Outbreak,* to name only a few. Computers, which have revolutionized all sectors of society, are themselves both vehicles for entertainment and ubiquitous and valuable tools. Even the movies reflect our fascination with and fear of technology: *Dr. Strangelove, On the Beach*, and *Failsafe.* Concern about nuclear testing produced *Godzilla, Spiderman*, and the giant ants of *Them.*

Looking Back at 1900-1949

At the end of the nineteenth century scientists were beginning to believe that there was little left to learn. The basic laws of nature had been determined and all that remained was to tie up a few niggling loose ends and then refine the calculation of some physical constants more precisely. As it turned out, tying up those loose ends led to a revolution in physics, new theories about atomic structure, a better understanding of the Earth's age, knowledge of how stars produce energy, how atoms and molecules bond to form chemicals, and much more. It is safe to say that the scientific discoveries of the first part of the twentieth century made possible the incredible technological advances of the century's second half. It is entirely possible that the scientific discoveries of the first half of the century are more remarkable, in the context of their times, than those of the second half because, in many instances, they were fundamental discoveries that completely changed the way we view the universe. On the other hand, these discoveries were not fully appreciated at the time, and it was left to more recent scientists to explain, understand, and capitalize on the discoveries made between 1900 and 1949.

1950-present: Understanding Ourselves, Our Planet and the Universe

If early twentieth-century science produced new scientific and conceptual tools, they were put to work in the second half of the century. These tools were used to design new experiments and techniques with which to probe ever deeper into the science that underlies our world. In other cases, the scientific concepts themselves helped forge a better understanding of the universe. In these explorations, we looked both inward and outward, and what we saw in either direction has had a profound and indelible impact on our society.

Looking Inward

Early in the century, researchers realized that genes explained the patterns in inherited traits. At first, however, they believed that nucleic acids were not sufficiently complex to convey this bewildering amount of information from generation to generation. James Watson and Francis Crick, by showing how DNA was organized, proved that the nucleic acids could, and did, carry this information. The paper they wrote announcing their discovery was not even two pages in length, but it was sufficient to win the Nobel Prize, and it set the stage for everything that was to follow.

Since then, molecular biology, genetics, and molecular genetics have given rise to the field of biotechnology and a far better understanding of how the living world works. Scientists can now add new genes to bacteria, allowing them to make drugs such as insulin or interferon cheaply and efficiently. Genetic engineering is beginning to change agriculture. Studies of the human genome have led to a deeper and more detailed understanding of certain diseases.

These amazing advances are not universally welcomed, however. Some people worry about the safety of genetically engineered foods because scientists have mixed genes from different organisms in ways that were not previously possible, and the ecological impact of these manipulations is not always certain. Progress in human genetics has raised fears that insurance companies might refuse to insure people with genetic markers for certain diseases, citing them as preexisting conditions. Other remarkable developments include synthetic hormones, artificial genes, DNA "fingerprinting" techniques, and an understanding of how cells generate energy. All influence our search to understand life on Earth. The implications of each innovation are hotly debated, as are issues surrounding cloning, gene therapy, and other advances made possible by our increasing knowledge of genetics and biology.

Medical advances have been equally significant. New surgical techniques let us transplant organs from one person to another and, in some cases, from one species to another. Surgical lasers help diagnose and treat diseases from hyperthyroidism to cancer. People routinely receive artificial replacement parts when their bones, joints, and heart valves wear out. The development of oral contraceptives gave us not only control of our population but the sexual revolution, which, in turn, may have contributed to a resurgence in sexually transmitted diseases, including AIDS, herpes, and others.

Looking Outward

Science also made tremendous strides toward understanding the Earth and the universe during this time. The discovery of plate tectonics led to a grand synthesis and explanation of many puzzling discoveries in geology, paleontology, and evolutionary studies, letting us see the Earth as a dynamic, living planet that is constantly changing and evolving. Exploration of the solar system by orbital telescopes and space probes showed us planets and satellites much different from the small, blurred, featureless images seen by ground-based instruments. Meanwhile, astronomers discovered new worlds around other stars. We now have a reasonably good understanding of how stars are born, evolve, and die, and have seen how galaxies form. The COBE orbital observatory has seen echoes of the birth of the universe, confirming that everything we see was formed in a Big Bang billions of years ago. Studies of the Big Bang, in turn, lead us back to particle physics in a sort of physics "great circle," and advances in this arena have been equally profound.

Threatening the World

Hans Bethe's discovery of stellar energy sources also helped design more efficient fusion bombs. Albert Einstein's famous equation, $E = mc^2$ helped explain certain facts about the universe, but it also helped us build an atomic bomb. Similarly, the discovery and increasing use of fossil fuels made life immeasurably easier, but it also led to oil spills, fears of global warming, and environmental damage. Virtually every scientific discovery is examined by the military to see if it could become a new weapon or improve an existing one. For the first time in history, man has the ability to wipe out every person on Earth and, at the same time, render the planet uninhabitable for any life more complex than a lichen. This realization has led to increasingly vocal environmentalism, a great deal of legislation, many international agreements, and a fervor in the media. Some wish to use the technology that created these problems to fix them; others would rather turn back the clock to a simpler and presumably better time. The solution to these problems will almost certainly be found in more technological discovery, not less.

Where Do We Go from Here?

The rapid progress in science and technology has led to some remarkable rifts, many that need not exist. There is no need, for example, to have to choose between religious belief and scientific fact. Indeed, most religious leaders and scientists manage to believe in both. Similarly, while technology currently seems to exacerbate the differences between the rich and the poor, there is no reason it cannot be used to help the poor advance economically. And, while many of our environmental problems may be due to technology, the same technology can be used to both extend human life and to make that life richer and more meaningful. Technology is morally neutral. Only the purposes to which it is put can be good or evil. Perhaps the largest challenge the future holds lies in improving humans, not in restricting technology.

P. ANDREW KARAM

Chronology: 1950–present

1950-53 A U.S.-led United Nations (UN) force fights combined Chinese and North Korean armies in the Korean War, which ends with a border established between North and South Korea at the 38th parallel.

1951 The introduction of the first successful oral contraceptive, based on discoveries by American biologist Gregory Pincus, sparks a social revolution with its ability to divorce the sex act from the consequences of impregnation.

1952 The United States explodes the first thermonuclear weapon, a hydrogen bomb, at the Eniwetok Atoll in the South Pacific.

1953 British climber Sir Edmund Hillary leads the first team to reach the summit of Mt. Everest.

1954 American surgeon Joseph Murray conducts the first successful organ transplant when he transfers the kidney of one twin to another.

1956 Soviet troops crush an uprising in Hungary, signaling the end of a "thaw" in the Cold War that followed Josef Stalin's death three years earlier.

1957 The Soviet Union launches *Sputnik 1*, the first man-made Earth satellite, thus inaugurating the space age—and the space race between the U.S. and the U.S.S.R.

1961 East Germany, supported by the Soviet Union, builds the Berlin Wall.

1961 Meteorologist Edward N. Lorenz discovers what comes to be called the butterfly effect—that small initial changes can result in large, completely random changes—thus forming the basis for chaos theory.

1962 *Silent Spring*, a book by American biologist Rachel Carson, raises international awareness concerning pollutants and spawns the environmental movement.

1963 U.S. President John F. Kennedy assassinated in Dallas on November 22.

1964 Murray Gell-Mann, an American physicist, first postulates the existence of unusual particles—which he dubs "quarks"—that carry fractional electrical charges.

1960s Rapid lifestyle changes in the West, particularly among young people, are manifested in rock music, the drug culture, the sexual revolution, and other movements.

1966 China's Chairman Mao Zedong launches the "Great Proletarian Cultural Revolution," which lasts 10 years and claims millions of lives.

1969 U.S. astronaut Neil Armstrong becomes the first person to walk on the surface of the Moon.

1969 The U.S. Department of Defense establishes the first packet-switched network, ARPANET—out of which will develop the Internet more than two decades later—to link computers in research facilities.

1969 U.S. troop strength in Vietnam peaks at 543,400; first sent in 1954, American forces will pull out after a 1973 peace treaty, and South Vietnam will fall to Communists in 1975.

1973 Organization of Petroleum Exporting Countries (OPEC) raises oil prices, spawning energy crisis and recession in the West.

1970s The last European colonies in Africa gain their independence; Marxist regimes seize power in several countries, and the continent is torn by ethnic and tribal clashes.

1975 The user-assembled Altair 8800 microcomputer makes its appearance, thus inaugurating the personal computer revolution; two years later, Commodore introduces the Personal Electronic Transactor (PET), the first personal computer designed for the mass market, and Apple debuts its highly popular Apple II.

1977 Two homosexual men in New York City, diagnosed as suffering from Kaposi's sarcoma, are the first reported cases of AIDS (acquired immune deficiency syndrome).

1979 Islamic fundamentalism grips the Middle East, as a Shi'ite regime seizes control in Iran and holds Americans hostage while *mujahideen* ("holy warriors") lead resistance to the Soviet takeover of Afghanistan.

1980 Luis and Walter Alvarez, father-and-son American physicists, speculate that a giant asteroid collided with Earth, causing a prolonged dust blackout and mass extinctions—including the disappearance of the dinosaurs.

1980-88 Iran and Iraq undergo the largest armed conflict since World War II, which ends in a stalemate.

1989 Communist regimes in Eastern Europe collapse, a fact symbolized by the opening of the Berlin Wall; three years later, the Soviet Union comes to an end.

1991 Following the Iraqi invasion of Kuwait, a U.S.-led UN force launches the brief Persian Gulf War; though Iraq is defeated and sanctions are imposed, Saddam Hussein's regime remains in power.

1993 English mathematician Andrew Wiles announces that he has proved Fermat's last theorem, a 325-year problem that many mathematicians had declared unsolvable; other mathematicians find fault with aspects of his proof, and a year later he presents a corrected version.

1997 At the Roslin Institute in Scotland, a lamb named "Dolly" is the result of the first successful effort to produce an exact genetic duplicate, or clone, from the genetic material of a mature mammal.

Exploration and Discovery

Chronology

1952 Michael G. F. Ventris deciphers Linear B, the written language of the Minoan civilization on Crete.

1953 British climber Sir Edmund Hillary leads the first team to reach the summit of Mt. Everest.

1960 Jacques Piccard and Donald Walsh in the *Trieste* descend to a record depth of 35,800 feet (10,912 m) in the Marianas Trench of the Pacific Ocean.

1961 Soviet cosmonaut Yuri Gagarin becomes the first human being in space.

1966 A Soviet craft lands on Venus, becoming the first man-made spacecraft to land on another planet.

1969 U.S. astronaut Neil Armstrong becomes the first person to walk on the surface of the Moon.

1971 The Soviet Union launches the first manned space station, *Salyut 1*.

1983 U.S. space probe *Pioneer 10* becomes the first man-made craft to travel beyond the orbit of the solar system's furthest planet.

1974 Chinese peasants uncover the tomb of Qin Shi Huang, China's first emperor, which contains some 7,000 life-size terracotta soldiers.

1981 The U.S. launches the space shuttle *Columbia,* the world's first winged, reusable spacecraft.

1985 A joint French-U.S. expedition uncovers the shipwrecked remains of the luxury liner *Titanic,* which sank in the north Atlantic in 1912.

1999 Brian Jones and Bertrand Piccard become the first human beings to fly around the world nonstop in a balloon, the *Breitling Orbiter 3*.

Overview: Exploration and Discovery 1950-present

The human motivation for exploration has always been clear: hope for national and individual profit or acclaim, the simple gratification of geographical curiosity, and the discovery and identification of the unknown. While these motives were still present in the twentieth century, fundamental developments in technology began to change the character of exploration.

Two significant discoveries propelled twentieth-century exploration to new heights—literally. In 1903 the historic flight of the Wright brothers ushered in a new era of technology, and with it new possibilities in exploration. Around the same time, American inventor Robert Goddard (1882-1945) began experimenting with rocket propulsion. In a 1920 technical report for the Smithsonian, Goddard outlined how a rocket might reach the moon. The scientific community labeled him a crackpot, but his report became the foundation for the early rocket program of the Nazi military, which made further advancements in rocket science during World War II. Goddard's rocketry research led to numerous patents and paved the way for modern rocket technology that would launch the first man-made satellites—and ultimately the first humans—into space in the second half of the twentieth century.

Expeditions into the skies above Earth became more than just science fiction in the second half of the twentieth century. Space, the ultimate mystery, the final frontier, became a little more familiar with the launching of unmanned probes, satellites, and manned space flights. From *Sputnik* (the first satellite rocketed into space in 1957) to *Apollo 11* (the first manned space flight to land on the moon in 1969), and *Salyut 1* (the first space station, inhabited in 1971), man proved he was not content with exploring Earth's surface and oceans.

The achievements of scientists, astronauts, and technicians toward solving the mysteries of outer space were extensive in the later twentieth century. The space race, set off by the Cold War between the United States and the Soviet Union, witnessed the development of satellites, the first man in space—Yuri Gagarin (1934-1968), launched into orbit in April 1961—and a 14-year experimental space station, *Mir,* launched in 1986 and scheduled for decommissioning in early 2000. And in yet another triumph of technology, the U.S. Space Shuttle program, in operation since 1981, proved reliable space transportation was feasible.

While scientists and astronauts explored space, other men and women were conquering some of the last known frontiers on Earth—its mountains, oceans, and atmosphere. Two noteworthy exploration firsts occurred in the skies over Earth. In 1986 the first nonstop, unrefueled aerial circumnavigation of the world was completed in the *Voyager* aircraft, piloted by Americans Dick Rutan (1939-) and Jeana Yeager (1952-). In 1999, the first nonstop, unrefueled balloon circumnavigation of the world was completed in the *Breitling Orbiter 3,* piloted by British aviator Brian Jones (1947-) and Swiss aviator Bertrand Piccard (1958-). On Earth's surface, in 1953, the world's highest peak, Mount Everest, was finally conquered by New Zealander Sir Edmund Hillary (1919-) and a Nepalese sherpa named Tenzing Norgay (1916-1986).

Mountains and landmasses comprise only 30 percent of the Earth's surface. The oceans cover the other seven-tenths. Deep-sea exploration requires mastery of the same skills used in geographical exploration; knowledge of the principles of biology, chemistry, geology, and physics; as well as extensive assistance from the technological realms of engineering and shipbuilding. In the later half of the twentieth century, ocean exploration was conducted for both knowledge and wealth. In 1960 Jacques Piccard (1922-)—father of Bertrand Piccard—and U.S. Navy Lieutenant Donald Walsh (1931-) piloted the bathyscaphe *Trieste* to a record depth of 35,800 feet in the Mariana Trench, nearly seven miles below the ocean's surface. Other underwater adventures were undertaken by submarines—notably the 1958 journey beneath the ice of the North Pole by the U.S.S. *Nautilus* and the 1960 submerged circumnavigation of the globe by the U.S.S. *Triton.*

With the assistance of new technological tools, twentieth-century explorers were able to make more detailed surveys of Earth's surface, explore the depths of the ocean and Earth's interior, and voyage to the Moon and stars, as the quest for the unknown extended beyond Earth. While these expeditions to discover and catalog the last unknowns of Earth's physical attributes were conducted, other explorers of a different nature—namely anthropologists, archaeologists, and even treasure hunters—continued their investigations into the origins of humans, examining past civilizations and their cultural distinctions. The wealth and culture of former civilizations were more fascinating to some twentieth-century discoverers than the land or sea itself.

While some ocean adventures had been undertaken for science and national pride, other deep-sea expeditions were motivated by fascination with maritime history, particularly the search for ships and cargoes that sank long ago. In the 1970s Dutch East India Company vessels were discovered, yielding priceless historical artifacts as well as silver, porcelain, and other relics. Ships from the Spanish Armada have also been found, including the warship *Girona* in 1967, and the galleon *Atocha* in 1985. In 1984, the discovery of the pirate ship *Whydah* yielded over 200,000 artifacts. In 1985 the *Titanic* was located and in the 1990s salvage missions were undertaken to her resting place. The fortunes aboard these vessels were valuable in terms of financial and cultural wealth, making such expeditions into the ocean deep a worthwhile enterprise for deep-sea explorers.

Similarly, a number of significant discoveries pertaining to ancient civilizations were made on land in the second half of the twentieth century. From artifacts of ancient man of the Paleolithic age to ancient Greek and Central American civilizations, scientists and explorers in the late twentieth century brought to light hundreds of thousands of cultural relics. In 1994 cave surveyors discovered paintings on the walls of the Chauvet-Pont-d'Arc cave near Avignon, France. The paintings were radiocarbon-dated between 30,300 and 32,000 years old. In 1974 Chinese peasants unearthed a site containing 7,000 life-size terra-cotta soldier figures near the tomb of China's first emperor, Qin Shi Huang (259?-210? BC).

Less haphazard and more formal excavations of early civilizations have been conducted in search of wealth and knowledge for several centuries. In the early part of the twentieth century, archaeologists digging in countries such as Egypt and Greece uncovered artifacts whose study returned vital truths of their origins. The discovery by Sir Arthur Evans (1851-1941) of the ancient Greek civilization of King Minos on the island of Crete and of the mysterious writings used by its people led to the decipherment in 1952 of Mycenaean Linear B script by Michael Ventris (1922-1956), assisted by John Chadwick (1920-1998). In the 1980s and early 1990s, Linda Schele (1942-1998) and Peter Mathews (1951-), among others, decoded other ancient hieroglyphic writings from Mayan ruins in Guatemala and other Central American countries.

As mankind enters the twenty-first century, our explorations of Earth and its skies will be ever influenced by the technologies that make them possible. Further space exploration, for example—which may include human exploration of the planets and celestial bodies closest to Earth, such as Mars, Venus, and Jupiter's moons—will be tied to scientific experimentation and studies at the International Space Station, which is scheduled to be completed in 2004. Space study has also drawn new attention to the fragility of Earth itself, our only known habitable planet. Global environmental research and exploration, therefore, will naturally be important to the survival of mankind. Scientific studies of worldwide phenomena such as deforestation, desertification, acid rain, land degradation, and water and energy deficiencies will rely on developing technological tools, as will space pioneers and explorers of the last mysterious regions on Earth.

ANN T. MARSDEN

The Decoding of Linear B Sheds New Light on Mycenaean Civilization

Overview

The Mycenaean civilization flourished in Greece and the surrounding islands in the Aegean Sea around 1400 B.C., during the era Homer depicted in his epics the *Iliad* and the *Odyssey*. The Mycenaean language was written in a script known as Linear B. Sir Arthur Evans first discovered specimens of the Linear B script in 1900 in

Crete, and Michael Ventris deciphered them about 50 years later. The ability to read the Mycenaean texts shed new light on this important culture.

Background

The Bronze Age civilization of Crete was the first society in Europe to be capable of fine craftsmanship, public architecture, and writing. It is often called the Minoan civilization, after the legendary King Minos, who was said to have ruled the Cretan city of Knossos. The Minoans spoke a local language about which little is known, but which may have been related to languages spoken in southwestern Turkey. They wrote using a script known as Linear A for four centuries beginning about 1850 B.C. Before that, they had employed a type of *hieroglyphic* script, using symbols to represent words. The famous Phaistos Disc, dating from about 1700 B.C., is stamped with a series of 45 hieroglyphs of yet another type, arranged in a spiral. The significance of this unique artifact remains a mystery. Its date, determined from the pottery with which it was found, suggests that older hieroglyphs may have been used for ceremonial purposes alongside the more mundane linear scripts.

Just as the Romans were later to borrow much of the basis of their civilization from the classical Greeks, the early civilization of mainland Greece, arising about 1600 B.C., was based upon that of the Minoans. Since one of its main centers was at Mycenae, it is called Mycenaean. Sometimes the term is used in a more general sense to refer to the civilizations in the area of the Aegean Sea from about 1400 B.C..

The Mycenaean Greeks modified the Minoan Linear A script to fit their own language, eliminating some signs and adding others. The result was a new script known as Linear B, which soon replaced Linear A in Crete as well. In fact, some scholars believe that the script was developed by Mycenaeans living in Crete. The first specimens of Linear B, scratched into about 4500 clay tablets, were discovered in 1900 by the British archaeologist Sir Arthur Evans (1851-1941) during his excavations at Knossos. The writing seemed to be used mainly as a way to keep royal, military, religious, and commercial records. A few hundred additional samples were found in Crete and several sites on the Greek mainland, including more clay tablets as well as short inscriptions on pots, jars, and vases.

The British cryptologist Michael Ventris (1922-1956) first became fascinated by the Linear B script as a teenager, when he heard Evans lecture on his finds. In 1949, after serving in the Royal Air Force during World War II, Ventris began working seriously on deciphering the script. His method involved assuming that the Mycenaean language was an archaic form of Greek and then employing statistical analysis. In June 1952 he announced on British radio that he had deciphered the script and confirmed that the language was the earliest known form of Greek.

Together with John Chadwick, a classical scholar and linguist at Cambridge University, Ventris published the seminal paper, "Evidence for Greek Dialect in the Mycenaean Archives," in 1953. The pair's book *Documents in Mycenaean Greek* was published in 1956, a few weeks after Ventris had died in an automobile accident. Chadwick wrote an account of their joint effort, *The Decipherment of Linear B,* in 1958.

Linear B consisted of about 90 signs made with straight or curved linear strokes. It was a *syllabic* script; that is, each symbol represented an individual syllable, such as *ma* or *ti*. In terms of our own phonetic alphabet, we would generally say that a syllable consists of a consonant followed by a vowel. However, vowel sounds alone may form the first syllable of a word; for example, "Athens." So Linear B had signs for *a, e, i, o,* and *u*. But the script did not distinguish between syllables beginning with *r* and those beginning with *l,* nor did it acknowledge a difference between *b* and *p*. It omitted final consonants, and if two consonants appeared together, as in the syllable *spe,* it either omitted the first or turned one syllable into two by reusing the vowel. In our example the result would be *se-pe*. These eccentricities often led to ambiguous spellings, which made the script more difficult to decipher.

Both Linear A and Linear B contained a number of ideograms, or symbols for words or concepts, in addition to their syllabaries. Interestingly, while both scripts use the same signs for the basic agricultural commodities and livestock, Linear B has many more signs for military equipment, furniture, and ritual objects.

Impact

The work of Ventris and Chadwick proved that the Mycenaeans on the Greek mainland during the period of the events in the Homeric epics, roughly 1400-1200 B.C., spoke Greek. Although little is actually known about Homer, he is thought to have lived about 500 years later.

Only hints of the ancient dialect appear in Homer's language, preserved by a long oral tradition. So the Mycenaean texts, terse and businesslike as they are, represent the oldest known Greek dialect and shed light on an important era in the development of the early Greek language and civilization.

The clay tablets found by Evans were now understood to show that Greek was also spoken in Knossos at the time of its destruction by fire in 1380 B.C.. The writings were inventories and similar records on unbaked clay tablets that were reused or replaced each year when new records were made. Ironically, the clay tablets with the records from the final year were baked and preserved by the fire that destroyed everything else. The specific cause of the fire is unknown. However, the city had been rebuilt before after previous accidental blazes. There was no sign of an earthquake or other natural causes in 1380 B.C. This leaves arson or invasion as the likely cause.

The fact that Mycenaean Greek was the scribal language of Knossos during this period has led most scholars to assume that Mycenaeans had taken over in Crete. This may have happened peacefully—by assimilation or dynastic marriage—or via conquest in the disarray after the volcano Thera erupted in about 1500 B.C. The effects of this eruption may also have led to the centralization of the Cretan bureaucracy at Knossos; no comparable records have been found in other Minoan population centers. The possibility has also been raised that Mycenaean Greek was not the local vernacular but rather was used for official records as the Aegean lingua franca, much as English is used worldwide today for commercial and scientific communications.

Because they are ledgers rather than literature, the Linear B texts revealed little about the souls of the Mycenaeans, their loves or their hates. However, the texts did provide a number of details about the society that could only be guessed at from the archaeological evidence or from memories preserved in the literature of early classical Greece.

For example, they included inventories of livestock and agricultural produce, textiles, vessels, furniture, military personnel, weapons, and chariots. This gave scholars an idea of what types of supplies they used and what was considered important enough to keep track of. Scientists could gain an understanding of the wool industry and farming practices in general. Military technology had apparently advanced to include tunics reinforced with bronze, and the shape of the lightweight Minoan chariots was shown in an ideogram.

Landholding records were important in terms of the evolution of real estate law into the classical Greek period. They also allowed comparison of the Mycenaean system with those of surrounding areas; for example, the Hittite Law Code. Records of religious tribute indicated that most of the classical Greek gods and goddesses were already worshipped in the Mycenaean era. Most scholars had already believed this, but now they had proof. Traditional Minoan deities such as the goddess Eleuthia were also included in the Knossos pantheon.

There remains much to be learned from the language itself. Understanding the linguistic forms and the meaning of the symbols may help in studying earlier forms of Cretan writing. Knowing what adjustments were made to the Minoan script in order to write Greek can provide hints about the unknown language that was written in Linear A. This is especially important because there are only one-tenth as many known existing Linear A inscriptions as there are for Linear B. In addition, a script related to Linear B was used on the island of Cyprus from the eleventh to the third centuries B.C. Greek as well as the native language Eteocypriot was written in this script.

There is still some controversy about the Mycenaean language and the Ventris decipherment. Although Ventris's theory has been widely accepted, a minority of scholars believe it is not entirely correct. A few even question whether the Mycenaeans spoke Greek at all. The records do contain many non-Greek proper names and technical terms. Others agree with Ventris's reading and believe that Mycenaean was a dialect of Greek, but one that was an evolutionary dead end. In this view, held by a relatively small number of scholars, the later forms of Greek were spread around the Aegean from other Greek-speaking areas and superseded the earlier dialect.

The additional inscriptions found after the Knossos excavations also make sense when interpreted as Greek, lending credence to Ventris's view. If longer passages of prose or poetry are found in the future, the decipherment could be tested conclusively. Archaeologists have found inked inscriptions on clay cups, suggesting that longer documents may have been written in ink on parchment or papyrus.

SHERRI CHASIN CALVO

Further Reading

Castelden, Rodney. *Minoans: Life in Bronze-Age Crete.* London: Routledge, 1990.

Chadwick, John. *Linear B and Related Scripts.* Berkeley: University of California Press, 1987.

Davies, Anna Morpurgo, and Duhoux, Yves, eds. *Linear B, a 1984 Survey: Proceedings of the Mycenaean Colloquium of the 8th Congress of the International Federation of the Societies of Classical Studies.* Louvain-la-Neuve: Cabay, 1985.

Levin, Saul. *The Linear B Decipherment Controversy Re-examined.* Albany: State University of New York, 1964.

Ventris, Michael, and Chadwick, John. *Documents in Mycenaean Greek.* Cambridge: Cambridge University Press, 1973.

Sir Edmund Hillary Leads the First Team to Reach the Summit of Mt. Everest

Overview

In 1953 Edmund Hillary (1919-) of Britain and Tenzing Norgay (1914-1986) of Nepal became the first individuals known to have reached the highest point on Earth, the summit of Mount Everest. Since that time, reaching Mount Everest's summit has become a matter of pride, both national and individual, and has led to a variety of expeditions sponsored by nations and private organizations and has even resulted in guided tours. This situation, in turn, has produced a steadily mounting death toll, culminating in the disastrous 1996 climbing season, in which eight climbers, many of them with paid guides, died during a single storm.

Background

In 1852 a worker with the British Governmental Survey of India was calculating the heights of a number of mountains in the Himalayas based on information gathered over the past few years. According to the story, he completed his calculations and, paper in hand, went to his supervisor to announce that he had just located the highest mountain in the world. Named Chomolunga (Goddess Mother of the World) by the local Sherpas, Peak XV (as it appeared on the British maps) was renamed Mount Everest in honor of Sir George Everest, the Indian Surveyor General from 1830 through 1843.

The first serious attempts to climb Mount Everest began in the 1920s, when Tibet opened its borders to outsiders and gave access to the mountain. In 1924 climbers George Mallory (1886-1924) and Andrew Irvine disappeared during an attempt on the summit. Although Mallory's body was found in 1999, his camera was not located, so whether they reached the summit is not known. As Edmund Hillary, however, pointed out when asked about the possibility he was not the first to reach Everest's summit, "The point of climbing Everest should not be just to reach the summit. I'm rather inclined to think that maybe it's quite important, the getting down."

At least thirteen climbers perished attempting to climb Everest before Hillary and Norgay succeeded. The early climbers set out with (by current standards) woefully inadequate clothing, equipment, and preparation. Mallory and Irvine decided to climb with oxygen during their fatal climb in 1924 but had no synthetic fibers to keep them warm, no modern climbing gear, and little in the way of training to climb in the extreme conditions that prevail in the Himalayas. Others were little better prepared.

Hillary succeeded because, unlike most of his predecessors, he attacked the mountain as a logistical challenge as well as a problem in climbing and endurance. Hundreds of support personnel, most of them Sherpas, carried tons of supplies to establish a base camp and seven subsequent camps progressively up the mountain. Hillary and Norgay set out from the highest of these camps to reach the summit on their historic climb. With few exceptions, all subsequent expeditions have used a similar strategy: take plenty of supplies and establish several camps at successively higher elevations. The most notable exception to this approach was the solo, single-day climb by the Italian Reinhold Messner (1944-) on August 20, 1980. Other exceptions include the elimination of some of Hillary's camps (most expeditions now use four camps plus the base camp) and the approximately 60 climbers who have reached the summit without the use of supplemental oxygen (at an altitude that commercial airliners frequent).

Mount Everest. *(Keren Su/Corbis. Reproduced by permission.)*

Impact

The most immediate impact of Hillary and Norgay's ascent was the knowledge that yet another extreme part of our planet had been conquered; human feet had trod yet another place. Everest was called the "third pole" and was perhaps even more difficult to reach than the North or South Poles. Its status as the highest point on Earth gave a certain amount of prestige to the climbers and their countries. Edmund Hillary was knighted and immediately became both national hero and international celebrity while Tenzing Norgay achieved similar acclaim among the Sherpas.

The conquest of Everest was perhaps among the first enterprises that depended as much on technology as on human perseverance and courage because without oxygen and modern equipment and clothing, Hillary and Norgay's expedition would likely have failed. From this perspective, the large number of subsequent "firsts" that have relied heavily on technology are interesting to note, perhaps because humans have reached the limits of what can be done without technology. For example, oxygen levels at Everest's peak are so low that they will not sustain life for longer than a few days, and even that duration is impossible without extensive preparation and conditioning. Other environments require even more sophisticated equipment: space suits for lunar landings, bathyscaphes for deep-ocean exploration, pressure suits and aircraft for altitude records, and so forth. Everest may well represent the limit of what humans can do without near-total reliance on technology. Or, as Peter Lloyd put it in 1984, "Were it 1000 feet lower it would have been climbed in 1924. Were it 1000 feet higher it would have been an engineering problem."

Between 1922 and 1953, 13 people died attempting to climb Everest. Between the first successful ascent and 1996, a total of 167 successful expeditions had placed 676 climbers atop Everest and, between 1922 and 1996, 148 people died on the mountain. Technology, experience, and repetition are obviously making Everest easier to climb, something being done with increasing regularity. This fact is also making death on Everest a more common event. Climbers also talk about the "world's highest garbage dump," where hundreds of abandoned and exhausted oxygen bottles lie, littering the slopes. They also talk matter-of-factly about climbing past the corpses of previous climbers who died attempting the summit. At high altitude with tight climbing schedules there is no time for adventurers to recover either bottles or bodies to return them to the bottom of the mountain.

As noted above, many of these factors have combined to make Mount Everest the world's most inaccessible tourist attraction. With the exceptions of the Kumbu Icefall and the Hillary Step, most of the climb is described as not being technically challenging, just terribly difficult because of the altitude, cold, and winds. These conditions have led to the growth of a small industry in which paying customers are guided to the summit. This option is still limited, of course, to those who are in adequate physical shape and who can pay tens of thousands of dollars for the trip, but the fact remains that you can reach the summit of Mount Everest by paying a guide to take you there. This development, in turn, has led to an increase in both the number of people reaching the summit of Everest and in the numbers of deaths on Everest's slopes. Between 1953 and 1973, a total of 38 people reached Everest's summit and 28 died trying to do so. Between 1973 and 1996, a further 638 people reached the summit and there were an additional 120 deaths. In 1985, however, the first amateur climber made the first commercial ascent and, since that date, more than 600 people have reached the summit while more than 75 have died trying to do so.

The statistics mentioned above are not meant to be a simple recitation of success and death. Rather, they demonstrate convincingly that Mount Everest, even nearly 50 years after it was first climbed, continues to compel people to climb it, even in the face of steadily mounting death tolls. In fact, the ability of inexperienced but driven people to sign up on expeditions has led to an explosion in deaths as well as successful climbs.

Lastly, it must be noted that Everest's pull on the imagination has been subject to politics. Serious attempts to climb Everest were impossible until Tibet opened its borders, because many of the best routes to reach the mountain went through there. Later, after the Chinese invasion of Tibet, these routes (and climbing routes from the north) were again closed to any who lacked permission from the Chinese government. The Tibetan routes have again been opened but only to those able to pay a hefty climbing fee, and climbers taking the favored Nepalese route must also pay a substantial fee for the privilege. These fees are in the vicinity of $10,000 per person to climb from the Nepalese side of Everest, with similar fees to climb from the Tibetan side. Add to this cost the supplies that must be purchased and the substantial numbers of Sherpas who are hired for these expeditions and the economic impact of Everest expeditions to the local governments becomes substantial. In fact, in 1996 nearly 200 climbers paid to attempt an Everest ascent.

People have been drawn to extremes for all of recorded history. Whether evidenced as exploring space, traveling to the South Pole, or climbing the world's highest peak, many are compelled to seek novelty continually. This urge often becomes a compulsion, which led Jon Krakauer to note "... attempting to climb Everest is an intrinsically irrational act—a triumph of desire over sensibility. Any person who would seriously consider it is almost by definition beyond the sway of reasoned argument." Identifying Mount Everest as the highest point on Earth guaranteed that many would try to climb it and that someone would succeed. And, the feat once accomplished, more knew it was possible and this knowledge led them to try.

P. ANDREW KARAM

Further Reading

Coburn, Broughton. *Everest, Mountain without Mercy.* Washington, DC: National Geographic Books, 1997.

Dyhrenfurth, G.O.*To the Third Pole: The History of the High Himalaya.* London: W. Laurie, 1955.

Hornbein, Thomas. *Everest: The West Ridge.* Seattle, WA: The Mountaineers, 1980.

Krakauer, Jon. *Into Thin Air.* New York: Villard Books, 1997.

Unsworth, Walt. *Everest, a Mountaineering History.* Seattle, WA: The Mountaineers, 1981.

Around the World *Beneath* the Sea: The USS *Triton* Retraces Magellan's Historic Circumnavigation of the Globe

Overview

The first known submarine was designed, but never built, by William Borne in 1578. From its early adventures (and misadventures) through the end of the twentieth century, the submarine played a vital role in both the exploration of the deep sea as well as the conquering of the globe. From the Revolutionary War to the Cold War, submarines made maritime history. In 1960 the USS *Triton* retraced the course of Ferdinand Magellan (c. 1480-1521) in a historic submerged circumnavigation of the globe.

Background

In 1620 Dutch inventor Cornelius van Drebbel (1572-1634) designed and constructed an oared submersible, recognized as the first submarine. By 1775, when Yale graduate David Bushnell (c. 1742-1824) built the *Turtle,* a one-man, human-powered submarine, man's desire to explore the ocean depths combined with his desire for naval superiority. History's first submarine attack came in 1776 when the *Turtle* was used by the Americans to attempt a break of the British blockade of New York Harbor during the Revolutionary War. From the *Turtle,* Bushnell attempted to attach a torpedo to the hull of the HMS *Eagle* but was unsuccessful.

Using the same principles developed by Bushnell, American steamboat inventor Robert Fulton (1765-1815) built the *Nautilus* and successfully submerged and operated it on the Seine in France in 1801. Technological developments continued, and in 1864 the Confederate submarine *H.L. Hunley* was the first to sink an enemy ship in combat when it rammed its spar torpedo into the hull of the Union sloop USS *Housatonic* off Charleston, South Carolina. In 1870, shortly after the Civil War, the U.S. Navy purchased its first submarine—a human-powered submarine called the *Intelligent Whale,* which failed during performance testing at sea and was never put into service.

Five years later John Philip Holland (1841-1914) submitted his first submarine design to the U.S. Navy, which rejected it as fantasy. Not discouraged, Holland went on to design and build a steam-powered submarine, the *Plunger,* according to Navy specifications, which also failed to pass tests. In 1900 Holland's John P. Holland Torpedo Boat Company completed his *Holland VI*, an internal combustion, gasoline-powered submarine, and after extensive trials, sold it to the U.S. Navy, which renamed it the USS *Holland* (SS-1), giving birth to the U.S. Navy's submarine force. (The idea for a submarine force came from the Assistant Secretary of the Navy, one Theodore Roosevelt [1858-1919], later President of the United States, who had seen its potential during the Spanish-American War.)

With its first seaworthy submarine, the U.S. Navy began focusing on design improvements. Following the lead of the French, who in 1904 built the *Aigrette*, the first submarine with a diesel engine for surface propulsion and electric engine for submerged operations, the U.S. debuted its first diesel-engine submarines in 1912. In 1916 the USS *Skipjack* (SS-24) became the first diesel-powered submarine to cross the Atlantic Ocean. During World War I submarines were put into service by both sides—and the superior German U-boats inflicted heavy damage on Allied ships. Following the war new design concepts were initiated when the U.S. had the opportunity to inspect conquered German submarines. Around this time, in 1917, the first passive sound navigation and ranging (sonar) technology was developed.

By 1941 new designs and technologies such as sonar and radar helped U.S. submarines in operations against the Japanese. Approximately five million tons of Japanese naval and merchant shipping were sunk, crippling that nation's economy and ultimately leading to her defeat. In fact, the U.S. submarine force caused 55 percent of Japan's maritime losses. Following the war, German U-boat technology again provided the U.S. Navy with technological improvements, including a snorkel mast that allowed for diesel operations at a shallow depth and battery charging while submerged. In the 1940s and early 1950s the U.S. Navy continued enhancing its underwater vessels, developing the teardrop-shaped hull that influenced all later U.S. submarines.

In 1951 the U.S. Navy signed a contract with Westinghouse and Electric Boat to build the first nuclear-powered submarine, the USS

Nautilus (SSN-571), which was completed in 1954 and launched in 1955. "Underway on nuclear power," the first message from the *Nautilus*, signaled a defining point in the history of the U.S. naval submarine force. (In 1958 the *Nautilus* was the first ship to pass beneath the North Pole on a four-day, 1830-mile voyage from the Pacific to the Atlantic.) Nuclear power allowed for a dramatic increase in range and operational flexibility. Nuclear submarines could remain submerged for nearly unlimited periods of time and, with the 1959 launching of the USS *George Washington* (SSN-598), could fire cruise or ballistic missiles at enemy land targets from a submerged position. The USS *Triton* (SSRN-586), the first (and only) dual-reactor submarine in the U.S. Navy, was also commissioned in 1959.

First launched in August 1958, the *Triton* carried a forward reactor that supplied steam to the forward engine room and drove the starboard propeller; a second reactor powered the after-engine room and port propeller. Packed with new technologies—from a periscope for navigating via the altitude of celestial bodies that was as accurate as through a sextant, to a Precision Depth Recorder, which would take soundings of the ocean floor and record them graphically to show its virtual shape—the *Triton* was a masterpiece of technology and innovation. Designed for high speed on the surface as well as below it, the *Triton* was 447.5 feet (136.4 meters) long—in her day, the longest submarine in the world—and the fifth nuclear submarine built for the U.S. Navy. The *Triton* began sea trials in September 1959, by which time some of her crew, including her captain, Edward Latimer Beach (1918-), had been involved in a rigorous nuclear submarine training program for a period of a year or more. In November 1959 the *Triton* was officially commissioned into the U.S. Navy.

After commissioning, the *Triton* began torpedo trials and other special tests. Then, on February 4, 1960, in a secret Pentagon conference, Beach learned that the *Triton*'s "shakedown" cruise, the final test of a commissioned naval vessel, would be to circumnavigate the globe in a voyage called "Operation Sandblast" (Beach's code name was "Sand"). To follow the track of Ferdinand Magellan (c. 1480-1521) and his crew's globe-circling voyage from 1519-22, the *Triton* would remain submerged for the journey, a feat never before attempted. With less than two weeks to make final preparations and under top secret conditions—the only men told of the planned operation were *Triton*'s officers and one enlisted man, the navigational assistant—provisions and equipment for 120 days were loaded onto the *Triton*, including 77,613 pounds (35,236 kg) of food. On February 15, 1960, a 24-hour, preshakedown cruise run turned up a number of small malfunctions, which were quickly fixed before her February 16 departure.

On February 24 the *Triton* reached St. Peter and St. Paul's Rocks, the official departing and terminating point of her circumnavigation voyage. She also made her first of four crossings of the equator. The crew occupied themselves with daily drills and exercises, and a doctor was aboard to study the psychological effects of long cruises, with volunteers completing daily questionnaires regarding their habits and other matters such as their general feelings and moods. No real problems surfaced until March 1, when the fathometer, vital to the soundings being taken in the uncharted waters through which the *Triton* was voyaging, experienced difficulties, a reactor problem was noted, and the chief radarman was diagnosed with a kidney stone (on March 5, the *Triton* partially surfaced—she remained 99 percent submerged—to transfer him to the USS *Macon* off Montevideo, Uruguay). The problems were fixed and on March 7, 1960, the *Triton* passed Cape Horn off the coast of South America (and went back and forth five times to allow all crewmembers the chance to view it through her periscope). On March 12 additional fathometer problems were discovered, and the *Triton* was forced to rely on her search sonar and a gravity-metering device being tested for the remainder of the voyage.

Part of the *Triton*'s assignment during her voyage was to conduct undetected photo reconnaissance—which she accomplished on March 13 off Easter Island and later on March 28 off Guam, where she observed navy planes taking off and landing. On April 1, while at periscope depth in Magellan Bay, the *Triton*'s periscope was sighted by a young man in a small dugout. This was the only unauthorized person to spot the *Triton* during her voyage, a 19-year-old Filipino named Rufino Baring who was convinced he had seen a sea monster.

Triton's course took her from the mid-Atlantic around Cape Horn, through the Philippine and Indonesian archipelagoes, and across the Indian Ocean. She rounded the Cape of Good Hope on April 17, 1960, and arrived back at St. Peter and St. Paul's Rocks on April 25, following her fourth crossing of the equator. With the circumnavigation voyage of 60 days, 21 hours, and 26,723 nautical miles (49,491 km) behind her, the *Triton*'s shakedown cruise wasn't yet over. Following an April 30 photo reconnaissance of the city of Santa Cruz on Tenerife, Canary Islands, and a May 2 transfer to the USS *Weeks* of the official mission photographer (and the boarding of a medical officer), the *Triton* finally surfaced on May 10 off the coast of Delaware, having been submerged 83 days, 10 hours and 36,014 miles (57,959 km). On May 11 she arrived in Connecticut after a journey of 36,335.1 nautical miles (67,293 km) and 84 days, 19 hours, 8 minutes—having accomplished a spectacular submerged retracing of Magellan's historic circumnavigation.

Impact

Application of nuclear power to submarines reinforced the image of the United States as a superpower and leader in technology. Many of the Triton's innovations and technological advances in naval nuclear power—as well as in the design and construction of submarines—were subsequently used in other industries. Civilian as well as naval submarines became an essential part of the science community; there were numerous expeditions in the oceans of the world where submarines participated in studies of marine life, collected oceanographic data, and made detailed studies of the ocean floor. The achievement of the USS *Triton* was an important part of that scientific milieu. It may also have improved American morale, which suffered after a U-2 spy plane was downed by a Russian missile on May 1, 1960. This disaster that cancelled a summit conference between the U.S. and Russia, delaying the cause of world peace for years.

ANN T. MARSDEN

Further Reading

Books

Beach, Edward Latimer. *Around the World Submerged: The Voyage of the Triton.* New York: Holt, Rinehard and Winston, 1962.

Beach, Edward Latimer. *Salt and Steel: Reflections of a Submariner.* Annapolis, MD: Naval Institute Press, 1999.

Parr, Charles McKew. *Ferdinand Magellan, Circumnavigator.* New York: Crowell, 1964.

Internet Sites

New Horizons. http://www.schurmann.com.br/site1ano/ing/press/i_press15.htm.

U.S.S. Triton (SSRN-586). http://www.subnet.com/fleet/ssn586.htm.

Deep-Sea Diving: Jacques Piccard and Donald Walsh Pilot the *Trieste* to a Record Depth of 35,800 Feet in the Mariana Trench in the Pacific Ocean

Overview

The greatest ocean depth yet located is the Challenger Deep, a part of the Mariana Trench that descends to a depth of 36,201 feet—almost seven miles down. While this great depth has not yet been reached, on January 23, 1960, the *Trieste* descended to 35,800 feet (10,912 meters), the greatest depth yet reached by man. This descent showed that the technology had been developed to take people virtually anywhere on Earth and that, just seven years after Mount Everest had been scaled, the depths of the sea had been conquered too. The technology that went into designing and building *Trieste* was later used for other research vessels and military submarines. It spurred developments that led to the remotely operated vehicles that discovered the *Titanic* and recovered the treasure of the *Central America* in the 1990s.

Background

Slightly over 70 percent of the Earth's surface is covered by ocean. Until the last decade of the twentieth century, however, more was known about the surfaces of Venus and Mars than was known about what lay beneath the oceans. In fact, for most of history men sailed on the surface of the ocean and submarines navigated the topmost few hundred meters, but the only ships that visited the ocean floor were those that sank, never to return.

Part of the reason for this lack of direct knowledge is sea pressure. The weight of a column of seawater increases by 44 pounds per square inch (psi) (3.1 kg/m^2) for every 100 feet (30.5 m) of depth. A mere 100 feet of seawater, then, will exert a pressure of 44 psi over each of the 144 square inches (929 cm^2) in one square foot (0.9 m$^{2)}$ for a pressure of 6,336 pounds (2,877 kg). In other words, a vessel with only one square foot of hull would have three tons (2.72 tonnes) of force acting against it at a depth of only 100 feet. The Challenger Deep, at a depth of about 36,000 feet (10,973 m) experiences a pressure of almost eight tons per square inch (11,249,112 kg/m$^{2)}$. Pressure alone is sufficient to keep people from venturing to these depths without taking extraordinary measures. Add to this equation the necessity to breathe, maneuver, and return to the surface and one begins to understand why the ocean depths were not visited until 1960 and why, even today, they are known chiefly only by indirect means.

In the 1950s a number of advances came together that began to make human visitation of the sea floor possible. Science gave us high-strength metals capable of withstanding the intense pressures that exist at great depths, while other advances helped make life-support systems that could keep people alive underwater for the many hours required to make a round trip to great depths. The engineering that went into designing better submarines in the post-World War II era also helped make deep-diving submersibles that could be steered, while advances in electrical engineering went into designing the lighting systems that allowed occupants to see during their dives. Finally, global politics spurred the International Geophysical Year (1957-1958), giving further impetus to explore the sea while the emerging possibilities of submarine warfare, seafloor ballistic missiles, and other military uses of the ocean gave navies a vested interest in learning more about the ocean and its floor. All of these trends intersected in the 1950s, leading to the design of the *Trieste,* the first submersible designed to travel to and return from the deepest parts of the ocean.

In 1953 Swiss oceanographer Jacques Piccard (1922-) helped his father Auguste Piccard (1884-1963) build the *Trieste,* which they dove to a depth of 10,168 feet off the Mediterranean island of Ponza. In 1956, under contract with the U.S. Navy, the Piccards redesigned the *Trieste* to withstand the pressure of any known sea depth; they sold the *Trieste* to the navy two years later. In 1960, accompanied by U.S. Navy Lieutenant Don Walsh, Jacques Piccard took the *Trieste* to the bottom of one of the deepest parts of the Mariana Trench, the Challenger Deep, where they touched bottom at a depth of 35,800 feet (10,912m), just 400 feet (122m) less than the deepest sounding recorded.

Impact

The *Trieste*'s visit to the bottom of the Mariana Trench resulted in a number of effects on science, engineering, and society. Some of the more important of these are:

Opening the ocean depths to direct exploration

Development of deep-sea technology used in a number of areas

Exciting the public interest in oceanographic exploration and marine biology at great depths

Each of these areas will be explored in greater detail in the remainder of this essay.

Before the *Trieste*'s descent, man's direct exploration of the oceans was limited to the uppermost few thousand feet, whereas the average depth of the oceans is over 20,000 feet (6,096 m). The continental shelves and areas near some islands could be observed directly, but very little else. All other deep-sea exploration was done by casting nets or dredges over the side of a vessel, dragging them along the ocean floor, and hauling them back to the surface. Because of such crude methods, the deep sea floor was thought to be lifeless.

This perception began to change in the 1950s when Jacques Cousteau (1910-1997) and Harold Edgerton (a professor at the Massachusetts Institute of Technology) developed the technology to take pictures at great depths. These photos showed evidence of life at virtually all depths and locations, gradually convincing marine biologists that life could exist even under the crushing pressures of the abyssal plains. *Trieste*'s visit showed life existed even at the deepest point on the planet; exploration by other vessels has confirmed that living communities inhabit most parts of the ocean bottom. This discovery, particularly the recent discovery of thriving communities around deep-sea hydrothermal vents, has caused biologists to reconsider questions of where life might have evolved and whether or not life may exist elsewhere in the solar system. The first few decades of the twenty-first century may see a submersible probe explore oceans thought to underlie the icy surface of Jupiter's moon, Europa, in search of extraterrestrial deep-sea life.

Trieste also helped to consolidate many advances in submersible design and to inspire other designers. As noted above, many advances came together to create *Trieste*. Her success encouraged others to design and build other vessels to explore the ocean. Jacques Piccard went on to invent the mesoscaph (in which "meso" means "middle"), a vessel for exploring intermediate ocean depths; the United States built the FLIP (floating instrument platform) to study near-surface oceanography and marine biology. In addition to these vessels, *Alvin, Deepstar,* and the navy's deep submergence rescue vehicle (designed to rescue crews from sunken submarines) were designed using lessons from *Trieste*. Some features of modern deep-diving nuclear submarines are the result of work that went into *Trieste*'s design as well.

In addition to the engineering and scientific advances represented by the *Trieste,* she and other deep-sea exploratory vessels excited the public's interest in oceanography, an interest that has carried on for several decades. The interest shown for most of the last half of the twentieth century was probably due mainly to the efforts of Jacques Cousteau, but the bizarre nature of deep-sea life has been sufficiently interesting to capture public attention in and of itself. In fact, deep-sea exploration often provokes newspaper headlines, stories in the nightly news, or feature articles in popular magazines. In addition to scientific discoveries, events such as the recovery of gold from the sunken ship *Central America,* the live broadcast from the wreck of the *Titanic,* and other events routinely command large television audiences. As with so many other oceanographic exploits, the technology that makes such deep submergence possible is a direct outgrowth of lessons learned while designing, building, and operating *Trieste,* including her dive into the Challenger Deep.

Finally, in a related vein, deep-sea exploration became important to the United States in the late 1950s and early 1960s as compensation of a sort for the Soviet Union's successes in space. The U.S.S.R. launched the first satellite and the first manned spaceflight as well as conducted the first spacewalk, all of which dealt temporary blows to the idea of the United States as a leading technological power and innovator. Moreover, while trying to catch the Soviet Union in space, the United States suffered a number of embarrassing rocket failures. At times, the only consolation for the United States seemed to be the American mastery of deep-sea technology and exploration.

In spite of the *Trieste*'s success and that of other manned and unmanned deep-sea exploration vessels, the bottom of the sea remains largely a mystery to science. The release of data from military satellites has been a tremendous boon to mapping the seafloor but provides no information about the organisms that exist there and how they live. Research on these communities of organisms is providing important information that will likely lead to a better understanding of the origins of life on earth and how that early life existed. These questions are of widespread scientific and popular interest, especially given the strides taken in the late 1990s in the search for life on other planets. In addition, rich deposits of metal nodules—mostly manganese and related metals—exist on the ocean's abyssal plains but, in spite of their economic potential, currently remain untouched. For these reasons, the *Trieste*'s 1960 dive to a depth of nearly seven miles ranks as a high accomplishment as well as sets the stage for even more dramatic achievements to come.

P. ANDREW KARAM

Further Reading

Piccard, Jacques. "Man's Deepest Dive." *National Geographic* (May 1960).

The Space Race and the Cold War

Overview

At the end of World War II, the United States and the Soviet Union began a decades-long battle for political, military, and technological superiority. In the absence of any real fighting, space exploration provided a focus for the competition between the two superpowers. From the 1950s to the 1970s the United States and the Soviet Union raced to conquer space, but when tensions eased between the two nations in the 1970s, the urgency of winning the race declined and the race ended with the superpowers cooperating on several projects.

Background

The United States and the Soviet Union emerged from World War II as adversaries in the Cold War—an open rivalry in which the two nations vied for political power and standing in the world without ever fighting an actual battle. Instead, they fought with propaganda and scientific and technological achievements.

Much of the technology that led to space exploration had military beginnings. World War I and World War II resulted in the development of government scientific research facilities charged with designing military airplanes. World War II had provided the motivation for rocket development in the United States, the Soviet Union, Great Britain, France, and other countries. But the Germans were by far the most advanced rocket designers: their V-2, a liquid-propellant-fueled rocket, was the ancestor of the rockets that would eventually reach space. Recognizing this, the United States brought several V-2s back for research after the war, and launched "Operation Paperclip," an effort to recruit as many top German scientists as possible to the United States to continue their research.

At the end of the war, it appeared that the United States was the clear technological giant in the world—they had detonated the first atomic bomb in 1945 and the first hydrogen bomb in 1952. Despite this advantage and the presence of German scientists in the United States, the Soviet Union quickly made great advances in rocketry. During the International Geophysical Year (1957-58) both countries announced plans to launch satellites into space. But the United States was still working on a launch vehicle when the Soviet Union stunned the world by announcing that it had successfully placed a satellite, *Sputnik I,* in orbit on October 4, 1957.

A month later, on November 3, 1957, the Soviet Union launched *Sputnik II,* carrying a dog named Laika. The United States tried to catch up, but its first attempt at a launch, on December 6, 1957, failed when the *Vanguard* rocket rose four feet and crashed back to the launch pad. It was instantly called "Flopnik," or "Kaputnik." Finally on January 31, 1958, the United States launched its first satellite, *Explorer I.* The space race had officially begun.

Impact

The early Soviet successes in space dealt a blow to American pride and confidence. Serious attempts to reach space had been neglected in the United States, where military officials preferred to concentrate on weapons development, and where the Eisenhower administration had been so concerned with keeping the nation's budget balanced that it had cut funding to all scientific efforts.

The launch of *Sputnik* was a wake-up call. Americans feared that the world would see the Soviet system as superior, and many questioned whether the free and open society of 1950s America was as dominant as they had thought. The U.S. space program, previously a concern only among scientists and engineers, was suddenly important to everyday people as well. Military experts, meanwhile, took the satellite launch as proof that the Soviet Union was probably ahead in ballistic missile development as well. The feeling was that if the Russians could get a satellite into space, then they could probably land a warhead on American soil as well.

With this fear spurring them on, U.S. officials scrambled to piece together a space program in an attempt to salvage some national pride and international prestige. President Eisenhower established the National Aeronautics and Space Administration (NASA) in 1958 to oversee the space program and to make sure the United States caught up to the Soviet Union. The space race continued though the 1950s and 1960s, with the United States and the Soviet Union competing for each progressive step of space exploration.

Having lost the initial leg of the race, the United States aimed to be the first to reach the moon. But the first attempt to launch, in August 1958, failed when the rocket carrying the *Pioneer 0* moon probe exploded on the launch pad. That same year the launches of *Pioneer* probes 1, 2, and 3 were also unsuccessful. Meanwhile, the Soviets were also working on a moon launch. As in the United States, the first attempt failed when the *Luna 1* probe launched but did not reach the moon in early 1959. But the Luna program soon got off the ground, and the Soviets racked up more firsts—the first solar orbit, the first impact on the moon, and the first photographs of the moon from a lunar orbit (which allowed the Russians to name many of the moon's geological features).

American pride was at a low. The nation that had emerged from World War II as the most powerful on earth was being humbled and technologically crippled by its enemy. In the face of this seeming defeat, the United States decided to aim for the ultimate prize—a man on the moon. With that in mind, Project Mercury was begun

Sputnik 1. (UPI/Corbis-Bettmann. Reproduced by permission.)

in 1958 with the goals of orbiting a manned spacecraft around the earth, studying man's ability to function in space, and recovering both man and spacecraft safely. But once again, the Soviet Union did it first. On April 12, 1961, Yuri Gagarin (1934-1968), a Russian cosmonaut, became the first man in space. This time, the United States was not so far behind. On May 5, 1961, Commander Alan Shepard (1923-1998) of the U.S. Navy became the first American in space, orbiting earth in the *Mercury* 7 capsule.

American officials scrambled to find a way to catch up. President John F. Kennedy met with advisers who felt that the only way to win the space race was to get a man to the moon first. So in a speech given on May 25, 1961, Kennedy rallied the nation around the space program. "If we are to win the battle that is now going on around the world between freedom and tyranny," he said, "now it is the time to take longer strides—time for a great new American enterprise—time for this nation to take a clearly leading role in space achievement, which in many ways may hold the key to our future on earth." Then he issued his famous challenge: "I believe that this nation should commit itself to achieving the goal, before this decade is out, of landing a man on the moon and returning him safely to earth."

Kennedy's challenge restored national interest in space. The U.S. space program accelerated, and the race to space with the Soviets intensified. On August 6, 1961, the Soviets struck again. Cosmonaut Gherman Titov (1935-) and the *Vostok* 2 capsule spent more than 25 hours in space, orbiting the earth 17 times. The next year, on February 20, 1962, John Glenn (1921-) became the first American in orbit. For the next seven years, the United States and the Soviet Union raced to get to the ultimate prize first. The Soviets put the first woman, Valentina Tereshkova (1937-), in space in 1963, and a cosmonaut took the first spacewalk in 1965. The first American spacewalk came just a few months later, but then the Soviets racked up a series of other firsts—the first impact on Venus, the first soft landing on the moon, and the first orbit of the moon with a safe return.

For all its earlier second-place finishes, the Unites States managed to cross the finish line first when it counted. The first man on the moon was an American, Neil Armstrong (1930-), and he walked on the moon before the end of the 1960s, just as Kennedy had promised. But soon after this victory, in the early 1970s, the United States' interest in conquering space waned, as sociopolitical issues preoccupied the nation's interest.

Simultaneously, the Soviet program began to falter. In 1971 the Soviet Union announced that it was shifting the focus of its space program to long-term living in space; later that year the Salyut program began, launching a number of stations that conducted experiments in space and hosted astronauts from other nations. Not to be outdone, the United States sent up the space station *Skylab* in 1973. But by this time, further détente between the Unites States and the Soviet Union cooled any chance of starting a new space race. The Cold War was coming to an end and the hostilities of the 1950s were being forgotten.

Some experts consider the official end of the space race to be 1975, when the Soviet *Soyuz* craft docked with the American *Apollo 18,* the first-ever international space rendezvous. The Cold War also ended peacefully, with the United States and Soviet Union never actually going to war—except to compete for the patriotism of their respective people and the international prestige of conquering space.

GERI CLARK

Further Reading

Burrows, William E. *This New Ocean.* New York: Random House, 1999.

Collins, Martin J. *Space Race: The U.S.–U.S.S.R. Competition to Reach the Moon.* New York: Pomegranate Press, 1999.

Crouch, Tom D. *Aiming for the Stars: The Dreamers and Doers of the Space Age.* Washington, DC: Smithsonian Institution Press, 1999.

Schefter, James. *The Race: The Uncensored Story of How America Beat Russia to the Moon.* New York: Doubleday, 1999.

Women in Space

Overview

In 1903, the historic flight of the Wright brothers ushered in a new era, not just in transportation, but also in lifestyle, adventure, and science. When American Bessica Raiche made a solo flight in 1910 using the aircraft she and husband François built, she opened the skies for future women aviators. By the time the National Aeronautics and Space Administration was chartered in the United States in July 1958, women were a fixture in aeronautical circles—not just in support roles, but as pioneers in astronomics, engineering, and mathematics. In 1963, when Russian cosmonaut Valentina Tereshkova (1937-) left Earth aboard the *Vostok 6*, she became the first woman in space, forever changing the destiny of women.

Background

Before the end of the nineteenth century, three women astronomers had made significant contributions to the science that would eventually lead mankind into space. The first, Maria Mitchell (1818-1889), discovered a comet in 1847 and became a professor of astronomy and director of the Vassar College observatory in 1865. The second, Henrietta Swan Leavitt (1868-1921), devised a method to measure the distances of stars from the Earth with stars in other galaxies. Her photographic measurements, key to determining astronomical distances, were known as the Harvard Standard and were accepted among the world's astronomers. The third, Annie Jump Cannon (1863-1941), a physicist, joined the staff of the Harvard College Observatory in 1897. In her 40-plus years on staff, Jump Cannon named and catalogued over 300,000 stars, perfected a universal system of stellar classification, and compiled the largest accumulation of astronomical information ever assembled by a single researcher.

While women astronomers were searching the far reaches of the galaxy via telescope, women aviators were exploring the skies closer to the Earth. Less than a decade after Orville Wright's (1871-1948) first successful flight, Harriet Quimby (1884-1912) became the first American woman to earn a pilot's license (1911). In 1912, she was the first woman to fly across the English Channel. The war effort expanded a flight school started in 1915 by Katherine (1891-1977) and Marjorie Stinson, who trained American and Royal Canadian pilots. In addition to her flight school achievements, Marjorie Stinson was appointed the first female airmail pilot in 1918. Another first was accomplished by Bessie Coleman (1896-1926), who became the first African-American (male or female) to earn a pilot's license (in 1921).

By the 1930s, women aviators had made their mark as stunt pilots, entertainers, and adventurers, and began making significant contributions in other areas of aviation. In 1931, Anne Morrow Lindbergh (1906-) earned her private pilot's license and went on to become the first female glider pilot and the first female navigator who, with her husband Charles (1902-1974), flew the world mapping transcontinental air routes for commercial aviation. Their pioneering routes were still in use in the late 1990s. In 1932, Amelia Earhart (1897-1937) was the first woman to make a transatlantic solo flight. (An aviation adventurer, she disappeared during her historic around-the-world flight in 1937.) In the same year Olive Ann Beech (1903-1993) co-founded Beech Aircraft with her husband Walter (1891-1950). She became President and CEO after his death and eventually transformed the company into a multimillion-dollar, international aerospace corporation.

Beech Aircraft wasn't the only environment for successful women. Other female professionals were influential during the development of aeronautics leading up to the space race. By 1943, in the midst of World War II, half a million women were working in the aviation industry, representing 36 percent of its workforce. In the United States, the National Advisory Committee for Aeronautics (NACA), the predecessor to NASA, welcomed female engineers, physicists, and computer specialists during the 1940s and 1950s. By 1945, the last year of World War II, nearly 1,000 women were working at NACA in technical positions.

When the National Aeronautics and Space Administration (NASA) was created in 1958. many of NACA's female engineers, mathematicians, scientists, and technicians, remained an integral part of the new organization. Women like Marcia Neugebauer (1932-), who served as the senior research scientist for NASA's Jet Propulsion Laboratory from 1956 to 1996, were critical to the involvement of women in the space age. When the Russian government launched cosmonaut Valentina Tereshkova (1937-) into space in 1963 aboard the *Vostok* 6, the final barriers to women in the aerospace arena were eradicated.

Impact

In 1961, when Russian cosmonaut Yuri Gagarin (1934-1968) became the first human in space, Valentina Tereshkova, an accomplished parachutist with over 125 jumps on her record, was employed as a cotton-spinning technology expert in a textile mill. Gagarin's achievement inspired Tereshkova, and she was selected for the Soviet space program in 1962. In June 1963, Tereshkova made her groundbreaking space flight when she was launched into orbit around Earth on *Vostok 6*. After 48 orbits of the Earth and more than 70 hours in space, Tereshkova guided the *Vostok* 6 back into the Earth's atmosphere, parachuted from the craft, and landed in central Asia. Although she never made another space flight, her space exploration launched new opportunities for other female astronauts, including those from the United States.

Before NASA appointed American women to fly in space, two U.S. Department of Defense divisions selected female pilots. In 1974, the U.S. Navy selected its first noncombatant female pilots and, in the same year, the U.S. Army trained its first female pilot, Lt. Sally Murphy. In 1976, women were admitted to American military academies. By the end of the decade, women comprised nearly 21 percent of NASA's workforce, and the first female astronauts had been selected. In 1978, astronaut candidates Rhea Seddon (1947-), Kathryn D. Sullivan (1951-), Judith A. Resnick (1949-1986), Sally K. Ride (1951-), Anna L. Fisher (1949-), and Shannon W. Lucid (1943-) became the first women chosen as part of NASA's space exploration program.

NASA's 1978 female astronaut candidates distinguished themselves as exemplary astronauts. In 1983, astrophysicist Sally Ride became the first American woman to be launched into space as a member of shuttle mission STS-7. (She was also a crew member of STS-41G in 1984.) In 1984, engineer Judy Resnick flew on the first Space Shuttle orbiter flight, operating its remote manipulator arm. (Sadly, Dr. Resnick lost her life in the tragic *Challenger* explosion in 1986.) Also in 1984, geologist Kathy Sullivan became the first American woman to walk in space during shuttle mission STS-41G. In 1985, Dr. Anna Fisher was the first American astronaut (male or female) to retrieve a malfunctioning satellite during NASA's first space salvage mission. The most extraordinary female astronaut of NASA's 1978 candidate class was Shannon Lucid, who was a veteran of four Space Shuttle missions before setting the record for the longest time spent in space by an American (188 days) during an assignment aboard the Russian space station *Mir* in 1996.

The Russian space program, highlighted by its many cosmonaut firsts, including Tereshkova's historic flight and that of Svetlana Savitskaya (1948-), the first female to walk in space (1984), made substantial contributions to the aerospace industry through its 14-year *Mir* space station mission, launched in 1986 (and scheduled to be decommissioned in early 2000). *Mir* housed international cosmonauts and astronauts who performed experiments of historical and scientific significance focusing on life in space and observational sciences. (It also served as the home base for the initial space construction of the International Space Station.) In 1991, Helen Patricia Sharman (1963-) became the first female British cosmonaut aboard Russia's *Soyuz TM-8* flight to *Mir*. From late 1994 to early 1995, Russian cosmonaut Yelena Kondakova (1957-) lived for 169 days aboard *Mir*, the second-longest female mission (after Shannon Lucid) on the station. While docked at *Mir* in 1996, female NASA astronaut Linda Godwin (1952-) and her male counterpart Michael Richard Clifford (1952-) made the first American spacewalk at an orbiting space station. Also in 1996, Claudie Andre-Deshays (1957-) became the first female French cosmonaut during a space flight aboard Russia's *Soyuz TM-24* trip to Mir.

Female astronauts from other nations also made significant strides in the space age. One of the six original Canadian astronauts selected in 1983 was Roberta Bondar (1945-). She was appointed the prime payload specialist on NASA's STS-42 for the first International Microgravity Laboratory mission (1992). The first female Japanese astronaut, Chiaki Mukai (1952-), was selected in 1985. In 1994, Dr. Mukai served as a payload mission specialist on NASA's STS-65, the second International Microgravity Laboratory mission. She also flew with ex-astronaut Senator John Glenn (1921-) and the first Spanish cosmonaut Pedro Duque (1963-) in 1998 on STS-95. A second female Canadian astronaut, Julie Payette (1963-), became a technical advisor for the International Space Station project after her selection and training in 1992. In 1999, she flew on NASA's STS-96 mission, which docked with the International Space Station to transfer equipment to the interior of the station.

As the international aerospace industry continued to expand in the 1990s, women at NASA made further contributions to the exploration of space—both in space and on the Earth. Women represented 25 percent of NASA's astronauts, 16 percent of its scientists, and one-third of its workforce. Opportunities for women at NASA were wide-ranging. In 1992, Mae Jemison (1956-) became the first African-American woman to fly in space and the first science mission specialist (male or female) on STS-47. In 1993, Ellen Ochoa (1958-) became the first Hispanic woman to fly in space on STS-56. Then, in 1997, Kalpana Chawla (1961-) became the first Indian-born American woman to fly in space on STS-87. The most significant achievements by a woman astronaut at NASA were made by Eileen Collins (1956-), who was the first female pilot selected by NASA (1990), the first female space shuttle pilot (1995), the first female pilot to dock with the Russian space station *Mir* (1997), and the first female space shuttle commander (1999). On the ground, NASA women made history in 1998 when nearly two-thirds of the flight control team for STS-95, including the flight director, launch commentator, ascent commentator, and CapCom (the communication between mission control and the shuttle crew), were female.

As adventurers, explorers, pioneers, and groundbreakers, women firmly established their place in aviation and aerospace history. In 2000 and beyond, women will follow in the footsteps of these innovators and help turn dreams, visions, and science fiction into reality. New computer technologies, a new field of study called aerospace bioengineering, and new medical advances will be a part of a new world in the space above the Earth. Other technologies will surely alter the way humankind lives—revolutionizing aviation and all aspects of aeronautics—and women will play a significant role at all levels of research, development, and implementation.

ANN T. MARSDEN

Further Reading

Books

Haynsworth, Leslie, and David M. Toomey. *Amelia Earhart's Daughters: The Wild and Glorious Story of American Women Aviators from World War II to the Dawn of the Space Age.* New York: William Morrow, 1998.

Russo, Carolyn. *Women and Flight: Portraits of Contemporary Women Pilots.* Boston: Bullfinch Press, 1997.

Other

Walley, Ellen C. and Terri Hudkins. "Women's Contributions to Aeronautics and Space." http://www.nasa.gov/women/ milestones.html. National Aeronautics and Space Administration, 1999.

The 1969 Moon Landing: First Humans to Walk on Another World

Overview

On July 20, 1969, Neil Armstrong (1930-) and Edwin "Buzz" Aldrin (1930-) landed an ungainly spacecraft named *Eagle* on the moon and spent two hours exploring the lunar surface. They left the next day, rendezvousing in lunar orbit with the command ship *Columbia* and returning safely to Earth. The *Apollo 11* landing ended a decade of competition between the Soviet and American space programs, helped to restore the nation's self-confidence, and began an intensive program of exploration that transformed scientists' understanding of the Moon.

Background

The dream of traveling to the moon was already centuries old when the Second World War ended in 1945. It had inspired Robert Goddard (1882-1945), who built and flew the first modern rockets in the New Mexico desert during the 1930s, and captivated Wernher von Braun (1912-1977), leader of a team that gave Nazi Germany the world's first guided missiles in 1944-45. Postwar Soviet and American leaders, recognizing the military potential of such missiles, clamored for bigger, more powerful versions. By 1957 the arms race had produced rockets strong enough to carry a nuclear bomb halfway around the world or a small satellite into Earth orbit. The Soviet Union launched such a satellite, *Sputnik I*, in October 1957. The success of *Sputnik* opened the Space Age and added a new dimension to the superpowers' already intense rivalry.

Soviet achievements in space overshadowed American ones from 1957 through April 1961, when Major Yuri Gagarin (1934-1968) of the Red Air Force became the first human to orbit Earth. America's seemingly permanent second-place status in space stung the pride and undermined the Cold War foreign policies of the newly inaugurated president, John F. Kennedy. He proposed, in a May 1961 address to Congress, that the United States take a bold step: committing itself to landing a man on the Moon and returning him safely to Earth by the end of the decade.

The engineering and organizational challenges involved in meeting Kennedy's goal were immense. Project Apollo (as the moon-landing program came to be known) would involve flights a half-million miles long, taking as much as two weeks to complete. It would require boosters more powerful, guidance systems more accurate, and spacecraft more complex than any then in existence. It would also require the command ship and the lander to rendezvous and dock twice: once in Earth orbit, and once in lunar orbit. No such maneuver had even been planned, much less carried out, in 1961.

Designing, building, and testing the *Apollo* spacecraft and its massive Saturn V booster took six years, millions of government dollars and the combined efforts of America's leading aerospace manufacturers. Simultaneously, the National Aeronautics and Space Administration (NASA) conducted preparatory flights designed to lay the groundwork for *Apollo*. The ten flights of Project Gemini (1964-66) tested rendezvous techniques and crew endurance in Earth orbit. Three series of robot probes—Ranger, Surveyor, and Orbiter—returned detailed information about the lunar surface, allowing NASA planners to select possible landing sites.

In January 1967, only weeks before the first manned test flight, Project Apollo suffered a tragic setback. Faulty wiring ignited a flash fire in the spacecraft during a routine launch simulation, killing astronauts Gus Grissom (1926-1967), Ed White (1930-1967), and Roger Chaffee (1935-1967). Extensively redesigned after the fire, the *Apollo* spacecraft would not fly with a human crew until late 1968. Once operational, however, it performed flawlessly. Two test flights in Earth orbit (*Apollo 7* and *9*) and two round trips to the moon (*Apollo 8* and *10*) proved its reliability, and gave NASA confidence to designate *Apollo 11* as the first lunar landing mission.

The July 20, 1969, lunar landing confirmed NASA's confidence in the *Apollo* spacecraft. Neil Armstrong's words as he jumped onto the surface of the Moon were heard by millions of Americans and have since become the stuff of legend: "That's one small step for a man, one giant leap for mankind."

Impact

Many of the technologies developed for Project Apollo eventually found their way onto the con-

Neil Armstrong was the first human to set foot on the moon's surface. *(NASA. Reproduced by permission.)*

sumer market: nonstick coatings, dehydrated foods, and miniaturized electronic components. NASA publicity often focused on such products in an effort to suggest that the space program provided taxpayers with tangible returns on their investment. These consumer spin-offs are, however, only the smallest part of Project Apollo's impact. The most significant results of *Apollo 11*, in particular, were intangible rather than tangible—scientific and social rather than technological.

The successful landing and return of *Apollo 11* ended the Soviet-American space race that had begun with *Sputnik* in 1957. No subsequent lunar landing could be as impressive as the first, Soviet planners recognized, and no other space achievement then within reach could have the same luster. A successful attempt to land a Soviet crew on the Moon would bring only modest benefits; a failed attempt, on the heels of America's success, would be disastrous. The longstanding political and military rivalry between the superpowers was also diminishing at that time, making a continuation of the space race even more unlikely. New leaders and new diplomatic initiatives such as arms-control treaties created a temporary thaw in the Cold War. With competition giving way to a new spirit of superpower coexistence (known as détente), the space race seemed to belong to another era.

The words and symbols connected with the *Apollo 11* landing dramatized this shift in attitudes. They reflected little of the intense superpower rivalry that gave birth to Project Apollo in 1961. Instead, they embodied the new ideal of superpower coexistence. Armstrong and Aldrin had ample cause to gloat and to celebrate as they set foot on the Moon, but they did neither. They planted their nation's flag where they landed but did not claim the land beneath it for their nation or their leaders. After stepping onto the Moon for the first time, Neil Armstrong's words were those of a human, not an American. A metal plaque left behind to commemorate the landing expressed the idea even more clearly. "Here men from the planet Earth first set foot upon the moon ... We came in peace, for all mankind."

Although the official symbols of *Apollo 11* did not define it as a specifically American triumph, most Americans saw it in just those terms. The year before the landing, 1968, had been one of the most turbulent in the nation's history. American forces suffered major setbacks in Vietnam; incumbent president Lyndon Johnson ended his bid for reelection; civil rights leader Martin Luther King was assassinated in April, and presidential hopeful Senator Robert Kennedy in June; protests against the Vietnam War grew increasingly angry and divisive; demonstrators and police fought in the streets of Chicago during the Democratic national convention. The series of successful *Apollo* missions that culminated in the landing of *Apollo 11* was welcome good news amid this string of national catastrophes. It was also proof, for those whose faith had begun to waver, that big government (NASA) and the American military (most of the astronauts) could still rise to greatness as they had during World War II.

Apollo 11, in particular, also boosted Americans' confidence in their ability to solve society's problems. The moon landing became proof of American competence and achievement. "If we can send a man to the moon," a popular expression asked, "why can't we cure cancer, clean up the air, end poverty, etc.?"

NASA promoted the *Apollo 11* landing as the climax of a decade of hard work and as the fulfillment of the late President Kennedy's 1961 challenge. News commentators called it epoch-making and compared it to the European discovery of the New World. These attitudes encouraged Americans to see the first moon landing as a triumph for the human race in general and America in particular. The same attitudes, however, made the flight of *Apollo 11* a nearly impossible act for NASA to follow. Public interest in Project Apollo diminished sharply after the first landing, as did Congressional support. Three projected lunar landing missions—*Apollo 18, 19,* and *20*—were cancelled for lack of such support. NASA undertook a variety of ambitious, successful missions in the three decades after *Apollo 11*, but few even came close to generating the same public interest or nationwide high spirits. NASA's desire to recapture the public confidence and substantial budgets it enjoyed in 1969 has, some critics charge, distorted its mission. Too often, they argue, the space agency neglects scientific research in order to fly missions that will draw public interest.

These criticisms, while valid to some extent, are also ironic. The *Apollo 11* landing itself made possible some of the most important science ever done in outer space. Neil Armstrong and Buzz Aldrin spent only a few hours on the lunar surface, deployed only a few scientific experiments, and collected only modest samples of lunar rock and soil. Because they were the *first* humans to walk on the Moon, however, even these limited contributions vastly expanded scientists' understanding of it. The robot orbiters and landers that preceded *Apollo 11* provided close-up pictures of the lunar surface, but they could not assess its texture or chemical makeup. Pictures allowed Earth-bound geologists to form hypotheses about the Moon but not to test them. Tests, and a clearer understanding of the Moon's structure, composition, and age, required samples. The *Apollo 11* landing provided those samples and began a revolution in the earth sciences.

Equally important, *Apollo 11* demonstrated that humans could make a soft landing on the moon, do useful work, and return safely to Earth. Premission concerns about possible hazards evaporated as the mission went on. Neither lander nor astronauts sank, as some had feared they would, into a thick layer of dust. Lunar soil did not burst into flames upon contact with oxygen. No alien microbes infected the returning astronauts. *Apollo 11* showed that the exploration of the Moon was well within NASA's capabilities. Its success opened the door for later Apollo missions to concentrate on science, and as long as its budget allowed, NASA took full advantage of the opportunity. A generation after Neil Armstrong took his "one small step," the legacy of *Apollo 11* remains very much alive. Scientists' understanding of the Moon is built almost entirely on data collected by the crews of *Apollo 11* and the five landing missions that followed. The landing remains a symbol of American greatness, and images of it were fixtures of century's-end retrospectives. And—for better or worse—NASA is still best remembered as the agency that put a man on the Moon.

A. BOWDOIN VAN RIPER

Further Reading

Armstrong, Neil, Michael Collins, and Edwin Aldrin. *First on the Moon.* Boston: Little, Brown, and Company, 1970.

Chaikin, Andrew. *A Man on the Moon: The Voyages of the Apollo Astronauts.* New York: Viking Penguin, 1994.

Collins, Michael. *Carrying the Fire: An Astronaut's Odyssey.* New York: Farrar, Straus and Giroux, 1974.

Lewis, Richard S. *Appointment on the Moon.* New York: Ballantine, 1969.

MacKinnon, Douglas, and Joseph Baldanza. *Footprints.* Washington: Acropolis Books, 1989.

Murray, Charles, and Catherine Bly Cox. *Apollo: The Race to the Moon.* New York: Touchstone/Simon & Schuster, 1989.

Wilford, John Noble. *We Reach the Moon.* New York: Bantam, 1969.

Wilhelms, Don E. *To a Rocky Moon: A Geologist's History of Lunar Exploration.* Tucson: University of Arizona Press, 1993.

Space Stations

Overview

The advent of space stations allowed humans to spend extended periods of time in space. They have provided a wealth of information about the challenges humans will face, and must overcome, if they are to survive outside Earth's life-supporting atmosphere while traveling to distant planets or one day inhabiting other worlds. The *Salyut 1,* launched in 1971 by the Soviet Union, became the first manned space station. The United States followed two years later with its version, called *Skylab.* As the end of the twentieth century neared, the United States, Russia (part of the former Soviet Union), Canada, Japan, Brazil, and the 11 nations of the European Space Agency combined efforts to plan construction of the International Space Station (ISS). The ISS is scheduled to be completed in 2004.

Background

The space race between the United States and the Soviet Union began when the world's first satellite, *Sputnik,* went into orbit around Earth on October 5, 1957. The launch date marked the 100th year after the birth of Russian rocketry pioneer Konstantin Tsiolkovsky (1857-1935), who in 1903 proved that a missile could escape Earth's atmosphere using a staged rocket design and liquid propellants. The 185-pound Russian *Sputnik*—the name means traveler—orbited the planet some 1,400 times during its 96-day mission.

Sputnik's successful mission came at a time when the Cold War raged between the United States and Soviet Union. Many Americans felt the satellite gave the Soviets the military and technological upper hand. In 1961 U.S. President John F. Kennedy announced the country's intention to put a man on the moon. Although he subsequently supported a joint American-Soviet effort, it didn't materialize. After Kennedy's assassination in 1963, the American government pursued a solely American mission, and in 1969 American astronauts put the first human footprints on the lunar surface.

The Russians, then, were the first to put a satellite into orbit and the Americans were the first to put a human on the moon. In late 1969, shortly after the lunar landing, the Soviet Union announced plans to build, launch, and man the first space station. Construction began in 1960, and on April 19, 1971, the Earth-orbiting space station named *Salyut 1* was launched. The 14.5-meter-long (48 m), 20-ton (18.16 tonne) *Salyut 1* was basically a series of three pressurized cylinders for use by cosmonauts and one nonpressurized cylinder for propellant storage. The pressurized cylinders included living and working quarters, various station controls and communications, and an ac-

Russian space station *Mir. (Corbis. Reproduced by permission.)*

cess module. The access module was designed to connect with *Soyuz* spacecraft, which would ferry cosmonauts to and from the station.

On June 7, 1971, the *Soyuz 11* brought the first cosmonauts to the *Salyut 1.* The two craft linked successfully, and three cosmonauts moved into the station for a 23-day stay. The team of cosmonauts was comprised of test engineer Viktor I. Patsayev, Lieutenant Colonel Georgi T. Dobrovolsky, and flight engineer Vladislav N. Volkov. This was Volkov's second Soyuz mission, and the first for Dobrovolsky and Patsayev. Before becoming a cosmonaut, Dobrovolsky was a fighter pilot and Patsayev was a design engineer. During the next three weeks and two days, the cosmonauts conducted a variety of equipment checks, performed medical and biological studies on plants and animals, and completed astronomical observation work.

During what became the longest manned space mission to date—the former record was held by cosmonauts aboard the *Soyuz 9*—the three cosmonauts took on hero status back at home. On June 30 the mission ended and the crew reboarded the *Soyuz 11* for the return trip. As the Soviet Union was finishing plans to celebrate the mission's success, tragedy struck. The return trip appeared to go as planned, and the *Soyuz 11* descent module landed as expected. When the recovery group reached the craft and opened its hatch, however, the group found all three men dead. Ensuing investigations revealed that a valve connecting the *Soyuz 11*'s descent and orbital modules had become unfastened, and the descent module's pressurized atmosphere escaped into space. The limited space inside the descent module precluded the cosmonauts from wearing their pressurized suits, and without that protection, the men died where they sat.

Following this sad turn of events, the Soviets fired the *Salyut 1*'s propulsion system for the last time on October 11, 1971, and allowed the space station's orbit to decay until the *Salyut 1* incinerated in Earth's atmosphere. Later Salyut stations and missions were unsuccessful or short-lived.

On May 14, 1973, the United States National Aeronautics and Space Administration (NASA) launched its version of a space station with the 36-meter-long *Skylab*. The station held an orbital workshop, airlock module, multiple docking adapter, and the Apollo Telescope mount. Despite some initial problems, which began with the loss of a meteoroid shield just 63 seconds after liftoff, the *Skylab* soon became functional. The initial crew to visit the station launched on May 25, 1973, for a 28-day stay. Two subsequent crews launched on July 28 and November 16, 1973, for missions of 59 and 84 days, respectively. After the manned missions, NASA conducted engineering tests of the station from Earth to learn more about long-term space flight and about the initial problems the station faced. Unexpectedly high solar activity ultimately affected the station's orbiting altitude, and on July 11, 1979, the station succumbed to Earth's gravity and collided with the Earth's surface over portions of the Indian Ocean and western Australia.

The next major space station effort on the part of the Soviet Union came with *Mir,* meaning "peace" or "world." Launched on February 20, 1986, the 20-ton *Mir* is similar to the *Salyut* series of space stations, but uses some of the interior area previously occupied by experimental equipment to create small, private crew cabins and a gymnasium. The *Mir* also has four docking ports for the attachment of scientific and experimental modules. The first cosmonaut crew to *Mir* spent 125 days in space. The station extended that record with its second mission, which brought two cosmonauts for the station. One stayed on board for six months, while the other spent 326 days in orbit.

In June 1995 the *Mir* was involved in another milestone when the U.S. space shuttle *Atlantis* docked with the station. A string of U.S.-Russian missions followed, ending with the return in June 1998 of the last American mission to *Mir.*

The *Mir* experienced a variety of problems that grew in number and severity through the years, including fires in 1994 and 1997. The final manned mission was planned for spring 2000, followed by the station's demise.

Plans for a new space station became official in 1994 when the United States Congress approved the International Space Station. Hailed by NASA as "the largest international scientific and technological endeavor ever undertaken," the joint effort includes contributions from the United States, Russia, Canada, Japan, Brazil, and the 11 nations of the European Space Agency. In 1995 NASA estimated the station would be up and running by 1998. Due primarily to economic and political turmoil in the nations involved, assembly didn't begin until 1998. Estimates in late 1999 pushed back the completion date for the ISS to 2004. Upon completion, the station will measure about 100 meters long and weigh some 475 tons.

Impact

If humans are to one day explore the solar system, much less the universe, we must learn how to survive extended periods beyond Earth's natural life-support systems and in microgravity conditions. In 1971 the first manned space station, *Salyut 1,* opened the doors to long-term space travel by allowing its first visiting cosmonauts to live in space for more than three weeks. Although the mission ultimately ended in tragedy for the three cosmonauts, the station itself was a success.

With a scientific payload of more than 2,640 pounds, the station carried two telescopes to gather astronomical information, and a number of cameras for studies of the Earth and the cosmos. Primarily, the *Salyut* crew conducted medical experiments relating to the ability of the human body to survive the rigors of long-term space travel. The cosmonauts also adopted a physical regimen to counter the deteriorating effects of the microgravity environment on their bones and muscles.

In addition, cosmonauts on that first manned space station performed a variety of plant and animal experiments, such as the hatching of frog eggs into tadpoles, and the growth of insects and small plants.

Like the *Salyut* missions, the *Skylab* and *Mir* station missions had a focus on medical experimentation to determine the effects of weightlessness on the human body. American astronaut Jerry Linenger, M.D., summed up the impact of microgravity during a press conference after his return from a 1997 mission aboard the *Mir* space station. He explained that the bones begin losing calcium and mass in a weightless environment. "You're just floating in space. Medically it's worse than bed rest." He added that bone loss would pose a perpetual problem during long periods in space. The *Mir* program confirmed that bone loss in microgravity continues at a rate of 1.2 percent in the lower hip and spine per month. Linenger reported that he experienced about 12 percent hip/lower spine bone mass loss while aboard *Mir*, and after a year back on Earth he was still down by 3-4 percent despite a rigorous exercise routine.

In addition to the medical experiments, astronauts aboard *Skylab* took nearly 175,000 pictures of Earth and the Sun with its telescopes and specially designed cameras, providing information for a variety of solar studies and investigations of Earth resources and weather patterns. Its solar studies produced images of coronal holes and observations of the comet Kohoutek, which neared Earth during the third *Skylab* mission.

On the non-human, biological side, cosmonauts and astronauts aboard the *Mir* were able to grow plants from seeds and harvest the second-generation seeds. These techniques—essentially farming in space—are a necessity for long-term space flights that require the crew to grow their own food. Technologically, each success—and failure—aboard the *Mir* station provided abundant information about space station design and systems. For example, the many successful dockings of spacecraft with the station helped designers understand the intricacies of joining two space vehicles; the 1977 fire led to a modification of the station's software so that the ventilation system, which fanned the fire, could be shut down with one command.

The International Space Station is the next step in the space station endeavor. From the amount of interior area that astronauts and cosmonauts will need to live and work, to exercise equipment that will help them prevent microgravity-induced bone loss, to new fail-safe systems to protect the crew from the unexpected, the ISS will draw on the histories of the *Salyut*, *Skylab* and *Mir* to create a more advanced facility for long-term space habitation.

LESLIE A. MERTZ

Further Reading

Books

Bernards, Neal. *Mir Space Station*. Mankato, MN: Smart Apple Media, 1999.

Bizony, Piers. *Island in the Sky: Building the International Space Station*. London: Aurum Press, 1996.

Bond, Peter. *Heroes in Space*. New York: Basil Blackwell, 1987.

Clark, Phillip. *The Soviet Manned Space Program*. New York: Orion Books, 1988.

Dyson, Marianne J. *Space Station Science: Life in Free Fall*. New York: Scholastic, 1999.

Heppenheimer, T. A. *Countdown: A History of Space Flight*. New York: John Wiley & Sons, 1999.

Periodical Articles

Kernan, Michael. "The Space Race." *Smithsonian* 28 (August 1997): 22.

Mandate from Heaven: The Tomb of Qin Shi Huang

Overview

In 1974, while digging a well, Chinese peasant Yang Zhifa uncovered bronze projectile points and pottery shards. He had no idea that he had found one of the largest and most spectacular archaeological sites in the world. This treasure, known as the Tomb of Qin Shi Huang, has given researchers insight into the historical, political, philosophical, and artistic life of people in what is now known as China. It has also given us an opportunity to appreciate the achievements and legacy of Chinese culture.

Life-size terra-cotta soldiers in tomb. *(WolfgangKaehler/ Corbis. Reproduced by permission.)*

Background

In 1974, a terra-cotta clay army of some 7,000 life-sized figures, equipped with actual chariots and bronze weapons, was discovered near the tomb of the first Emperor of Qin. Created nearly 2,000 years ago to accompany the dead emperor on his celestial journey, they were buried in an underground vault near a subterranean palace containing the dragon-shaped sarcophagus of the Emperor Qin Shi Huang.

The layout of the burial area is an imitation of the emperor's Xinnyang palace. The tomb is situated under a 15-story earth mound, Mt. Li. Covering approximately 500 acres, two protective walls enclose the central part of the mausoleum. The emperor's tomb was originally built in the center of an enclosed "spirit city." It contained sacred stone tablets, inscribed towers, and prayer temples. All of these structures were part of the "inner city" within a walled square more than a quarter mile long on each side. Beyond lay an "outer city" guarded by a high, rectangular stone wall, 23 feet (7 m) wide at its base with watchtowers at the corners.

Today archaeologists have unearthed more than 2,000 of the clay warriors, dozens of bronze horses, and a cache of 30,000 bronze swords, spears, crossbows, and other weapons. In 1980, a ceremonial procession containing the oldest bronze chariots and horses were found. Although the practice of burying live people and animals had been abandoned during the Shang Dynasty (1700-1100 B.C.), it appears that Emperor Qin Shi Huang revived it symbolically.

Ranks of clay warriors were carefully arranged four abreast in long tunnels paved with tightly fitting green bricks and then covered with wooden planks and a layer of preserving clay. The Qin craftsmen gave each warrior a different hairstyle, and each of the faces are painted differently. The specificity of the uniforms, armor, and weapons, as well as the arrangement of the figures into distinct battle units, offers an unusual glimpse of the Qin Dynasty and people's lives at the time.

Each figure has been individually shaped from coiled clay. Once formed, the figures were fired, cooled, painted, and placed in position. The horses and charioteers were first modeled in clay, and then cast in bronze with hand-crafted overlays and painted. All of the faces have distinct characteristics, indicating that the artists had been ordered to model after live soldiers. In addition, bronze weapons found at the site, together with molded clay bricks and tiles, give insight into the physical environment. The material goods excavated from the tombs depict native beliefs regarding the afterlife, and the artistic and ornate objects speak of high sophistication and craftsmanship in Chinese culture at the time.

Excavation work on the site is moving ahead, however, there is no timetable for working on the actual tomb of the emperor. A Chinese historical record written a century after his death describes the tomb as filled with precious jewels and roofed with pearl replicas of the stars, sun, and moon. These same records suggest that the bodies of the tomb artisans and conscripted work force may have been buried alive to protect the secret of Qin Shi Huang sarcophagus.

Impact

By about 1500 B.C. the population of what is now known as China had grown to a substantial level and along with it the desire to consolidate the provinces into a unified territory. A pattern of warfare, slavery, and ideology was set into motion during the years prior to Qin, known as the period of "Spring and Autumn" Period and the "Warring States" Period. Though marked by civil strife and disunity, this epoch, also known as the Golden Age of China, is testimony to an unprecedented era of cultural richness and economic prosperity. This era had a profound political and philosophical impact on China for the next 2,000 years.

This period was marked by constant warfare as the feudal system weakened and crumbled. Confucianism and Taoism, two powerful systems of philosophy and thought, emerged during this time and contributed to political and social transformation. Confucius believed that the only way society could work was for each person to act according to prescribed relationships. To Confucius, the government functioned to sustain ethical values.

To the Taoist school of thought, life focused on the individual finding balance and harmony adjusted to the rhythm of the natural and supernatural world. The early rulers of China also believed that a mandate from Heaven guided social and political authority. This mandate determined that the emperor governed by divine right; if dethroned, he had lost divine favor. These philosophies had profound implications in the way the Qin empire developed.

Also during this time, commerce was stimulated through the introduction of coins. Iron came into general use, making irrigation and canal projects feasible. Enormous walls were built around cities and along the broad northern frontier. Warring regional lords competed in amassing strong and loyal armies and increasing economic production for a broader tax base.

In 246 B.C., a 13-year-old prince inherited the throne of the Qin Kingdom. The next 25 years of his life were spent in battle, which resulted in the unification of China. Ruling over this feudal empire, he proclaimed himself the first emperor of a unified China. He asserted that his dynasty would rule for 10,000 years, but it turned out to be the shortest in the imperial system. There would be a long line of imperial dynasties until the fall of the lineage in 1912.

Although Qin Shi Huang ruled with an iron hand, often with barbaric methods, he was a leader with considerable foresight and talent. Under his 14-year rule, sweeping changes made his dynasty a turning point in Chinese history. Forty provinces were organized into a unified empire. Each was administered by non-hereditary officials appointed by the emperor. A system of education was developed in order to fill the burgeoning civil service positions. A hierarchical chain of authority from provincial governors to the emperor was initiated.

Under his rule, coins and other forms of money, weights, and measures conformed to a standard. Wide highways were constructed, all leading to the capital city, with bridges and lined with shade trees. These thorough fares allowed taxes to be collected and couriers to deliver messages to the emperor. A uniform system of writing was also developed.

His success in unifying the country is symbolized by the consolidation of the disparate, pre-existing segments of the northern frontier into the famed 1,500-mile (2,413 km) long Great Wall. Emperor Qin Shi Huang mobilized more than 500,000 soldiers and conscripted laborers to connect the wall segments into a single fortified defense line.

Several years before his death in 210 B.C., Emperor Qin Shi Huang provided himself with a massive tomb and clay army to protect him in the afterlife. The emperor had declared a Mandate from Heaven, whereby he declared himself a god as well as the Emperor of China. The construction of the tomb lasted for 40 years until the end of the Qin Dynasty. Nearly 750,000 laborers were conscripted to build the tomb and army, located near present-day Xian in Shaanxi province in the Yellow River valley. The ability of the Emperor to assemble this workforce speaks of his enormous political power and resources.

However, the measures taken by Qin Shi Huang were to prove too severe. The people of the eastern region found the restraint imposed upon them unacceptable. The huge work projects of building the Great Wall, the Imperial Palace, and the Emperor's Tomb exhausted the manpower and drained the treasury. To pay for the extravagances of the Emperor, taxes were raised and the people faced poverty or was forced to work at hard labor for the remainder of their lives. Furthermore, following advice given to him, the Emperor ordered books burned and libraries destroyed.

The first emperor failed in founding a lasting dynasty, however, a unified China continued, proving to be the world's most durable geopolitical unit. The remarkable systems of philosophy and thought, as well as the development of art, literature, medicine, and science, continue to be a reminder of the remarkable culture that evolved prior to that of the West. The unification of China over 2,200 years ago remains one of the most momentous events in the history of the world.

LESLIE HUTCHINSON

Further Reading

Periodical Articles

Carlson, Bobby. *National Geographic* vol. 153 (April 1978).

Hearn, Maxwell. *Smithsonian Magazine* (November 1979).

Other

Davis, B. K. "The Chinese Emperor's Eternal Armies." www.jadedragon.com/archives/feb98/emperors.html

"Qin ShiHuang's Terracotta Army." www.mc.maricopa.edu/anthro/asb_china/qin/slide1.html

The Unmanned Exploration of the Solar System: *Mariner, Viking, Pioneer,* and *Voyager*

Overview

On January 2, 1959, the Soviet Union launched *Luna 1,* the first manmade object designed to explore another celestial body. *Luna 1* passed about 3,600 miles above the moon's surface, performing some basic scientific observations before entering a solar orbit. Since then, space probes—primarily American and Soviet/Russian—have explored every major body in the solar system except Pluto. They've landed on the Moon, Mars, and Venus; flown through the erupting dust and gases of Halley's comet; mapped asteroids; and dropped probes into the atmospheres of Jupiter, Venus, and (soon) one of Saturn's moons, Titan. In the latter half of the twentieth century, our vision of the solar system changed from that of small, blurred dots in a telescope to a collection of real, unique planets. As William Burrows said in *Exploring Space,* "It was the interplanetary exploring machines . . . simultaneously drawing in the edges of the world and expanding them."

Background

Throughout history, mankind has looked toward the skies. Even before recorded history, people noticed that while most stars seemed fixed in the sky, some of the brightest ones moved. These were called *planets* (meaning wanderers) by the ancient Greeks. Powers were attributed to them, and elaborate mythologies arose to explain their origins and characters. Until just a few centuries ago, our understanding of the planets stayed at about that level of sophistication. Then Galileo's telescope began to put a face to the planets, while Copernicus, Kepler, and Newton explained their motions. But even in 1960 man's knowledge of the planets was limited to blurry telescopic images seen through a thick and dirty atmosphere from millions to billions of miles away. Although our theorizing grew ever-more elaborate, we had yet to see what lay outside—or inside, for that matter—Earth's orbit.

Our close-up exploration of the solar system began with the Soviet probe *Luna 1,* but that gave only fleeting views of our nearest neighbor. The first probe designed to explore beyond Earth's orbit was launched in July 1962. Unfortunately, instead of traveling to Venus, *Mariner 1* ended up at the bottom of the Atlantic Ocean. In December 1962 *Mariner 2* became the first probe to send back information from another planet, Venus. Throughout the decade the United States and Soviet Union sent probes to Mercury, Venus, and Mars, sending back photos and data from each of these planets, largely on fly-by missions that scooted past, never to return. But, by the end of the 1960s, the inner solar system was becoming known, if only in passing. This culminated with the landing of the *Viking* landers on the surface of Mars in 1976.

In 1972 the first missions were launched to explore the outer solar system. Sent as pathfinders to prove contemporary technology and navigation, *Pioneers 10* and *11* were launched a year apart to explore Jupiter and Saturn, and to pave the way for the follow-up Voyager missions. The Pioneer spacecraft performed superbly, returning a tremendous amount of data. They were followed by the more sophisticated Voyager probes, which were able to visit all four major planets in a single "Grand Tour" of the solar system, a once-in-175-years opportunity. The *Galileo, Cassini,* and *Magellan* probes followed, each dedicated to studying a single planetary system for several years. At the same time, orbital observatories launched by NASA, Europe, and Japan helped expand our knowledge of the rest of the universe even more dramatically; other missions visited comets, asteroids, and observed the Sun.

View of Mars from *Viking I. (NASA. Reproduced by permission.)*

Key to this entire process was the development of technology that could send large space probes to any point in the solar system and return scientifically valuable data. These missions required powerful rockets, compact instruments, navigational software, onboard computers, electrical generators, and communications equipment that could function for decades over billions of miles. Even missions closer to home required care, as evidenced by the loss of the billion-dollar *Observer.*

Unmanned exploration of the solar system has been largely successful, returning enormous amounts of information and tens of thousands of images. The success was hard-won, however. The unmanned space program continually fought for funding against the space shuttle and space station programs, and was nearly abandoned on a number of occasions. Despite the public acclaim that accompanied each new set of spectacular photos, the memory of failure lingered.

Impact

If scientific and technological advances made exploration of the solar system possible, space programs helped drive earthbound technology, too. The lack of abundant sunlight in the outer solar system led to the development of radioisotopic thermal generators for electrical power, new imaging techniques for photography, and ways to stabilize cameras while a spacecraft sped by a planet or moon at tens of thousands of miles per hour. Problems such as *Galileo*'s jammed high-gain antenna or *Voyager*'s flighty tape recorder forced engineers and scientists to devise clever diagnostic and repair procedures for a spacecraft several light-hours away. Weight and power limitations forced them to discover ways of transmitting data using no more power than is needed by a refrigerator lightbulb. Finally, although more prosaically, this data had to be plucked from background noise, stored, analyzed, and cataloged so that scientists decades later could locate, retrieve, and read it.

In addition to overcoming these technological obstacles, NASA had to surmount political and economic challenges as well. In the closing days of the Apollo program, the American public lost interest in space exploration as their attention turned toward the Vietnam War and domestic problems. NASA's funds were cut, forcing the cancellation of at least two planned lunar landings. To maintain public interest and government funding, NASA promoted their man-in-space program, the centerpiece of which was the space shuttle. It worked—both the public and legislature were intrigued. Congress may well have been attracted by the number of jobs the program would create; the fact that many of those jobs would be in vote-rich California appealed to both presidents and presidential aspirants alike.

Unfortunately, time and cost overruns began to tarnish the space shuttle's image and, by association, all of NASA. Unreliable schedules, cancelled launches, and the *Challenger* explosion all diminished NASA in the eyes of both Congress and the public. The subsequent loss of the *Observer,* early problems with the Hubble Space Telescope, loss of the Mars *Climate Orbiter,* and the *Galileo* antenna problems only added to the public's skepticism about NASA. Fortunately, the agency had successes as well: the Mars *Pathfinder,* the repaired Hubble Space Telescope and other orbital observatories, the *Magellan* mission to map Venus, and *Galileo* all performed extraordinarily well.

The public and Congress alike, however, seemed to take success for granted while castigating NASA for failures. At the same time, increased public and political pressure to balance the U.S. budget placed even more financial stress on NASA. As of this writing, NASA continues to operate under financial pressure, although the future seems a little brighter. The heady Apollo days of virtually unlimited budgets are unlikely to return, forcing NASA to continue to budget and consider carefully all space exploration for the foreseeable future. Their new motto: "faster, better, cheaper."

Other nations are also launching their own spacecraft. Despite efforts by the Japanese, Europeans, and Russians, however, the United States remains the only nation to send spacecraft beyond Mars—and the only one to have both the capability and plans to continue solar system exploration for the near future.

In spite of its problems, space exploration has captured public imagination and changed the perception of man's place in the universe more than any other enterprise in history. As we learn about our solar system and the universe, we develop a better appreciation for Earth's value and fragility. We've seen that the universe is awe inspiring, beautiful, vast—and largely empty. A photo of Earth rising beyond the desolate lunar surface taken by the crew of *Apollo 8* is a dramatic emphasis of this fact.

The discovery of other planetary systems suggests we might not be alone in the universe, although no planet hospitable to life has yet been found. Given this situation, we can continue to hope and dream, but we must still consider the possibility that Earth and its inhabitants represent the only outpost of life we will know. This realization may prove to be the most lasting legacy of space exploration.

P. ANDREW KARAM

NASA'S WORK ON ADVANCED SPACECRAFT PROPULSION SYSTEMS

Since Robert Goddard's work in the early part of the twentieth century, most rockets have been fueled by chemical reactions, burning either solid or liquid propellants. In the last few decades of the twentieth century, NASA began experimenting with alternate spacecraft propulsion systems, searching for smaller, lighter, faster, and more efficient ways to move spacecraft around the Solar System. These included solar sails, ion drives, and an electromagnetic drive that uses a large cloud of ionized gas to pull a spacecraft with it away from the Sun. Of these systems, the most efficient would be the solar sail. Constructed of a huge sheet of mylar or some other light plastic, a solar sail would be driven by the pressure of photons hitting it, driving it through the solar system. It is most efficient because, once beyond the atmosphere, it would sail forever, requiring no further fuel or propellant. Ion drives, already demonstrated on some small spacecraft, use electricity to generate a stream of ions, or charged atoms. These are driven from the ship in a stream, thrusting the craft in the opposite direction. Ion drives are more efficient than chemical rockets but still require a fuel supply. NASA's hope is that one of these drives, or something not yet conceived, might prove the key to opening the Solar System to manned exploration by making interplanetary trips faster and much less expensive. Until then, NASA is likely to continue developing and testing advanced propulsion systems on their unmanned spacecraft.

Further Reading

Burrows, William. *Exploring Space.* New York: Random House, 1990.

Chaikin, Andrew. *A Man in the Moon.* New York: Penguin Books, 1994.

Kippenhahn, Rudolf. *Bound to the Sun.* New York: WH Freeman and Company, 1990.

Miner, Ellis. *Uranus: The Planet, Rings, and Satellites.* New York: Ellis Horwood, Ltd., 1990.

Morrison, David. *Exploring Planetary Worlds.* New York: Scientific American Library, 1993.

Space Shuttles

Overview

At the United States National Aeronautics and Space Administration (NASA), the space shuttle program goes by the name Space Transportation System (STS). The shuttle program, as it is commonly referred to, has enabled humans to travel to space on reusable craft, providing transportation to and from space stations, and allowing scientists and the public to begin thinking about regular transportation to moons and planets. The shuttle was originally envisioned as a fully reusable craft that would be both efficient and economical; the latter benefit is still argued by some politicians and others. As designed, the space shuttle is semi-reusable. The shuttle basically comprises two solid rocket boosters, an external fuel tank and three main engines for launching purposes, and the reusable airplane-like orbiter spacecraft. In operation since 1981, the space shuttle has become the foundation of NASA's space program.

Background

In fall 1969, shortly after Neil Armstrong (1930-) became the first human to walk on the Moon, a special Space Task Group, appointed by the president, recommended a plan for the future of the U.S. space program. NASA had lobbied the group during the preceding months, and the plan included NASA's requests for the development of a reusable space shuttle. President Richard M. Nixon accepted the group's report, but did not put a high national priority on the space program.

More than two years after the task group report, following meetings between Nixon and NASA Administrator James C. Fletcher, an announcement was made unveiling plans to build "an entirely new type of space transportation system designed to help transform the space frontier of the 1970s into familiar territory, easily accessible for human endeavor in the 1980s and '90s." By 1975, NASA had demonstrated that a shuttle-like craft could return from orbit and land safely on a runway, a key part of the Space Transportation System.

Tests continued, and on February 18, 1977, the first space shuttle was attached to the top of a Boeing 747 ferrying aircraft at NASA's Dryden Flight Research Center in southern California for its first flight test. The 747 jet still serves as the ferrying vehicle for shuttles, transporting them when necessary from the landing runway to the launch pad between missions. Named *Enterprise* after the spacecraft in the fictional television series "Star Trek," the space shuttle made its maiden free-flight tests six months later. Those tests demonstrated the craft's ability to glide to a safe landing.

By spring 1981, the shuttle was ready for its first extended flight. On April 12, astronauts John W. Young and Robert L. Crippin took the space shuttle *Columbia* into orbit for a flight lasting more than two days, and landed the craft safely at Edwards Air Force Base in southern California. Upon landing, the *Columbia*—named after one of the first U.S. Navy ships to circumnavigate the globe—was heralded as the first airplane-like craft to return from an orbital mission for reuse.

The next in NASA's series of shuttles was the *Challenger,* which had its first mission from April 4-9, 1983. In this flight and previous *Columbia* flights, the shuttles deployed a variety of communications and other satellites. Astronauts Crippin and Young continued to fly shuttle missions. In June 1983 Crippin was joined by an astronaut team that included Sally K. Ride (1951-), who became the nation's first female astronaut. On November 28, 1983, Young was part of a *Columbia* crew that included the U.S. space program's first non-American astronaut, Ulf Merbold of West Germany. The flight also made history by carrying *Spacelab 1,* an onboard research laboratory that allowed astronauts to conduct scientific experiments while in orbit. Unlike the U.S. space station *Skylab,* which was also used for scientific experiments, *Spacelab 1* was not a space station since it was never deployed free of the orbiter.

Other "firsts" in the shuttle program included: the first black American astronaut, Guion S. Bluford (1942-), who flew aboard the *Challenger* after its August 30, 1983, launch; the first use of the Manned Maneuvering Unit (MMU), which allowed American astronauts to perform spacewalks without being tethered to the shuttle, in February 1984; the first "on-orbit" satellite repair mission, which occurred during the April 1984 *Challenger* mission; and the inaugural flights of the space shuttles *Discovery* and *At-*

lantis, launched on August 30, 1984, and August 8, 1985, respectively.

NASA's success with its shuttle program laid the groundwork for its decision to invite civilian school teacher Christa McAuliffe (1948-1986) to join its *Challenger* mission, which launched on January 28, 1986. The crew of seven also included Francis R. (Dick) Scobee, Michael J. Smith, Judith A. Resnik, Ronald E. McNair, Ellison S. Onizuka, and Gregory B. Jarvis. School children across the nation watched this highly anticipated launch live from their classrooms. However, the excitement gave way to disbelief and sorrow when the shuttle exploded just 73 seconds into the flight, killing all aboard.

Subsequent investigations concluded that the explosion was prompted by a faulty seal in one of the two solid rocket boosters. A review of photographic data showed a puff of smoke coming from the seal area. Several other puffs followed over the next two seconds. The smoke puffs indicated that hot propellant gases were burning and eroding the seal, including its rubber O-rings. A flame ignited on the solid rocket booster at 59 seconds into the flight, and five seconds later breached the external tank, which began to leak its liquid hydrogen fuel. At 73 seconds, various structural failures occurred, and the shuttle exploded into several sections that fell to Earth.

After the explosion, NASA ceased further shuttle missions and conducted internal investigations. These led to a redesign of the solid rocket booster seal, and several management changes, including the reappointment of Fletcher as NASA administrator (he had held the title from 1971-77) and the selection of astronaut Richard H. Truly to lead the shuttle program. Truly and Daniel C. Brandstein piloted the *Challenger* on the August 1983 flight mentioned above. In addition, NASA initiated an Office of Safety, Reliability, Maintainability, and Quality Assurance.

More than two years passed after the ill-fated *Challenger* flight before NASA launched another shuttle mission. On September 29, 1988, the *Discovery* marked the return of the shuttle program and the 26th shuttle mission.

Impact

The space shuttle program has provided a reliable transportation vehicle for not only the U.S. space program, but the programs of other countries. Although the program experienced a major catastrophe and setback with the 1986 explosion, the shuttle has continued to be the workhorse driving the U.S. space program. Since the first orbital shuttle flight in 1981, shuttles have flown dozens of missions and have had major direct and indirect impacts on scientific endeavors as well as political efforts.

One of the shuttle program's major milestones came in April 1990 when the *Discovery* carried and launched the Hubble Space Telescope (HST). This Earth-orbiting observatory, which revolves around the planet once every 93 minutes, is able to collect data unavailable to telescopes on Earth because the HST is located beyond Earth's atmospheric distortion. A flaw in the Hubble's optics prompted a December 1993 flight by the shuttle *Endeavour,* a new shuttle that made its inaugural flight in May 1992. The *Endeavour's* astronauts were able to repair the optics, perform routine maintenance work, and release the telescope back into its orbit.

As the ferrying, and later the servicing, craft for the telescope, the shuttles are indirectly responsible for the wide variety of spectacular images that have been made available by the Hubble Space Telescope. In the decade since its launch, the Hubble has provided images and data about black holes, the birth of stars, previously unseen galaxies, and planets in our own solar system.

The shuttle was also an important part of space history when its American crew was joined by a Russian cosmonaut in February 1994. That initial U.S. space mission with a Russian on board opened the doors to later U.S.-Russian missions, including the maiden link-up between the U.S. shuttle and the Russian space station *Mir* in the summer of 1995. Carrying two cosmonauts to replace the crew on *Mir,* the shuttle *Atlantis* docked with the space station. After several days of activities and experiments, the American crew, along with an American astronaut and two cosmonauts who had been stationed on the *Mir* since March, boarded *Atlantis* for a return voyage to Earth. Later missions to *Mir* ferried several astronauts to the station, including Shannon Lucid, the first American woman on the station. Lucid remained at the station for five months. United States involvement with the station ended in 1998, and the station is scheduled to be discontinued in early summer of 2000.

NASA points to numerous applications that have arisen from the myriad scientific research projects conducted during the shuttle-*Mir* missions. A cardio-muscular conditioner, for example, is a piece of exercise equipment designed hand-in-hand with experiments performed in a microgravity environment. The conditioner, which employs elastic cords and a kick plate, has potential uses for "athletes and heart disease patients by promoting cardiovascular fitness and muscular strength development," according to NASA's Shuttle-*Mir* web site. Other developments with potential Earth-bound applications are a heart rate monitor, a cardiac imaging system for use by cardiologists, and implantable and external pumps to help people with diabetes maintain insulin levels.

On the political side, the shuttle program has helped to cement a cooperative working arrangement between Russia and the United States. The shuttle program will also be a major part of the scheduled construction of an International Space Station. The station will involve cooperation between 15 countries in addition to the United States. Plans call for the shuttles to ferry component parts to the station at least through its completion, and later deliver food, water, and other supplies to station crews. With involvement in the International Space Station, NASA's shuttle program is likely to extend its role as the foundation of the U.S. space program and become a cornerstone of the combined world space program.

LESLIE A. MERTZ

Further Reading

Clark, Phillip. *The Soviet Manned Space Program.* New York: Orion Books, 1988.

Harland, David M. *The Space Shuttle: Roles, Missions, and Accomplishments.* New York: Wiley, 1998.

Heppenheimer, T. A. *Countdown: A History of Space Flight.* New York: John Wiley & Sons, 1999.

Richardson, Adele D. *Space Shuttle.* Mankato, MN: Smart Apple Media, 1999.

Remains of the RMS *Titanic* Discovered

Overview

In 1985 a joint French and American team found the submerged wreckage of the *Titanic,* the famed luxury liner that struck an iceberg and sank on April 14, 1912. More than 1,500 passengers and crew perished in the wreck. Using a revolutionary sonar vehicle system and a submersible camera-outfitted robot called *Argo* during the expedition, representatives of the U.S. Woods Hole Oceanographic Institution and the French Research Institute for the Exploration of the Sea (INFREMER) located the remains of the Titanic on September 1, 1985, after a 56-day search. The *Titanic* rested some 2.5 miles (4 km) beneath the ocean surface and about 350 miles (563 km) from the coast of Nova Scotia in the North Atlantic. Return trips have yielded ghostly images of rust-encrusted wreckage along with information about that starry night in 1912 when the vessel sank.

Background

> Oh, they built the ship *Titanic* to sail the ocean blue. / They thought they had a ship that the water would never go through, / But the good Lord raised his hand, and said the ship would never land. / It was sad when the great ship went down.

The words and tune to this childhood song vary slightly from place to place, but the story remains the same: The supposedly unsinkable R.M.S. (Royal Mail Ship) *Titanic* slipped beneath the sea, and hundreds of people on board died in the icy waters of the Atlantic Ocean on April 14, 1912.

The luxury liner set out on its maiden voyage only five days earlier. When compared with her contemporaries, the ship was huge, nearly 883 feet in length. The voyage went smoothly until the night of April 13th. On that night, a nearby ship, the *Californian,* radioed that it had come upon impassable ice to the north of the *Titanic*'s location. Following the transmission, the *Californian*'s crew shut down the radio for a period of time.

Aboard the *Titanic* about 40 minutes later, the lookout realized that the ship was heading straight for an iceberg just a quarter mile ahead. The first officer ordered the engines reversed and the wheel hard to the starboard. The evasive action was too late, and the iceberg grazed the *Titanic,* slicing a catastrophic 300-foot-long (91 m) gash into the ship's hull. The navigator quickly radioed a distress call along with the doomed ship's position, but the message went unnoticed by the *Californian*'s crew, whose radio was still turned off.

After midnight, the orders were given to abandon ship. Lax safety regulations allowed the *Titanic* to carry only enough lifeboats for about half of the passengers, and in the confusion many were launched before they were completely filled. Some of the crew went below to try to load additional passengers onto lowering lifeboats from openings on the sides of the ship. Still, many lifeboats floated away from the *Titanic* at less than full capacity. Although the *Titanic* had some 2,200 passengers and crew on board, its 20 lifeboats and rafts only carried 705 of them. Those 705 were the only survivors.

While the lifeboats were boarding and launching, the *Titanic* shot signal rockets into the air. Although some of the *Californian*'s crew saw them and reportedly passed along the sightings to the ship's captain, the *Californian* never responded to this plea for help.

At 2:20 a.m., the mighty *Titanic* sank. The lifeboats floated for another two hours before another liner, the *Carpathia,* arrived at the scene. The *Carpathia,* which had heard the *Titanic*'s first radio distress call, pulled the *Titanic*'s survivors to safety. In the end, more than 1,500 passengers and crew had died, and only about 300 bodies were recovered.

The tragedy and drama of the historical shipwreck caught the attention of Robert D. Ballard, a marine geologist at the Woods Hole Oceanographic Institution and head of its Deep Submergence Laboratory. He spent 13 years searching for the shipwreck before finding the *Titanic* during a joint expedition with INFREMER in 1985. Ballard and Jean-Louis Michel of INFREMER were the chief scientists of the expedition.

The scientists had a difficult time finding the *Titanic* for several reasons. The exact location of the *Titanic* was unknown. When the *Carpathia* began picking up survivors, the *Titanic* had already sunk and the lifeboats had likely floated a considerable distance from the shipwreck site. In addition, the sea in the area was more than two miles (3.2 km) deep, making impossible anything but dives with highly advanced equipment that could survive the great pressures of the depths.

In 1985 the American component of the joint expedition embarked aboard the U.S. Navy research vessel *Knorr* to scan a 150-square-mile (389 sq km) area in the north Atlantic Ocean for the shipwreck. The French team, which had arrived earlier, was aboard another research vessel, the *Le Suroit.* Combined, the crew used both sonar devices and remote TV cameras to view the ocean floor. Crew members aboard the *Knorr* rotated turns watching the images relayed from the cameras. Fifty-six days into the expedition—at 1 a.m. on September 1, 1985—the crew member watching the images stopped and said, "Wreckage," followed a second later by shouting, "Bingo!" The crew on watch erupted in cheers.

Impact

Both the search for the *Titanic* and its findings have important ramifications. They have demonstrated advancements in undersea exploration, while shining some light on the final moments of the *Titanic.*

The joint expedition had several key components. One was the *Argo,* a robot search vehicle that contained cameras, sonar devices, timing systems and other electronic equipment. The expedition was actually a testing ground for the *Argo,* which was developed in Ballard's lab. The Navy was supporting the work on *Argo* as a tool to find lost submarines and to perform deep-sea intelligence missions. On the *Titanic* expedition, however, *Argo*'s primary duty was to locate the shipwreck. The *Argo,* towed by the *Knorr,* skimmed the ocean floor, taking pictures, collecting data, and sending specimens up to the *Knorr.* On the surface, scientists watched and waited. It was the *Argo* that transmitted the first images of the *Titanic.*

The French team relied mostly on SAR, a sonar search vehicle that on every pass scanned a swath of ocean floor a half-mile (0.8 km) wide. As it turned out, the SAR had come within 900 feet (274 m) of the *Titanic* before the French ship left the expedition in early August, a month before the *Titanic* was discovered.

Following the discovery of the ocean liner's resting place, the joint U.S.-French team launched a remote camera capable of taking high-resolution, still photographs. Those photographs provided detailed images of the so-called debris field, which held the remains of the *Titanic.* The photographs showed torn and twisted pieces of the ship alongside scattered anchor chains, plates, and possibly intact bottles of wine.

Nearly a year after the discovery of the *Titanic,* Ballard returned to the shipwreck site. This time he brought *Alvin,* a manned, deep-sea submersible. *Alvin* includes a titanium crew compartment that is 7 feet (2.13 m) in diameter, and can hold up to three people and a variety of equipment. Ballard also brought the much smaller *Jason Jr.,* a robot submarine developed in his lab for the Navy. Unlike the *Argo, Alvin* and *Jason Jr.* are physically unconnected from the "mother ship" at the surface. *Jason Jr.* is an unmanned vehicle controlled by a scientist aboard *Alvin*. *Jason Jr.* can travel up to 200 feet (61 m) from the *Alvin* during operation. While scientists could get close to the *Titanic* in *Alvin*, the shipwreck's wires, railings, and other obstacles made those types of ventures dangerous. *Jason Jr.*, however, was small enough to negotiate around the obstacles and even enter the ship, and its single camera could relay pictures directly to the scientist controlling it.

During the *Alvin* and *Jason Jr.* missions, both experienced problems and taught the scientists more about deep-sea research. *Alvin* took its first trip to the *Titanic* without *Jason Jr.* After a 2.5-hour free fall through the water to the ocean floor—the free-fall descent saves energy for sea-bottom travel—*Alvin's* batteries developed leaks and its sonar failed. The *Alvin's* crew brought the submersible back to the surface for a night of repairs. The following day, *Alvin* worked and the deep-sea mission went smoothly until the final ascent to the ship. Attached to *Alvin* during the ascent, *Jason Jr.* came loose at the ocean surface, and divers had to jump in and save the little robot, which was no longer under *Alvin's* control. Again, the scientists and technicians had to make quick repairs. Subsequent dives with *Alvin* and *Jason Jr.* were successful and provided close-up and "aerial" views of the wreckage.

Many of the photographs from these expeditions to the *Titanic* provided clues to the ship's demise. For instance, photographs indicate that one of the ship's four giant stacks was ripped from its foundation and fell across the bridge as the ship sank. The collapsing stack also pulled the mast backward. Survivors reported seeing multimillionaire businessman John Jacob Astor for the last time standing approximately where the stack fell. Astor, who had purchased passage on the *Titanic* for himself and his new wife as part of their honeymoon, likely died there on the deck.

Other photographs of the shipwreck show that the entire debris field astern of the *Titanic* is filled with remnants of one area of the hull. From these photographs and other evidence, the research team concluded that the hull had been torn asunder. The photographs and data clearly show two major sections of the *Titanic* some 1,800 feet (549 m) apart, which supports survivor accounts that the ship split in half as it sank.

After analyzing the research team's evidence, Dr. Ballard concluded that these initial violent events at the surface were followed by a slow fall to the depths of the ocean, where the *Titanic* settled without much further damage. Other authorities had previously hypothesized that the *Titanic* underwent a 100-mile-per-hour (161 kph) collision with the sea floor.

RMS *Titanic* before its ill-fated voyage in 1912. *(Archive Photos/Popperfoto. Reproduced by permission.)*

In addition to providing insight into events on the night the *Titanic* sank, the expedition demonstrated that an unmanned submarine, and *Jason Jr.* in particular, could be a boon for deep-sea oceanographic research. Ballard believed that if control of these submarines could be extended well beyond the 200-foot (61-m) limit for *Jason Jr.*, eventually technicians would be able to remain aboard the research vessel and control the robot vehicles from there. Dangerous manned missions to the ocean depths, then, would become obsolete. In 1995 Ballard was part of the team that discovered the underwater wreckage of the aircraft carrier U.S.S. *Yorktown*. For that expedition, the team used a remotely operated submersible called the Advanced Tethered Vehicle to reach the wreckage on the floor of the north-central Pacific, some 16,650 feet (26,796 m) down.

In the 1990s, recovery plans led to a series of salvage missions to the *Titanic*. These missions were joint operations between IFREMER and R.M.S. Titanic Inc., which "owns" the wreck.

LESLIE A. MERTZ

Further Reading

Books

Ballard, Robert. *The Discovery of the Titanic.* New York: Warner Books, 1987.

Carter, J., and J. Hirschhorn. *Titanic Adventure.* New Jersey: New Horizon Press, 1999.

Periodical Articles

Ballard, Robert. "A Long Last Look at Titanic." *National Geographic* 170 (December 1986): 697-727.

Ballard, Robert. "How We Found the Titanic." *National Geographic* 168 (December 1985): 696-719.

Cook, William J. "Beneath the Waves, Traces of World War II." *U.S. News & World Report* (15 June 1998): 26.

Holden, C. "Americans and French Find the Titanic." *Science* (27 September 1985): 1368-9.

Kentley, Eric. "Diving to the Wreck." *USA Today* 123 (March 1995): 67.

Wilson-Smith, Anthony. "That Sinking Feeling." *Maclean's* (9 September 1996): 16-7.

Dick Rutan and Jeana Yeager Pilot the First Aircraft to Fly around the World Nonstop

Overview

In December 1986 two pilots, Dick Rutan (1939-) and Jeana Yeager (1952-), landed an odd-looking aircraft called *Voyager* in the California desert after making the first nonstop flight around the world without refueling. The *Voyager* pilots spent 9 days, 3 minutes, and 44 seconds aloft in a cabin the size of a phone booth. The 25,012-mile (40,244 km) flight was the last major milestone left in aviation and was the result of six years of work. Pilot Dick Rutan and his brother Burt, *Voyager's* designer, intended the plane and the round-the-world flight to usher in a new era in aviation that would take advantage of novel materials and designs.

Background

The concept of an extremely efficient plane capable of flying around the world began as a sketch on a restaurant napkin when Dick Rutan first proposed the idea to his brother. The original design was of a flying fuel tank that would accommodate as much fuel as possible. Burt determined from the original napkin drawing that a typical aluminum construction for the aircraft would be too big to build, so the decision was made to use composite resin materials that would be lighter and tougher than metal and more fuel efficient.

The $2 million *Voyager* was built by hand over 18 months in a hangar at Mojave Airport. The money to finance the project was raised from corporations in exchange for commercial endorsements by the pilots after the flight. Money was also raised through the sale of T-shirts and other merchandise promoting the project. Volunteers helped in the construction, and materials used to build the plane were donated by corporate sponsors. The composite material that was used in the plane was a combination of carbon graphite fibers in epoxy set over a layer of woven fiberglass honeycomb. This construction was 20% lighter than aluminum and seven times as tough.

The finished plane had a 110-foot (33.5 m) wingspan and two engines, one on each end of the main fuselage. The rear engine provided most of the propulsion while the front engine was used for extra energy in climbs. The flexible wings were designed to flop by as much as 30 feet (9 m).

The Rutans developed serious doubts about the *Voyager* when test flights showed that it became unstable and difficult to pilot when weighted down with fuel. They did not voice their concerns to the public or aviation officials for fear their doubts could hold back the project's progress. During 350 hours of tests, the aircraft suffered seven major failures. The *Voyager* never flew with a full load of fuel until it took off for its flight around the world in 1986. In a July 1986 test flight along the California coast *Voyager* flew 11,600 miles (18,664 km) in 111 hours to break the record for the longest flight without refueling on a course beginning and ending at the same point. That record was set by pilot Bill Stephenson in 1962 when he flew a U.S. Air Force B-52 11,337 miles (18,241 km).

Dick Rutan, who was 49 years old at the time of the flight, was a highly decorated Air Force officer who had flown over 300 missions during the Vietnam War. Jeana Yeager was an accomplished pilot who held several aviation

The *Voyager*, piloted by Dick Rutan and Jeana Yeager. *(The Library of Congress. Reproduced by permission.)*

records for flight distance and speed, some of which had been held by her co-pilot, Dick Rutan. Rutan and Yeager were romantically linked during the development phase of the *Voyager* project, but split up several months before they piloted the flight. The two pilots had different approaches to the mission. Rutan was reportedly determined to set a milestone by completing the flight, while Yeager said that she would have been satisfied at having broken ground in aviation, even if the flight around the world was not successful.

Voyager took off on its round-the-world flight from Edwards Air Force base early in the morning of December 14, 1986. The tips of *Voyager's* flexible wings were damaged during takeoff as the heavy load of the full fuel tanks made them drag along the runway. More than two feet (61 cm) of the composite skin covering the wings had been torn off at takeoff. After circling the airport a few times after takeoff the team shook off the damaged wing tips and decided the aircraft could continue with the flight as planned. It was not until *Voyager* landed nine days later that anyone could see the extent of the damage done to the wings at takeoff. The plane climbed slowly with its full load of 1,200 gallons (545 kg) of fuel weighing 9,500 pounds (4,313 kg).

The flight was helped by unseasonably favorable tailwinds all along the flight. Those tailwinds helped *Voyager* conserve fuel through deviations in their course and the added wind resistance from the damaged wing tips. *Voyager* cruised at an altitude between 7,000 (2,134 m) and 11,000 (2,253 m) most of the time, at one point getting as high as 20,000 (6,096 m) to avoid bad weather over Africa and another time falling 3,500 feet (1,067 m) over Mexico after the rear engine stopped running.

Other problems plagued the flight. On one occasion the pilots were sickened by oxygen deprivation at high altitude and two other times they were overcome by fumes from fuel that had spilled into the cockpit. Within the tiny cabin the noise from the two engines exceeded 100 decibels, louder than what would be heard at the first row of a hard rock concert. This required the pilots to use ear plugs and an electronic noise dampening device for the cabin. To sustain the pilots, the plane carried 90 pounds (41 kg) of drinking water. Rutan and Yeager ate prepared foods, which they warmed on a heating duct inside the cabin, and also consumed liquid meals.

The team had originally planned to make their flight through the southern hemisphere, but weather conditions forced them to stay north of the equator for the trip. A team of meteorologists on the ground guided *Voyager* to avoid storms and take advantage of good tailwinds through the flight. Rutan flew the *Voyager* for about 40 of the flight's first 48 hours. After those two days the pilots adopted the planned schedule to alternate flying and rest. 12,532 miles (20,164 km) into the flight, over the Indian Ocean, Rutan and Yeager broke the record for the longest un-refueled flight in a straight line that was set in 1962 by pilot Clyde Evely in a specially designed U.S. Air Force B-52.

While flying over Africa faulty gas gauge left the pilots wondering how much fuel was left in the *Voyager* Fuel level readings had to be double-checked by chase planes that calculated the weight of the fuel on board *Voyager* by watching how the plane performed aerial maneuvers. The tests determined that *Voyager* had enough fuel to return to California.

Towards the end of the flight the plane was caught in bad weather over South America, and at one point *Voyager* was flying sideways at a 90-degree angle bank. The main engine at the rear of the aircraft shut off after crossing Central America and approaching California from the Pacific Ocean, but Rutan, who was flying at the time, was able to start it again, and the team decided to keep both engines running for the rest of the flight.

The tailwinds that helped throughout the flight allowed the plane to arrive back in California a day early, and *Voyager* touched down at Edwards Air Force Base on December 23, 1986, after nine days, three minutes, and 44 seconds aloft. The average speed for the 25,012-mile (40,244 km) flight was 115.8 miles (186 km) per hour, relatively slow by modern aviation standards. The two pilots were lifted out of their tiny cabin and taken to the base hospital for examination after posing for photographers with *Voyager*. After landing *Voyager*'s tanks were found to have only 18 (8 kg) of the original 1,200 gallons (545 kg) of fuel left.

Impact

The aircraft now hangs in the Smithsonian's Air and Space Museum in Washington, D.C. *Voyager*'s flight marked the first time since the Second World War that a major aviation milestone was set by a non-military aircraft and with private funding. After the flight the two pilots, Rutan and Yeager, went their separate ways, though they appeared together to fulfill their commitment to endorse the companies that made contributions to the *Voyager* project. Rutan made an unsuccessful attempt to run for the House of Representatives, during which Yeager endorsed his opponent.

Rutan meant for the development of *Voyager*, as well as its flight, to inaugurate a new era in aviation. The flight demonstrated the possibilities of non-conventional design and construction to satisfy fields of aviation. Rutan compared the construction of *Voyager* to the revolution that occurred in the 1930s when aluminum replaced wood and canvas as the dominant construction material for airplanes. The applications of long-duration flight are now being explored in communications technology. Experiments with unmanned computer-controlled aircraft powered by solar energy are examining the possibilities of using high-altitude aircraft to replace space-based satellites in orbit. Such planes would maintain a constant circling pattern over their service area to relay telecommunication signals at less cost than orbiting satellites. An experimental version of these planes over Hawaii has set the record for the most time aloft of any aircraft ever.

A California company hopes to bring composite-built jets to consumers. It is conducting tests of a vertical takeoff and landing jet designed for general use. The company hopes someday it will be used with a computer-guided flight system to make the planes easy enough to fly so that they could replace the family car.

GEORGE SUAREZ

Further Reading

Books

Yeager, Jeana and Dick Rutan with Phil Patton. *Voyager.* New York : Knopf, 1987.

Norris, Jack. *Voyager: The World Flight: The Official Log, Flight Analysis and Narrative Explanation of the Record around the World Flight of the Voyager Aircraft.* Northridge, CA.: J. Norris, 1988.

The Legacy of Cave Paintings

Overview

What did early humans do with their time? Evidence discovered since the mid-nineteenth century in Western Europe suggests that they did a lot of drawing on the walls of caves. France and Spain have been the centers of an extraordinary number of cave painting discoveries. What is notable and significant about these cave paintings are the archaic depictions of animal life in pre-historic times, which our pre-human ancestors apparently encountered. While the first discoveries were made in the nineteenth century, the most spectacular discoveries were made in the last half of the twentieth century. The discovery of these cave paintings suggests to modern researchers a sophistication and artistic sensibility of prehistoric humanity, which was once unthinkable among anthropologists. This pre-historic art suggests that our ancestors were not only aware of their environment, but were probably very articulate.

Background

Marcellino de Sautuola, at Altamira, Spain, discovered the first significant cave paintings in 1875. The findings were so extraordinary and contrary to popular thought about early and pre-humans that most experts refused to believe they were Paleolithic. Later, around 1900, similar discoveries at Les Eyzies, France, were finally accepted and recognized as one of the most surprising and exciting archaeological discoveries of all time. A gradual succession of similar finds has continued throughout the twentieth century. Arguably the most famous of these was discovered in 1940 at Lascaux, France.

The Lascaux cave was discovered by four teenage boys in September 1940, and was first studied by the French archaeologist Henri-Edouard-Prosper Breuil. Some of the most compelling and informative cave paintings have been documented at this site. The layout and dimensions are notable in and of themselves. Consisting of a main cavern measuring approximately 66 feet (20 m) wide and 16 feet (4.9 m) high, there are many very steep galleries. All of the walls are amazingly decorated with engraved, drawn, and painted figures. Archaeologists have found some 600 painted and drawn animals and symbols, along with nearly 1,500 engravings.

The paintings appear to have been done on a light background in various shades of yellow, red, brown, and black. Among the most remarkable pictures are four huge (over 16 feet [4.9 m] long) aurochs (a now extinct species of wild ox). The paintings also depict a mysterious two-horned animal (misleadingly nicknamed the "unicorn"), which some researchers suspect was intended to depict a mythical creature. Several other species are scattered about the walls, including red deer, great herds of horses and bison, the heads and necks of several stags, which appear to be swimming across a river, a series of six big cats (possibly lions or panthers), and a rare narrative-like illustration. Dots, geometric motifs of unknown significance, accompany many of these animal figures.

The cave paintings are thought to date from about 20,000-15,000 B.C. Their vivid pigments have most likely been preserved by a natural process caused by rainwater seeping through the limestone rocks to produce a preservative-acting saturated bicarbonate. The colors appear to have been originally rubbed across the rock walls and ceilings with hard, sharpened lumps of dirt (probably yellow, red, and brown ochre). Outlines were drawn with black sticks of wood charcoal. The discovery of mixing dishes suggests that liquid pigment mixed with fat was also used and smeared on the walls by hand.

When the cave was discovered in 1940, the paintings were very well preserved. The stable levels of moisture and temperature within the cave provide an ideal environment for the preservation of pigments over thousands of years. After the cave was opened to the public in 1948, as many as 100,000 visitors a year came to view the famous paintings. The presence of so many people soon disturbed the delicate environment of the cave, and the paintings began to deteriorate. The colors faded and a green fungus grew over the pigments. The cave was closed to the public in 1963. A replica of the cave, known as Lascaux II, was constructed using the same pigments and methods believed to have been used by the original artists. Lascaux II opened in 1983 and now receives about 300,000 visitors a year.

In addition to the caves in Spain and France, there are signs that people were painting and creating art objects in other parts of the world at least 18,000 years ago. Some caves have been discovered in Australia and South Africa, and there seems to be a similar pattern to the designs and motifs used. There are many questions which researchers have been trying to answer with regards to this ancient art. One fundamental question is "why did human beings suddenly start to paint on cave walls?" Unfortunately, there is little intelligent speculation about this.

Impact

Since the discovery of these first caves in Spain and then subsequently in France, there has been a virtual flood of discoveries. It would appear that vast regions of these countries are rife with incredible caves, which were elaborately and enigmatically decorated by our distant ancestors. Lascaux was not to be a solitary phenomenon.

Others discoveries include the Chauvet cave and over 150 archaeological sites, including in the Eyzies-de-Tayac caves located in the Vézère Valley. These sites were collectively designated an UNESCO World Heritage site in 1979. As late as 1994, the Vallon-Pont-d'Arc cave in the Ardèche region of France was discovered to incorporate several very large galleries depicting a diverse menagerie, including rhinoceroses, felines, bears, owls, and mammoths.

As discoveries are made almost annually, one of the greatest contributions these prehistoric cave paintings have made to science has been the clear renderings of the wildlife that inhabited the plains of Europe. As a few of the more notable examples, the paintings of the horses found on the cave walls do not resemble any horse still remaining in the West. In fact, they more resemble a species of horse that is considered endangered and that is now only found in the high plains of Mongolia and Western China, the Przewalski's (per-zhe-vall-skeez) horse.

Impressively, the large, hairy, elephant-like animals known as the mammoth that lived during the last Ice Age, some 20,000-30,000 years ago and are now extinct, are rendered on the cave walls. This lends clear evidence that prehistoric man was very familiar with the mammoth.

Scientists have also discovered paintings of hyenas, rhinoceroses, lions, and panthers, which are now all extinct in the wild throughout Europe and only live in European zoos. One of the most startling discoveries was an entire hallway in the Lascaux Caves filled with paintings of bison. This clearly indicted that bison were once extremely common in Europe—this great species is now also extinct on the continent.

Archaeologists have theorized that the cave served over a long period of time as a center for the performance of hunting and magical rites—a theory supported by the depiction of a number of arrows and traps on or near the animals. In fact, several theories have developed over the years in an attempt to explain what these paintings might have meant to Paleolithic peoples. There are researchers that have suggested that the ancient artists created the art simply for art's sake. This supposes that they had the same attitudes about art as we do today. Did the artists decorate the caves simply to entertain themselves or an audience? Possibly, but the evidence researchers have discovered about living hunter-gatherer societies suggests that there were much more profound motives.

Other scientists have embraced theories of "sympathetic hunting magic." These theories suggest that, since most of the images are of animals, then the art was probably connected with the fertility of animals and successful hunting. The art is most certainly connected with the animal world, but the relationship between the animals and humans seems far more complex than merely one of hunting magic.

The latest theories attempting to hypothesize on the meaning of these paintings are based on a comparative study. Referred to as "symbolism theories," scientists are examining the symbolism of modern hunter-gatherers and their art and relating that to the art left by Paleolithic humans. When researchers began studying San art (the San were an African group who until the 1970s comprised one of the last hunter-gatherer societies in the world), it was generally interpreted as representing simple, schematic images of everyday San life. Recently, however, researchers have begun to understand that the images were not realistic, but instead were shamanistic art, which has a different kind of reality, the reality of another world.

In light of this evidence, however, it is still unlikely that we will ever know the true meaning of the images in the caves. Whatever the motivation of our ancestors, researchers will continue to investigate these marvelous relics.

LESLIE HUTCHINSON

Further Reading

Books

Chauvet, Jean-Marie, Christian Hillaire, and Eliette Brunel Deschamps. *Dawn of Art: The Chauvet Cave:*

The Oldest Known Paintings in the World. Harry N. Abrams Inc., 1996.

Clottes, Jean. *The Cave beneath the Sea: Paleolithic Images at Cosquer.* Harry N. Abrams Inc., 1996.

Clottes, Jean, David Lewis-Williams, and Sophie Hawkes (Translator). *The Shamans of Prehistory: Trance and Magic in the Painted Caves.* Harry N. Abrams Inc., 1998.

Conkey, Margaret W. *Beyond Art: Pleistocene Image and Symbol.* California Academy of Sciences, 1996.

Hinshaw, Dorothy. *Mystery of the Lascaux Cave.* Benchmark Books, 1998.

Lauber, Patricia. *Painter of the Caves.* National Geographic Society, 1998.

Perez-Seoane, Matilde M., Antonio B. Martinez, and Pedro A. Saura Ramos. *The Cave of Altamira.* Harry N. Abrams, Inc., 1999.

The Circumnavigation of the Earth by Balloon

Overview

On March 21, 1999, Swiss psychiatrist Bertrand Piccard and English balloon instructor Brian Jones became the first team to fly around the world by balloon, nonstop and without refueling, setting records for distance and duration, and winning a million-dollar purse staked by Anheuser-Busch. The three-week adventure, beginning in Switzerland and ending in Egypt, was an accomplishment in the history of exploration. Using new balloon designs and taking advantage of jet-stream developments, the team ended a 20-year quest and set a new milestone in the 200-year history of ballooning.

Orbiter III over the Egyptian desert. *(AFP/Corbis. Reproduced by permission.)*

Background

The history of ballooning began in 1783 with a 25-minute ride at 3,000 feet (914 meters) above the city of Paris. In the years that followed, balloons developed in size and structure, and greater distances of flight and altitudes were achieved. In 1978 a three-man crew aboard the *Double Eagle II,* became the first to cross the Atlantic by balloon, from Presque Isle, Maine, to Miserey, France. The quest to push the limit of the balloon had begun.

In 1980 a father and son piloted their balloon, the *Kitty Hawk,* from Fort Baker, California, to Quebec in four days, becoming the first to cross a continent by balloon. In 1981 four crew members aboard the *Double Eagle V* crossed the Pacific from Nagashima, Japan, to Covelo, California, a trip taking just over 84 hours. In 1984 a lone pilot in the *Rosie O'Grady* launched from Caribou, Maine, and crossed the Atlantic, landing in Savona, Italy. This pilot's second solo effort in 1997, aboard the *Solo Spirit,* set an endurance record of six days, two hours, and 44 minutes, as well as a distance record of 10,361 miles (16,674 km), travelling from St. Louis, Missouri, to Sultanpur, India. During a third effort in 1998, aboard the *Solo Spirit II,* the same pilot further increased the balloon distance record by travelling from Mendoza, Argentina, to a spot 500 miles (805 km) off the coast of Australia, where he was forced to land after his balloon ruptured in a thunderstorm. The total distance of this trip was 14,236 miles (22,911 km). These events, especially attempts at global circumnavigation during the last 20 years, set the stage for Jones and Piccard's record-setting flight in their balloon the *Breitling Orbiter 3.*

Two failed attempts by Piccard preceded the *Breitling Orbiter 3* voyage. In 1997, just six hours into the flight, Piccard was forced to land when a loose clip caused fuel to leak into the gondola. The following year, aboard the *Breitling Orbiter 2,* Piccard flew from Switzerland to Myanmar, where the Chinese government refused to allow the balloon into its airspace. In preparation for a third attempt, Swiss diplomats worked with the Chinese to make the trip possible.

New designs made the trip possible as well. The 10-ton *Breitling Orbiter 3* was designed to hold 15% more helium than its predecessor. *Orbiter 2* had used an experimental kerosene fuel that proved inefficient, prompting a switch to propane for *Orbiter 3.* Also, due to the large consumption of kerosene by *Orbiter 2* to maintain altitude, the balloon was redesigned with solar-powered fans to keep the helium balloon cool in the day. Solar panels, suspended below the gondola, charged five lead batteries and supplied energy for equipment on board. Propane burners provided the heated air needed to keep the balloon aloft; the burners were turned on at night. In case of an emergency, the lower part of the balloon was detachable, leaving the remaining upper half attached and functioning as a parachute. The gondola in which Piccard and Jones lived was equipped with a toilet, writing desks, sleeping bunks, satellite telephones, and a fax machine.

The balloon, with all of these improvements, was ready for the journey by the end of November 1998. Circumstances, however, were far from ideal. Iraq, possibly in the flight path, was again under NATO air attack. In addition, because a British balloon had recently drifted over forbidden parts of China, the Chinese government halted Jones and Piccard's flight plans. By the time Swiss diplomats had received permission for the team to fly over China, the optimal round-the-world-ballooning season was virtually over. The wait stretched over the winter months and finally on March 1, 1999, the team was ready to take off.

Impact

Piccard and Jones began their journey aboard the *Breitling 3* in the Swiss village of Chateau-d'Oex, ascending into the morning air amid the cheers of thousands of spectators. During that first day of flight Piccard and Jones had beautiful views of Mont Blanc and the Matterhorn. They enjoyed a meal of emu steaks, rice, and vegetables, reheated in plastic bags that evening. The two piloted the balloon in rotating eight-hour shifts; while one piloted, the other slept. The rest of the time was spent preparing food and planning the flight.

On the second day they passed Italy, then flew south and west over Mauritania to catch the jet stream eastward. By the fifth day they were over Libya. With the assistance of their flight team back in Switzerland, they raised and lowered their altitude to ride the optimum jet streams at the right speeds so that their path eastward was as controlled and on schedule as possible. This was be crucial, especially as they neared China, since the Chinese government had agreed to let the team fly through Chinese airspace, but only south of the 26th parallel. The *Breitling 3* drifted toward southern Egypt, over Sudan, Saudi Arabia, then further eastward over India and Bangladesh.

On the ninth day, Piccard and Jones determined that they were following the narrow course that would take them over the approved path designated by the Chinese government. Once over China, however, the *Breitling 3* drifted to within 25 miles (40 km) of the forbidden 26th parallel. The Chinese government radioed the team and asked them to prepare to land. Suddenly, a wind blew them away from this sensitive area—a close call indeed. The balloon sped on toward the next hurdle, the Pacific Ocean, a 10,000-mile (16,093 km) expanse of water.

Using computer models to predict when and where optimum jet streams would develop, the control center in Geneva instructed Piccard and Jones to let themselves be pushed to the equator, where a jet stream was expected to develop in a few days. One drawback of flying so close to the equator, though, was that the balloon's aluminum covering blocked the satellite signal being broadcast many miles directly above. As a result, Piccard and Jones temporarily lost contact with the control center in Geneva.

After six days of drifting lazily over the Pacific, the powerful jet stream that had been predicted finally materialized. The *Breitling 3* was propelled toward Mexico at 115 miles per hour (186 kph). Unfortunately, the team was at such a high altitude that the cold forced their propane burners to use more fuel than planned. Suddenly, their speed dropped, they somehow lost the jet stream, and began drifting southeast toward Venezuela.

Morale aboard the balloon fell. In a last effort to salvage the trip, Piccard and Jones decided to ascend as high as 35,000 feet (10,668 m) to catch a jet stream that the control center indicated was there. Indeed, after reaching that altitude the direction of flight corrected and speed eventually increased. After passing over Jamaica, the balloon was back on course. At this point, 17 days into the adventure, only four of the original 32 fuel tanks remained. Despite concern that they lacked enough fuel to cross the Atlantic Ocean, Piccard and Jones decided to press on.

Halfway across the Atlantic, the balloon was being carried in the center of a jet stream at 105 miles an hour (169 kph). By sunrise on March 20th, the team was only hours away from Mauritania, the point from which they began their journey eastward. They flew over Mauritania on March 21st, after a total of 19 days, 21 hours, and 47 minutes of flight. The *Breitling Orbiter 3* landed in Egypt, bringing an end to their journey. They had flown 26,050 miles (41,923 km).

The successful balloon circumnavigation by Piccard and Jones stands as an impressive technological accomplishment and an inspiring testament to human will. However, while they drifted high above the world and its problems, they had cause for serious reflection. The impact of their adventure is perhaps best expressed in the words of Piccard: "During our three-week flight, protected by our high-tech cocoon, we have flown over millions of people suffering on this Earth, which we were looking at with such admiration. Why are we so lucky? At this moment it occurs to me that we could use the largest portion of the Budweiser Cup million-dollar prize to create a humanitarian foundation, the Winds of Hope, to promote respect for man and nature."

MICHAEL T. YANCEY

Further Reading

Books

Piccard, Bertrand. *Around the World in 20 Days: The Story of Our History-Making Flight.* New York: John Wiley & Sons, 1999.

Periodicals

Nadya, Labi. "Around the World in 20 Days." *Time* 153, No. 12 (March 29, 1999): 44-7.

Piccard, Bertrand. "Around at Last!" *National Geographic* 196, No. 3 (September 1999): 30-51.

Scott, Phil. "The Balloon That Flew Round the World." *Scientific American* 281, No. 5 (November 1999): 111-13.

Future Space Exploration: New Research, Developments in Space Exploration, and the Search for Extraterrestrial Life

Overview

By the end of the twentieth century, space probes—some of them conducting long-term studies of planets and their satellite systems—had visited every planet in the solar system except Pluto. In addition to these exploratory missions, a number of large orbital observatories were launched to better study the cosmos. Most of these missions were expensive and took many years of preparation and construction. Their cost siphoned money and time away from other projects and, in the event of a failure, left no backup. In the 1990s NASA began to replace large, expensive missions that tried to do everything with smaller, cheaper, more limited missions that could be built and launched more quickly. This philosophy promises to return a great deal of information about our solar system. While these missions go out to explore the solar system directly, observatories promise to help us learn more about the origins of the universe, its ultimate fate, and about the existence of other solar systems or life on other worlds.

Background

The first artificial satellite, the Soviet Union's unmanned *Sputnik 1,* was launched into orbit in 1957. Later that year *Sputnik 2* had a canine passenger, Laika, the first living creature launched into space.

The first U.S. satellites, the *Explorer* and *Vanguard* probes, were launched in 1958, discovering (respectively) the Van Allen radiation belts that surround the Earth and the fact that the planet is more pear shaped than spherical.

Over the next few years, these launches were followed by other probes to the Moon, Venus, and Mars, most of which were launched by the Soviet Union. Many were unsuccessful and all were crude by current standards. The first stage of planetary exploration, however, had begun.

During the 1960s the unmanned exploration of space was part of the race to land men on the Moon, although efforts to gather direct information from Mars and Venus continued.

THE SETI PROJECT

Ever since radio astronomers tuned into the skies, scientists have listened for an elusive radio signal that would confirm the existence of extraterrestrial life. One of the major efforts in the last quarter of the twentieth century was a project termed the Search for Extraterrestrial Intelligence (SETI). On August 15, 1977, astronomer Jerry Ehman was going through the computer printouts of an earlier SETI-like project run by Ohio State University that was dubbed "Big Ear" when he discovered the reception of what remained throughout the twentieth century as the best candidate for a signal that might have as its origin an extraterrestrial intelligence. Excited, Ehman scribbled "WOW!" on the printout, and forever after the signal became known as the WOW! signal.

Despite repeated attempts to reacquire the signal, the fact that the signal was never again recorded makes many astronomers, including Ehman, skeptical about the origins of the WOW! signal. If it were an intentional signal, astronomers argue, the sending civilization would repeat it—or something like it—many times. Some scientists explain the WOW! signal as a mere reflection of a signal from Earth off of a satellite. Although government funding fluctuates, SETI projects continue, sometimes as private research efforts.

Some probes were designed to gather information about the interplanetary environment, such as radiation levels, dust concentrations, and other information that might affect future lunar-landing missions. Other probes attempted to land on the moon (some of which succeeded), to map it from orbit, or to gain information about its surface characteristics.

The golden age of big-budget planetary exploration occurred during the last three decades of the twentieth century, when several sophisticated and expensive space missions, many of which cost over one billion dollars, were launched. These missions began with an extended Mars orbital mission in 1971 and continued through the Pioneer missions to Jupiter and Saturn. Later operations included the *Voyager* flyby missions to the outer solar system, the *Magellan* mission to map Venus, the *Galileo* mission to Jupiter, the *Cassini* mission to Saturn, the Mars *Observer,* the Hubble Space Telescope, the Compton Gamma Ray Observatory, and more. In addition, the European Space Agency (ESA) launched a mission to Halley's Comet in 1986 and the Japanese launched lunar missions in 1990. Of these, the *Observer* was lost and the *Cassini* mission is still en route to Saturn as of this writing. The other missions (including the Hubble, despite of initial problems) have returned an incredible amount of information, increasing our knowledge of the solar system, our galaxy, and the rest of the universe.

At the same time, NASA and ESA began to launch sophisticated orbital observatories. These were designed to look into the universe in a variety of bands of light with unprecedented detail, and were supplemented by some Japanese satellites and a few European, non-ESA missions (such as the Dutch-Italian *Beppo-Sax* satellite). These missions looked at microwave, visible-light, x-ray, gamma-ray, ultraviolet, and infrared wavelengths; others measured the distances to nearby stars with incredible accuracy. Space-based observations were complemented by new generations of large and precise Earth-based radio and visible-light telescopes. Out of all these observations a more accurate picture of the solar system and universe began to emerge and the first extra-solar planetary systems were discovered. This spurred the search for extraterrestrial life, a search that took on a more serious cast when reputed nanofossils were discovered in a martian meteorite.

In the late 1990s, however, government funds began to tighten, forcing both NASA and ESA to scale back on the size and scope of their missions. The last large, multipurpose planetary mission launched in the twentieth century was the *Cassini* mission to study the Saturn system; the last to begin its work was the *Galileo* mission to Jupiter.

While these missions were being completed before launch, the first of the newer, smaller missions was built, the *Clementine* lunar probe. This was followed by the Mars *Pathfinder* and a bevy of follow-up martian expeditions. NASA also plans to send a fleet of smaller, cheaper, simpler craft to comets, asteroids, and other planets. The disadvantage of such craft is that they will each likely gather only a limited amount of information, so a single mission is unlikely ever to produce the flood of data experienced during the *Voyager, Galileo,* and *Magellan* missions. On the positive side, standardized craft built with "off-the-shelf" components are significantly less expensive to build and launch, making each one less indispensable. The loss of the *Observer,* for example, cost over $1 billion and resulted in a total loss of data. A smaller, more limited probe, however, would contain only a fraction of the total data. In other words, putting scientific eggs in a multitude of smaller baskets, reduces the effects of any single failure.

Impact

While the golden age of large-scale space exploration may have ended with *Cassini,* the golden age of data return may be yet to come, and its impact promises to be significant. Currently planned missions include the continuing study of Mars and a possible study of the jovian moon Europa. Unmanned missions to Mars will eventually bring back samples from its surface; these will be studied for signs of life and to learn more about martian geology. Such exploration would also shed light on the origins of life on Earth. In addition, martian exploration may help prepare for future manned missions to the red planet, assuming that both the money and technology necessary materialize. After Mars, the next-most-likely place to find life in the solar system is thought to be Europa, which may be covered by a global ocean capped with ice.

Other missions will try to locate extra-solar planetary systems that resemble our own solar system. Thus far, our techniques can only locate very large planets orbiting close to their stars, and those are unlikely to harbor any sort of life we are familiar with. Future missions should be able to detect earthlike planets in what are called "Goldilocks orbits," orbits not so close to the stars that the planets are boiled dry ,or so far from the star that it's ice -bound. The direct observation of planets may also be possible at some point. If this possibility comes to pass, scientists will look for the presence of free oxygen in the planetary atmospheres because, to the best of our knowledge, oxygen can only exist if there is plant life to produce it. The other major efforts in this area are the SETI (Search for Extra-Terrestrial Intelligence) projects, including the SETI at Home project that uses millions of home computers to sift through radio data, looking for signals of intelligent origin. The discovery of other intelligent life in the universe, even if we cannot communicate with it, would be one of the most significant events in our species' history. This discovery would also raise a host of ethical, philosophical, moral, and religious questions, many of which have already appeared in science fiction novels and stories. The answers to these questions will likely have await that event.

The discovery of even simple life forms (or evidence of past life) elsewhere in the solar system would be one of the single-most significant scientific discoveries ever made, eclipsed only by the discovery of extraterrestrial intelligent life.

Finally, continued multi-wavelength observation of the cosmos has begun to address questions of nearly equal significance. The COBE (Cosmic Background Explorer) satellite conclusively showed that the universe began with the Big Bang and data from the Hubble Space Telescope seems to show that the event happened between 12 and 15 billion years ago. Observations from *Beppo-Sax,* the Compton Gamma Ray Observatory, and other satellites are showing that gamma-ray bursts (flashes of high-energy radiation) are more frequent that had been suspected, and may have played an important role in the development of life on Earth and elsewhere in the universe. Other observations are suggesting that the universe may contain too little matter to be "closed," meaning it will continue to expand forever, eventually fading away as the last stars exhaust their fuel and burn out. Future observations by these satellites and planned observatories (such as the Next Generation Space Telescope), supplemented by earth-based observations, will continue to add scientific data to help answer these important questions about the universe and solar system in which we live and our place in them.

P. ANDREW KARAM

Further Reading

Burrows, William. *Exploring Space.* New York: Random House, 1990

Morrison, David. *Exploring Planetary Worlds.* New York: The Scientific American Library, 1993.

NASA Homepage. www.nasa.gov.

Biographical Sketches

Neil Alden Armstrong
1930-
American Astronaut, Test Pilot, and Engineer

Neil Armstrong enjoyed a distinguished career as a research test pilot before becoming a NASA astronaut in 1962. After leaving NASA he served on two Presidential commissions that helped to define the agency's future. He is best known, however, for leaving the first human footprints on another world.

Armstrong was born on a farm near Wapakoneta, Ohio, on August 5, 1930. In love with flying from boyhood, he wanted to study aeronautical engineering in college. A U. S. Navy scholarship gave him the money he needed, but a call to active duty interrupted his studies at Purdue University. Armstrong flew jet fighters for the navy from 1949 to 1952, flying 78 combat missions from the aircraft carrier *Essex* during the Korean War. After leaving the service he returned to his studies, receiving a B.S. in aeronautical engineering in 1955.

Neil Armstrong. *(NASA. Reproduced by permission.)*

Now both an engineer and an experienced jet pilot, Armstrong went to work for the National Advisory Committee on Aeronautics as an aeronautical research pilot. NACA, the forerunner of NASA, used jet fighters and experimental rocket planes to test new techniques for designing high performance aircraft. The experience Armstrong gained on such flights made him an ideal astronaut candidate. He joined the manned space program in 1962, one of the first two civilians to do so. He commanded the *Gemini 8* mission in March 1966 and, on the ground, played supporting roles in four other Gemini missions.

Armstrong's selection as commander of the *Apollo 11* mission—the first lunar landing—was not foreordained. It was the result of a complex, unpredictable set of events that shaped, and reshaped, Apollo crew assignments in 1967-69. Once selected, however, Armstrong proved to be an ideal choice. When the planned landing site on the Moon turned out to be strewn with large boulders, Armstrong coolly overrode onboard computers and manually guided the lunar module *Eagle* to a safer place. He landed it so gently that its shock absorbers barely compressed, with only seconds of fuel to spare. During his two-hour, 14-minute moonwalk on July 20, 1969, Armstrong adapted easily to work on the lunar surface. The data and samples that he and crewmate Buzz Aldrin (1930-) collected were modest by the standards of later missions, but they gave Earth-bound scientists a priceless first-hand glimpse of another world.

The landing of *Apollo 11* was more than a triumph of science and engineering: it was a public event. Millions followed its progress on television, eager for the chance to see one of the great events of the century unfold. Neil Armstrong's voice was, for millions who watched live and millions more who have since watched the films, the soundtrack for the landing. Knowing, perhaps, that this would be so, Armstrong rose to the occasion. His carefully low-key announcement of the landing let the event speak for itself: "Houston,Tranquility Base here, the *Eagle* has landed." His one-line speech after stepping into the lunar dust expressed a soaring sentiment in simple words. "That's one small step for [a] man, one giant leap for mankind." (The "a" was not heard by listeners due to the lack of sensitivity in Armstrong's microphone.) His running commentary during the moonwalk perfectly echoed the emotions that viewers on Earth felt: pride, exuberance, and endless curiosity.

Armstrong left NASA for business and university teaching in 1971, and he has remained the most private of private citizens since then. He did, however, serve on two presidential commissions in the mid-1980s. The first, the National Commission on Space, developed long-term goals for America's space program in 1984-85. The second, generally known as the Rogers Commission, investigated the explosion of the space shuttle *Challenger* in 1986. Neil Armstrong brought to both commissions what he had brought to the space program: razor-sharp intelligence, cool judgement, and an air of absolute dedication.

A. BOWDOIN VAN RIPER

Jacques Yves Cousteau
1910-1997

French Oceanographer and Documentary Filmmaker

Jacques Cousteau. *(Corbis-Bettmann.Reproduced by permission.)*

Traveling around the world aboard his ship the *Calypso,* oceanographer and filmmaker Jacques-Yves Cousteau made underwater exploration popular with his fascinating research and documentary films. He also collaborated on the invention of the first scuba (self-contained underwater breathing apparatus) device in 1943.

Born in 1910 in the village of St. Andre-de-Cubzac, France, Cousteau was the son of Daniel and Elizabeth Cousteau. His first seven years were plagued by ill health serious enough to prevent his participation in strenuous activities. As a result, he turned to reading and found pleasure in books—particularly those that dealt with life at sea and the exploits of treasure-hungry pirates.

A few years later, his father's employer suggested that an ongoing swimming program might strengthen the young boy's physique. This proved a good recommendation, especially when the family moved to New York in 1920 and Cousteau began swimming in earnest. He learned to hold his breath while diving and swimming in local lakes. When the family returned to France, Cousteau saved enough money to purchase a movie camera. He used it to produce numerous small films, each crediting him as producer, director, and cinematographer.

Cousteau went on to join the French navy, hoping to become an aviator. Three weeks before the final exam, he was in a serious automobile accident that nearly cost him the use of his arms. As part of his rehabilitation, Cousteau once again took to a heavy swimming program, this time in the Mediterranean. During his recovery, he began to experiment with devices that would permit underwater breathing for prolonged periods. Over the next few years, he worked with Frederic Dumas and Emil Gagnan to produce the Aqua-Lung, the first underwater enabler or scuba (self-contained underwater breathing apparatus) device.

When World War II was over, Cousteau persuaded his government to sponsor the Undersea Research Group. In 1950 he acquired an American minesweeper that he christened *Calypso,* after the Greek nymph who took Ulysses captive. Cousteau's *Calypso* would become more recognizable than any luxury liner or cruise ship on the high seas during its time.

Cousteau's fame as an oceanographer was nearly eclipsed by his secondary career as an independent film producer. His films were based on his prolific writings about his expeditions and adventures. His first 18-minute film was highly praised at the Cannes Film Festival in France. When his next film—based on his best-seller *The Silent World*—was shown at the Cannes Film Festival, it won the coveted Palme d'Or award. His popular television series, *The Undersea World of Jacques Cousteau,* was televised over an eight-year period by American Broadcasting Corporation, bringing him much fame and recognition.

Though Cousteau became famous worldwide for his scientific and marine biology research, he never held an actual degree in any field of science. Through his work in conserving the bounties and beauty of the oceans, he visited, lectured, and educated in all parts of the world. He received numerous international awards, medals, and honorary degrees. When he died in 1997 at the age of 87, he was honored by a memorial service at Notre Dame Cathedral in Paris. His final book, *Man, the Octopus, and the Orchid,* was published posthumously.

BROOK HALL

Melvin A. Fisher

1922-1998

American Treasure Hunter

Famed treasure hunter Mel Fisher caught the fever as a young boy when he read Robert Louis Stevenson's *Treasure Island*. His dreams of gold and other underwater riches led him to pursue a diving business and ultimately a career in historic underwater salvage. In 1985, his team made the discovery of a lifetime when they located the treasure cargo of the lost Spanish galleon *Atocha*.

On August 21, 1922, Mel Fisher was born in Hobart, Indiana, where, as a young boy, he dreamed of pirates and treasure after reading *Treasure Island*. At the time, news stories chronicled the adventures of deep-sea diving pioneers such as Dr. William Beebe (1877-1962) who were exploring the depths of the ocean using "hard hat" suits and bathyspheres. Inspired, eleven-year-old Mel invented his own dive helmet using a bucket, some hose, and a bicycle pump so, in the absence of nearby oceans, he could explore a mud-bottomed lagoon in his hometown.

After graduation from high school in Glen Park, Indiana, Fisher studied engineering at Purdue University. With the commencement of World War II, he joined the U.S. Army and trained with its Corps of Engineers. Before being shipped to Europe, Fisher pursued further engineering studies at the University of Alabama, which later awarded him an honorary doctorate. During his European tour with the Corps of Engineers, Fisher traveled in France and Germany making repairs to damaged structures and building his repertory of engineering skills.

By the end of World War II, developments in the sport and technology of diving prompted Fisher to open California's first dive shop on his family's chicken ranch in Torrance. His small operation offered dive lessons and equipment, including Fisher's modified snorkel gear. He supported his personal diving activities by making some of the first underwater movies, encouraging others to learn how to dive. In 1953, Fisher decided to sell the ranch and concentrate on his dive business. While negotiating with the potential buyers, he met their daughter, Dolores (known as "Deo"), whom he introduced to diving and quickly married. Their honeymoon was spent diving on shipwrecks in Florida and its Keys, while Fisher shot one of his best-known diving films, *The Other End of the Line*. Upon returning to California, the Fishers dove commercially for spiny lobsters while they built their own diving business, eventually opening Mel's Aqua Shop in Redondo Beach.

Early business for Mel's Aqua Shop included introducing scuba diving for gold in California's rivers, Fisher's first successful treasure hunting. The family explored the California coast and the Caribbean for shipwrecks, while Fisher continued to experiment with new equipment, developing new wet suits, underwater cameras, and spear guns. In 1962, treasure fever took hold of the Fishers, and the family abandoned its business in California to move to Florida's East Coast in search of the remains of the 1715 Spanish Plate Fleet—and to work for no pay while searching for treasure. Nearly a year went by without success when one of Fisher's salvage inventions, a device called the "mailbox," uncovered over a thousand gold coins, the first of several significant discoveries by the team.

By the late 1960s, Fisher's focus shifted to other treasures of the Spanish Main, and he learned about a Spanish galleon called the *Nuestra Senora de Atocha*, which had gone down in 1622 in the Florida Keys. The search for the *Atocha* consumed Fisher for nearly two decades. Along the way, tragedy struck the family when one of the salvage boats capsized in 1975 and his oldest son, Dirk, his daughter-in-law, Angel, and another diver were lost. There were successes, however, including the 1980 discovery of the *Atocha*'s sister ship, the *Santa Margarita*, which yielded more than 20 million dollars worth of gold and other riches. In 1985, the *Atocha* was finally located—its cargo represented the richest treasure found since the opening of King Tut's tomb in the 1930s. Estimated value put it at nearly half a billion dollars. More valuable to cultural archaeologists were *Atocha*'s thousands of historical artifacts, including rare seventeenth-century navigational instruments.

The monumental discoveries made by Mel Fisher and his team of divers, archaeologists, and salvage experts were only a small part of his legacy after his death in 1998 from cancer. He also left behind the Mel Fisher Maritime Heritage Society, founded in 1982, and the Mel Fisher Center, Inc., opened in the 1990s to provide conservation and exhibition facilities for many of the treasures located by his salvage expeditions. Society archaeologists continue to search the world's oceans for shipwrecks, extending Fisher's legacy by adding to the artifacts in the collection housed by the Society.

ANN T. MARSDEN

Sir Vivian Ernest Fuchs
1908-
English Arctic Explorer

British geologist and explorer Vivian Fuchs lead the first coast-to-coast land crossing of Antarctica in 1957. Fuchs's historic trek, undertaken in collaboration with a New Zealand team lead by Edmund Hillary (1919-), was both a personal and technological victory over the severe Antarctic environment.

Fuchs was born on the Isle of Wight in 1908. In 1929 he earned a master's degree from Cambridge University and took part in the Cambridge Greenland Expedition as a field geologist. This venture gave him the necessary credentials to participate in two field trips to east Africa the following year. His next assignment was to lead an expedition to the Lake Rudolf-Rift Valley, where he was to survey 40,000 square miles of the Ethiopia-Kenya portion of Africa. The work he did on this expedition earned him a Ph.D. in geology from Cambridge in 1935.

During World War II, Fuchs entered the British army as a second lieutenant and, by war's end, had risen to the rank of major. In 1947 he was selected to take charge of a survey project in the Falkland Islands. It was during this tour of duty that he first became seriously interested in Antarctica.

For many years, the North and South Poles presented a dual attraction for researchers and explorers—first, scientific interest in the physical magnetics of the poles; and second, the lure of historic immortality for discovering the geographic and magnetic points. The magnetic North Pole was discovered by James Clark Ross

Vivian Fuchs. *(UPI Bettmann Newsphotos. Reproduced by permission.)*

(1800-1862) in 1831, and Robert Peary (1856-1920) was the first to reach the geographic North Pole in 1909. The magnetic South Pole was located on Antarctica by Tannatt W. E. David (1858-1934) and Douglas Mawson (1882-1958) in 1909. And finally, after an unsuccessful try by Ernest Shackleton (1874-1922) in 1908, two teams reached the geographic South Pole within one month—Roald Amundsen (1872-1928) of Norway in December 1911, followed by Robert Falcon Scott (1868-1912) and his British team in January 1912.

Finding the poles was one phase of the explorers' dreams. The dramatic sequel was an overland crossing of the Antarctic continent. Earlier, in 1914, when Shackleton attempted an inland crossing, his ship, the *Endurance,* was crushed in the heavy ice of the Weddell Sea and the attempt failed. The plan to cross the deadly expanse of snow and ice was shelved for some years until Vivian Fuchs revived it in 1957.

Fuchs had an excellent collaborator who knew much about snow and ice—Sir Edmund Hillary, the world-famous explorer who had already left his mark at the top of Mount Everest in the Himalayas. With the aid of aerial penetration and tracked vehicles, Fuchs and the British Commonwealth Trans-Antarctica Expedition were able to accomplish the Antarctic crossing after so many had failed. They took off from the

Shackleton Base on the Filchner Ice Shelf on November 24, 1957, and made their way across the South Pole and on to New Zealand Scott Base. They arrived at Ross Island on March 2, 1958.

That same year, the British crown knighted Fuchs and named him director of the British Antarctic Survey, a position he held from 1958 until his retirement in 1975. Fuchs and Hillary also published a well-received book about their expedition, *The Crossing of Antarctica* (1958).

Sir Vivian Fuchs received additional honors, including election as president of the International Glaciological Society and the Royal Geographical Society (of which he is still honorary vice president). His later years were devoted to an autobiographical account of his adventures, *A Time to Speak,* published in 1990.

BROOK HALL

Yuri Alekseyevich Gagarin
1934-1968
Soviet Cosmonaut

Yuri Gagarin. *(Archive Photos. Reproduced by permission.)*

When Yuri Alekseyevich Gagarin was born in 1934, his parents had no way of knowing that he was destined for a unique place in world history: the first human to travel in space. His father, a carpenter by trade, left the collective farm where the family lived in Gzhatsk, about 100 miles from Moscow. Gagarin senior chose to join the Russian army when his not-yet-famous son was seven years old. He wanted to protect his family from the German invaders and insisted that his wife take Yuri and his two older siblings away from the battlefront. There was no place near Moscow where Yuri could be accepted in elementary school, so instead he enrolled in a trade school in nearby Lyubertsy. When he graduated in 1951, he had a degree in metal working but had become interested in aviation.

While he was continuing his education at an industrial school in Saratov, he began to take flying lessons. His teachers were quick to spot his unusual abilities, and he was soon moved to the cadet school in Orenburg, which would lead to a post in the Soviet air force. He completed his flight education in 1957 and immediately joined the air force, where other opportunities soon appeared for him.

The Soviets were deep into research and developments that would allow man to leave Earth and its gravitational force to explore space. When Gagarin became aware of this program, he was quick to volunteer as an astronaut (a word found formerly only in science-fiction writings). The following year, the space program chose Gagarin and 19 other pilots for advanced flight training. When money became scarce, the total number was reduced to six.

On April 11, 1961, the Soviet government announced to the waiting world that Yuri Gagarin had been selected to pilot their spacecraft—*Vostok I*—and be the first human to enter outer space on the following day.

Stories are told and repeated with increasing drama about how all of Russia stayed up that night and worried about their new "favorite son." However, it was reported that the future space traveler slept well and, when questioned about his ability to relax, he was said to have replied: "Would it be right to take off if I were not rested? It was my duty to sleep, so I slept."

At 9:07 a.m. on April 12, 1961, the *Vostok I* took flight. The launch site was in Baikonur, an isolated plain in the southern part of the Soviet Union. *Vostok* took 1 hour 29 minutes or 1 hour 48 minutes (conflicting records) to orbit Earth and reached a top speed of 17,000 mph (27,359 kph). At the peak of its orbit, its altitude was calculated to be 187 miles (301 km) above Earth.

The Russian plan was to eject Gagarin from the *Vostok* at 4 miles (6.4 km) from Earth's surface—a plan they kept secret for some years since an unmanned landing would not qualify for various world records. Gagarin did, in fact, make it safely to Earth and his subsequent immortality. He appeared on television, and his journey was chronicled in newspapers all over the world. The Soviets raised monuments to honor him, named streets in various cities after him, and, after his death, renamed the town of Gzhatsk after its most renowned citizen, Gagarin.

His celebrity did not deter Yuri from taking an active part in training future Soviet cosmonauts. In 1968 he was flying with another pilot in a two-seat jet aircraft on a routine training flight when the plane crashed and killed both men. His ashes were placed in a special niche in the Kremlin wall. His life—though short—will be remembered for its achievement. He inspired Americans (political and otherwise) to actively join in the exploration of space and its possibilities of interplanetary travel and other resources.

BROOK ELLEN HALL

John Glenn. *(NASA. Reproduced by permission.)*

John Herschel Glenn Jr.

1921-

American Astronaut

John Glenn achieved two astronautic milestones at different points in his career. In 1962 he was the first American to orbit around Earth, and 36 years later he was the oldest astronaut to fly on a space mission. The first accomplishment bolstered a fledgling space program, reassuring the American public and validating scientists' and engineers' innovations. Glenn's second success renewed Americans' flagging interest in the possibilities of space travel, especially as the National Aeronautics and Space Administration (NASA) budget was severely slashed. This flight also recognized that American demographics were reflecting a growing population of senior citizens.

Born on July 18, 1921, in Cambridge, Ohio, Glenn grew up in New Concord. His parents, John Hershel and Clara (Sproat) Glenn, emphasized values of community, patriotism, and duty. Glenn graduated from New Concord High School, which was later renamed John Glenn High School in his honor. In 1939 he enrolled at Muskingum College and studied chemistry. In addition to classes, Glenn took flying lessons and earned a private pilot's license. When the United States entered World War II, Glenn joined the Naval Aviation Cadet Program at the University of Iowa and then was commissioned in the Marine Corps. He married his childhood sweetheart, Anna Margaret Castor, in 1943.

As a member of Marine Fighter Squadron 155, Glenn flew 59 combat missions in the war zone around the Marshall Islands. After World War II, Glenn performed reconnaissance missions over northern China and Guam. During the Korean War, Glenn completed 90 combat missions and shot down three enemy jets. Glenn then trained at the Naval Test Pilot School before being assigned to the Navy Bureau of Aeronautics in Washington, D.C. He flew a jet on the first supersonic transcontinental trip from Los Angeles to New York City in three hours 23 minutes on July 16, 1957.

When the Soviet satellite *Sputnik* orbited Earth in 1957, the National Advisory Committee on Aeronautics (the predecessor of NASA) and the Navy focused on developing manned spacecraft, and Glenn volunteered as a test subject for the high-speed centrifuge to gauge the stresses humans might encounter in space. In April 1959 Glenn joined six other men to become America's first astronauts as part of the Mercury program. Although he hoped to become the first American in space, Glenn instead served as the backup astronaut for Alan Shepard and Gus Grissom when they underwent their suborbital flights. On February 20, 1962, Glenn was the first American to orbit Earth. His spacecraft, the

Friendship 7, completed three orbits in five hours. Glenn tested communications with tracking stations and system controls. Physicians had compiled detailed reports about his physical status and doctors assessed his vital signs during the flight.

Glenn became an international hero because television and radio coverage increased awareness of his mission. He received such honors as the Congressional Space Medal of Honor. Glenn assisted in the development of spacecraft for NASA's Apollo program, but President John F. Kennedy, Jr., ordered NASA officials not to assign Glenn to another flight because he considered Glenn a viable political candidate. Glenn retired from NASA in January 1964 and began campaigning. He won a Senate seat on his second attempt and served a total of four terms. He was a presidential candidate in 1983 but withdrew before the party's national convention.

Wanting to return to space, Glenn initiated talks with NASA administrator Daniel Goldin about conducting geriatric investigations aboard the space shuttle, suggesting that researchers could compare data collected on him in 1961 and 1998 and evaluate how space travel affects aging. Glenn devised experiments and NASA assigned him to a shuttle flight as a payload specialist. That mission flew in October 1998, reinforcing Glenn's role as an American icon, one who demonstrated technological prowess through mastery of space travel. Glenn published his autobiography, *John Glenn: A Memoir,* in 1999.

ELIZABETH D. SCHAFER

Walter William Herbert

1934-

English Explorer and Writer

Walter William Herbert, also known as Wally Herbert, led the first surface crossing of the Arctic Ocean from Alaska, via the North Pole, to Spitzbergen, Norway. In addition to that historic journey, which took place in the late 1960s, he has led and participated in numerous exploratory trips in the Arctic and Antarctic regions. Among these was the first circumnavigation of Greenland by dog sledge and skin boat, a feat undertaken during the late 1970s.

Herbert was born on October 24, 1934, the son of Walter and Helen Herbert, in York, England. Educated in Lichfield, Staffordshire, England, he served in the Royal Engineers of the British army from 1952 to 1955. After his military service ended, he embarked on the first of his many trips to the far reaches of the planet, as a surveyor on the Falkland Islands Dependencies Survey from 1955 to 1958.

In 1960 Herbert was part of an expedition to Lapland and Spitzbergen, and from 1960 to 1962 worked as a surveyor with the New Zealand Antarctic Expedition. He spent the years that followed planning and preparing for his trans-Arctic expedition, and from 1966 to 1967 tested equipment and techniques in northwestern Greenland. Then came his great Arctic trek, which Britain's Prince Philip—Herbert's patron—described as "among the greatest triumphs of human skill and endurance."

In 1969, after he returned to England, Herbert married Marie McGaughey, a writer, and they had two children. In 1971 he was back in the frozen north, leading an expedition to northwestern Greenland to make a film about polar Eskimos. This project lasted nearly two years; then, in 1975, he led an expedition to Lapland. During the following years, he prepared for his circumnavigation of Greenland by dog sledge and skin boat, which he completed in the period from 1977 to 1979. He also led several short expeditions to Greenland in the early 1980s.

Herbert published his first book, *A World of Men,* in 1969, and has produced a number of works since, including *Across the Top of the World* (1969), *The Last Great Journey on Earth* (1971), and the novel *A Noose of Laurels* (1989). Both a mountain range and a plateau in the Antarctic have been named for him, as has a range of mountains in the Arctic.

JUDSON KNIGHT

Sir Edmund Percival Hillary

1919-

New Zealand Mountaineer and Explorer

Born in Auckland, New Zealand, in 1919, this future world figure started his adult life as a beekeeper. His interest in climbing began in the New Zealand Alps and, in 1951, he joined a group that was headed for the Himalayan range and a reconnaissance of the south face of Mount Everest. In 1953 he and his Sherpa guide, Tenzing Norgay, became the first persons to reach the summit of this famed mountain.

Mount Everest had been attracting climbers since 1920, and in the subsequent 32 years seven major expeditions failed to reach the summit—

some individuals losing their lives along the way. The peak stands at 29,028 feet (8,848 m)—higher than any other solid form on Earth. (It is more than twice the height of Mt. Whitney in California, which is the highest point in the continental United States.) Everest rises on the border between Tibet and Nepal, and its summit is calculated as reaching two-thirds to the top of Earth's atmosphere, where oxygen levels become dangerously low. Absolutely nothing grows on the upper slopes of Everest. When scientists tried to discover what lay beneath the top layers of snow and ice, all they found was much more snow and ice. Auxiliary oxygen, which has since become standard practice for most climbers attempting Mt. Everest, was not an option at that time.

The 1953 ascent that brought enduring fame to Edmund Hillary and Tenzing Norgay (1914-1986), was actually under the command of Sir John Hunt, a British army officer with heavy experience in exploration and mountaineering. All who have studied the first successful climb have been quick to give much of the credit to Tenzing Norgay. He had been personally involved with assaults on Everest since 1935, and in the next few years took part in more Everest expeditions than any other person. He was at Hillary's side when they reached the summit on May, 29, 1953, and true to his Buddhist beliefs, he left an offering of food at the highest point of the mountain. He received many honors from the British government and collaborated with several authors in recounting his adventures.

Hillary's triumph was heralded around the world. It came on the eve of the British coronation of the young Queen Elizabeth II and gave the entire war-weary nation a double reason to celebrate. Upon his return to England, he was immediately knighted by the newly crowned queen.

Between 1955 and 1958 Hillary and Vivian Fuchs (1908-) led a hardy New Zealand group of adventurers on the British Commonwealth Trans-Antarctic Expedition. On January 4, 1958, they reached the South Pole (riding on a tractor). The pair collaborated on two books detailing the exploration: *The Crossing of Antarctica* (1958) and a sequel called *No Latitude for Error* (1961).

Hillary went on to scale Mount Herschel (10,941 ft.) on a follow-up expedition to Antarctica. It was the first time this peak had been targeted for a climb. Hillary opted for warmer weather on his next adventure, a 1977 jet boat excursion—the first such journey—up the religiously revered Ganges River in India. When the navigable portion had been exhausted, he got off the boat and climbed to the river's source high in the Himalayas.

Sir Edmund Hillary. *(UPI/Corbis-Bettmann. Reproduced by permission.)*

Along with his exciting and interesting life, Hillary took the time to form a Himalayan trust fund that funds humanitarian work among the Sherpas, building clinics, hospitals, and numerous schools. He became an environmentalist before the word existed and persuaded the Nepalese government to enact legislation protecting their forests and to set aside the area around Everest as a Nepalese National Park. His autobiography *Nothing Venture, Nothing Win* was published in 1975.

BROOK ELLEN HALL

Mae Carol Jemison
1956-
African-American Astronaut

Mae Jemison was the first black woman to fly in space. Her flight reinforced the inclusion of minorities as professionals in the National Aeronautics and Space Administration (NASA). Although Jemison was not the first American woman or African American assigned to a spaceflight, her mission represented the possibilities for a racially diverse astronaut corps to explore space. Since Jemison's selection as an astronaut, many female minorities have joined—as space travelers, scientists, and engineers—NASA's efforts to study the universe.

Born on October 17, 1956, in Decatur, Alabama, Jemison was born to Charlie and Dorothy Jemison. When Jemison was a toddler, the family moved to Chicago, Illinois. Jemison watched the television series *Star Trek,* admiring the character Lieutenant Uhura, who was a black female astronaut. Reading space books and having watched the televised lunar landing, Jemison planned to become an astronaut even though her teachers at Morgan Park High School tried to deter her. Jemison attended Stanford University, where she suffered from racism from professors who ignored or belittled her. Graduating with dual chemical engineering and African-American studies degrees in 1977, Jemison entered medical school at Cornell University and performed volunteer medical work in Thailand, Cuba, and Kenya between semesters. After earning her diploma, Jemison joined the Peace Corps as a medical officer in Sierra Leone and Liberia. During this time, she focused on improving vaccines for both hepatitis and rabies. In 1985 Jemison returned to America to work as a general practitioner for CIGNA Health Plans in Los Angeles, California, and also took graduate engineering courses.

At the same time, NASA was seeking qualified female astronauts because civil rights legislation in the early 1970s had forbidden federal agencies from discriminating on the basis of gender. By January 1978 NASA had selected six female astronauts. Jemison was encouraged by the acceptance of women and African Americans in the astronaut corps. She asked black astronaut Ron McNair for advice regarding astronaut evaluation policies. After not being picked on her first try, Jemison was selected by NASA as an astronaut candidate in June 1987. She was the fifth African-American astronaut chosen and the first black female.

Jemison devoted the next year to training and preparing for her flight. She flew on one shuttle mission that had initially been scheduled for launch in August 1988 but that had been postponed after the space shuttle *Challenger* exploded. In August 1992 Jemison joined crew members on the shuttle *Endeavour.* This flight, known as Spacelab J, was significant because it was the first cooperative mission with Japan. Because of her engineering and medical background, Jemison was designated to monitor such scientific experiments as chronicling how hornets, fish, and frogs behaved in microgravity. Jemison also investigated how body fluids shift into astronauts' chests while living in microgravity, and explored possible ways to move the fluid back to their legs before the shuttle landed. She showed how the Autogenic Feedback Training Vestibular Symptomatology Suit could assess astronauts' vital signs and enable wearers to initiate biofeedback in order to mitigate space sickness.

Mae Jemison. *(AP/World Wide Photos. Reproduced by permission.)*

When Jemison returned to Earth, she spoke to groups of children and adults, urging minorities to seek space-related careers because of the equal opportunities offered by NASA. Jemison resigned from NASA and began teaching at Dartmouth College, examining developing countries and space age technology. She created the Jemison Group in Houston, Texas, as a means to advance West African health care. Having retained her aerospace interests, Jemison played a guest role as Lieutenant Palmer on *Star Trek: The Next Generation* and wrote the afterword for Doris L. Rich's biography of aviator Bessie Coleman, entitled *Queen Bess,* which was published in 1993.

ELIZABETH D. SCHAFER

Brian Jones
1947-
English Balloonist

Brian Jones and Bertrand Piccard (1958-) became the first men to circumnavigate the globe in a balloon when they completed their around-the-world flight in the *Breitling Orbiter 3* on March 21, 1999. The two had set out from the Swiss Alps 20 days before, and except for perils over the Pacific Ocean, their flight had gone amazingly well.

Jones was born on March 27, 1947, in Bristol, England. He learned to fly at the age of 16, and served 13 years in the Royal Air Force (RAF). He first became involved in ballooning in 1986, when he was 39 years old, and soon acquired a commercial ballooning license. By 1989, he had been licensed as an instructor, and was later certified as an examiner for balloon flight licenses by the British Civil Aviation Authority.

His experience as a ballooning enthusiast put him into the same circles as Bertrand Piccard, a Swiss psychiatrist who came from a distinguished family of adventurers. In 1992, Piccard and Wim Verstraeten won the Chrysler Transatlantic Challenge, and when the two men decided to attempt an around-the-world flight, they chose Jones as their back-up pilot. As a result, Jones underwent all the training necessary for the flight, but was not aboard the *Breitling Orbiter* when it took off in January 1997. (Breitling, a Swiss watch company, sponsored the flight.)

Because of a fuel leak, Piccard and Verstraeten had to bail out, but in January 1998, they tried again on the *Breitling Orbiter 2*. This time Jones helped organize the flight, and assisted Alan Noble at mission control in Geneva, Switzerland, as Piccard, Verstraeten, and Andy Elson took off. Once again the attempt failed, due to a fuel leak and problems gaining permission to cross Chinese airspace.

As Piccard prepared for a third attempt on *Breitling Orbiter 3*, Jones initially signed on as project manager, in which capacity he was responsible for the construction of the gondola and flight systems on this ultra-high-tech, 180-foot-tall (55 m) craft. The gondola was 18 feet (5.5 m) long, and *National Geographic* (which published Piccard's account of their trip in its September 1999 issue) described it as "about the size of a minivan." The pressurized cabin had one bunk in which one pilot could sleep while the other flew, and up front was a computerized control panel, which allowed the pilot to operate the burners, switch propane tanks, and release empty ones.

As it turned out, Jones was Piccard's copilot, and the two took off at 9:05 a.m. local time on March 1, 1999, from the village of Château-d'Oex, Switzerland. Piccard later described their first meal aboard the craft as "emu steaks, rice, and vegetables, reheated in plastic bags in the kettle." Although their ultimate direction was easterly, their initial flight plan took them in a southwestward direction, to a point over the Sahara at which they began catching a jet stream that propelled them eastward.

Jones and Piccard crossed Libya at almost 90 miles (145 km) per hour, then headed southeastward to avoid no-fly zones over Egypt and Yemen on March 6. On the following day, they learned that Elson and Colin Prescott, who were also attempting an around-the-world balloon flight, had been forced to ditch their craft off Japan.

By March 11, Jones and Piccard had reached the eastern edge of the Asian continent, and began sailing over the Pacific Ocean. They encountered several problems in their six days over the Pacific, including a loss of contact with mission control in Geneva, as well as a broken heater that lowered the cabin temperature; but on March 16, they passed the previous distance record of 14,236 miles (22,906 km) set by balloonist Steve Fossett (1945-). On March 19, by now on the other side of the Americas headed toward Africa, they surpassed the duration record of 17 days, 17 hours, 41 minutes. The following day, March 20, marked the end of their global circumnavigation, as they crossed the point on the Sahara where they had turned east, but they did not land until March 21, when they set down in the Egyptian desert.

Following the completion of their historic flight, the two men were feted and honored around the world. Anheuser-Busch Companies, Inc. awarded them its Budweiser Cup and a million-dollar prize, most of which they donated to charity.

Jones is married and a father of two, with three grandchildren.

JUDSON KNIGHT

Dervla Murphy
1931-
Irish Writer and Traveler

Dervla Murphy is a fearless, female traveler who writes about her adventures with a straightforward punch. She travels simplistically by bicycle, by pony, or on foot, allowing her to truly connect with the people she encounters.

Murphy has been writing and traveling for over 30 years with no plans of stopping. Her first travel book, *Full Tilt*, describes her 1963 experience on a solo bicycle journey over 4,000 miles (6,436 km) from Dublin to New Delhi. She traveled with only one change of clothes, a toothbrush, and a .25 automatic pistol in her backpack. Luckily, she only had to use the gun on a pack of wild wolves she encountered while riding through Turkey.

Born in Lismore, County Waterford, Ireland, on November 28, 1931, Murphy was an only child who dreamed of far away places. Her father, the county librarian, brought home picture books and Murphy would plot journeys through each page. Murphy got her own atlas and bicycle at age 10 and mapped out her first journey to India. However this journey would have to wait. Four years later, she left boarding school to keep house for her father and take care of her invalid mother. Later, after both parents died, she prepared for her long bicycle ride to New Delhi, which began a life filled with incredible travels.

While in India, she worked with Tibetan refugees and wrote her second book, *The Waiting Land: A Spell in Nepal*. In 1968, her daughter, Rachel, was born. Married life did not agree with Murphy, but she readily took on the role as a single parent. After a five-year hiatus, Murphy returned to India with her daughter, who became her companion through many of her future adventures. Together they roamed through southern India, described in her 1976 book *On a Shoestring to Coorg*, trekked across 1,300 miles (2,092 km) of the Andes (*Eight Feet in the Andes*, 1983), and more recently crossed Cameroon on a horse (*Cameroon with Egbert*, 1990). In the later book, Rachel, who was 18 at the time, accompanied her mother by choice and traveled more as Murphy's adult companion than her daughter.

Although she has written of her many adventures in Madagascar, Nepal, and Africa, she has also written about less exotic destinations, some of which were politically important to Murphy. In *A Race to the Finish* (1982), Murphy describes the nuclear arms struggle in Northern Ireland. She also described accounts of the Handsworth riot in 1985 and the troubles of ethnic minorities living in Northern Ireland.

Her interest in these causes and others around the world gave Murphy another reason to travel and write about what she encountered. Since her early 20s she has been interested in the old South African apartheid regime. Visiting there on three separate occasions, twice before the 1994 election that brought about a democratic government, and once since, Murphy has given her readers a serious look at world travel. While traveling the second time through South Africa on bike, she wrote about her encounters with the serious disease of AIDS, or "ukimwi." Although her journey from Kenya to Zimbabwe was intended to be lighthearted, Murphy knew that "ukimwi" would follow her throughout her trip. Even at the age of 62, she bicycled 40 to 80 miles (64-129 km) a day. She also contracted malaria, starved, and was beaten up during this visit.

Dervla Murphy. *(John Reeves. Reproduced by permission.)*

Although she has survived many unpleasant experiences, Murphy is no stranger to fear. In 1969, when she was riding a mule through Ethiopia, four bandits robbed her and debated whether to kill her. In her latest travelogue, *One Foot in Laos*, both brakes on Murphy's bike gave out while riding down a steep track with hairpin turns and a 500-foot (152 m) drop. Ironically, although she is fearless on her treks, Murphy continues to have a phobia of spiders.

Murphy's books serve as a guide for other travel/adventure writers who look to her for inspiration. Riding alone on a bike through scenic landscapes and tucked-away villages allows her to get close to people. She is an honored guest in many villages, where she sleeps on the floors of huts, eats insect delicacies, and enjoys hearty portions of rat stew. She is interested in and fascinated with every detail of her journeys, which makes for excellent travel writing.

KATHERINE BATCHELOR

Bertrand Piccard

1958-

Swiss Psychiatrist and Balloonist

On March 21, 1999, Swiss psychiatrist Bertrand Piccard and British balloon instructor Brian Jones (1947-) became the first men to circumnavigate the globe in a balloon. Their voyage aboard *Breitling Orbiter 3*, a high-tech hot- air balloon, took 20 days.

Born in 1958, Piccard comes from a distinguished family of adventurers. His grandfather was Auguste Piccard (1884-1962), a Swiss physician whose inventions included the pressurized cabin. Aboard a pressurized balloon in 1931, Auguste Piccard and Paul Kipfer become the first men to enter the stratosphere. Jacques Piccard (1922-), son of Auguste and father of Bertrand, invented the bathyscaphe, a type of submarine made for extreme depths. Aboard the bathyscaphe *Trieste* in 1960, he and Donald Walsh descended to a record depth of 35,800 feet (10,916 meters) in the Marianas Trench of the Pacific Ocean.

Bertrand Piccard has pursued a dual career as a psychiatrist and adventurer in various forms of flight. At the age of 16 in 1974, he was one of the world's first notable hang- gliding enthusiasts, and he became the European aerobatics champion. He also invented several new aerobatic figures and set a world altitude record for hang-gliding. As an adult, Piccard was not only a hang-gliding instructor and paraglider, but a qualified ultra-light flyer as well.

Piccard made his first hot-air balloon flight in 1979. Later, with balloonist Wim Verstraeten, he won the Chrysler Transatlantic Challenge in 1992. In January 1997 Piccard made his first attempted around-the-world balloon flight, with Verstraeten aboard the *Breitling Orbiter.* (Breitling, a Swiss watch company, sponsored this and Piccard's two later flights.) A fuel leak forced them to bail out over the Mediterranean, just hours after beginning their flight. Again in January 1998 Piccard, Verstraeten, and Andy Elson got as far as Myanmar (Burma) in the *Breitling Orbiter 2* before a loss of fuel, combined with China's refusal to let them cross its air space, forced them down. They had, however, set a record for the longest amount of time aloft in a balloon: nine days, 18 hours.

Piccard and his companions were certainly not the only balloonists who had unsuccessfully attempted the circumnavigation of the globe. Starting in 1981, more than 20 crews had tried to circle the Earth, and just a week after Piccard and Jones began their historic flight, another crew failed: Elson, this time with Colin Prescot aboard the *Cable & Wireless,* had been plunged into the Pacific Ocean after a thunderstorm tore their balloon apart. (The two men survived.)

Piccard's and Jones's flight had already been delayed for several weeks as Swiss officials worked to obtain permission for them to cross China; finally, at 9:05 a.m. local time on March 1, 1999, they set sail from the village of Château-d'Oex, in the Swiss Alps. Initially they backtracked, flying in a southwesterly direction, though in fact their entire route had been carefully planned. Thus they sailed toward Morocco, then headed northeast to catch a jet stream, high-altitude winds that typically move from west to east.

Moving at a speed of 60 mph (97 kph), *Breitling Orbiter 3* and its two-man crew hastened toward India by way of the Red Sea and the southern tip of the Arabian Peninsula. On March 10 they spent 15 hours over Chinese airspace, crossing that nation in the extreme south due to an agreement whereby they would not stray north of the 26th parallel. Over the Pacific Ocean they encountered their most serious difficulties, when they lost contact with mission control for four days. Also during this time, a heater broke, dropping the cabin temperature to just 46 degrees Fahrenheit (8 degrees Celsius). But after crossing Central America, they caught a 100-mph (161 kph) jet stream over the Atlantic Ocean, which sped them toward Africa. On March 21, 1999, they landed safely in Egypt at 5:52 a.m. Greenwich Mean Time (GMT).

Piccard, who wrote about his adventure in the September 1999 issue of *National Geographic,* is married and has three daughters.

JUDSON KNIGHT

Jacques Ernest-Jean Piccard

1922-

Swiss Oceanic Engineer and Physicist

Jacques Piccard, born on July 28,1922, is an oceanic engineer and physicist who assisted his father, Auguste Piccard, in building the bathyscaph for deep-sea exploration. This submarine-like vessel allowed scientists to explore the deepest depths of the ocean and has provided information on ocean temperature, animal life, and useful geophysical information.

Jacques, at an early age, was introduced to both air and sea by both his father and uncle. Auguste and Jean-Felix Piccard were intensely interested in science and became fascinated with lighter-than-air balloons. In 1931, Auguste gained worldwide attention by making the first balloon ascent into the stratosphere at over 51,000 feet (15,545 m) to study cosmic rays from distant stars. He flew in a spherical gondola that could carry two people above 40,000 feet (12,192 m) without the need for pressurized suits.

Jacques spent most of his early childhood participating in the adventures and explorations of his family. In 1943, he studied physics at the University of Geneva, taking a year off in 1944-45 to serve in the French army. On completion of his studies, he taught at the University of Geneva for two years. During this time, his father became interested in undersea exploration and the design of a submersible capsule. They worked together designing the bathyscaphs, a deep-sea submarine vessel engineered to operate at great depths.

The first vessel, the *FNRS 2*, was designed and built in 1947, and was capable of submerging to 13,125 feet (4,000 m) and operated under water pressure of 0.42 metric ton/sq.cm. This first bathyscaph contained two main components: a steel cabin, heavier than water and resistant to sea pressure, to accommodate the observers and a light container called a float, filled with gasoline, which being lighter than water, provided the lifting power. On the surface, one or more of the ballast tanks were filled with air to keep the bathyscaph afloat. When the ballast tanks valves were opened, air escaped and was replaced by water, and as the vessel became increasingly heavy, it descended. The gasoline was in direct contact with the seawater and was compressed at a rate nearly identical with the depth of the water. The bathyscaph gradually lost buoyancy as it descended. To slow down or to begin the ascent, iron shot stored as ballast was released, which was held in place by electromagnets.

A second vessel, the *Trieste*, was reconfigured to withstand the demands of deeper descents. After several successful dives the United States Navy acquired the bathyscaph. The Navy equipped the craft with a new cabin designed to enable it to reach the seabed of the great oceanic trenches. These Grand Canyon-like trenches were little understood, and no one had ever ventured to the sea bottom of these underwater canyons.

Captained by Jacques and U.S. Navy officer John Walsh, they set a world record on January 23, 1960. The submarine-like craft descended 35,810 feet (10,915 m) to the bottom of the Mariana Trench, the deepest known point in the ocean southwest of Guam in the South Pacific Ocean. Jacques described this expedition in great detail for *National Geographic Magazine* in 1960. The expedition was successful and yielded previously unknown information about deep sea trenches as well as abundant data on how the bathyscaph handled at great depth.

The *Trieste* completed nearly 64 missions before being retired in the late 1960s. In 1963, Jacques and his father designed and built a mesoscaphe, capable of carrying 40 people for underwater observation.

LESLIE HUTCHINSON

Valery Vladimirovich Polyakov

1942-

Russian Cosmonaut

Dr. Valery Vladimirovich Polyakov—physician, researcher, and cosmonaut—holds the record for human longevity in space. In 1994-95 Polyakov spent 437 days and 17 hours in space, the longest uninterrupted time by an individual human being. His total space time (678.7 days) is second only to fellow cosmonaut Sergei Vasilyevich Avdeyev (1956-).

Polyakov was born April 27, 1942, in the town of Tula, located in the Russian Republic of the former Soviet Union. After completing high school, Polyakov went to Moscow, where he received a medical degree in 1965 from the I. M. Sechenov Medical Institute. He has taught and conducted research in leading Russian medical and biological institutes, as well as the Russian Mission Control Center in Moscow. In 1972 Polyakov was selected to become a cosmonaut and began a rigorous training program at the Yuri Gagarin Cosmonaut Training Center. At the same time he continued his work as a medical researcher, expanding his focus to the biological effects of space travel and emergency medicine.

Polyakov participated in two space flights, each of which involved long-term stays aboard *Mir,* the orbital space station launched by the Soviet Union on February 20, 1986. His main responsibilities were to conduct life science experiments and monitor the health of the other cosmonauts. Polyakov first went into space in 1988 as a research cosmonaut and flight engineer for the *Soyuz* TM-6 mission. Previously he had been a member of the backup crew for the *Soyuz* T-3 and *Soyuz* T-10 space flights in 1980 and 1984.

On August 29, 1988, Polyakov and two others were transported to *Mir,* where he remained for 240.94 days, returning April 27, 1989. While on board *Mir,* Polyakov joined with French astronaut Jean-Loup Chretien (1938-) to conduct 22 days of joint French-Soviet science and medical experiments. Polyakov also participated in the first successful Ham Radio transmissions, which enabled Soviet cosmonauts to make contacts with amateur radio operators all over the world.

During his second trip into space, Polyakov shattered the old space-flight record of 365 days set in 1988. This time he lived on *Mir* for 437.7 days, from January 8, 1994, to March 22, 1995. The aim of his extended stay in space was to investigate the effects of long-term weightlessness on the human body, particularly the loss of bone and muscle density. Polyakov directed experiments involving his fellow cosmonauts that measured physiological reactions to exercise and the effect of microgravity on vision. Memorable moments of his second mission included the first rendezvous of the U.S. Space Shuttle *Discovery* with *Mir,* which occurred in February 1995, and the arrival on March 14, 1995, of Norman Thagard (1943-), the first American to board the Russian space station.

Polyakov's capacity to endure a long period in space has significant implications for future aerospace projects, including manned voyages to Mars. Polyakov's successful experiments proved that humans could endure a two- or three-year mission to Mars, and be able to function on the planet after the long flight, estimated to be at least 160 days. Polyakov maintained a strict exercise regimen, which ranged from 90 minutes to three hours daily. At the conclusion of his mission, the 52-year-old cosmonaut was in excellent condition. Remarkably, rather than having to be carried from the ship as was often the case for those returning from space, upon landing Polyakov was able to walk with help to a chair a few feet from the spacecraft, and within one day he was jogging.

Polyakov is a published author and the recipient of numerous awards both at home and abroad. In 1989 he was made a Hero of the Soviet Union. In 1995 he was honored for his contribution to the field of aerospace by the *Aviation Week and Space Technology* annual Aerospace Laurels. In 1996 Polyakov was inducted into the International Space Hall of Fame.

Polyakov is married with two children and resides in Russia. Though no longer active in the Russian space program, he has been an ardent spokesman and supporter of the financially-strapped Russian space program and the space station *Mir,* as well as space travel in general.

ELAINE M. MACKINNON

Sally Kristen Ride
1951-
American Astronaut

Sally Ride was the first American woman to fly in space, her accomplishment symbolizing the equal opportunities for women offered by astronautical careers. Although a Soviet woman had orbited the earth two decades previously, Ride's mission represented the National Aeronautics and Space Administration's (NASA) commitment for all individuals to have access to space exploration as well as reinvigorated public enthusiasm for the space program. While the Soviet cosmonaut's pioneering flight was a publicity stunt to boost the Soviet Union's political prestige during the Cold War, Ride's achievement and vision of future goals in space represent the possibilities offered by the inclusion of a diverse group of space travelers to advance aerospace knowledge.

Born on May 26, 1951, at Los Angeles, California, Ride was born to Dale and Joyce (Anderson) Ride. Ride benefited from her education at Beverly Hills's Westlake School for Girls, where her scientific curiosity was nurtured. She went on to study astronomy, physics, and English at Stanford University, completing two bachelor's degrees in 1973. She continued with graduate studies at Stanford, earning a doctorate five years later. Ride planned to be a space researcher because she saw that American women were excluded from the astronaut corps at the time.

In the early 1970s civil rights legislation required federal agencies to implement equitable hiring policies. NASA accordingly initiated efforts to select female astronauts. Ride underwent interviews and physical examinations to assess her abilities, research interests, and potential before she was chosen as an astronaut candidate in January 1978 with five other women. Training as a mission specialist at the Lyndon B. Johnson Space Center in Houston, Texas, Ride practiced on simulators for every possible aspect of a shuttle flight. Ride assisted Canadian scientists in creating the Remote Manipulator System (RMS), which is a robotic arm used to maneuver space hardware. During the first shuttle mission in April 1981, Ride flew in the T-38 chase plane to photograph the landing spacecraft, and she was the first female capsule communicator at Houston's Mission Control, relaying information for the second and third shuttle flights.

Because of her consistent engineering ingenuity, even when under pressure, Ride was selected to become the first American woman in space. The shuttle on which she traveled was launched from Cape Canaveral on June 18, 1983. Ride observed experiments and used the RMS to capture a satellite to transport it to Earth for repairs. When she landed at Edwards Air Force Base a week later, Ride became a national hero. She also became the first American female astronaut assigned to a second shuttle mission, flying in October 1984. Chosen to participate in a third mission, Ride ceased training after the *Challenger* explosion and was the only active astronaut on the Rogers Commission, which investigated the accident. At that time, she stressed that astronauts should be hired for some NASA management positions in order to oversee safety considerations.

Ride then became a special assistant to NASA's administrator and wrote *Leadership and America's Future in Space*. Known as the "Ride Report," this document proposed future space exploration, especially "Mission to Planet Earth," which relied on spacecraft observations of Earth. Ride also wanted to build moon outposts, launch unmanned probes to explore the solar system, and initiate manned missions to Mars in an effort to regain America's space leadership and increase public awareness of the universe. Ride resigned from NASA in 1987 to conduct research at Stanford's Center for International Security and Arms Control before serving as director at the California Space Studies Institute. In 1999 Ride became president of Space.com, an internet-based company. Ride has received many awards and written several books about space as well as promoted opportunities for women in space research and exploration.

ELIZABETH D. SCHAFER

Sally Ride. *(NASA. Reproduced by permission.)*

Dick Rutan

1939-

American Pilot

While many Americans devote their lives to breaking existing records in their chosen fields, an elite circle goes about setting those remarkable firsts in unusual arenas. One of these history-making events bears the names Dick Rutan and Jeana Yeager (1952-): the first pilots to circumnavigate the globe without refueling. They accomplished this in December of 1986 in the now-famous aircraft *Voyager*, which was a privately funded endeavor, designed and built by enthusiastic volunteers and manufacturers who believed in the project.

Dick Rutan—the "favorite son" of Mojave, California—brought a wealth of experience and training to the flight. He had earned and received both his driver's and pilot's licenses on his 16th birthday and, when he was 19 years of age, was accepted in the USAF Aviation Cadet Program. Upon graduation from the Academy, he was commissioned with the rank of Lieutenant.

During his 20 years in the Air Force, Rutan flew 325 combat missions in Vietnam while he was a Tactical Air Command fighter pilot. While on a mission in September of 1968, he was forced to eject from his burning aircraft (an F-100) but was successful in evading capture until he was rescued by an army helicopter team. When Rutan retired from the USAF in 1978, he had earned the Silver Star, five Distinguished Flying Crosses, 16 Air Medals, and the Purple Heart.

While this illustrious career would be enough for most people, Rutan began new quests by joining his brother, Burt, in his Rutan Aircraft factory. As Production Manager and Chief Test Pilot, Dick Rutan set numerous speed/distance records in the company's LongEZ, a home-built craft which sold well. His flights received international recognition and earned him the Louis Bleriot Medal in Brussels, Belgium.

In 1981, Dick Rutan left his brother's company and formed Voyager Aircraft, Inc. Much of the proposed craft was designed by Burt Rutan, whose experience with lightweight but durable units was put to good use. The body was composed of layered pieces of fiber tape and a special paper which had been saturated with epoxy and glued together with an epoxy resin. The craft weighed only 2,500 pounds (1,135 kg) empty, with space for the two occupants only as big as a phone booth. The estimated fuel load (which was stored in the fuselage, wings, and other frame parts of the craft) brought the full weight up to 10,500 pounds (4,767 kg).

Early in the morning of December 14, 1986, Rutan and Yeager left the desert floor of Edwards Air Force Base, which is located 60 miles (96 km) northeast of Los Angeles. For the next nine days, the intrepid pair braved obstacles (like a 600-mph [965 kph] typhoon) by turning south to avoid the storm and continually monitoring their fuel supply. History notes that upon their return to Edwards nine days, three minutes, and 44 second later, there were only a few gallons of fuel remaining in the *Voyager*.

Since the famous flight, Rutan has traveled the world, speaking to various organizations. However, his appetite for adventure was still strong and he has since taken on new challenges for lengthy world flights. As a qualified and experienced balloon pilot and instructor, he has teamed up with Mike Melvill for additional quests. In 1997, they completed the Spirit of EAA Friendship World Tour in two small LongEZ craft which they built over 16 years before. The tour was called "Around the World in 80 Nights" and was well received wherever they appeared.

As of this writing, Rutan holds not only the *Voyager* flight record but 20 other world speed and distance records, three of which are absolute world records. The craft *Voyager* is proudly displayed at the Smithsonian Air and Space Museum's "Milestones of Flight" gallery.

BROOK ELLEN HALL

Alan Barlett Shepard Jr.

1923-1998

American Astronaut

A distinguished career as a U.S. Navy pilot and officer resulted in Alan Shepard's selec-

Alan Shepard. *(Archive Photos. Reproduced by permission.)*

tion for NASA's first class of astronaut candidates. When he was blasted into space in *Mercury 3*'s *Freedom 7* capsule on May 5, 1961, he became the first American in space, launching several decades of intensive NASA space exploration. In 1971, Shepard became the fifth (and oldest, at 47) man to walk on the moon. Before his death from cancer in 1998, his accomplishments as an astronaut were equaled by his success in business.

Born in East Derry, New Hampshire, on November 18, 1923, the son of a retired U.S. Army colonel and businessman, Alan Shepard grew up on a farm and went to school in a one-room schoolhouse. His boyhood hero was pioneering aviator Charles Lindbergh (1902 - 1974), the first to fly solo across the Atlantic, who inspired him to become a pilot. He attended high school in Derry at the Pinkerton Academy and completed a post-graduate year at Admiral Farragut Academy in anticipation of an education at the U.S. Naval Academy and an eventual career as a naval pilot.

In 1944, Shepard earned his B.S. from the U.S. Naval Academy. Upon graduation, he married Louise Brewer whom he met while at Annapolis. He began his distinguished naval career as an ensign on the destroyer *Cogswell* in the Pacific during the last year of World War II. Eager to become a pilot, he completed flight training and earned his wings in 1947 before serving several tours with the 42nd Fighter Squadron on aircraft carriers in the Mediterranean. In 1950, Shepard was admitted to the Navy Test Pilot School. He went on to become one of the Navy's top test pilots, completing two tours testing experimental fighters, conducting high-altitude tests, and perfecting landing techniques for the Navy's new angled-deck aircraft carriers. In 1958, Shepard graduated from additional studies at the Naval War College in Newport, Rhode Island.

The pivotal year in Shepard's career was 1959 when he was selected as one of the United States's original seven Mercury astronauts. On May 5, 1961, less than a month after the Russians launched the first human, Yuri Gagarin (1934-1968), into space, Shepard made his historic space flight aboard *Freedom 7*, the *Mercury 3* capsule. His short yet pioneering space flight, seen on live television by millions around the world, succeeded in laying the foundation for the United States to become a great innovator in space exploration and spurred President John F. Kennedy (1917-1963) to declare the United States's intention to land a man on the moon before the decade was over.

Ironically, it would be nearly a decade before Shepard would return to space, as he was grounded from 1963 to 1969 by an inner ear problem called Meuniere's Syndrome. Before surgery corrected the problem and he was once again cleared for space flight, he was made Chief of the Astronaut Office at NASA. Successful surgery put Shepard back on the active astronaut roster and on January 31, 1971, he was launched into space in command of the *Apollo 14* lunar mission. Shepard and fellow astronaut Edgar Mitchell (1930-) landed their lunar module *Antares* from which they explored the lunar surface, gathering soil and rock samples and deploying scientific equipment. Before departing the Moon's surface, Shepard also conducted a memorable, non-scientific test of hitting two golf balls—one chipped into a crater and the other a long drive. (It was said that as many people remembered Shepard's golfing on the moon as those who recalled Neil Armstrong's [1930-] first steps on its surface.)

In 1974, Shepard retired from both NASA and the U.S. Navy, where he had risen to the rank of Rear Admiral (and logged over 8,000 hours of flight time, including 4,000 hours in jet aircraft). A millionaire before his retirement, Shepard chose to devote more time to his business activities, including his Seven Fourteen Enterprises, named for his *Freedom 7* and *Apollo 14* missions. In 1984, he and other surviving Mercury astronauts, along with the widow of astronaut Virgil (Gus) Grissom (1926-1967), founded the Mercury Seven Foundation to raise money for college science and engineering scholarships (the organization was renamed the Astronaut Scholarship Foundation in 1995). Shepard remained active in his business and philanthropic activities until his death from leukemia (diagnosed in 1996) on July 22, 1998.

ANN T. MARSDEN

Valentina Vladimirovna Tereshkova

1937-

Soviet Cosmonaut

Valentina Tereshkova was the first woman to fly in space. Although her flight was a publicity stunt staged by the Soviet Union in an effort to seize control of the space race during the Cold War, Tereshkova's flight benefited space exploration. Her orbits around Earth reinforced public awareness that space travel was possible for women as well as men. American women cited Tereshkova's successful mission as justification for the National Aeronautics and Space Administration (NASA) to include women in aerospace careers.

Born on March 6, 1937, at Masslenikovo in the Yaroslavl region of the Soviet Union, Tereshkova was born to Vladimir Aksenovich and Yelena Fedorovna Tereshkova. Tereshkova grew up on a collective farm, worked at a mill, and took correspondence classes from the Yaroslavl Technical School of Light Industry to master cotton-spinning technology. Tereshkova joined the Komsomol, the Young Communist League. She also became affiliated with the mill's parachute jump club, earning a certificate for being a proficient parachutist.

After cosmonaut Gherman Titov's flight in August 1961, Tereshkova wrote the Soviet space center, asking how to become a cosmonaut. Nikolai Kamanin, who was in charge of cosmonaut training, invited Tereshkova to travel to Moscow, where she passed interviews and medical tests. In March 1962 Tereshkova and four other women moved to the Yuri A. Gagarin Cosmonaut Training Center at Star City. The male cosmonauts were aloof toward the women, disdaining them for their lack of flying experience. Because the Soviet air force was the sole source of cosmonauts in the 1960s, the women were enlisted as privates then commissioned as junior lieutenants.

Tereshkova studied rocket technology, geophysics, and navigational theory. She learned to fly jets, was spun on a centrifuge, and placed in an isolation chamber to prepare for the solitude of a space capsule. Tereshkova rode in planes designed to produce temporary zero gravity conditions in order to experience the weightlessness of space. Soviet officials designed experiments to determine how females were affected by space travel. Soviet premier Nikita Khrushchev decided that Tereshkova should be the first woman in space because her family consisted of workers representative of the Soviet Union's political ethic.

On June 16, 1963, Tereshkova was launched into space in *Vostok 6*. Most cosmonaut officials considered her mission successful. Ground controllers, however, complained that Tereshkova fell asleep often and was difficult to awaken. Tereshkova completed 48 orbits in three days. Her call name in flight was "Chaika," which means "seagull." Images of Tereshkova in space were broadcast to television viewers in the Soviet Union. Also, her conversations with Valery Bykovsky, who was orbiting at the same time in *Vostok 5,* were transmitted by radio. On June 19, Tereshkova ejected from the capsule and touched down at Kazakh. The Soviets feted Tereshkova at a Red Square celebration. She spoke about her flight to audiences and traveled to foreign nations. Tereshkova married cosmonaut Andrian Nikolayev in November 1963 in a public ceremony. They had one daughter, Yelena, and divorced 19 years later. Citing Tereshkova's successful flight, Khrushchev criticized western countries for forbidding women the opportunity to fly in space.

From 1965 to 1966, Tereshkova trained for potential Vostok and Soyuz flights that were later canceled. She attended the Zhukovsky Air Force Engineering Academy and graduated in 1969, the same year that the female cosmonaut program ended. Tereshkova earned a technical sciences degree in 1976 and also received an honorary doctorate from the University of Edinburgh. She rose to the rank of major-general before retiring in March 1997. Tereshkova held Communist Party offices, was elected to the Congress of People's Deputies, and was honored with various awards. Some countries have produced stamps and coins bearing Tereshkova's likeness and a moon crater has been named for her. In addition, Tereshkova has penned an autobiography.

ELIZABETH D. SCHAFER

Valentina Tereshkova. *(The Library of Congress. Reproduced by permission.)*

Helen Thayer

1938-

American Arctic Adventurer

In 1988 Helen Thayer became the first woman explorer to plan and carry out an unsupported solo expedition to the magnetic North Pole. No one has successfully soloed to the North Pole by foot since her successful expedition.

Thayer was born in 1938 in Whangerei, New Zealand. Her parents were both amateur mountain climbers and instilled in her a lifelong love of outdoor adventure. At age nine she climbed her first mountain, Mount Egmont (8,258 feet) in New Zealand.

Thayer was educated in New Zealand and enjoyed an exciting career as a multitalented international athlete. As a discus thrower, she representing teams from New Zealand, Guatemala (where she lived for four years), and the United States. In 1975 she won the United States luge championship and represented the United States in Europe. Thayer also excelled as a Nordic ski racer and instructor, a kayak racer, a mountain climber, and climbing instructor. She has climbed some of the highest mountains in New Zealand, North America, Mexico, China, and the former Soviet Union.

In 1988, at the age of 50, Thayer decided to walk alone to the magnetic North Pole without the support of aircraft resupply, dog teams, or a snow mobile. She had dreamed of traveling to one of the world's poles since 1958, when Sir Edmund Hillary (1919-) traveled to the South Pole. She had always been fascinated by the barren islands, the hardy plant and animals adapted to one of the world's harshest climate, and the treacherous sea ice. She was especially fascinated by polar bears, of which the North Pole had the largest population in the world. This expedition would be the first of many trips conducted through her environmental education program for students throughout the United States.

The trip was to take approximately one month. Thayer would travel by foot or ski and carry all of her supplies on a sled that she would pull. There would be no resupply, and she would check in by radio once a day. Four months before departure, she traveled to Resolute Bay to test her equipment. The Inuit (local Eskimo people) felt that she needed to take a dog team for protection.

In March 1988 she returned to Resolute Bay to begin her expedition. Again, the local Inuit encouraged her to take a dog team. Just days before leaving, she decided to take a local dog on the trip as her companion. Charlie was a large black Husky, trained to warn villagers of polar bears near the Inuit village. Charlie's job was to walk at her side and protect her from polar bears. Thayer later credited Charlie for alerting her to polar bears at least nine times, and actually saving her life early in the expedition.

Thayer and Charlie departed on March 30, beginning a month-long 365-mile journey. Since her expedition was the only excursion to the magnetic North Pole in 1988, she had no advance warning of the treacherous ice conditions that she would encounter. Because the magnetic North Pole is constantly in motion, she traveled to the approximate area of the Pole. The actual location of the Pole is in the center of this large area and it wanders daily in an elliptical path and may fluctuate much as 50 miles (80 km.)

She traveled most of the expedition across the sea ice, photographing and taking notes as she went. She recounted her journey in a book, *Polar Dreams.* In 1988 Thayer and her husband established an environmental education program, Environmental Expeditions, for students in kindergarten through the 12th grade. Through Internet communications, specially designed curricula, and summer opportunities for older students, this program was designed to allow students to "travel" with her on expeditions to Antarctica, the Sahara Desert, and the Amazon rain forest.

LESLIE HUTCHINSON

Michael George Francis Ventris
1922-1956

British Architect, Archaeologist, and Cryptographer

Although educated as an architect, Michael Ventris won acclaim for solving one of the genuine mysteries of archaeology. An early fascination with languages and scripts was the basis for his success in his 1953 decipherment of *Linear B*, regarded as the greatest of all archaeological decipherments. Tragically, Ventris was killed in a 1956 car accident shortly before his only book, *Documents in Mycenaean Greek*, was published.

Born on July 12, 1922, Michael Ventris was the son of a British Army officer who served in India and a mother whose father was a native of Poland and had settled in England. As a young boy, his mother introduced him to archaeology through regular trips to London's British Museum and its objects of antiquity. Ventris began his schooling in Switzerland, where he became fluent in French and German. He had a talent for languages—even teaching himself Polish at age six. After school in Gstaad, Ventris attended the Stowe School where he majored in the classics and, in 1940, went on to the Architectural Association school in London. In March 1942, he married Lois Elizabeth Knox-Niven, a fellow architecture student, with whom he eventually had a son and daughter.

With the outbreak of World War II, Ventris's education was interrupted by a brief stint as a navigator in the Royal Air Force, but he received his diploma, with honors, in 1948. Ventris went on to work as an architect for Britain's Ministry of Education, where he designed new schools. In 1956, he was awarded the first *Architects' Journal* Research Fellowship; his research subject was "Information for the Architect."

Those who had seen Ventris's work as an architectural student had predicted a brilliant future for him as an architect, but it was his early interest in archaeology that brought him distinction. In 1936, Ventris was part of a school group that visited a London exhibition organized to mark the 50th anniversary of the British School of Archaeology at Athens. The group attended a lecture by Sir Arthur Evans (1851-1941) who detailed his discovery of the ancient Greek civilization of King Minos on the island of Crete and of the mysterious writings used by its people. It was during this lecture that Ventris learned about the supposed "Minoan" tablets of baked clay covered with inscriptions which could not be deciphered, and he decided to be the one to solve the puzzle of their cryptic writings, which Evans later termed *Linear B*.

Between 1940, when he published his first theory regarding the language of the clay tablets in the *American Journal of Archaeology*, and 1953, when he published his solution to Linear B in the *Journal of Hellenic Studies*, Ventris spent considerable time researching the possibilities of the puzzle of Linear B and the clay tablets, corresponding with scholars around the world, including fellow Linear B researcher John Chadwick (1920-1998), who eventually assisted him in finalizing his solution. Ventris's background in architecture may have given him a unique ability to see the language of Linear B at multiple layers. His eventual solution included an elaborate grid system showing the relationships of the symbols and writings on the tablets. Although initially regarded with skepticism, Ventris's solution was confirmed by the analysis of a new tablet, found in 1952.

In 1955, Ventris received the Order of the British Empire for services to Mycenaean paleography. He was also made an honorary research associate at University College, London, and received an honorary doctorate of philosophy from the University of Uppsala. On September 6, 1956, as his only book, *Documents of Mycenaean Greek*, which gave a detailed account of his search for the solution to Linear B, was about to be published, Ventris was killed when his car collided with a truck in the early hours of the morning.

ANN T. MARSDEN

Jeana Yeager
1952-
American Piolot

Jeana Yeager, with co-pilot Dick Rutan (1939-), piloted the experimental *Voyager* airplane they had developed on the first non-stop flight around the world. The flight was the result of a six-year effort that combined advanced aircraft construction and design with expert piloting to achieve a major milestone in aviation.

Jeana Yeager was born and raised in Fort Worth, Texas. Her interest in aviation began with the desire to fly helicopters. After learning to fly fixed-wing airplanes as a step towards gaining a license to fly helicopters, Yeager maintained an interest in aviation while working as a draftsperson. In California she found work on a private project to build a rocket for launch into space.

In 1980 she met Dick Rutan and went on to set new aviation records for speed and distance in Rutan's airplanes. The following year they formed the Voyager Corporation to design and build an airplane that would fly around the world without landing or refueling. The two started with a design drawn on a restaurant napkin by Rutan's brother Burt, an aircraft designer.

Yeager was the quiet half of the pair, in contrast to the wild and gregarious Rutan. The two were romantically involved during the development of the *Voyager* project, but split up before the flight. After a series of test flights, one of which set a new record for distance flight without refueling, the stability of the aircraft was still in doubt. The *Voyager* team kept their reservations about the plane's stability to themselves to keep aviation officials from delaying the flight.

Voyager took off on its nine-day flight from an airfield in California's Mojave Desert on December 14, 1986, with Rutan and Yeager sharing a cabin the size of a phone booth. The plane headed out over the Pacific Ocean while a team of meteorologists on the ground guided its course. Rutan did most of the piloting during the first two days of the flight before the two pilots began to follow a schedule to allow one pilot to rest while the other flew.

The pilots encountered various problems during the long flight. Noise from the two engines at the front and rear of the fuselage required the pilots to wear earplugs during the flight. Yeager and Rutan experienced oxygen deprivation at high altitude and at other times were overcome by fumes from fuel that leaked into the tiny cabin. A faulty fuel gauge left the team wondering if there was enough fuel to fly back to California. A storm over South America tossed the hard-to-maneuver *Voyager* through the air. And an engine failed and had to be restarted as the plane approached California for its historic landing after 9 days, 3 minutes, and 44 seconds aloft.

After landing at Edwards Air Force Base Yeager received the Presidential Citizens Medal by President Ronald Reagan as well as the Gold Medal from the Royal Aero Club of Great Britain. She also became the first woman to receive the Robert J. Collier Trophy for achievement in aeronautics. After the flight Yeager and

Rutan continued to appear together to fulfill their endorsement obligations to the corporations that supported the six-year *Voyager* project and co-authored a book about the experience.

GEORGE SUAREZ

Biographical Mentions

Ben Abruzzo
1930-1985

American balloonist who took part in two historic ocean crossings. In 1978, along with Maxie Anderson and Larry Newman, Abruzzo flew from Maine to France aboard the *Double Eagle II* to make the first Atlantic crossing by ballon. In 1981 as captain of the helium-filled *Double Eagle V* he crash landed in California after making the first balloon flight across the Pacific Ocean with Newman and two other crew members.

Eugine Edwin Aldrin
1930-

American engineer and astronaut who walked in space during *Gemini 12* and explored the moon during *Apollo 11*. After a 1951 graduation from the U.S. Military Academy in West Point, "Buzz" Aldrin served in the U.S. Air Force, where he earned his pilot wings (1952) and flew combat missions during the Korean War. In 1963 NASA selected Aldrin for the third group of U.S. astronauts. In 1969 he and Neil Armstrong (1930-) became the first humans to set foot on the surface of the Moon.

Bryan Allen
1952-

American aviator who developed human-powered-flight techniques. A pilot and a bicyclist, Allen used pedals to power the *Gossamer Condor.* He won the first Kremer Prize in 1977. Allen next flew the *Gossamer Albatross* across the English Channel. His aircraft have also run on stored energy that he physically produces. He has set distance, duration, and speed records for human-powered aircraft and blimps. The *Gossamer Condor* is now displayed at the National Air and Space Museum.

Maxie Anderson
1934-1983

American balloonist who completed the first balloon crossing of the Atlantic—from Presque Isle, Maine, to Miserey, France—on August 13, 1978, flying 3,107 miles (5,000 km) in 137 hours. In 1980 Anderson and his son Kristian made the first crossing of the North American continent, flying a gas balloon from San Francisco to Sainte-Félicité, Quebec, Canada—2,823 miles (4,500 km) in 100 hours.

William A. Anders
1933-

American astronaut who was among the first to orbit the moon. Anders graduated from the U.S. Naval Academy and earned a master's degree in nuclear engineering from the Air Force Institute of Engineering. His studies focused the effect of radiation exposure on astronauts. In December 1968 he and his crewmates completed 10 lunar orbits on *Apollo 8*. Anders was also selected as backup pilot for *Apollo 11*. He later monitored space research for civilian organizations and was an aerospace industry executive.

Edward Latimer Beach
1918-

American naval officer and author whose experiences as a submarine commander were celebrated on the pages of best-selling fiction and non-fiction books. After graduating second in his U.S. Naval Academy class (1939), Beach was among the first to attend submarine school (1941). His first book, *Submarine!*, was published in 1952, and his novel *Run Silent, Run Deep* (1955) became a popular motion picture. In 1960 he commanded the nuclear submarine *Triton* on a record underwater circumnavigation of the globe.

Alan LaVern Bean
1932-

American astronaut who walked on the Moon and spent two months in space aboard the *Skylab* space station. Bean served in supporting roles on three Gemini missions and one Apollo mission before becoming the fourth person to walk on the Moon during the *Apollo 12* mission in November 1969. In 1973, as commander of the second crew to work aboard *Skylab,* he oversaw scientific experiments and crucial repairs to the station.

Charles F. Blair
1909-1978

American aviator who advanced navigational science. A mechanical engineer and U.S. Naval Flying School graduate, Blair was a military and commercial airline pilot, testing new aircraft and routes. He completed the first trans-Atlantic nonstop commercial flight in 1951, assessing the jet stream's effect on aircraft. Next, he became the first pilot to fly solo over the Arctic and North Pole. He devised and tested a polar navigational system and received numerous aviation awards acknowledging his achievements.

Guion S. Bluford
1942-

American astronaut who was the first African American in space. Bluford graduated from Pennsylvania State University and then served as a U.S. Air Force pilot in Vietnam. He earned an aerospace engineering doctorate from the Air Force Institute of Technology, and served as a mission specialist on a 1983 space shuttle flight. Two years later, he conducted scientific experiments aboard *Spacelab*. He flew twice again in the 1990s, deploying military and commercial satellites. After retiring from NASA, Bluford became an engineering executive.

Frank Borman
1928-

American astronaut who commanded *Apollo 8*, the first manned spacecraft to orbit the Moon. *Apollo 8* was the second manned flight of the *Apollo* spacecraft and the first to use the massive Saturn V booster rocket. The success of the December 1968 flight ended the Soviet-American "space race" by pre-empting a similar Soviet mission planned for early 1969, and it paved the way for the *Apollo 11* lunar landing eight months later.

Richard Branson
1950-

British entrepreneur and adventurer who with Per Lindstrand made the first transatlantic (1987) and transpacific (1991) hot-air balloon flights. Perhaps best known for his position as the founder and chairman of The Virgin Group of companies (founded in 1970), which include interests from music to an international airline, Branson is also well-known for his worldwide adventures breaking land, water, and air speed and distance records.

Richard Evelyn Byrd
1888-1957

American naval explorer and aviator who led five expeditions to Antarctica between 1929 and 1957. Several of these were privately funded; the remainder were official government enterprises. During his 1929 trip, he flew over the South Pole with three others, and he ultimately mapped extensive areas of the continent. All five of Byrd's Antarctic expeditions achieved vitally important geographic, climatic, and meteorological results. The inventor of vital devices used in his expeditions, including the aerial sextant and wind-drift instruments, Byrd also journeyed to the North Pole and with Floyd Bennett was the first to fly over the North Pole.

Eugene Cernan
1934-

American astronaut who left his spacecraft for more than two hours of extravehicular activity during the *Gemini 9* mission in 1966. Cernan joined the U.S. Navy in 1956, became a test pilot, and earned a master's degree in aeronautical engineering. In 1963 he was selected by NASA to serve as an astronaut. Cernan also participated in the *Apollo 10* and *Apollo 17* moon missions. He resigned from the navy and the space program in 1976 to enter private business.

Nicholas Clapp

American filmmaker credited with finding the lost Arabian city of Ubar. Clapp began researching ancient maps and literature for possible locations of the legendary city. Radar images of ancient caravan routes from NASA led Clapp and others to a site in Oman where they uncovered a large fortress. Greek and Roman artifacts found there indicated the site had been a major trading center. The team also found evidence of a sinkhole that may be the basis of the legend that Ubar was swallowed by the desert sands.

Eugenie Clark
1922-

American ichthyologist known for her research on poisonous fish and shark behavior. In 1949 the United States Office of Naval Research sent her to the South Seas to collect species of poisonous fish. From that experience she wrote *Lady with a Spear* (1953). In the early 1960s she turned her attention to shark behavior, the subject of *The Lady and the Sharks* (1969). In addition to her worldwide travels and studies, Clark led a series of deep-sea expeditions in the late-1980s and is an advocate for ocean conservation.

Barry Clifford
1946-

American deep-sea explorer and treasure hunter whose discovery of the pirate ship *Whydah* in 1984 signaled the beginning of a celebrated career of underwater salvage. (Over 200,000 artifacts were recovered from the wreck.) In 1987 alone Clifford helped discover seven seventeenth- and eighteenth-century shipwrecks in the Indian Ocean. During the late 1980s his team made significant discoveries in New York's East

River and Boston's inner harbor. Clifford also discovered up to 18 elaborate French warships and pirate vessels off Venezuela's coast (1998).

Michael Clifford
1952-

American astronaut who made the first spacewalk from a spacecraft docked to an operating space station. This spacewalk, six hours in duration, was conducted to mount experiment packages on the docking module of the Russian space station *Mir.* A former Army officer, Clifford also served as mission specialist on two other space shuttle flights and worked as a military liaison with NASA prior to selection as an astronaut.

Jacqueline Cochran
1910?-1980

American aviator who held numerous speed records and in 1935 became the first woman to enter—and win—the Bendix Transcontinental Air Race. Born around 1910 in Pensacola, Florida (training base for American navy pilots), Cochran was orphaned early and reared in poverty by foster parents. Following a series of menial jobs, she ultimately learned to fly in 1932 and married an industrialist/banker named Floyd Odlum. During World War II she was a captain in the British Air Force Auxiliary and headed the women pilots who ferried aircraft all over Europe. When the United States entered the conflict, Cochran was made director of the Women's Air Force Service Pilots. She eventually retired as a full colonel.

Eileen Marie Collins
1956-

American pilot and astronaut who accomplished many significant firsts for women involved in the space race. A colonel in the U.S. Air Force, Collins was the first female pilot selected by NASA (1990), the first female Space Shuttle pilot on STS-63 (1995), the pilot of the first Space Shuttle flight to dock with the Russian *Mir* space station on STS-84 (1997), as well as the first female Space Shuttle commander on STS-93 (1999).

Michael Collins
1930-

American astronaut who orbited the Moon aboard the *Apollo 11* command module *Columbia* during the first manned lunar landing in July 1969. Collins's skill at rendezvous and docking, developed on the *Gemini 10* mission in 1966, made it possible for moon-walkers Neil Armstrong and Buzz Aldrin to return safely to Earth. He went on to direct the National Air and Space Museum and to write a widely praised memoir, *Carrying the Fire* (1973).

Charles Conrad Jr.
1930-1999

American astronaut who walked on the Moon and commanded the first crew to live aboard the *Skylab* space station. Conrad's handling of the lunar module *Intrepid* during the *Apollo 12* mission of November 1969 showed that pinpoint lunar landings were possible—a prerequisite for the science-oriented missions that followed. His three-member crew spent a month aboard *Skylab* in 1973, carrying out scientific experiments and repairing damage that the space station had suffered during launch.

Leroy Gordon Cooper
1927-

American astronaut whose flights generated crucial data on the challenges of long-duration space missions. Cooper, one of the seven original NASA astronauts, orbited Earth for 34 hours on the last of the Project Mercury missions and for eight days as commander of the *Gemini 5* mission. Scheduled to command a lunar landing flight, Cooper was eased out of the NASA flight rotation because of his reputation as a maverick and a daredevil.

Robert Laurel Crippen
1937-

American engineer and astronaut who had a highly decorated career with the U.S. Air Force and NASA that included being part of the three-man crew of the first Space Shuttle mission (1981). Crippen became a NASA astronaut in 1969 and served on the support crews of the *Skylab 2, 3,* and *4* missions in 1973 before flying aboard the Space Shuttle on STS-1 (1981), STS-7 (1983), STS-41-C (1984), and STS-41-G (1984). He logged over 565 cumulative hours of space flight.

Gerard D'Aboville
1946?-

French sailor who is the only person to have rowed across both the Atlantic and Pacific oceans without assistance. D'Aboville rowed across the Atlantic Ocean in 72 days in 1980, rowing a distance of 2,735 miles (4,400 km) from west to east. In 1991 he rowed from Japan to the United States in just under 4½ months, making history in the process.

Georgy Timiofeyevich Dobrovolsky
1928-1971

Soviet cosmonaut who commanded the first crew to live aboard a space station. Dobrovolsky acted as communications officer for manned *Soyuz* flights between 1967-69, and trained for the abortive Soviet moon landing program. He transferred to the *Salyut* space station program in 1970 and, with Viktor Patsayev and Vladislav Volkov, spent 23 days aboard the *Salyut 1* station in 1971. He died, along with his crew, when the spacecraft returning them to Earth depressurized.

Charles M. Duke, Jr.
1935-

American astronaut who was the tenth person to walk on the moon. Duke graduated from the U.S. Naval Academy and completed a master's degree in aeronautics and astronautics at the Massachusetts Institute of Technology. He served as an Air Force test pilot. As the lunar module pilot of *Apollo 16,* Duke spent several days on the moon in 1972. He was the backup pilot for *Apollo 13* and *17.* After retiring from NASA, he became an entrepreneur and minister. He published his autobiography, *Moonwalk,* in 1990.

Sylvia Earle
1945-

American marine scientist and undersea explorer who holds the world's depth record for solo diving—1,000 meters (3281 feet). In 1970 she led the Tektite II Mission, in which she and four other women lived 50 feet (15 meters) below the ocean for two weeks. In the 1980s she founded Deep Ocean Engineering, a company that designs and builds undersea vehicles. She took a leave from Deep Ocean in the early 1990s to serve as chief scientist of the National Oceanic and Atmospheric Administration, where she was responsible for overseeing the health of the nation's waters.

Linda Finch
1951?-

American aviator who made history on March 17, 1997, flying from Oakland, California, in an original Lockheed Electra 10E airplane to recreate Amelia Earhart's 1937 ill-fated journey of circumnavigating the globe. After 73 days and 26,000 miles (41,834 km), Finch returned to Oakland on May 28. Finch is an experienced pilot with more than 20 year's experience in flying historic aircrafts, as well as a millionaire businesswoman who runs two nursing homes and a retirement community in Texas.

Steve Fossett
1945-

American yachtsman, balloonist, and endurance sportsman who made a career of setting world sports records during his worldwide adventures. Fossett swam the English Channel in 1985. In 1992 he participated in the Idatarod Dogsled Race. In 1995 he made the first solo transpacific balloon flight. He was also the first balloonist to cross Asia, the South Atlantic Ocean, and the Indian Ocean. By the end of the 1990s Fossett had set over 10 World Sailing records, including the 24-hour record (580.23 miles), and eight yacht race records.

Birute Galdikas
1948-

Canadian primatologist known for her work with orangutans. Along with Dian Fossey and Jane Goodall, Galdikas is one of the three pre-emminent researchers of large apes. Her pioneering studies in the 1970s required her to spend weeks on end tracking individual orangutans through the rainforest. Since then she has worked to preserve orangutan habiat with the Indonesian government by establishing national parks and has adopted orphaned orangutans to prepare them for re-release into the wild.

Linda Godwin
1952-

American astronaut who logged more than 633 hours in space. She received her Ph.D. in physics in 1980 from the University of Missouri and joined NASA as an astronaut in 1985. In 1996 she and a crew of six joined and docked with the Russian space station Mir, and delivered 4800 pounds (2,177 kg) of science equipment and hardware, food, water, and air supplies to the crew. She performed a six-hour space walk, the first while docked to an orbiting space station.

Richard Francis Gordon
1929-

American astronaut who made several space flights in the 1960s. Born in Seattle, Washington, Gordon completed his initial schooling by graduating from the nearby University of Washington in 1951. He was accepted into the naval aviation program and, six years later, became a test pilot. He first came to public attention when he won the Bendix Trophy Race in 1961 and was chosen for astronaut training in 1963. His first flight into space was in September 1966 aboard the *Gemini II* with Pete Conrad, during which he took a 45-minute space walk between *Gemini* and an Agena target vessel. In November 1969, along with Alan Bean and Conrad, Gordon piloted the command module on the *Apollo 12* mission. He remained in orbit while Bean and Conrad conducted the lunar exploration. He retired from the Navy and space program in 1972.

Frederick Drew Gregory
1941-

American astronaut who was the first African-American shuttle commander. A U.S. Air Force Academy alumnus, Gregory completed flight training at the Naval Test Pilot School. He served in Vietnam, commanding helicopter rescue crews that evacuated refugees from the American embassy in Saigon. Gregory piloted a 1985 shuttle flight. In 1989 and 1991 he was commander of two shuttle missions that deployed satellites. By 1992 Gregory became the associate administrator for safety at NASA headquarters. He has also served as chief of astronaut training.

Virgil Ivan Grissom
1926-1967

American engineer and astronaut who was the second American in space during a 1961 suborbital Mercury test flight in the *Liberty Bell 7* spacecraft, which sank in the Atlantic after splashdown (the capsule was recovered in 1999). After earning his Air Force wings (in 1951) and flying over 100 combat missions in Korea, "Gus" Grissom was the command pilot on the first manned *Gemini* flight (1965). He died tragically with two other astronauts in the *Apollo 1* launch pad fire (1967) during a countdown simulation test.

Thor Heyerdahl
1914-

Norwegian anthropologist and explorer who led several transoceanic voyages aboard primitive vessels to prove the possibility of ancient sea migrations. During his 1947 *Kon-Tiki* expedition, Heyerdahl sailed on a small raft from the Pacific coast of South America to Polynesia. In 1969 he crossed the Atlantic Ocean from Morocco to the coast of Central America in the *Ra II,* a replica of an ancient Egyptian reed boat. Since then Heyerdahl has made similar voyages down the Tigris River, to Easter Island, and Peru.

James B. Irwin
1930-1991

American astronaut who was one of the first to drive the lunar rover on the moon. Irwin graduated from the U.S. Naval Academy and earned an engineering master's degree from the University of Michigan before completing Air Force test pilot training. Irwin served as the lunar module pilot for *Apollo 15* in 1971 and explored the moon's surface. The next year he established the High Flight Foundation. His autobiography, *To Rule the Night,* discusses his spiritual epiphanies on the moon.

Josepf P. Kerwin
1932-

American astronaut who was the first U.S. physician in space. Kerwin earned a medical degree from Northwestern University Medical School and then graduated from the Naval School of Aviation Medicine. For 28 days in 1973 on *Skylab 2,* Kerwin assessed the crew's physical condition and determined that it was medically possible for astronauts to participate in sustained missions. He served as director of life sciences for NASA, evaluated biological aspects of shuttle flights, and worked as an aerospace industry consultant.

Vladimir Mikhailovich Komarov
1927-1967

Soviet engineer, pilot, and cosmonaut who was among the first to complete a mission in a spaceship built for more than one occupant. On October 12, 1964, Komarov, a veteran of World War II and one of the original 18 selected to train as cosmonauts, piloted the three-man spacecraft *Voskhod* around the earth 16 times. In 1967 Komarov became the first human fatality of space travel when, after a problematic one-day test flight aboard the *Soyuz-1,* the main parachute on his descent craft failed to open. A crater on the far side of the moon is named after him.

Alexei Arkhipovich Leonov
1934-

Soviet cosmonaut who made the world's first space walk and participated in the first international space mission. Leonov left *Voskhod 2* for ten minutes during its March 1965 flight, floating at the end of a 17-foot tether. Ten years later, during the Apollo-Soyuz Test Project flight, Leonov and Valery Kubasov docked their spacecraft with a three-man American ship in Earth orbit. The two-day rendezvous symbolically ended the Soviet-American "space race" of the 1960s.

Per Lindstrand
1950-

Swedish balloonist who completed a balloon flight across the Atlantic and set balloon altitude and distance records. Lindstrand served for a short time as an engineering officer in the

Swedish Air Force, at which time he began ballooning as a hobby. After making his first solo flight in a homemade balloon, he decided to set up his own balloon manufacturing business. Lindstrand is best known for his Atlantic crossing, which he completed with Richard Branson, owner of Virgin Atlantic Airlines and Records. Lindstrand also set the world altitude record, taking his Stratoquest balloon to 65,000 feet (19,800 m) over Laredo, Texas, 10,000 feet (3,000 m) higher than the previous record. In January 1991 Lindstrand and Branson completed the longest balloon flight—6761 miles (11,000 km) from Japan to Northern Canada.

James A. Lovell, Jr.
1928-

American astronaut who was the first individual to complete four space missions. Lovell graduated from the U.S. Naval Academy and the Naval Test Pilot School. He piloted *Gemini 7* for two weeks in 1965 and then commanded *Gemini 12*. On *Apollo 8* in 1968, Lovell orbited the moon in a pioneering spaceflight. Two years later, he survived the *Apollo 13* explosion. Lovell next served as the deputy director of science and applications at the Johnson Space Center. He published his autobiography, *Lost Moon*, in 1993.

Shannon Wells Lucid
1943-

American biochemist and astronaut who in 1996 spent 188 days in space on the U.S. shuttle and Russian *Mir* space station, for which she received the Congressional Space Medal of Honor. In 1979, following a teaching and research career, Dr. Lucid became an astronaut and qualified as a Space Shuttle mission specialist. In addition to her flight time on *Mir*, she was a veteran of four other space missions—STS-51G (1985), STS-34 (1989), STS-43 (1991), STS-58 (1993)—and two additional Space Shuttle flights in 1996, STS-76 and STS-79.

Peter Mathews
1951-

Australian archaeologist and epigrapher who specialized in Mayan hieroglyphics, writing systems, and linguistics. Dr. Mathews, a world-renowned Mayan archaeologist, made significant discoveries in Mesoamerica. Especially noteworthy was an altar discovery in El Cayo, Chiapas, Mexico, which almost led to his death when local bandits and looters attacked and robbed his team in 1997. He authored numerous essays, articles, and books on Mayan art and iconography and hieroglyphic inscriptions. With Linda Schele (1942-1998), Mathews was considered one of the first Mayan glyph decoders.

Thomas K. Mattingly
1936-

American astronaut who piloted three missions. Mattingly, a 1958 Auburn University aerospace engineering graduate, became a Navy pilot then an astronaut. Mattingly was removed from the *Apollo 13* crew because of measles exposure only to monitor the stricken flight from mission control. In 1972 Mattingly served as pilot of *Apollo 16*'s command module. He later assisted space shuttle development and was commander of two shuttle missions carrying Department of Defense payloads. Mattingly has also been an administrator for both governmental and commercial aerospace interests.

Sharon Christa McAuliffe
1948-1986

American teacher and astronaut who captured the heart of a nation after her selection by NASA as the first teacher to fly in space. Upon her college graduation, McAuliffe began teaching, specializing in American history and social studies. In 1984, she was selected by NASA to communicate with students from space during a Space Shuttle mission. Tragically, the shuttle *Challenger* exploded shortly after takeoff in January 1986, killing all astronauts aboard. McAuliffe's popularity made the explosion especially horrific to Americans and the world.

Bruce McCandless
1937-

American astronaut who was the first individual to spacewalk untethered. McCandless graduated from the U.S. Naval Academy and Stanford University, specializing in electrical engineering. McCandless contributed to the development of the manned maneuvering unit, testing it during a 1984 shuttle flight. He demonstrated how extravehicular activity free of connecting lines could be useful in repairing satellites. In 1990 McCandless and crewmates deployed the Hubble Space Telescope. After that mission he applied his knowledge to the international space station's construction.

James A. McDivitt
1929-

American astronaut who photographed the first American spacewalk. After high school, McDivitt joined the U.S. Air Force, serving in Korea. He earned an aeronautical engineering degree from the University of Michigan and completed test pilot school. McDivitt piloted the 1965 *Gemini 4* flight, documenting Edward White's historic spacewalk. In 1969 McDivitt commanded *Apollo 9,* testing the lunar module for the moon landing. Afterwards, he served as manager of lunar landing operations and the Apollo spacecraft program. McDivitt then became an industrial executive.

Edgar Dean Mitchell
1930-

American astronaut who was the sixth person to walk on the Moon. An expert on the Apollo lunar module, Mitchell played supporting roles on the *Apollo 9* and *10* missions before going to the Moon as lunar module pilot for *Apollo 14* in February 1971. He spent a total of nine hours on the lunar surface, helping to carry out the most ambitious scientific program yet attempted on an Apollo mission.

Larry M. Newman
1947-

American balloonist who in 1978 participated in the first nonstop balloon journey across the Atlantic Ocean. Newman, Maxie Anderson, and Ben Abruzzo flew their balloon, the *Double Eagle II,* at a high altitude across the Atlantic in six days, taking advantage of decent weather and favorable winds to cross the Atlantic from west to east. Newman was added to the crew after the previous year's attempt had been forced down by a combination of poor weather and fatigue.

Tenzing Norgay
1914-1986

Nepalese Sherpa mountaineer who, with Sir Edmund Hillary, was the first to reach the summit of Mount Everest, the highest peak in the world. Norgay was born and spent his early years in Solo Khumbu, Nepal, just south of Mount Everest. As a boy, he ran away from home and settled in Darjeeling, West Bengal, India. In 1935 he served as a porter to Sir Eric Shipton on a reconnaissance expedition to Everest. Several other Everest expeditions followed, and Norgay progressed to the position of sirdar, an organizer of porters. In 1953 he accompanied Hillary to the summit of Everest, making history as the first to set foot on the mountain's 29,028-foot (8,850 m) peak. Norgay was a hero to many Nepalese and Indians, and was awarded several honors for his mountaineering.

Viktor Patsayev
1933-1971

Soviet engineer and cosmonaut who was test engineer of the first crew to live aboard a space station. Patsayev joined the Korolev spacecraft design bureau as an engineer in 1957. He worked extensively on the design of *Salyut 1*, the world's first station, and spent 23 days in space as a member of its first crew. He died, along with flight engineer Vladislav Volkov and mission commander Georgy Dobrovolsky, when the spacecraft returning them to Earth depressurized.

Steve R. Ptacek
1953?-

American aviator who piloted the first solar-powered airplane, the *Solar Challenger,* in 1981. Ptacek's plane, designed and built by Paul MacCready, was based on MacCready's previous designs for the first human-powered airplanes to fly a figure-8 course (winning the Kremer Prize) and to cross the English Channel. The *Solar Challenger* weighed about 210 pounds (95 kg) and was powered by over 16,000 solar cells. On his record-breaking flight, Ptacek flew 160 miles (257 km) at an average speed of nearly 30 mph (48 kph) at an altitude of 11,000 feet (3,350 m).

Yuri V. Romanenko
1944-

Russian cosmonaut who set an endurance record for manned space flight by remaining aboard the *Mir* space station for 326 days. A veteran of three other space flights, Romanenko showed that it is possible for people to live in space for prolonged periods and still return to function normally on Earth. The data collected during his stay in orbit and subsequent return to Earth is expected to prove important for the planning of future missions to Mars.

Stuart Allen Roosa
1933-

American astronaut who orbited the Moon as command module pilot of *Apollo 14*. The successful flight of *Apollo 14,* in February 1971, restored public confidence in NASA after the aborted *Apollo 13* mission. After his return from the Moon, Roosa filled the vital-but-thankless role of backup command module pilot for the last two lunar landing missions, *Apollo 16* and *Apollo 17*. He also contributed to the early development of the space shuttle.

Valery V. Ryumin
1939-

Russian astronaut who served as flight director of the *Salyut-7* and *Mir* space stations from 1981-89. Ryumin spent a total of 362 days in space, primarily on two missions to the *Salyut-6* space station—175 days in 1977 and 185 days in 1980. In 1998 Ryumin was selected by NASA as a crew member of STS-91, the final Shuttle-*Mir* docking mission and the last phase of the U.S.-Russian joint space program, to be supplanted by the construction of the International Space Station.

Linda Schele
1942-1998

American art historian who was world-renowned for her work in decoding the hieroglyphic language of the Maya. After an early career as an artist and studio art professor, Schele became interested in the Mayan culture after a 1970 trip to the Yucatan; she earned her Ph.D. in Latin American Studies in 1980. From her ensuing years of research, Dr. Schele developed an extensive knowledge of Mayan culture and history, and she published numerous essays, articles, and award-winning books about the subject.

Harrison Hagan Schmitt
1935-

American geologist and astronaut who was the first scientist-astronaut in space and who landed on the Moon during *Apollo 17* (1972), the last space flight to the Moon. Dr. Schmitt worked as a geologist until NASA selected him in June 1965 in its first group of scientist-astronauts. He attended a flight training course to gain pilot's wings before training for his lunar mission. Schmitt served as a U.S. Senator for his home state of New Mexico from 1976 to 1983.

Russell L. Schweikart
1935-

American astronaut who contributed to the knowledge of space adaptation syndrome. Schweikart graduated from the Massachusetts Institute of Technology. In 1969 he flew on *Apollo 9* to perform a two-hour spacewalk to test the lunar spacesuit. Because he became dizzy, Schweikart shortened his spacewalk to half an hour, which was sufficient to assess the spacesuit's capabilities. For *Skylab* preparations, he served as a medical subject for doctors studying how space affected astronauts. Schweikart later founded the Association of Space Explorers.

David Randolph Scott
1932-

American engineer and astronaut who piloted *Gemini 8* (1966), served as command mobile pilot of *Apollo 9* (1969), and landed on the Moon as spacecraft commander of *Apollo 15* (1971). Following graduation from West Point in 1954, Scott entered the U.S. Air Force, where he attended the Experimental Test Pilot and Aerospace Research Pilot schools. After selection in 1963 for NASA's third astronaut group, he flew on three space flights and served as backup crew on three other missions.

Sir Peter Markham Scott
1909-1989

English ornithologist and painter who became a well-known painter of birds, especially waterfowl, exhibiting his work in London and New York. He mapped the previously unexplored Perry River region in the Canadian Arctic in 1949 and subsequently led naturalists' expeditions to Antarctica, Iceland, the Seychelles, the Galapagos, and Australasia. A founder and long-time officer of both the World Wildlife Fund and Wildfowl Trust, he was also president of a number of other international conservation organizations. Author of 18 books on ornithology and conservation, he also illustrated 20 volumes written by other individuals. Scott received many honors for his work in conservation and ornithology, including 8 honorary degrees, and was the first Englishman to be knighted (in 1973) for his contributions to conservation.

Thomas P. Stafford
1930-

American astronaut who was the first U.S. general in space. A Naval Academy alumnus, Stafford completed flight training and became an instructor at Edwards Air Force Base. He piloted *Gemini 6* and *9* in 1965 and 1966, respectively. As commander of the 1969 *Apollo 10* flight, he tested the lunar module. In 1975 Stafford was the American commander for the Apollo-Soyuz Test Project, the first international spaceflight. He served as chief of the astronaut office and deputy director of flight crew operations.

Freya Stark
1893-1993

British writer well known for her books describing local history, culture, and everyday life of countries she visited. Many of her trips were to remote places where no woman had ever traveled. Her travels in Iran were the inspiration for her first book, *The Valley of the Assassins* (1934). Stark traveled extensively throughout the Middle East, Turkey, Greece, Italy, and Asia, and learned to speak Arabic, Iranian, and Turkish (she already spoke French, German, and Italian). She wrote 24 books, several volumes of collected letters, and four volumes of memoirs. Stark died at the age of 100 at her home in Italy.

Will Steger
1945

American explorer who has completed a number of risky—and record-setting—solo trips through the Arctic and Antarctic. Steger has explored and traveled his entire life, beginning with a raft trip down the Mississippi River at age 15. Since then, he has traveled in the Arctic and Antarctic, making the longest unsupported dogsled trip (1,600 miles [2,575 km], south to north across Greenland) in 1988, and the first dogsled trip across Antarctica (3,471 miles [5,586 km]) in 1989-90. He also made the first unsupported dogsled journey to the North Pole in 1986.

M. Robert Stenuit
1933-

Belgian diver and deep-sea explorer whose experiments in a submersible decompression chamber in 1964 provided invaluable scientific data for underwater divers and scientists. Stenuit also spent years collecting data on shipwrecks and sifting accounts of famous ships, notably the Spanish Armada and Dutch East India Company vessels, in Europe's libraries and archives. His most significant discoveries included the Armada warship *Girona* (1967), Dutch East India Company silver bars found on the *Slotter Hooge* (1974), and a treasure trove of rare Ming porcelain, insulated and preserved by tons of pepper from the spice cargo of the *Witte Leeuw* (1977).

Karen Thorndike
1942-

American transoceanic sailor born in Snohomish, Washington. In 1996 Thorndike sailed from San Diego, California, to attempt a solo ocean voyage around the world on her 36-foot (11 m) sloop, *Amelia.* Two years later, in August 1998, she completed her journey. This marked the first time that an American woman successfully solo-navigated around the five "Great Capes": Cape Horn (South America), Cape of Good Hope (South Africa), Cape Leeuwin (Australia), South East Cape (Tasmania), and Southwest Cape (New Zealand). Thorndike was 56 when she completed her voyage, making her the oldest woman to sail solo around the world.

Gherman Stepanovich Titov
1935-

Soviet cosmonaut who was the first human to spend an entire day in orbit. Titov's flight, originally scheduled for three orbits, was extended to seventeen in order to break the 24-hour mark. Titov made the first television broadcasts from space, giving a "tour" of the cabin and commenting on the sights below. Severely affected by space sickness, Titov never made another flight but remained active as a spokesman and administrator in the Soviet space program.

Vladislav Nikolayevich Volkov
1935-1971

Soviet engineer and cosmonaut who was flight engineer of the first crew to live aboard a space station. Volkov designed spacecraft and trained cosmonauts before becoming a cosmonaut himself in 1966. He flew aboard *Soyuz 7* in 1969 and spent 23 days on *Salyut 1*, the world's first station, in 1971. Volkov died, along with test engineer Viktor Patsayev and mission commander Georgy Dobrovolsky, when the spacecraft returning them to Earth depressurized.

Donald Walsh
1931-

American naval officer who, with Jacques Piccard, made the first and (to date) only descent to the bottom of the Marianas Trench. Descending in a submersible, the *Trieste,* to a depth of nearly 36,000 feet (11,000 m), Walsh and Piccard became the first people to visit and explore the deepest spot on the planet. This trip was not only record-setting, it also showed that, contrary to accepted scientific thought at the time, life existed and flourished in the ocean's furthest depths.

Paul J. Weitz
1932-

American astronaut who flew in the original space shuttle missions and served as pilot of the *Skylab* space station project. He received his M.S. degree in aeronautical engineering in 1964 and was chosen by NASA in 1966 to be an astronaut. He served as pilot aboard the *Skylab 2*, the first piloted mission of the space station project. The crew returned to Earth after 28 days in space, setting what was then a record for a single space mission.

Edward Higgins White
1930-1967

American engineer and astronaut who was the first U.S. astronaut to walk in space—during the *Gemini 4* mission (1965). Following graduation from West Point (1952), White flew in the U.S. Air Force, service that included time as an experimental test pilot with the Aeronautical System Division. He was selected by NASA in 1962. In 1967 White and fellow *Apollo 1* astronauts died tragically in a launch pad fire during a countdown simulation test.

Alfred M. Worden
1932-

American astronaut who piloted the command module *Endeavor* during the *Apollo 15* lunar mission. In 1955 he received his B.S. degree from the U.S. Military Academy at West Point, and in 1966 he was chosen by NASA to join the astronaut program. As pilot of *Endeavor,* he conducted scientific experiments to determine the chemical properties of the Moon, measured the lunar atmosphere, and mapped the Moon's surface. On the return flight to Earth, he performed a 38-minute space walk.

John W. Young
1930-

American astronaut who was the first astronaut to complete six flights. Young piloted the 1965 *Gemini 3* flight and commanded *Gemini 10* the next year. In 1969 Young was the command module pilot for *Apollo 10.* He landed on the moon in 1972 as commander of *Apollo 16.* Young flew on the space shuttle's 1981 inaugural flight and commanded the 1983 shuttle mission that first carried *Spacelab* experiments. Because he questioned safety procedures concerning the *Challenger* disaster, Young was replaced as chief of the astronaut office.

Bibliography of Primary Sources

Books

Fuchs, Vivian, and Edmund Hillary. *The Crossing of Antarctica* (1958). An account of the first overland crossing of Antarctica, led by Fuchs and Hillary during the British Commonwealth Trans-Antarctic Expedition in 1957-58. They also published a sequel, *No Latitude for Error* (1961).

Hillary, Edmund. *High Adventure* (1955). Detailed Hillary's 1953 feat in becoming the first person—along with his Sherpa guide, Tenzing Norgay—to reach the summit of Mount Everest.

Periodical Articles

Piccard, Jacques. "Man's Deepest Dive." *National Geographic* (May 1960). Piccard recounted his 1960 deep-sea dive with Donald Walsh, during which the pair piloted the *Trieste* to a record depth of 35,800 feet in the Mariana Trench of the Pacific Ocean.

Ventris, Michael, and John Chadwick. "Evidence for Greek Dialect in the Mycenaean Archives" (1953). A seminal paper on the decipherment of Mycenaean writing, called Linear B, the oldest known form of the Greek language. Mycenaean civilization flourished in Greece and the surrounding islands in the Aegean Sea around 1400 B.C.

Life Sciences

Chronology

1952 English physiologists Alan Hodgkin and Andrew Huxley first delineate the mechanism of nerve-impulse transmission, showing that "sodium pump" system carries impulses.

1953 James Watson and Francis Crick discover the double-helix structure of DNA.

1957 American psycholinguist Noam Chomsky first claims, in his book *Syntactic Structure,* that the human brain is born programmed to learn language.

1961 Louis and Mary Leakey discover the first fossil remains of *Homo habilis,* the earliest-known member of the genus *Homo,* in Tanzania.

1962 *Silent Spring,* a book by American biologist Rachel Carson, raises international awareness of pollutants and spawns the environmental movement.

1966 Chinese-American chemist and endocrinologist Choh Hao Li becomes the first to describes the structure of human growth hormone, which he later synthesizes.

1970 Har Gobind Khorana and colleagues construct the first artificial gene.

1972 Gerald M. Edelman and Rodney R. Porter are the first to determine the chemical structure of the human antibody immunoglobulin.

1977 Floyd Bloom, an American neurobiologist, pinpoints the location of endorphins, discovered the previous year by Roger Guillemin, in the pituitary gland.

1985 Alec Jeffreys first develops a method of "fingerprinting" with DNA and shows that there are certain core sequences of DNA unique to each individual.

1990 The Human Genome Project, whose goal is to locate all the genes within a human cell's 46 chromosomes and to determine the precise order of the 3 billion gene subparts or nucleotides that form the genetic code, begins in the United States.

1995 Scientists at the University of Copenhagen discover a new phylum, *Cycliophora,* in which they place a newly discovered creature named *Symbion pandora,* which lives in the mouths of Norwegian lobsters.

1997 At the Roslin Institute in Scotland, a lamb named "Dolly" is the result of the first successful effort to produce an exact genetic duplicate, or clone, from the genetic material of a mature mammal.

Overview: Life Sciences 1950-present

Background: Advances between 1900-1949

Early in the twentieth century the rediscovery of Gregor Mendel's (1822-1884) work on heredity in pea plants led to the development of the field of genetics. In the 1940s Oswald Avery (1877-1947) and his research associates at the Rockefeller Institute in New York found that DNA was the genetic material, the chemical information passed on from one generation to the next, that determines traits in all species. At about the same time, a group of researchers from a number of different areas of biology—zoology, ecology, genetics, and paleontology—developed what came to be called the Modern Synthesis, the updating of Charles Darwin's (1809-1882) theory of evolution to incorporate the new discoveries in genetics and other areas of biology. It was also during the first half of the century that ecologists developed mathematical methods for describing changes in populations over time, techniques for sampling populations and for studying interactions between species, and concepts such as food chains and webs. The second half of the twentieth century saw further development in these fields, with greater knowledge of DNA influencing almost all areas of biology.

1950-present: Molecular Biology and Genetics

Certainly the most important event in biology in the post-1950 period was the discovery of the structure of DNA by James Watson (1928-) and Francis Crick (1916-) in 1953. This opened the way for a thorough investigation of precisely how a gene, a piece of DNA containing the information to make one cellular substance, is controlled, and how the information in the genes is used to make proteins, the molecules that are key to the functioning of cells. Through the 1950s and 1960s biologists worked out the steps in this process, called protein synthesis, and learned a great deal more about cellular structures including the fundamentals of gene regulation. Much of this research was done on bacteria and viruses, the simplest forms of life. It was only in the 1970s that techniques were developed for studying the genes of plants and animals, leading ultimately to the Human Genome Project, a massive effort to discover the complete human genetic makeup.

With recombinant-DNA techniques, it became possible to isolate specific genes and to insert them into the DNA of other organisms. This type of genetic engineering has blossomed, making possible the large-scale production of insulin and many other human proteins. It has also revolutionized agriculture and the food industry, with the insertion of genes for accelerated growth into animals and for insect and herbicide resistance into plants. In addition, new techniques developed since the 1960s make it much easier to analyze genetic differences between individuals; studies have unearthed a great deal of genetic diversity within species and races, calling into question the whole concept of race as a biological entity. Finally, in 1997 the first clone of an adult mammal, a sheep named Dolly, was born. Since then a number of other mammalian species, including mice, monkeys, and cattle, have been cloned, leading to a heated debate on the ethics of cloning humans.

Ecology and Environmental Science

Since the 1950s ecologists have become more aware of the complexity of ecosystems and the difficulty of predicting change. They continue to debate whether species diversity makes ecosystems more or less stable and to search for better mathematical models to describe the structure of ecosystems, of how the organisms within an ecosystem interact with each other. Environmental science developed in the 1960s in response to an increasing awareness of environmental deterioration, sparked in part by the writings of such activists as Rachel Carson (1907-1964), whose *Silent Spring* (1961) dealt with the dangers of pesticides. Late in the century "biodiversity" became an important concept, marking a shift away from efforts to save individual species from extinction to saving habitats and thus a diversity of species. Since the 1980s concerns have also grown about the destruction of the ozone layer in the upper atmosphere that screens out ultraviolet radiation and about global warming, the apparent increase in average temperatures that seems to be leading to climatic changes that influence the species composition of ecosystems.

There has also been mounting worry over the continued destruction of rainforests and wetlands and the relationship between environmental destruction and the emergence of new infectious diseases. With the destruction of their habitats, animals are forced into more contact with humans and thus are more likely to transmit animal diseases to them. There is also growing concern that genetically altered crops will have significant environmental impact as genes move from crop plants into related wild species. At the same time, many farmers are attempting to reduce the environmental impact of farming by using fewer pesticides and herbicides and by adopting methods that conserve soil and lessen water pollution.

Evolution

At the end of the twentieth century evolution remains the unifying concept that underlies all of biology. While the Modern Synthesis is still central to most biologists' thinking on evolution, a number of newer ideas have caused some modification of it. These include punctuated equilibrium, the concept that change in species does not necessarily occur gradually, but that a period of rapid change may be followed by long intervals with little modification. Debates among biologists about the mechanisms of evolution, though not about the fact of its occurrence, have made it easier for creationists to challenge the theory. Creationists see the idea of the creation of living things by a divine being as one that should be given at least equal standing with evolution in the schools, despite the fact that the past 50 years have seen the discovery of much new evidence for evolution. Rich fossil beds in China and Mongolia have revealed the existence of many previously unknown dinosaur species as well as much new information on the origin of mammals and birds. A great deal of evidence has been discovered supporting the theory that the final extinction of the dinosaurs was due to a comet hitting Earth, causing massive climate changes that the dinosaurs could not survive. Another important trend has been the discovery of many new fossils of early humans and their ancestors, making it obvious that the human family tree has had many branches; modern humans are only one of several lines and the only one to survive to the present day.

Neurobiology and Behavior

Over the past 50 years the study of primates, the closest living relatives to humans, has grown significantly, with a number of researchers including Jane Goodall (1934-) and Dian Fossey (1932-1985) dedicating themselves to behavior studies of specific ape species. In the 1950s Noam Chomsky (1928-) developed the idea that human language acquisition is biologically encoded; this sparked research on the language abilities of apes. In terms of humans, Alfred Kinsey (1894-1956) published landmark studies on sexual behavior that others have since added to and amended. Neurobiology has also flourished, as major discoveries have been made on brain chemistry and the functioning of nerve cells. New imaging techniques such as PET scans have made it possible to study brain activity of normal subjects carrying out everyday tasks such as reading and puzzle-solving.

The Future

In the next half century it is clear that continued exploration of the genes and how they are controlled will lead to new drugs, new foods, and new knowledge about the human body and its diseases. There will also be increased efforts to understand and preserve the great diversity of life on Earth, especially as threats to that diversity intensify. The exploration of our evolutionary past will continue, as biologists search for new fossil evidence while at the same time using genetic analysis to discover relationships among present-day species. In the past 50 years biology has become a much more complex science with a growing impact on our lives, and that trend is likely to continue in the near future.

ROBERT HENDRICK

Evolution and Creationism in American Public Schools

Overview

The history of the teaching of evolution in America cannot be addressed without analyzing the origins of contemporary creationism. The Scopes Trial of 1925 is often regarded as a landmark in the battle to teach evolution in American public schools, but the U.S. Supreme Court did not overturn state laws that banned the teaching of evolution until 1968. In response, creationists adopted a new strategy, calling for "balanced treatment" in the teaching of evolution and the Creation. That is, they demanded that "Special Creation" be taught as science and that evolution should be described as "merely" a theory. "Creation Science" has traditionally embraced religious tenets, most notably that of divine creation "from nothing," distinct "kinds" of plants and animals, a worldwide flood, and a relatively recent origin of the universe. In 1987 the Supreme Court ruled that state laws requiring equal time for creationism were unconstitutional. In the 1990s some states removed evolution from their curricular mandates and gave local school boards the right to decide whether or not to teach evolution. Even though the National Academy of Science and the National Science Foundation have identified evolution as the unifying concept of modern biology, not all states require the inclusion of evolution in high school biology courses.

Background

Scientists and philosophers were interested in the concept of evolution long before Charles Darwin (1809-1882) published *The Origin of Species* in 1859, but the theory of evolution remained incomplete because no convincing mechanism for it had been proposed. Darwin's theory of evolution states that species have evolved from ancestral forms over the course of millions of years and that the mechanism of evolution is natural selection. *The Origin of Species* was both a popular and controversial book. Theologians as well as scientists were divided about Darwin's work, especially his ideas about human origins. Some supporters of evolutionary theory attempted to retain a role for divine intervention, at least in the case of the intellectual and spiritual faculties of human beings.

The most revolutionary aspect of the Darwinian revolution was its rejection of the comforting belief that the world had been designed and created especially for human life. Nevertheless, many theologians were able to abandon the idea of the fixity of species in order to explore new concepts regarding the relationship between God and the natural world. Others, however, held to a literal interpretation of the Bible and rejected any alternatives. When *The Origin of Species* was published, various theories of the Creation co-existed within the Christian community. Some theories managed to achieve a measure of harmony between theology and science. For example, advocates of the "day-age theory" taught that the "days" in scripture referred to ages rather than 24-hour days. The "gap theory," in turn, suggested that the time period described by the book of Genesis could encompass a gap of millions of years between the creation of the earth and that of mankind. During the 1960s literal beliefs about the account of the Creation in Genesis became dominant among American anti-evolutionists.

In America, Darwinism was often used to justify business and political practices. Many states, however, excluded evolution from biology courses. The Scopes Trial is the best known example of the conflict surrounding the teaching of evolution. Even in the 1920s, most scientists were sure that teaching biology without references to evolution was impossible, but Christian Fundamentalists saw evolution as a threat to religious belief. Anti-evolutionary literature equated Darwinism with agnosticism and atheism. Fundamentalists tried to legislate the theory of evolution out of the classroom. In Kentucky in 1922 attempts to pass a law prohibiting the teaching of "Darwinism, Atheism, Agnosticism, or the theory of Evolution as it pertains to man" were narrowly defeated. In 1925, however, a Tennessee law made it illegal to teach "any theory that denies the story of the Divine Creation of man as taught in the Bible, and to teach instead that man has descended from a lower order of animals." John Thomas Scopes was tried in Dayton, Tennessee, for teaching "the theory of the simian descent of man, in violation of a lately passed state law." As recorded on a marker established by the Tennessee Historical Society on the courthouse grounds, Scopes was convicted. In 1927 a Tennessee court reversed the decision on a technicality. The Scopes Trial became the focus of a play called *Inherit the Wind,* which became a popular film in 1960.

Impact

While the Scopes Trial created a national sensation, the outcome was far from clear. The publicity may have stopped some states from enacting similar laws, but anti-evolutionary laws were passed in many other states. Teachers in Tennessee in the 1960s were still required to sign a pledge that they would not teach evolution. The teaching of evolutionary science at the high school level actually declined after the Scopes trial. Many states passed laws similar to the Tennessee anti-evolutionary law and most biology textbooks omitted any mention of Darwin or evolution. Evolution did not regain a significant place in biology textbooks until the Soviet Union's launch of *Sputnik* in 1957 led to a new interest in science education in the United States.

In 1968 the U.S. Supreme Court finally overturned the laws that had banned the teaching of evolution. The Court ruled that banning the teaching of evolution "for the sole reason that it is deemed in conflict with a particular religious doctrine" was unconstitutional. Scientists assumed that the issue had been settled, but the rise of Protestant Fundamentalism and "televangelists" (television preachers) led to a vigorous new campaign against the teaching of evolution. Opponents of evolution launched a campaign to require "equal time" for the teaching of Special Creationism whenever Darwinian evolution was discussed, demanding that teachers refer to Special Creationism as science and describe evolution as a theory. The success of this approach was clear when Texas rejected biology textbooks that included evolution. Publishers rushed to eliminate or water down references to evolution.

Challenging state "Equal Time" laws, scientists and philosophers of science have tried to explain what science is and how one can distinguish science from religion. Science has the following characteristics: 1) science works upon a foundation of natural law—it cannot explain away anomalies as "miracles"; 2) science makes predictions by inference about what might occur; and 3) scientists acknowledge the fact that their predictions could be wrong. After considering such arguments in a case brought to a U.S. district court in Arkansas, Judge William Overton on January 5, 1982, ruled against the Arkansas Equal Time Creation Science law on the following grounds: 1) Creation Science is religion; 2) to teach religion in the public schools is illegal; and 3) Creation Science is not science.

The Louisiana Balanced Treatment for Creation-Science and Evolution-Science Act was examined by the U. S. Supreme Court in 1986. Again, the court ruled against Creation Science. Scientists testified that the case was crucial to the future of basic science education in the United States. They argued that science education should accurately portray the premises and processes of science and the current state of scientific knowledge. Science education, scientists maintained, was damaged where the curriculum called for teaching religious ideas as science. Teaching creationism, in turn, established a false conflict between science and religion and created confusion about the nature of scientific inquiry. Future progress in science, medicine, and technology, they continued, would suffer if Americans were not encouraged to discriminate between natural phenomena and supernatural articles of faith. Moreover, the Louisiana act created confusion about the way in which scientists used the terms "fact" and "theory." A "fact" to scientists is a property of a natural phenomenon and a "theory" is a system of knowledge that explains a body of facts. These definitions permeate all fields of scientific endeavor and are no less relevant to discussions of the origin of the universe and life than to any other area of research. A review of the process and vocabulary of science will confirm that the essence of a scientific "theory" does not vary from discipline to discipline. The Louisiana act's "fact-theory" distinction reflected the belief system of certain Fundamentalist sects.

The Court held that the terms "creation" and "creation-science" embody the principles of a particular religious sect or group of sects. It concluded that the legislature intended Louisiana's public school teachers to offer students evidence that mankind and the universe were brought into existence by a divine Creator. In other words, the statute called for teachers to balance evolution against a particular religious belief. The biology textbooks recommended by advocates of Creation-Science included statements such as "the age of the earth can be measured in thousands rather than millions or billions of years" and "the dinosaurs were directly created at the same time as men, so that the humans and dinosaurs did live together for many years. However...the dinosaurs died in the great Flood."

The Court concluded that Creation Science as presented in the Louisiana case was religion, but it suggested that the teaching of any scientific views on origins would be acceptable. Therefore, Special Creationists saw the decision as opening up the possibility of getting creationism into the biology classroom as a "science" if creationist texts were carefully rewritten to provide scientific evidence. For example, the creationist claim for a young earth leads to the conclusion that dinosaurs and humans must have co-existed. In support of this assertion, creationists often claim to have found footprints of man and dinosaurs together.

Most polls say that the majority of Americans support the position that schools should "teach both sides." Scientists object to this and in reply ask whether we should also teach Aristotle's physics, in which the earth is located at the center of the universe. Community pressure generally favors Creation Science and "balanced treatment," but some of the major opponents of creationism are actually theologians from mainstream religious denominations. They say that because Special Creation is religion and not science it should be taught in comparative religion courses and not in biology classes. For example, Catholic schools have accepted the teaching of evolution since the 1950s. Moreover, in 1996, after many years of deliberation, Pope John Paul II declared that evolution did not conflict with Catholic doctrine.

Anti-evolutionary activism has grown in the United States since 1980. Theological and social factors contributed to the phenomenal popularity of scientific creationism in the late twentieth century as Americans became increasingly suspicious of science. The battles between evolutionism and creationism have raised many questions about the separation of church and state, the teaching of controversial subjects in public schools, and the ability of scientists to communicate with the public. Many scientists find believing that the theory of evolution still needs defending to be impossible, but public opinions polls show that the majority of Americans do not accept the science of evolution, especially with respect to human origins. Many people do not like to believe that they are the product of a long series of natural events rather than the products of wise design.

While details regarding the theory of evolution have been refined in light of new scientific discoveries, evolution's general principles have remained unaltered. Evolution, moreover, remains unsurpassed in its ability to provide a detailed explanation for both the diversity and the history of life on Earth.

The controversy over evolution proves more clearly than virtually any other episode in the history of science that science has had a profound impact on society and human thought. In no other case have the offshoots of a science created more social challenges than evolution. At

THE CREATION RESEARCH SOCIETY

The Creation Research Society was founded in 1963 in Ann Arbor, Michigan. The society is "a professional organization of trained scientists and interested laypersons who are firmly committed to scientific special creation." The primary functions of the society are: (1) Publication of a quarterly peer-reviewed journal; (2) Conducting research to develop and test creation models; (3) Providing research grants and facilities to creation scientists. Voting members of the organization must have a postgraduate degree in science; all members must agree with the following statement of principle: (1) The Bible is the written Word of God...all of its assertions are historically and scientifically true in all of the original autographs...the account of origins in Genesis is a factual presentation of simple historical truth. (2) All basic types of living things, including man, were made by direct creative acts of God during Creation Week as described in Genesis. Whatever biological changes have occurred since Creation have accomplished only changes within the original created kinds. (3) The great Flood in Genesis, commonly referred to as the Noachian Deluge, was an historical event, worldwide in its extent and effect. (4) Finally, we are an organization of Christian men of science, who accept Jesus Christ as our Lord and Savior. The account of the special creation of Adam and Eve as one man and one woman, and their subsequence Fall into sin, is the basis for our belief in the necessity of a Savior for all mankind. Therefore, salvation can come only through accepting Jesus Christ as our Savior.

the end of the twentieth century, many science educators are convinced that the debate over the teaching of evolution will likely take many forms in the future.

LOIS N. MAGNER

Further Reading

Awbrey, Frank and William Thwaites, eds. *Evolutionists Confront Creationists.* San Francisco, CA: American Association for the Advancement of Science, 1984.

Kitcher, P. *Abusing Science: The Case Against Creationism.* Cambridge, MA: Massachusetts Institute of Technology, 1983.

Morris, Henry M., Duane T. Gish, and George M. Hillestad, eds. *Creation: Acts, Facts, Impacts.* San Diego, CA: Creation-Life Publishers, 1974.

Morris, Henry M., and Gary E. Parker. *What Is Creation Science?* El Cajon, CA: Master Books, 1987.

Nelkin, Dorothy. *The Creation Controversy: Science or Scripture in the Schools.* New York: W. W. Norton & Company, 1982.

Numbers, Ronald L. *The Creationists: The Evolution of Scientific Creationism.* New York: Alfred A. Knopf, 1992.

Numbers, Ronald L., ed. *Creation-Evolution Debates.* New York: Garland Pub., 1995.

Ruse, Michael, ed. *But Is It Science? The Philosophical Question in the Creation/Evolution Controversy.* Amherst, NY: Prometheus Books, 1996.

Webb, George E. *The Evolution Controversy in America.* Lexington, KY: The University Press of Kentucky, 1994.

Wilder-Smith, A. E. *Man's Origin, Man's Destiny: A Critical Survey of the Principles of Evolution and Christianity.* Minneapolis, MN: Bethany Fellowship, 1975.

The Rise of Environmental Science

Overview

Environmental science is the study of the natural processes that occur in the environment and how humans affect them. The ideas of environmental science are closely related to ecology, the branch of science that deals with the interrelationships of plants, animals, and the environment. At times, the two words have been used interchangeably, especially during the last part of the twentieth century.

The rise of environmental science as a discipline occurred in the last three decades of the twentieth century. Researchers had studied plants and animals, but concepts tended to stay in the academic realms of pure botany or zoology. People thus had little knowledge or interest in the environment.

A group of writers were responsible for the eventual change in this situation. The turning point was the publication of a book by Rachel Carson called *Silent Spring.* The book highlighted the damage done to the environment by pesticides. The vibrations of Carson's work resounded not only in academia but in the mind of the public as well. Like many movements of the counter-culture during the 1960s and 1970s, the environmental movement was driven by the public.

The Environmental Protection Agency (EPA) was created to be the overseer of this new movement. Such legislation as the Clean Air Act of 1970 and Clean Water Act of 1972 was driven by public demands for change. The movement expanded worldwide with the United Nations establishing an environmental arm in 1972.

Background

Environmental issues focus on three major areas:

> Resource use. Resources are anything in the environment used by people. Renewable resources are those that may be replaced in a short time, such as trees, wind, or sunlight. Non-renewable resources are not replaceable and include coal and oil.
>
> Population growth. Up until about the year 1650, the world's population grew slowly. The explosion of the world's population is a major environmental concern.
>
> Pollution. Any problem in the environment that has a negative effect is considered pollution. A problem's source may be something that greatly benefits humans.

Arriving at a classification of these environmental issues has been gradual. In the past, various elements of the natural sciences were not connected. Several scientists were involved in zoology and botany. Europeans such as Charles Darwin traveled the world studying exotic plants and animals. Thomas Malthus (1766-1834) wrote *Essay on the Principle of Population,* in which he described how human growth fostered grave problems. The industrial revolution was a double-edged sword; while it created products to make life better at the same time it dumped waste into water and air.

American naturalists began a back-to-nature movement. John James Audubon (1785-1851), John Muir (1838-1914), William Bartram (1739-1823), and Meriwether Lewis (1774-1809) and William Clark (1770-1838) in their famous expeditions from 1804 to 1806 evoked interest in the importance of nature.

No one group figured out that these living and non-living things were interconnected. Many factors, beginning in the 1960s, were necessary to conceive of the biotic whole.

Impact

Although several studies relating to what was happening in the environment were available, they were highly technical and gathered dust on university book shelves. A group of perceptive journalists, however, became interested and plowed through the "dull" science to bring the studies to life.

Marjorie Stoneman Douglas (1890-1998) was one of these writers. Douglas, daughter of the editor of the *Miami Herald,* investigated the Miami River for a book assignment. She discovered that the river was part of the wilderness called the Everglades. With the real estate and population booms in Florida, the Everglades were being drained and exploited—a fact that would soon result in the cutting off of the water supply to South Florida. With a hydrologist, Gerry Parker, she devoured scientific studies, conducted interviews, and presented her findings in a book, *The River of Grass,* in 1947. The book was immediately a bestseller. Douglas worked through the last part of this century to promote environmental causes. When she died at the age of 108, she was still fighting to save the Everglades. Douglas was the first to write eloquent and descriptive prose about environmental concerns.

Rachel Carson (1912-1964) was both a scientist and writer and served as editor-in-chief of the U.S. Fish and Wildlife Services publications. To supplement her income, she wrote articles and books, translating her scientific knowledge into beautiful prose. When she received a letter from Olga Huckins of Duxbury, Massachusetts, describing how community spraying of DDT to kill mosquitoes had also killed the songbirds in her yard, Carson became interested. Her research into pesticides and their industry were published in the book *Silent Spring,* whose very title evokes the emotion of a spring without the sound of birds. She described how indiscriminate spraying of pesticides was poisoning our food and water and offered an outline how to stop this irresponsible use. Immediately, the book became a bestseller. Although she was dying of cancer, Carson threw herself into a campaign to influence legislation. The controversy that she created around the use of pesticides is still alive and well.

In 1962 the word "environment" was not in anyone's political vocabulary. People had complained about the terrible "smog" in cities such as Los Angeles, but the only mention of conservation in the 1960 Republican and Democratic Conventions was related to national parks and natural resources.

The public had no knowledge that their water was being slowly poisoned by pesticides. Criticism of *Silent Spring* and Carson was predictable. Major chemical companies branded Carson as a hysterical extremist. Claiming she used "emotion-fanning words," the industry attacked her credibility as a scientist. The book, however, still had great support. The message was out.

President John Kennedy asked for a special sub-committee to be created. When Senator Abraham Ribicoff heard Carson testify before the Senate committee, he recalled the words that Abraham Lincoln had used when he met Harriet Beecher Stowe, author of *Uncle Tom's Cabin*: "So you are the lady who started all this." Carson was soon dubbed the lady who had started the environmental movement.

Due to concerns Carson had raised, the Environmental Protection Agency (EPA) was created in 1970. States had begun to pass their own confusing and often contradictory protection laws. President Richard Nixon established the EPA to digest these new laws and monitor them. Several responsibilities were spread out under different governmental departments, including the Departments of the Interior, Agriculture, and

Trees destroyed by acid rain in the Great Smoky Mountains. *(JLM Visuals. Reproduced by permission.)*

Health Education and Welfare. This act brought environmental areas under one agency.

Environmental science as a discipline had not been previously known, the problems of the environment being complicated and involving complex scientific principles and techniques. Rene Dubos (1901-1982), a French-born biologist and ecologist, became interested in the "total environment" and by 1964 was a leading spokesperson for the fledgling environmental movement. He was an outspoken critic of what he considered a short-sighted outlook by most biologists. He won a Pulitzer Prize for his book *Only One Earth: The Care and Maintenance of a Small Planet.*

Another activist, Claire Patterson (1922-), an American geochemist, alerted the public to the dangers of lead. While studying ocean plankton, he was shocked to find a dramatic increase in lead levels in the ocean. He set out to measure lead in the atmosphere and the polar ice caps as well as the oceans. From 1930 to 1960, gasoline from automobiles had thrown tons of lead particles into the air. These high levels were dangerous to all. Armed with Patterson's research, environmentalists successfully lobbied for the Clean Air Act of 1970.

The 1970 Clean Air Act sought to combat pollution from industry in addition to motor vehicles. The responsibilities for enforcing the act were given to the states and each state had to submit an implementation plan. An early success in this endeavor was getting automobile manufacturers to install catalytic converters, thereby reducing emissions 85%. From 1970 to 1990, air pollution in the United States declined by one-third. During the 1980s, the pollution standards index improved.

The 1990 Clean Air Act amended the 1970 act and covered interstate as well as international problems involving Mexico and Canada. This act gave new enforcement power to the EPA. The newly flexible programs were market based, offering choices and incentives. In 1980 the Comprehensive Environmental Response, Compensation, and Liability Act, also called Superfund, provided billions of dollars for cleaning up abandoned waste dumps.

The creation of environmental science led to the development of a highly technical discipline. Its vocabulary is specific and laden with bureaucratic jargon. For example, "criteria air pollutants" are those pollutants that have a quantitative measurement set for a geographical region. There also are sets of primary standards to protect health as well as secondary standards to prevent property damage, such as that caused by acid rain. Overall, the 1990 act included 189 hazardous air pollutants. In 1997 the EPA changed the air quality standards for smog, called ground level ozone, as well as for particulate matter (PM), including soot, dust, and smoke.

Following on the heels of the Clean Air Act of 1970 came the Clean Water Act of 1972. The Ohio's Cuyahoga River, laden with toxic pollutants, had burst into flames, incensing the public. This act was designed to protect lakes, rivers, wetlands, and coasts from pollution. The Clean Water Act's requirements have prevented more than 900 million pounds of sewage and chemicals from entering the nation's water. Nevertheless, more than 2,000 beaches were closed in 1994, and warnings about fish contaminated with mercury have been issued in over 1,500 areas. Polluted run-off and the destruction of wet lands continue to be problems. The complex monitoring system needed to enforce the Clean Water Act requires scientific applications in such new fields as limnology (the study of lakes) and oceanography.

Numerous groups have also arisen in support of environmental interests, including the Audubon Society, the Sierra Club, and Greenpeace. Some of these organizations have gained notoriety through their tactics.

While the U.S. conscience was piqued for action, the United Nations organized an environmental conference in Stockholm in 1972, leading to the U.N. Environmental Programme (UNEP). Projects such as cleaning the Mediterranean, protecting water, combating deforestation, and banning ozone-depleting materials were started. Many countries, however, questioned the scientific basis for environmental concerns. The largest intergovernmental conference in history met in Rio de Janiero in 1992 to discuss the preservation of natural resources.

Much of the resulting debate and publicity has focused on the destruction of the rainforests in Central and South America. Drawing attention to the rain forests has been an amazing success, tugging at the heart strings of young and old alike. Environmental problems, however, also affect farmlands and cities. South America, for example, has had many such environmental problems. Among these problems are the struggles of indigenous peoples against petroleum groups in Equador and rubber tappers in Brazil.

Environmental pollution is also a legacy of the Soviet Union, which collapsed in 1991. Shortly before its fall, 45 million protesters in Eastern Europe demanded the end of Communist rule and mismanagement of the environment. The area, ranging from Poland to Romania to the Czech Republic, is the most polluted in the world. Adding to this debacle was the Chernobyl nuclear explosion of 1986. Eastern Europeans began to organize in 1995 with hundreds of "green" (environmental) organizations working to reverse decades of neglect. Elsewhere, in 1997 a group of biologists founded the Green Belt Movement to encourage restoring forests in

Kenya and other African nations. Other nations are following suit.

The idea of planet Earth as a total environment continues as part of the environmental movement. Finding appropriate policies to address the worldwide issues of resource use, population growth, and pollution will be the challenge of the twenty-first century.

EVELYN B. KELLY

Further Reading

Archer, Julie. *To Save the Earth.* New York: Viking Penquin, 1998.

Brooks, Paul. *Rachel Carson at Work.* San Francisco: Sierra Club Books, 1998.

Carson, Rachel. *Silent Spring.* New York: Houghton-Mifflin, 1962.

Collinson, Helen, editor. *Green Guerillas: Environmental Conflicts and Initiative in Latin American and the Caribbean. A Reader.* New York: Monthly Review Press, 1996.

Douglas, Marjorie Stoneman. *The River of Grass.* Sarasota, FL: Pineapple Press, 1985.

Dubos, Rene. *Only One Earth: The Care and Maintenance of a Small Planet.* New York: Norton, 1972.

Lear, Linda. *Rachel Carson: Witness for Nature.* New York: Henry Holt, 1998.

Trends in the Environmental Sciences since 1950

Overview

In the 1960s the environmental movement gained strength and direction, first in the United States and then in many other nations. Increasing knowledge of the environment and the impact of humans on it led to increasing clamor for regulations protecting areas such as wetlands, rainforests, the oceans, waterways, endangered species, and the atmosphere. In addition to marked environmental changes, these regulations and other measures have changed many aspects of industry, and environmental concerns have repeatedly been raised as issues in international trade discussions and similar venues.

Background

Some say that environmentalism began with Henry David Thoreau's (1817-1862) books, written in the 1840s and 1850s. Others claim that John Muir (1838-1914) in the latter decades of the nineteenth century helped to start the modern environmental movement, preserving lands for posterity rather than simply writing about them. Still others feel that Rachel Carson's (1907-1964) 1962 book, *Silent Spring* launched the environmental movement. Each of these claims has some justification, and it may be most accurate to point out that the environmental movement has been born in stages. Thoreau taught us to see and appreciate nature, Muir helped us to realize that nature is not endless, and Carson showed us that humans can destroy nature.

The first part of the twentieth century was a time of human civilization exerting dominion over the earth. Giant dams, spreading cities, bridges, railroads, airports, skyscrapers, canals, tunnels, and more were built with enthusiasm and endless optimism. By taming the planet we were making it better, safer, more productive, and more comfortable for people. This phase of boundless optimism in technology and relative lack of concern about its potential ill effects began to wane in the wake of World War II, but any concerns were more like nagging second thoughts rather than full-fledged worries. Part of the reason for this was political and economic; post-war Europe was struggling for survival and did not have the luxury of worrying about the environment, Japan was occupied and had virtually no economy, the Soviet Union was under totalitarian control, and the United States was too flush with new-found power to care much about the environment.

This started to change, however, in the 1960s, as Europe recovered, Japan built its economy, and the U.S. was caught in turmoil and doubt. With an unpopular war in Vietnam, activists in the U.S. were ready to believe that other governmental policies might be equally ill-considered. *Silent Spring* raised the specter of DDT and other insecticides poisoning the planet, and significant degradations of water quality and air quality led Richard Nixon (1913-1994) to form the Environmental Protection Agency in response. The EPA, in turn, helped to enforce a number of new laws, including the Clean Water Act, the Clean Air Act, and the Wetlands Protection Act.

As awareness of the importance of biodiversity grew, the Endangered Species Act was passed, offering protection to many plants and animals that were threatened with extinction. This also led to the realization that many plants and animals in other parts of the world were endangered due to deforestation, primarily from slash-and-burn agriculture that is practiced in many less-developed nations. With this realization came an understanding that we did not fully understand the impact on an entire ecosystem of removing some individual species through extinction. E.O. Wilson's (1929-) studies of island biogeography spurred the work of others who helped to elucidate some of these dependencies, and efforts have been made to encourage nations hosting large tracts of rain forest or threatened species to take measures to conserve them for the future.

In the 1980s global warming and ozone depletion were recognized as potential problems. Man's burning of coal, petroleum, and wood added carbon dioxide to the atmosphere, helping to trap the Sun's heat and threatening to turn Earth into a sauna. Global warming activists painted pictures of melting ice caps, violent weather, and spreading deserts driving people from the coasts and causing crops to fail. It remains to be seen whether these scenarios will come to pass, however, because there is still considerable debate over whether or not Earth is actually warming and, if so, if pollution is the actual cause. The retreat of many mountain glaciers and the partial collapse of parts of the Antarctic ice sheet suggest that global warming may have arrived, but it must be remembered that Earth is currently in an interglacial period of an ice age and, as such, these changes may be expected.

The case for depletion of the ozone layer is stronger, because it seems definitive that the chemicals found in the stratosphere are capable of causing ozone destruction and are created only by man-made products. In this case the concern is that loss of ozone over heavily inhabited regions could lead to an epidemic of skin cancer and could severely impact native plants and animals not adapted to survive in high-UV (ultraviolet radiation) environments.

The other major environmental concerns raised in 1970s and carried forward through the 1990s include those of nuclear power, nuclear war, and radioactive waste. Originally billed as a great advance, nuclear power started to lose its luster as concerns about nuclear war heated up. The accidents at Three Mile Island (which actually resulted in very low radiation doses to any members of the public) and Chernobyl (which had a significant impact on the local environment and health) only served to confirm fears of nuclear power as an unsafe technology. In addition, research into the comet impact that likely killed the dinosaurs led to the concept of a global winter sparked by massive dust clouds. This, in turn, led to the idea of a "nuclear winter" that, in the wake of a nuclear war, could further cripple humanity. Finally, many people were led to believe that there was simply no way to safely store nuclear waste. These three factors led many to raise heartfelt concerns about the safety of all things radioactive.

Impact

The impact of the environmental movement has been tremendous and has been tremendously varied. This movement has raised an enormous amount of controversy because so many of its proposals deal with issues that could cost billions of dollars to clean up and/or prevent, and that address phenomena that are sometimes subject to continuing scientific debate. On the one hand, we can point to cleaner skies and water and see that environmental regulations have led to considerable good. On the other hand, we can also point to global warming and ask if the money spent to reduce greenhouse gas emissions is well-spent when there is still no clear scientific consensus as to whether there is a warming trend and, if so, if it is man-made.

Equally problematic is the net effect of many civil engineering projects from earlier in the twentieth century. For example, the Hoover and Glen Canyon dams in the United States have substantially affected the Colorado River and parts of the Grand Canyon. However, they provide electricity and water to a sizable population, and the reservoirs impounded by these dams provide recreation to many people and a home to many fish.

Another impact of the explosive growth in environmental concern is the accountability of governments to their citizens in such issues. As a case in point, the U.S. Department of Energy has been taken to task on many occasions as the nation became informed of a number of environmental transgressions growing out of the nuclear weapons program. Environmental release, both

The endangered Giant Panda. *(JLM Visuals. Reproduced by permission.)*

accidental and deliberate, of both toxic and radioactive materials led to massive efforts to characterize and remediate these sites, as well as paying restitution to the most affected people in nearby communities. This same level of awareness has spread to other governmental sites, including former military bases. While these environmental restoration projects have been very expensive, they have also led to the growth of the environmental restoration industry and the development of many new tools for such remediation projects.

Environmental issues have been added to many seemingly unrelated political agendas. The North American Free Trade Agreement was delayed because of environmental concerns, and a 1999 meeting of the World Trade Organization in Seattle was disrupted by environmentalists (and others) concerned about the possible impact of global free trade on the environment. Other issues, such as attempts to ban tuna fish from certain nations that do not practice "dolphin-safe" netting practices, have led to international trade disputes. Meanwhile, the World Bank has often made a priority of requiring some environmental concessions of nations hoping to borrow funds.

At this point, it is probably safe to say that the environment is measurably better in most of the developed world because of the hue and cry raised by the environmentalists over the past 40 years. The environmental movement has, without a doubt, played an important role in addressing the most serious pollution problems that faced us in the last half of the twentieth century. It is also true that some environmental initiatives may simply be premature, enacted without clear scientific justification and without the consensus of the scientific community. In any case, as we move into the twenty-first century, humans will continue to be faced with the problems of how best to protect and preserve the planet's natural resources while also accommodating the needs of a technology-based society.

P. ANDREW KARAM

Further Reading

Adams, Douglas. *Last Chance to See*. New York: Ballantine Books, 1990.

Carson, Rachel. *Silent Spring*. Boston: Houghton Mifflin, 1962.

Harr, Johathan. *A Civil Action*. New York: Random House, 1995.

Lyons, Janet and Sandra Jordan. *Walking the Wetlands*. New York: John Wiley & Sons, 1989.

Quammen, David. *The Song of the Dodo*. New York: Scribner, 1996.

The Emergence of Biodiversity as an Issue of Importance

Overview

Biodiversity, sometimes measured by the total number of different plant and animal species in a given area, emerged as an issue of great importance in the later part of the twentieth century. As an agenda item in many national and global political arenas, it was pushed by both environmental groups and scientists, capturing the attention of the public and politicians alike. Awareness of the potential problem grew in the 1970s and 1980s, while the 1990s saw the first concrete steps taken to halt the decline. This process tends to pit developed nations against less-developed ones to some extent because the greatest amount of diversity tends to be in poorer tropical nations. With much at stake for all parties, this issue is likely to continue to be hotly debated for years or decades to come.

Background

Until relatively recently, man's relationship with the rest of terrestrial life was best summarized by the biblical dictate: "Be fruitful and multiply, and fill the earth and subdue it; and have dominion over the fish of the sea and over the birds of the air and over every living thing that moves upon the earth." Virtually all cultures, religions, nationalities, and groups exercised this degree of dominion at some time or another during their history through farming, hunting, and other activities. Some North American animals were almost certainly exterminated by early Native American hunters, who likely burned clearings for farming and hunting. European settlers were no better, adding advanced technology to make the clearing, planting, and hunting more efficient. Asians have also hunted and, in some places, the terrain is so heavily terraced for rice production that the original contours of the land are impossible to ascertain. The Aztec, Maya, and Inca nations cleared fields, planted crops, and hunted extensively, as did many African nations and the New Guinea highlanders. In Australia many large native mammals disappeared coincidentally with the first human settlements. In short, it seems to be human nature to exploit

Deforestation of the Brazilian jungle. *(National Audubon Society Collection/Photo Researchers. Reproduced by permission.)*

our environment for our gain, even when doing so is injurious to the environment. However, it must also be pointed out that this same trait is what propelled humanity forward, for it was not until we mastered some aspects of our environment that we could begin to develop and exploit our natural tools of intelligence, an opposable thumb, upright posture, and adaptability.

In the nineteenth century, scientists become aware that species were not a permanent part of nature, that species could become extinct. With this realization and the growing acceptance of evolutionary theory came the unavoidable corollary that species could also be driven into extinction. This fact was witnessed with the documented extinction of the Carolina parakeet, the passenger pigeon, and, perhaps most famously, the dodo. At the same time, however, colonial powers continued planting colonies for economic gain and newly-independent nations, most of them in tropical regions, began to assert their own economic interests.

By the 1960s most of the former colonies had vanished, leaving many resource-rich and cash-poor young nations. Most of these nations had little to no industry because, as colonies, their mission was to send raw materials to the factories of the imperial country. With a profound lack of skilled workers and manufacturing plants in which to work, these nations either turned to agriculture, continued selling their natural resources, or tried to settle their frontiers. All of these options led to clearing forests to free land for commercial and economic purposes.

With the growing impact of the environmental movement in the 1960s came an increasing awareness of what had been lost in the developed nations and what was potentially at stake in the new tropical states. This was a time, too, when scientists began to discover an increasing number of useful compounds that could be created from plants and animals, including new types of drugs and foods. Other scientific studies, originally begun on islands, showed that patches of rainforest left standing between clear-cut areas came to resemble islands, whose size, determined the species diversity that could be supported. Finally, climatologists pointed out the tremendous impact of rain forests on global climate, particularly their role in the removal of pollutants from the atmosphere, the generation of oxygen, regulation of the global carbon cycle, and the moderation of rainfall patterns. Thus, to some extent, pharmaceutical companies, climate scientists, and island biogeographers came together to point out that the continued exploitation of the rain forests could have long-term effects on the quality of human life, as well as the livability of the Earth as a whole.

There are a great many areas in which biodiversity is threatened or has been severely depleted. Indonesia suffered from extensive forest fires in 1997 and 1998, the nearly universal forest of eastern North America has all but vanished, and coral reefs are under attack in Australia and near Florida. However, it may be instructive to focus attention on the condition of the Amazon rain forest, as it is the largest, possibly the most threatened, and the best known of such examples. It is also safe to say that the example of the Amazon is fairly typical.

Impact

The recent appreciation of the importance of biodiversity has led to a number of significant developments in the modern world, particularly in terms of scientific and medical research, ecological effects, and political and economic effects.

There is a long history of using natural compounds for medical care. In fact, until very recently, all medical treatments were found in natural substances because, until the advent of modern chemistry, there was no way to reliably synthesize compounds. Although traditional treatments were often dismissed by scientists, awareness has grown that many natural compounds are far superior to synthetic alternatives for the treatment of disease. This led to a growing appreciation of the role the Amazon rain forest can play in the quest for more effective medicines and other treatments, as the Amazon contains a large fraction of the world's genetic diversity.

At the same time, the global food supply is now being viewed as somewhat tenuous because of the widespread use of just a few genetic varieties of the major crops. This raises the specter of widespread famine should a disease or parasite emerge that attacks one or more major crops. Such a pestilence, if spread worldwide by modern transportation, could eliminate a large part of global grain production in a short period of time. The Irish potato famine of 1845-47 resulted in the deaths of over 1 million people and precipitated the emigration of another 1 or 2 million from a single small nation that, even today, has a population of only about 4 million. If the world's wheat or rice crops were to suffer a similar fate, the effect would be global and catastrophic. For this reason, scientists are turning to the tropics for genetic variability and potential new crops, hoping that new species may be amenable to domestication or that genetic advantages of some species may be transferable to existing crops to help make them more resistant to disease.

The ecological effects of rain forest loss are briefly described in the preceding section. Of these, the effects on terrestrial climate are still poorly understood. However, the aspects of island biogeography that apply to loss of rain forest are fairly well understood and are significant.

Simply put, small areas hold fewer species than do large areas. Furthermore, a number of small areas will, in composite, hold fewer species than a single large area of the same area as the combined smaller areas. In the case of the Amazon, clear-cutting or burning large swathes of rain forest for lumber, agriculture, or population expansion typically leaves stands of trees that act as islands of diversity in a sea of farmland or bare soil. As time passes, the genetic diversity of these stands decreases because species that have large ranges cannot support a stable breeding population and die out. Then, the species that depend on them die out, followed by the species that depend on them, and so forth until the "islands" are left with a greatly simplified ecosystem containing only a fraction of its original diversity. Given that some species have extremely limited ranges (some are limited to individual mountains, hills, or valleys), this "island effect" can rapidly lead to the extinction of a large number of species in a relatively short period of time.

In the recent past, with the growing awareness of the importance and relative fragility of the Amazon and other rain forests, many of the developed nations have been pressuring developing nations to preserve their remaining wilderness regions. In many cases, this pressure is the direct or indirect result of political pressure from ecologically concerned constituents. In most cases, this unwanted outside pressure is not warmly received.

Most developing nations point out that the developed world reached that status by exploiting their own natural resources and, when they were exhausted, by exploiting those of their colonies. They then point out the hypocrisy of such a stance on the part of the developed nations, most of whom continue to use resources at a rate far in excess of the rest of the world. In addition, the developing nations note that they are simply trying to improve their economies and standards of living through the same paradigm used by the developed nations. They argue that refusing to use what raw materials they have places them and their citizens at a significant disadvantage in terms of quality and length of life, national wealth, and other factors. Finally, it is often suggested that the developed nations are simply trying to hold less developed nations in an economically and politically disadvantaged position, the better to continue practicing political and economic domination. The developed nations, on the other hand, contend that they have learned much in the past century and are trying to save new countries from making the same mistakes. This debate is far from being settled and will likely continue for the foreseeable future, with its ultimate resolution difficult to predict.

P. ANDREW KARAM

Further Reading

Brin, David. *Earth.* New York: Bantam, 1990.

Diamond, Jared. *Guns, Germs, and Steel: The Fate of Human Societies.* New York: W. W. Norton, 1999.

Quammen, David. *The Song of the Dodo.* New York: Scribner, 1996.

Terborgh, John. *Diversity and the Rain Forest.* New York: Scientific American Library, 1992.

Wilson, E. O. *The Diversity of Life.* Cambridge: Harvard University Press, 1992.

Advances in Ecological Theory

Overview

A number of scientific disciplines have come together in the last few decades, combining with each other to give a much more complete understanding of the way that organisms interact with each other and their surroundings to form an ecosystem. The science of ecology has grown vastly more sophisticated in the recent past, sparked in part by the growing ecological awareness of the public and their insistence on ecological studies. Among the subdisciplines now included in ecological theory are population genetics, population and ecological computer modeling, and hydrogeology. It is somewhat ironic that previous ecological errors have actually helped fuel current ecological theories, as efforts to correct these earlier mistakes require a deeper understanding of what a healthy ecology should look like.

Background

Humans have long had an impact on their environment through hunting, agriculture, and city building, even in earliest times. Until recently, however, very little was known about the environment that was being changed. To be sure, hunter-gatherer cultures developed empirical knowledge of their surroundings because it was necessary for survival. But there is little, if any, extant evidence of such early observations that relates the way in which the various parts of the environment interacted in their time. As a result, most cultures changed their environment with little regard for future impacts, sometimes suffering as a result. For example, it is thought that the Mayan civilization fell, in part, because of food shortages due to poor land management.

The Industrial Revolution, technological and medical advances, and the concomitant increase in human life-span and population have resulted in a situation in which humanity can have a dramatic and long-lasting effect on large areas. While the disappearance of species and other ecological effects have been noted for at least a few centuries, it is only recently that humans have realized that each relatively minor change may cause far-reaching effects through a network of interconnections that we still see only dimly. The study of these interconnections between various forms of life, their environment, and each other is the study of ecology.

Ecological science became important in the 1960s with the explosive growth of the environmental movement. Sparked in part by atmospheric nuclear weapons testing and industrial emissions that became unbearable, the early environmentalists began to develop tools to study the interactions and interrelationships that comprise an ecology. What began as empirical observations led to conceptual models and, from there, to mathematical formulations and models. In some cases, once it was felt that a system was somewhat understood, attempts were made to take corrective or preventative actions. One example is the population management of deer herds near urban areas. An absence of predators can lead to an explosion in deer populations. This, in turn, results in loss of food supply for the deer, disease, and starvation. To redress this imbalance, it is often decided that humans must take the place of other natural predators by extending the hunting season for deer or, in some cases, by employing marksmen to cull the deer herds. Other, perhaps less dramatic examples of interventions include attempts to construct artificial wetlands to replace those lost to development, environmental restoration projects, and attempts to save species in the Amazon by mandating that stands of forest be left intact even in areas otherwise clear-cut for agriculture.

Impact

Our deepening understanding of ecological dynamics is due largely to the wider array of scientific and conceptual tools that can be brought to bear now as compared to previously. In addition, there was something of a paradigm shift in human thought, whereupon we finally realized that, not only were we changing isolated bits of the environment, but that these bits interacted with each other, sometimes over large distances, causing unforeseen consequences with potentially far-reaching effects. This deeper understanding stands as the primary impact that ecological studies and tools have had on society. In turn, it has sparked secondary effects on society, primarily social and political in nature.

Before proceeding, it is necessary to define the terms "conceptual model" and "conceptual tool." For the purposes of this essay, a conceptual model refers to an idealized and simplified system that provides the "big picture" idea of processes in an ecosystem. The "circle of life" made popular by recent movies is a conceptual model. A conceptual tool is a way of thinking of or approaching a problem that helps to make it solvable. For example, evolutionary theory explained many observations quite well, but lacked a credible method for transmitting changed characteristics to offspring. The conceptual tool of genes helped explain how traits could evolve, even though the method by which genes work, or even where they appeared in an organism was not discovered for several decades.

In the case of ecological studies, the conceptual tools that were brought to bear on the subject included advances in population genetics and advanced statistical and mathematical methods to help untangle and quantify the relationships observed in the field. These led researchers to the basic conceptual model of most ecosystems, the now cliché concept of the "web of life." It must also be pointed out that, cliche or not, the idea of a web is an accurate one from at least two perspectives. First, of course, it suggests that any single point is connected to multiple other points, each of which is, in turn, connected to many other points. In addition, more concretely, tugging on any part of a web causes the entire web to flex and deform, even if only slightly. This serves as an excellent conceptual model for one of the most important points that has been learned—nothing exists in a vacuum. In other words, traditional scientific methods of examining each small bit minutely and in isolation will not work because of the high degree of interconnectivity involved in ecological webs. Studying each part in isolation and assembling the pieces will not work because the whole system depends on the connections as much as it does on the individual pieces.

The web analogy is also accurate and instructive in another way. For example, the removal of a single strand from a web may diminish its strength slightly, but the web will remain. By judiciously plucking strands, a great deal of a web may be removed without excessive impact. However, there comes a point at which the web suddenly begins to lose cohesion and utility because it begins to unravel. At this point, the loss of each successive strand places excessive strain on the remaining structure. We are beginning to understand that this, too, bears a resemblance to ecosystems, some of which originate with fewer "strands" than others.

Ecological studies have had a social impact that continues to grow. As ecological understanding increased in depth and detail, so too did awareness that individual actions are important. This, in turn, has led to increases in recycling, "green" products and processes, growth in "organic" products, more fuel-efficient internal combustion engines, and measures designed to reduce the stress on local and regional ecologies. The environmental restoration projects at contaminated sites have launched a major industry devoted to remedying ecological damage of the past. Included in many of these projects is funding for detailed ecological studies of the sites before, during, and after remedial activities. Those now espousing environmentalist beliefs or studying the ecological impact of activities, many of whom were once on the fringes of science and society in the early 1960s, are now considered mainstream. This, in and of itself demonstrates the profound impact that advances in the field of ecology have had on society in the last half of the twentieth century.

Because of the growing social awareness of ecological concerns, the environmental movement is growing more politically powerful. Part of this increase is due to the fact that early environmentalists are now assuming positions of authority in society, while part is no doubt due to the increasing political strength of the environmental movement in many countries. Politicians respond by approving additional funding for ecological studies, passing environmental legislation, authorizing environmental restoration projects, and attempting to place constraints on activities by other nations that are perceived to be "ecologically incorrect." However, each of these activities, in turn, has consequences.

Despite these apparently positive developments, it is worth noting that attention (whether monetary, regulatory, or political) won by environmental and ecological causes is often at the expense of other important causes. Therefore, it is possible that, in addressing one high-profile concern, resources may be taken from other, higher-risk concerns that are not as politically popular. This may result in a paradoxical situation where, for instance, funding for a badly needed environmental restoration project is lost to a high profile but less immediate concern. Similarly, political pressure brought to bear on allies or trading partners can strain international relationships, resulting in unforeseen adverse consequences. This makes it even more important to develop and use the scientific tools necessary to ensure that we understand ecological relationships sufficiently well that, when recommending corrective actions, we are confident that the actions to be taken will:

1. Improve the perceived problem
2. Not place added strain on existing ecological relationships
3. Result in an actual reduction of risk to society when all risk factors are considered (including the risk of any excavation, construction, and transportation activities that would be required)
4. Be at least as effective as other possible actions that would cost less to implement.

Only in this way can we ensure that our activities actually result in a net gain to society, rather than simply satisfying our urge to do what seems on the surface to be the right thing to do. Here, again, the ecological tools that led us to an awareness of the problems should also help us to understand and to correct them.

P. ANDREW KARAM

Further Reading

Brown, James, and Leslie Real, eds. *Foundations of Ecology: Classic Papers with Commentary.* Chicago: University of Chicago Press, 1991.

Carson, Rachel. *Silent Spring.* Boston: Houghton Mifflin, 1982.

Leopold, Aldo. *A Sand County Almanac.* New York: Ballantine, 1966.

Lyons, Janet, and Sandra Jordan. *Walking the Wetlands.* New York: John Wiley & Sons, 1989.

Peterson, David L., and V. T. Parker. *Ecological Scale, Theory, and Applications.* New York: Columbia University Press, 1998.

Pimm, Stuart L. *The Balance of Nature? Ecological Issues in the Conservation of Species and Communities.* Chicago: University of Chicago Press, 1991.

Ray, Dixy Lee, and Lou Guzzo. *Environmental Overkill: Whatever Happened to Common Sense?* New York: HarperPerennial, 1994.

Ray, Dixy Lee, and Lou Guzzo. *Trashing the Planet: How Science Can Help Us Deal With Acid Rain, Depletion of the Ozone, and Nuclear Waste (Among Other Things).* Washington, DC: Regnery Gateway, 1990.

The Rise of Biotechnology as Big Business

Overview

Biotechnology emerged as a big business in the last quarter of the twentieth century. The year of its birth was 1977. The promise of biotechnology has been compared to that of alchemy, which in the Middle Ages sought not only to turn lead into gold but to find the secrets of life itself. Biotechnology-related research is mushrooming not only in the field of medicine but in such areas as agriculture, fuels, plastics, the environment, and mining.

The meshing of science with business, however, is not without its problems. While scientific investigations may be done within an academic setting, the problems involved in taking these findings and making practical use of them are legion. This goal requires investment of capital in firms that will produce and package a product. The question of patents also arises. With an invention such as a car device, for example, there are no ethical problems, but with genes and organisms the situation is completely different. A legal and ethical debate has arisen over who can own genes and over issues of technology transfer. In the meanwhile, the business of biotechnology has added new phrases to the English vocabulary, including "research and development" (R & D), "technology transfer," and "venture capital."

Background

Scientific discoveries often wait many years before being used. For example, when a new drug is created it must undergo lengthy testing by the U.S. Food and Drug Administration (FDA) before it is approved. If the work takes place at a university, the drug development program must pass through a scrutinizing R & D committee before it can even get off the ground. The lengthy process of FDA approval, in turn, is a nightmare of bureaucratic red tape. Researchers must go through pre-clinical tests using both in vitro (test tube) experiments and in vivo (living animal) studies. At this point, researchers file for an "initial new drug" (IND). All work is highly regulated and must be supported with massive amounts of data. The drug then enters Phase I testing, involving a small number of humans to look for side effects. If approved for Phase II, researchers must recruit more subjects. Phase III involves even larger numbers of test subjects. An investigation may be ended at any point; some drugs have reached phase III and not been approved. Other Western countries have similar procedures, although the time taken may be shorter. Prospects of the business of biotechnology are therefore usually considered with a long-term outlook. Biotechnology has become a great economic opportunity in the twentieth century and is expected to loom even larger in the twenty-first.

Impact

In a laboratory at the University of California, San Francisco, Herbert Boyer inserted a synthetic insulin gene into a bacterium, *E. coli.* Boyer later convinced Robert Swanson, a venture capitalist, to invest in a company called Genentech. The term "venture capitalist" would soon become very important. Venture capitalists are individuals who invest in a company to start it up and support it. In 1977 Genentech reported the production of somatostatin, a human growth hormone created by recombinant DNA technology. That moment began a slow trickle of developments that soon became a downpour, involving diagnostic tools and techniques. Universities and fledgling companies entered the biotechnology race. Many consider 1977 the dawn of the "Age of Biotechnology."

Biotechnology's birth, however, was not without labor pains. Reacting to fears, Congress had sixteen bills introduced to regulate recombinant DNA research, but none of the sixteen bills passed.

Other developments came that year. Bill Rutter and Howard Goodman isolated the gene for rat insulin. Walter Gilbert (1932-) and Allan Maxin likewise initiated a new epoch in the study of genetics when they devised a method for sequencing DNA using chemicals rather than enzymes. In 1978 Biogen, SA of Geneva was founded by a consortium of businessmen and scientists that included Gilbert. A pharmaceutical company, Schering-Plough, Inc. was a major investor.

The small company Genentech began to grow and signed research agreements with several large pharmaceutical houses. Bitter legal battles, however, soon erupted. The University of California sued Hoffman-LaRoche and Genentech, claiming that a line of cells used to produce interferon was created under their auspices. Another squabble ensued over royalties regarding the human growth hormone. Genentech in the process lost much money and the possibility of future royalties. These disputes, however, would lead to collaboration agreements between universities and industry to benefit both financially.

Cetus Corporation was founded by a physician, a biochemist, and a physicist. The corporation's initial work involved using genetically engineered organisms to produce industrial chemicals, such as ethylene oxide for making chemicals and plastics. In 1980 the corporation's Kary Mullis invented a technique for replicating DNA sequences in vitro, called polymerase chain reaction, or PCR, revolutionizing molecular biology. Cetus patented PCR and in 1991 sold the patent to Hoffman-LaRoche, Inc. for 300 million dollars.

In the past, diabetics had to purchase insulin made by extractions from the pancreases of cows and pigs. About five percent, however, suffered allergic reactions. In 1982 Genentech received the FDA's approval to market genetically engineered human insulin, in which every atom is identical to human insulin. In 1983 Eli Lilly presented this product to the market.

In 1980 the U.S. Supreme Court ruled that genetically altered life forms could be patented. The case involved the Exxon oil company, which wanted to patent an oil-eating microorganism. This landmark ruling opened enormous possibilities for the commercial growth of genetic engineering.

Collaborations between industry and academics began to build. In 1981 Hoechst AG, a German chemical company, gave Massachusetts General Hospital, a Harvard Medical School teaching facility, 70 million dollars to build a new department for molecular biology. The only "catch" was that the company would have the rights to license any technology emerging from the facility.

These new agreements prompted a series of hearings held by Congressman Al Gore on the relationship between the academic world and industry. The hearing focused on the potential for profit from the intellectual property as well as on patent rights the universities could own. Jonathan King of the Massachusetts Institute of Technology, however, reminded the biotechnology industry that the most important long-term goal of research is for the good of mankind. One of the problems with a marriage between academia and industry is that few businessmen understand the nature of science and research and few scientists know business.

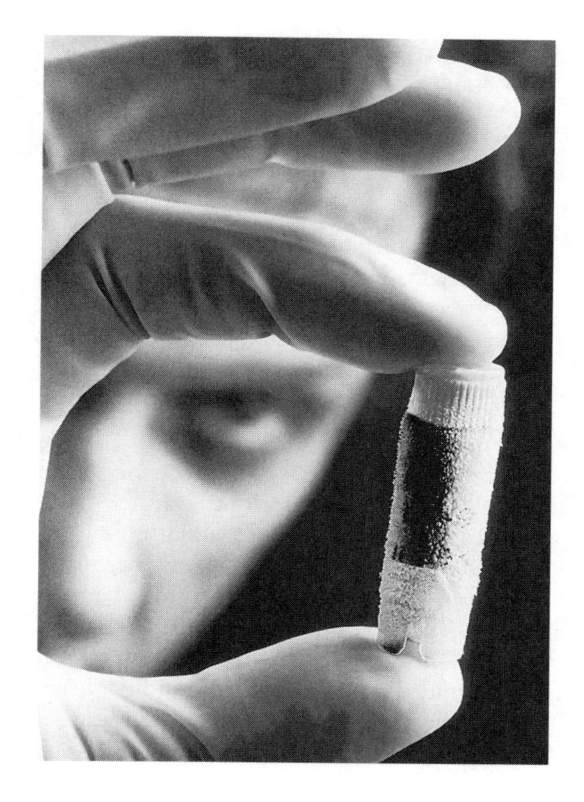

A biotechnology researcher examines a vial. *(Bill Varie/Corbis. Reproduced by permission.)*

In 1983 patents were granted to companies for genetically engineered plants, and Syntex Corporation received FDA approval for a monoclonal antibody diagnostic test for chlamydia.

In 1984 English researcher Alec Jeffreys (1950-) hit upon the idea of using differences in DNA for identification. He coined the term "DNA fingerprinting." Jeffreys used the technique in March of 1985 to prove that a boy was the son of a British citizen. The use of DNA fingerprinting in forensics grew into big business, receiving major publicity in the O.J. Simpson trial of 1995.

In 1988 Harvard investigators Philip Leder and Timothy Stewart were awarded the first patent for a genetically altered animal, a mouse that is susceptible to breast cancer. In the same year SyStemix Inc. received a patent for the SCIDHU Mouse, a mouse engineered for AIDS research. The value of "transgenic animals" for research should not be understated. These animals are engineered to display human traits such as hemophilia or produce factors such as insulin. In the summer of 1999 a company's transgenic mice, programmed to express Alzheimer's disease, showed significant results after receiving a vaccine for the disease. (Sprague-Dawley rats are raised in special clinical conditions for the study of heart disease, high cholesterol, and even obesity.

The 1990 novel *Jurassic Park* evoked a lot of interest in biotechnology in its presentation of a genetic engineering experiment with dinosaurs going awry. This year also saw the launch of the Human Genome Project, an international effort to map all the genes in the human genome. In fact, much of the work on the genome project has been farmed out to small biotechnology companies.

"Biotech" companies and discoveries began to explode. From genetically engineered tissue to frost-resistant fruits, investigations multiplied. Universities and emerging companies negotiated complex agreements that even cross national borders. The Biotechnology Industry Organization (BIO) was formed in 1993 to promote growth and development in the industry. The BIO emphasizes that it takes 10 to 15 years of research to develop just one product. The venture capital needed to bring such a product to market can exceed 100 million dollars.

As the twentieth century ended, there were 1,274 biotechnology companies in the United States. There were also 300 biotechnology-related products and vaccines in human trials and hundreds more in development. During 1998, 385 new products were introduced and 364 new companies were formed in the industry.

Federally sponsored research in universities combined with the idea of technology transfer has paid off. "Technology transfer" is the patenting and licensing of discoveries made in academic areas and taken from the laboratory to the commercial sector. One example is the work conducted by Florida State University chemist Robert Holton, who began work in the 1970s on the Pacific yew tree bark. He received a grant from the NIH and in the early 1990s synthesized paclitaxel (TaxolTM). Taxol was first marketed by Bristol-Myers-Squibb in 1992 as a treatment for ovarian cancer (but used only as a last resort). In 1998 the FDA approved Taxol for the treatment of both ovarian and breast cancer. Taxol had sales in 1998 of 1.2 billion dollars. Time and investment eventually paid off for Taxol's owner.

The debate over patents, however, has continued. In 1997 the U.S. Patent and Trademark Office announced that it would allow patents on expressed sequence tags (ESTs), which are short sequences of human DNA that are used in genome mapping. Many scientists strongly objected to this announcement. Agreements with other countries, including emerging and developing ones, also involve legal and ethical hurdles yet to be addressed.

At the beginning of the twenty-first century the biotechnology industry holds great promise. The 50-billion-dollar figure shown by the industry at the end of the twentieth century should pale in comparison to future prospects. The business of biotechnology, moreover, is expected to expand into such areas as the environment, gene therapy, vaccines, manufacturing, food safety, and bioprocessing.

EVELYN B. KELLY

Further Reading

Crow, James, and William F. Dove. *Perspectives on Genetics: Anecdotal, Historical, and Critical Commentaries.* Madison: University of Wisconsin Press, 2000.

Landau, Ralph, ed. *Pharmaceutical Revolution: Revolutionizing Human Health.* Philadelphia: Heritage Press, 1999.

Oliver, Richard W. *The Coming Biotech Age: The Business of Bio-materials.* New York: McGraw-Hill, 1999.

Sterck, Sigrid. *Biotechnology, Patents, and Morality.* Brookfield, VT: Ashgate Publishing, 2000.

Population Genetics and the Problem of Diversity

Overview

The extent to which genes differ between individuals, races, and species has been the central theme of population genetics in the twentieth century. New experimental methods developed in the 1960s allowed the first estimates of the degree of genetic variation in natural populations of human and non-human species. The unexpected finding was that on average, at least one out of every three genes in a species had more than one molecular form, revealing substantial genetic variation among members of the same species. In humans related studies revealed that the genetic variation between individuals of the same race was much more pronounced than that between races. These findings called for a re-thinking of the role of natural selection in evolution and brought a deeper understanding of the close symbiosis between genes and the environment. The inadequacy of racial classifications in humans became clear, with far-reaching implications for the use of racial distinctions in human society.

Background

Natural selection, which is sometimes referred to as "survival of the fittest," formed the cornerstone of the theory of evolution formulated by Charles Darwin (1809-1882). Species evolve due to hereditary variations that favor their survival and reproduction. It is an organism's phenotype, or outwardly expressed characteristics like physiology and behavior, that is subject to selection by the environment. By contrast, what is actually transmitted from one generation to the next is the genotype, the collection of genes that is inherited by the organism. The phenotype of an organism is determined both by its genotype as well as the environment in which it grows. Human skin color, for instance, varies with the amount of exposure to ultraviolet sunlight, with darker skin having a selective advantage for people who live in lower latitudes.

Each gene can have one or more molecular forms, giving rise to the notion of genetic *polymorphism.* Population genetics deals with characterizing the frequency with which one or other form of a gene prevails in a population. A polymorphic gene is one that has more than one variant present in significant amounts—usually a few percent—in natural populations. The earliest example of genetic polymorphism in humans was found in studies of blood types. Different forms of a gene gave rise to different types of blood cells that were incompatible with one another.

For many years, it was thought that most genes were not polymorphic. This was in accordance with the assumption that natural selection would weed out unfit variants of a gene, leaving a relatively homogenous gene pool in each species. This view was challenged in 1966 by Richard Lewontin (1929-) and Jack Hubby, who worked with fruit flies, and by Henry Harris, who worked with humans. Their method relied on the known link between proteins and genes—that the sequence of molecules making up a protein was a translated copy of the gene sequence. By studying the protein sequence, they could make inferences about the gene sequence and distinguish one form of a gene from another. They used a technique called *gel electrophoresis,* in which proteins of different electric charge migrated different distances in a gel. Even a single variation in the protein sequence could be detected in this way. The main advantage of this method over blood group studies was that a nearly random selection of genes, not affiliated with a particular trait, could be analyzed for polymorphism.

Lewontin and Hubby found that 7 out of 18 genes in their fruit fly samples were polymorphic, while Harris found that 3 out of 10 genes in his samples taken from the human population were polymorphic. Subsequent studies in other species, including other primates, amphibians, rodents, and birds, reached the same conclusion: on average, at least one out of every three genes in a species harbored polymorphic variants. In humans, this meant that any two individuals taken at random from a population are likely to have no more than two-thirds of their genetic material in common. It is natural to ask whether it matters if these individuals are chosen according to their racial or ethnic affiliation.

Although individual traits such as blood type or disease susceptibility were known to show marked racial variation, no single trait could be used to determine a person's race with certainty, and no two traits were seen to yield similar racial groupings. What was needed was a random sampling of the human genome to determine the extent to which humans of different races diverged from one another genetically. Harris, Lewontin, and others used electrophoretic studies in the early 1970s to estimate racial differences in humans. These studies relied on a measure of genetic variation that was more precise than polymorphism, called *heterozygosity.* In sexually reproducing organisms, genes occur in pairs, one of each pair inherited from each parent. Heterozygosity measures the probability that a gene pair has dissimilar genes for an individual taken at random from the population. While polymorphism does not indicate the number of variant forms of a gene (other than to say that such variation exists), heterozygosity would be higher for a gene with more variants.

A famous study by Lewontin in 1972 used 18 polymorphic genes in 7 races, including Africans, Caucasians, and Mongoloids. He found that 85% of the heterozygosity in the human species was already present in a single nation or tribe. Race contributed about 7% of the decrease in total heterozygosity, while nationality contributed the remaining 8% of the decrease. In other words, genetic diversity in humans was most pronounced at the individual leve, rather than at the national or racial level.

Impact

The discovery of large amounts of genetic variation in humans and non-human species revolutionized the understanding of the evolutionary theory of natural selection. Polymorphism is introduced into a species by mutation, the random errors that occur in the replication and transmission of genes from parents to offspring. Once introduced, it was thought that these mutant genes would either become prevalent or die out under the forces of natural selection, in either case moving toward a reduction in variation. The continued maintenance of so much variation in a species called for new evolutionary assumptions, including some that departed from traditional Darwinian evolution.

In 1968 Motoo Kimura (1924-) introduced the idea that perhaps a large fraction of the genetic variants observed were selectively neutral—that is, they were immune to the forces of natural selection. He assumed that these variations were introduced into a population by mutation and maintained by random genetic drift. The latter refers to the statistical fluctuation in the frequencies of gene variants transmitted from one generation to the next. This implied a less deterministic view of evolution and one that was less adaptive to the environment.

Advocates of natural selection, on the other hand, considered mechanisms of selection that maintained, rather than reduced, genetic variation. Selection could be frequency-dependent, favoring high frequencies of polymorphism in the species. Heterozygosis could also have a selective advantage, as in the case of the sickle-cell gene in humans. This gene causes anemia but offers resistance against malaria, so having one member of the gene pair of the sickle-cell type optimizes the chances of survival in areas where malaria is prevalent. That genetic variation in a species could be preserved by natural selection seems contrary to Darwinism, although Lewontin takes the view that Darwinian evolution is essentially the conversion of variation within populations to variations between populations in space and time.

The neutralist-selectionist controversy of the 1970s left its mark on evolutionary theory, with elements of both views gaining currency with new experimental evidence. In the end, it was an effort to come to terms with the immense genetic diversity inherent in every species. Various factors were identified that could sustain this diversity and still allow for deterministic evolution. Genetic diversity implied a variation in the observed characteristics of an organism, and this link between the genotype and the phenotype was a necessary ingredient in determining the effects of natural selection. Apportioning causes to genes and the environment in determining the phenotype of an organism was at the heart of another controversy that has had a long history in science, the problem of nature versus nurture. Apart from the genetic contribution, the immense range of environments available to an organism further added to the diversity of phenotypes observed.

In some sense, the discovery of genetic diversity among humans came as no surprise, since it has often been said that "no two individuals look exactly alike." By the same token, the very features that are used to distinguish individuals—skin color, hair form, eye shape, nose shape, etc.—are the basis for the perception of distinct races in human society. The discovery in the 1970s was that, at the biochemical level, races were not nearly as homogeneous as we perceived them to be, and racial distinctions were, at best, superficial. Racial groupings continue to be used in anthropological studies, but more as a means of describing the findings rather than explaining them. Population biologists continue to study the variation of human traits across populations, defined as groups of interbreeding individuals, without inferring any racial taxonomy per se.

Although the race concept has lost its utility in these fields of study, it still informs many issues in public policy, especially in multi-racial societies. This is because race can be associated with social inequalities among people that are attributable to their perceived racial differences in the society. Even genetic differences between races continue to raise complex public policy concerns. One issue is whether companies should be allowed to screen individuals for race-specific genetic diseases such as sickle-cell anemia. Another is whether law enforcement agencies can rely on deoxyribonucleic acid (DNA) fingerprinting for incriminating individuals, given the possibility that such techniques may not be racially unbiased. It is noteworthy that the average genetic differences between races that were studied in the 1970s do not obviate such concerns.

New developments in DNA technology in the 1980s made it possible to map individual DNA sequences in each gene directly. This provided the impetus for mapping the entire human genome, an international ongoing effort known as the Human Genome Project. An offshoot of this project was the Human Genome Diversity Project (HGDP), a proposed survey of the genetic variation in humans from all racial and ethnic groups in the world. The urgency of this project is based on the assumption that the genetic isolation of the world's populations is fast disappearing today due to the rapid merging of peoples through migration. Still in its infancy, the HGDP continues the trend of studying the racial diversity and unity preserved in the human genome that was inaugurated by the electrophoretic studies in the 1970s.

ASHOK MUTHUKRISHNAN

Further Reading

Books

Cavalli-Sforza, Luigi L., Paolo Menozzi, and Alberto Piazza. *The History and Geography of Human Genes.* Princeton, NJ: Princeton University Press, 1994.

Committee on Human Genome Diversity, Commission on Life Sciences, National Research Council. *Evaluating Human Genome Diversity*. Washington, DC: National Academy Press, 1997.

Hedrick, Philip W. *Genetics of Populations.* Boston: Science Books International, 1983.

Kimura, Motoo. *The Neutral Theory of Molecular Evolution.* Cambridge: Cambridge University Press, 1983.

Lewontin, Richard C. *The Genetic Basis of Evolutionary Change.* New York: Columbia University Press, 1974.

Montagu, Ashley. *Man's Most Dangerous Myth: The Fallacy of Race.* 6th ed. London: SAGE Publications Ltd., 1997.

Periodical Articles

Lewontin, Richard C. "The Apportionment of Human Diversity." *Evolutionary Biology* 6 (1972): 381-98.

The Human Genome Project

Overview

The Human Genome Project is a massive scientific effort to identify all human genes and determine the sequence of the chemical bases (nucleotides) of the entire human genome. The human genome has perhaps 50,000-100,000 genes, which are constructed from about 3 billion base pairs of nucleotides. There are four different nucleotides—adenine, cytosine, guanine, and thymine—and the arrangement of these four bases in specific sequences acts as a hereditary code. These 3 billion nucleotides are located on DNA molecules within 23 pairs of human chromosomes in the nucleus of cells in the human body. The Human Genome Project began in 1990 and was funded at $200 million per year by the United States National Institute of Health and the Department of Energy. At its inception it was expected to last 15 years and cost $3 billion, making it one of the largest single scientific projects funded by the United States government.

Background

In 1900 the experiments on heredity by the Austrian monk Gregor Mendel (1822-1884) were rediscovered, giving rise to Mendelian genetics. About the same time biologists discovered that the chromosome is the part of the cell involved in passing on heredity. The field of genetics took a great leap forward in 1953 with the discovery of the double helix model for DNA—the molecule in the chromosome responsible for transmitting genetic information from one generation to the next—by James Watson (1928-) and Francis Crick (1916-). This capped a long search by biologists for the chemical basis for heredity. It also opened up exciting new vistas for research into the nature of heredity and hereditary illnesses.

Impetus to begin the Human Genome Project came largely from health professionals who believed that the knowledge gained through the project could dramatically improve diagnosis and treatment of a wide range of genetic illnesses. For the past couple of decades, medical geneticists have been searching for various genes that cause major illnesses and have already identified some. In 1987, for example, a major breakthrough came when the gene causing cystic fibrosis was discovered. Finding the specific gene behind a particular illness cannot only greatly improve diagnosis—even in advance of the onset of the disease—but can also help scientists understand the root causes of the disease.

In 1987 the Department of Energy's Biological and Environmental Research Advisory Committee recommended the implementation of a large-scale research project to map the entire human genome. The Department of Energy had already funded research on the effects of various kinds of radiation on human health, and it had a good track record of managing large scientific research projects. The following year the National Research Council issued its own recommendation that the United States government, specifically the National Institute of Health, take the lead in human gene sequencing. To further this research, three national genome research centers were established in 1998-99 at Lawrence Berkeley National Laboratory, Lawrence Livermore National Laboratory, and Los Alamos National Laboratory.

In 1990 the project officially began, and James Watson, co-discoverer of the double helix model for DNA in 1953, became the first director of the National Center for Human Genome Research (its name changed to National Human Genome Research Institute in 1997). After Watson resigned in 1992, Francis Collins, a leading genetics researcher at the University of Michigan, took over leadership of the project in 1993.

The actual gene sequencing is being carried out at many different research centers in the United States, Japan, England, France, Germany, and China. Research teams at these various sites work on sequencing a discrete segment of a chromosome, and the results are synthesized. In addition to working on mapping and sequencing the human genome, part of the project include sequencing other organisms, partly to gain experience and refine research techniques. These include several microorganisms, the fruit fly, and the mouse. The latter two organisms have been two of the most important organisms used by scientists for research in genetics. The data from these organisms can provide useful comparisons between human heredity and the heredity of other animals. In 1995 the first whole genome for an organism other than a virus was sequenced when scientists completed sequencing the genome of a relatively simple bacterium, *Haemophilus influenzae*. By sequencing the microorganism *Methanococcus jannaschii*, scientists confirmed in 1996 that it belongs to a third major branch of life called Archaea (the other two are prokaryotes [bacteria] and eukaryotes [most other organisms]).

Producing physical maps of chromosomes involves cutting the DNA molecules using restriction enzymes—special chemicals that slice DNA at specific locations. Once the DNA molecule is cut into fragments, the fragments are cloned or copied. When enough overlapping DNA fragments are analyzed, maps of the chromosomes can be constructed. Maps are the first step in sequencing all the nucleotides in the DNA. Data obtained from the sequencing projects is stored in the Genome Database, established in 1991 at Johns Hopkins University. Because data are made available over the Internet, scientists throughout the world can take advantage of the information as soon as it is processed.

Impact

With the information generated by the Human Genome Project, scientists hope to be able to locate the genes that cause specific illnesses, as well as those responsible for normal functions. By knowing the nucleotide sequences they can figure out the ultimate physical basis for hereditary illnesses and for normal growth and development. Geneticists hope that this will not only aid diagnosis, but will ultimately lead to new treatments for genetic illnesses. While this aspect of the Human Genome Project is relatively unremarkable, more controversial is the possibility of using the information for genetic engineering. Treatments that involve manipulating the DNA are already being used on some illnesses, and such methods will likely increase as our understanding of human genetics expands.

The Human Genome Project has aroused many ethical concerns. The concerns were so pronounced that 5% of the project's budget was set aside to fund discussion of the ethical, social, and legal implications of the projected knowledge. This makes the project the first large scientific project to address the ethical dimensions of the knowledge generated. Some fear the Human Genome Project may presage "designer genes" and may mark the advent of a reinvigorated eugenics (conscious attempts to "improve" human heredity).

However, the most persistent concern seems to be privacy of information. Once we have the knowledge to locate many hereditary illnesses, perhaps years before they manifest themselves, how will we use this knowledge? Could knowledge about one's genetic makeup be passed on to insurance companies, employers, or prospective mates? There seems to be broad agreement that people should have right to privacy regarding their own genetic makeup. The Human Genome Project's Ethical, Legal, and Social Issues branch in 1994 drafted legislation, the Genetic Privacy Act, to protect people against the misuse of genetic information, though Congress has not yet enacted it (as of late 1999).

As of December 1999, the Human Genome Project was ahead of schedule, partly because of advances in technology and sequencing techniques. It had already completed sequencing 15% of the human genome and also had some more in draft form. At the same time the project announced the completion of sequencing the first full human chromosome—chromosome 22. This chromosome consists of 33.5 million nucleotides and has at least 545 genes (scientists estimate there may be another 200-300 that are still not identified). The mean size of the genes on this chromosome is 190,000 nucleotides. Despite this remarkable accomplishment, there are still some gaps in the sequencing that present technology is unable to fill in, but even the gaps are identified.

HGP anticipates completing the rough draft of the entire genome in spring 2000 and completing the entire project sometime in 2003. However, a commercial competitor, Celera Genomics, was formed in 1998, and is racing to beat the Human Genome Project. Celera uses less precise sequencing methods, but boasts that its methods are faster and will produce a complete sequence of the genome by 2001.

RICHARD WEIKART

Further Reading

Books

Cooper, Necia Grant. *The Human Genome Project: Deciphering the Blueprint of Heredity.* Mill Valley, CA: University Science Books, 1994.

Kevles, Daniel. *The Code of Codes: Scientific and Social Issues in the Human Genome Project.* Cambridge: Harvard University Press, 1992.

Nelson, J. Robert. *On the New Frontiers of Genetics and Religion.* Grand Rapids: Eerdmans, 1994.

Rothman, Barbara. *Genetic Maps and Human Imaginations: The Limits of Science in Understanding Who We Are.* New York: Norton, 1998.

Other

Human Genome Project Information Website. www.ornl.gov/TechResources/Human_Genome/home.html

National Human Genome Research Institute's Website. www.nhgri.nih.gov

Current Trends in Gene Manipulation

Overview

Biotechnology involves the alteration or use of cells or molecules for specific applications. The biotechnology that relates to the manipulation of genetic material is called genetic engineering. The result of combining the genetic material of one species with another is called recombinant DNA.

Recombinant DNA technology had its roots in 1975, when 140 microbiologists met to discuss the possibilities of a new type of experiment. Technology had developed the tools for recombining genes from different species. Called restriction enzymes, these tools cut pieces of DNA so that other pieces can be added. These restriction enzymes allow scientists to recombine the genes of different species. The process is also known as genetic engineering. A transgenic organism develops from a gamete that has been altered by the recombination of genes. Transgenic animals are used to "pharm" (produce) useful human proteins or to model human diseases. Cloning is accomplished by inserting material from a somatic, or body, cell into a germ cell (egg).

With the development of procedures that manipulate living things and combine the traits of uniquely different species, ethical questions arise. Fear of the unknown also comes to play a part. Legal and political challenges have emerged from many corners.

Background

Biotechnology in a broad sense has been around since ancient times. Adding yeast to make bread or wine, using pectin to make jelly, or using enzymes to remove stains from dirty clothes are all examples of using biological materials. Molecular biology, however, has its roots in 1938, when Warren Weaver, director of the Rockefeller

Foundation, spoke to his trustees about a new branch of science called "molecular biology" that was beginning to uncover the secrets of the cell. This moment was the first time the term "molecular biology" appeared in print. The term was again used in 1950, when a British x-ray crystallographer used it in a Harvey lecture. "Molecular biology" was off and running.

Impact

In February 1975 Asilomar, a conference center in California, was the site of a meeting of 140 molecular biologists. Researchers had found a simple way to combine the traits of two species and were concerned about where the new field of biotechnology was heading. They discussed safety implications and restrictions that should be in place to prevent the escape of a recombinant organism from the laboratory. Guidelines were also drawn up for laboratory features, such as hoods, and a general tightening of procedures. When the group met ten years later in 1985, they were amazed that the technology had proved safer than predicted and had moved from the research lab to industry much faster than they could have imagined.

In 1977 the age of biotechnology was born. That year a manmade gene was used to manufacture a human protein in a bacterium. Herbert Boyer at the University of California, San Francisco inserted a synthetic version of the human insulin gene into the bacterium *Escherichia coli*.

Manipulating the DNA molecule required biochemical scissors called restriction enzymes. Scientists use these enzymes to cut a gene from its normal location, insert it into a circular piece of DNA (called a plasmid), then transfer these circles into the cells of another species. The value of plasmids is that they can pass from one cell to another. For example, a human sequence called cDNA can be inserted into the plasmid ring, which is then inserted into a microbe's DNA. Plasmids thus are used as vectors to transfer genes between organisms. Bacterial cells divide approximately every 20 seconds, so millions of copies, or clones, containing the human gene can rapidly result. This procedure allows a genome to be manipulated more readily than does the natural breeding process in addition to being able to combine the traits of species. The potential applications of genetic engineering in such fields as food technology, agriculture, forensics, and medicine are staggering.

With such a technology, there are naturally legal and ethical conflicts. In 1980 the U.S. Supreme Court ruled that genetically altered life forms can be patented. Although the decision specifically allowed the Exxon oil company to patent an oil-eating bacteria, the door was opened for commercial genetic engineering. In 1981 Congress held a series of hearings on intellectual property and patent rights.

In 1990 the Human Genome Project was begun to map all human genes, and the first gene therapy trial was carried out on a four-year-old girl with a genetic disorder, called ADA deficiency. While the therapy worked, it set off a firestorm of ethical debate that exploded in the media. Michael Crichton's novel *Jurassic Park* described a theme park where bioengineered dinosaurs roam and frightening results occur when the experiment goes awry. The book's dramatization into a film fanned the flames even more. In 1995 a coalition of religious groups sought to overturn current laws allowing the patenting of genes used for medical work and research. Jeremy Rifkin, an outspoken critic of biotechnology, led a media charge. A survey taken in 1996 showed most people to be suspicious of gene therapy.

The prospects of genetic testing also led to ethical questions. The U.S. Defense Department created a kind of "DNA dogtag" file of genetic fingerprints that could help identify the bodily remains of military personnel. These "prints" would be stored in a repository. The question emerged as to who would have access to the DNA repository—for example, the police or insurance companies that might use the information to refuse coverage if a person carried a gene for a debilitating disease. The questions over the potential benefit and damage of such a program have still not been answered.

The area of genetic screening has raised many questions. Surveys have revealed that the public knows little about genetic testing and even less about gene therapy. Even more frightening is that people who knew nothing about the subject approved of using genetic engineering to improve their children's physical appearance or intelligence. Television programs promoting "perfect people" also added to the confusion.

The year 1997 was rife with discovery and controversy. That year, the public first heard about cloned sheep and monkeys, babies from frozen eggs, headless frog embryos, a 63-year-old mother, and sperm taken from dead men. Debates over bioethics grew but in no way could keep up with the pace of science.

When Dolly, the first cloned sheep, nuzzled her way into headlines in 1997, Ian Wilmut of the Roslin Institute in Edinburgh, Scotland, became a media sensation. He and his staff had cloned Dolly from a cell derived from the mammary gland of a six-year-old sheep. The procedure in theory was simple. An unfertilized egg (oocyte) was removed from the ovary of an adult ewe. The nucleus was replaced with the nucleus of a cell from an adult sheep's mammary gland, and then the egg was implanted into another ewe.

In July the same researchers cloned Polly, the first sheep bearing human genes. Polly could be good news for hemophiliacs, who rely on expensive protein therapy for their condition. On March 4 of the same year, a group of researchers in Oregon cloned two rhesus monkeys from early stage embryos, using nuclear transfer methods. Donald Wolf, the senior scientist on the project, said such animals could be useful in the design of studies for cancer and AIDS.

This cloning research, however, set off alarm bells in Washington, D.C., as visions of cloning humans came closer to reality. The media unleashed a furor centered on pictures of cloned armies, tanks of living organisms for transplantation, and bringing back the dead. President Clinton banned human cloning and told the National Bioethics Commission to review the ethical and legal implications of cloning. In addition, Congress submitted bills banning human cloning and revived old bills on genetic screening, or "snooping."

The research that created Dolly was not new and had been in the making for many years. Nuclear transfer in frogs began in 1950. In 1986 a team of scientists from the Medical College of Pennsylvania transferred nuclei from red blood cells (unlike humans, frogs have nuclei in their red blood cells) to enucleated frog eggs. ("Enucleated" refers to a cell whose nucleus has been taken out.) The clones, however, did not live past the tadpole stage.

Nancy, Ethyl, and Herman were transgenic pioneers. Nancy is a sheep that produces human alpha-1-trypsin (ATT), a material necessary to inflate lungs properly. To engineer Nancy, scientists gave a ewe a drug to produce mature eggs then artificially inseminated her. The fertilized eggs were washed out and micro-injected with copies of the ATT gene. These eggs were then inserted into a surrogate ewe. While the process may sound simple, it was not. Many genes did not go into the correct part of the host's genome, where they were to be expressed. Of 152 implanted eggs, only Nancy had the proper transgene. Ethyl, from Scotland, was engineered to make factor VIII, a protein necessary for blood clotting. At Virginia Tech, genetically engineered piglets were developed to produce the factor in their milk, so that 300 to 600 milking sows could meet the needs of the world's hemophiliacs. Herman, a transgenic dairy calf, has a gene that codes for lactoferon, to be added to infant formula to prevent infection. Dolly was special because she was cloned from the cell of an adult ewe. In 1998 Japanese scientists cloned eight identical calves using cells from one adult.

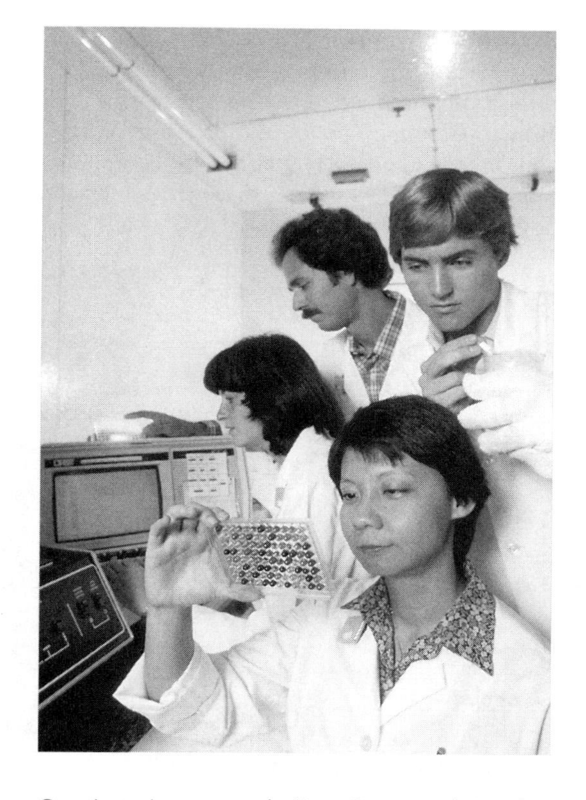

Genetic engineers at work. *(RogerRessmeyer/CORBIS. Reproduced by permission.)*

By the end of the twentieth century, discoveries had grown so numerous they could not be counted. Techniques for analysis and study are likewise becoming sophisticated and complex. For example, a technique may combine PCR, DNA chips, and computer programming to search the genome. Also, high throughput technology allows a multitude of analyses in a short period of time and is valuable in drug discovery. A December 1999 release from Ian Dunham of Cambridge University stated that his team had cloned the genetic sequence for all of chromosome 22, the smallest of the 23 human chromosomes. Also in December, Clyde Hutchison of the University of North Carolina published his discovery of the minimal number of genes required to produce a living organism. The Human Genome Project, moreover, is on time and under budget. A target year for completion was set at 2003. As genetic discoveries increase, however, so will the ethical and legal challenges.

EVELYN B. KELLY

Further Reading

Caplan, Arthur. *Moral Matters: Ethical Issues in Medicine and the Life Sciences.* New York: John Wiley and Sons, 1995.

Heinberg, Richard. *Cloning the Buddha: The Moral Impact of Biotechnology.* Theosophical Publishing House, 1999.

Knoppers, Bartha Maria. *Socio-ethical Issues in Human Genetics.* Cowansville: Ed. Yvon Blais, 1998.

Rifkin, Jeremy. *The Biotech Century.* New York: Penquin, 1999.

U.S. Congress. *Genetic Testing in the New Millinium: Advances, Standards, Applications.* Washington, DC: USGPO, 1999.

Wilmut, Ian. *The Second Creation: Dolly and the Age of Biological Control.* New York: Farrar, Straus, and Giroux, 1999.

Agricultural Science since 1950

Overview

Agriculture underwent a dramatic transformation after 1950. Farming became industrialized, and governments and corporations, for the most part, dictated what crops would be grown and where they could be sold. The stereotypical independent farmer, providing food for his family and selling surpluses at local markets for extra cash, began to vanish almost as quickly as fertile rural acreage was appropriated for development near expanding urban areas. While the amount of farmland and labor declined, populations worldwide increased rapidly. Production of adequate, nutritional food supplies was essential. Agricultural scientists and engineers focused on research to provide practical solutions for utilizing available resources efficiently. Several of these answers, however, provoked environmentally aware people to criticize scientific and engineered responses and suggest alternative agricultural methods.

Background

Some historians consider agriculture to have been static until the mid-twentieth century, when an agricultural revolution occurred because farmers embraced scientific techniques. During the Second World War an extreme shortage of farm workers resulted when people departed for military service, and mechanization was considered essential as a form of substitute labor. Such large machinery permitted one vehicle operator to perform a task in a few hours in comparison to many laborers requiring days. Although for several centuries some farmers had adopted tools of varying technological sophistication, few early twentieth-century agriculturists had access to machinery. The war was the catalyst for the transition to agricultural reliance on technology and science to meet production demands.

The foundation for agricultural science enabling such efficiency was initially developed during the Scientific Revolution of the 1600s-1800s. Although earlier agriculturists might have intuitively bred the best plants and animals, practiced crop rotation, or slanted a plow blade, systematic experimentation to understand agricultural processes was not pursued until later. As early as the seventeenth century, scientists, curious about the natural world, wanted to devise techniques to improve agriculture. Gregor Mendel's (1822-1884) nineteenth-century pea investigations inspired agriculturists to conduct similar hereditary experiments with other plants.

Legislation aided the advancement of agricultural science and engineering. The 1862 and 1890 Morrill Acts enabled the establishment of land-grant colleges in the United States and set professional standards for agricultural education and research. The United States Department of Agriculture, created during the Civil War (1861-65), also bettered scientific research, financing investigations, collecting data, and publishing yearbooks. Politicians passed additional legislation in the late nineteenth and early twentieth centuries, expanding funding for scientific agriculture. Federally supported agricultural extension services and experiment stations assisted farmers and implemented vocational education programs that incorporated scientific agriculture.

In the early 1900s agriculturists devised high-yielding hybrid seeds. By 1908 hybrid corn was the United States' most valuable crop, adding $1,615,000,000 yearly to the economy. Many farmers, however, were still reluctant to utilize educated experts' advice and considered machinery too luxurious to purchase. Statistics show that by 1920 the United States' urban population exceeded rural residents. This demographic shift necessitated the ultimate industrialization of agriculture worldwide because there were fewer farmers to feed more consumers. The 1930s economic depression and World War II interrupted development of scientific agriculture.

Impact

Many agricultural scientists temporarily abandoned their research for World War II service. American scientists returned to laboratories, readjusting their hypotheses based on observations of foreign agriculture and conversations with peers they met in the military. European and Asian agricultural scientists suffered the loss of research facilities and experimental plots due to enemy bombardment. Enhanced American agricultural production was necessary to rebuild Europe and Asia after the war and to reinforce democracies during the Cold War. Also, the Korean War (1950-53) created new demands for military agricultural supplies. At the same time, the global population boomed. In the United States some veterans used benefits from the GI Bill to study agricultural sciences, ranging from agronomy to veterinary medicine. As interstates crossed the country and city limits expanded, arable agricultural land was swallowed by urban development, reinforcing the urgency for scientific agriculture to feed the world.

Scientific and technological applications to agriculture since 1950 have been primarily beneficial. These ideas were developed to speed up agricultural processes or to perform tasks in new ways to produce more food and fiber. Science and technology increased farmers' earning potential and created employment opportunities in agriculturally related industries. Perhaps science and technology's most valuable achievement was saving agriculturists time. In 1945 a farmer spent an average of 19 hours per acre of corn harvested. Fifty years later an agriculturist relying on technology invested 1 hour per acre of corn.

By the 1950s farmers were more willing to adopt technology because of familiarity with mechanical vehicles such as automobiles. After World War II steel was no longer rationed for wartime use, and more tractors were produced. Newly tapped oil fields assured ample, inexpensive supplies of tractor fuel. Engineers refined machinery design for specific duties and crops such as cotton pickers. Commercial farms began to dominate agriculture, relying on science to meet domestic and international market demand. Scientists developed agricultural technology such as freeze-drying and refrigerated containers to ship goods to distant markets. Engineers designed agricultural buildings to protect livestock and crops from pests and predators, sometimes incorporating solar panels to collect renewable energy. Automatic devices dispensed food, water, and insect repellents to large quantities of livestock at what are referred to as factory farms because of their size and high productivity. Agricultural scientists sought ways to reduce livestock odors, especially important as urban and rural areas merged. They improved soil conservation methods to protect topsoil from wind and precipitation. Scientists also developed nonagricultural applications of crops, including ethanol made from corn as fuel and newspaper ink and motor oil from soybeans.

Scientists created polymers to lubricate plows to reduce friction, saving 1,500 million gallons of gasoline annually. Global Positioning Satellites (GPS) monitored agricultural fields, and computer models permitted engineers to test agricultural machinery designs without risking crops. Automatic controls, computerized sensors, and electronically based guidance systems enabled unmanned machinery to perform mundane tasks. Soil dynamics, the scientific study of relationships between machinery and soil, was first defined in the 1920s by Mark L. Nichols (1888-1971) in cooperation with the Alabama Agricultural Experiment Station. Federally funded by New Deal programs in the 1930s, soil dynamics emerged as a significant agricultural science in the 1950s, guiding implement manufacturers to create improved machinery with better traction, thus conserving fuel and enhancing yields. Within decades, soil dynamics knowledge aided the application of artificial intelligence and robotics to sophisticated agricultural machinery.

Bioengineering also dominated agricultural science after 1950. Scientists manipulated organisms' genes regarding such factors as growth and biochemical activity, in an attempt to create perfect specimens through identification of genes for certain qualities and excision of those for negative traits. During recombination, specific genes were isolated from one DNA strand and spliced into another. Some bioengineered plants were genetically altered to repel and kill harmful insects and to manufacture fertilizer, saving expensive losses from damaged crops and application of chemical sprays. These plants were programmed to withstand environmental stresses. Bioengineered livestock were genetically altered to be immune to contagious diseases, reducing the need for vaccinations. A bovine growth hormone increased the size of beef cattle so that they required less feed and produced more meat.

Major corporations invested millions of dollars in bioengineering, hastening the transformation of agriculture from subsistence to commercial production. Researchers discovered that they could move genes between species, creating animals such as the geep, a goat-sheep hybrid. Biotechnology was also used to manufacture agriculturally based pharmaceuticals as new means to combat human diseases. Consumer goods such as blue jeans and fast food also incorporated genetically altered agricultural resources. In 1998 an estimated 69.5 million acres globally were cultivated with bioengineered crops.

The Green Revolution introduced bioengineering methods to the Third World in an attempt to ease hunger. Norman E. Borlaug (1914-), a plant pathologist, created a strain of high-yield dwarf spring wheat that had sufficient protein and calories to alleviate malnourishment. Beginning in the 1960s Borlaug distributed his seeds in Asia and educated farmers about their use. In addition to the seeds providing needed nutrients, Borlaug argued that his seeds would prevent habitats from being slashed and burned to clear more farmland. Within one year yields increased 70%, and Pakistan became agriculturally self-sufficient by 1968. India produced surplus wheat that was exported for profit. Borlaug was credited with saving millions of people from dying of starvation. He won a Nobel Prize, an honor that was also awarded in the late twentieth century to other notable agricultural scientists including Barbara McClintock (1902-1992).

Scientists developed irradiation in the 1990s to combat toxins and bacteria such as *E. coli* in meat, vegetables, and fruits. Thousands of people die annually from contaminated foodstuffs, and millions more become sick. Irradiation is a process in which foods are exposed to gamma rays or an electron beam. Promoted as a measure for food safety, irradiation also keeps meat from spoiling and prevents pest infestation. Many consumers are reluctant to eat irradiated foods, which advocates compare to the initial reaction people expressed toward milk pasteurization. Activists boycott irradiated goods and demand that agricultural process facilities be sanitized to remove the toxic threats that irradiation counters.

A backlash protesting the potentially harmful effects of agricultural science techniques to human consumers and the environment emerged in the 1960s, spurred by Rachel Carson's (1907-1964) writings. The organic farming movement and sustainable agriculture have gradually gained strength. Organic farmers strive to farm naturally without chemical substances. Sustainable agriculture encourages farmers to adopt agricultural methods that will assure plentiful food supplies for present and future generations while maintaining environmental stability and earning sufficient profits Each farmer encounters differing conditions and is encouraged to act as an amateur scientists to experiment with varying crops and ground cover to replenish soil, prevent erosion, and maintain a symbiotic balance between humans, earth, and organisms. Both organic and sustainable agriculturists feel a responsibility to protect nature. Organic agriculture comprises almost 10% of crops in some European countries. Organic farming is a thriving agribusiness catering to agriculturists nostalgic for less-scientific times; the Seed Savers Exchange at Decorah, Iowa, sells heirloom seeds descended from stock originally grown centuries ago, and living history farms reenact pioneer agricultural methods.

Green activists decry the loss of employment for blue-collar agricultural laborers, such as migrant workers, replaced by machinery. They also rail against the rise of government and corporate interference such as the 1980 grain embargo against the Soviet Union, and socioeconomic conditions that led to the decline of family farms during the 1980s. Protestors call bioengineered produce "Frankenfood"—after Mary Shelley's famous monster, Frankenstein—and seek to ban all genetically modified foods. They also criticize agricultural science for damaging the environment, citing such problems as pesticides polluting groundwater. Protestors warn that agricultural chemicals pose health hazards and are potentially carcinogenic. In reaction, some agriculturists have sued protestors for slander, including television personality Oprah Winfrey, who expressed her concerns about the safety of meat.

Both agricultural science proponents and protestors use computers and the Internet to express their opinions and share information. Computers serve as tools to record farm accounts and analyze yields. They provide isolated farmers access to the Internet and on-line marketplaces. E-mail, listservs, and electronic bulletin boards connect agriculturists and provide access to experts. Many agricultural colleges post news about ongoing research on Web pages (see Further Reading section below). As the twenty-first century dawned, it was clear that the benefits and potential drawbacks of the post-1950 revolution in the agricultural sciences were still being debated.

ELIZABETH D. SCHAFER

Further Reading

Books

Berardi, Gigi M., and Charles C. Geisler, eds. *The Social Consequences and Challenges of New Agricultural Technologies*. Boulder, CO: Westview Press, 1984.

Crabb, Richard A. *The Hybrid-Corn Makers: Prophets of Plenty*. Revised ed. New Brunswick, NJ: Rutgers University Press, 1993.

Doyle, Jake. *Altered Harvest: Agriculture, Genetics, and the Fate of the World's Food Supply*. New York: Viking, 1985.

Fabry, Judith. "Agricultural Science and Technology in the West." In *The Rural West since World War II*, ed. by R. Douglas Hurt. Lawrence: University Press of Kansas, 1998, 169-189.

Fite, Gilbert C. *American Farmers: The New Minority*. Bloomington: Indiana University Press, 1981.

Hurt, R. Douglas. *Agricultural Technology in the Twentieth Century*. Manhattan, KS: Sunflower University Press, 1991.

Hurt, R. Douglas, and Mary Ellen Hurt. *The History of Agricultural Science and Technology: An International Annotated Bibliography*. New York: Garland Publishing, 1994.

Internet Sites

Council for Agricultural Science and Technology. http://www.cast-science.org

Food and Drug Administration. http://www.fda.gov

Iowa State University's College of Agriculture. http://www.ag.iastate.edu

United States Department of Agriculture. http://www.usda.gov

Advances in Understanding Nonhuman Primate Behavior

Overview

The close relationship between humans and other primates continues to spark intense interest in nonhuman primate behavior. Because of the common evolutionary roots between humans and other primates, many scientists believe that the behaviors of these animals can provide glimpses of our collective past, as well as correlations to the current actions of individuals and to human society as a whole. The understanding of nonhuman primate behavior has grown enormously since the 1960s when famous paleontologist Louis S. B. Leakey (1903-1972) selected several women to study different primate species in the wild. That work led to such findings as tool use in chimpanzees, never-before-seen social behavior in orangutans, and the peaceful nature of gorillas.

Background

The study of nonhuman primates in the wild began in earnest in 1960 when British paleontologist Louis S.B. Leakey began an experiment of sorts. His studies confirmed that early humans and other primates were connected, and he felt that additional studies of the great apes would provide important insights into human evolution.

The great apes include gorillas, chimpanzees, and orangutans. (Some classifications include the gibbons as well.) Like other primates, the great apes have well developed ears and eyes, a large cranium, and hands with long fingers and an opposable thumb. Like humans, they have binocular, three-dimensional vision. Their cranium holds a brain that is large for the body size, compared to other animals, and their long fingers and opposable thumbs enable them to manipulate food, sticks, and other objects.

The women Leakey chose to study different primates included Dian Fossey (1932-1985) for gorillas, Jane Goodall (1934-) for chimpanzees, and Biruté M.F. Galdikas for orangutans. He believed that women would be less threatening to the animals than men, which would allow closer and more natural observation of behaviors. He also suspected that these women would have the patience to perform the lengthy, detailed observations needed to learn about the many aspects of the animals' behaviors.

In 1960, Goodall became the first of the three to set out on her studies. She had been working in Africa with Leakey as an assistant secretary for three years when he sent her with essentially no training to observe chimpanzees in an area now called the Gombe National Park in Tanzania. Her work with the chimpanzees has led to the longest study of wild animal communities ever conducted. The study is still ongoing as other researchers continue her work. Although many scientists questioned the wisdom of sending an untrained person into the field, Goodall made a string of stunning discoveries, including two important findings in her first year: chimpanzees are carnivorous, and they use tools.

Fossey conducted her work with gorillas in Rwanda after establishing the Karisoke Research Center there in 1967. Over the years, she found the gorillas to be peaceful creatures, a far cry from the vicious, often predatory animal usually imagined. In 1983 she reported altruistic behavior among the gorillas, when she observed one give himself up to poachers evidently to save the others in his troop from death. Fossey continued to study the gorillas until she was murdered in her field cabin on Dec. 26, 1985, the apparent victim of those who opposed her attempts to ban gorilla poaching.

In 1971 Galdikas and her husband Rod arrived at Tanjung Puting, a governmental reserve in Borneo, to study orangutans. Because of the reclusive and truly arboreal nature of the orangutans, her work progressed much more slowly than Goodall's studies of chimpanzees. In addition, Galdikas remarked in a *Discover* article, "Compared with humans, chimpanzees and most other primates, orangutans seem to operate in slow motion. (Once,) Jane Goodall commented to me that it took me two years to observe as much orangutan activity as she observed with chimpanzees in two hours!" Nonetheless, Galdikas was able to view some limited social interactions among the animals, which were previously thought to pursue almost exclusively solitary lives.

Impact

The studies of Goodall, Fossey, and Galdikas helped illustrate the many similarities and differences between humans and their primate relatives, possibly providing new views of human behavior. The work also led to the formation of individual preserves as well as worldwide conservation efforts.

Goodall's work is perhaps the best known. When she went into the field in 1960, neither she nor Leakey—and certainly not the many skeptical scientists—had any idea how important her observations would become. Before she was in the field a year, she had already observed the chimps eating meat, and manufacturing and using tools. The first discovery was amazing in itself: chimps had previously been considered complete vegetarians. Her second observation set the scientific community on its ear: she saw a male chimpanzee trimming a grass blade and using it to extract termites from a nest. Before this report, no other animal in the wild had ever been observed making even rudimentary tools. This formerly unique human characteristic was now shared with wild animals. Since this discovery of tool use, chimps have been observed using stone tools and sticks in food gathering, employing leaves to soak up water for drinking, and even using sticks to make crude "soles" for the bottoms of their feet.

Four years later, Goodall also found evidence of planning in chimpanzees. She observed a male running off with a female's baby apparently in an attempt to get the mother and other chimps to follow him to a new site. The ploy succeeded. This finding shattered conceptions that only humans had the level of intelligence necessary to engage in planning.

During her early years in the field, Goodall believed that chimpanzees were peaceful animals that lacked the violent tendencies present in humans. That notion changed in 1974 when she witnessed one group of chimpanzees savagely attack another, beginning what she called the "Four-Year War." She and other researchers saw the attacking chimps sneak through the forest in organized raiding parties to attack members of the other group, ripping them apart. Over the four years, the attacking chimps killed 10 adults—three of them females—and their young. (Some scientists questioned whether the war was a natural occurrence or was spurred by artificial feeding provided at the Gombe research site. They, along with Leakey on some occasions, felt that providing artificial food altered the chimpanzees' behaviors.) These attacks, along with others reported in different areas of Africa, led to considerable discussion among scientists about whether aggressive behavior in nonhuman primates has any correlation to violence among humans. Those discussions continue today.

Goodall no longer considers herself a field researcher, campaigning instead to protect chimpanzee habitat and to stop what she sees as inhumane laboratory use of animals. She is still involved in chimpanzee studies, however. In 1999, she and eight other primatologists compiled a comprehensive view of wild chimpanzee behavior, including the suggestion that cultural habits, like tool-making methods and specific courtship rituals, are passed from generation to generation through a learning process. By reviewing data gathered in a collective 150 years of observational studies, the researchers identified what they believe are definite behavioral differences between groups of chimpanzees. These cultural habits, they say, are another shared characteristic between humans and chimpanzees.

The discovery of altruistic behavior in gorillas, which was reported by Fossey two years before her death, also brought the attention of other scientists. This behavior, in which an individual gives itself up to harm or even death to protect other members of its group, is interesting to scientists who are trying to understand the reason the sacrificed individual engages in such behavior. Fossey observed a male gorilla apparently give himself up to poachers so that the members of his troop could escape. Many scientists now believe that such behavior promotes the passage of the individual's genes to subsequent generations, not through the individual's own offspring, but through the offspring of close relatives. Such behavior falls under the relatively new heading of "inclusive fitness."

Additional primate studies appear to indicate the presence of emotion in primates. Although difficult to document, Goodall has noted emotions in her study subjects on numerous occasions. Other scientists concur. For example, primatologist Takayoshi Kano reported an apparently emotional relationship between pygmy chimpanzees, which are also called bonobos. When the sister died, her older brother picked up her body and, according to Kano, "He carried his little sister's body with all four limbs hanging down lifelessly; one of his arms pressed her against his chest; and he walked slowly in the tree apparently in deep thought." On the following day, Kano observed the mother sitting with the corpse and shooing away flies. Although Kano said he may have been a bit anthropomorphic in his description of the scene, he believes scientists should consider the possibility that humans may not be the only living beings with emotions.

Jane Goodall with a baby chimpanzee. *(The Library of Congress. Reproduced by permission.)*

Scientists also study primates to learn more about male-female relationships. While male chimpanzees and bonobos are both larger and stronger than their female counterparts, their societal structures are quite different. Males dominate chimp society, occasionally mating forcibly with females. In bonobo, society, however, females band together against the individually stronger males, which effectively eliminates forcible mating. Some scientists believe these findings may have correlations to the increase in rape among humans over the last few decades. They hypothesize that rape was less common when women lived near their families, because a woman could turn to this strong support structure to defend against sexual violence. As families become separated geographically and their support structures disintegrate, scientists suggest, women become more vulnerable to sexual assaults, leading to the increase in rapes.

Through the years, scientists have continued to make remarkable discoveries about nonhuman primates, but few have drawn as much attention as those made by Fossey, Galdikas, and Goodall. Their accomplishments, and those of numerous researchers who have followed, spurred numerous conservation efforts to protect the animals and their habitats, and to promote a sense of stewardship of the environment. Goodall's Roots and Shoots children's program, for example, promotes environmental education and compassion toward the Earth's living things. She originally began the program in the hopes of influencing young people in Africa. The program quickly spread and now has chapters worldwide. Goodall has also been influential in establishing wildlife sanctuaries in Africa, including one in Congo and another in Uganda. Fossey's research and educational efforts were instrumental in gaining governmental protection for mountain gorillas in Rwanda.

These sanctuaries, along with habitat protection and educational efforts, will help ensure that primatologists can continue to study the Great Apes for years to come, and to learn whether humans and other primates are as closely related behaviorally as they are genetically.

LESLIE A. MERTZ

Further Reading

Books

Fossey, D. *Gorillas in the Mist.* Boston: Houghton Mifflin, 1983.

Goodall, J. *The Chimpanzees of Gombe: Patterns of Behavior.* Cambridge: Harvard University Press, 1986.

Stone, L. *Kinship and Gender.* Boulder: Westview Press, Harper-Collins, 1997.

Periodical Articles

Bower, B. "Monkeys Provide Models of Child Abuse." *Science News* (23 May 1998): 324.

de Waal , F. "Bonobo Sex and Society." *Scientific American* (March 1995): 82-88.

Galdikas, B. "Waiting for Orangutans." *Discover* (December 1994): 100-106.

Miller, P. "Crusading for Chimps and Humans... Jane Goodall." *National Geographic* (December 1995): 102-129.

Other

McCarthy, S. "Jane Goodall: The Hopeful Messenger." http://www.salon.com/people/feature/1999/10/27/reason/

Theories of the Origin and Early Evolution of Life

Overview

After the theory of spontaneous generation was discredited, only religious explanations were offered to explain the origin of life. Alexander Oparin (1894-1980), an atheist, suggested that natural chemical reactions produced biological molecules that came together to form the first living thing. Later, Stanley Miller tested this hypothesis and produced chemical "building blocks" but not life itself. In spite of much progress, there is still no clear consensus on how life originated on Earth. Some scientists are even looking to outer space for the origin of life.

Background

The first scientist to synthesize a molecule normally produced by living organisms was Friedrich Wöhler (1800-1882). In 1828, he accidentally made urea by heating ammonium cyanate. This finding helped dispel a theory known as "vitalism," which taught that living things and their components possessed a "vital force." At the time, scientists believed that living things consisted of "organic matter" driven by that vital force which separated them from non-living things. Wöhler's discovery suggested that life forms, like non-living forms, were composed of molecules that obey the laws of chemistry and physics. Further, it might be possible to produce other molecules of life by experimental or natural means.

Around the time that Charles Darwin (1809-1882) proposed his theory of evolution by natural selection (1859), two other scientists, Rudolf Virchow (1821-1902) and Louis Pasteur (1822-1895) showed that another commonly held theory was false. Spontaneous generation is the term that describes the formation of living things from non-living starting material. Scientists believed that worms, insects, mice, and microscopic organisms simply "arose" from decaying meat, grain, broth, or even dirty underwear. The theory seemed reasonable at the time because no one had any idea of the complexity or the multitude of interactive molecules that make up even the simplest bacteria. For them, a cell was only "protoplasm," not much more complicated than gelatin.

Pasteur and Virchow discredited spontaneous generation and laid the groundwork for the biogenic law which asserts that life comes only from life. This principle became a key component of the cell theory: every cell is made from a pre-existing cell. The implication of their work was that only God could have created the first life that would subsequently reproduce. Their demonstration was so effective that it virtually prevented any research on the origin of life for decades.

In the 1920s, a Russian chemist named Alexander Oparin coined the term "primordial soup" and suggested that the building blocks of life could spontaneously form and then coalesce together to form the first living cell. In his view, the basic components of cells (lipids, carbohydrates, amino acids) aggregate together, forming what he called "coacervates." Presumably, these coacervates would eventually carry out rudimentary metabolism and some would reproduce. Oparin also proposed that the atmosphere of the early Earth differed from the present one by having reducing gases such as hydrogen, methane, and ammonia in abundance. The British physiologist J.B.S. Haldane (1892-1964) independently concurred with Oparin, proposing that oxygen was absent during the origin of life because it would have prevented the formation of important organic molecules. This assumption about the atmosphere was not based on experimental evidence but on an understanding of the requirements for producing the desired molecules.

Oparin's hypothesis was tested in the early 1950s by Stanley Miller, a graduate student in Harold Urey's (1893-1981) laboratory, at the University of Chicago. Miller designed an apparatus that would simulate a reducing atmosphere and the presumed conditions of the early Earth. He used a spark discharge to mimic lightning and provide the energy required for the organic synthesis reactions. Miller's chamber lacked oxygen because this gas would prevent the formation of the desired molecules. In a short time, Miller found that the chamber produced 13 of the 20 amino acids found in proteins. Variations of this type of experiment were later shown to produce carbohydrates and the nitrogen-containing bases of nucleotides found in DNA and RNA. The work of Urey and Miller was hailed as producing the "building blocks of life."

However, producing such "building blocks" is not the same as producing life and was not qualitatively different than Wohler synthesizing urea. Chemical synthesis of building blocks is complicated by several factors. When amino acids are synthesized, a mixture of right- and left-handed molecules are produced. However, only the left-handed version is found in proteins. When carbohydrates are produced, many different sugars are made. However, the ribose and deoxyribose found in nucleotides are not made in appreciable amounts. Polymerization of amino acids into proteins and nucleotides into RNA and DNA is also a problem. Even then, these molecules are not living—they cannot reproduce themselves, carry out metabolism, and lack a boundary.

Later, Sydney Fox heated amino acids and they reacted together to form "proteinoids." Unlike normal proteins which are linear polymers of amino acids linked by peptide bonds, the proteinoids were branched polymers with both peptide and non-peptide bonds. The proteinoids could aggregate into microspheres and absorb various molecules. The aggregates were observed to enlarge and split into smaller fragments, although this could hardly be called reproduction.

As scientists began to unravel how the amino acid sequence of proteins is coded for in DNA and how DNA is replicated, there arose a paradox. The sequence of amino acids in a protein is not random but determined by the exact sequence of nucleotides in DNA. Therefore, a meaningful DNA sequence is required to produce a functional protein. However, proteins and enzymes are necessary in the replication of DNA, the transcription of mRNA, and the production of the nucleotides themselves. A conceptual difficulty arose because one could not start life with either proteins or DNA since each is so dependent on the other.

The conundrum was apparently resolved by Walter Gilbert (1932-), who proposed that life originated in an "RNA world." He suggested that the first living things consisted solely of RNA—that proteins and DNA were later developments. This was based on the observation that proteins are translated from mRNA with the help of tRNA and rRNA. Scientists also found that RNA could be reverse transcribed into DNA, a process carried out by the HIV virus. Further, certain RNA called ribozymes carry out limited catalytic activities like enzymes. RNA, then, appears to have the perfect combination of features to be the first molecules of life. However, the relative instability of nucleotides at high temperatures, the lack of appreciable ribose, and the inability for RNA to replicate itself pose serious problems for this hypothesis. Therefore, some scientists are suggesting a pre-RNA world that would later give rise to the RNA world. They have proposed clay to serve in this role.

However the first living cell arose, it must have done so very quickly. Many scientists believe that the Earth is about 4.5 billion years old and that the Earth would be much too hot to support life until about 4 billion years ago. J. William Schopf described fossil bacteria found in structures called stromatolites that according to radiometric dating were 3.5-3.8 billion years old. These microfossils, apparently a type of filamentous blue-green algae found in pre-Cambrian rocks, are supposedly the oldest fossil evidence of life on Earth. If these assumptions are correct, this would imply that life appeared on Earth as soon as it possibly could since a considerable amount of time would seem necessary between the origin of life and the formation of the complex cells in the stromatolites.

Early scientists classified living things into basically two categories: plant and animal. As more types of organisms were discovered it became clear that this type of classification was inadequate. Robert Whittaker developed the five kingdom classification system. Monera is the kingdom for bacteria and prokaryotic cells. Protista consists mostly of single-celled eukaryotic organisms with some colonial forms included. The remaining three kingdoms, plant, fungi, and animal, are all multicellular eukaryotes.

Such a system appears to reflect evolution since bacteria are the simplest organisms followed by protists. Plants, fungi, and animals are more complicated and arguably equidistant from the other two kingdoms. The discovery of Archeans has complicated this scenario. These cells, found in harsh conditions such as high salt and very high temperatures, were initially believed to be the first cells and led to true bacteria later. But upon further study, they are in many ways more similar to the eukaryotic kingdoms than they are to true bacteria, except that they are prokaryotes.

Eukaryotic cells share many features in common in spite of their differences. They have membrane-bound organelles such as the nucleus and mitochondria while prokaryotic cells lack them. The origin of these subcellular structures is unknown, since it has been established that a cell cannot simply "create" them once they have been lost. During cell division, the components of the organelles or the organelles themselves are divided between the two daughter cells. Lynn Margulis (1938-) proposed the endosymbiont hypothesis to explain the origin of organelles, including the mitochondria. According to this view, the mitochondria and other organelles were once bacteria that were internalized by another larger cell. Since the mitochondria, for example could use oxygen to produce energy, this gave an advantage to the cell that protected it. The endosymbiont hypothesis has been widely accepted, although recent data on protein targeting suggests the origin of organelles may not be so simple.

Impact

Once Pasteur and Virchow discredited the theory of spontaneous generation it became difficult to discuss the origin of life outside a theological context. Scientists conducted research on evolution but not on the origin of life until Oparin reopened the field. Because Oparin was in Soviet Russia, a nation committed to atheism, he was able to develop a naturalistic theory for origin of life. One could also argue that his commitment to atheism forced him to devise an origin of life consistent with that view. Nonetheless, Oparin's work paved the way for Stanley Miller.

The elegance and simplicity of the work by Stanley Miller producing amino acids from a gaseous mixture has dominated the field of origin of life research for decades. Although scientists now question his choice of starting material and debate the conditions of the early Earth, they have been slow to offer a better alternative. Therefore, Miller's experiment continues to play a prominent role in textbooks in spite of the difficulties with it. Some have suggested life arose in deep sea vents in the ocean or in lagoons near volcanoes instead of in the atmosphere. The most radical suggestion is that the molecules of life, or even life itself, was carried to Earth from outer space, a theory called panspermia.

If life could arise by natural processes on Earth, then some suggest the same conditions and processes may occur elsewhere as well. In 1969, a meteorite was found to contain organic compounds including the same amino acids in a similar ratio to Miller's experiment. This observation has provided hope to the possibility of finding life on another planet. The desire to understand the origin of life has helped to fuel the SETI (Search for Extraterrestrial Intelligence) project in spite of the current lack of evidence for extraterrestrial life.

Theories of the origin of life are likely to remain controversial because uncertainty will always remain. If scientists do create life in the lab, it would still not prove that such a process occurred in the past.

DAVID A. DEWITT

Further Reading

Books

Margulis, Lynn and Dorian Sagan. *What is Life?* New York: Simon & Schuster, 1995.

Miller, Stanley L. and Leslie E. Orgel. *The Origins of Life on Earth.* New Jersey: Prentice-Hall, 1974.

Schopf, J.W. "Archean Microfossils: New Evidence of Ancient Microbes," in *Earth's Earliest Biosphere.* New Jersey: Princeton University Press, 1983.

Periodical Articles

Duve, Christian de. "The Beginnings of Life on Earth." *American Scientist* 83, no. 5 (September/October 1995).

Joyce, Gerald F. "RNA Evolution and the Origins of Life." *Nature* 338 (1989): 217-224.

McClendon, John H. "The Origin of Life." *Earth-Science Reviews* 47 (1999): 71-93.

Miller, S. L. "A Production of Amino Acids under Possible Primitive Earth Conditions." *Science* 117 (1953): 528-529.

Orgel, Leslie. "The Origin of Life: A Review of Facts and Speculations." *Trends in Biochemical Sciences* (1998): 491-495.

Shapiro, Robert. "Prebiotic Ribose Synthesis: A Critical Analysis." *Origins of Life and Evolution of the Biosphere* 18 (1988): 71-85.

Shapiro, Robert. "Prebiotic Cytosine Synthesis: A Critical Analysis and Implications for the Origin of Life." *Proceedings of the National Academy of Sciences* 96 (1999): 4396-4401.

Woese, C.R. "Archaebacteria." *Scientific American* 244 (1981): 98-122.

Cracking the Genetic Code

Overview

"Cracking" the genetic code was one of the most exciting discoveries of the twentieth century. Although philosophers and early scientists had long pondered the nature of inheritance, it was not until 1953 that James Watson (1928-) and Francis Crick (1916-) announced that they had determined that the code for life resides in the molecular structure of deoxyribonucleic acid (DNA). This announcement began a frenzy of investigation that still continues today. One of the hottest topics in science at the end of the twentieth century is molecular biology.

Many scientists have added to the knowledge of the genetic code. In 1955 Mahlon B. Hoagland (1921-) isolated transfer ribonucleic acid (tRNA) while Robert Holley (1922-1993) described the complete structure of tRNA in 1965. In 1956 George Palade (1912-), working with the small structures (organelles) within the cytoplasm of the cell, discovered ribosomes, the protein factories of the cell. In 1967 Charles Yanofsky (1927-) and Sydney Brenner (1927-) described the organization of base groups that make up a protein. Marshall Nirenberg (1912-) and his team cracked the genetic code with a description of how the base pairs are related to twenty amino acids. These scientists laid the foundation for biotechnology and genetic engineering.

Background

A few scientists in the 1800s argued that the nature of living organisms could be reduced to basic chemistry and physics. Most were resigned to the prospect that the mystery of life and its mechanisms would never be solved. While a Swiss scientist in 1869 isolated the chemical DNA from pus cells, he did not recognize the importance of his finding.

At the beginning of the twentieth century, scientists had determined that nucleic acids were in all cells. Likewise, they knew that cells had three key ingredients: a ribose or deoxyribose sugar, a phosphate, and bases made of nitrogen and carbon. In 1938 Warren Weaver used the term "molecular biology" for the first time in an annual report to the Trustees of the Rockefeller Foundation. The foundation was supporting research into x-ray crystallography, which became instrumental in cracking the genetic code.

The 1940s, including the events of World War II, encouraged a new frenzy of scientific thinking that led to exciting discoveries in many fields, ranging from nuclear physics to biochemistry. In 1944 O.T. Avery (1877-1955) and his colleagues identified a substance, named deoxyribonucleic acid, that was able to change one strain of bacteria into another. The science of molecular biology was built on the work of sci-

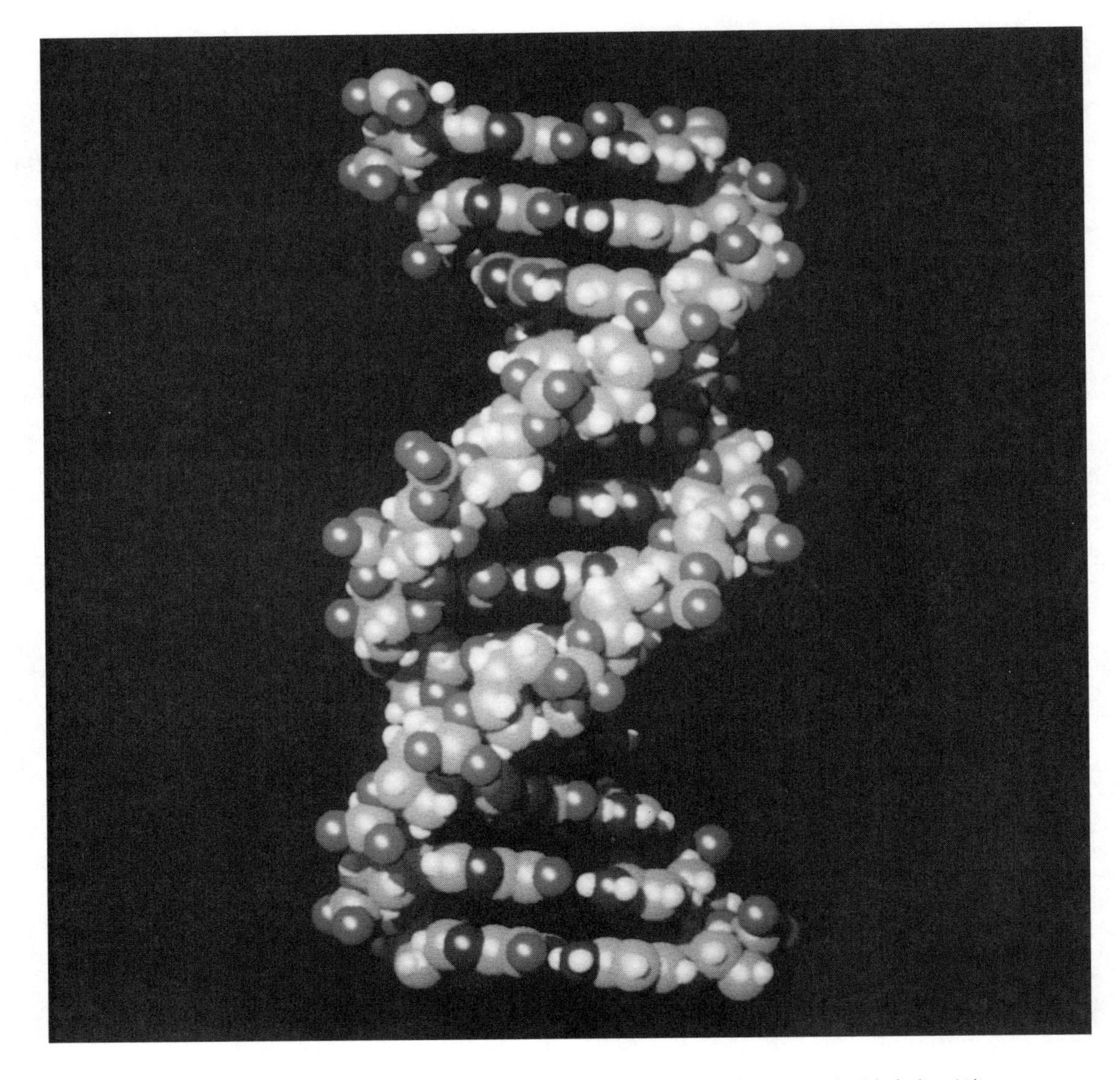

James Watson and Francis Crick discovered that DNA is shaped like a spiral staircase, or double helix. *(Clive Freeman/Photo Researchers, Inc. Reproduced by permission.)*

entists such as Walther Flemming (1843-1905), who studied cell processes, and Hugo de Vries (1848-1935), who investigated mutations. While the discovery of DNA was eclipsed by developments in nuclear technology, the study of reproduction, inheritance, and growth was taking off.

Impact

In 1951 a young post-doctoral fellow named James Watson came to the Cavendish Laboratories in Cambridge, England, where he met physicist-turned-biologist Francis H.C. Crick. The two immediately became focused on the problem of DNA structure. Using x-ray diffraction, a technique that shows chemical structure, they studied the protein coat of the tobacco mosaic virus but made no great progress.

At another laboratory at King's College, London, Rosalind Franklin (1920-1958) and Maurice Wilkins (1916-) had taken great diffraction pictures of DNA. All, however, was not well between the two colleagues, who disliked each other. When Wilkins showed Watson one of the pictures they had made of the double helix, Watson immediately recognized the parameters he needed to establish the structure of DNA.

Franklin died of cancer in 1958 at the age of 37, but her work helped illuminate the way toward the double helix. Crick praised her key experimental work and acknowledged that he and Watson had benefited from her criticism, which, while it offended them, caused them to rethink their procedures.

While the existence of four DNA bases—adenine (A), thymine (T), cytosine (C), and guanine (G)—had already been established, Watson looked at how the pairs fit together. Fiddling with models, he showed that adenine and thymine are held together by a hydrogen bond and that this pair was identical in shape to the guanine-cytosine pair. From looking at the x-ray pictures of the structure, he deduced that the pairs were the like the rungs of a ladder. The base pairs that unzip and rezip form the key to DNA replication. For example, rung by rung, C-A-T would have a complementary strand, G-T-A. When these pairs unzip, each of the two separate helixes forms a template upon which is built a new double helix.

On February 28, 1953, Crick strolled into the Eagle pub in Cambridge, England, and casually announced he and Watson had just completed a model for the secret of life. What caught the attention of scientists as well as the public was the model of the double helix, which presented to the world a tangible picture of what the molecule of life looked like. Although the two scientists wrote an account that was published in the journal *Nature,* it was the picture of the twisted ladder made of colored balls (representing atoms) that excited the world. In 1962 Watson, Crick, and Wilkins were awarded a Nobel Prize for their work.

In 1954 George Gamow (1904-1968) published the first theoretical consideration of a genetic code, suggesting that amino acids fit into the "holes" on the DNA. Other scientists were also investigating the relationship.

With the double helix established, Crick turned his attention to finding out how nucleotide sequences in DNA make up the sequences of amino acids in proteins. Amino acids have a nitrogen base or NH_2 radical. In 1954 he founded the ribonucleic acid (RNA) Tie Club, a group of scientists determined to find out how RNA coded for the amino acid sequences. Crick suggested that amino acids are attached to adapter molecules before being attracted to a nucleic acid molecule. In 1957 Crick and Gamov worked out the "central dogma," explaining how DNA makes a protein. The central theme of this idea is the "sequence hypothesis," which states that the order of pairs on the DNA determine a specific amino acid. They also suggested information goes only in one direction: from DNA to RNA to protein.

By the early 1950s, researchers had realized that RNA carries information from structures called codons to build proteins. Codons consist of three base pairs (for example, C-A-T) located on the DNA within the cell nucleus. RNA carries the codon instructions from the cell's nucleus into the cytoplasm, where ribosomes use this information to assemble amino acids into proteins. Biochemist Paul Berg (1926-) had determined that the combining of amino acids into proteins involved adenosine triphosphate (ATP), but it was unknown how the amino acids recognized the coding.

While Crick and Watson were working in England, other scientists on the other side of the Atlantic Ocean were also conducting research. In 1956 Mahlon Hoagland (1921-) and a team at Harvard stumbled onto the idea that amino acids attach not to RNA on the ribosome but instead to locations on small, soluble molecules called transfer RNA (tRNA), which, in turn, attach to the ribosomal RNA. This discovery fit Crick's theory that amino acids attach to an adapter molecule.

Examining how the genetic code controls synthesis of proteins, American biochemist Robert Holley began his research on RNA after studying with James F. Bonner at the California Institute of Technology in 1956. By 1960 Holley and colleagues had shown that tRNA provides instructions for the assembly of amino acids into proteins. This discovery was independent of Hoagland. Holley's team developed techniques to separate tRNA from the cell. By 1965 Holley had established how tRNA incorporates the amino acid alanine to form specific proteins. First, he determined the sequence of the nucleotides by digesting the molecule with enzymes, identifying the resulting pieces, and then figuring out how these pieces fit back together. All tRNA molecules have been determined to have similar structures. Marshall Nirenberg built a strand of tRNA comprised only of the base uracil. Calling the strand "poly U," he discovered that UUU is the codon for the amino acid phenylalanine. This finding was the first step in setting up the code for other amino acids. In 1966 the genetic code was cracked when Nirenberg and his team announced that a specific sequence of three nucleotide bases (a codon) determined each of twenty amino acids. Holley, along with Nirenberg and Har Gobind Khorana (1922-), shared the Nobel Prize in physiology or medicine in 1968 for their studies of amino acids and proteins.

Charles Yanofsky, an American geneticist, worked with a bacterium, *Esherichia coli (E. coli),* to show that the sequence of nitrogen-containing bases of the genetic structure has a linear correspondence to the amino acid sequence of proteins. (An amino acid always has an NH_2 radical.) His establishment of the co-linearity of gene and protein structures was built upon by other scientists to establish the genetic code.

From 1960 to 1964, South-African born biologist Sydney Brenner teamed up with Crick to study the genetics of bacterial viruses called bacteriophages. Studying carefully chosen mutations, they found information on the number of nucleotides that form the cores of amino acids. Using sophisticated detection methods, they found that a particular type of nucleotide forms a codon, which specifies an amino acid. Brenner was also a member of the first scientific team to introduce messenger RNA (mRNA), which carries the information that specifies a particular protein product.

From the Watson-Crick model of the DNA double helix in 1953, the gene emerged as a continuous string of information. The gene is organized so that three base pairs, or codons, hold the information for amino acids, which form proteins. By 1969 molecular biologists thought that they had found all the major players in the genetic code, but their discoveries were only the beginning.

EVELYN B. KELLY

Further Reading

Clarke, B.C., et al., eds. *The Evolution of DNA Sequences.* Scholium Int., 1986.

Crick, Francis. *Of Molecules and Men.* 1967.

Crick, Francis. *What Mad Pursuit: A Personal View of Scientific Discovery.* New York: Basic Books, 1988.

Olby, Robert. *The Path to the Double Helix: Discovery of DNA.* New York: Dover, 1994.

Sayre, Ann. *Rosalind Franklin and DNA.* New York: Norton, 1990.

Watson, James. *The Double Helix: A Personal Account of the Discovery and Structure of DNA.* New York: Penguin, 1976.

Watson, James. *Passion for DNA: Genes, Genomes, and Society.* New York: Cold Harbor Press, 2000.

Advances in Gene Regulation, Gene Expression, and Developmental Genetics

Overview

The "central dogma of molecular biology," elaborated shortly after James Watson (1928-) and Francis Crick (1916-) proposed their model of the DNA double helix, states that genetic information is encoded in the primary structure of nucleic acids and that this information is transferred to proteins. In summary, the path of information transfer is DNA→RNA→protein. That is, information in DNA, the genetic material, is transcribed into an RNA intermediate called messenger RNA, which is then translated into proteins. For many years, the central dogma served as a stimulus and a framework guiding investigations of genetic mechanisms at the molecular level. In order for scientists to understand how the genetic material found in the nuclei of all the cells in the body directs the special characteristics of highly differentiated cells, as well as how all of these cells develop from a single cell, they had to unravel the complex mechanisms that control gene regulation, gene expression, and developmental genetics.

Background

At the beginning of the twentieth century, scientists had just rediscovered Mendelian genetics and were starting to search for the physical basis for Mendel's genes. Although several studies conducted in the 1940s suggested that DNA might be the hereditary material, this possibility was dismissed because nucleic acids were thought to be simple linear tetranucleotide polymers. Most chemists, physicists, and geneticists thought that the genetic material must be a protein. Between 1946 and 1950, however, Erwin Chargaff (1905-) produced chemical studies that revolutionized attitudes towards DNA as the genetic material. In 1953 James Watson and Francis Crick described a model of DNA structure that immediately suggested explanations for its biological activity. The double helix could theoretically explain how DNA reproduces itself and makes more copies of DNA exactly like the original. Watson and Crick noted that the specific base-pairing that they had postulated suggested a possible copying mechanism for the genetic

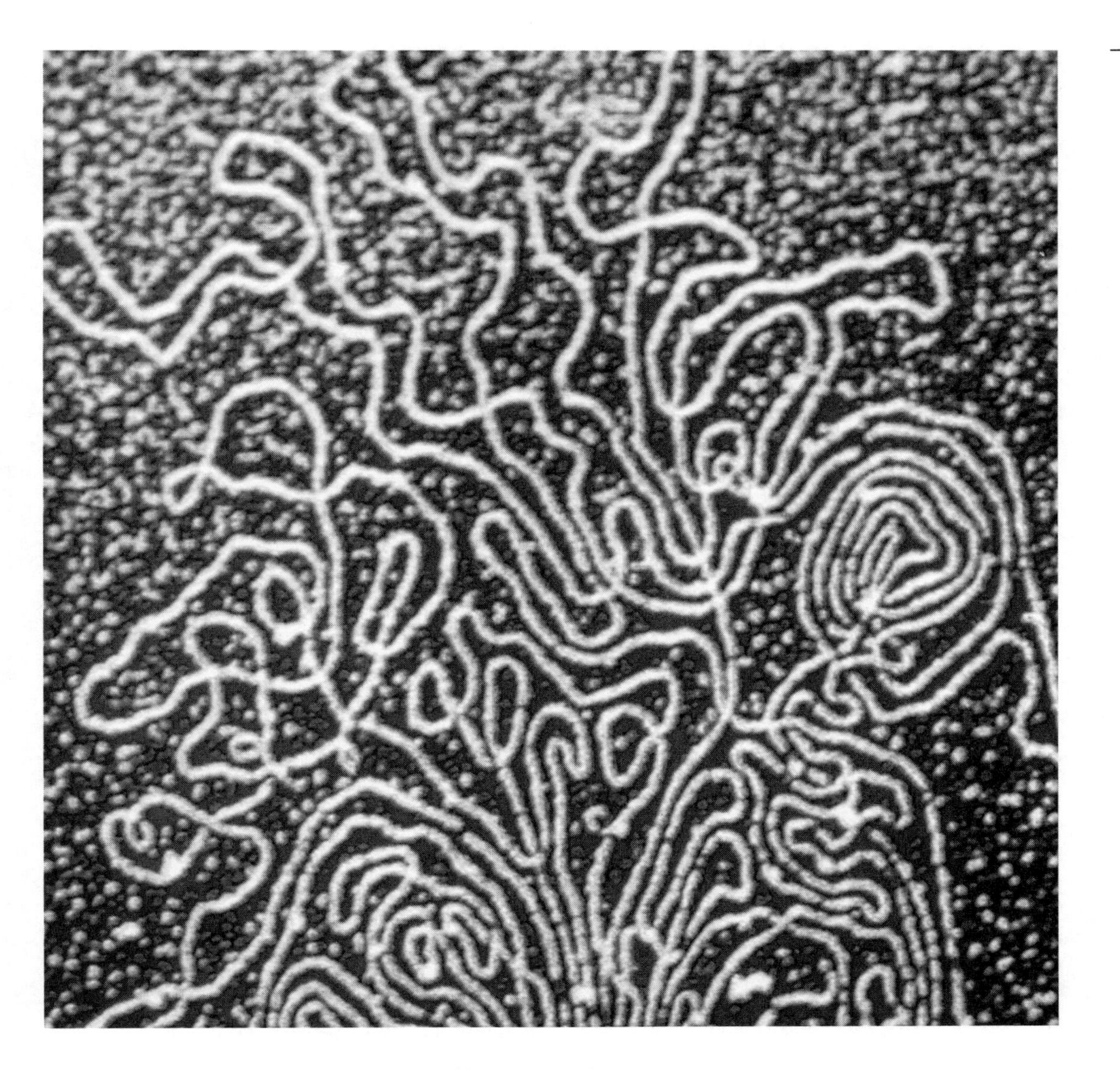

Strands of DNA. *(The Stock Market. Reproduced by permission.)*

material. Five years were needed for researchers to prove the essential validity of the DNA replication scheme proposed in 1953. Once the chemical nature of the gene was established, scientists were able to link knowledge of what a gene was to studies of how genes worked.

For cells to function and respond to changes in their environment, the transcription of genes must be tightly regulated. In prokaryotes, only a small percentage (about 3%) of the genes undergo transcription at any given time but, in the differentiated cells of eukaryotes, the percentage of genes being transcribed is even lower (about 0.01%). Although much of our fundamental knowledge about gene regulation and gene expression derives from studies of prokaryotes, the genetic apparatus of eukaryotic cells is much more complex and diverse as well as larger. Mammalian cells, for example, contain about 15,000 times as much DNA as *E. coli.* In prokaryotes, transcription and translation occur in close association. In eukaryotes, however, these processes are both spatially and temporally separated; transcription occurs within the nucleus while translation occurs in the cytoplasm. Thus, in eukaryotes, post-transcriptional processing of mRNA is another important aspect of gene regulation.

Transcriptional regulation is fundamental to coordinating metabolic activity and cell division as well as the complex patterns of embryological development and cell differentiation. Specific sequences on DNA serve as promoters that determine where transcription begins. In prokaryotes, a single kind of DNA-dependent RNA polymerase is responsible for virtually all RNA synthesis. Eukaryotic cells, in contrast, rely on three specialized classes of RNA polymerase, which are large, complex enzymes found in the nucleus. A variety of transcription factors are involved in determining the specificity of interactions between the three classes of RNA polymerases and the DNA. Transcription factors are DNA-binding proteins that can recognize and initiate transcription at specific promoter sites. The activity of RNA polymerase II is of special importance in allowing the cell to modulate gene expression in accordance with changing patterns of growth, differentiation, and metabolic conditions.

Not surprisingly, considering the complex needs of the cells of eukaryotes, various control mechanisms are involved in orchestrating transcriptional regulation. The fundamental organizing principles of gene regulation involve DNA-protein interactions, protein-protein interactions, and environmental cues that act as signals so that the regulator proteins can convey environmental information to the genome. For transcription to occur, a transcriptional activator protein must bind to the DNA and make protein-protein contacts with RNA polymerase. Transcriptional activation is essentially proportional to the strength of the protein-protein interaction. DNA-binding proteins that activate transcription have an activation domain that interacts with RNA polymerase and a DNA-binding domain. Proteins recognize nucleic acid sequences in terms of the complementarity of their three-dimensional shape and the surface of a DNA sequence. Most of the regulatory proteins that bind to specific DNA sequences have distinctive structural motifs that belong to three categories: the helix-turn-helix, the zinc finger, and the leucine zipper-basic region.

In 1995 the Nobel Prize for physiology or medicine was awarded to three developmental biologists: Edward B. Lewis (1918-), Eric F. Wieschaus (1947-), and Christiane Nüsslein-Volhard (1942-). It was the first Nobel Prize for research in developmental biology since 1935. Their work, done between the 1940s and the 1970s, led to the discovery of a family of genes that are critical in determining the architecture of the body during embryological development in the fruit fly *Drosophila.* Others researchers later determined that what Lewis, Wieschaus, and Nüsslein-Volhard had discovered in *Drosophila* also applies to humans.

By studying mutations that cause abnormal development in *Drosophila,* Lewis discovered a cluster of genes that control the development of individual body segments. These genes were later named "homeotic selector genes." "Homeotic" is a functional description that refers to genes that cause transformations in defined segments of the body. Wieschaus and Nüsslein-Volhard, in turn, focused on earlier developmental stages in order to discover the genetic events that activate the homeotic genes as a fertilized *Drosophila* egg develops into a segmented embryo. They created mutants with abnormal body segments and identified a small number of genes that are critical for determining the body plan.

Homeobox genes play an important role in the temporal and spatial aspects of genome expression during development. Genes that encode homeodomain (homeobox domain) proteins were first found among the *Drosophila* homeotic genes, from which the name "homeobox" was derived. Since the discovery of *Drosophila* homeobox genes, hundreds of homeobox genes have been found throughout a wide distribution of species, from yeasts and plants to invertebrates and vertebrates. Mutations in homologous genes found in vertebrates can also produce changes in the structure of body parts. Further investigations proved that not all homeobox genes are homeotic genes. Therefore, the homeobox is more generally understood as the DNA sequence motif that encodes the homeobox domain. Almost all eukaryotes, from yeasts to humans, that have been studied to date contain such DNA motifs. The homeobox domain contains a helix-turn-helix (HTH) motif.

Homeobox genes and the homeodomain proteins they encode play important roles in the developmental processes of many multicellular organisms. Homeobox genes have been shown to be involved in developmental mechanisms in fruit flies from the early steps in embryogenesis to crucial steps in cell differentiation. Originally, the homeobox was described as a conserved DNA motif of about 180 base pairs that encoded a protein domain (the homeodomain) about 60 amino acids long. In addition to the typical homeodomains, however, there are various atypical homeodomains, which may have more or less than 60 amino acids. In general, the homeobox portion of the protein contains about 10 percent of its mass. The rest of the protein is involved in protein-protein interactions that are essential aspects of transcription regulation. The homeodomain is a DNA-binding domain; homeobox domain proteins serve as sequence-specific transcription factors.. Homeodomain proteins are, therefore, basically transcription factors that usually play a role in development.

Some scientists think that the homeobox might serve as a "Rosetta stone" for the study of animal development. Such domains may be very ancient and may have been present in the common ancestor of all modern animals. Complex interactions among homeobox genes are the basis of fundamental genetic networks. Studies of homeoboxes and genetic networks may provide key insights into development, phylogeny, and evolution. Gene duplication, gene deletions, and chromosomal rearrangements may have been involved in the evolution of homeoboxes. Such changes may alter the expression of genes and could result in the activation of different sets of genes. A general outline of gene origin involves: gene duplication and fixation of the gene in the population through selection or drift; maintenance of gene function by selection; and gene evolution through mutation, transpositions, and selection. Transposable elements (transposons and retroposons) appear to be important in the establishment of new genes, changes in gene expression during development, and the genesis of major genomic rearrangements. Indeed, in 1984 Barbara McClintock (1902-1992) called these genetic phenomena "genomic shock."

Impact

Scientists believe that an understanding of gene regulation will ultimately provide crucial insights into the areas of human development, health, and disease. The identification and mapping of the entire human genome will illuminate gene networks and regulatory mechanisms that affect these areas. The Humane Genome Project, therefore, may result in major advances in medicine, such as new ways to treat or prevent birth defects, and innovative approaches to diseases, including cancer, schizophrenia, and AIDS. The Human Genome Project, an international effort to sequence the entire genome, began in 1994. Initially, scientists concentrated on developing methods that would increase the speed and accuracy of sequencing. Government scientists openly publish their results, but many private laboratories conduct sequencing in order to patent gene sequences and develop commercial products.

At least one-third of the human genome—one billion of the three billion base pairs—was identified, sequenced, and published by the year 2000. Essentially all of human chromosome 22 was completely mapped. Chromosome 22, which is a small chromosome, contains about 700 closely packed and very active genes, or a little more than 1% of all human genes. Scientists have now mapped the order of about 545 of those genes. Mutations in genes on chromosome 22 have been associated with cancers, heart defects, immune system disorders, schizophrenia, and mental retardation. Scientists predicted that a working draft of the entire human genome would be completed within the year, but verifying the entire sequence might take several additional years.

Although the Human Genome Project will certainly lead to novel biological therapies, gene therapy is still an experimental procedure that involves significant risks, as indicated by the death of a 17-year-old male who died during a gene therapy experiment. Jesse Gelsinger of Tucson, Arizona, was being treated for ornithine transcarbamylase deficiency. Although Gelsinger did have a severe genetic disorder, at the time of the experiment he was in fairly good health and was controlling his disorder with drugs and diet. He died on September 17, 1999, four days after he received an infusion of engineered corrective genes encased in a weakened cold virus. Other patients had gone through the same procedure without severe side-effects or adverse reactions. Gelsinger appears to have been the first person to die as a result of experimental gene therapy. Shortly after Gelsinger's death, a National Institutes of Health advisory panel proposed more stringent rules for gene therapy research.

Gene therapy and genetic engineering raise complex social, ethical, moral, and perhaps even long-term evolutionary questions. Some critics argue that gene therapy should only be used as a last resort for patients with severe genetic disorders that cannot be controlled by conventional treatments. Genetic engineering, however, could still have a profound impact on human health if attention were focused on using it to change plants and animals in order to increase food production and to provide more nutritious foods that include higher-quality proteins, lower levels of saturated fat and cholesterol, and higher vitamin contents. By improving nutritional standards throughout the world, genetic manipulation of domesticated plants and animals might have profound effects on twenty-first century patterns of morbidity and mortality. Additionally, genetically engineered plants might provide a new generation of pharmacologically active and commercially valuable products.

LOIS N. MAGNER

Further Reading

Books

Bürglin, T.R. "Homeodomain Proteins." In *Encyclopedia of Molecular Biology and Molecular Medicine*. VCH Verlagsgesellschaft mbH, Weinheim, 1996.

Duboule, Denis, ed. *Guidebook to the Homeobox Genes*. New York: Oxford University Press, 1994.

Gehring, Walter J. *Master Control Genes in Development and Evolution: The Homeobox Story*. New Haven: Yale University Press, 1998.

Kirsch, Ilan R., ed. *The Causes and Consequences of Chromosomal Aberrations*. Boca Raton, FL: CRC Press, 1993.

Schwartz, Jeffrey H. *Sudden Origins: Fossils, Genes, and the Emergence of Species*. New York: Wiley, 1999.

Periodicals

Erickson, Deborah. "Genes to Order." *Scientific American* 266 (June 1992): 112-14.

Lucas, P.C., and D.K. Granner. "Hormone Response Domains in Gene Transcription." *Annual Review of Biochemistry 61* (1992): 1131-73.

McClintock B. "The Significance of Responses of the Genome to Challenge." *Science 226* (1984): 792-801.

Steitz, T.A. "Structural Studies of Protein-nucleic Acid Interaction: The Sources of Sequence-specific Binding." *Quarterly Review of Biophysics* 23 (1990): 205-80.

Scientists Learn More about the Evolution and Acquisition of Human Language

Overview

Of all of the behaviors that human beings engage in, probably none is so complex and yet so commonplace as speaking and listening to language. Human language is what makes communication with others possible and gives order to our physical and social environments. The question, however, of how language is acquired by children has been and continues to be an interesting and lively debate. The inquiry into how much of language is present at birth is far from settled. Research is transforming traditional views of how language may have evolved, how the human brain works, and how children acquire language.

Background

Attempting to reconstruct the evolution of human language has been extremely difficult. Whereas the physical remains of our ancestors have endured for millions of years, we have no record of early speech. Anthropologists hypothesize that human beings as we know them have existed for only 100,000 years, but the earliest written records are barely 6,000 years old. These records appear so late in the history of the development of language that they provide no clue at all as to the origin of verbal language. The language or languages spoken by our earliest ancestors are irretrievably lost.

The idea that language was divinely bestowed upon humankind is found in myth and religion throughout the world. In the Judeo-Christian heritage, God gave Adam the power to name all things. In Egypt the creator of speech was the god Thoth. The Babylonians believed language was given by the god Nabu. In the Hindu creation story, the goddess Sarasvasti gave language to the people.

The assumption of the divine origin of language stimulated "experiments" in which children were isolated in the belief that their first words would reveal an original language. The Egyptian Pharaoh Psammetichus, James IV of Scotland, the Holy Roman Emperor Frederick II, and German J. G Becanus all conducted experiments to determine an "original language." All were unsuccessful.

Other early theories suggested that language is a human invention. The early Greeks believed that a "legislator" gave the true names of things. The naturalists argued that there was a natural connection between the forms of language and the essence of things. In other words, language may have begun as a vocalization of the sounds heard by early humans in their environment.

Later, evolutionary theories would oppose the divine-origin and the invention theory of human language. Charles Darwin's (1809-1882) theory of natural selection, arrived at in the 1800s, suggested that there must have been genetic variation among individuals in their ability to communicate with one another. There would have been a series of steps leading from no language at all to language, as we currently know it, with enough evolutionary time and genetic variation separating humans from our non-linguistic ancestors.

More recently, studies of the evolution of the human vocal tract to fit the function of speech have been conducted. Studies of early hominids have shown genetic variation that may have led to the ability to speak. Using fossil evidence and computer modeling, the research concludes that the Cro-Magnon—early modern human beings who lived in Europe between 10,000 and 35,000 years ago—appeared to have an important advantage in their favor: a vocal tract capable of producing all of the sounds of human speech.

Study of the evolution of the human vocal tract led scientists to study the evolution of the human brain. It is likely that the throat and mouth would not have evolved the way they did to facilitate language while compromising eating and breathing if the brain had not been capable of producing and comprehending language.

By 1950 the debate of language as evolution or divine miracle continued. The question of how children acquire language became the inquiry that captivated researchers. The acquisition of language may not appear initially to be very different from the other things that children learn, but it became a pivotal research question because of the on-going developments of evolutionary theory.

Contrary to both theories, B. F. Skinner (1904-1990), an American behavioral psychologist, proposed in 1952 that speech is a learned behavior, one reinforced by the stimulus; response and reinforcement can explain how and why language is learned. By receiving positive reinforcement for sounds, words, and sentences, a child is able to gradually shape his/her verbal skills over time to approximate the language of the community.

Impact

In 1957 the publication of *Syntactic Structures* by American linguist Noam Chomsky (1928-) revolutionized the inquiry into language acquisition. He proposed a theory that would account for both linguistic structure and for the creativity of language—the idea that human beings can create entirely original sentences and understand sentences never spoken before.

Chomsky suggested that children are born with the ability to understand the formal principles of grammatical structure, in marked contrast to the idea that language is essentially a system of grammatical habits established by training and experience. Children cannot possibly learn the full rules and structure of languages strictly by imitating what they hear. Instead, nature gives human children a head start by wiring them from birth with the ability to acquire their parent's native language; they can fit what they hear into a pre-existing template for the basic structure shared by all languages.

Current research within the last 10 years supports Chomsky's theory. Linguists have conducted research that suggests that within 48 hours of birth infants show a preference for the language of their parents. Infants can perceive the entire range of phonemes (sounds that make up words), and by the time they are 8 months of age they have begun to distinguish between nouns and verbs. By the age of 16 months most children know where a word belongs in a sentence, and by the age of 3 years most children have the essential patterns of grammar sorted out. All normal children by the age of 4 years, regardless of intelligence, are proficient speakers of their native language.

When Chomsky first voiced his idea that language is hardwired in the brain, he didn't have the benefit of the current findings in cognitive science, which has begun to pry open the human mind with sophisticated research and computer modeling. Until recently, linguists could only marvel at how quickly children master the abstract rules of language that give every human being who can speak (or sign) the power to express an infinite number of ideas from a finite number of words.

Chomsky proposed that children are born with the help of a "language acquisition device," preprogrammed circuits in the brain. Ninety percent of the sentences spoken by a three-year old are grammatically correct. It appears that the baby's brain is humming with activity. Neurobiologists once assumed that the wiring in a baby's brain was set at birth. After that, the brain, like arms and legs, just grew larger. Images made using brain-scanning technique have revealed that when a baby is eight or nine months old, the part of the brain that stores and indexes many kinds of memory becomes totally functional. This is precisely when babies appear to attach meaning to words.

Other leaps in a child's language development coincide with remarkable changes in the brain. For example, an adult listener can recognize eleph as elephant within 400 milliseconds because of an ability called *fast mapping,* which demands that the brain process speech sounds very quickly. To understand strings of words, humans have to identify individual words rapidly. At 15 months a child needs more than a second to recognize a familiar word, like dog. At 18 months the child can understand the word almost immediately, and by two years of age she knows the word in 600 milliseconds, as soon as the word has been spoken.

This ability coincides with a dramatic reorganization of the child's brain, in which language-related operations, particularly grammar, shift from both sides of the brain into the left hemisphere. While infants and toddlers work with language in both hemispheres of the brain, most adult brains process grammar almost entirely in the left temporal lobe.

Although the ability to acquire language appears to be innate in the child's brain, they need human interaction to attach meaning to words. They need to hear people speaking and conversing for this remarkable ability to emerge. Hearing more than one language in infancy and early childhood makes it easier for a child to hear the distinctions between phemones of another language later on. It also appears that the window of opportunity begins to narrow around age six. Children who do not learn a second language by age 12-15 will have difficulty becoming fluent in another language. This is a compelling argument for teaching a second language in elementary school, when the brain is primed for acquiring language.

Linguists have a long way to go before they can say exactly how a child goes from babbling to talking, what the first languages might have been like, or how the brain transforms thoughts into words, sentences, and complex ideas. However, it appears that innate brain wiring plus experience and practice equal language knowledge and ability. Continued research into human evolution and brain development will perhaps lead towards an answer of how human beings acquire language. In the meantime it is accepted that language is one of the great wonders of the natural world.

LESLIE HUTCHINSON

Further Reading

Brownlee, Shannon. "Baby Talk." *U.S. News and World Report* (June 15, 1998).

Chomsky, Noam. *Reflections on Language.* New York: Pantheon Books, 1975.

Eastman, Carol. *Aspects of Language and Culture.* Chandler and Sharp, 1992.

Fromkin, Victoria. *An Introduction to Language.* New York: Holt, Rinehart and Winston, 1983.

The Advent of Sociobiology Sheds New Light on Animal Societies

Overview

Sociobiology is the attempt to understand the biological origins and development of animal societies—including human society—using the Darwinian theory of natural selection. It is based in part on the observation that certain behaviors in animal societies are universal, meaning they must be based on hereditary instincts. This includes behaviors that make social life possible, such as cooperation, division of labor, social hierarchy, etc. Sociobiology explains the origins of these social instincts as the product of natural selection and explains hereditary behaviors as traits that help animals survive and reproduce.

Background

Charles Darwin (1809-1882) himself had grappled with the problem of social behavior and ethics in human society. In *The Descent of Man* (1871) he expressed the view that human ethics was similar to the social instincts of other animals, such as wolves or apes, but more highly developed because of the human capacity to reason. He believed that the social instincts were hereditary in humans, just as they are in other animals.

A difficulty Darwin faced was how to explain the origin of ethical behavior in naturalistic terms. After all, his theory of natural selection through the struggle for existence implied that all organisms were competing for scarce resources. How could social instincts or ethical behavior develop if the individual practicing selfless behavior would perish at a greater rate than more selfish individuals? Wouldn't the more ruthless individuals in any species triumph over the more meek, humble, or loving individuals in the struggle for existence? The explanation Darwin provided to explain this puzzling situation is often called group selection, though Darwin never used that term. Group selection means that even though selfless behavior (altruism), such as sacrificing one's life for the sake of another, does not benefit the individual in the struggle for existence, it benefits the individual's group (family, tribe, or nation). The group that has the greatest amount of altruism or self-sacrifice would thus supplant groups that cooperate less. Thus the groups with greater social instincts would pass on their traits to the next generation.

By the mid-twentieth century most biologists rejected the idea of group selection, insisting that selection operated only on individuals. Most scholars in the humanities and social sciences considered human ethics environmentally rather than genetically determined, placing human ethics beyond biological explanation. These factors militated against Darwinian explanations of social behavior, especially in human society.

Nevertheless, sociobiology exploded on the scene as a new scientific research field in the 1960s and 1970s. Konrad Lorenz (1903-1989) and Nikolaus Tinbergen (1907-1988) laid the groundwork by arousing interest in the biological roots of animal behavior. They founded a new branch of biology—ethology—to rigorously analyze and explain animal behaviors and the hereditary instincts on which behavior is based. Lorenz and Tinbergen shared the 1973 Nobel Prize in medicine for their work in ethology.

However, it was W. D. Hamilton's seminal paper on kin selection in 1964 that made sociobiology plausible and won over its strongest advocate, the Harvard biologist Edward O. Wilson (1929-). Hamilton grappled with the same issue as Darwin—the origin of altruistic behavior in social animals. He argued that natural selection can indeed account for the origin of hereditary altruistic behaviors in social animals. He believed that by furthering the interests of close kin, who share many of the same genes, a self-sacrificing individual is actually helping promote the reproduction of his or her own genes, even if the individual does not leave any direct offspring. For example, among social insects there are often neuter castes of workers or soldiers, who cannot reproduce, but who work to support the queen; only the queen can reproduce. If natural selection operates only at the individual level, it is hard to explain the origin of these neuter castes. However, Hamilton pointed out that because all individuals in the colony are descended from the queen, they all share genes with the queen. Thus helping the queen survive and reproduce means they can pass on their own genes (including the biological social instincts) to the next generation.

Impact

Other works in the 1960s promoted the idea that animal behaviors, including human ethical behavior, can be understood as instincts produced through Darwinian evolution. In the mid-1960s Lorenz, Desmond Morris, and Robert Ardrey wrote popular works arguing that human aggression and territoriality are hereditary instincts based on evolutionary competition. The idea that human behavior is conditioned by hereditary instincts began to gain some currency in psychology and the social sciences as well.

Wilson, building on Lorenz and Hamilton's ideas, became the most prominent proponent of sociobiology and brought it to the attention of the wider public. His own specialization was ant biology, and in 1971 he published *The Insect Societies*, applying sociobiology to his field of expertise. Then in 1975 he expanded his treatment of sociobiology to include all social animals, including humans, in *Sociobiology: The New Synthesis*. In this work he argued that instincts producing social behavior are genetically determined and are produced through natural selection. He identified various behavioral traits that are universal within social animal species and showed the survival and reproductive advantage such traits endow on their possessors.

Wilson's *Sociobiology* spawned a public debate, as some scientists and many social scientists vigorously rejected his genetic determinism, especially as applied to human society. In response to the harsh criticism he received, Wilson expanded his treatment of human sociobiology in *On Human Nature* (1978), which won him a Pulitzer Prize. In this work he analyzed behaviors and ethical standards that are universal or almost universal in human societies. He examined human tendencies toward division of labor between the sexes, altruism toward kin, territorial aggression, incest avoidance, tribalism, male dominance, etc. He claimed humans had genetic predispositions for these behaviors.

Two of his colleagues at Harvard, the geneticist Richard Lewontin and the paleontologist Stephen Jay Gould (1941-), both staunch Darwinists, led the opposition against sociobiology, and they are still Wilson's strongest critics. They argued that human social and ethical behavior is not biologically determined. They considered Wilson's views politically dangerous, since it seems to suggest that hierarchy, male dominance, and tribalism are natural and unavoidable in human society. Wilson was accused of racism, sexism, and justifying the oppressive status quo, charges which spawned public protests against Wilson. At the American Association for the Advancement of Science meeting in Washington in February 1978, Wilson was doused with ice water during his presentation by anti-racist demonstrators.

Wilson responded to his critics, who, he claimed, misunderstood his position. He asserted that he was not strictly a genetic determinist, because in his view genes do not compel specific behaviors, but only make humans predisposed toward certain behaviors. He also acknowledged that culture, even though linked to genetic tendencies, plays a significant role in the development of specific behaviors and moral codes. Even cultural traits, in his view, evolve through natural selection in a process he termed "gene-culture coevolution," since cultural traits that promote survival and reproduction will be selected, just as biological traits are.

Despite widespread criticism, Wilson also gained important allies and wielded considerable influence, especially among biologists. Richard Dawkins in his famous book, *The Selfish Gene* (1976), helped promote sociobiology, as have many of his students and followers, such as Helena Cronin. In the 1980s and 1990s there was a huge outpouring of works by biologists, anthropologists, psychologists, and other scholars promoting sociobiology and evolutionary ethics. Scientists are studying and explaining ever more animal and human behaviors within the perspective of sociobiology. Some prefer to call this new field evolutionary psychology, especially when dealing with human behavior.

Sociobiology has become a hot topic in the popular media. In 1977 *Time* magazine carried a cover story on sociobiology. In 1995 Robert Wright, a popularizer of evolutionary psychology in his *The Moral Animal: Evolutionary Psychology and Everyday Life*, wrote a cover story for *Time* magazine, in which he explained the human male's alleged proclivity for adultery as a genetic predisposition explicable through Darwinian theory. Sociobiologists and evolutionary psychologists have tried to explain everything from infanticide to homosexuality to President Clinton's sexual escapades as genetic tendencies produced through Darwinian evolution. One of the most prominent recent advocates of evolutionary psychology is MIT psychology professor Steven Pinker, whose book, *How the Mind Works* (1998), reached a popular audience.

Sociobiology and evolutionary psychology, especially as applied to humans, remain highly contested topics today. They have many supporters, but also many detractors. Adherents consider sociobiology the key to understanding who we are as humans, the solution to the riddle of human nature. However, many still object to the genetic determinism and the just-so stories that are so often used to support human sociobiology. One prominent biologist, Steve Jones, has asserted that "the attempt to explain the modern world in terms of the sex life of the Stone Age is the biggest load of hogwash ever foisted on to an unsuspecting public."

RICHARD WEIKART

Further Reading

Books

Caplan, Arthur L., ed. *The Sociobiology Debate: Readings on Ethical and Scientific Issues.* New York: Harper and Row, 1978.

Degler, Carl. *In Search of Human Nature: The Decline and Revival of Darwinism in American Social Thought.* New York: Oxford University Press, 1991.

Kaye, Howard L. *The Social Meaning of Modern Biology: From Social Darwinism to Sociobiology.* New Haven: Yale University Press, 1986.

Kitcher, Philip. *Vaulting Ambition: Sociobiology and the Quest for Human Nature.* Cambridge: MIT Press, 1985.

Wilson, E. O. *Sociobiology: The New Synthesis.* Cambridge: Belknap Press of Harvard University Press, 1975.

Human Ancestors: The Search Continues

Overview

The search for human ancestors and origins made great strides in the second half of the twentieth century. During the nineteenth and early twentieth century few actual fossils were known, it was thought that there were only a few types of archaic (early) people, and the age of those fossils was in dispute. Today, sophisticated dating techniques and DNA analysis have been brought to bear upon the question of age. Also, far more fossil material has been found showing that the Neanderthals and Cro-Magnons were not alone. The evidence proves that there were many early hominids (human-like creatures). Paleoanthropologists (those scientists who study ancient humans) agree that the cradle of mankind was not Central Asia, as was once thought, but Africa—an idea that suggests we are all of African descent regardless of our skin color. While more is known of our ancestors than ever before, questions and controversies still rage over just how humans evolved to the stage we are at today.

Background

Evidence that the first human ancestors originated in Africa was discovered by Raymond Dart (1893-1988) as early as 1924, though few at that time believed such proof was conclusive. In fact, Charles Darwin (1809-1882), in his *Descent of Man* (1873), had hypothesized that Africa would prove the cradle of man, but he was ignored. Dart was a young professor of anatomy at the University of Witwatersrand in Johannesburg, South Africa. Some interesting primate fossils had been discovered by workers in a limestone quarry in the nearby region of Taung, and Dart arranged to have them sent to him. In a box of these fossils Dart noticed the broken remains of a tiny skull that was neither a baboon nor other known primate. Over time he pieced the fragments together to reveal the skull of a child. He dubbed the diminutive relic *Australopithecus africanus* (the South African Ape) and decided it was an intermediate stage between anthropoids (monkeys) and man.

Dart published a description of the fossil shortly thereafter, but few scientists seemed interested in what Dart thought to be the earliest known human ancestor ever found. Those who were interested found it so because it suggested that the primates had ventured much further south during their evolution than previously thought. They did not think it a human ancestor, however. Dart argued that it was a human ancestor because the Foramen Magnum—the hole in the skull that the spinal cord passes through—was in the bottom like a human, not in the back like a primate. Therefore, he reasoned, the creature must have walked upright on two feet. Analysis of the fossil specimen proved problematic because it was an infant (organisms often change their morphology, or physical structure, as they grow to adults) and there were no other fossil parts to go with it. Also, the skull had both primate and human characteristics. As such "Taung Child," as Dart called it, was dispensed by the scientific world to a strange limbo of existence.

Convincing evidence that Africa was the cradle of human origins was discovered in the 1950s and 1960s by the husband and wife team of Louis and Mary Leakey. Louis Leakey (1903-1972) was born in Kenya and was raised with the children of the local Kikuyu tribe. He studied anthropology at Cambridge University, England, and became obsessed with discovering the oldest humans, and convinced that those humans would be found in Africa. Mary Leakey (1913-1996) was Louis's second wife and shared his fascination with human ancestry. They made their discoveries in a wasteland area of Tanzania known as Olduvai Gorge. The area is excellent for fossil hunting because many layers of rock, or strata, are exposed, enabling fossils found at that location to be dated. The Leakey's originally came to Olduvai Gorge in the early 1930s. They discovered numerous broken stones that they believed were actually the oldest known tools made by humans. The problem was that they could find no human fossil remains to go with those tools. By the 1950s the Leakeys were able to return to the area to do a more systematic search.

One day in 1959, while Louis lay in bed with a fever, Mary went prospecting and discovered a set of fossil hominid teeth set in a jaw. Electrified by the find, the team scoured the area for more and eventually pieced together an entire skull. They also discovered more stone tools. Chemical dating later showed the skull specimen to be 1.75 million years old. Dubbed *Zinjanthropus boisei* (East African Man) by Mary

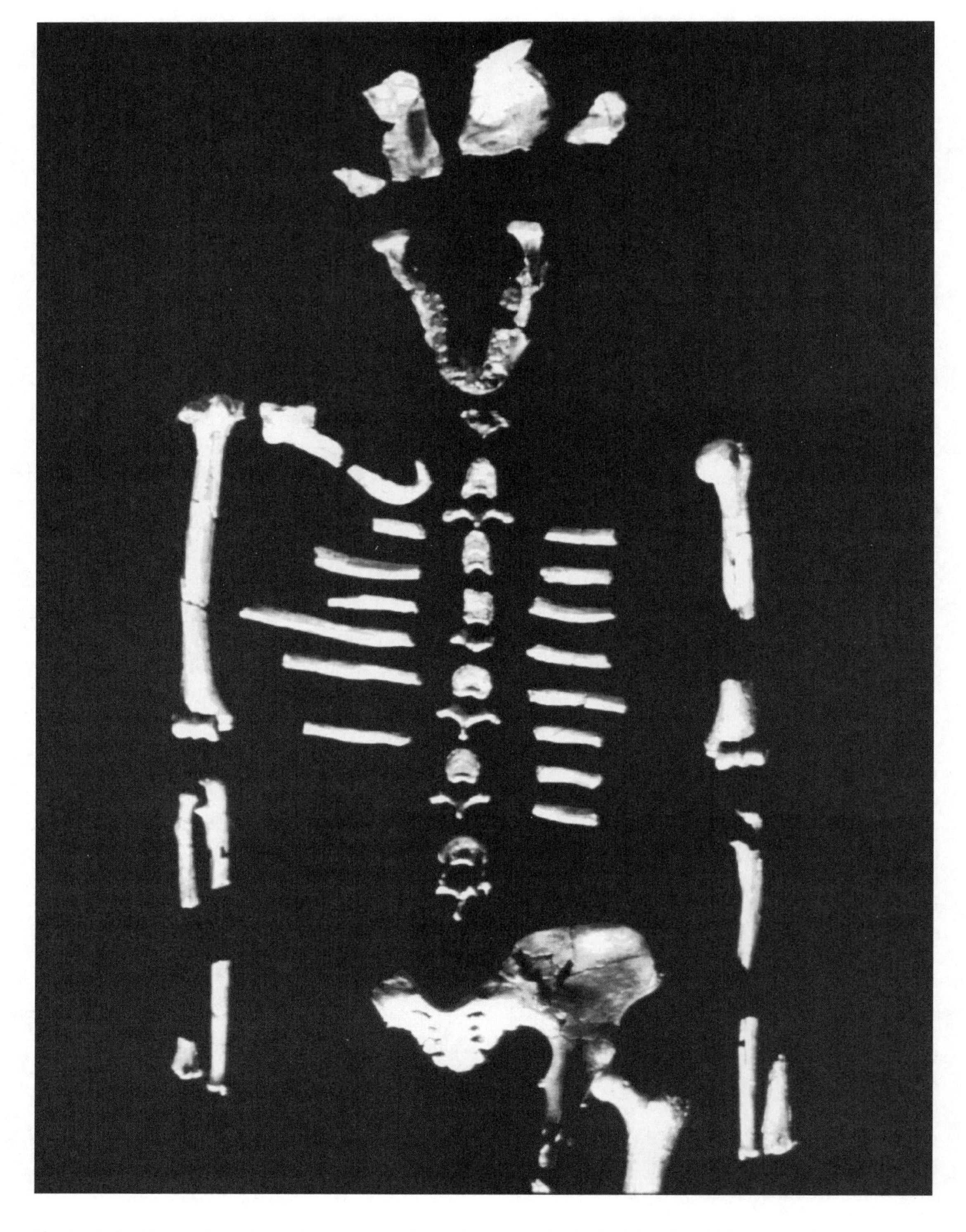

The *Australopithicus afarensis* remains, known as "Lucy," discovered in Ethiopia by Donald Johanson. *(Photo Researchers. Reproduced by permission.)*

Leakey, the skull represented a creature directly ancestral to modern humans.

Louis Leakey had never thought Raymond Dart's *Australopithecus* to be ancestral to man, but in fact *Zinjanthropus* was an australopithecine. Leakey saw his creature as more modern than Dart's and directly related to humans. He refused to accept Zinj, as Leakey called it, as anything other than a human ancestor. This caused a debate over just what group the skull fell into.

Eventually, Leakey came around to the fact that his Zinj was an *Australopithecus.*

With their success at Olduvai Gorge the Leakey's became famous and thus had less trouble getting the money necessary to continue their work. Louis became curator of the Coryndon Museum, while Mary was able to take over the day to day control of the dig site. Their son Jonathan assisted his mother in discovering a different type of hominid that was clearly different from Zinj. Soon dubbed *Homo Erectus* (the upright man), this creature stood fully on two legs like a modern human. Louis did return to Olduvai when he could. In 1962 he found hominid parts that were later ascribed to the earliest known human toolmaker, which Leakey called *Homo habilus* (the handy man). The term *Homo* designates a creature as being fully in the same family as modern humans.

However, they discovered more than just fossils and tools. In May 1970 Mary Leakey uncovered a series of fossil footprints near Laetoli, Tanzania, not far from Olduvai Girge. The trackway was especially significant because, while fossilized bones are representations of a dead creature, trackways are representative of a creature that is alive and moving about. The footprints, made in volcanic ash, show what some argue was a group, possibly a family, of archaic hominids—a male, a female, and a child—walking together, perhaps holding hands. They were walking upright 3.75 million years ago.

The entire Leakey family took part in the discovery of human ancestry. Their youngest son, Richard Leakey (1944-), also made important contributions to the study of paleoanthropology. Richard Leakey made his discoveries around Lake Turkana, Kenya. This fossil-laden area is known as the Koobi Fora. In 1972 his team (particularly his field assistant Bernard Ngeneo) discovered several hundred fragments that were carefully pieced together to create a skull designated 1470. It was a remarkable three million years old and had a relatively large brain. Neither Richard Leakey nor his colleagues could agree whether 1470 was an *Australopithecus* or *Homo.* Skull 1470 may have represented a different hominid line, suggesting that several groups had lived in the same area at the same time—a possibility not generally accepted within scientific circles of the day. If there were more than one type of hominid in Africa at the same time, the entire history of human ancestry, and how scientists viewed that ancestry, would have to be reconsidered.

The next major discovery in the search for human ancestors was made in Hadar, Ethiopia, in 1974. A team led by American paleoanthropologist Donald Johanson discovered the most complete *Australopithecus* skeleton ever found. Nicknamed "Lucy"—after the Beatles's song "Lucy in the Sky With Diamonds"—the creature was a true biped, weighing about 60 pounds and standing about three and a half feet tall. Fossils can not be directly dated. Because their approximate age is established by dating the matrix, or surrounding rock, in which they are found. In this manner Lucy was found to be roughly 3.18 million years old.

After Lucy was uncovered, Johanson's team found the remains of over a dozen individuals. Though they seemed similar to Lucy, they were older and physically larger. As they were found together, they were dubbed the "First Family." It was later realized that these creatures did not die as a group, but were washed into the location by drowning at different times. Johanson made his discoveries public in 1978. He argued that the group of creatures Lucy was a part of (not her as an individual) represented the earliest human ancestors. In fact, he argued that humans as a group find their point of origin in these creatures. With that he created a new hominid category—something only rarely done and with much serious consideration and evidentiary support—called *Australopithecus aferensis.* This made the human line almost four million years old.

Impact

Fossils represent facts that are generally firm. Interpretation of those facts is another question entirely. It is in the interpretation of these fossils that problems arise. The central question is one of relationship: how does one group of fossils relate to another in time and space? Just because one genus of human fossils is older than another does not necessarily mean that one gave rise to the other. The old linear model of human evolution asserted that one group appeared and evolved into another (with the old group dying off), which then evolved into another, and so on to modern humans. It is a simple straightforward view: One group at a time inhabiting the world to be pushed out by the next group. Because of the finds discussed above, this simplistic model is no longer accepted.

It is clear now that different groups of archaic humans inhabited the earth along side other archaic humans, as well as advanced primates. Groups overlapped each other. Some groups gave rise to new variations, while some groups died off as evolutionary dead ends. Just what that progression was remains a major question. There are several theories that are put forward to explain the progression of human evolution. One popular idea is the Out of Africa theory. This theory states that a group of advanced hominids walked out of Africa several million years ago. As they spread out they evolved into other forms until the appearance of modern *Homo sapiens* (the smart man) about 150,000 years ago. There is also the Out of Africa Two theory, which states that one group did leave Africa to evolve into more complex forms, but was followed by a second wave that left Africa later and competed with those hominids already out in the world.

Another version of evolution is the single species theory, which claims that all humans, whether extinct or extant, are of one species, but variations on a general theme. There is also the Eve hypothesis, which states that a tiny isolated group of early hominids, probably in Africa, were the original ancestral stock of all modern humans. A more controversial view is the Multi-Regional hypothesis, which argues that modern humans appeared in several different locations at different times, completely independently of one another, and gave rise to the different modern human groups. In contrast, there is the Diversity hypothesis, which contends that there were many different human species living side by side at one time, but that for one reason or another only one species survived. That surviving species gave rise to all modern humans, making us all the same species regardless of superficial differences.

At this point there is no consensus among scientists about the relationship of one group of archaic humans and hominids to another, or their relationship to modern humans. The search for answers to these puzzling questions continues, as does the debate about what it all means.

BRIAN REGAL

Further Reading

Howells, William. *Getting Here: The Story of Human Evolution.* Washington, DC: Compass Press, 1997.

Lewin, Roger. *Bones of Contention.* New York: Simon & Schuster, 1988.

Shreeve, James. *The Neanderthal Enigma.* New York: William Morrow, 1995.

Tattersall, Ian. *The Fossil Trail: How We Know What We Think We Know about Human Evolution.* New York: Oxford University Press, 1995.

A *Tyrannosaurus Rex* Named Sue

Overview

Vertebrate paleontology has always inspired scientific and political debates. Humans have been fascinated with discoveries of ancient, and sometimes bizarre, animals that reflect Earth's history. No recent event has produced as much controversy as the discovery of a *Tyrannosaurus rex* dinosaur in South Dakota, initially called "Sue."

Background

The saga began in 1990 when a field collector for a private company, the Black Hills Institute of Geological Research, discovered the most complete fossil of a *Tyrannosaurus rex* ever found. The research institute allegedly paid the Cheyenne Sioux landowner, Maurice Williams, $5,000 for the fossil and began their excavations. The amazing specimen was placed in protective plaster jackets and transported to the company's facilities for removal from its 67 million-year-old earthly encasement. Some of the bones were partially prepared for making replicas and display at the institute's headquarters.

News of the rare fossil's discovery spread like wildfire and soon scientists, museums, and people from all over the world were made aware of the stunning specimen. Controversy immediately followed. The academic world protested the sale of the fossil to the institute, claiming the primary goal of the company was the sale, not research, of vertebrate fossils. It was feared that the invaluable specimen would be sold to buyers whose main interest was not science and disappear from study forever. The worry also included fears of improper preparation. Vertebrate fossils often contain a great deal of information of interest to much of the scientific world, especially those interested in paleontology. Dinosaurs are especially fascinating to people of all ages,

A reconstructed dinosaur skeleton. *(Paul A. Souders/Corbis. Reproduced by permission.)*

and this T. rex was certainly the most complete specimen of its kind. It was thought that if the fossil were improperly removed and preserved, a wealth of potential scientific information it contained might be lost. The resulting battle that ensued over the specimen lasted for many years.

On May 14, 1992, the federal government executed a search and seizure warrant issued by the acting U.S. Attorney, Kevin Schieffer. Nine FBI agents, four National Park rangers, two agents from the Department of the Interior office of the Inspector General, one Bureau of Indian Affairs agent, one South Dakota Highway Patrolman, one Pennington County Deputy Sheriff, and many paleontologists arrived at the institute to confiscate the fossil from its owner, Peter Larsen. The National Guard arrived to help in the removal of the giant dinosaur. The fossil was taken to the Rapid City Museum of Natural History for safekeeping until the legal battle for custody of the rare specimen could be determined.

Years of legal battles followed. The Tyrannosaurus, which had been named "Sue" by Mr. Larsen and the institute, became an international celebrity. Schoolchildren all over the country knew about this legendary animal fossil. Many opinions about what should be done with the now famous dinosaur bones were published in magazines and newspapers. Legal judgments were overturned and then reestablished by several levels of the United States court system. There was even a request to have the case heard by the Supreme Court, though it was denied.

Impact

The central controversy of the legal battle is one that, in general, has not yet been decided. Who really owns national treasures like this amazing T. rex? Is it of such value that it should belong to the people of the country. Many foreign countries, like China, have protective laws regarding the discovery of important paleontological and archaeological finds. In the United States there are laws governing antiquities such as archaeological discoveries. But the laws regarding animal fossils are weak and ambiguous. The South Dakota T. rex became the focus of this ongoing debate. Since the fossil was found on private land, did it belong to the rancher to sell? Did the Black Hills Institute have the right to buy it and do whatever they wanted with the fossil? Do the people of America have the right to claim such remarkable finds as national treasures? Many additional legal questions were proposed in addition to these.

The argument proposed by the federal government was that the land owned by Mr. Williams was actually held in trust for the Cheyenne people by the United States government. This was the basis for seizing the fossil from the Black Hills Institute. Eventually that decision was overturned.

It was decided that the fossil did, indeed, belong to Mr. Williams. He chose to put the T. rex up for auction at Sotheby's New York. This unprecedented sale occurred on October 2, 1997. The bidding started at $500,000 and, in a few minutes, closed at $8.36 million, the largest sum ever paid for a fossil in history. The new owner of the spectacular fossil is the Chicago Field Museum, with a consortium of investors including McDonald's, Walt Disney World, the California State University system, and private individuals. It will have its public unveiling in the year 2000 at both the museum and Disney World, where replicas made from the bones of the giant animal will be placed on display.

A great deal of information about the fossil has been published. However, the name "Sue," provided at the Black Hills Institute, can no longer be used because Peter Larsen copyrighted it. A museum sponsored contest for schoolchildren will find a replacement name.

The characteristics and history of the famous dinosaur specimen have undergone much study. The huge 2000-pound (908 kg), 5-foot-long (1.5 m) skull was scanned at a Boeing lab in California in 1998. After 500 hours of x-raying the very thin sections of the skull and digitally reconstructing the skull, some interesting details have come to light. The Tyrannosaurus could both see and hear quite well. However, its greatest sense appears to be its ability to smell. Huge olfactory bulbs and canals for nerves support the hypothesis that a T. rex could sniff out food very easily.

Other yet to be confirmed data on the 67 million-year-old dinosaur indicates that it had a femur, or thigh bone, 54 (137 cm) inches long. It appears that the animal incurred some injuries before its death such as a broken tail and leg bone. It may have been bitten in the head by another T. rex since it seems to have a broken facial bone. Its brain was about the size of a grapefruit. Big for a dinosaur, but not very big in relation to its large body.

There is information that can never be known for certain, such as the gender of the animal or whether it was warm- or cold-blooded. Was it a predator or scavenger? And how or why did it actually die? What did its skin look like? How did it reproduce? Although the preserved skeleton provides much information about the specimen and the species, it is still just a skeleton. Without tissues and other samples these are questions that cannot be answered at this time, or perhaps ever.

The debate and legal battles over this amazing fossil have focused attention on the problem of what to do with fossils. Who has the rights to them? Should they be protected by federal laws? Should they be allowed to be sold outside of their country of origin?

Recent discoveries of important fossils and the information they provide has stirred a great deal of interest in Earth's history. New fossils of dinosaurs from China indicate that some dinosaurs may have had feathers, supporting the idea that birds are closely related to dinosaurs. What would happen if these fossils had been sold to private collectors? Dinosaur eggs are unique in that they reveal a great deal of information about the embryology or juvenile growth of certain dinosaurs. They even indicate some dinosaur behavior since they are often found in grouped nests. This grouping tends to support the idea of herding or social grouping between dinosaurs. Scientists argue that this type of information should be protected for study and that fossil hunters should not be allowed to simply collect and sell fossils.

On the other hand, the rights of landowners and the support of free markets are part of the basic foundations of society in the United States. Some people argue that to interfere with these rights may be unconstitutional.

Whatever the sentiment, it is a debate that is not likely to end very soon. Even with federal protection, important archaeological sites are often raided by poachers and valuable artifacts are stolen for sale on the black market. Other countries face the same problems. There are not enough finances to protect all the sites. Even if fossils are protected by law, it is believed that the illegal sale of fossils will continue.

Not long after the Black Hills Institute discovered "Sue," they found another T. rex they named "Stan." North America and Canada are regions where many dinosaurs lived and died. The more fossils that are discovered, the more science will understand the natural history of these animals.

In addition to dinosaurs, there are thousands of other fossil animals being unearthed every day. These fossils add to the knowledge of how life evolved. Patterns of evolution are recognized throughout the fossil record. Even human history is recorded in fossils. The search is always on to find the relatives of ancestral humans. What the world does with these fossils has yet to be determined.

BROOK ELLEN HALL

Further Reading

Fiffer, Steve. *Tyrannosaurus Sue: The Extraordinary Saga of the Largest, Most Fought Over T. Rex Ever Found.* New York: W. H. Freeman, 2000.

Horner, John R., and Don Lessem. *The Complete T. Rex.* New York: Simon & Schuster, 1993.

Horner, John R., and Don Lessem. *Digging Up Tyrannosaurus Rex.* New York: Crown, 1992.

Lindsay, William. *American Museum of Natural History: Tyrannosaurus.* London and New York: Dorling Kindersley, 1992.

Advances in Neurobiology and Brain Function

Overview

The deepest unknowns are not in outer space but behind the eyes, in the brain. This inner space is the newest frontier of investigation. Neuroscience, the study of the brain and nervous system, seeks not only to understand brain structure but also human thought and behavior. From the 1950s on, steady progress in research has been made—so much that President George Bush declared the 1990s to be the "decade of the brain." Using tools ranging from computers to special dyes to high throughput technology, neuroscientists study neurons, neural networks, the brain, and behavior. Learning how the nervous system functions normally is important in understanding complex neurological disorders, such as schizophrenia and Alzheimer's disease.

Background

For thousands of years people have tried to figure out why they act and behave in a certain way and in the process have assigned certain roles to the brain. While the ancient Egyptians preserved other organs but discarded the brain, the Greeks in the sixth century B.C. decided that the brain was the "organ of the mind." Aristotle taught that the brain was a cooling system for the body's blood. During the seventeenth century, anatomical examinations determined the brain's general outline.

In the early 1800s a pseudoscience called phrenology taught that personality traits had specific locations in the brain and that a trained person could analyze personality by measuring lumps on the head. While the fad of phrenology was ill-founded, it became an important step in understanding the brain's role in human behavior.

At the beginning of the twentieth century, people still thought that dissection tampered with the soul. As the century unfolded, however, knowledge about the brain slowly evolved. In 1906 the Spanish anatomist Ramon y Cajal won a Nobel Prize for determining that the brain was built from independent nerve cells, or neurons. In the 1920s researchers discovered that neurons used a chemical signal or neurotransmitter. Not until the electron microscope was developed in the 1930s, however, could one picture the brain's neuronal structure. Pioneers in the study of neurons included Otto Loewi, Charles Sherrington, Alan Hodgkin, Andrew Huxley, John Eccles, Bernard Katz, and Julius Axelrod.

Much of neuroscience research has focused on the nerve cell. Each person is born with about 100 billion neurons—as many as he will ever have. The neurons never touch each other but rather communicate by sending electrical signals across a synapse, or gap. These signals are assisted by chemicals called neurotransmitters. Drugs may affect this transmission, and certain mental illness are caused by imbalances in the neurotransmitters. The brain, in turn, is composed of the cerebrum, the thinking and largest part of the brain; the cerebellum, which controls motion and balance; and the brain stem, which is the seat of involuntary processes such as breathing and digestion.

Sir Charles Sherrington (1857-1952) is known as the father of modern neurology. He won a Nobel prize for physiology or medicine in 1932. He described the brain as "an enchanted loom, where millions of flashing shuttles weave a dissolving pattern, always a meaning pattern, through never an abiding one." This quotation foreshadowed the current theory of brain plasticity.

Impact

In the last half of the twentieth century, neuroscience integrated biology, chemistry, and physics with the studies of physiology, structure, and behavior. Scientists began to determine the electrical and chemical processes involved in how neurons communicate with each other across synapses. Nerve impulses involve the opening and closing of ion channels, which are tunnels that allow charged atoms, or ions, to pass through. The voltage resulting from the movement of ions triggers the release of neurotransmitters, special chemicals that relay the impulse from one neuron to the next. Acetylcholine, found in the 1920s, was the first neurotransmitter to be identified. Other neurotransmitters include dopamine, norepinephrine, serotonin, and chemicals called opioids. The actions of these chemicals play important roles in key body functions, including learning, memory, movement, and emotion. Understanding neurotransmitters is one of the top challenges of neuroscience.

Roger Wolcott Sperry (1913-1994), an American neurobiologist, shared the 1981 Nobel Prize for physiology or medicine for his investigation into brain function. While his early work was on the regeneration of nerve fibers, he eventually became interested in split-brain research. Using animals at first, he later began to work with epileptics whose brains had been divided surgically to cure their epilepsy. The human brain has two halves, called hemispheres, that are joined by a thick band of nerve fibers, the corpus callosum. Sperry's studies demonstrated that the left half has definite functions for analytical and verbal tasks, while the right half controls such functions as spatial tasks and musical abilities.

Sperry's research in the early 1950s laid the foundation for other work. Memory research was reinforced in 1953 when an individual, named H.M., had surgery in the medial area of the temporal lobe to cure epilepsy. H.M. could remember things that pre-dated the surgery but could not remember for more than 15 minutes any events that occurred after the surgery. This result indicated different kinds of memory, dubbed "short term" and "long term." At one time, the brain was thought to operate like a computer, with memory patterns laid down in units. More exhaustive research is showing that memories are stored throughout the cerebral cortex in glial cells, the connective tissue of the brain. The process of memory is still a mystery.

Another 1981 Nobel Prize winner, David H. Hubel (1926-), analyzed the flow of nerve impulses from the retina to the sensory and motor centers of the brain. He used tiny electrodes to track the electrical discharges that occur as light passes from the retina to the brain. Other scientists would show that electrical impulses from any of the senses are the same and that the destination area in the brain determines whether the sense is sight, sound, touch, taste, or smell. For example, when a person hits his finger with a hammer, what matters is how the message registers in the brain. At the site of the injury, the body produces prostaglandins, increasing pain sensations. Aspirin works by blocking this production; acetaminophen, in turn, blocks pain reception in the brain, while opiates, such as morphine, block pain signals in the spinal column. Electrode implants may also block pain as well as stimulate vision or hearing.

Research on electrical stimulation of the brain has yielded important findings. Spanish researcher Jose Delgado brought dramatic attention to his work by donning a matador's cape and standing in front of a charging bull. When the bull reached full speed, Delgado pressed a remote control that controlled a hair-like electrode in the bull's brain, freezing the bull in its tracks. In the late 1950s electrode experiments with rats accidentally uncovered "pleasure centers." The rats learned to press a bar that stimulated the pleasure center, pressing until they dropped from exhaustion. Researchers now relate these pleasure centers to various addictions.

While recent revelations in neuroscience are legion, the following may be pursued and refined in future years:

1) At one time, people thought the brain was static and inevitably declined with age. Research is finding out that what distinguishes the brain from a computer is its adaptability, called by scientists "plasticity," meaning that the brain is flexible and dynamic. The brain, therefore, is not just a rigid box programmed up to the age of two or three but is forever adapting to experience, hormones, and injury as well as countless learning experiences.

2) Knowledge about neurotransmitters and molecular chemistry have made new drug investigations, called "drug discovery," possible. Drugs are now being developed through a process known as "high throughput technology," which combines mass micropipetting with spectrographic analysis using computer data—a sophisticated and expensive development process.

3) New gene-related techniques and the Human Genome Project are finding the chromosomal locations for genetic defects, such as Huntington's disease, Down syndrome, and

Alzheimer's disease. Conditions that are carried on a single gene pair are candidates for gene therapy, which entails replacing a defective gene so that its characteristics will not be expressed.

4) The development of "transgenic animals," such as mice that have been bred to express genes suspected in human Alzheimer's disease, are used to study diseases and develop drugs. Novel treatments are being tried that once went against all dogma. For example, in 1999 a vaccine was developed for Alzheimer's that was found to be successful in treating the transgenic rats that had been bred to show symptoms of human Alzheimer's. Embryonic and fetal tissue transplantations for spinal cord regeneration are also in trials.

5) More and more medical instruments are less damaging and invasive. For example, the stereostatic linear accelerator, a precision radiosurgical instrument that delivers a high dose of radiation without destroying normal brain tissue, has been developed.

Much of the work in neuroscience in the 1950s and 1960s resulted from animal studies, but techniques have since evolved that allow better observations of the human brain. The electroencephalogram (EEG) records trace activity of the brain by means of electrodes on the head; computer-processed information allows brain waves to be studied for abnormal patterns. Positron emission tomography (PET) involves the injection of a radioactive material that goes to active brain areas, which are then shown as three-dimensional color-coded computer pictures. PET is used to help doctors determine how well a drug or treatment is working on a neurological disorder. Single photon emission computer tomography (SPECT) is similar to PET but less expensive. Magnetic resonance imaging (MRI)uses magnetic coils to detect signals from tissue and give an accurate picture of brain anatomy. Magnetic source imaging (MSI), the newest scanner, reveals a weak electronic field from neurons and is useful in pinpointing the origins of epilepsy. The magnetic resonance spectroscope (MRS) is related to MRI but scans for chemistry rather than anatomy.

Exciting new breakthroughs in neuroscience are forthcoming. One expected breakthrough is transplanting pieces of one brain into another. Another possibility includes treating children with mental retardation with drugs. While more understood now than they were ten or twenty years ago, the brain and nervous system are still the least known of all the body systems and remain a relatively unexplored frontier.

EVELYN B. KELLY

Further Reading

Dennett, Daniel C. *Brainchildren: Essays on Designing Minds.* Bradford, CT: MIT Press, 1998.

Gazzaniga, Michael S. *The Mind's Past.* Berkeley, CA: University of California Press, 1998.

Gillick, Muriel R. *Tangled Minds: Understanding Alzheimer's Disease and Other Dementias.* New York: Dutton, 1998.

Katz, Lawrence. *Keep Your Brain Alive: 83 Neurobic Exercises.* New York: Workman, 1999.

Wyden, Peter. *Conquering Schizophrenia: A Father, His Son, and a Medical Breakthrough.* New York: Knopf, 1998.

New Directions in Evolutionary Theory

Overview

Darwinian evolution, that is, evolution through the accumulation of almost undetectable changes over millions of years, has been the mainstay of evolutionary theory since its formulation by Charles Darwin (1809-1882) and Alfred Wallace (1823-1913) in the mid-nineteenth century. In the last half of the twentieth century, this view of evolution was further developed to accommodate the realization that natural selection can operate in alternate ways, that evolution does not always operate with geologic slowness, and that the fittest do not always survive. These revisions to evolutionary theory have had a profound impact on how we view the natural world and our place in it, as they serve to emphasize that there is no evolutionary "reason" for us to exist. Instead, it is becoming more obvious that humans, like any other species, are the result of a number of factors, including random chance and contingency.

Background

Evolutionary theory was born in the Galapagos Islands and the Indonesian Archipelago, the intellectual child of Darwin and Wallace. Both men, working independently, reached the conclusion that species must be able to change, or evolve over time, giving rise to new species and that natural selection, or "survival of the fittest," was the primary factor in determining which traits became the basis for new species. Although Darwin and Wallace acknowledged that natural selection was not likely to be the sole motive force for evolution, and that evolution did not necessarily proceed in a gradual and smooth manner, they did express a preference for gradual, smooth transitions between species. This gradualist ("Darwinian") viewpoint held sway from 1859, the year that Darwin published his landmark work *On the Origin of Species,* until 1972, when Niles Eldredge and Stephen Jay Gould (1941-) suggested that evolution consisted of long periods of evolutionary stasis punctuated with bursts of extremely rapid evolutionary change and speciation. This view of evolutionary theory was called "punctuated equilibrium" and it rapidly gained adherents.

At the same time, genetic research performed by Motoo Kimura (1924-) suggested that not all genetic changes (mutations) resulted in a change in a creature's fitness to survive. Many mutations were neither good nor bad but were, instead, neutral in their effects. This gave scientists the equivalent of a clock, based on neutral mutations, that could be used to date the amount of genetic change accumulated in different organisms; a clock that could be used to determine when two lineages diverged in the geologic past.

In addition to Kimura's work, Gould and others pointed out that natural selection need not always choose the organisms best suited for survival. Modern species may, instead, simply be those that were lucky enough to survive catastrophic events. For example, an asteroid striking the Amazon would likely kill every organism living in the region of the impact. Not considering the climatic effects this event would have, Amazonian organisms would become extinct, not because they were less fit to survive, but because they happened to have evolved in an area that was struck by the asteroid. After the dust settled the region would be full of ecological niches with no native organisms to fill them. This would invite rapid speciation by the first organisms to arrive, and the fossil record would show the sudden appearance of a multitude of new species. By conventional, Darwinian logic, these new organisms would be found in the fossil record and would be assumed to have been better adapted. Using the tools of punctuated equilibrium and contingency, a researcher may now suggest otherwise. In reality, both mechanisms affect evolution, but contingency and punctuated equilibrium are proving to be far more significant than previously thought.

Finally, there is no reason to suppose that evolution leads inevitably to intelligence. In other words, humans are not the natural endpoint of evolution. Bacteria have dominated life since its first appearance and continue to do so today. Gould posits that the apparent "direction" of evolution toward increasingly complex organisms simply records the fact that evolution works randomly, towards both increasing simplicity and increasing complexity. However, there is a "wall," below which simpler life cannot exist. Therefore, random change will tend to be towards the direction of complexity simply because that is the only direction to go. Yet, he also points out that complex organisms tend to simplify with time as frequently as they become more complex.

To summarize, evolutionary theories have changed dramatically since the 1970s. While paleontologists and evolutionary biologists affirm the important role that natural selection plays in evolution, most also agree that other factors, including luck, have had equally important roles in determining the shape of modern life on earth. Put together into what is called the Modern Synthesis, these concepts have had an important impact on the way we view life and evolution in general, as well as our place in the world.

Impact

From a strictly scientific standpoint, viewing evolution in terms of punctuated equilibrium has opened the field up tremendously. In many cases, scientists struggled to explain some changes in species in terms of perceived adaptive benefit, often constructing hypotheses that seemed far-fetched at best. Understanding that factors other than strict Darwinian fitness can govern evolutionary selection has coincided with the realization that strict uniformitarianism is also an oversimplification. Uniformitarianism states that evolutionary forces have worked with the same intensity and the same frequency since life emerged. Punctuated equilibrium reflects the fact that geologic change takes the form of episodic catastrophe as well as slow, steady accumulation of minuscule effects. In both cases, there is a uniformity of physical laws and processes, some processes just happen to occur randomly and lead to very rapid change as opposed to the majority of processes that operate almost invisibly over millions of years. In this way, these two fields of study have tended to reinforce each other, each leading to a fuller and more dynamic view of the earth's history.

The effects of these advances in evolutionary theory have largely been confined to the realm of science and scientists. However, this contingent and dynamic view of life has inescapable philosophical implications. One of the most significant social impacts of this revised thinking involves the continuing debate between those who believe in the evolution of species and those who believe in the literal truth of the Bible. Many biblical literalists (often referred to as religious fundamentalists) view the new scientific debate surrounding evolution, as an indication that the theory is discredited and falling into disfavor among scientists. Drawing some strength from this argument, biblical literalists have managed to reopen arguments concerning the teaching of evolution in American public schools, an issue many presumed was settled once and for all with the famous Scopes trial in 1925. It should be noted, incidentally, that many major religions accept that evolution occurs. Pope John Paul II conceded in a 1996 papal encyclical that "the tehory of evolution [is] more than hypothesis."

Many biblical literalists state that evolution's status as a "theory" also undermines its scientific legitimacy. This, however, is based on a fallacious interpretation of the term "theory." In science, "theory" refers to an idea that is firmly accepted as true by the scientific community, even if some details still remain to be worked out. Thus, we have plate tectonic theory, genetic theory, gravitational theory, and evolutionary theory, to name a few. Science does not doubt that these occur—tectonic plates move, genes exist and transmit information between generations, mass attracts other mass, and species evolve.

Finally, scientific revisions of evolutionary theory have challenged some philosophical views regarding man's place in the universe. For millennia men and women have found solace and strength in the view that humans enjoy a special, privileged relationship with God. Darwin and Wallace challenged this belief, relegating mankind to a place above the ape, but an animal nonetheless, counting beasts in our lineage. Yet, even in this view it was still possible to find consolation in the assumption that evolution's progress led inexorably to humans as

SCIENCE WARS

Starting in the 1960s there developed an increasing tension between scientists and those who study the philosophy and sociology of science. In 1962 Thomas Kuhn set forth the argument that scientific knowledge—far from being objective—reflected the cultures that produced it. Postmodernists and others involved in cultural studies subsequently began to reassert airy and discarded philosophical viewpoints that properties associated with the physical world were influenced by human psychology or cultural perspectives. Scientists maintained that science was not simply another form of cultural criticism and that it remained the most accurate and productive way to understand a real and knowable world. The debate was not new to scientists. During the Scientific Revolution Bishop Berkeley, a contemporary clerical critic of Sir Isaac Newton, once claimed that gravity and matter were "imaginary" constructions of the human mind. Legend has it that Berkeley's assertion was promptly refuted by one Royal Society scientist's stubbing of his toe upon a rock as he pronounced, "I refute it thus!"

The modern debate also often proved so heated and acrimonious that many have dubbed it the "Science Wars." In 1996 Alan Sokal, a respected physicist at New York University, brought the issue to the attention of the general public with his celebrated hoax of the cultural studies journal, *Social Text.* Sokal deliberately slipped past the editors of an issue devoted to anti-science viewpoints a scholarly sounding but obtuse article titled, "Transgressing the Boundaries: Toward a Transformative Hermeneutics of Quantum Gravity" that repeatedly offered gibberish prose to support sweeping philosophical, cultural, and political conclusions based on bizarre representations of mathematical and physical theories. Sokal perpetrated the hoax to reveal what he maintained was a lack of scholarly rigor on the part of those who viewed science as a cultural expression. The hoax went unnoticed by the editors and reviewers of *Social Text* until Sokal revealed his deception in the journal *Lingua Franca.*

the most complex, intelligent, and best-adapted animal yet to appear on Earth. Whether guided by divine intervention or by the laws of the universe, life progressed to form our own "advanced" species.

According to Mark Twain, this was similar to saying that, by analogy, the whole of the Eiffel Tower existed for the purpose of the coat of paint at the very top, and current thinking would not disagree. Instead, we find ourselves lucky to be here, for there was no way to predict that intelligent life would come to pass on Earth at all, let alone life that looked like us. In a sense, we have lost the ability to think of ourselves as the logical outcome of billions of years of evolution, just as we have lost the ability to reason that the laws of nature had to produce a thinking, bipedal creature. From the standpoint of evolution, we are not special. However, this view has proven liberating, too, for accepting that we are the product of chance, contingency, and random process also means that we can revel in our good fortune to be here at all as a species. Also, if we are not created for any specific purpose, neither are we destined to fill any purpose other than what we set for ourselves. Thus, as a species, we have the opportunity and the responsibility to use our existence wisely. These philosophical and ethical implications, in addition to the obvious scientific significance, are among the most profound consequences of the new thinking about evolution.

P. ANDREW KARAM

Further Reading

Darwin, Charles. *On the Origin of Species by Means of Natural Selection or the Preservation of Favoured Races in the Struggle for Life.* 1859. Reprint. New York: Modern Library, 1993.

Eldredge, Niles. *Fossils: The Evolution and Extinction of Species.* New York: H. N. Abrams, 1991.

Gould, Stephen Jay. *Full House: The Spread of Excellence from Plato to Darwin.* New York: Harmony Books, 1996.

Gould, Stephen Jay. *Rocks of Ages: Science and Religion in the Fullness of Life.* New York: Ballantine, 1999.

Gould, Stephen Jay. *Wonderful Life: The Burgess Shale and the Nature of History.* New York: W. W. Norton, 1989.

Skelton, Peter, ed. *Evolution: A Biological and Palaeontological Approach.* Reading, MA: Addison Wesley, 1993.

Advances in Plant Biology since 1950

Overview

Like many other areas of biology, the study of plants—also called botany—underwent dramatic changes over the past 50 years. Two major and related trends were largely responsible for these changes. First was the intensification of research begun earlier in the century to look for the molecular basis of plant structures and functions; in other words, scientists investigated the chemical reactions that are involved in everything from how plants harness the Sun's energy to how they respond to invasion by disease organisms. The second major trend was the exploration of the genetic basis for plant characteristics and processes. The discovery of the structure of DNA, the molecule that contains genetic information, led to a great deal of research on genes, the pieces of DNA responsible for specific traits or activities in all living things, including plants. This essay will review some of the results produced by both these lines of research.

Background

During the first half of the twentieth century plant biologists discovered a number of plant hormones, chemical signals that coordinate activities such as plant stems bending toward the light or roots growing against the force of gravity. But it has only been more recently that researchers have determined more precisely how these hormones produce their effects. For example, the simplest plant hormone, ethylene, uses the same signaling system in plant cells to create such diverse effects as cell elongation, root development, and fruit ripening. In other research on plant chemistry, biologists have recently discovered that the production of salicylic acid (a cousin of the chemical in aspirin) causes a number of other chemical changes in the plant that make it more resistant to damage by viruses, fungi, and bacteria. Studying how plants resist damage is important in agriculture where the organisms just cited, along with parasitical worms, are responsible for a great deal of crop damage.

Many plants are sensitive to day length and will only develop flowers when the days start to become longer in the spring. In the 1950s phytochrome, the light-sensitive chemical responsible for this effect, was discovered. But as with the effects of plant hormones, it has taken researchers years to decipher the precise molecular processes that phytochrome triggers. Recently, they found the cellular molecule, the protein, that activates phytochrome and the mechanism by which the activated form in turn changes gene expression, leading to the wide-ranging changes involved in flower development.

It was at mid-century that Melvin Calvin (1911-1997) and his associates at the University of California-Berkeley discovered a key sequence of chemical reactions involved in plants' harnessing the Sun's energy in photosynthesis. Called the Calvin cycle, these reactions convert carbon dioxide into sugar using the light energy trapped by the green pigment chlorophyll. Calvin received the Nobel Prize in chemistry for this work in 1961, but his research hardly provided a complete understanding of photosynthesis. Researchers are still investigating how the process is controlled and have discovered many processes related to the Calvin cycle. These include photorespiration, in which oxygen replaces carbon dioxide in the cycle and reduces sugar production. This is most likely to occur on hot, dry days, and some plants, including those adapted to desert life, use variations on the Calvin cycle that prevent much of the energy drain of photorespiration.

In the 1970s recombinant DNA techniques were developed, making possible the transfer of genes from one species to another. At first it was difficult to use these techniques on plants, because the methods to transfer genes into bacteria and into animal cells did not work on plants. When an efficient transfer method was developed, genetic engineering of plants moved ahead quickly with the focus on economically important food plants. Ultimately, the development of cloning made it easier to genetically engineer plants than animals.

In the 1950s it was first demonstrated that whole plants could be regenerated from cells taken from a fully developed carrot plant and grown in culture on an artificial medium containing nutrients and hormones. This meant that by teasing apart the cells of one plant and growing them separately, researchers could produce large numbers of genetically identical plants, again with the focus on economically important species. It also meant that single plant cells could be genetically engineered, and then whole plants with altered properties could be grown from these cells.

Impact

An obstacle that slowed analysis of plant genetics was the absence of a model organism about which a great deal was known and which was easy to grow and manipulate, in other words, a plant comparable to the mice and rats used in animal research and the bacterium *E. coli* employed so frequently in the early days of molecular biology. In the 1980s attention focused on *Arabidopsis thaliana,* a tiny weed in the mustard family; it is easy to grow even in a test tube, has a short life cycle of only six weeks, and has a compact genome, that is, a relatively small number of genes. Research on this plant increased tremendously in the last decade of the century as many biologists used it to work out the genetic basis of everything from flower formation to cell-wall synthesis. In late 1999 the complete sequence for one *Arabidopsis* chromosome, a segment of its DNA, was reported, with the assumption that the sequence for the complete genome would be available within a few years.

But *Arabidopsis* can't be used to investigate all the issues important in plant biology. For example, it is not one of the legumes, the group of plants that have nodules filled with bacteria growing on their roots. Nitrogen is an essential plant nutrient, but though the air is over three-quarters nitrogen, it is not in a chemical form that can be used directly by plants. The bacteria in legume root nodules can convert nitrogen into a useable form by a process called nitrogen fixation; this means that legumes can grow well even in nitrogen-poor soil. Since the 1970s researchers have found more than 20 bacterial genes required for nitrogen fixation. At first, they thought that they could make non-legumes such as corn into nitrogen fixers by transferring the bacterial genes into these plants. Having little success with this tack, they are now focusing on creating conditions in the roots of non-legumes that would make it possible for the nitrogen-fixing bacteria to live there. Along with this work, there has been a steady rise in plant biologists' interest in roots in general, with increasing estimates of the extent and significance of roots to plant growth. Researchers have found that most plants rely heavily on fungi living in and around their roots to make it easier for the plants to absorb soil nutrients.

Work on root bacteria and fungi indicate how important other species can be to plants. There have also been investigations of how crucial animals are as pollinators, carrying pollen grains that contain the male sex cell from one flower to another of the same species, where the male cell can fertilize the female sex cell to create a seed. In many plant species the male and female sex organs are in the same flower, but pollen from that flower, or that plant in general, can't fertilize the egg of the same plant. Using *Arabidopsis,* researchers have begun to work out the molecular basis for this phenomenon, discovering that there is a complex series of molecular interactions involved in fertilization and that these interactions can fail at any of a number of different points, thus preventing self-fertilization. In any case this means that pollen must be transported from one plant to another in order for pollination and fertilization to be successful. For some species wind is the chief means of transport, but for many species animals—especially insects and birds—serve as pollen carriers. In some cases the relationship between plant and pollinator is very specific, with a particular animal species responsible for pollination. Plant biologists are increasingly interested in how both species in such relationships have evolved adaptations to make such interactions more effective, and a debate has developed over whether or not the diversity of insect species was key in the evolution of flowering plants.

Paleontologists, those who study fossils of past life, have estimated that flowering plants first appeared about 90 million years ago. Now those estimates are being pushed back with the recent discovery in China of 90-million-year-old fossils of flowers with characteristics that are not at all primitive, indicating that by this time flowering plants had probably been in existence for tens of millions of years. As to which present-day species are most closely related to these first flowering plants, the magnolia had been considered the most likely candidate because of its primitive flower structure. But recent genetic analyses comparing the same gene sequence in a large number of flowering plants indicate that a little-known plant called *Amborella* is the closest living relative of the first flowering plant.

This research was part of a project begun in the 1990s called Deep Green, which coordinated the work of a large number of plant biologists looking at the classification of plants and how they are related to each other. The project's results also called into question another long-held botanical concept, that flowering plants belong to two fundamentally different categories: monocots, with one seed leaf and usually narrow leaves like grasses, and dicots, with two seed leaves and usually broad leaves like tomatoes and oaks. But Deep Green results indicate that some monocots are more closely related genetically to some dicots than to any other monocot. And other findings suggest that some green algae, water plants, are more like land plants than they are like other algae.

The Deep Green project suggests how plant science will be carried out in the future. The stress on molecular research and genetic analysis that developed in the last 50 years is likely to continue and intensify in the next 50. There will also be an intensification of efforts to identify, describe, and classify new plants in an effort to document Earth's dwindling biodiversity.

MAURA C. FLANNERY

Further Reading

Books

Attenborough, David. *The Private Life of Plants: A Natural History of Plant Behavior.* Princeton, NJ: Princeton University Press, 1995.

Barth, Friedrich. *Insects and Flowers.* Princeton, NJ: Princeton University Press, 1985.

Bernhardt, Peter. *Natural Affairs.* New York: Random House, 1993.

Bernhardt, Peter. *The Rose's Kiss: A Natural History of Flowers.* Washington, DC: Island Press, 1999.

Calvin, Melvin. *Following the Trail of Light: A Scientific Odyssey.* Washington, DC: American Chemical Society, 1992.

Huxley, Anthony. *Plant and Planet.* New York: Penguin, 1987.

Rissler, Jane, and Margaret Mellon. *The Ecological Risks of Engineered Crops.* Cambridge, MA: MIT Press, 1996.

Swain, Roger. *Earthly Pleasures.* New York: Scribner's, 1981.

Swain, Roger. *Field Days.* New York: Scribner's, 1983.

Periodical Articles

Brown, Kathryn. "Deep Green Rewrites Evolutionary History of Plants." *Science* 285 (1999): 990-1.

Kenrick, Paul, and Peter R. Crane. "The Origin and Early Evolution of Plants on Land." *Nature* 389 (1997): 33-9.

The Study of Human Sexuality

Overview

The study of human sexual behavior is a relatively new science compared to other scientific disciplines. While disciplines such as cell biology were limited by the technology of the day, serious investigations into human sexual form and function were hindered by ethical constraints. The groundbreaking studies of Alfred Kinsey (1894-1956), whose systematic research reported the sexual behaviors of Americans, laid the foundation for the sexual revolution of the 1960s and 1970s.

Background

Historically, scientists have explored physical and biological phenomena through careful observation and methodical investigation. Sound science is rooted in a researcher's ability to remain objective about the subject of his or her investigation. Presumably, a researcher's objectivity is easily maintained when cultural mores, the fundamental moral views of a group, are not called into question. Objectivity is almost never a confounding factor in most physical and biological sciences.

However, prior to the 1930s, the assumption of objectivity failed for investigations involving human sexual behavior. Early sex scientists, or sexologists, were physicians and psychiatrists unschooled in the scientific method, the systematic approach to solving problems. The results were early sexuality studies fraught with inaccurate information and personal bias.

During the eighteenth century, the guardianship of sexual study began to shift. What had been almost entirely a moral issue became the focus of discussions concerning sexual ethics. Although few physicians had any specialized knowledge of human sexuality and behavior outside the treatment of sexually transmitted diseases, physicians were viewed as authorities in the infant field of sexology. Discussions of sexuality gave rise to the first programs of public and private sex education as well as new classifications and documentation of sexual behaviors. Samuel Tissot's published 1760 warning against masturbation, in order to prevent "masturbatory insanity," became a dominant theme in adolescent sex education.

By the nineteenth century, the medical view of sexual behavior was expanded to include sexual behaviors classified as mental diseases. Heinrich Kaan's *Psychopathia sexualis* (1843) introduced the concept of deviance, behavior that diverges from the accepted norm, and perversion, behavior caused by a determination not to do that which is expected, both regarded as functions of mental illness. Less than 30 years later, as a result of the publication of case histories by prominent psychiatrists, homosexuality became viewed by the medical community—and, therefore, by society—as a mental illness.

In 1873 Anthony Comstock (1844-1915) lead a campaign to regulate sexual behavior, resulting in the passage of the Comstock Laws, which further limited sexual freedom, particularly that of women. These laws made the distribution of information about contraception illegal. Physicians were prohibited from providing patients with contraceptive information and were imprisoned when found in violation of the law.

Iwan Bloch (1872-1922) is credited with founding the modern study of sexuality. Dissatisfied with the medical view of sexual behavior at the turn of the twentieth century, Bloch challenged conventional views of sexuality. He proposed a reexamination of perceived pathological and degenerative behaviors, such as prostitution and homosexuality, from both a historical perspective and on a global scale. Bloch co-founded the *Journal for Sexology* with his colleague, Magnus Hirschfield. Hirschfield pioneered the first gay rights organization and opened the first Institute for Sexology in 1919.

Establishment of the first Institute for Sexology and funding provided by the Rockefeller Foundation for the purpose of studying American sexual behavior lent an air of credibility to the study of sexuality. The missing piece in the puzzle was a researcher with the qualifications and desire to carry out studies of actual human sexual behaviors. Until the mid-1930s, sex research was based primarily on field observations of animals, historical data, and poorly constructed questionnaires. Due to the sensitive nature of the topic, few scientists were prepared or willing to embark on a study of actual human sexual behaviors. All of that changed when Alfred Kinsey signed on to teach a course on marriage and family in 1938.

Alfred Kinsey. *(The Library of Congress. Reproduced by permission.)*

Kinsey was an ideal candidate for conducting the types of research studies that the Institute had in mind. Unlike previous physicians and psychiatrists who studied the topic, Kinsey was an objective researcher. As an accomplished author of several biology textbooks and an experienced researcher well versed in the scientific method, Kinsey found the information that was to be used in the marriage course lacking. Dismayed by the thin veil of science that underscored published sex studies, Kinsey set himself to the task of building a new sex behavior knowledge base, employing the scientific method during the process.

Kinsey began his research into the sexual experiences of others using questionnaires. However, concerns over the validity of the responses recorded through this technique caused Kinsey—and, subsequently, his colleagues—to switch to face-to-face interviews. Kinsey devised an elaborate series of carefully constructed questions and questioning techniques designed to elicit extremely intimate details, while at the same time maintaining objectivity and anonymity. Through this interview technique, Kinsey set a new benchmark in sex research.

Based on empirical data gathered through thousands of interviews, Kinsey and his colleagues, Wardell Pomeroy, Clyde Martin, and Paul Gebhard, published *Sexual Behavior in the Human Male* in 1948 and *Sexual Behavior in the Human Female* in 1953. Information contained in these two volumes turned conventional perceptions of human sexuality on its head, as old myths were dispelled. The contents of these two controversial publications forever changed precepts (accepted conditions of moral behavior) regarding such topics as masturbation, homosexuality, premarital and extramarital intercourse, and the role of sex in the lives of women.

Impact

In 1948 Kinsey reported that 37 percent of American males had at least one homosexual experience during their lifetime. By 1950 homophobia swept the nation, as police and government agencies attempted to rid public positions of "deviants" and "sex perverts."

Evelyn Hooker became one of the most influential figures in the highly successful movement to convince the American people that homosexuality is a "normal variant" of human sexual behavior. Her 1957 study *The Adjustment of the Male Overt Homosexual* picked up where Kinsey and his colleagues left off. Hooker assembled a panel of experienced psychologists for the purpose of evaluating a series of psychological tests administered to 60 men. The inability of the psychologists to discriminate between homosexual and heterosexual responses provided evidence that homosexual behavior was normal human sexual behavior. Replication of Hooker's research results by other researchers and through other means of evaluation caused the American Psychiatric Academy to reevaluate its classification of homosexuality as a mental illness. In 1973 the Academy removed homosexuality from its list of psychiatric disorders.

In 1950 Margaret Sanger (1879-1966) lead efforts to solicit research money for the purpose of finding an effective oral contraceptive that would give women more control over their childbearing decisions. The right to use birth control was limited to married women, a nineteenth-century law that would not be repealed until 1972. Through the collaborative work of Gregory Pincus (1903-1967) and John Rock (1890-1984), Enovid—the Pill—was approved by the Food and Drug Administration (FDA) in 1960. By the mid 1960s, more than 80 percent of all married college graduates under the age of 25, and half of all married women under age 20, were "on the Pill." The sexual revolution had begun.

With the control of reproduction in the hands of women through the availability of oral contraception, and armed with the knowledge that females were as sexual as men, compliments of *Sexual Behavior in the Human Female,* the stage was set for the next major advances in the study of human sexuality. If the 1950s had been guided largely by the work of Kinsey, then the 1960s and 1970s belonged to William Masters and Virginia Johnson.

Their investigations into the physiological aspects of sexuality produced some of the first reliable data in the field of human sexuality. Masters and Johnson's first book, *Human Sexual Response,* recorded the results and conclusions of detailed laboratory studies on the physical aspects of sexual arousal and orgasm in a large number of men and women. Published for the medical community, it quickly became a best-seller purchased by the general public. The Masters & Johnson Institute opened in 1964, providing sex counseling and therapy for individuals and couples experiencing sexual difficulties. Masters and Johnson pioneered the field of sex therapy and trained other therapists in clinical counseling. The Masters & Johnson sex therapy program became a model for clinics elsewhere.

Sex research conducted during the last 100 years significantly impacted social and cultural mores of Americans and others around the world. With the debunking of nineteenth-century myths surrounding human sexuality, and a more sophisticated understanding of the role sexuality plays in people's lives, future sexual behavior research is likely to involve studies into disease risk prevention, as well as assessments of how sexuality changes as a function of age. Society will continue to reap the benefits of sound scientific investigations, increasingly focused on social and physiologic factors that shape sexuality and the development of sexually "healthy" adults.

MICHELLE ROSE

Further Reading

Kinsey, Alfred C., et al. *Sexual Behavior in the Human Female.* 1953. Reprint. Bloomington: Indiana University Press, 1998.

Kinsey, Alfred C., Wardell Baxter Pomeroy, and Clyde E. Martin. *Sexual Behavior in the Human Male.* 1948. Reprint. Bloomington: Indiana University Press, 1998.

Masters, William H., and Virginia E. Johnson. *Human Sexual Response.* Boston: Little, Brown, 1966.

Masters, William H., Virginia E. Johnson, and Robert C. Kolodny. *Human Sexuality.* New York: HarperCollins College Publishers, 1995.

Petersen, J. R. *The Century of Sex: Playboy's History of the Sexual Revolution.* New York: Grove Press, 1999.

Robinson, Paul A. *The Modernization of Sex: Havelock Ellis, Alfred Kinsey, William Masters, and Virginia Johnson.* New York: Harper & Row, 1976.

Zgourides, George. *Human Sexuality: Contemporary Perspectives.* New York: HarperCollins College Publishers, 1996.

The Emergence of Biotechnology

Overview

James Watson (1928-) and Francis Crick's (1916-) publication on April 25, 1953, of the double helix model for deoxyribonucleic acid (DNA) propelled the biological sciences and biotechnology, the use of microorganisms to produce specific chemical compounds, into the modern age. Today, biotechnology and molecular biological techniques, once strictly confined to the realm of geneticists and molecular biologists, are finding applications in fields ranging from medicine to species conservation.

Background

Microorganisms, such as yeast, have enhanced the quality of human life for thousands of years. Primitive civilizations used simple forms of biotechnology as they fermented juices to produce alcoholic beverages. By the mid-1700s, microorganisms had been incorporated into cheese, bread, and beverage production. In the 1920s, Alexander Fleming (1881-1955) serendipitously discovered that a mold produced the chemical substance penicillin, which became the first antibiotic used to fight infection. During the next 25 years, as more antibiotics were isolated from microorganisms, early immunology was limited by the amount of the desired chemical substance that these microorganisms could produce naturally.

While immunologists studied microorganisms, their products, and the effects of these products on disease, biologists in other fields were working with other experimental systems to unravel biological mysteries. Experimental, or model, systems were often selected by each scientific discipline for both practical and research purposes. On the practical side, as a body of knowledge grew with regard to one model system, continuing to study that system rather than starting at the beginning again with another species was more efficient. Animal, plant, and bacterial systems were selected on the basis of short life cycles and their similarities to other more complex systems as well as how well these organisms lent themselves to experimental manipulation.

Initially, embryologists utilized frogs, *Xenopus laevis* and *Rana pipen,* as model systems for studies of vertebrate development. Fertilized eggs were easily obtained and manipulated. Frog eggs and embryos proved particularly hardy and well-suited for translocation, the movement of cells from one location to another. Later, the mouse, chick, and—most recently—zebrafish were added to the list of model organisms. With a life cycle of nine weeks from fertilization to maturity, the mouse, however, remains the most efficient model system for studying mammalian development.

The fruit fly, *Drosophila melanogaster,* has been and remains one of the premier tools of the geneticist's trade. *Drosophila*'s attractiveness as a model system stems from its diploid (having two copies of each chromosome) nature—a characteristic it shares with other animals, including mice and humans—and its short life cycle. Even in the absence of modern molecular techniques, *Drosophila* became invaluable in the study of the effects of mutation because of the fly's many phenotypic markers (readily observable physical characteristics) and the large number of offspring generated quickly from specifically designed genetic crosses. These features permitted early geneticists to map chromosomal mutations and, more specifically, X-linked mutations, from which were extrapolated the implications for human X-linked diseases. Full exploration of mutations as the result of gene expression, however, had to wait until the molecular details of gene expression were uncovered by François Jacob (1920-) and Jacques Monod (1910-1976) in 1961.

While the contributions of animal and plant models to the fields of embryology, genetics, and molecular biology cannot be overstated, knowledge of these systems (and the systems they represent) have been tremendously affected by that which is not so easily seen—bacteria, viruses, and plasmids. The structural simplicity of these model systems provided the drawing board on which the modern story of molecular biology has been written.

As a result of a series of genetic and biochemical experiments using the common intestinal bacterium, *Escherichia coli,* Jacob and Monod proposed that clusters of genes with a related function belonged to a single regulatory unit, an operon, in which all of the genes within the cluster are turned on and off together. Thus, these clusters operate as units of transcription and regulation.

While Jacob and Monod were ferreting out the mechanism for gene expression, Francis Crick and Sydney Brenner (1927-) worked on deducing the nature of the genetic code. Charles Yanofsky's (1925-) studies of tryptophan sythetase in *E. coli* confirmed that a sequence relationship existed between DNA and the proteins it encodes. Crick and Brenner set out to figure out how many nucleotides were necessary to specify each amino acid. Through careful studies of the effects of the mutagen proflavin on a bacteriophage, a virus that only infects bacteria, Crick and Brenner concluded that nucleotides in DNA must be read in groups of three. Within five years of Crick and Brenner's publication of the triplet nature of the genetic code, the genetic code was cracked. Many advances in molecular biology during the 1960s came from studies of *E. coli* and of bacteriophages and plasmids that use it as a host.

The 1970s were host to an avalanche of technical discoveries that led to dramatic advances in molecular cell biology. The discovery of restriction enzymes, enzymes found in most bacterial cells that protect the cell from foreign DNA, in 1971 was a boon to molecular biologists. Restriction enzymes enabled biologist to cut DNA from any organism at specific sequences thereby generating a reproducible set of fragments. The isolation or determination of fragment lengths was accomplished using gel electrophoresis. Analysis of variability in the lengths of fragments was then added to the molecular biologist's tool box. Restriction fragment length polymorphism (RFLP) analysis proved to be particularly useful in the study of genetic variability within and between species.

By 1973, Stanley Cohen (1922-) and Herbert Boyer had used DNA fragments and ligases, enzymes normally involved in DNA replication and repair, to produce the first recombinant DNA organism. The insertion of a restriction fragment containing specific genetic information into a bacterial genome gave birth to the new science of recombinant DNA technology, which now includes cloning and genetic engineering.

While there have been major biotechnological advances made since 1973, none have so profoundly affected molecular biology as the development of the polymerase chain reaction (PCR), developed in 1985. Using PCR techniques, very small samples of genomic DNA can be amplified, resulting in sample sizes large enough to carry out standard DNA analysis protocols. This technology has made significant contributions to many fields, including law enforcement, immunology, and conservation biology.

Impact

Medicine is at the forefront of the biotechnology revolution. After his groundbreaking work with Cohen, Boyer co-founded Genentech, the first biological engineering company. Genentech pioneered the bioengineering industry, producing the first bioengineered human protein, insulin, and human growth hormone, which is used in the treatment of children. By 1982, Genentech had begun marketing genetically engineered insulin, thereby changing the pharmaceutical industry. Today, through the aid of bioengineering, the volume of natural antibiotics produced by microorganisms no longer limits the work of immunologists as they now have synthetic and recombinant weaponry in their arsenal against disease. Additionally, innovative treatments for age-old diseases such as cystic fibrosis and muscular dystrophy are giving some patients afflicted with these diseases a new lease on life through gene therapy.

As science gains a fuller understanding of genes—how they operate and how they can be manipulated—science endeavors to attack disease at its foundation. Since many diseases are rooted in missing or flawed genes, physicians and molecular biologists are striving together to correct genomic problems through gene therapy. Healthy copies of genes that compensate for faulty genes are the focus of the gene therapy industry.

Second only to the focus on biotechnological applications in medicine is the impact of advances in molecular techniques on agriculture. As agriculturists endeavor to improve crop yield and quality while at the same time reducing production costs, bioengineers are increasingly called upon to accomplish these tasks. Bioengineered plants are expected to reduce the amount of fertilizer and pesticide needed to ensure high crop yields.

In addition to the health care and agricultural industries, advances in molecular techniques have significantly influenced other social and natural sciences. For example, the discovery of restriction enzymes and the fragments that they generate has been used to examine phylogenetic relationships within and between species. Southern, Western, and Northern Blot techniques as well as RFLP analysis, DNA amplification using PCR technology, and DNA sequencing represent only a short list of the molecular techniques employed to investigate species relatedness (the time since the evolutionary divergence between two species) and levels of genetic variability (a measure used to evaluate species health and viability) in captive and wild populations.

The twentieth century witnessed rapid and remarkable advances in science and medicine. Life expectancy in developed nations at the time of Watson and Crick's 1953 publication was approximately 66 years. As a result of monumental advances in both biological knowledge and biotechnology—less than 50 years after Watson and Crick's historic discovery—many individuals living in industrialized nations can expect to live well into their 70s or beyond. Similarly, through better health care and reproductive technology, infant mortality in developed nations is at its lowest level in history. The results of the Human Genome Project, slated for release in 2003, will mark the fiftieth anniversary of the publication of the double helix model of DNA. With information on every gene sequence in the human genome expected to be available, the twenty-first century is already being referred to as the "biology century."

MICHELLE ROSE

Further Reading

Books and Periodicals

Becker, W. M., and D. W. Deamer. *The World of the Cell.* Second edition. New York: The Benjamin/Cummings Publishing Company, Inc., 1991.

Lodish, H., D. Baltimore, A. Berk, S. L. Zipursky, P. Matsudaira, and J. Darnell. *Molecular Cell Biology.* Third edition. New York: W. H. Freeman and Company, 1996.

O'Brien, S. J. "A Role for Molecular Genetics in Biological Conservation." *Proc. National Academy of Sci. 91* (June 1994): 5748-5755.

Internet Sites

ABS Global. "Cloning and Other Biotechnology Applications to Benefit Production Agriculture." http://www.absglobal.com/pr1222.htm

The Human Genome Project. http://www.ornl.gov/TechResources/Human_Genome/home.html

"What is Gene Therapy?" http://www.med.upenn.edu/ihgt/info/whatisgt.html

"Barnyard 101: An Introduction to Transgenic Farm Animals." http://www.accessexcellence.org/AB/BA/casestudy3.html

Biographical Sketches

George Wells Beadle
1903-1989
American Geneticist

In 1958 George Wells Beadle shared the Nobel Prize in medicine with Joshua Lederberg (1925-) and Edward Lawrie Tatum (1909-1975) for their discoveries that demonstrated the relationship between genes and the proteins they controlled. Beadle and Tatum demonstrated that genes act by regulating specific chemical events. Exploiting the potential of the bread mold *Neurospora* as a genetic and biochemical tool, they established the "one gene–one enzyme" theory.

Beadle was born in Wahoo, Nebraska, on a small family farm. One of his high school teachers recognized his potential and urged him to continue his education. He became interested in genetics while earning his bachelor's and master's degrees from the College of Agriculture of the University of Nebraska in 1926 and 1927, respectively. A professor of agronomy at Nebraska suggested that he continue graduate work with Rollins A. Emerson, an early advocate of Mendelian genetics and an eminent maize geneticist, at Cornell University. After earning his Ph.D. in 1931, Beadle worked as a postdoctoral fellow in the laboratory of Thomas Hunt Morgan (1866-1945) at the California Institute of Technology. He taught at Caltech for two years before spending a year in Paris researching the biochemical genetics of eye color mutations in *Drosophila* (fruit flies).

At the Institut de Biologie Physico-Chimique, he worked with Boris Ephrussi, who was trained in embryology and tissue culture and had studied *Drosophila* genetics in Morgan's laboratory. Working together, Beadle and Ephrussi analyzed the biochemical basis of heredity by studying how genes controlled the insects' eye-color. Their tests with various eye color mutants proved that the steps involved in the sequential synthesis of eye-pigments were controlled by different genes.

Beadle became assistant professor of genetics at Harvard University in 1936, but moved to Stanford as professor of biology one year later. When Morgan died in 1945, Beadle became chairman of the Division of Biology at Caltech, where geneticists were moving from classical genetics to studies of the biochemistry of gene action. Classical genetics could address the question of how the gene was transmitted, but could not answer the question of how the gene worked or determine its chemical nature. Edward Tatum, a microbiologist and biochemist, joined Beadle to study the substances responsible for eye color in *Drosophila*.

Frustrated by the work with fruit flies, Beadle and Tatum decided that the bread mold *Neurospora crassa* would be more practical for studying the relationship between genes and the enzymes that controlled particular processes. *Neurospora* had a fairly short life cycle, techniques for genetic analysis had already been worked out, and its biochemical pathways were already well known. It was also easier to work with than the organisms that had traditionally been used by geneticists, and could be grown in a synthetic culture medium consisting of sugar, salts, and a simple growth factor (biotin).

Beadle and Tatum decided to reverse the procedures generally used to identify specific genes with particular chemical reactions. Instead of taking a mutant as their starting point and then searching for the chemical reaction it controlled, they decided to begin with known chemical reactions and look for the genes that controlled them. Mutations were induced by using x rays, then mutants that lost the ability to synthesize certain organic substances were selected. The induced mutations exhibited Mendelian patterns of inheritance. Biochemical and genetic analyses proved that the mutant strains were genetically different from the parental type.

When Beadle enunciated the "one gene–one enzyme" hypothesis in 1945, the chemical identity of the gene was still unknown. However, the work of Beadle and Tatum provided a valuable approach to discovering how genes work and became one of the foundations of modern genetics. Their success stimulated efforts to use simple organisms to solve fundamental questions about the nature of the gene.

LOIS N. MAGNER

Rachel Louise Carson
1907-1964
American Biologist and Writer

Rachel Carson was a scientist and writer who first revealed the residual hazards of indiscriminate pesticide use, drawing attention to their serious ill effects on animals and humans. Carson also won recognition and lasting esteem as a conservationist and important early leader of the environmental movement.

Born May 27, 1907, on a farm in Springdale, Pennsylvania, Carson credited her mother with encouraging her love of nature. She displayed an early talent for writing and went to Pennsylvania College for Women to major in English, with the aim of becoming a writer. When a biology teacher piqued her interest in marine biology, she changed her major. She received a scholarship to Johns Hopkins University in Baltimore, graduating with a master's degree in 1932. She went on to complete postgraduate work at Woods Hole Marine Laboratory in Massachusetts, and then worked on the zoology staff at the University of Maryland.

In order to support her mother and nieces, Carson took a position as an aquatic biologist with the Bureau of Fisheries in 1936, one of the first women to be hired by the bureau for a professional position. She supplemented her income by writing magazine articles on natural history and conservation. All her writing was about science and the sea.

In 1940 the Bureau of Fisheries merged with the Biological Survey to form the U.S. Fish and Wildlife Service, whose major focus was conservation. Carson was still writing for magazines and published a popular essay in the *Atlantic* on life in the sea. She decided to turn the article into a book, and 1941 published *Under the Sea Wind.* In it she emphasized that man is part of the natural world, not separate from it, and not its master. The book won critical acclaim, but sales were interrupted by World War II. In 1947 she became editor-in-chief of publications at the bureau and promoted a national policy of conserving

Rachel Carson. *(AP/Wide World Photos. Reproduced by permission.)*

natural resources in a series of twelve booklets called "Conservation in Action."

In 1951 she wrote about the sea again in the book *The Sea Around Us.* It was an immediate success, winning several awards, including the National Book Award. It was translated into 30 languages and led to a Guggenheim Foundation fellowship that allowed her to take a year's absence from the bureau. The next year she resigned from her job and bought a cottage by the sea in Maine to continue her writing.

A turning point came when a friend wrote a letter telling her how DDT had been sprayed in their community to kill mosquitoes. She found seven songbirds dead in her yard. The woman had scrubbed and scrubbed her birdbath but three more birds died. Carson began to research the use of pesticides. Focusing on DDT, she outlined the trail of poison from soil and water to humans. She described how the toxins accumulate from the lower parts of the food chain, such as plankton, to the next. She emphasized how environmental health problems are created by radiation and the never-ending stream of pesticides.

Her book *Silent Spring* was published in 1962 and was an immediate bestseller. The title suggested the silence after the birds are decimated by pesticides. The book aroused passion and excitement, as well as controversy. But public awareness of the subject had been raised, and people began to look at the environment. Ecology, a subject that had been ignored and relegated to the back of textbooks was now front and center.

Carson had never considered herself a crusader, but she launched a campaign to influence legislation. By 1962 bills to stop the use of pesticides were introduced in several states. She testified before Congress and appeared on television. President Kennedy set up a commission to study the issue.

On April 14, 1964, Carson died of breast cancer, but she had set in motion efforts to protect the environment that continue to this day. Regarded as the mother of the environmental movement, Carson was one of the first to make Americans aware of the delicate ecological balance and the devastating effects of pollution and unregulated exploitation of the earth's resources.

EVELYN B. KELLY

Avram Noam Chomsky
1928-
American Linguist and Philosopher

Noam Chomsky has had a profound and lasting effect on the study of linguistics and language acquisition in this century. Prior to his work, the focus in linguistics (the study of human speech patterns) was based on classification, rather than on exploration for the universal and biological basis of language.

Chomsky was born December 7, 1928, in Philadelphia, Pennsylvania. His father fled from Russia in 1913 to avoid being drafted into the Czarist army. In America the elder Chomsky supported himself by working in sweatshops, eventually he graduating from John Hopkins University. He wrote many books on the Hebrew language and was an influential teacher, administrator, and champion of the establishment of a Jewish state.

Chomsky's parents had an enormous impact on their son. From an early age Noam and his brother were immersed in the revival of the Jewish culture and the Hebrew language. Young Noam studied Hebrew literature with his father. He spent time in Hebrew school and later became a Hebrew teacher himself.

Chomsky was introduced to linguistics by his father. He studied under the linguist Zellig Harris at the University of Pennsylvania and earned his bachelor's (1949), master's, and Ph.D. degrees there. He taught modern languages and linguistics at the Massachusetts Institute of Technology (MIT) in 1955 at the age of 33.

As professor of linguistics, Chomsky pioneered new theories of language acquisition and how speakers of specific languages recognize and utilize grammar and words in order to be understood. Since the beginning of the 1960s, research on language acquisition has been influenced by Chomsky's innovative ideas. Although his work is controversial, it spawned many exciting theories and ideas about how we acquire speech and language.

All normal children everywhere learn language. This ability is not dependent on race, social class, geography, or intelligence. Before the 1960s, scientists thought that language was acquired by structured learning: training, and repetition of words, sentence structure, and grammatical guidelines.

Chomsky rejected that view, proposing instead that human beings have an innate ability to understand their particular language and do not need to be taught the underlying grammatic structure. This innate capacity explains how children are able to understand and construct complex sounds, words, and sentences that they may have never heard before.

Chomsky pioneered the idea that the logical structure of language may be universal, reflecting an unconscious structure of the mind. He described this as *transformational-generative grammar,* a system of language analysis that recognizes the relationships among the various elements of a sentence.

People recognize the grammatical sentences of their language and know how various words must be arranged to make sense to the listener or reader. Everyone is capable of producing and understanding an unlimited number of new sentences never before spoken or heard. In analyzing this innate ability to construct generative grammars, Chomsky distinguished between two levels of structure in sentences: *surface structures,* which are the actual words and structures used, and *deep structures,* which carry a sentence's underlying meaning.

Listeners or readers are able to create and interpret sentences by generating the words of surface structures from deep structures according to a specific protocol Chomsky called *transformational rules.* He argued that these are universal in all languages and correspond to innate, genetically inherited patterns in the human brain.

Chomsky has published more than 70 books and more than 1,000 articles covering a broad range of topics including linguistics, philosophy, politics, and psychology. In 1988 he was awarded the Kyoto Prize, the Japanese equivalent of the Nobel Prize. Chomsky is also well respected as a social activist and critic. He spoke out publicly against the Vietnam War. He continues to teach and write on the interface of human beings, science, and technology.

LESLIE HUTCHINSON

Francis Harry Compton Crick
1916-
English Physicist and Molecular Biologist

In 1962 Francis Crick shared the Nobel Prize in Medicine with James Watson (1928-)and Maurice Wilkins (1916-) for their discoveries concerning the molecular structure of deoxyribonucleic acid (DNA) and its significance for the transmission of genetic information. Molecular biologists have called the discovery of the double-helical structure of DNA one of the most important developments in twentieth-century biology. The structure of DNA proposed by Crick and Watson in 1953 immediately suggested insights into the nature of the gene, the genetic code, and mechanism by which information stored in DNA was transmitted from generation to generation.

Francis Crick was born in Northampton, England, where his father ran a shoe factory. He attended Northampton Grammar School and entered Mill Hill School in London at the age of 14. Even as a child Crick was extremely curious and interested in scientific discoveries. He confided to his mother that he was afraid that by the time he grew up everything important would have been discovered. He received a degree in physics from University College, London, in 1937. Crick's initial Ph.D. program was interrupted by the outbreak of World War II. During the war, he served as a scientist for the British Admiralty, working on magnetic and acoustic mines.

In his autobiographical memoir *What Mad Pursuit: A Personal View of Scientific Discovery,* Crick notes that after the war he realized that his qualifications for research appeared to be rather limited. He decided, however, that he could turn his deficiency into an advantage. That is, because he had no particular expertise, he was free to chose entirely new fields of inquiry. Like many physicists of his generation, he became interested biology after reading *What is Life?* by Erwin Schrödinger (1887-1961). Crick believed that biological research could provide scientific explanations for areas that were regarded as mysteries, such as the borderline between the living and the nonliving and the problems of consciousness, areas now known as molecular biology and neurobiology. He left the Admiralty in 1947 to study biophysics at the Strangeways Research Laboratory. In 1949 he joined the Medical Research Council Unit headed by Max Perutz (1914-) in Cambridge, where scientists were attempting to determine the structure of proteins by X-ray crystallography. At the time, many scientists still believed that proteins must serve as the chemical basis of the gene, but Crick was open to the possibility that DNA might be involved in gene structure and gene replication. In 1954, after the discovery of the double helix, Crick finally earned his Ph.D. for a thesis entitled "X-ray Diffraction: Polypeptides and Proteins."

In 1951 Crick met James Watson, an American postdoctoral fellow with a background in genetics. Despite differences in personality and scientific training, the two instantly discovered that they shared a passion for discovering the "secret of the gene." Moreover, they were both convinced that DNA, rather than protein, would prove to be the macromolecule responsible for passing genetic information from generation to generation. A solution to the structure of DNA should, therefore, lead to an explanation of the replication of genes. After numerous false starts, in 1953 Watson and Crick arrived at a solution to the three-dimensional structure of DNA on the basis of model building, data from X-ray crystallography studies by Rosalind Franklin (1920-1958), and general knowledge about the chemistry of DNA. After proposing a general scheme for the transmission of genetic information, which became known as the "central dogma," Crick suggested approaches to working out the details of the genetic code and the mechanisms by which information in DNA was copied into RNA and then used in the biosynthesis of proteins.

By the mid-1960s Crick thought that the foundations of molecular biology had been established and he began to explore other intractable problems, such as embryology and the workings of the brain. In 1976 he left the Medical Research Council and joined the Salk Institute in La Jolla, California, where he devoted himself to theoretical studies of the brain, the problem of consciousness, the nature of dreams, and neural networks. Rather than confine himself to the molecular aspects of these problems, he has attempted to incorporate psychological aspects, neuroanatomy, neurophysiology, and related philosophical issues.

LOIS N. MAGNER

Dian Fossey

1932-1985

American Zoologist

American zoologist Dian Fossey is best known for her field studies of mountain gorillas in the Virunga Mountains of Rwanda and Zaire, which served to dispel many myths about the violent and aggressive nature of gorillas. Her dedicated work combined research and conservation to ensure the survival of these elusive and endangered animals.

Born in San Francisco, Fossey graduated from San Jose State College in 1954 with a degree in occupational therapy; she then worked at a children's hospital in Kentucky for several years. During this time she read and studied all that she could about African primates. Inspired by the writings of American zoologist George B. Schaller, including *Life of the Gorilla,* Fossey traveled to Africa on holiday in 1963.

In letters home, she described her trip in colorful detail. She observed mountain gorillas in their home habitat, the mist-shrouded volcanoes in central Africa. It was there that she met British anthropologist Louis Leakey (1903-1972). Leakey, believing that studies of great apes would shed light on the subject of human evolution, encouraged Fossey to undertake a long-term field study of gorillas. He felt that the gorilla study, in conjunction with the studies of chimpanzees by Jane Goodall (1934-), would generate data supporting an evolutionary link between humans and primates.

Although Fossey had no formal training in animal behavior or zoology, Leakey felt that her excitement and interest in the gorillas would be an important and valuable asset. Fossey returned to the United States and, in 1966, resigned from her job, sold her possessions, and traveled to Rwanda's Virunga Mountains, the last bastion of the endangered mountain gorilla. Upon her arrival, Goodall gave Fossey a two-day crash course in data collection and observation methods. She was then on her own, following gorillas up and down the steep mountainous terrain. The local people called her "the woman who lives alone in the mountains."

For the next 22 years, Fossey was an astute and patient observer of gorilla behavior. Her field methods were unorthodox, gentle, and simple. She quietly and sensitively allowed the gorillas to accept her into their world. Unarmed, she sat within a few feet of them every day for years. She knew each individual in her study area by the names she had given them, and came to regard the gorillas as gentle, social animals, not violent and aggressive as was popularly thought at the time. Fossey received a Ph.D. in zoology from Cambridge University in 1974 on the basis of her fieldwork.

Fossey established Karisoke Research Center for gorilla research and conservation in 1967. She understood that the survival and well-being of the mountain gorillas was dependent on their human neighbors. Poaching and the export trade of gorilla infants for zoos and medical research were taking a serious toll on the gorilla population. Research was not enough, she asserted. Without a strong conservation program in place, the gorilla population would become unable to survive.

Fossey worked diligently to encourage the formation of National Parks to protect the gorillas and their habitat. She attempted to work with the local people to gain support for protecting the gorillas. Her position, however, was not popular and, in 1985, she was found murdered at her cabin at Karisoke. Some authorities believe she was murdered in retaliation for her efforts to stop the poaching of gorillas and other animals in Africa. Her murder has yet to be solved.

Fossey's book, *Gorillas in the Mist* (1983), recounts observations from her years of field research. The book was subsequently made into a popular movie starring Sigourney Weaver, introducing millions of viewers to the plight of the mountain gorilla.

Due largely to Fossey's research and conservation work, mountain gorillas are now protected by the government of Rwanda and by the international conservation and scientific communities.

LESLIE HUTCHINSON

Rosalind Elsie Franklin

1920-1958

English Physical Chemist and Molecular Biologist

Rosalind Franklin made important studies of the physical chemistry of coal and played a significant role in the determination of the structure of deoxyribonucleic acid, a role which was not adequately acknowledged until a number of years after her death.

Franklin was born in London. She completed her undergraduate studies in physical chemistry at Newnham College of Cambridge University in 1941 and, in 1942, began work in the laboratories of the British Coal Utilization Research Association. The structural studies of coal and coke that she carried out there produced significant results that found important applications in industrial processes. In 1945, she returned to Cambridge to receive her Ph.D.

In 1947, she left England to accept a position at the Laboratoire Centrale des Services Chimiques de l'Etat in Paris. During her continued studies of the structure of carbon in France, she began to learn and apply x-ray crystallographic methods to crystal structure determination.

When she returned to England in 1951, she undertook the study of the structure of crystalline deoxyribonucleic acid (DNA) as a research fellow in biophysics at King's College in London. DNA is among the most important molecules of nature. It is found in the nuclei of virtually all cells and functions in the synthesis of proteins as well as containing the genetic material that functions as the carrier of the information of heredity.

Franklin applied the x-ray crystallographic methods that she had learned in Paris to the DNA structure problem and invented new techniques for the application of x-ray crystallography that could be applied specifically to this type of study. She determined that the phosphate sugar groups are located on the outer part of DNA and was continuing her analysis of x-ray photographs of DNA crystals in her attempt to learn more about its structure.

Her co-worker on this project at King's College was Maurice Wilkins (1916-). Unfortunately, they did not work well together. In the midst of her studies, Wilkins showed her data, without her knowledge, to James D. Watson (1928-) and Francis H.C. Crick (1916-) who were working on the DNA structure problem at Cambridge. Watson and Crick were able to use Franklin's data to support their conclusion that the DNA molecule is shaped like a double helix. They published their proposal in the scientific journal *Nature* in January 1953.

In 1953, Franklin moved to the laboratory of crystallographer J.D. Bernal at Birkbeck College in London where her subsequent studies of DNA provided additional support for the double helix model for the structure of DNA. She also undertook the determination of the structure of plant viruses, including tobacco mosaic virus. She demonstrated a single-stranded helical structure for the ribonucleic acid in this virus. She was studying the structure of live polio virus when she died of cancer at the age of 37 in 1958. This study was regarded as so dangerous that it was discontinued after her death.

In 1962, Watson, Crick, and Wilkins shared the Nobel Prize for the determination of the structure of DNA. The Nobel is only given to living individuals, and Franklin, therefore, would have been ineligible to share the Prize. Her significant contributions, however, were largely ignored at the time. The importance of Rosalind Franklin's role has since been brought to light and widely accepted, largely as the result of the efforts of individuals such as Anne Colquhoun Sayer, who published *Rosalind Franklin and DNA* in 1975. Franklin's case has become an important example in the study of sexism in science, the ethics of science, and the sociology of science.

J. WILLIAM MONCRIEF

Jane Goodall
1934-
English Primatologist

Primatologist Jane Goodall is best known for her long-term field studies of chimpanzee life and behavior. A leading expert on the subject, many Americans came to know her through a public television series chronicling her life and ongoing research on chimpanzees in Tanzania.

Born in England in 1934, Goodall moved with her family to France when she was five years old. When the Nazis threatened to subjugate France during World War II, the Goodalls returned to England and lived at the family estate known as the Birches, located in Barnemouth and managed by Goodall's maternal grandmother. Goodall remained there until she graduated from high school.

Goodall subsequently attended secretarial school, followed by an assortment of jobs at Oxford University. She later worked briefly for an independent film company specializing in documentaries. Her life brightened considerably when an old school friend invited her to Kenya, where she was living at the time. Since Goodall had long been interested in the African continent, the invitation was a welcome opportunity to see and experience what she had read about for many years.

She secured passage on a liner called the *Kenya Castle* and sailed at the earliest opportunity.

Goodall was in Africa only two months before she was introduced to Louis Leakey (1903-1972), the world-famous anthropologist. He gave her a job on his research staff and, after nine months in Africa, she had saved enough to invite her mother for a visit. During this visit, Leakey suggested that Goodall take a companion with her and begin a comprehensive study of chimpanzees in the wild. The project interested Goodall and, when her mother agreed to accompany her, she began making plans for the study.

After a series of short delays, Goodall and her mother took off for the boat trip to Gombe on July 16, 1960. They were accompanied by two scouts from the Gombe National Park and their cook, Dominic.

Until that time, the observation and study of animals in their natural habitats fell under several different scientific disciplines. However, it was then given its own classification—Ethology. When Leakey suggested that Goodall study chimpanzees, he did so because of their genetic similarity to humans. Since chimp DNA is only one percent different from that of humans, he believed that satisfactory data could be gathered in months rather than years of research.

However, the months that Leakey anticipated turned into more than 30 years of work. During this time, Goodall earned a Ph.D. from Cambridge University (1965) and international recognition for her fieldwork and popular books. She also married (and later divorced) Dutch photographer Hugo van Lawick, with whom she collaborated on several books and films, and gave birth to a son.

Goodall encountered a series of adventures and challenges. She and her mother contracted malaria early in their stay in Gombe, but recovered and remained to see the first evidence of a chimp using a tool. In the ensuing years, Goodall corrected many erroneous misunderstandings about her subjects. For instance, it was generally believed that all primates were vegetarians. However, Goodall observed chimps in Gombe eating meats. They were also filmed making and using various implements to improve their subsistence.

All these events, and many more, were effectively chronicled in her books, including *In the Shadow of Man* (1971) and *The Chimpanzees of Gombe: Patterns of Behavior.* (1986). Goodall has been honored for her work in conservation and is the recipient of the Albert Schweitzer Award (1987), the Encyclopedia Britannica Award (1989), and the prestigious Kyoto Prize for Science (1990). She remains personally involved in the longest, unbroken field study of any group of animals in the wild.

BROOK HALL

Stephen Jay Gould
1941-
American Paleontologist and Educator

American paleontologist Stephen Jay Gould has contributed important insights into the nature of life and evolutionary science. His theory of punctuated equilibrium posed a compelling emendation to Darwinian theories of evolution. As both a respected researcher and science popularizer, he has published many books, demonstrating a unique ability to make complex scientific data clear and entertaining to average readers and students.

Born in New York City, Gould attended Antioch College in Yellow Springs, Ohio, where he graduated with an A.B. He remained at Antioch and taught geology through 1966. His deepening interest in paleontology led him to pursue graduate work. He received his Ph.D. in evolutionary biology and paleontology from Columbia University in 1967.

Gould's next teaching assignment was an assistant professorship in the Geology Department at Harvard University, followed by an associate professorship. In 1973 he became a full professor at Harvard, where he has remained until the present.

During the 1970s, Gould developed the theory of punctuated equilibrium, a revised version of the Darwinian belief that species evolve over long periods of time. He proposed that most evolutionary change takes place in much shorter time frames—thousands instead of millions of years, and also in fairly rapid succession instead of gradual, miniscule developments.

His conclusions resulted from a research project involving West Indian land snails (traditionally slow movers). Gould moved on to other species and was particularly devoted to the theory of exaptation, which proceeds much more rapidly than adaptation. It promotes the premise that instead of slow growth changes in structures, some species used existing structures for new and constructive purposes. He continues to acknowledge that, for the most part, evolution is dependent on fundamental processes like direct inheritance, varying DNA throughout millions of years, and natural selection that includes competition for basic elements of survival: plants, water, predators, limited space (as in oceanic or continental islands), floral and fauna collapse, and inevitable ecosystem decay.

One of Gould's best-known examples of exaptation is chronicled in *The Panda's Thumb,* published in 1980. It graphically illustrates how the panda's wrist-bone modification, which must have been in place for some time, "suddenly" enabled the panda to strip leaves from its primary food supply: bamboo shoots. In subsequent publications, Gould used other illustrations of exaptation, such as the swim bladders in fish that kept them buoyant in water, but were readily converted into lungs for land species because of the thin flap of tissue that permitted an exchange of gases both in and out of water.

Gould's prolific writings brought him numerous awards and continuing assignments. *The Panda's Thumb* earned him the Notable Book Citation from the American Library Association and the American Book Award in *Science* in 1981. Other books that have contributed to the popularization of evolutionary biology include *Ontogeny and Phylogeny* (1977), *The Mismeasure of Man* (1981), *Time's Arrow, Time's Cycle* (1987), and *Wonderful Life* (1989).

Gould has been a regular contributor to *Natural History* magazine and many of his essays were later collected and published together. His ability to clarify difficult conceptual theories into understandable, plausible arguments has made him popular with readers from all walks of life.

In addition to his many awards and prizes, Gould is a member of the American Association for the Advancement of Science, the American Society of Naturalists, the Paleontological Society, the Society for the Study of Evolution, the Society of Systematic Zoology, and Sigma Xi. He continues to work as a prominent paleontologist, writer, and educator in his field.

BROOK HALL

Sir Alan Lloyd Hodgkin
1914-1998
English Biophysicist and Physiologist

Sir Alan Lloyd Hodgkin was an English biophysicist and physiologist who was awarded the 1963 Nobel Prize in Physiology or Medicine with Andrew Huxley (1917-) for their pioneering

Alan Lloyd Hodgkin. *(The Library of Congress. Reproduced by permission.)*

research in the electrical and chemical events involved with nerve cell impulses. They shared their prize with Sir John Eccles (1903-1997). Their use of the "squid giant axon" to explain nerve behavior provided needed information that demonstrated the precise inner workings of nerve cells. He was knighted for his scientific efforts in 1972.

Sir Alan Lloyd Hodgkin was born on February 5, 1914, in Oxfordshire, England. He attended Trinity College in Cambridge (1932-1936). While at Trinity College, Hodgkin contemplated studying history because of family tradition, but because of his passion for science, he chose to concentrate on biology and chemistry instead. When he began his studies, he was advised to learn as much mathematics and physics as he possibly could so that he could reach his fullest scientific potential. This recommendation proved to be extremely valuable to Hodgkin throughout his life as he excelled in all areas of science.

Another aspect of his education that proved to be extremely valuable was the significant number of eminent professors he studied under at Trinity College. Many had a significant impact on his work and molded his early thinking and training. In fact, it was the famous physiologist A. V. Hill (1886-1977) who chaired his thesis work and helped him get his first research position in the United States. Hodgkin spent two years at the Rockefeller Institute in New York

(1937-1938) and during this time, he was introduced to the technique of dissecting a large squid nerve for scientific study. He later returned to Cambridge with the intent of using this technique as a research model, but before he could begin in earnest with a student of his, Andrew Huxley, World War II pulled him away.

During the war, Hodgkin worked primarily in the areas of aviation medicine and radar research for the British Air Ministry (1939-1945). He returned to a teaching post at Cambridge after the war to continue his association with Huxley. They were primarily interested in the ionic mechanisms in living cells. Hodgkin's research efforts were helped by a reduction in teaching load and monetary grants. His most significant contribution to the field consisted of measuring the electrical and chemical activity on squid nerve fibers (*Loligo forbesi*). Hodgkin and Huxley used microelectrodes to show that the electrical voltage within a nerve fiber during a nerve impulse exceeds the electrical voltage of that fiber at rest. This idea went against conventional theory at the time, which postulated that the cell membrane actually broke down during an impulse. They further reported in 1947 that the activity of a nerve fiber is dependent upon the concentrations of certain chemicals both inside and outside of the nerve cell. This work provided specific experimental data on the mechanisms of nerve conduction, and it was because of these experiments that they won their Noble Prize.

Hodgkin married Marion Rous, daughter of distinguished American pathologist Peyton Rous (1879-1970), in 1944 during a brief visit to the United States near the end of World War II. Hodgkin had previously met his wife while at the Rockefeller Institute in 1938.

Professor Hodgkin's distinguished career included election into the Royal Society. He served on many of the councils that set policy for the Royal Society throughout his life and was elected president in 1970. He also accepted the position of Chancellor at Leicester University in 1971. During his esteemed career, he received numerous commendations and honorary awards. Sir Alan Lloyd Hodgkin died on December 20, 1998, at age 84.

JAMES J. HOFFMANN

Dorothy Crowfoot Hodgkin

1910-1994

English Chemist and Crystallographer

Dorothy Crowfoot Hodgkin was a pioneer in the use of x-ray crystallographic methods for the determination of crystal and molecular structures and is widely regarded as the founder of protein crystallography. She both developed the x-ray crystallographic methodology and used it to solve the molecular structures of a number of complex biologically important molecules. She also served as a much-admired mentor and role model for several generations of x-ray crystallographers throughout the world.

Dorothy Crowfoot was born in Cairo, Egypt. Her father was an archeologist and her mother an artist. She was educated in England and received two degrees from Somerville College of Oxford University, a B.A. in 1931 and a B.Sc. in 1932. She studied chemistry at Oxford and did her first crystallographic studies as an undergraduate. In 1933, she continued her studies at Cambridge University under the direction of x-ray crystallographer John D. Bernal. During her graduate studies, she took the first x-ray diffraction photograph of a protein (pepsin). She was awarded her doctorate by Cambridge in 1937. In 1935, she returned to Oxford University, where she became a member of the faculty. In 1937 she married Thomas L. Hodgkin, an historian whose specialty was Africa. They subsequently became the parents of three children.

During the 1940s, she used x-ray crystallographic methods, many of which she developed herself, to determine the molecular structures of cholesterol, penicillin, and hemoglobin. X-ray crystallographic methods require numerous repetitive calculations involving very large sets of data. Today, such calculations are performed quickly with the use of computers. There were, however, no high-speed computers available at the time Dorothy Hodgkin undertook her work, and each structure determination required lengthy, tedious calculations to be done by hand. Even the determination of the structure of a relatively small molecule was a lengthy enterprise; those of the complex molecules that she chose to study each required a number of years to complete.

The dramatic success of her research led to her election as a fellow of the Royal Society in 1947, and she served as the Society's Wolfson Professor during 1960-77. She was awarded the Nobel Prize for chemistry in 1964 for the determination of the structure of vitamin B_{12} by x-ray crystallographic analysis. Vitamin B_{12} is used to prevent and to treat pernicious anemia. In 1965, she became the second woman ever to receive the British Order of Merit; the only other woman so honored had been Florence Nightingale in 1907. Hodgkin was further honored with the position of Chancellor at Bristol University, and she served in this role from 1970-88.

One of her greatest successes came in 1969 when she announced the successful determination of the molecular structure of insulin, a protein used in the treatment of diabetes. It had taken her a total of 34 years to complete this major work.

In addition to her scientific undertakings, Hodgkin remained actively dedicated to the cause of world peace throughout her life. In 1957, she was a founder of the Pugwash Conference on Science and World Affairs and used every opportunity to speak or otherwise lend her support to the movement.

Hodgkin was not only a pioneer in x-ray crystallography and in the determination of the molecular structures of proteins, she was also a pioneer as a woman in the scientific research and academic establishments. Her distinguished success not only led to her own personal acceptance but made it easier for women who followed her. It is also worth noting that her accomplishments came in spite of the fact that she was crippled by rheumatoid arthritis for much of her life.

J. WILLIAM MONCRIEF

Robert William Holley
1922-1993
American Biochemist

In 1968 Robert W. Holley shared the Nobel Prize in Medicine or Physiology with Har Gobind Khorana (1922-) and Marshall W. Nirenberg (1927-) "for their interpretation of the genetic code and its function in protein synthesis." Holley's work provided insight into the mechanism the cell uses to translate the information in the genetic code into essential proteins. Holley was one of the discoverers of the special type of nucleic acid called transfer-RNA. Having developed techniques for determining the structure of nucleic acids, Holley isolated alanine transfer RNA and determined the total sequence of nucleotides in this polynucleotide chain. This work represented the first determination of the complete chemical structure of a biologically active nucleic acid.

Holley was born in Urbana, Illinois, but he also lived in California and Idaho during his youth and developed a love of the outdoors and an interest in living things. Both of his parents were teachers. He majored in chemistry at the University of Illinois and received his B. A. degree in 1942. His wife, Anne Dworkin, was a chemist and mathematics teacher. In 1947 he was awarded the Ph.D. degree in organic chemistry from Cornell University. World War II interrupted his graduate work. From 1944 to 1946, he carried out research for the United States Office of Research and Development at Cornell University Medical College, where he participated in the first chemical synthesis of penicillin. Holley spent two years as an instructor and American Chemical Society Postdoctoral Fellow at Washington State University before returning to Cornell in 1948 as assistant professor of organic chemistry at the Geneva Experiment Station. During a sabbatical year (1955-1956), he was a Guggenheim Memorial Fellow at the California Institute of Technology. After returning to Cornell, he took the position of Research Chemist at the United States Department of Agriculture's Plant, Soil, and Nutrition Laboratory at Cornell. He was Professor of Biochemistry and Molecular Biology at Cornell until 1966. Although he maintained an affiliation with Cornell, he moved to the Salk Institute for Biological Studies in La Jolla, California, in 1967, where he became a resident fellow and professor. He also held the position of Adjunct Professor at the University of California at San Diego.

Robert Holley. *(The Library of Congress. Reproduced by permission.)*

Holley's initial research topics concerned the organic chemistry of natural products and gradually turned to more biological subjects, including work on amino acids and peptides, and eventually work on the biosynthesis of proteins. Holley did his landmark RNA work while at Cornell University. His classic paper on deciphering the genetic code for RNA, "Sequences in Yeast Alanine Transfer Ribonucleic Acid," appeared in the *Journal of Biological Chemistry* in 1965. During his studies of the biosynthesis of proteins, Holley discovered alanine transfer RNA. Isolating and determining the structure of this RNA with the available techniques took about 10 years. The laborious work of sequencing the polynucleotide sequence of the alanine transfer RNA was completed in 1964. The nucleotide sequence of alanine transfer RNA was the first nucleic acid for which a complete structure had been established. Moreover, Holley's work helped explain how transfer RNAs are involved in reading the genetic code and transforming the information in the nucleic acid of the gene into the amino acids of the proteins.

In the 1970s, Holley became more involved in studies of the factors that control cell division in mammalian cells. He concluded that the factors that controlled growth and development were probably polypeptide hormones, hormone-like substances, and various low molecular weight nutrients.

In addition to the Nobel Prize, Holley won many other honors, including the Albert Lasker Award in Basic Medical Research, the Distinguished Service Award of the U. S. Department of Agriculture, and the U. S. Steel Foundation Award in Molecular Biology of the National Academy of Sciences. Holley was a member of the National Academy of Sciences, the American Academy of Arts and Sciences, the American Association for the Advancement of Science, the American Society of Biological Chemists, and the American Chemical Society.

LOIS N. MAGNER

Sir Andrew Fielding Huxley

1917-

English Physiologist

Sir Andrew Fielding Huxley received the Nobel Prize in medicine for his work on the chemical properties of nerve and muscle impulses. This work has influenced and furthered research into nerve disorders and brought insight into how the brain works.

Andrew Fielding Huxley. *(The Library of Congress. Reproduced by permission.)*

Andrew Huxley was born in London on November 27, 1917. He had two brothers, Julian (1887-1975), a well-known scientist, and Aldous (1894-1963), author of the classic novel *Brave New World.* Andrew came from a long lineage of scientists and writers. Both his father and grandfather had strong science backgrounds and introduced the young Huxley to their enthusiasm for creative thinking, scientific research, and writing. Being mechanically minded, he enjoyed building and using microscopes throughout his secondary school years.

Interestingly, Huxley entered Cambridge University's Trinity College with the intent to study literature. He changed his focus from literature to the sciences in 1932. He assumed that he would concentrate on the physical sciences, physics in particular. However, students were required to take another science course at Trinity. After some contemplation, he chose physiology, on the recommendation of friends. The department had exciting professors, he enjoyed the subject matter, and at the completion of the course decided to study for entrance into medical school.

While studying anatomy and physiology he joined his colleague Alan Hodgkin (1914-) at the Marine Biology Laboratory for his first introduction to research and the first of their collaborative projects. Hodgkin had been conducting research into the nerve fibers of crabs and the recently discovered giant nerve fibers in squids. He wanted to understand the mechanism by which the action potential travels along the nerve fiber. The action potential is the electrical change that activates the nerve impulse. Together Huxley and Hodgkin recorded the first electrical impulses from the nerve cells of the giant squid. The large size of the cells permitted the insertion of electrodes directly into the interior of the cells. Before Huxley and Hodgkin's experiment, no one had succeeded in recording the electrical impulse from inside the nerve fiber. They had conclusive evidence that the electrical impulse is caused by selective changes in the permeability of the cell membrane.

He returned to Trinity to continue his medical training, but the entrance of Britain into World War II curtailed all clinical training. Because all medical teaching was suspended due to air attacks, he spent the rest of the war working on gunnery research for various branches of the armed services. After the war he completed his science studies and was appointed to the faculty of Trinity College in the Department of Physiology. There Huxley and Hodgkin resumed their research collaboration, focusing on nerve conduction and nerve fibers.

In 1963 Huxley, Hodgkin, and John Eccles (1903-1997) received the Nobel Prize in medicine for continued research into the transmission of nerve impulses. This information, transmitted by action potentials, is specified by the frequency of transmission and by the connections each nerve cell has with neighboring cells. During the rising phase of the impulse, the membrane becomes permeable, and this permeability is highly specific for sodium ions. The pores in the membrane open up as a result of the potential change and let sodium ions diffuse in, bringing the potential very nearly to the equilibrium of the sodium ions. During the falling phase of the impulse, potassium ions diffuse out.

With a judicious combination of electrochemistry, modern electronics, and mathematical modeling, Huxley's team was able to show that the exchange of sodium and potassium ions causes a brief reversal in a nerve cell's electrical polarization; this phenomenon, known as an action potential, results in the transmission of an impulse along a nerve fiber. The techniques that they developed have been applied with minor modifications to other excitable tissues for all subsequent research to understand the function of the central nervous system. Huxley was knighted in 1974.

LESLIE HUTCHINSON

Sir Alec John Jeffreys

1950-

British Geneticist

Sir Alec John Jeffreys developed a groundbreaking new technique to identify different genetic patterns found in each individual person, except identical twins. He coined the term DNA fingerprinting, and his procedure revolutionized criminal investigations by enabling forensic scientists to identify suspects based on scant DNA evidence found in blood, tissue, and body fluids.

Jeffreys was born on January 9, 1950, in Oxford, England. When he was eight years old his father, Sidney Victor Jeffreys, a designer and engineer in the car industry, gave him a microscope and chemistry set. He became hooked on both biology and chemistry, and found the two subjects fit together in biochemistry. In 1975 Jeffreys received a Ph.D. in human genetics from the University of Oxford and went to Amsterdam to work on a project. While in Amsterdam, an interesting question occurred to him: If genes can be detected in DNA, can inherited differences among people be detected in their genes? When this was later found to be possible, a new area of molecular genetics research was opened up.

Finding how bits of human DNA are different came unexpectedly while working on another project concerning genes and the course of evolution. Jeffreys noted the same genetic sequences are repeated over and over, an occurrence he called "stuttered DNA." There are different numbers of stutters between people. These stutters or variable numbers of tandem repeats (VNTRs) he found could be cut with restriction enzymes and then separated into fragments by a process called polyacrylamide gel electrophoresis, or PAGE. The process uses agarose gel, a jelly-like material, in a device that creates an electrical field. The groups migrate according to size, creating a pattern that represents a distinct DNA fingerprint. The procedure is also called restriction fragment length polymorphism analysis (RFLP).

Jeffreys realized its potential for solving problems of identification. His first opportunity to use DNA fingerprinting occurred in March 1985, when he proved a boy was the son of a British citizen, allowing him to enter the country.

The first case of human DNA used in crime detection occurred the next year. A teenage girl was raped and strangled in a village near Jeffreys's laboratory in Leicester. Body fluids were recovered, but no suspect was found. Three years later it happened again. Another teenage girl was strangled in the same way as the first. A 19-year-old caterer confessed to the second murder but not to the first. However, the DNA evidence from both murders was the same. The man had falsely confessed to the murder of the second girl. The DNA evidence did eventually reveal the real killer when blood samples from 4582 village men were taken.

DNA evidence was first used in an American court to convict Tommy Lee Andrews of a Florida rape. The use of DNA evidence in criminal investigations soon spread worldwide.

In 1988 Kary Mullis (1944-) discovered polymerase chain reaction (PCR), a technique that allows small amounts of DNA to be copied or amplified in test tubes. In 1990 Jeffreys used DNA analysis, with the help of PCR, to identify the skeletal remains of Joseph Mengele, the infamous Nazi doctor who performed barbaric experiments on Jewish prisoners at Auschwitz.

The technique developed by Jeffreys has been refined to identify small sequences called short tandem repeats (STRs). One of the most famous cases involving the use of DNA was the O.J. Simpson investigation in 1995. Jeffreys's work has provided an important tool for solving crimes.

Jeffreys teaches human genetics at the University of Leicester and has shifted his research focus to the study of DNA mutations from one generation to the next. He is especially interested in the effects of environment on DNA, including the genetic consequences of the Russian nuclear power plant accident in Chernobyl.

In 1994 Jeffreys was knighted by the Queen for his research in genetics and his contribution to the field of forensic DNA. He has received over 30 honors and prizes for his work.

EVELYN B. KELLY

Donald C. Johanson

1943-

American Paleoanthropologist

Donald Johanson, born June 28, 1943, is an American paleoanthropologist specializing in the study of human evolution. His discovery in 1974 of the fossil skeleton Lucy dramatically changed our understanding of how human beings may have evolved. He has dedicated the last 25 years to looking for clues to questions that have puzzled scientists since Charles Darwin (1809-1882): What made us human? When and why did we begin to walk upright? Why did we develop such intellectual prowess? By approaching these questions from a variety of directions, incorporating techniques borrowed from molecular biology, archeological excavation, and sociobiological studies of primates and hunter-gatherer societies, Johanson has provided new insight into our human origins.

Johanson was born in Chicago, Illinois, the son of Swedish immigrants. His father died when he was two years old, and his mother moved to Hartford, Connecticut. He developed an early interest in anthropology from a neighbor who taught at a nearby seminary. When he was in high school, he was told by his guidance counselor to forget going to college because of poor scores on the Scholastic Aptitude Test. Deciding not to be discouraged, he applied and was accepted at Illinois State University.

Johanson initially studied chemistry but switched over to anthropology and decided to specialize in the study of human origins. He completed his Ph.D. with a comprehensive study on chimpanzee dentition in 1966. He later taught anthropology at Case Western University, Kent State University, Stanford, and the Cleveland Museum of Natural History, but his reputation is based on his fieldwork in Africa.

In 1973 Johanson began to search for fossil remains of hominids, the primitive ancestors of modern humans in the Great Rift Valley in Africa. Exposed sedimentary layers, often millions of years old, contain buried and fossilized animal remains, including those of our direct ancestors. In 1974 in the Afar region of Ethiopia, he found the fossilized remains of a 3 million-year-old female skeleton, named Lucy by field crew. The following year Johanson's crew discovered the fossilized remains of 13 individuals believed to be the oldest evidence of human ancestors living in-groups. In 1986 at Olduvi Gorge, a 1.8 million-year-old partial skeleton was found that was believed to be the first tool maker.

As we travel back in time, our ancestors look less and less like us and begin to resemble our closest ancestor, the African apes. The previously accepted explanation of human evolution suggested that a line of primates with larger brains had evolved, became capable of making tools, and began walking upright to free up their hands. However, it appeared that Lucy and other hominids found at the site were walking upright, although their brains were only slightly larger than the chimpanzee. No stone tools were found at the site, so it may be inferred that our ancestors walked upright for another reason.

Scientists studying human origins had long attempted to find a missing link, the shared ancestor between human beings and apes. After years of studying the fossil remains found at the Hadar site, Johanson in 1978 shocked the scientific community with the assertion that the remains belonged to a species that could indeed be a missing link between humans and apes. He named this new species *Australopithecus afarensis*. In 1990 Johanson returned to Africa and discovered a large portion of another *A. afarensis* skull. The reconstruction of the skull convinced most of his critics that the species is the ancestor of both *Australopithecus africanus* and modern humans, *Homo sapiens*.

Director of the Institute of Human Origins at Arizona State University, Donald Johanson continues his work in researching human evolution. In 1978 his ideas on the origins of humankind were presented at a Nobel symposium on human origins in Sweden.

LESLIE HUTCHINSON

Har Gobind Khorana
1922-
Indian-born American Organic Chemist and Biochemist

In 1968 Har Gobind Khorana shared the Nobel Prize in Medicine or Physiology with Robert W. Holley (1922-1993) and Marshall W. Nirenberg (1927-) "for their interpretation of the genetic code and its function in protein synthesis." As a result of their independent, but interrelated nucleic acid researches, they were able to break the genetic code and prove that the universal language of nucleic acids is spelled out in three-letter words. Each codon, or triplet, codes for a specific amino acid. Specifically designed oligonucleotides, which can be thought of as artificial genes, became essential tools in research and biotechnology for sequencing, cloning, and bioengineering new plants and animals.

Khorana was the youngest of five children born to Hindu parents in Raipur, a village of about 100 people in Punjab, India, which is now part of West Pakistan. His father, who was dedicated to his children's education, was an agricultural taxation clerk in the British government. Khorana attended D.A.V. High School in Multan, and earned his B.Sc. and M.Sc. from the Punjab University in Lahore in 1943 and 1945, respectively. In 1945 a Government of India Fellowship allowed him to go to the University of Liverpool, England, where he was awarded the Ph.D. in 1948. For further training, Khorana spent a year in the laboratory of Vladimir Prelog (1906-) at the Federal Institute of Technology in Zurich, Switzerland. After a brief stay in India, Khorana returned to England from 1950 to 1952, where he obtained a fellowship to work with G. W. Kenner and Alexander Todd (1907-1997). During this period, Khorana became interested in both proteins and nucleic acids.

In 1952 he received an offer of a research position in the organic chemistry section of the British Columbia Research Council and the University of British Columbia, Canada. Although Khorana first received international recognition during this period for the synthesis of coenzyme A, he generally focused his research group on biologically interesting phosphate esters and nucleic acids. In 1960 Khorana moved to the Institute for Enzyme Research at the University of Wisconsin, Madison, where he became Professor of Biochemistry and co-director of the Institute. In 1970 he accepted the position of Alfred P. Sloan Professor of Biology and Chemistry at Massachusetts Institute of Technology. He has also served as visiting professor at Stanford University, Harvard Medical School, Cornell University, and Rockefeller University.

While at the University of Wisconsin, Khorana devised methods for synthesizing specific oligonucleotides, that is, long chains of RNA in which the base sequences were precisely known. Khorana was able to replicate each of the 64 possible codons (triplets) and create RNA-like molecules. He could then demonstrate, for example, that an oligonucleotide containing alternating triplets of CUC and UCU directed the synthesis of a polypeptide in which leucine alternated with serine. Using this approach, Khorana proved that the code consisted of three-letter, non-overlapping words, read in a specific linear fashion. The code was also shown to have "punctuation marks," that is, certain triplets dictated the beginning and the ending of polypeptide chain synthesis. Khorana and his associates worked out methods for the chemical synthesis of both RNA and DNA polynucleotides. His announcement in 1970 of the complete synthesis of the gene for alanine transfer RNA, the first wholly artificial gene, was greeted as a major landmark in molecular biology.

In the 1980s Khorana turned to research on the chemistry and molecular biology of rhodopsin, the light transducing pigment of the retina, and bacteriorhodopsin, (a form of rhodopsin found in bacteria). Khorana's group synthesized the gene for rhodopsin and studied its mechanisms of action and expression. In the 1990s Khorana's work on vision led to the discovery that the misfolding of defective rhodopsin might be responsible for the inherited form of blindness known as retinitis pigmentosa. Studies of these mutant forms of rhodopsin may help explain certain clinical findings associated with retinitis pigmentosa.

LOIS N. MAGNER

Alfred Charles Kinsey

1894-1956

American Zoologist and Sexologist

In the 1950s, American zoologist and sex researcher Alfred Kinsey established himself as the preeminent authority on the sexual behavior of men and women in America. The publication of *Sexual Behavior in the Human Male* (1948) and *Sexual Behavior in the Human Female* (1953) blazed new trails in the field of sex research, leading to reassessments of research practices and of medical, psychiatric, and public attitudes towards sex. Before achieving notoriety as a sexologist, Kinsey was a trained entomologist and the world's leading expert on the American gall wasp. Regardless of the focus of his research, throughout his career Kinsey sought to uncover details and amass evidence, thereby making significant contributions to the fields that he explored.

Alfred Charles Kinsey was born on June 23, 1894, in Hoboken, New Jersey, to Alfred and Sarah Ann Kinsey. His mother and father, a self-educated teacher at the Stevens Institute of Technology, were strict and deeply religious. In spite of being a sickly child, Kinsey joined the newly formed Boy Scouts of America and was one of the first persons to earn the rank of Eagle Scout. A natural born taxonomist, Kinsey filled his boyhood days collecting plant and animal specimens.

After high school, Kinsey studied mechanical engineering at the Stevens Institute for two years before switching colleges and career direction. With little support from his father, Kinsey enrolled in the biology program at Bowdoin College, graduating in the top of his class in 1916.

A scholarship from Harvard financed Kinsey's graduate studies. Between 1916 and 1920, Kinsey collected thousands of ant-sized insects during the course of his study of gall wasps, an insect that lays its eggs inside plant stems. At the completion of his graduate studies, he accepted a teaching position at Indiana University, where he continued his gall wasps studies while carrying a full teaching load and completing two high school textbooks. The results of his gall wasp research were published in two separate texts—*The Gall Wasp Genus Cynips: A Study in the Origin of the Species* (1930) and *The Origin of Higher Categories of Cynips* (1936).

Kinsey's research interests changed dramatically in 1938 when he was invited to coordinate a new interdisciplinary course on marriage and family. Dissatisfied with the available information on human sexuality, Kinsey set out to bring the study of sexuality up to date, conducting new research based on the objective principles of sound scientific investigation rather than moralistic guidelines and outdated laws.

In 1938 Kinsey's early study of human sexual behavior was based on responses to questionnaires completed by students enrolled in his course. As the validity of these responses came into question, Kinsey changed his data gathering technique to face to face interviews. Over time, his questioning technique evolved into an elaborate series of questions designed to tweeze out the most intimate details of an individual's sexual history, while at the same time reducing or eliminating the opportunity for the individual to provide false information. His questioning technique and that of his three other colleagues stood up under rigorous scrutiny by his peers.

In December 1948 Kinsey and his colleagues published *Sexual Behavior in the Human Male,* the first of two highly controversial texts. This and his subsequent book, *Sexual Behavior in the Human Female* (1953), came to be known as the Kinsey Reports. Both volumes were filled with frank discussions of biological functions in nonjudgmental contexts. Information derived from thousands of American men and women challenged established perceptions of homosexuality, masturbation, premarital and extramarital relationships, and the role of sex in women's lives.

Before his death on August 25, 1956, resulting from pneumonia and heart complications, Kinsey had personally conducted 7,935 interviews. Additionally, a $23,000 grant awarded by the Rockefeller Foundation in 1943, with continued funding until 1954, enabled Kinsey and his colleagues to found the Institute of Sex Research of Indiana University.

MICHELLE ROSE

Louis Seymour Bazett Leakey

1903-1972

British Kenyan Archaeologist and Anthropologist

Whenever a scientist makes a discovery that challenges the established information of his or her time, the result is predictable—controversy. This was certainly the case during the illustrious career of anthropologist Louis Leakey.

Born on August 7, 1903, to British missionaries stationed in Kabete, just outside Nairobi in Kenya, Leakey grew up with the Kikuyu people of that area. He played with their children, learned their customs and culture, and received his early schooling from his parents.

He later returned to England, where he attended Cambridge University and majored in anthropology. When his initial schooling there was completed in 1926, he applied for and received a position to join an archaeological mission in Tanzania, where he put his childhood experiences to good use. When the assignment was completed, he returned to Cambridge for additional studies and in 1930 received his Ph.D. in African prehistory. He was also elected a fellow of St. John's in Cambridge.

When this phase of his education was completed, Leakey returned to Tanzania and entered into what would be the future site of his important discoveries. He focused on the Olduvai Gorge, where he uncovered numerous animal fossils and primitive stone tools. He had always believed that Africa was the home of the earliest men on earth, and the artifacts and bones he found confirmed that opinion. When he published his first book, *The Stone Age Cultures of Kenya Colony,* St. John's College gave him a grant that enabled him to continue his studies in Olduvai Gorge. He subsequently discovered the oldest-known skeletal remains of *Homo sapiens.* Although his academic associates refuted his claim, Leakey eventually uncovered even older skulls whose age could be verified.

Returning to England, Leakey was disappointed to learn that his reputation had suffered because of the controversial results of his early finds. He held a conference at Cambridge soon after his return, and the solid evidence he produced caused the objectors to revise their opinions and accept his discoveries as genuine.

Louis Leakey. *(AP/Wide World Photos. Reproduced by permission.)*

Although Leakey's academic career was flourishing, his personal life became increasingly strained. While married to Frida Avern (a fellow Briton whom he had met in Africa), Leakey fell in love with Mary Nicol (1913-1996), a scientific illustrator. He pursued this affair even though he had a son and his wife was pregnant with their second child. In spite of these deterrents, Mary went to Africa with Leakey and, when he returned to England in 1935, she also returned and lived openly with him. Frida filed for divorce in 1936, enabling Louis and Mary to marry.

In 1945 Leakey accepted the curatorship of the Coryndon Museum in Kenya. The pay was very low but it meant he could continue his research in Kenya. As a team, the Leakeys were remarkably successful. They found an important Miocene ape fossil in 1948—important enough for them to secure funding for additional research.

Later, Leakey was instrumental in starting both Jane Goodall (1934-) and Dian Fossey (1932-1985) on their impressive research projects in Africa. He also became involved with a primate research center, an Ethiopian dig, and a dig in California where there were rumors of ancient human remains at Calico Hills. However, Leakey is best remembered for his great discoveries of hominid fossils in his favorite site—the Olduvai Gorge.

Leakey's last years were spent traveling, mostly in America, where he was a popular speaker and public personality. He died in England in October 1972.

A few days prior to Leakey's death, his son, Richard Leakey (1944-), from whom he had been long estranged, showed him a new fossil skull (ER 1470). This fossil reinforced Louis Leakey's early beliefs that the genus *Homo* had its own long history and was not a descendant of the so-called "missing link." Although his last years had been difficult, this reconciliation with his son made his final days especially meaningful.

BROOK HALL

Mary Douglas Leakey
1913-1996
British Archaeologist and Anthropologist

Mary Leakey made several major archaeological discoveries that shed light on the origins and prehistory of humans. Working both individually and with her husband Louis Leakey (1903-1972) in Africa, her ceaseless excavations resulted in significant contributions to archaeology, geology, and paleontology.

Born Mary Nicol in London, England, Leakey had an interesting childhood. Since her father was a painter by profession, the family moved from town to town and country to country. Her future began to take shape when she was 11 years old and living in the Dordogne in France. It was there that she met Abbe Lemozi, who was in the midst of an archaeological dig at the Cabrerets. She found the adventure exciting and began to read and study the importance of recovering ancient artifacts. When her father died in 1926, she was determined to pursue studies in prehistory and eventually attended lectures on the subject at the University of London.

Mary's talent for drawing and sketching came to the attention of Dr. Gertrude Canton-Thompson, who invited her to illustrate her book, *The Desert Fayoum*. At that time, Louis Leakey was scheduled to give a lecture at the Royal Anthropologist Institute. Dr. Canton-Thompson introduced Mary to him and, after looking over her superb drawings, Leakey asked her to illustrate his book, *Adam's Ancestors* (1934). She agreed and over the next few months their professional relationship became personal (leading to Leakey's divorce from wife Frida Leakey in 1936).

Mary Leakey holding casts of footprints made by human ancestors 3.6 million years ago. *(Bettmann/Corbis. Reproduced by permission.)*

Prior to her marriage to Leakey, Mary worked with Dorothy Liddell on an important dig at Hembury Fort in Devon, England. Liddell was an expert in the techniques of proper excavation and Mary was a gifted student. Four months later, Mary undertook her own project, an excavation at Jaywick Sands in Essex that led to the publication of her first scientific paper.

Mary Leakey and her new husband moved to Kenya in 1935, where they remained until 1959. They worked at Olduvai Gorge in the Serengeti Plains of northern Tanzania, documenting aspects of Stone Age culture as revealed by the primitive tools, bones, and artifacts they uncovered.

On one of their trips to Rusinga Island in Lake Victoria (1947), Mary discovered a *Proconsul africanus* skull that was later found to be 20 million years old. To this date, only three others are extant. Mary Leakey's discovery merited the Stopes Medal from the Geological Association, which was presented to her and her husband jointly. She is also credited with finding the now-famous primate fossil known as *Jinjanthropus* in 1959. This work and its documentation is considered an important factor in the study of human origins.

When Louis Leakey began spending more time abroad, lecturing and trying to secure funding for their work in Africa, Mary took over the research project and spent most of the next 25 years digging, cataloging, and writing about their findings.

She made her first trip to America in 1962. Mary and Louis were the guests of the National Geographic Society in Washington, D.C., where they were jointly awarded the society's Gold Hubbard Medal. Seven years later, Mary Leakey earned her first honorary degree from the University of Witwatersrand in Johannesburg, South Africa.

After the death of her husband in 1972, Mary Leakey continued to work at Olduvai and another site called Laetoli. The latter site became well known when Mary found hominid fossils (more than 3.75 million years old), 15 new species, and a single new genus. Although Mary Leakey is often referred to as an archaeologist, most scientific writings refer to her as a physical anthropologist.

Along with her awards and degrees, Mary Leakey has two books to her credit—*Olduvai Gorge: My Search for Early Man* (1979) and the autobiography *Disclosing the Past* (1984). Her son, Richard Leakey (1944-), followed his parents' path and distinguished himself in the science of paleontology. Mary Leakey died in Nairobi in 1996.

BROOK HALL

Joshua Lederberg
1925-
American Geneticist and Microbiologist

In 1958 Joshua Lederberg shared the Nobel Prize in Medicine with George Wells Beadle (1903-1989) and Edward Lawrie Tatum (1909-1975) for his discovery of sexual reproduction and genetic recombination in bacteria. This work was fundamental to overcoming skepticism about the value of microbes as model systems for research in genetics. Previously, some classical geneticists had dismissed the possibility of using bacteria because these simple organisms were thought to have "no genes, nuclei, or sex." Lederberg and Edward Tatum announced their discovery on the sex life of bacteria at the 1946 Cold Spring Harbor Symposium on Microbial Genetics. This announcement convinced many biologists that, like the traditional subjects of genetic research, bacteria had chromosomes and mutable genes that could undergo recombination and replication processes. New approaches to studying the nature of the gene in bacteria and viruses led to the establishment of molecular genetics and genetic engineering, and to fundamental insights into processes common to all forms of life.

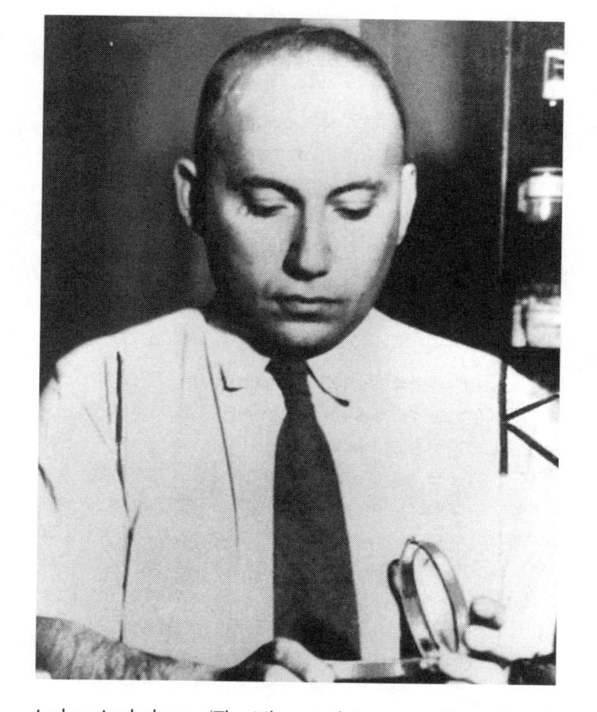

Joshua Lederberg. *(The Library of Congress. Reproduced by permission.)*

Lederberg was born in Montclair, New Jersey, and raised in New York City. At age 16 he entered Columbia University, where he planned to study chemistry in preparation for medical school. Professor Francis J. Ryan, who had worked with George Beadle and Edward Tatum, aroused his interest in microbial genetics. In 1944 Lederberg earned his B.A. in zoology and began medical studies at Columbia. While a medical student, he carried out genetic research with Ryan and served in the U.S. Navy Hospital Corps.

In 1945, after learning about work in the laboratory of Oswald Avery (1877-1955) that suggested DNA might be the genetic material in pneumococcus bacteria, Lederberg decided to switch from *Neurospora* to bacteria. Ryan suggested that Lederberg work with Tatum, who was investigating the genetics of *Escherichia coli* and establishing a new program in microbiology at Yale University. Lederberg transferred from Columbia to Yale in 1946, and was awarded a Ph.D. in 1948. Tatum and Lederberg worked with easily identifiable nutritional mutants of *E. coli* strain K-12. Different mutant strains of the bacteria were grown separately and together, and the characteristics of the mutations generated in such cultures were tested. Lederberg and Tatum were able to isolate a nutritionally independent strain that could only be explained as the result of a sexual mating process that led to the segregation and reassortment of the genetic material of the parental types. Stimulated by the work of Lederberg and Tatum, other geneticists soon confirmed their results and discovered other methods of genetic transformation in bacteria. Joshua Lederberg and Esther Lederberg, his first wife, demonstrated in 1953 that bacterial viruses could transfer bacterial genes in *Salmonella*.

After receiving his degree from Yale, Lederberg became professor of genetics at the University of Wisconsin. Although he won the Nobel Prize for his work in bacterial genetics, his research interests were very broad, including computer science and artificial intelligence. During a 1957 sabbatical at Melbourne University in Australia, he carried out studies of monoclonal antibodies. When he returned to Wisconsin, he established a new Department of Medical Genetics, but in 1959 he moved to Stanford University to organize a Department of Genetics. At Stanford he served as chairman of the Department of Genetics, professor of biology, and professor of computer science. He became director of the Kennedy Laboratories for Molecular Medicine in 1962. In 1978 he became president of Rockefeller University.

In addition to his scientific work, Lederberg was very interested in public service and communicating scientific concepts to the public. He served as an advisor to the Arms Control and Disarmament Agency, where his expertise was important to negotiating a biological weapons disarmament treaty. He also served as an advisor to the U.S. Defense Science Board and the Mariner and Viking missions to Mars.

LOIS N. MAGNER

Lynn Margulis
1938-
American Geneticist

American geneticist Lynn Margulis is noted for her investigations of the intricate, fundamental systems by which life creates and maintains itself on Earth. She is often regarded as the co-creator, with James Lovelock (1919-), of the Gaia hypothesis, an idea suggesting that the earth is an ecosystem, or the sum of many ecosystems.

Margulis was born in Chicago in 1938, the eldest of four daughters. At age 14 she enrolled in an undergraduate program at the University of Chicago, where she was introduced to the natural sciences. After graduating, she pursued an M.S. in zoology and genetics at the University of Wisconsin. In 1965 she received her Ph.D. in genetics from the University of California at Berkeley. In 1970 she moved to Massachusetts, where over the next 22 years she raised four children and taught at Boston University. She now teaches at the University of Massachusetts.

Margulis is considered an expert on the biological kingdom Protoctista, which includes an estimated 250,000 algae, seaweeds, amoebas, and other little known life forms. In the 1960s she began looking for DNA where no one had thought it could be found—outside the nucleus of the algae cell. She found it, and her discovery supported a revolutionary theory of symbiosis in the origin of the cell.

Margulis is also considered the co-author of the Gaia hypothesis. The Gaia hypothesis suggests that certain conditions that sustain life are regulated by life itself. More specifically, she contends, the atmosphere and all life on Earth act as a single integrated physiological system.

The strongest evidence for the Gaia hypothesis comes from the study of atmospheric chemistry. Her work in reconstructing early life on the planet revealed that bacteria produce and remove all types of atmospheric gases. The composition of the earth's atmosphere differs from our nearest neighbors, Mars and Venus. Both of these planets have carbon dioxide-rich, steady-state atmospheres that remain in equilibrium. However, our atmosphere on Earth is very different. Loaded with reactive gases, our atmosphere contains oxygen, nitrogen, and methane, among others, which are violently reactive to one another. There is no way to explain this by chemistry alone.

James Lovelock, a British atmospheric chemist, felt that the presence of reactive gases is evidence that atmospheric gases on Earth are actively regulated. Margulis agreed and suggested the atmosphere is an extension of life itself. If the surface of Earth were not covered with oxygen-emitting algae and plants, methane producing bacteria, hydrogen-producing fermenters, and countless other life forms, its atmosphere would long ago have reached the same steady state of Mars and Venus.

Magulis suggests that another argument for Gaia comes from the study of astronomy. According to accepted models, the sun is 30 to 70 percent hotter today than it was in the early days of the earth's history. If the earth's temperature was consistent with this increase in solar radiation, we would now be at a boiling point. However, the temperature of Earth has remained stable and conducive to life for all of this time. She concludes that growing populations of gas-producing organisms have actively maintained surface temperatures within a range suitable for life.

Ultimately, Margulis notes, "it doesn't matter whether the Gaia hypotheses is supported or not." The fact that it has generated new thoughts and new work is the best evidence of its value. She also points out that although soil erosion, loss of nutrients, methane production, ozone depletion, deforestation, and the loss of species diversification may all be Gaian processes, our human habits and behaviors have accentuated them to the point of near catastrophe. It is possible that our lack of environmental policies and our overpopulation will stress the system to such an extent that the earth will roll over into another steady-state regime, which may or may not include human life.

LESLIE HUTCHINSON

Barbara McClintock. (UPI/Corbis-Bettmann. Reproduced by permission.)

Barbara McClintock
1902-1992
American Geneticist

Barbara McClintock discovered that certain types of genes, called "jumping genes," can move from one place on a chromosome to another, from one generation to the next. For this discovery she was awarded the 1983 Nobel Prize in Physiology or Medicine.

Born in Hartford, Connecticut, on June 16, 1902, McClintock attended Erasmus Hall High School in Brooklyn, then went on to Cornell University, where she earned her bachelor's degree in 1923. As an undergraduate she was very active in social activity, playing the banjo in a jazz band and serving as president of the freshmen women's class. She soon became absorbed in her studies and retreated from social activities to pursue academics. At that time genetics was still a relatively new field, as it had been only 21 years since the rediscovery of Mendel's principles of heredity. However, interest in genetic research was growing. In January 1922 McClintock was invited by Dr. Hutcheson to attend his class on genetics, the only course of its type then offered at Cornell. This proved a momentous event, as McClintock found her life's work in genetics.

As McClintock worked on her master's degree, she became fascinated by the structure of chromosomes and the role of mitosis and meiosis. She completed her Ph.D. in 1927 and remained at Cornell to study the 10 chromosomes of maize. At that time Cornell did not appoint women as faculty professors. After 11 years as an instructor and researcher, she accepted a position as assistant professor of botany at the University of Missouri, where she remained for five years. That was the last time she taught, for she began to do research exclusively. In 1942 she moved to Cold Springs Harbor, New York, where she continued to work long hours, seven days a week, until just before her death. She even had a small apartment on the grounds of the laboratory.

Concentrating her efforts on the genetics and cellular composition of maize, or corn, McClintock observed variations among kernels of corn and found that genetic information does not stay in one place. Tracing pigmentation changes and examining the large chromosomes under a microscope, she found two genes were "controlling" elements. These genes actually told the other genes what to do by moving along a gene at a different site. She dubbed these "jumping genes" and suggested these transposable genes were responsible for mutations. This research, initially conducted in 1944, was not recognized until the 1960s, when scientists learned the genetic material was DNA and two French scientists discovered genetic anomalies in bacteria similar to those McClintock earlier observed in corn.

McClintock's work subsequently won recognition for its great contribution to the understanding of genetic function and organization. In 1970 she received the National Medal of Science. She received many additional medals and prizes, including the prestigious Nobel Prize in 1983.

McClintock had many personal and professional challenges. Her independence, originality, and accomplishments were intimidating to many of her colleagues. Living at a time when genetics was beginning to come into its own, she saw her work go unrecognized for many years. However, her groundbreaking research established her as a leader in the field of genetics. She died in Huntington, Long Island, on September 2, 1992.

EVELYN B. KELLY

Jacques Monod. *(The Library of Congress. Reproduced by permission.)*

Jacques Lucien Monod
1910-1976
French Geneticist

In 1965 Jacques Lucien Monod shared the Nobel Prize in Medicine with François Jacob (1920-) and André Lwoff (1902-) for their contributions to discoveries concerning the genetic regulation of enzymes and viral synthesis. In 1961 Monod and Jacob proposed the concept of messenger RNA and the operon theory.

Monod was born in Paris. His father was a French artist and his mother was an American. Monod studied zoology at the Université de Paris (Sorbonne). He earned his B.Sc. in 1931 and his doctorate in 1941. In 1936 he worked in the laboratory of Thomas Hunt Morgan (1866-1945) at Caltech. During World War II, Monod was involved in the French Resistance. He served as a professor at the Université de Paris until 1945, when he moved to the Pasteur Institute, where he spent the rest of his scientific career. He became head of the Department of Cellular Biochemistry in 1953 and director of the Institute in 1971.

As a student, Monod was fascinated by the new science of genetics, but his early researches involved the problem of growth in bacteria. His doctoral dissertation involved quantitative studies of the factors that govern the rate of growth of bacterial cultures. Monod noticed that when bacteria were grown with certain sugars in the medium, the initial growth phase was followed by a lag phase, and then another growth spurt. In 1940 Lwoff suggested that this growth pattern might be the result of enzyme adaptation, or induction. Monod began to think that his studies of inducible enzymes and bacterial growth might be linked to the new work on bacterial mutants reported by Salvador Luria (1912-1991) and Max Delbruck (1906-1981). Studies of the enzyme ß-galactosidase in a mutant strain of bacteria indicated that enzyme induction involved the synthesis of new protein molecules. Monod speculated that a mutation in the adaptive function of an enzyme might be the result of prior mutations in the gene. Experiments carried out by Monod and his colleagues suggested that the biosynthesis of enzymes was stimulated by the presence of an "inducer." In most cases, the inducer seemed to be the substrate of the enzyme.

A series of experiments carried out by Jacob and his colleagues between 1958 and 1963 led to the "operon theory," a theoretical framework linking gene expression and the induction of enzyme synthesis. The operon was assumed to be a fundamental unit of bacterial gene expression and regulation, consisting of structural genes, regulator genes, and control elements. Genes were assumed to play two essential roles: encoding protein structure and regulating protein synthesis. According to Jacob and Monod, gene expression is made possible when an inhibitor, called the "repressor," is removed from the gene. This leads to the biosynthesis of an inducible enzyme. The classic experiment that supported the operon theory became known as the "PaJaMo experiment," in honor of Arthur Pardee (1921-), Jacob, and Monod, who published the results of the experiment in 1959 in the *Journal of Molecular Biology.*

Another important contribution to enzyme control mechanisms made by Monod is known as the theory of allosteric regulation. In 1961 Monod suggested that interactions between enzymes and small molecules, such as substrates, activators, and inhibitors, might change the shape of the enzyme in ways that would affect its activity and affinity for its substrates.

Both Jacob and Monod wrote eloquently about science, life, and philosophy. In *Chance and Necessity* (1970) Monod noted that biology was sometimes accorded only a marginal place among the sciences, because the study of living things on earth did not seem to lead to universal, cosmic laws. He argued that the ultimate aim of science was to clarify the relationship between human beings and the universe; therefore, biology deserved a place at the center of the sciences. The main thesis of his philosophical essay *Chance and Necessity* was that life arose by chance and that all species of life, including human beings, were ultimately the products of random genetic mutations. Monod was a gifted musician, a sailor and rock-climber, an opponent of Lysenkoism and other forms of pseudoscience, and a supporter of women's rights and educational reform.

LOIS N. MAGNER

Linus Carl Pauling
1901-1994
American Chemist

In 1954 Linus Pauling was awarded the Nobel Prize in Chemistry for his research into the nature of the chemical bond and its application to the elucidation of the structure of complex substances. Pauling's concepts and experiments made it possible for chemists to understand the forces that held proteins together. His classic text *The Nature of the Chemical Bond* (1939) remains a model of clarity. In 1962 he was awarded the Nobel Peace Prize for his campaign against the use, testing, and proliferation of nuclear weapons and the very idea of waging war as a way of solving international conflicts.

Linus Carl Pauling. *(The Library of Congress. Reproduced by permission.)*

Using the prestige inherent in his status as a Nobel laureate, Pauling was able to attract attention to the dangers of the radioactive fallout that was produced by the testing of nuclear weapons. The Peace Prize recognized the role Pauling played in establishing the banning of atmospheric testing of nuclear weapons by the United States and the Soviet Union.

Reflecting on his life and work, Linus Pauling said that he had known many people who might have been smarter than he was, but they had generally gone into theoretical physics. His consolation was that even though such physicists might have been deeper thinkers, he had enjoyed broader interests. Pauling's remarkably varied body of scientific work evolved from his interest in the nature of the chemical bond. His experimental work encompassed diverse aspects of physical chemistry, such as the use of X-ray diffraction to elucidate the structure of crystals, the use of electron diffraction to determine the structure of gas molecules, and studies of the magnetic properties of various substances. Pauling was also an ingenious theoretical chemist, as demonstrated by his attempts to apply quantum mechanics to the structure of molecules and the nature of the chemical bond, the extension of the theory of valence to include metals and intermetallic compounds, and the development of a theory of the structure of atomic nuclei and the nature of the process of nuclear fission. Indeed, as James Watson (1928-) and Francis Crick (1916-) struggled to discover the secret of the gene, they deliberately attempted to emulate Pauling's remarkable innovation of building three-dimensional models as a means of determining the structure of a molecule with only a minimum of experimental evidence. Pauling had used this approach when he suggested that the X-ray diffraction pattern of the protein keratin could be attributed to alpha-helices coiled round each other. Unfortunately, Pauling's preliminary speculations about the structure of DNA were based on obsolete and incomplete information. Pauling could not assess the new data assembled by Rosalind Franklin (1920-1958) at King's College because he was unable to travel freely. The State Department had taken away his passport for allegedly subversive activities that primarily involved participation in the Ban the Bomb movement.

Pauling's immense curiosity about the chemical nature of complex biological compounds led to fruitful investigations of the structure of proteins, the molecular basis of general anesthesia, and the nature of serological systems, and the structure of antibodies. Fundamental insights into understanding disease processes at the molecular level grew out of Pauling's studies of the relationship between abnormal hemoglobin molecules and hereditary hemolytic anemias, including sickle-cell anemia. In 1949 Pauling proved that the sickling phenomenon was caused by an alteration in hemoglobin. By analyzing the abnormal hemoglobin produced by patients with sickle cell anemia, Pauling and Vernon M. Ingram established the existence of a specific molecular disease. Pauling also investigated the relationship between abnormal enzymes and mental disease. These studies eventually led to his interest in the relationship between the chemistry of nutrition and medical problems.

Pauling was born in Portland, Oregon. He earned his B.S. in Chemical Engineering from Oregon State College, and his Ph.D. from the California Institute of Technology. In 1923 Pauling married Ava Helen Miller, who was an important influence on his campaigns against nuclear weapons. Pauling was a member of the faculty of Caltech for 41 years. He held academic appointments and visiting professorships at many other institutions, including Stanford, Cornell, MIT, Harvard, Princeton, Madras, Oxford University, and the Center for the Study of Democratic Institutions in Santa Barbara, California. He established the Linus Pauling Institute of Science and Medicine, where he served as Research Professor. At the Institute, Pauling served as an advocate of "orthomolecular medicine" and megadose vitamin C to prevent and treat colds, schizophrenia, and cancer.

Ranked along with Isaac Newton (1642-1727) and Albert Einstein (1879-1955) as one of the twenty most important scientists of all time, Pauling's willingness to tackle a wide array of scientific, social, political, and medical questions has also earned him a reputation as one of the twentieth-century's most controversial scientists.

LOIS N. MAGNER

Esther May Sternberg
1951-

Canadian-American Physician, Neuroendocrinologist, and Administrator

Esther M. Sternberg has made many contributions to the study of rheumatology, neuroendocrinology, stress and neurological disorders, and the relationship between emotions and disease. Her research was fundamental to elucidating the etiology of a puzzling 1989 epidemic of eosinophilia-myalgia syndrome. Sternberg's work on the new field of research known as "mind-body interactions" has been instrumental in explaining how the immune system and the nervous system communicate with each other.

Sternberg was born in Montreal in 1951. In 1991 she became a citizen of the United States. Her choice of a career was greatly influenced by her family background. Joseph Sternberg, her father, was a physician-scientist and a pioneer in the fields of radiation biology and nuclear medicine.

In 1972 Sternberg received a B.Sc., with Great Distinction, from McGill University, located in Montreal. Two years later, she was awarded an M.D. from McGill. During her post-graduate medical training at the Royal Victoria Hospital, she selected rheumatology as her area of clinical and research specialization. Between 1981 and 1986, she was a research associate in the Division of Allergy and Clinical Immunology at the Washington University School of Medicine in St. Louis, Missouri. In 1987 she accepted a research position in clinical neurosciences at the National Institute of Mental Health (NIMH) in Bethesda, Maryland. This job has led to a series of increasingly prestigious positions at the NIMH. Her current position there is Chief of the Section on Neuroendocrine Immunology and Behavior. She is also a research full professor at the American University in Washington, D.C.

Sternberg has published over 60 articles and books. These writings explore many medical problems, including rheumatoid arthritis, scleroderma, fibromyalgia, multiple sclerosis, the effect of serotonin on the immune response, the design of drugs to inhibit HIV receptor binding, the relationship between L-tryptophan and human eosinophilia-myalgia syndrome, neuroendocrinology and the immune response, and the stress response and regulation of inflammatory disease. Other works by Sternberg consider hyperimmune fatigue syndrome, tamoxifen, lymphokines, the relationship between exercise and the immune system, neuroendocrine aspects of autoimmunity, neuroimmune stress interactions, neuroendocrine factors in susceptibility to inflammatory disease, emotions and disease, and other aspects of mind-body interactions. Sternberg has received many awards and honors for her research on rheumatology, EMS, and the relationship between emotions and disease. Her expertise has been sought in areas as diverse as scleroderma, asthma, mind-body interactions, influenza and pneumoccocal vaccines, stress in neurological disorders, the health of deployed U.S. military forces, and military nutrition research. In 1989 a previously unknown disorder that came to be called eosinophilia-myalgia syndrome (EMS) appeared in an epidemic pattern in the United States. About 1,500 cases and 40 deaths were reported. Case studies, epidemiological data, and laboratory research on animals linked the syndrome to the amino acid L-tryptophan, sold by a particular manufacturer as a dietary supplement. Sternberg's research was fundamental in elucidating the etiology (origin) of EMS and in drawing attention to the pressure that the manufacturer of the suspected L-tryptophan was bringing against researchers trying to investigate the disorder.

For much of the twentieth century, medical science and even medical practice had become, according to many critics, increasingly specialized and mechanistic. The field of neuroimmune interactions, in contract, is an intensely interdisciplinary field that encompasses immunology, neurobiology, neuroendocrinology, and the behavioral sciences. Beliefs about the relationship between emotions and disease that go back to the writings of Hippocrates are now being investigated in terms of molecules, cells, and nerve signals. Biomedical scientists and physicians involved in this challenging area have been able to integrate research results from molecular biology with clinical observations of behaviors, emotions, and disease. As a leader in the study of mind-body interactions, Sternberg has called for research that is precise, focused, and integrative. For example, vague references to "stress" are giving way to precise definitions and measurements of both external stressors and internal responses (e.g., neural, neuroendocrine, and immune factors). The task of future research, according to Sternberg, will be to provide rigorous evidence that neuroimmune interactions play a role in susceptibility and resistance to inflammatory and infectious diseases and to find ways to apply these new scientific insights to human health and healing.

LOIS N. MAGNER

Emmanuel Bandele Thompson
1928-
African-American Pharmacologist

Emmanuel B. Thompson is a pharmacologist and educator known for his research leading toward treatments for high blood pressure and sickle-cell anemia. Both diseases, particularly the latter, have a high rate of incidence among African Americans, and Thompson has published numerous studies regarding these conditions. In addition, he has conducted important research on drug screening, and published a textbook on the subject.

The oldest of five children, Thompson was born on March 15, 1928, in the town of Zaria in the northern part of what became Nigeria after the British decolonized that nation two decades later. His father was a successful merchant for the United Africa Company, which purchased the agricultural products of the area and sold manufactured goods from outside to the locals. Thompson attended a Roman Catholic school, where he took an early interest in biology—an interest that was piqued when renowned African-American scientist George Washington Carver (1860-1943) visited the school.

After earning his high school diploma from a prestigious boarding school in the Nigerian capital of Lagos, Thompson worked at a variety of jobs before entering the Yaba School of Pharmacy in Lagos in 1951. Late in 1952, he received a diploma as a technician, and soon afterward earned a scholarship, established by a Quaker missionary in Nigeria, to Rockhurst College in McPherson, Kansas. Thompson finished his biology degree at Rockhurst in three years, graduating with a B.S. in 1955. He then entered the University of Missouri in Kansas City, where he met and married Nova Garner, with whom he later had two daughters. In 1959 he earned his pharmacy certification, but chose to continue his education and in 1961 began work on his M.S. in pharmacology at the University of Nebraska in Lincoln. Earning his M.S. in 1963, he went on to the University of Washington in Seattle, where he earned his Ph.D. in pharmacology in 1966. His dissertation dealt with the treatment of high blood pressure with synthetic compounds to inhibit nerve messages to the heart.

Thompson went to work as a senior research pharmacist at Baxter Laboratories in Morton Grove, Illinois, near Chicago. In this capacity he developed a variety of general anesthetics, one of which gained wide use among surgeons in Japan. Following a three-year stint with Baxter, Thompson became an assistant professor at the University of Illinois Medical Center's College of Pharmacy. In 1971 he began working part-time at West Side Veteran's Administration Hospital in Chicago as its principal researcher and consultant. In 1973 Thompson became an associate professor, and, though his base remained in the College of Pharmacy, he also taught classes in the schools of Public Health and Associated Medical Sciences, and in the College of Nursing. An occasional lecturer at the Illinois College of Pediatric Medicine and the Chicago State University School of Nursing, he published a textbook called *Drug Bioscreening* in 1985. A revised edition saw print in 1990.

In addition to his work on high blood pressure, in the 1980s and 1990s Thompson researched treatments for sickle-cell anemia. As of the late 1990s there was no cure for sickle-cell anemia, a disease that primarily affects African Americans. However, in dozens of papers on this disease and high blood pressure, Thompson suggested methods of treating it. He has also conducted vital research in the area of screening tropical plants for their medicinal uses.

Thompson retired from teaching in 1997, but continues to consult with hospitals. He is a member of the New York Academy of Science, the American Association of Colleges of Pharmacy, and the American Pharmaceutical Association. He lives in Chicago.

JUDSON KNIGHT

James Dewey Watson

1928-

American Molecular Biologist

In 1962 James Dewey Watson shared the Nobel Prize in Medicine with Francis Crick (1916-) and Maurice Wilkins (1916-) for their discoveries concerning the molecular structure of deoxyribonucleic acid (DNA) and its significance for the transmission of genetic information. Molecular biologists have called the discovery of the double-helical structure of DNA one of the most important developments in twentieth-century biology. The structure of DNA proposed by Crick and Watson in 1953 immediately suggested insights into the nature of the gene, the genetic code, and mechanism by which information stored in DNA was transmitted from generation to generation.

Watson was born in Chicago, Illinois. He was a bright and precocious child who appeared as a Quiz Kid on a national radio program. Only 15 when he entered the University of Chicago, Watson was attracted to the natural sciences, especially ornithology. He later claimed that he became obsessed with finding the secret of the gene after reading *What is Life?* by Erwin Schrödinger (1887-1961). Rejected by Harvard and Caltech, Watson accepted a fellowship at Indiana University and worked with Salvador Luria (1912-1991), a charter member of the "phage group." At the age of 22, Watson was awarded his Ph.D. for a dissertation on the effects of X-rays on phage replication. Luria suggested that Watson go to Europe to study biochemistry. After a disappointing year in Copenhagen, Watson joined a group of researchers headed by Max Perutz (1914-) at the Cavendish Laboratory to work on the molecular structure of nucleic acids extracted from plant viruses.

At Cambridge in 1951, Watson met Francis Crick, who was 12 years his senior but still a graduate student. Despite differences in personality and scientific background, the two discovered that they shared a passion for discovering the "secret of the gene." Moreover, they were both convinced that DNA, rather than protein, would prove to be the macromolecule responsible for passing genetic information from generation to generation. A solution to the structure of DNA should, therefore, lead to an explanation of the replication of genes. Both attributed their success to their special relationship: their ability to complement, criticize, and stimulate each other. The Watson-Crick collaboration was a fortunate one, for Crick doubted that either he or Watson could have discovered the structure of DNA alone. According to Crick, the structure might have been solved by Rosalind Franklin (1920-1958), Maurice Wilkins, Linus Pauling (1901-1994), or by further refinements of biochemistry. After numerous false starts, in 1953 Watson and Crick arrived at a solution to the

James Watson. *(The Library of Congress. Reproduced by permission.)*

three-dimensional structure of DNA on the basis of model building, data from Rosalind Franklin's X-ray crystallography work, and general knowledge about the chemistry of DNA. After proposing a general scheme for the transmission of genetic information, which became known as the "central dogma," Watson pinned a cryptic note above his desk that said "DNA→RNA→protein." Watson and Crick played an important role in formulating the general principles that explain how information stored in DNA is replicated and passed on to daughter molecules, and how information encoded in DNA was used in the production of proteins.

In their first *Nature* paper of 1953, Watson and Crick described their model for the structure of deoxyribose nucleic acid. The novel feature of their double helix was the way in which the two chains were held together by the purine and pyrimidine bases. A purine on one chain always paired with a pyrimidine on the other chain; adenine always paired with thymine, and guanine with cytosine. Although any sequence of bases was possible on one chain, the rules of base pairing automatically determined the sequence of bases on the other chain. Thus, the DNA double helix immediately explained data produced by Erwin Chargaff (1905-) concerning the molar ratios of purines to pyrimidines and how DNA exhibits order and stability, as well as variety and mutability. Shortly after their first paper in *Nature,* Watson and Crick elaborated upon the genetic implications of their model. Watson's best-selling books include *The Double Helix* and *The Molecular Biology of the Gene.* In 1968 Watson became director of the Cold Spring Harbor Laboratories. He served as head of the United States human genome research program from 1989 to 1992.

LOIS N. MAGNER

Ian Wilmut
1944-
English Embryologist

In February 1997 Ian Wilmut made international headlines when he announced that he and other researchers at the Roslin Institute in Scotland had successfully cloned the first mammal, a sheep named Dolly, from an adult animal. This meant that in the future animals could be cloned to produce proteins for the manufacture of certain pharmaceuticals. However, the experiment was not without controversy. Many speculated about what cloning could mean if applied to human beings, though Wilmut emphasized the fact that his work was intended purely for animals.

Wilmut was born on July 7, 1944, in Hampton Lucey, Warwick, England. His interest in embryology began as a student at the University of Nottingham, where he met renowned reproduction expert G. Eric Lamming. Wilmut graduated from Nottingham in 1967 with a degree in agricultural sciences, and went on to pursue his doctoral studies at Darwin College, Cambridge. There he performed his dissertation, completed in 1973, on techniques for freezing boar semen.

Upon receiving his degree, Wilmut went to work at what was then called the Animal Breeding Research Station, a government facility in Roslin, near Edinburgh, Scotland. Later it would be renamed the Roslin Institute. In 1973 he produced the first calf, Frosty, born from a frozen embryo implanted in a surrogate mother. He went on to pursue research in cloning, despite the fact that during the 1980s scientists became increasingly skeptical that cloning could become a reality. However, Wilmut was convinced that it could, and when he heard that Steen M. Willadsen of Grenada Genetics in the United States had cloned a lamb using a differentiated cell from an already developing embryo, he stepped up the pace of his research.

Keith Campbell, a biologist at the Roslin Institute, had noted that the cycles of adult and

Ian Wilmut. *(AFP/CORBIS.Reproduced by permission.)*

embryo cells were not synchronized, and for this reason, embryos were less likely to accept the genetic material from a transplanted adult cell. Therefore, he developed a technique for slowing down adult cells by depriving them of nutrients. Using this method, Campbell and Wilmut cloned two sheep, Megan and Morag, from developing embryo cells.

Next, they turned their attention to cloning an adult sheep. In order to do this, Wilmut and Campbell harvested mammary, or udder, cells from a six-year-old ewe and preserved them in test tubes, starving them by reducing their serum concentration over a five-day period. Success was not immediate: they made 276 attempts at cloning before an embryo survived. They implanted the embryo into a surrogate mother, and on July 5, 1996, Dolly—named for country singer Dolly Parton—was born. The Roslin Institute did not publish the results of its experiment until it had secured the patent for the cloning process.

Wilmut and his wife Vivian live in a small village near Edinburgh with their three children, Helen, Naomi, and Dean. An honorary fellow at the Institute of Ecology and Resource Management at the University of Edinburgh, Wilmut lives quietly, in spite of the controversy his work has raised. He continues to conduct cloning research in the hope that cloning and genetic engineering can produce proteins such as the clotting factor lacking in the genetic makeup of hemophiliacs.

JUDSON KNIGHT

Edward Osborne Wilson

1929-

American Biologist

Edward O. Wilson is a specialist in ant biology at Harvard University. He first gained renown among biologists for his discovery of ants' ability to communicate using chemicals called pheromones. He gained even greater fame as one of the key figures in the founding of sociobiology (also known today as evolutionary psychology). His book, *Sociobiology: The New Synthesis* (1975) stirred up considerable controversy. Wilson's sociobiology is the attempt to explain animal societies—including humans—as the product of evolutionary development.

As a boy growing up in the southeastern United States, Wilson was enthralled with nature and began collecting insects and carefully observing animals in their habitat. By high school he had decided to pursue the vocation of biology and had even settled on a specialization—ants. He attended the University of Alabama in the late 1940s, where he enthusiastically embraced the neo-Darwinian synthesis by reading Ernst Mayr's (1904-) *Systematics and the Origin of Species.* From that time on, evolutionary theory was a central theme not only for his biology, but for his whole world view. Though he had previously embraced Christianity as a teenager, thereafter he usually called himself a scientific materialist, explaining religion and ethics as the product of material processes, especially biological evolution.

After beginning doctoral studies at the University of Tennessee, he transferred to Harvard University in 1951, where he received his Ph.D. In 1956 he was appointed professor at Harvard, and soon thereafter he began studying ant communication. He was the first to discover that animals can communicate through their sense of smell by using chemical signals called pheromones. For example, an ant in distress can emit a certain chemical to alert other ants in its colony to come to its aid.

Wilson's interest in sociobiology was stimulated partly through his research on ants as social insects, but also through his encounters with Konrad Lorenz (1903-1989) and Nikolaus Tinbergen (1907-1988), both of whom contributed to the founding of ethology, the study of animal behavior. Another important influence on the development of Wilson's sociobiology was W. D. Hamilton's idea of kin selection, advanced in a 1964 paper that Wilson read the following year. Hamilton's notion of kin selection tried to explain the puzzle of altruistic (selfless) behavior within a Darwinian framework. Hamilton argued that altruism could benefit the kin of the altruistic individual, and since the kin shared many of the same genes, altruism could thus promote survival of one's genes.

Wilson adopted Hamilton's idea and applied it first to ant societies and later to other social animals. By applying sociobiology to humans, he aroused a storm of controversy in the mid-1970s. Other prominent scientists and especially social scientists began protesting his view that many human behaviors and ethical systems are determined by or at least heavily influenced by genetic predispositions. Wilson did not claim that all specific behaviors or specific moral standards are genetically determined, but he did believe that many tendencies, such as division of labor between sexes, altruism toward kin, tribalism, male dominance, and territorial aggression, were biological instincts produced through evolution. Critics claimed that Wilson was politically motivated, justifying racism and sexism, but Wilson claimed this was a misunderstanding.

Despite the opposition, Wilson won many influential followers, some of whom preferred the term evolutionary psychology. In 1977 *Time* magazine carried a cover story on sociobiology, and that same year Wilson won the National Medal of Science for his work in sociobiology. The following year he published his Pulitzer Prize-winning *On Human Nature,* providing greater detail on human sociobiology. Responding to widespread criticisms of sociobiology as genetic determinism, Wilson worked out his views on the interaction of heredity and culture, a view he called "gene-culture coevolution."

In the 1980s Wilson became an environmental activist because of his concern about the extinction of many species. He lamented the rapid decline in biodiversity, which he considered the product of eons of evolutionary development. He also began warning about human overpopulation.

Wilson's entire world view is laid out in *Consilience: The Unity of Knowledge* (1998). He considers the empirical scientific method the only valid method for attaining knowledge about anything, and thus wants to infuse the humanities and social sciences with the scientific method. This is because he considers all of nature, including every aspect of humans, the product of mindless evolutionary development. According to Wilson, ethics and religion are merely genetic predispositions that helped our ancestors survive, but only science can impart real truth. Wilson's sociobiology and his philosophy of consilience are still highly controversial.

RICHARD WEIKART

Rosalyn Sussman Yalow

1921-

American Physicist

Rosalyn Sussman Yalow won the 1977 Nobel Prize for the development of radioimmunoassays for peptide hormones. The revolutionary radioimmunoassay (RAI) developed by Yalow and her colleague Solomon Berson (1918-1972) made possible the measurement of extremely minute amounts of almost any substance in blood and body tissues.

Yalow was born in New York City and has lived and worked there ever since, except for her graduate work at the University of Illinois. As a high school student, Yalow was interested in mathematics and chemistry, but at Hunter College she became attracted to nuclear physics. Like many other women scientists, Yalow was touched by Eve Curie's biography of her mother, *Madame Curie,* a book that Yalow still considers essential reading for aspiring female scientists.

Although Yalow hoped to become a physicist, in keeping with prevailing views of proper roles for women, her parents thought that a career as an elementary school teacher was a more practical goal. Indeed, few graduate programs in science were willing to accept a woman. During her senior year at Hunter, one of her physics professors suggested that she accept a position as secretary to Rudolf Schoenheimer at Columbia University's College of Physicians and Surgeons. Hopeful that this job would provide a "backdoor" entráe into graduate courses, Yalow agreed to take a course in stenography. Soon after she received her B.A. degree from Hunter College in 1941, Yalow was offered a teaching assistantship in physics from the University of Illinois. At the first meeting of the Faculty of the College of Engineering, she discovered that she was the only woman out of about 400 faculty members and students. Yalow considered her being hired an important achievement, but she realized that the draft of young men into the armed forces made the college more willing to admit women. Here, she met Aaron Yalow, another new graduate student in physics, and married him in 1943. Even though Yalow earned As in most of her courses, the chairman of the physics department told her than her A- in one laboratory course proved "women do not do well at laboratory work." She was awarded her M.S. in physics in 1942 and her Ph.D. in nuclear physics in 1945.

From 1945 to 1950, Yalow was a physics teacher and researcher at Hunter College. She also worked as a part-time consultant to the Radiotherapy Service at the Bronx Veterans Administration (VA) Hospital from 1947 until 1950, when she established a radioisotope laboratory at the Bronx VA Hospital. (She conducted full-time research here until 1980.) Soon after she founded the laboratory, she began a fruitful collaboration with Solomon A. Berson, who was a resident in internal medicine at the VA Hospital. Yalow and Berson worked together until his death in 1972. At Yalow's request, her laboratory was later designated the Solomon A. Berson Research Laboratory.

Their first investigations were in the application of radioisotopes to blood volume determination, clinical diagnosis of thyroid diseases, and the kinetics of iodine metabolism. They extended these techniques to studies of the distribution of globin, serum proteins, and small peptide hormones, such as insulin. Yalow and Berson used insulin that was tagged with radioactive iodine and injected it into diabetic and normal subjects. After discovering that insulin-treated patients developed antibodies to animal insulins, they realized that they had a tool that could potentially be used to measure insulin in the blood. Eventually, the concept was transformed into a practical methodology for the measurement of plasma insulin. Yalow and Berson named their technique, which they developed in 1959, the radioimmunoassay (RIA). The RIA made possible the measurement of virtually any substance of biological interest, such as antibodies, enzymes, hormones, or drugs, in blood and body tissues. (Often, these important factors are present in amounts that are too small for conventional direct measurements.) The RIA was eventually turned into a diagnostic kit that is routinely used to measure hundreds of substances of biological interest.

LOIS N. MAGNER

Charles Yanofsky
1925-
American Geneticist and Microbiologist

Charles Yanofsky's most significant contributions to genetics and biochemistry developed from his studies of the genetics and biochemistry of tryptophan synthetase. Yanofsky's pioneering investigation of tryptophan synthetase was the first to demonstrate than an enzyme could contain two dissimilar subunits. *Eschericia coli* tryptophan synthetase catalyzes the final two sequential reactions in the biosynthesis of tryptophan. Yanofsky's work on the relationship between the genes controlling the enzyme and the synthesis and regulation of the enzyme contributed to a more sophisticated version of the "one gene—one enzyme" concept advanced by George Beadle (1903-1989) and Edward Tatum (1909-1975).

Yanofsky was born in New York City. He received his B.S. degree, with a major in biochemistry, from the City College of New York in 1948. He earned his M.S. and Ph.D. degrees in microbiology from Yale University in 1950 and 1951, respectively. From 1944 to 1946, he served with the Armed Forces of the United States. After spending two years as a research assistant in microbiology (1951-1953), he became Assistant Professor of Microbiology at Western Reserve University Medical School (1954-1958). In 1958 he accepted a professorship in the Department of Biological Sciences at Stanford University. In 1967 he was appointed Herzstein Professor of Biology. He was elected to the American Academy of Arts and Sciences in 1964, and the National Academy of Sciences in 1966.

One of the first biosynthetic pathways to be thoroughly elucidated by biochemical and genetic analyses in *Neurospora crassa* was the one leading to the amino acid tryptophan. Further research with typtophan-requiring mutants of *Eschericia coli* and *Salmonella typhimurium* confirmed the findings in *Neurospora*. Stimulated by the work of Beadle and Tatum, Yanofsky's advisor at Yale, David Bonner, attempted to investigate the relationship between genes and enzymes by examining the enzymes of *Neurospora* that appeared to be defective or missing in specific mutants. By the 1950s members of Bonner's group had chosen enzymes in *Neurospora* or *E. coli* for further enzymatic and genetic analyses, hoping to reveal the structural relationship between gene and protein. Because of his previous research experience, Yanofsky chose tryptophan synthetase. Work with this complex enzyme would provide valuable insights into the structural relationship between genes and enzymes, including such specific aspects as suppression, reaction mechanisms, active sites, protein folding, and the variability of enzymes from different microbial species.

In 1954 Yanofsky and his colleagues unequivocally proved that tryptophan synthetase in *Eschericia coli* consisted of two separable protein subunits. Yanofsky's group also determined the relationship between the protein subunits and the series of reactions catalyzed by the intact protein, the ability of the subunits to aggregate, and the location of the active sits for substrates.

In the 1980s Yanofsky carried out a series of experiments that illuminated the phenomenon of attenuation in the control of bacterial operons concerned with the biosynthesis of amino acids. According to these experiments, the operons in the bacterial chromosome that are responsible for the biosynthesis of amino acids contain a site called the attenuator. The translation product of the initial segment of these operons is a peptide that is rich in the amino acid whose synthesis is controlled by that operon. When the supply of that amino acid was very low, translation at the relevant codons of the transcript was inhibited. This process allowed RNA polymerase to proceed through a site that would terminate transcription when the supply of the amino acid in question is high. Attenuation provided a new mechanism for the regulation of gene expression based on the selective reduction of the transcription of distal portions of an operon. Yanofsky explained that the existence of two mechanisms—the repression system and attenuation—for regulating transcription of the tryptophan operon might be explained in terms of the various metabolic reactions that are involved in the biosynthesis and utilization of tryptophan. According to Yanofsky, the combination of the two regulatory mechanisms allowed the bacterium to recognize and respond efficiently to both external and internal events. Studies of tryptophan synthetase were so fruitful that Yanofsky has called it a "charmed enzyme."

LOIS N. MAGNER

Biographical Mentions

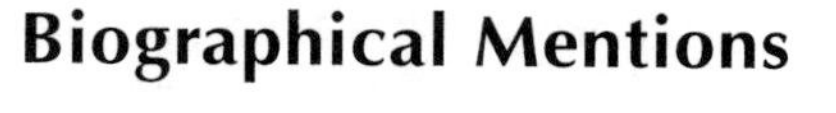

Jon E. Ahlquist

American ornithologist who used DNA to measure genetic similarities among different species of birds. In 1972 Ahlquist and Charles Silbey found that DNA hybridization applied to the classifications of birds. After studying over 1,700 species, they published *Phylogeny and Classification of Birds,* in which they introduced a revised taxonomy of the world's birds and discussed their evolutionary relationships. Ahlquist and Sibley received the Daniel Giraud Elliott Medal from the National Academy of Science in 1988 for their work.

Christian Boehmer Anfinson
1916-1995

American biochemist whose research into the properties of enzymes was recognized with the Nobel Prize in chemistry in 1972. Anfinson conducted and collaborated on research into many aspects of enzyme chemistry, seeking to learn more about how these vital molecules catalyze and mediate biological functions. Without enzymes, it is likely that nonbacterial life could not exist at all, and most bacterial life would cease as well. Anfinson's work led to a greater understanding of their importance and properties.

Werner Arber
1929-

Swiss microbiologist whose discovery that enzymes break large pieces of DNA into smaller, manageable pieces led to a revolution in genetics research. Building on Salvatore Luria's findings that bateriophages (viruses that infect bacteria) undergo changes themselves, Arber found that restriction enzymes helped bacteria defend themselves by cutting the DNA into pieces. For his work in molecular genetics, Arber received the 1978 Nobel Prize for physiology or medicine, which he shared with Daniel Nathans and Hamilton Smith.

Julius Axelrod
1912-

American biochemist and pharmacologist whose work with neurotransmitters provided important insight for the understanding of the nervous system. Building on the discoveries that chemicals like noradrenaline transmit nerve impulses, Axelrod found that certain enzymes could neutralize these neurotransmitters. This discovery showed how certain drugs may affect conditions like hypertension and schizophrenia. Axelrod served as the chief of pharmacology at the National Institute of Mental Health. He received the 1970 Nobel Prize for physiology or medicine, which he shared with Bernard Katz and Ulf von Euler.

David Baltimore
1938

American virologist whose research led to an understanding of the role of viruses in the development of cancer. While working on tumor-causing RNA viruses (now called retroviruses), Baltimore found an unusual enzyme, called reverse transcriptase, that copies DNA. In this way, viral DNA invades the cell, leading to the development of cancer. Baltimore received the 1975 Nobel Prize for physiology in medicine.

Georg von Békésy
1899-1972

Hungarian-born American physicist whose discovery of how sound waves affect the cochlea, a part of the inner ear, led to greater understanding of the ear and sensory perception. In 1947 Békésy took a position at the Harvard psycho-acoustic laboratory, where he developed a mechanical model of the ear that showed that cilia lining the basilar membrane in the cochlea functions as a receptor of pitch and loudness. He received the 1961 Nobel Prize for physiology or medicine.

Paul Berg
1926-

American biochemist who developed the technique of splicing DNA from different organisms and recombining it in a separate hybrid. Recombinant DNA technology became a fundamental advance in genetic research, giving scientists a valuable tool for studying chromosomes and genetic diseases. In 1985 Berg became the director of the Beckman Center for Molecular and Genetic Medicine. He received the 1980 Nobel Prize for chemistry, shared with Walter Gilbert and Frederick Sanger, for his recombinant DNA research.

Elizabeth Blackburn
1948-

American molecular biologist who has been a leader in the study of telomeres and telomerase. Telomeres are specialized structures containing short, repeating nucleotide sequences and are found at the ends of eukaryotic chromosomes. Telomeres seem to protect the integrity of the chromosome, but normal somatic cells lose portions of the ends of the telomeres during cell division. Eventually, the unprotected chromosomes become unstable and the cells die. These observations support the telomere theory of aging. In 1985 Blackburn and Carol Greider discovered the enzyme telomerase, an unusual RNA-containing DNA polymerase that can add to the ends of chromosomal DNA. Most normal somatic cells lack telomerase, but cancer cells display telomerase activity, which might explain their ability to multiply indefinitely.

Gunther Blobel
1936-

German-born naturalized-American biochemist who won a Nobel Prize in 1999 for discovering that proteins have intrinsic signals that govern their transport and localization within the cell. Blobel found that when proteins are first synthesized within the cell they carry signals, rather like ZIP codes, that indicate whether they should be exported from the cell or take their place in a specific compartment of the cell. Errors in these special codes may be involved in diseases like cystic fibrosis.

Konrad Emil Bloch
1912-

German-born American biochemist who received the 1964 Nobel Prize for physiology or medicine, shared with Feodor Lynen, for research into the metabolism of cholesterol and fatty acids. Early in his scientific studies Bloch found that acetic acid is a major precursor to cholesterol, a molecule that has 27 carbons in its base. Cholesterol, an essential compound in all animal cells, is necessary for the formation of bile and hormones, such as cortisone and the sex hormones. Using radioactive carbon-14, Block traced the process through which acetic acid is transformed by the body into cholesterol.

Floyd Elliott Bloom
1936-

American neurobiologist noted for his investigations of neuron structure and function, and for his studies concerning the effects of drugs on neural activity. Bloom received his doctor of medicine from Washington University in 1960. He is the author of various technical publications and books, and has received numerous awards. Since 1983 he has worked at the Scripps Research Institute in La Jolla, California.

Walter Fred Bodmer
1936-

German-born English geneticist noted for his research into the genetics and biology of colon and rectal cancer. Bodmer has also worked on the human immune response to cancer and on human genome analysis. He has been active in many cancer research societies, including the Imperial Cancer Research Fund, London, for which he served as its director beginning in 1991.

James Frederick Bonner
1910-1996

American geneticist who discovered how histones control gene activity. Attempting to understand chromosome control, Bonner isolated genes and found that they could be turned on or off by the action by a protein histone. Bonner received his Ph.D. in plant physiology and genetics from California Institute of Technology in 1934, where he taught for more than four decades. He contributed to several scientific fields, including 500 publications and 10 books in three dozen fields of inquiry.

Sydney Brenner
1927-

South African geneticist responsible for many significant advances in genetic research. Brenner confirmed in 1964 that genetic information is stored in sequence along chromosomes, and that the order corresponds to the sequence of amino acids that make up a protein. In 1967 he and Francis Crick decoded the DNA "stop" codon, one of the last to be deciphered. Later, Brenner and his colleagues set out to learn everything possible about an entire organism, *C. elegans* (a small nematode), setting the stage for the Human Genome Project.

Roger Cecil Burgus
1934-

American biochemist who with Nobel laureate Roger Guillemin isolated a hypothalamic hormone known as thyrotropin hormone releasing factor. This molecule regulates the function of the thyroid gland through its effect on the pituitary gland in the human brain. He received his undergraduate, master's, (1960) and Ph.D. (1962) degrees from Iowa State University. He has taught and conducted research at Iowa State University, Baylor University, Salk Institute, and Oral Roberts School of Medicine.

Melvin Calvin
1911-1997

American biochemist who won the 1961 Nobel Prize for chemistry for elucidating the process of photosynthesis in plants. Calvin began work on photosynthesis in the 1940s, studying algae to determine the biochemical interactions by which water and carbon dioxide are transformed into carbohydrates and oxygen. By using the radioactive isotope carbon-14, he was able to identify the reactions involved in the different stages of photosynthesis.

Timothy M. Casey
1946-

American ecologist whose work in comparative physiological ecology has given scientists new insight into evolution. He measured the energy expenditure and muscle function of insects during flight and compared this data to that of powered aircraft. This interdisciplinary approach to insect form and function may assist scientists and agriculturists in reducing economical losses due to migrating flying insects.

Thomas R. Cech
1947-

American chemist whose work on the structure and function of chromosomes was recognized by a number of prizes, including the 1989 Nobel Prize in Chemistry, which he shared with Sidney Altman. Cech's studies of RNA and nuclear enzymes helped to elucidate the process by which chromosomes are manipulated so that genetic information can be read, copied, and used by cells.

Martha Cowles Chase
1927-

American scientist who, with Alfred Hershey (1908–) made important discoveries about the nature of DNA, paving the way for modern molecular biology. In 1952 Chase and Hershey proved that viral DNA, not the protein capsule (called the capsid) that surrounds it, carries the genetic code in all living things. Their discovery spurred more research in the field, and was followed one year later by Watson and Crick's discovery of the double-helix structure of DNA.

Albert Claude
1898-1983

American cell biologist who received the 1974 Nobel Prize for Physiology or Medicine, shared with George Palade and Christian de Duve, for research toward defining the specialized functions of various cell structures. Claude developed a method for separating cell components into different sizes, shapes, and densities using a centrifuge, a device that spins at high speeds, distributing the cell material in layers. Claude pioneered the use the electron microscope. He also discovered the endoplastic reticulum, a structure that plays a part in the formation and transport of fats and protein, and the mitochondria, the center of the cell's energy production.

Stanley Cohen
1922-

American biochemist who shared the Nobel Prize in 1986 with Rita Levi-Montalcini for their work on epidermal growth factors. Cohen and Montalcini demonstrated that a factor released by certain mouse tumors stimulated the growth of specific tissues in chick embryos. Cohen later discovered a similar factor in extracts of mouse submaxillary glands. The term "epidermal growth factor" was first used in the 1960s to describe the origins and physiology of factors that altered the timing of specific developmental processes.

Elias James Corey
1928-

American chemist who was awarded the 1990 Nobel Prize for Chemistry for his work on retrosynthetic analysis, a technique for simplifying the synthesis of large complex molecules. Corey successfully synthesized many organic substances previously found only in nature. He developed the synthetic prostaglandins, a hormone-like compound used to treat infertility and induce labor, and ginkgolide B, an active chemical compound used to treat asthma and circulatory problems. The synthesis of these new biologically active molecules has contributed to understanding how diseases develop at the molecular level.

Christian René de Duve
1917-

Belgian biochemist and cell biologist who shared the 1974 Nobel Prize with Albert Claude and George E. Palade for their independent discoveries concerning the microscopic structural and functional organization of the cell. Duve identified the lysosome, a subcellular entity that contains digestive enzymes, and the peroxisome, the site of multiple oxidation reactions. Duve suggested that various diseases could be traced to defects in these cellular particles.

Johann Deisenhofer
1943-

German chemist who received the 1988 Nobel Prize for Chemistry, which he shared with Robert Huber and Hartmut Michel. Deisenhofer contributed to the understanding of photosynthesis, the transformation of sunlight into energy in plants. Using x-ray crystallography and advanced mathematical calculations to interpret the x-ray pattern, Deisenhofer, Huber, and Michel detailed the complete atomic structure of the photosynthetic reaction center. Their work provided new insight into the process of photosynthesis and revitalized its study.

Vincent Du Vigneaud
1901-1978

American biochemist who won the 1955 Nobel Prize in Chemistry for synthesizing the first hormone, oxytocin. Du Vigneaud also deciphered the chemical structure of biotin, one of the B vitamins, making synthesis possible. His work on vitamin and hormone synthesis presaged the now common and widespread use of such compounds. In addition to his work with oxytocin and biotin, du Vigneaud conducted important research into other compounds, including insulin and penicillin.

John Carew Eccles
1903-1997

Australian physiologist who demonstrated the electrical basis of neural inhibition in the polarization of the nerve cell membrane. In the 1950s Eccles described excitatory postsynaptic potential and inverse inhibitory reaction. Eccles believed that inhibition was essential in directing and maintaining neural function. His investigations of the patterns and organization of neural pathways led to his 1964 identification of synaptic inhibitory neurons in the brain. Sir John Eccles shared the 1963 Nobel Prize with Alan Hodgkin and Andrew Huxley.

Gerald Maurice Edelman
1929-

American biochemist whose contributions in understanding the chemical structure of antibodies earned him the Nobel Prize for medicine in 1972 with Rodney Porter. He received his medical degree in 1954 and his Ph.D. in 1960 from Rockefeller University. His Nobel Prize research focused on mapping the antibody molecule, discovering that it consisted of 1,300 molecules. His later research focused on morphogenesis—the formation of and differentiation of tissues and organs.

Anne Ehrlich
1935-

American biologist who has collaborated with her husband, Paul Ehrlich, on human ecological policy research. They are best known for their efforts to focus public attention on the connection between human population, resource exploitation, and the environment. Their research has focused on deforestation, over-fishing, global warming, depletion of topsoil, groundwater, and clean air and other environmental issues. She has co-authored with her husband *Extinc-*

tion, The Stork and the Plow, and *Betrayal of Science and Reason.*

Paul R. Ehrlich
1932-

American biologist who received the Crafoord Prize (awarded by the Swedish Academy of Sciences to support areas of science not covered by the Nobel Prize) for his work on the problems of population growth, resource exploitation, and the environment. Ehrlich received his Ph.D. in zoology from the University of Kansas. He joined the faculty of Stanford University in 1957 and specialized in entomology. His book, *The Population Bomb* (1968) brought overpopulation issues into the public arena.

Charles Ellington

British zoologist who made groundbreaking discoveries related to winged insects. In 1994 Ellington and his colleagues constructed an insect model—fairly large and a slow-motion type—to use in wind tunnel tests. They were rewarded when the tests revealed a microscale vortex that adhered to the wing's leading edge. The resulting swirling provided low pressure over the wings and an extraordinary amount of lift, helping to explain how winged insects can fly. Ellington has long taught at Cambridge University in England.

Heinz Fraenkel-Conrat
1910-1999

German-American biochemist whose research helped to reveal the structural components of the virus. In a series of experiments with the tobacco mosaic virus, he was able to disassemble the virus into its infectious and noninfectious parts and reveal how the infectious process functions in the cell. He received his medical degree in 1934 and his Ph.D. in biochemistry in 1936 from the University of Edinburgh. After becoming a U.S. citizen, he joined the faculty at University of California at Berkeley in 1958.

Peter Funch
1965-

Danish zoologist whose work in arctic biology has contributed toward understanding the coevolutionary nature of lifecycles and morphology of animals in arctic ecosystems. Funch's extensive research in Greenland led to the discovery of *Symbion pandora,* a new life form and new phylum. He has also pioneered new techniques of light microscopy and scanning electromicroscopy that have furthered understanding of the molecular level of arctic meiofauna.

Martin Gellert

American biologist whose work focuses on the molecular biology of the lymphatic and immune system, specifically T-cell genetics and immunoglobbins. The behavior of DNA in recombination is a primary research interest. He began his academic career at Harvard, where he received his Bachelor of Arts degree in 1950. He received his Ph.D. in molecular biology at Columbia University in 1956. He later became Metabolic Enzymes Section Chief of the Laboratory of Molecular Biology with the National Institutes of Health in Bethesda, Maryland.

Walter Gilbert
1932-

American molecular biologist who isolated the lactose repressor in *Escherichia coli,* the first time such a genetic control component had been identified. In 1980 he shared the Nobel Prize in chemistry with Frederick Sanger and Paul Berg for contributions concerning the determination of base sequences in nucleic acids. The sequencing techniques developed by Gilbert and Sanger, and the gene-splitting techniques of Berg, provided the basis for genetic engineering. In 1978 Gilbert helped found Biogen, a pioneering biotechnology firm, and served as its chairman and CEO until 1984.

Ragnar Arthur Granit
1900-1991

Finnish-born Swedish physiologist who was a corecipient (with George Wald and Haldan Hartline) of the 1967 Nobel Prize in medicine for their work on the biochemical and electrical processes of vision. Later in his life, his research focused on movement, specifically the role of muscle spindles and tendon organs. This work helped determine the neural pathways and processes in muscles. He received his M.D. degree in 1927 from the University of Helsinki.

Henry Harris
1925-

Australian cell biologist known for his work on true cell hybrids, in which the nuclei of the parent cells coalesce. This work brought him worldwide acclaim and sparked productive research in genetic characteristics of various cell types in Europe and the United States. His work on cell biology led to his appointment as Director of Research of the British Empire Cancer Campaign at the Sir William Dunn School of Pathology, Oxford.

Stephen Coplan Harrison
1946-

American biochemist and molecular biologist who was the first to work out the atom-by-atom structure of a virus. He also mapped the makeup of the receptor used by the AIDS virus to attach to human cells using x-ray crystallography. The technique utilizes purifying a virus, crystallizing it, and scanning it with x rays. He received his Ph.D. in 1967 and produced his first three-dimensional image of a virus in 1977. In 1997 he won the ICN International Prize in virology.

Haldan Keffer Hartline
1903-1983

American biophysicist who was corecipient (with Ragnar Granit and George Wald) of the 1967 Nobel Prize in medicine for their work on the biochemical and electrical processes of vision. Hartline was the first to determine the mechanics of how the visual system nerves receive information and transfer it to the brain. His work contributed to advances in night-vision, pattern recognition, and motion-detection devices. He earned his M.D. in 1927 from Johns Hopkins University.

Alfred Hershey
1908-1997

American geneticist and a founding member of the "phage group" who helped to establish that DNA carries the genetic code in all living things. In their famous "blender experiments" on radioactively labeled bacteriophage virus in 1952, Hershey and Martha Chase showed that viral DNA, and not the protein coat that surrounds it, contains genetic information. Hershey shared the 1969 Nobel Prize with Max Delbrück and Salvador Luria.

Mahlon Bush Hoagland
1921-

American biochemist who was the first to isolate transfer RNA, which plays an essential part in intracellular protein synthesis. In the late 1950s he isolated various types of RNA molecules from cytoplasm and demonstrated that each type can combine with only one specific amino acid. He obtained his M.D. degree in 1948 from Harvard University Medical School and conducted research as a fellow of the American Cancer Society, Dartmouth Medical School, and the Worcester Foundation for Experimental Biology.

Evelyn Hooker
1907-1996

American psychologist known for her 1950s' research showing that homosexuality is not a mental illness. In a highly controversial report in 1957, she challenged the then-prevailing beliefs about homosexuality. In 1973 the American Psychiatric Association removed homosexuality from its Diagnostic and Statistical Manual of Psychiatric Disorders. She earned her bachelor and master's degrees in psychology at the University of Colorado. She received her Ph.D. from Johns Hopkins University.

David Hunter Hubel
1926-

American neurobiologist who shared the 1981 Nobel Prize in Medicine with Torsten Wiesel and Roger Sperry. Hubel and Wiesel, colleagues at Johns Hopkins University, studied the mechanisms of the brain's visual cortex. In experiments with cats and monkeys, they were able to trace the exact path of amino acids in the brain, uncovering the transmission route for nerve impulses. They found that the visual cortex contains a variety of cells, ranging from simple, complex, and hypercomplex, arranged in vertical columns that allow for sight in animals and humans.

Robert Huber
1937-

German chemist and corecipient of the 1988 Nobel Prize in chemistry with Hartmut Michel and Johann Deisenhofer for their work in photosynthesis, the chemical process that turns sunlight into energy in plants. Huber's laboratory pioneered a technique using x-ray crystallography to determine the molecular structure of membrane-bound proteins and the process of photosynthesis. Huber received his Ph.D. degree in 1963 from Technical University in Munich, Germany, and later headed the Max Planck Institute for Biochemistry.

John Hughes
1942-

British biochemist who has devoted his academic life to the study of pharmacology, paying special interest to naturally compounds. Although his initial research centered on opiate-like chemicals, he has since centered his efforts on understanding the pharmacology biochemistry of naturally occurring neuroactive substances. Hughes received undergraduate and postgraduate degrees at the University of London but traveled to the United States to earn a Ph.D at Yale University. Hughes has been Director of the Parke-Davis Research Unit at Cambridge University, Senior Research Fellow at Wolfson College, and Honorary Professor of Neuropharmacology at Cambridge.

Libbie Henrietta Hyman
1888-1969

American zoologist who authored several widely used texts and reference works on invertebrate and vertebrate zoology during the 1920s and 1930s. Hyman received her doctorate in 1915 from the University of Chicago, where she worked for zoologist Charles Manning Child. Hyman's area of expertise was invertebrate zoology, and she became an authority on invertebrate taxonomy. She held an honorary research appointment at the American Museum of Natural History in New York City, and was highly regarded for her encyclopedic knowledge and elegant writing.

François Jacob
1920-

French biochemist who won the 1965 Nobel Prize for contributing to discoveries about the genetic regulation of enzymes. With Elie Wollman he proved that bacteria are able to exchange genetic material during a sexual reproductive process called conjugation. This discovery allowed the mapping of the bacterial chromosome. Work on bacteriophage viruses led to the development of the operon theory of genetic control, which predicted the existence of messenger RNA and the role of repressors in gene expression.

Sir Bernard Katz
1911-

German-born British biophysicist and corecipient of the 1970 Nobel Prize in medicine with Ulf von Euler and Julius Axelrod for their work identifying the activation, inactivation, and storage of neurotransmitters, chemicals that stimulate nerve and muscle cells. His research helped explain how the nervous system operates, because the storage, release, and inactivation of acetylcholine was determined to be identical to other neurotransmitters. Katz earned his M.D. degree in 1934 and his Ph.D. in physiology from the University of London in 1938.

Sir John Cowdery Kendrew
1917-1997

British molecular biochemist and corecipient with Max Perutz of the 1962 Nobel Prize in chemistry for their work in determining the structure of the protein molecule. Protein molecules are essential parts of all living cells. Kendrew's research pioneered the use of x-ray crystallography to determine the structure of proteins and their complex amino acid components. He received his Ph.D. in biochemistry in 1948 from Cavendish Laboratory in Cambridge, England.

Henry Bernard David Kettlewell
1907-1979

English geneticist known for his work on melanism—the occurrence of dark pigments in isolated groups of moths. Always an outstanding student, Kettlewell studied medicine and received his degrees at Gonville and Caius College at Cambridge University and St. Bartholomew's Hospital in London. Although Kettlewell practiced medicine during World War II, he moved on to other scholastic areas, holding positions in the genetics section at Oxford's zoology department. Acknowledged by his peers to be an exceptional naturalist, he devoted much of his time to the study of melanism. His attention was held by noticing that the melanism occurred primarily in industrial areas where pollution and atmospheric carbon were prevalent.

Georges J. F. Köhler
1946-1996

German immunologist and corecipient of the 1984 Nobel Prize in medicine with Niels Jerne and César Milstein for their work with antibodies and how they interact with the immune system. They developed a technique for producing monolocal antibodies that are used in medical and biological research throughout the world. He received his Ph.D. in biology from the University of Freiburg in 1974 and has served as director of the Max Planck Institute.

Arthur Kornberg
1918-

American biochemist and corecipient with Severo Ochoa of the 1959 Nobel Prize in medicine for artificially producing a chemically exact but inert molecule of deoxyribonucleic acid (DNA), a basic component of genes. In 1967 at Stanford University, he headed a team that built on the Nobel Prize work by synthesizing DNA in a biologically active state. He was educated at the University of Rochester, receiving his Ph.D. there in 1943.

Edwin G. Krebs
1918-

American biochemist who shared the 1992 Nobel Prize in medicine with Edmond H. Fischer for their discovery of how the body breaks glycogen down into glucose. This process, called reversible phosphorylation, activates and regulates various cellular processes. This discovery contributed to techniques that prevent the body from rejecting transplanted organs as well as links to investigative work in cancer, blood pressure, and inflammatory reactions research.

Reinhardt Kristensen

Danish zoologist whose research is focused on the phylogeny and systematics, or relationships, of controversial groups of microscopic animals called the tardigrades (water bears), lociferans, and cyliophorans. His groundbreaking work on these little-understood groups continues with numerous investigations from around the world. Kristensen is Professor and Curator of the Invertebrate Department of the Zoology Museum of the University of Copenhagen, Denmark.

Richard Erskine Leakey
1944-

American paleoanthropologist whose research has contributed to the study of human evolution. Leakey studied the fossilized remains of extinct human-like creatures called hominids, some of which are thought to be the ancestors of modern humans. He discovered the earliest skull of *Australopithecus robustus,* as well as a complete skeleton of "Turkana Boy," thought to be nearly 1.6 million years old. Leakey is also credited with finding the 17 million-year-old jaw, teeth, and skull fragments of *Sivapithecus,* a possible ancestor of both apes and humans.

Jérôme Jean Louis Marie Lejeune
1926-1994

French geneticist who discovered the chromosomal abnormality linked to Down's syndrome. During the 1950s, while at the National Center for Scientific Research (CNRS) in Paris, Lejeune conducted research concerning genetic predisposition for Down's syndrome, a congenital form of mental retardation. In 1959 he discovered that normal children were born with 46 chromosomes, or 23 pairs of chromosomes, while children with Down's syndrome exhibited an extra chromosome, which made one of the pairs a triplet. Lejeune was appointed director of research at the CNRS in 1963, and professor of fundamental genetics at the Faculty of Medicine in Paris in 1964.

Choh Hao Li
1913-

Chinese-American endocrinologist who was the first to synthesize human growth hormone. Li was also the first to determine the exact amino acid sequence and construction of the hormone ACTH, and the first to isolate human growth hormone. Because of the crucial regulatory functions played by hormones, these discoveries were of great importance to the fields of medicine in general and endocrinology (the study of the hormone-producing glands) in particular.

Konrad Lorenz
1903-1989

Austrian zoologist who was a cofounder of ethology, the science of animal behavior. He received a medical degree in 1928 and a Ph.D. in zoology in 1933 from the University of Vienna. Thereafter he studied instinctive animal behavior in natural settings. In 1963 he wrote a book on human aggression as an instinctual behavior. He was corecipient of the 1973 Nobel Prize in medicine for his work in ethology.

Andre Michel Lwoff
1902-1994

French microbiologist who was awarded the 1965 Nobel Prize for his contributions to discoveries concerning the genetic regulation of enzymes and viral replication. His research included studies of protozoa, patterns of biochemical evolution, growth factors in microorganisms, and the cancer-causing genes known as oncogenes. His studies of lysogenic bacteria, viruses, and his demonstration of the induction of the prophage provided the foundations for experiments in molecular biology.

Feodor Felix Konrad Lynen
1911-1979

German biochemist who specialized in the study of lipid metabolism. He shared the 1964 Nobel Prize with Konrad Bloch for discoveries concerning the mechanism and regulation of cholesterol and fatty acid metabolism. Lynen carried out research on the enzymatic steps of the fatty acid cycle, the role of biotin in lipid metabolism, fermentation, biological phosphorylation and oxidation, and the regulatory mechanisms of metabolic pathways. His work contributed to studies of the relationship between cholesterol and disease.

Robert Helmer MacArthur

Canadian-born American ecologist who developed quantitative models in ecology— measuring relationships between the number of species and the size of habitat. Born in Toronto, Canada, MacArthur moved to the United States when he was 17 to study mathematics. While working toward his Ph.D at Yale University, he made an unusual choice by changing his major to zoology. He later became professor of biology at Princeton in 1965.

Hartmut Michel
1948-

German biochemist who received the 1988 Nobel Prize for Chemistry, which he shared with Johann Deisenhofer and Robert Huber. Together, they determined the structure of certain proteins necessary for photosynthesis (the process by which plants convert sunlight into energy). From 1978-82 Michel documented the three-dimensional structure of a four-protein complex, called a photosynthetic reaction center, necessary for photosynthesis in certain bacteria. Michel was able to crystallize the membrane-bound protein complex into a pure crystalline form, making it possible to determine the protein structure atom-by-atom by means of x-ray diffraction.

Jacques F. A. Miller
1931-

French-Australian physician who discovered that the thymus gland is a part of the immune system. Miller made this discovery by removing the thymus in baby mice, subsequently noting that the mice failed to develop either white blood cells or lymph nodes. He also noted that the mice would then accept grafts that would otherwise have been rejected. This discovery helped advance understanding of the immune system greatly.

César Milstein
1927-

Argentinean molecular biologist who shared the 1984 Nobel Prize with Georges Köhler and Niels K. Jerne for their contributions to immunology. Their research led to the successful hybridoma technique for the production of pure antibodies in virtually unlimited amounts. The hybridoma technique fuses myeloma cells and sensitized lympocytes to establish a line of antibody-producing cells. In 1975 Milstein and Köhler predicted that hybridoma cultures would be used to provide "designer antibodies" for medical and industrial purposes.

Stanford Moore
1913-1982

American biochemist who received the 1972 Nobel Prize for Chemistry, along with William Stein and Christian Anfinsen, for his research on the structure of protein molecules. Moore and Stein developed an efficient way of analyzing each of the 20 common amino acids that make up proteins. Using a process called chromatography, they were able to determine the exact arrangements of all 124 amino acids in a molecule of ribonuclease, an enzyme important in biochemical reactions.

Andrew W. Murray

American biochemist who, with Jack Szostak, developed the first artificial chromosome in 1983. Murray and Szostak worked with yeast chromosomes, inserting a number of genes from another organism into the yeast chromosome to create a Yeast Artificial Chromosome (YAC). So-called "YAC libraries" are used to catalog the genetic information from a number of organisms, helping to organize the vast amount of genetic information that is discovered yearly.

Daniel Nathans
1928-1999

American molecular biologist who was corecipient of the 1978 Nobel Prize in medicine with Hamilton Smith and Werner Arber for the discovery and use of restriction enzymes. Restriction enzymes are proteins that cut DNA chains. Nathans's work allowed researchers to modify the DNA molecule and paved the way for new innovations in biotechnology. Biotechnology companies use his work to make synthetic versions of insulin and other hormones. He received his M.D. degree from the University of Delaware in 1954.

Christiane Nüsslein-Volhard
1942-

German geneticist who received the 1995 Nobel Prize in medicine with Edward Lewis and Eric Wieschaus for their groundbreaking research into the genetic blueprint of the fruit fly. This research may help explain birth defects and miscarriages in humans. By raising and studying the fruit fly, she identified specific genes for specific functions that correlated with human embryonic development. She received her Ph.D. in 1962 and later conducted research at the Max Planck Institute in Germany.

Severo Ochoa
1905-1993

Spanish-American biochemist who shared the 1959 Nobel Prize with Arthur Kornberg for their discovery of the mechanism of the biological synthesis of ribonucleic acid (RNA). Ochoa's American research career began in the laboratory of Carl and Gerty Cori. His studies of high-energy phosphates and the enzymatic processes involved in biological oxidation and the transfer of energy led to the discovery of the enzyme polynucleotide phosphorylase, which catalyzes the synthesis of RNA.

George Emil Palade
1912-

Romanian cell biologist who shared the 1974 Nobel Prize with Albert Claude and Christian R. de Duvé for their research on the fine microscopic structure and functions of the cells. Palade significantly improved Claude's methods of centrifugal fractionation and electron microscopy. By elucidating the modern map of the cell, Palade established the foundations of current cell biology. Palade provided a pioneering description of the mechanism of protein synthesis and the role of the subcellular particles known as ribosomes.

Candace Pert
1946-

American biochemist regarded as a codiscoverer of endorphins, which are peptide neurotransmitters that serve as natural opioids. Working with Solomon Snyder in 1973 she demonstrated the existence of opiate receptors in the brain. Pert contends that neuropeptides and their receptors are the biochemicals of emotions, transmitting information through a network that links the material world of molecules with the nonmaterial world of the psyche. Her book *Molecules of Emotion: Why You Feel the Way You Feel* (1997) illuminates the "mind-body-spirit revolution" in modern medical science.

Max Ferdinand Perutz
1914-

Austrian-born British biochemist who shared the 1962 Nobel Prize in chemistry with John Kendrew for their studies of the structure of the globular proteins hemoglobin and myoglobin. Perutz demonstrated that it was possible to determine the structure of proteins by x-ray diffraction, a technique previously used with much smaller molecules. The key to solving the structure of such a large molecule was developed in 1953 when Perutz succeeded in incorporating mercury atoms into specific positions in the hemoglobin molecule.

Rodney Robert Porter
1917-1985

British biochemist who received the 1972 Nobel Prize for Physiology or Medicine, which he shared with Gerald Edelman, for research concerning the chemical structure of antibodies. Antibodies (or immunoglobulins) are proteins in the blood that help to eliminate antigens, foreign substances such as bacteria and toxins. The antibody-antigen reaction is the primary mechanism in the body's immune system. The work of Porter and Edelman contributed toward greater understanding of how the body defends itself against millions of diverse antigens.

Mark Steven Ptashne
1940-

American molecular biologist who was the first to identify and describe the operation of repressor genes in an organism. This discovery, made simultaneously and independently of Walter Gilbert's work, helped to shed additional light on the workings of the genome. Ptashne demonstrated, for example, that an enzyme responsible for digesting lactose is turned off by a repressor gene when lactose is absent from the cell. This process is reversible when lactose exists in the cell.

Richard John Roberts
1943

British molecular biologist who received the 1993 Nobel Prize for Physiology or Medicine with Phillip A. Sharp for the discovery of "split genes." In 1977 Roberts and a team of researchers discovered that the genes of the adenovirus (which causes the common cold) were discontinuous, and that certain segments did not contain any genetic information. Biologists had previously believed that DNA consisted only of unbroken structures, all of which encoded protein structure. His finding influenced the study of genetic diseases, and gave scientists new clues to how evolution was aided by DNA's ability to generate new genetic combinations.

Martin Rodbell
1925-1998

American biochemist who received the 1994 Nobel Prize for Physiology or Medicine for the discovery of G-proteins, a class of proteins responsible for translating chemical and hormonal messages to cells. In the 1960s, Rodbell discovered that G-proteins, which bind with nucleotides guanosine diphosphate (GDP) and guanosine triphosphate (GTP), allow cells to communicate with one another. Rodbell's research led to great inroads in the understanding of diseases like cholera, diabetes, alcoholism, and cancer. Rodbell shared the Nobel Prize with pharmacologist Alfred Gilman, who later isolated the G-protein and provided its name.

Andrew Victor Schally
1926-

Polish-American endocrinologist who received the 1977 Nobel Prize for Physiology or Medicine, along with Roger Guillemin and Rosalyn Yalow, for his work in isolating and synthesizing the hormones that control other hormone-producing glands. Schally was responsible for synthesizing TRH (thyrotropin-releasing hormone) and isolating LH-RH (luteinizing hormone-releasing hormone), hormones secreted by the hypothalamus that control the activities of other hormone-producing glands.

Phillip Allen Sharp
1944-

American molecular biologist who discovered the presence of split genes. Sharp's studies revealed that genes are comprised of various DNA regions, set apart by introns, areas where no genetic information is stored. It was previously believed that genes consisted of uninterrupted strands of DNA. Introns, which are apparently unused genetic debris, are thought to play a role in some genetic diseases. Sharp was awarded the 1993 Nobel Prize in Medicine, which he shared with Richard Roberts, for his discovery.

Charles G. Silbey

American molecular biologist who during the 1980s adopted a controversial technique for comparing the DNA of different species to document ranges of genetic differences. Silbey applied this technique to the problem of human-chimp-gorilla phylogeny. He reported that humans were, genetically, closer relatives to chimpanzees than to gorillas; it was previously thought that humans were equally related to both. In a series of papers, Silbey claimed that his procedure was genetic, precise, replicable, and quantitative. However, his work was subsequently discredited.

Hamilton Othanel Smith
1931-

American molecular biologist who received the 1978 Nobel Prize for Physiology or Medicine, which he shared with Daniel Nathans and Werner Arber, for the discovery of restriction enzymes. Restriction enzymes are proteins that cut DNA chains and allow for modification of DNA molecules. This breakthrough opened up the field of biotechnology. Biotechnology companies now use hundreds of these restriction enzymes to make synthetic hormones and insulin, forming the basis for gene therapy, and for gene-mapping research, such as conducted by the Human Genome Project.

Solomon Halbert Snyder
1938-

American biologist whose research centered on the investigation of neurotransmitters found in different regions of the brain. His most important research focused on a special enzyme that may regulate RNA synthesis. This is linked to his work on the effects of opiates and similar substances on the brain. In 1963 he was successful in demonstrating that there are specialized targets (receptors) on the cell membranes of nervous tissues. Snyder earned his medical degree at Georgetown University. Trained in both psychiatry and pharmacology, he served his internship at Kaiser Foundation Hospital in San Francisco. During his brilliant career, he eventually held a number of professorships at Johns Hopkins Hospital in Baltimore.

Roger Sperry
1913-1994

American neurologist renowned for his research on the brain. Sperry and his colleagues, David Hubel and Torsten Wiesel, were awarded the 1981 Nobel Prize in Physiology or Medicine for their pioneering work in determining how information and functions are organized within the brain. Their research has contributed toward a better understanding of how the brain works, leading to advances in the treatment of brain injury, brain disease, and similar ailments.

William Howard Stein
1911-1980

American biochemist who received the 1972 Nobel Prize for Chemistry, shared with Stanford Moore and Christian Anfinsen, for research on the complex structure of protein molecules. Stein helped develop an efficient way of analyzing and identifying the 20 common amino acids that make up protein molecules. Using a special technique, chromatography, Stein and Moore were able to determine the exact arrangement of all 124 amino acids in RNA (ribonuclease). Their technique became the basis for modern protein analysis, which has been instrumental for medical advances, genetic engineering, and criminal investigations.

Earl Wilber Sutherland Jr.
1915-1974

American biochemist who won the Nobel Prize in 1971 for discoveries concerned with the operation of hormones. Sutherland discovered cyclic AMP and demonstrated that it converts the inactive form of the enzyme phosphorylase to an active form that converts glycogen to glucose. Because cyclic AMP proved to be essential to the actions of several hormones, and affected numerous cellular processes, Sutherland's work served as a unifying concept for the mechanism of hormone action.

Jack William Szostak
1952-

Canadian biochemist who, in collaboration with Andrew Murray, developed the first synthetic chromosome, the Yeast Artificial Chromosome (YAC). In addition to aiding the organization of genetic information that poured into laboratories during the 1980s and 1990s, the YAC helped researchers develop techniques for manipulating genes. These techniques were used with great success during the subsequent surge of genetic research.

Edward Lawrie Tatum
1909-1975

American microbiologist who shared the 1958 Nobel Prize with George Beadle and Joshua Lederberg for discoveries that demonstrated the relationship between genes and the proteins they controlled. Exploiting the potential of the bread mold *Neurospora* as a genetic and biochemical tool, they established the "one gene, one enzyme" theory. Beadle and Tatum irradiated the mold to create nutritional mutants and identified specific defects in the mutants. The induced mutations exhibited Mendelian patterns of inheritance.

Howard Martin Temin
1934-1994

American biochemist and virologist who shared the 1975 Nobel Prize with Renato Dulbecco and David Baltimore for their discoveries concerning the interaction between tumor viruses and the genetic material of the cell. Temin developed the first reproducible in vitro assay for a tumor virus. His research on Rous sarcoma virus, an RNA virus, led to the formulation of the "provirus hypothesis" and the demonstration of the enzyme known as reverse transcriptase.

Lars Y. Terenius

Swedish neurochemist who was the first to show that certain small peptides, produced naturally in the body, act on the brain's opiate receptors in a manner similar to that of other opiates. This discovery helped elucidate the pharmacological effects of opiates and other painkillers, widely used in the practice of medicine. It also gave important insights into the chemistry of the brain, helping researchers to better understand the origins of some mental disorders resulting from chemical imbalances.

Susumu Tonegawa
1939-

Japanese molecular biologist who made important discoveries relating to the immune system and its genetics. In 1976 Tonegawa reported that the genes responsible for antibody production moved physically closer to one another on their respective chromosomes. This suggested that the immune system is able to produce antibodies against new and novel invaders more quickly by forming efficient recombinations of existing genetic information and patterns. Tonegawa's research provided new insight into the workings of the immune system, for which he was awarded the 1987 Nobel Prize in Physiology or Medicine.

Ageneta Wahlstrom

Swedish biochemist who, with Lars Terenius, performed important research into the chemistry of the brain. Their research had two important effects: it lead to a better understanding of mental illness and contributed toward the design of improved drugs for the purpose of halting or correcting mental illness. Some of the techniques pioneered by Wahlstrom and Terenius have also proven valuable in other studies of the brain and its interaction with various compounds.

Robert A. Weinberg
1942

American biologist and cancer researcher at Massachusetts Institute of Technology who conducted important research involving the genetic manipulation of cells in mice. In 1999 Weinberg created a cancerous human cell by genetically altering three genes in a healthy cell. Oncologists (cancer experts) believe that this breakthrough could advance scientific understanding of the causes of different kinds of cancer, as well as promote additional research into new therapies that target specific genetic problems in humans.

Torsten N. Wiesel
1924-

Swedish neuroscientist who received the 1981 Nobel Prize for Medicine or Physiology, which he shared with Roger Sperry and David Hubel, for research on the organization and local functions of the brain. Their work, which remains of vital importance, provided a better understanding of which areas of the brain are called upon when performing certain mental and physical tasks. This knowledge has led toward advances in the treatment of various problems arising from brain trauma and disease.

Frederick Maurice Hugh Wilkins
1916-

New Zealand-born British molecular biologist who shared the 1962 Nobel Prize with James Watson and Francis Crick for their 1953 discovery of the double helix structure of DNA. Wilkins established methods for the preparation of crystalline DNA fibers and attempted to use x-ray diffraction analysis to determine the structure of DNA. Conflicts between Wilkins and Rosalind Franklin led Wilkins to collaborate with Watson and Crick. Based on Franklin's x-ray photographs and their own model-building work, Watson and Crick proposed the double helix.

Allan Charles Wilson
1934-1991

American biochemist who, with Russell Higuchi, was the first to successfully clone cells from an extinct animal. Taking cells from the skin of a quagga, an extinct zebra species, Wilson and Higuchi were able to force them to replicate in the laboratory. While this is a far cry from cloning an entire animal, their experiment showed that it might be possible to someday clone extinct animals, restoring them to the Earth.

Robert Burns Woodward
1917-1979

American biochemist who was awarded the 1965 Nobel Prize in Chemistry for synthesizing many organic and biologically important chemicals. Woodward was the first person to successfully synthesize chlorophyll, cholesterol, and vitamin B_{12}. He later synthesized tranquilizers, the antibiotic tetracycline, and steroids, including cortisone, widely used to control inflammation. Vitamin B_{12} is an essential nutrient that is added to many vitamin-enriched foods.

Norton David Zinder
1928-

American geneticist who, while working with Joshua Lederberg, discovered the phenomenon known as transduction, by which genetic information can be transferred from one bacterial strain to another by certain carrier viruses called bacteriophages. Bacteria infected in this way acquire new hereditary characteristics, such as drug resistance. Zinder and Lederberg published their landmark paper on genetic exchange in *Salmonella* in 1952. Zinder and Lederberg had developed an important technique for the rapid isolation of metabolic mutants of *Eschericia coli.*

Bibliography of Primary Sources

Books

Carson, Rachel. *Silent Spring* (1962). Groundbreaking book about pesticides and the industries that use and make them. Carson described how indiscriminate spraying of pesticides was poisoning our food and water and offered an outline of how to stop this irresponsible use. The controversy that she created around the use of pesticides is still alive and well.

Chomsky, Noam. *Syntactic Structures* (1957). The publication of this book revolutionized the inquiry into language acquisition. Chomsky suggested that children are born with the ability to understand the formal principles of grammatical structure, in marked contrast to the idea that language is essentially a system of grammatical habits established by training and experience. He proposed a theory that would account for both linguistic structure and for the creativity of language—the idea that human beings can create entirely original sentences and understand sentences never spoken before.

Dubos, René and Barbara Ward. *Only One Earth: The Care and Maintenance of a Small Planet* (1972). French-born biologist and ecologist Dubos shared a Pulitzer Prize with co-author Ward for this work in which they advocated for the "total environment." Dubos was an outspoken critic of what he considered a short-sighted outlook by most biologists.

Fossey, Dian. *Gorillas in the Mist* (1983). Recounts Fossey's observations of gorillas during her many years of field research. The book was subsequently made into a popular movie starring Sigourney Weaver, introducing millions of viewers to the plight of the mountain gorilla.

Kinsey, Alfred, et al. *Sexual Behavior in the Human Male* (1948) and *Sexual Behavior in the Human Female* (1953). Together, these books blazed new trails in the field of sex research, leading to reassessments of research practices and of medical, psychiatric, and public attitudes towards sex. Known collectively as the Kinsey Reports, both volumes were filled with frank discussions of biological functions in nonjudgmental contexts and challenged established perceptions of homosexuality, masturbation, premarital and extramarital relationships, and the role of sex in women's lives.

Leakey, Louis S. B. *Olduvai Gorge* (1952). In this work archeologist Leakey discussed his pioneering efforts at the famous Olduvai Gorge site in Africa, where he uncovered new evidence about the lineage of human beings, including that the species appeared to have originated in Africa, not Asia as had previously been thought.

Watson, James D. *The Double Helix: A Personal Account of the Discovery of the Structure of DNA* (1968). This book is Watson's account of the dramatic discovery of the double-helical nature of DNA, a discovery Watson made along with Francis Crick and Maurice Wilkins. The book became a bestseller despite controversy surrounding its publication, when Crick and Wilkins refused to sign release forms for the original publisher (Harvard University Press). Watson and Crick initially reported their findings in a 1953 article in the scientific journal *Nature*.

Wilson, Edward O. *Sociobiology: The New Synthesis* (1975). In this work acclaimed sociobiologist Wilson argued that instincts producing social behavior are genetically determined and are produced through natural selection. He identified various behavioral traits that are universal within social animal species and showed the survival and reproductive advantage such traits endow on their possessors.

Wilson, Edward O. *On Human Nature* (1978). In this Pulitzer Prize-winning work Wilson provided greater detail on his views of human sociobiology. He analyzed behaviors and ethical standards that are universal or almost universal in human societies. He examined human tendencies toward division of labor between the sexes, altruism toward kin, territorial aggression, incest avoidance, tribalism, male dominance, etc., and claimed that humans had genetic predispositions for these behaviors.

Periodicals

Holley, Robert. "Sequences in Yeast Alanine Transfer Ribonucleic Acid" (1965). Holley's classic paper on deciphering the genetic code for ribonucleic acid (RNA).

Mathematics

Chronology

1950 French mathematician L. Schwartz develops a new branch of analysis he calls the theory of distributions, which has applications in numerous areas of pure and applied analysis.

1951 The National Research Council of the U.S. National Academy of Sciences creates a new, separate division of mathematics in order to better represent its growing needs.

1955 *Homological Algebra* by Henri-Paul Cartan and Samuel Eilenberg offers a powerful cross between abstract algebra and algebraic topology, thus beginning a new field of study.

1961 Meteorologist Edward N. Lorenz discovers what comes to be called the butterfly effect—that small initial changes can result in large, completely random changes—thus forming the basis for chaos theory.

1971 In the first major use of a computer for sophisticated calculations other than to figure the value of π, American mathematicians John Billhart and Michael Morrison accomplish the factorization of the Fermat number F7.

1976 Using a combination of computer methods and theoretical reasoning, Kenneth Appel and Wolfgang Haken prove the conjecture posed more than 120 years before by English mathematician Francis Guthrie, that no more than four colors are needed to create a map in which no same-colored regions adjoin.

1980 Close cooperation among mathematicians in different institutions results in the complete classification of the finite simple groups, the basic building blocks for a major part of modern algebra.

1982 Polish-American mathematician Benoit Mandelbrot publishes *The Fractal Geometry of Nature* in which he founds a new branch of mathematics based on the study of figures that are self-similar at varying scales.

1985 Dutch mathematician Hendrik Lenstra devises a method of factoring based on so-called elliptic curves, thus making it possible to factor large numbers that had resisted all other methods.

1993 English mathematician Andrew Wiles announces that he has proved Fermat's last theorem, a 325-year problem that many mathematicians had declared unsolvable; other mathematicians find fault with aspects of his proof, and a year later he presents a corrected version.

1995 David and Gregory Chudnovsky at Columbia University calculate the value of π to more than four billion places, and devise a method to ensure that all the decimals are correct.

Overview: Mathematics 1950-present

Background: Mathematics Becomes the Language of Scientific, Philosophical, and Cultural Revolution

During the nineteenth century advances in mathematics pointed toward a universe not necessarily limited to three dimensions and not necessarily absolute in time and space. By developing new mathematical models and precise formulas with enormous predictive power, mathematicians profoundly shaped the understanding and application of twentieth-century relativity and quantum theories. In many cases, innovative mathematical models became the only means to describe profoundly revolutionary scientific and philosophical concepts regarding the structure and workings of nature.

Throughout the twentieth century there was a steady pace to the refinement and discovery of new applications for mathematical principles. In particular, advancements in differential equations (equations that relate the rates of change of physical quantities to the values of those quantities themselves) found continued application in astronomy and physics. Mathematicians and physicists labored to find mathematical formulas, expressions, and constants to that which, in essence, governed the cosmos. Along with the speed of light, Planck's constant, for example, was found to be a fundamental constant used in the mathematical expression of the Heisenberg uncertainty principle and, as a consequence, carried profound philosophical implications regarding limits on knowledge.

French mathematician Alexander Grothendieck (1928-) once wrote that "mathematical activity involves essentially three things: studying numbers, studying shapes and measuring distances." Grothendieck contended that all mathematical reasoning and divisions of study (e.g., number theory, calculus, probability, topology, or algebraic geometry) branched from one or a combination of these methodologies. Indeed, just as modern physicists have sought grand unification theories to reconcile quantum and relativity theory, during the last half of the twentieth century mathematicians sought, with varying degrees of success, to interrelate mathematical theories. The use of statistics, for example—beyond being just a mathematical convenience useful in describing the average workings of large systems—became the only way to describe some of the finer, quantum level workings of nature.

It may be fairly argued that in 1931 German mathematician Kurt Gödel's (1906-1978) theorem regarding the limitations of mathematical proofs was the assertion of a mathematical "uncertainty principle." Regardless, it became one of the most powerful and philosophically influential mathematical discoveries of the century—especially with regard to postmodern, existential, and abstract expressionist movements. Advancements in later twentieth-century mathematics, however, often refocused on classical mathematical theory to advance man's understanding of non-linearity and of complex or chaotic phenomena. Without question, English physicist Sir Isaac Newton's (1642-1727) classical mechanics and French mathematician Jules Henri Poincaré's (1854-1912) studies of the chaotic behavior of systems provided a path for the development of twentieth-century chaos theory.

As concepts regarding the dualism of mind and body underwent philosophical revision in the twentieth century, advances in both pure and applied mathematics worked their way into new and exciting concepts of physical and social order. Just as there was an increasing emphasis on the duality between the need for diversity and the need for interdependence of world-wide cultures, in the later half of the twentieth century mathematicians, scientists, and philosophers freely crossed blurred intellectual boundaries in an effort to more accurately describe an increasingly complex non-Newtonian world in which no classical, linear, God's-eye view of nature was possible.

Advances in Theory and Application

Especially for theorists, challenging mathematical terrain to scale was clearly mapped at the beginning the twentieth century with the posting of mathematician David Hilbert's (1862-1943) famous list of 23 problems. Throughout the century mathematicians wrestled with Hilbert's problems and, in some cases, only pinned down solutions or partial solutions in the last decades of the century. One of Hilbert's problems apparently finding resolution late in the 1990s, for example, included proof of Kepler's conjecture regarding the most efficient geometrical arrangement for stacked spheres. Although the "sphere-packing" problem seemed proved by everyday experience, a mathematical proof eluded mathematicians for nearly four centuries. The utilization of computers in providing proofs, however, spurred philosophical discussion about the nature and future of mathematical proofs.

Highlights of twentieth-century mathematical advancements would be incomplete without mention of four-color mapping theory, advances in understanding of Georg Cantor's (1845-1918) continuum hypothesis, and René Thom's (1923-) influential catastrophe theory. Moreover, during the later half of the twentieth century, mathematics often moved rapidly from theory to application. For example, American mathematician John Forbes Nash's (1918-) work in noncooperative games became influential in economic and social science. Although still controversial in theoretical aspects, game theory found application in the development of strategy for war, politics, and business.

Mathematics and Science

After 1950 advances in science, especially in physics and cosmology, became increasingly dependent upon advances and application of mathematics. English mathematician Sir Roger Penrose (1931-), for example, one of the leading mathematicians of the later half of the twentieth century, is perhaps best known for his collaborative work with fellow English physicist Stephen Hawking (1942-) regarding the calculation and prediction of the fundamental properties associated with black holes.

At the other end of the cosmic scale, English mathematician Simon Donaldson's (1957-) work in low-dimensional topological geometry has been used by particle physicists to describe short-lived subatomic particle-like wave packets called *instantons.* The development of chaos theory also fused scientific and mathematical efforts to seek order in complex and seemingly unpredictable systems. In the last decades of the twentieth century chaos theory became an important tool in the study of population trends, epidemiology (the study of the spread of disease), explosions, meteorology, and complex chemical reactions.

Mathematics and Emerging Technology

Almost all of the research and innovation in statistics during the last two decades of the twentieth century was a result of, or was deeply influenced by, the increasing availability and power of computers. Powerful computer-based techniques referred to by statisticians as "bootstrap statistics," for example, allow mathematicians, scientists, and scholars working with problems in statistics to determine with great accuracy the reliability of data. The techniques, invented in 1977 by Stanford University mathematician Bradley Efron, allow statisticians to analyze data and make predictions from small samples of data. Accordingly, bootstrap techniques have found wide use politics (e.g., polls), economics, biology, and astrophysics.

Using the emerging tools of computer graphics, Polish-born American mathematician Benoit Mandelbrot's (1924-) work in fractal geometry created a mathematical school with broad scope and application. Fractals seemed to be everywhere—a universality in nature—and were used by astrophysicists, for example, to construct computer simulations depicting the dynamics involved in the highly complex formation of galaxies and planetary systems.

In addition to igniting a world-wide microelectronics revolution, by the end of the twentieth century the invention of the hand-held pocket calculator and powerful computer software such as *Mathematica* placed at the fingertips of the average middle school student the most powerful and elegant of mathematical concepts.

Mathematics and Education

Although the tools of mathematics became cheaper and the mechanics of math more accessible, methods for teaching mathematics, especially in the United States, became mired in controversy. "New Math," for example, launched into American schools in the early 1960s, stressed conceptual understanding of the principles of mathematics and de-emphasized technical computing skills in an effort to teach children basic mathematical truths they could apply to more specific problems in a rapidly specializing scientific and technical world. New Math also, however, stirred controversy akin to a national strategic crisis and fostered sharply divided political opinions and passionate social debate regarding pedagogy (teaching methodologies) as schools sought to boost student's lagging mathematical skills.

Interest in bolstering mathematical skills was not, however, solely an American concern. Harvard professor Heisuke Hironaka (1931-), one of a number of influential Japanese-born mathematicians and executive director for the Japan Association for Mathematical Sciences, is often credited with providing the inspiration for the International Math Olympics competition for schoolchildren in an effort to encourage mathematical accuracy and speed.

Mathematics and Popular Culture

Just as games of mathematical logic became popular in Victorian England a century earlier, in the last decade of the twentieth century mathematics once again provided a source of popular entertainment. During the 1990s a number of biographies of mathematicians and books on mathematical theory soared to the top tiers of many best-seller lists. Movies using the complexities and subtleties of mathematics and the culture of mathematicians became box-office hits and, as an increasingly technological society sought deeper meanings behind the science and mathematics enabling the information age, books and articles explaining often difficult and abstract mathematical concepts in simple terms gained popularity.

Books, for example, ranged in topics from biographies of the Greek mathematician Archimedes (c. 287-212 B.C.) to the brilliantly eccentric Paul Erdös (1913-1996). Other works treated specialized areas of mathematics such as the history of π as fresh, exciting, and readable history. Regarding these popular works, however, none captured more public attention than the controversy and scholarly drama surrounding the proof of Fermat's last theorem by Princeton mathematician Andrew Wiles (1953-).

Mathematics and Twenty-First Century Society

From the darker—decidedly unpublic—worlds of political intrigue and espionage, advances in mathematics allowed cryptography to become a part of the everyday experience. Cryptography allows its users, whether governments, military, businesses, or individuals, to maintain privacy and confidentiality in their communications. Although the attempt to preserve the privacy of communications is an age-old quest, many cryptologists and communications specialists insist that a truly global electronic economy will be dependent on the development of cryptographic systems that allow transactions to carry the same legal weight as paper contracts. Development of such systems is highly dependent upon further advances in number theory.

Modern philosophers and social ethicists also assert that, in an expanding, electronically networked world, preservation of traditional notions of privacy may be dependent on cryptologic applications of higher mathematics. In a very real sense, mathematics developed during the later half of the twentieth century with the intent of helping us probe the innermost secrets of nature may, at the same time, provide the means to protect the sanctity of our innermost selves in the twenty-first century.

K. LEE LERNER

The Proof of Fermat's Last Theorem

Overview

> *But one cannot split a cube into two cubes, nor a fourth power into two fourth powers, nor in general any power* in infinitum *beyond the square into two like powers. I have uncovered a marvelous demonstration indeed of this, but the narrowness of the margin will not contain it.*

These words, written by Pierre de Fermat (1601-1665) in the margin of his copy of Diophantus's *Arithemetica,* have challenged and sometimes haunted mathematicians for more than 350 years. When a successful proof of Fermat's Last Theorem was finally found in 1993, it ended centuries of interesting and often controversial attempts to solve this famous problem.

Background

In modern algebraic terms, the theorem states that the equation

$$x^n + y^n = z^n$$

has no whole number solutions for $n > 2$. If $n = 2$, we have the famous Pythagorean Theorem:

$$x^2 + y^2 = z^2.$$

For instance, if $x = 3$, $y = 4$, and $z = 5$, we have,

$$3^2 + 4^2 = 5^2$$

$$9 + 16 = 25 .$$

This is only one of an infinite number of solutions, called Pythagorean triples. But Fermat claimed that if *n* were three or four or any other whole number larger than 2, then there were no solutions to the equation. For example, the equations

$$x^3 + y^3 = z^3$$

and

$$x^4 + y^4 = z^4$$

or any other equation with larger exponents, have no whole number solutions.

Why does this seemingly simple statement rank among the most famous theorems produced by the mind of a mathematician? The list of mathematicians who have tried to prove it reads like a "who's who" of mathematics: Leonhard Euler (1707-1783), Carl Friedrich Gauss (1777-1855), Niels Abel (1802-1829), Sophie Germain (1776-1831), Adrien-Marie Legendre (1752-1833), Lejeune Dirichlet (1805-1859), Henri Lebesgue (1875-1941), Joseph Liouville (1809-1882), Augustin Cauchy (1789-1857), Carl Jacobi (1804-1851), Gabriel Lamé (1795 1870), David Hilbert (1862-1943), Richard Dedekind (1831-1916), Ernst Kummer (1810-1893), and many, many others. This theorem, which has been tagged "Fermat's Last Theorem" because it was the last of Fermat's theorems to be proved, has played a pivotal role in the history of mathematics.

Fermat's Last Theorem is enigmatic in several ways. First of all, Fermat himself was not very interested in proving theorems and rarely found time to do so, claiming, "I am content to have discovered the truth and to know the means of proving it whenever I shall have the leisure to do so." It's ironic that proving one of his theorems should become the Holy Grail of mathematics. Secondly, although Fermat claimed to have a proof of the theorem, most mathematicians doubt this. The techniques used by mathematician Andrew Wiles (1953-) to finally prove Fermat's Last Theorem were not available to seventeenth-century mathematicians. That does not make Fermat a liar, however. He was probably only a little overconfident in his ability to provide a legitimate proof. Fermat used a technique called the method of infinite descent to prove his theorem for the case $n = 4$, and he may have

Andrew J. Wiles conquered the most famous unsolved problem in mathematics when he proved Fermat's Last Theorem. *(Photo Researchers, Inc. Reproduced by permission.)*

believed (incorrectly) that this method would work for the general case where *n* is any integer. Whether Fermat actually had a proof or not, his famous statement has contributed to the lore surrounding the theorem. It is reported that on a graffiti-covered wall at a New York train station the following words were found:

> $x^n + y^n = z^n$
>
> I have found a truly remarkable proof of this, but I can't write it out now because my train is coming.

Impact

Fermat's Last Theorem belongs to a branch of mathematics called number theory, a field in which few seventeenth-century mathematicians showed any interest. The theorem received little attention until almost one hundred years after Fermat's death, when the famous mathematician Leonhard Euler revived interest in number theory. Euler proved the theorem for $n = 3$, and later Sophie Germain, one of the first great women mathematicians, did important work in establishing the theorem for a certain class of prime numbers.

The nineteenth century saw still more advances in the search for a proof. Legendre proved the theorem for $n = 5$ and Lamé for $n =$

7. The most important advance came from the German mathematician Ernst Kummer, who proved the theorem for a class of numbers called *regular primes* in 1850. Kummer showed that the theorem was true for all prime exponents between 3 and 100 except for 37, 59, and 67. It had taken approximately two centuries to accumulate proofs for $n = 3, 4, 5$, and 7, and then, remarkably, Kummer proved the theorem for nearly all of the other values up to 100. Ironically, Kummer had presented a complete but erroneous proof of Fermat's Last Theorem a few years earlier; he had now made the biggest advance in the search for a general proof.

The twentieth century saw the eventual proof of Fermat's Last Theorem, a solution aided by the use of modern technology. Using long-established mathematical methods and high-speed computers, the theorem was proved for values of n up to 150,000 by 1987, 1 million by 1991, and 4 million by 1993. This strengthened the belief of many mathematicians who intuitively believed that Fermat's Last Theorem was true. To nonmathematicians, it seemed to settle the question entirely. If this theorem is true for exponents as large as we can realistically calculate, isn't that enough evidence to pronounce the theorem true for all values of n? A famous example refutes this type of thinking.

Euler stated the following theorem over 200 years ago: The equation

$$x^4 + y^4 + z^4 = k^4$$

has no whole number solutions for x, y, z, or k. Although never proved, this theorem was believed to be true by Euler and most subsequent mathematicians. Then, two centuries after the theorem was stated, a whole number solution was found to exist, nullifying the theorem. Why did it take so long for mathematicians to find a solution? Because the solutions (the problem actually has many solutions) involve extremely large numbers. The smallest such solution is $x = 95{,}800$, $y = 217{,}519$, $z = 414{,}560$, and $k = 422{,}481$. Examples like this are part of the reason mathematicians stubbornly refuse to accept the truth of a theorem until it is proved for all cases.

Although love of mathematics has been the main motivation for most mathematicians who searched for a proof of Fermat's Last Theorem, fame and financial awards also played a part. In 1908 the German mathematician Paul Wolfskehl left 100,000 marks in his will as a prize for a proof of Fermat's Last Theorem. At one time in his life, Wolfskehl had been on the verge of suicide. He had made his plans, written his farewell letters, and was awaiting the time he had appointed to take his own life. While waiting, he began to read a recent publication on Fermat's Last Theorem. He became so engrossed in the work that the time appointed for his suicide came and went. Wolfskehl had a change of heart and canceled his suicide plans. Many years later, he expressed his gratitude by leaving the prize money to the eventual conqueror of Fermat's Last Theorem.

Because of the prizes and notoriety attached to the problem, interest in proving Fermat's Last Theorem became intense among amateur mathematicians. Edmund Landau, a mathematician appointed to administer the Wolfskehl prize, had the following form letter made:

> Dear Sir or Madam: Your proof of Fermat's Last Theorem has been received. The first mistake is on page ________, line ________.

Landau's students were then given the job of reviewing the many incorrect proofs received and returning the postcards.

The search for a proof of Fermat's Last Theorem finally ended when Andrew Wiles announced his proof in the summer of 1993. Wiles, a Princeton mathematician, had worked in isolation on the problem for many years and his announcement came as a surprise to the mathematics community. Wiles's work combined two fields of mathematics, elliptical functions and modular forms, to solve the elusive problem.

In proving Fermat's Last Theorem, Wiles had actually solved another problem in mathematics, the Taniyama-Shimura Conjecture. Goro Shimura and Yutaka Taniyama were two Japanese mathematicians who, in the 1950s, conjectured that there was a relationship between elliptical equations and modular forms. Later, thanks to the work of mathematicians Gerhard Frey, Ken Ribet, and Barry Mazur, it was shown that if the Taniyama-Shimura Conjecture were true, then so was Fermat's Last Theorem.

In a dramatic series of lectures at a conference in Cambridge, England, in the summer of 1993, Wiles presented the proof to his colleagues. He did not announce the true intention of his lectures, but by the third day, it became apparent to the mathematicians attending the conference that Wiles's work was leading toward a proof of Fermat's Last Theorem. The excitement grew in the packed room where Wiles was presenting his lectures. Finally, he reached the conclusion of his proof, and put to rest the most famous problem in the history of modern mathematics.

Or had he? Not long after Wiles announced his discovery, an error was found in one section of his long and difficult proof. At first it seemed that Wiles would be able to fix the error and save the proof, but as time went on the "correction" became more and more difficult. Finally, with the help of one of his former students, Richard Taylor, Wiles was able to make the necessary corrections. These corrections took over a year to complete, however, illustrating the complexity of the proof that Wiles had constructed.

Although the search for a proof of Fermat's Theorem is over, its impact upon mathematics is not. Many mathematical advances are credited to mathematicians who were attempting to solve this problem. As in most theorems, finding one proof does not end useful work on the problem. Other proofs using other mathematic tools may be found, and of course the question remains: "Did Fermat really have a proof, and if so, what was it?"

TODD TIMMONS

Further Reading

Aczel, Amir D. *Fermat's Last Theorem: Unlocking the Secret of an Ancient Mathematical Problem.* New York: Dell Publishing, 1996.

Bell, E. T. *The Last Problem.* New York: Simon and Schuster, 1961.

Mahoney, Michael. *The Mathematical Career of Pierre de Fermat.* Princeton, NJ: Princeton University Press, 1994.

Ridenboim, Paulo. *Thirteen Lectures on Fermat's Last Theorem.* New York: Springer-Verlag, 1979.

Singh, Simon. *Fermat's Enigma.* New York: Doubleday Press, 1997.

The Development of Computer Assisted Mathematics

Overview

The development of programmable electronic computers placed a powerful new tool in the hands of mathematicians. The computer's ability to manipulate symbols allows it to perform the same sort of rearrangements used by humans to solve equations. Early artificial intelligence programs were able to discover simple proofs in symbolic logic and geometry, and to solve some problems in calculus. In 1976 two University of Illinois mathematicians announced the computer-assisted proof of the famous conjecture that any map drawn on a sheet of paper could be colored with just four colors. This particular result, which required calculations more extensive than could be checked within a human mathematician's lifetime, raised a number of important questions about the nature and role of proof in mathematics.

Background

The first electronic computers were number "crunchers." Built during the Second World War and the years immediately following, they were built to find approximate numerical solutions to the complex equations that described the behavior of explosives, high performance aircraft, and the weather. In these applications, they differed from the adding machines used by accountants, mainly by being able to store many numerical quantities in different locations, and to follow a program, a set of instructions, that indicated which arithmetic operations were to be performed and in which order.

British mathematician Alan Turing (1912-1954) had developed the basic theory of the programmable computer in the 1930s. Based on an analysis of the way humans did computations, Turing described a universal symbol-processing machine that could be realized using electronic components. The components available at first were rather bulky. One of the first computers, the ENIAC, built at the University of Pennsylvania using vacuum tubes, weighed 30 tons and occupied 16,200 cubic feet. It could perform about 5,000 additions per second.

By the 1950s a number of pioneering computer scientists, Alan Turing among them, were seriously exploring the possibility that computers could display intelligent behavior. The symbol rearrangement capability of the computer was used to find simple proofs in symbolic logic and to search for theorems in plane geometry. Programs were also written to solve some problems using the calculus. Overall, however, the development of such applications proceeded slowly.

In 1976 the majority of mathematicians regarded computers primarily as a labor saving device. Many were caught by surprise when two mathematicians at the University of Illinois, Kenneth Appel (1932-)and Wolfgang Haken (1928-), announced that they had found a proof of a long-standing conjecture, that all maps drawn in the plane can be colored using only four colors. This issue was first raised in 1852 by an Englishman, Frederick Guthrie, who noticed that he could color the counties on a map of England using only four colors and wondered if the same would be true of maps in general. The problem came to the attention of the great British mathematician Augustus DeMorgan (1806-1871) through Guthrie's younger brother, who was one of DeMorgan's students. DeMorgan quickly determined that the problem could not be easily solved and encouraged other mathematicians to look at the problem.

To prove that the four-color conjecture was in fact true, Haken and Appel first demonstrated that all possible maps consistent with some very simple rules could be classified as one of 1,936 possible cases and then used the computer to verify that each of the possible cases could be colored with four colors or less. Using one of the fastest computers available, their proof took over 1,200 hours. They had also used the computer to examine many sample cases while developing their strategy for the proof. Their published paper could only provide a sketch of the proof. The details were made available to mathematicians in a set of 400 microfiche cards, each card providing reduced-size images of dozens of computer-generated pages. Critics were quick to suggest that the proof could not be trusted. Electronic computers do make occasional errors and the fact that a single mathematician could not check the steps of the proof, even in an entire human lifetime, left one unsure that the proof could be trusted. The fact that the proof could be repeated on different computers afforded some confidence in the result, however. The search for a far simpler proof is still going on, even though mathematics affords no guarantee that even a simple true statement like the four-color conjecture will have a simple proof.

Over the second half of the twentieth century, the replacement of vacuum tubes by transistors and then by microprocessors allowed a phenomenal reduction of size and cost, while computational speed and memory increased many times. By 1990 the owner of an inexpensive desktop computer would have the equivalent of numerous ENIACs at his or her disposal. Over the same time period a number of advances in software and output devices made it possible for people who were not computer specialists to input information into the computer and to understand the results of extensive computations. By the 1960s computer languages like ALGOL and FORTRAN allowed mathematicians to specify the operations to be performed in a manner similar to that used in writing equations on paper. Graphical output devices allowed them to see the results plotted as graphs or as solid objects viewed in perspective. In the 1970s and 1980s computer applications like spreadsheet software, made it possible to do extensive calculations and visualization without writing a program at all.

The workaday world of mathematics was changed by the release in June 1988 of Mathematica, a computer software package designed by Stephen Wolfram (1959-) and a group of collaborators. Mathematica could run on an inexpensive desktop computer, accepting inputs in a form very close to that used by mathematicians and then manipulating the symbols to find an exact solution when possible. Mathematica could provide approximate numerical results to any number of decimal places and could plot graphs in two-dimensional form or in three-dimensional perspective. Furthermore, Mathematica included capabilities to draw a large number of three-dimensional mathematical objects in perspective. Mathematica and a number of competing software packages are being released in updated and more powerful versions every few years.

Impact

The uneasiness felt by many mathematicians at having to accept a proof that could not be checked in any realistic amount of time has died down, but not disappeared. Gone forever, however, are the days when mathematics professors would brag about the small number of computers in their department. Instead, the computer has become not only a faithful assistant for doing the more tedious manipulation of symbols but also as a tool for discovery. A computer can, for instance, answer questions about what will happen if a mathematical operation is repeated a million times. This has lead do the discovery of a number of those mathematical objects called fractals, many of which are defined as sets of points that have a certain property when a specified operation is continued indefinitely. The study of equations that are too complex to have an exact solution has remained important for both pure and applied mathematicians.

Computer software can also perform visualization tasks is spaces of more than three dimensions. The fundamental connection between algebra and geometry was established in the seventeenth century by the French Philosopher René Descartes (1596-1650). Using Descartes's analytic geometry, mathematicians are able to describe the solution to a set of equations as a set of points in a two- or three- or higher dimensional space. Often the points form a geometrical structure: a curve, a surface, or a related structure. While humans can draw curves in two-dimensional space and make models of surfaces in three-dimensional space, to understand the structure of objects in higher dimensions it is often useful to look at their "shadow" or projection in two dimensions. This is a task that is relatively easy for computers with modern display devices, and this makes it possible for a mathematician to "view" a higher dimensional object from any angle until he or she understands the structure well.

The development of computers has also served to define a number of new problems and several new areas of mathematics. The field of automata theory deals with the behavior of simple symbol processing machines that can exist in only a finite number of states. This field includes problems such as the so-called "busy-beaver" problem, which asks, for each number of states, what is the longest string of symbols that such an automaton can be made to print and then come to a stop, given a string of zeros as input. It has been found that only 13 symbols can be output by a four-state automaton, but that a five state automaton can output 4098 or more symbols before stopping.

Another mathematical field, called algorithmic complexity theory, is concerned with measuring the difficulty of mathematical tasks. It asks such questions as: What is the length of the shortest set of instructions needed to compute the digits of "pi" or the square root of two. Related to it is the field of computational complexity, which attempts to classify problems by the number of steps required to solve them. Given that computers can be expected to become faster and more powerful in the years to come, it is likely that the role played by computers in mathematics will be even more significant in the future.

DONALD R. FRANCESCHETTI

Further Reading

Books

Dewdney, A. K. *The New Turing Omnibus*. New York: Freeman, 1993.

Feigenbaum, E. A., and J. Feldman, eds. *Computers and Thought*. Menlo Park, CA: AAAI Press, 1995.

Peterson, Ivars. *The Mathematical Tourist*. New York: Freeman, 1998.

Wolfram, Stephen. *Mathematica: A System for Doing Mathematics by Computer*. 2nd ed. New York: Addison Wesley, 1991.

Periodical Articles

Appel, K., and W. Haken. "The Solution of the Four-Color-Map Problem." *Scientific American* 237 (October 1977): 108-21.

Gerd Faltings Proves Mordell's Conjecture (1983)

Overview

German mathematician Gerd Faltings (1954-) proved Mordell's conjecture in 1983, an accomplishment that earned him the prestigious Fields Medal, mathematics' highest honor. His method of altering a familiar geometric theorem into algebraic terms led him to solve the complex geometric theorem proposed by Louis Mordell in 1922. His success in proving this conjecture has contributed to the advancement of the studies of algebra and geometry.

Background

Faltings has been awarded many honors in his lifetime. Most notable is his receipt of the distinguished Fields Medal, which he received in 1986. Faltings earned this honor because he proved Mordell's conjecture using algebraic geometry. The conjecture that Louis Mordell (1888-1972) initiated in 1922 stated that a given set of algebraic equations with rational coefficients defining an algebraic curve of *n* greater than or equal to 2 must have only a finite number of rational solutions. To prove the conjecture, Faltings used a method initiated out of developing the Arakelov theorem in order to produce an arithmetic version of yet another theorem—the commonly used, geometrically based Riemann-Roch theorem.

Mordell in his era proved the finite generation in mathematics, and mathematicians have since built upon his findings and developed the crossover use of algebra in solving geometric problems utilizing the methods of one to enable proof of the other. André Weil (1906-1998), for example, extended Mordell's result of finiteness using algebraic geometry to prove finiteness in Abelian varieties over number fields using multiple geometric strategies. Geometrically, Mordell's conjecture translated well into algebraic terms, providing scholars with the opportunity to use functional algebraic methods to solve geometric problems. Mordell's conjecture, when translated algebraically, revealed that a group of curves based on his conjecture would result in only a finite number of sections if the group were not constant. Thus, Faltings was able to prove the accuracy of the conjecture.

Impact

Faltings's proof is important for several reasons, which are listed below and then discussed in greater detail in the paragraphs following:

1. It shows how the work of one mathematician can lead others to advance the knowledge of a given field.

2. It has led others to find new ways to solve age-old problems related to number theory.

3. It was instrumental in helping Andrew Wiles finally prove what was probably mathematics' greatest unsolved problem: Fermat's last theorem.

The historical context of Faltings's achievement is important because it sheds light on how the work of one mathematician can lead other scholars to advance knowledge in a given area. A mathematician named Gillet-Soule, for instance, expanded on the technique of utilizing many arithmetic surfaces to translate another well-known theorem, the Reimann-Roch theorem. Gillet-Soule also added varieties pertaining to arbitrary dimensions, complex differential geometry as it relates to the components at infinity, and real partial differential equations. He utilized this expansion of the Reimann-Roch theorem by using both the Hirzebruch-Grothendieck theorem and the original Riemann-Roch theorem. Another example of how scholars have built upon the accomplishments of others outside of their specialty in mathematics can be seen in Bismut's work. Bismut took Gillet-Soule's developments, which sprung from Faltings's work as it related specifically to real partial differential equations, and expanded them to function with the analogues of Green functions, particularly in higher dimensional cases. Following Faltings's proof of Mordell's conjecture, another mathematician—a man named P. Vojta—was later able to find another way to prove Mordell's conjecture. To accomplish this, he utilized the fundamental information that proved the Reimann-Roch theorem, and then he expounded upon particular aspects of it to create yet another way to prove Mordell's conjecture.

All of these various advancements from great scholars such as Faltings encompass the expansion of several areas of mathematics that have found their residence in the Riemann-Roch theorem. Throughout time, the Riemann-Roch theorem has been transformed into something that encompasses other arithmetic areas of study as a result of dedicated scholars who have contributed to the discovery of new elements in algebra and geometry, adding new dimensions to original theorems. Faltings has written a book (*Lectures on the Arithmetic Riemann-Roch Theorem*) discussing the Riemann-Roch theorem in algebraic terms.

The widespread use of contemporary methods to solve mathematical problems posed ages ago relative to number theory, as is the case with Faltings's proof of Mordell's conjecture, has led many scholars to discover new ways to incorporate several areas of mathematics to solve old problems. In his case, Faltings was able to utilize the general concepts of algebraic geometry along with complete objects, including the components of Arakelov's infinity. In using the knowledge of Arakelov's infinity theorem, Faltings illustrated the importance of building upon new discoveries. The Arakelov theory prepared the means for other mathematicians to create unification of intersection theory in algebraic fields as well as unification in classical intersection theory. Arakelov defined the infinity of intersection numbers as the values specifically of Green's functions.

Proving Mordell's conjecture led Faltings to further the study of Fermat's last theorem, long considered the greatest unsolved problem in mathematics until it was finally proven by Andrew Wiles (1953-) in 1994. Faltings illustrated that Fermat's last theorem, $x^n + y^n = z^n$, is only capable of a finite number of rational solutions in integers for which *n* is greater than 2 and, therefore, has no solutions. Faltings also used ideas of Vojta's to prove a conjecture created by Serge Lang that dealt with higher dimensional diophantine analogues for use in basic sub-varieties of Abelian varieties. Diophantine approximations are the formal relationships existing in the functions of a theory and the counting processes used to determine the number of functions in classical as well as specific asymptotic estimates.

Faltings's methods of algebraic geometry have also been used to examine the particular finiteness conjecture in Galois representations. The practicality of utilizing various areas in mathematics has given rise to many developments in the various fields. Interest in mathematics has grown tremendously with the advancements accomplished in number theory throughout time.

BROOKE COATES

Further Reading

Faltings, Gerd. *Lectures on the Arithmetic Riemann-Roch Theorem (Anal of Mathematics Studies, 127)*. Princeton, NJ: Princeton University Press, 1992.

Faltings, Gerd. *Rational Points*. Friedrich Vieweg & Sohn, 1992.

Soule, C., D. Abramovich, J. K. Kramer, and J. F. Burnol. *Lectures on Arakelov Geometry (Cambridge Studies in Advanced Mathematics, 33)*. New York: Cambridge University Press, 1995.

The Independence of the Continuum Hypothesis

Overview

One of the questions that accompanied the rigorous foundation of set theory at the end of the nineteenth century was the relationship of the relative sizes of the set of real numbers and the set of rationals. The axioms that had been laid down shortly thereafter were expected to provide an answer to the question of whether there was any infinite number between the sizes of those two sets. After an earlier proof that there might be a negative answer to the question, the work of Paul J. Cohen in the 1960s demonstrated that the question was not settled by the standard axioms. As a result, the notion of truth for statements about infinite sets was regarded as perhaps in need of revision.

Background

The notion of the infinite was one of the legacies of Greek philosophy, as it appeared in the work of Aristotle (384-322 B.C.). Aristotle, however, dismissed the idea of the "actual infinite" in favor of the "potential infinite," a sequence that could be continued indefinitely. In the centuries that followed various writers took a look at the notion of the infinite, but the subject seemed to be hedged about with paradoxes—arguments that seemed to give opposing answers to a single question. Galileo (1564-1642) showed that the number of even numbers was the same as that of the number of whole numbers, which violated the principle that the whole was greater than the part. Mathematicians left the subject of the infinite to the philosophers, who speculated about it in vague and grandiose terms.

An immense change took place with the work of Georg Cantor (1845-1918). Although Cantor had a mystical streak, his writings were genuinely mathematical and consisted of arguments and proofs about the actual infinite. Cantor demonstrated that the number of rational numbers was the same as the number of whole numbers, which was itself something of a surprise, since there seem to be so many more fractions than whole numbers. What Cantor did was to define the notion of a one-to-one correspondence between two sets in terms of matching up each element in one set with one element in the other, and then argue that being able to find such a correspondence between two sets amounted to showing that they had the same number of members. Cantor went ahead to show that the set of real numbers was larger than the set of whole numbers by proving that there could be no one-to-one correspondence between them.

A fundamental question that Cantor could not answer was whether there were any infinite sets whose number of members lay in between the whole numbers and the real numbers. The claim that no such intermediate set existed was called the *continuum hypothesis*. The name comes from referring to the collection of real numbers in terms of a continuous sequence of points on a line. It was regarded as of sufficient importance that David Hilbert (1862-1943), in an address in 1900 that stated a mathematical agenda for the twentieth century, put the continuum hypothesis at the top of his list of problems.

In an effort to work on the continuum hypothesis and other problems in the theory of infinite sets, Ernst Zermelo (1871-1956) came up with a set of axioms to try to avoid any concealed paradoxes. Other mathematicians had already run aground on some of the intricacies of dealing with the infinite, and Zermelo's axiomatization (known in a slightly altered form as ZF) was designed both to avoid paradoxes and to make further progress possible. He had a particular interest in the status of what has become known as the axiom of choice, but his system of axioms for set theory proved to be useful in addressing other issues as well.

The most important advance in the first half of the twentieth century with regard to the continuum hypothesis was the work of Kurt Gödel (1906-1978). He had already established some of the most important results in mathematical logic and assured the field of its status as an independent discipline. Then in 1938 he proved that the continuum hypothesis was consistent with the ZF axioms for set theory. This result suggested that there was good reason to keep working on the problem with the hope that the continuum hypothesis could be proved from the axioms. On the other hand, it did not establish that the continuum hypothesis was a consequence of the axioms, which would have finally answered the question posed by Hilbert.

Paul J. Cohen (1934-) was a mathematician who did not specialize in mathematical logic when he arrived at Stanford University as an assistant professor of mathematics in 1961. He did, however, have an impressive mathematical background, and he was looking for a problem of some importance on which to work. His attention was directed to the continuum hypothesis, and he undertook a thorough study of the literature that surrounded the earlier attempts to establish it on the basis of the axioms of set theory. Over the next few years he managed to create an entirely new technique in mathematical logic that enabled him to provide a kind of answer to Hilbert's question.

There are three possible relationships between a statement and a set of axioms: the statement can be provable from the axioms, its negation can be provable from the axioms, or the statement can be neither provable nor unprovable from the axioms. A good example is the parallel postulate included by Euclid (fl. 300 B.C.) in his list of axioms for geometry. For many years mathematicians tried to prove that statement on the basis of the other axioms provided by Euclid, but they were always unsuccessful. Not until the nineteenth century was it demonstrated that the parallel postulate could not be proved from the other axioms.

The way in which this was demonstrated was to come up with one model for the other axioms in which the parallel postulate was true and another in which the parallel postulate was false. A model is a specific collection of objects to which the axioms apply. If two different models for a set of axioms can give two different answers for the question of the truth of a statement, that statement is said to be independent of the axioms. Specific geometric models for the Euclidean axioms without the parallel postulate showed that the parallel postulate was indeed independent of the other axioms.

Gödel succeeded in showing that the continuum hypothesis was consistent with the axioms of set theory by constructing a model of set theory based on the axiom of constructibility. This is the assertion that every set is built up from other sets by certain well-defined processes. Within this model (called "the constructible universe") the continuum hypothesis could be proved. As a result, the continuum hypothesis had been shown to be consistent with the other axioms of set theory—in other words, no contradiction could arise from including the continuum hypothesis with the other axioms. There was some disagreement about whether the axiom of constructibility adequately captured mathematical intuition about the realm of infinite sets.

Cohen introduced the method of "forcing" to try to show the other side of independence for the continuum hypothesis. He needed a model for the other axioms of set theory in which the continuum hypothesis was false. From that it would follow that the continuum hypothesis was independent of the axioms of set theory. Forcing involves specifying which statements (Cohen started by working with statements about the positive whole numbers) are supposed to be true in the model being constructed. In particular, one introduces a kind of relationship that determines the truth of compound statements by the truth of the component statements of which it is made up. The only statements true in the model are those that are forced to be true by the forcing conditions. This kind of case-by-case analysis had also appeared in Gödel's proof that the continuum hypothesis was consistent with the axioms of set theory.

Impact

When the news of Cohen's result became known in the community of mathematical logicians, it was widely regarded as the most important development in set theory since it had first been axiomatized. He received the highest honor paid by the mathematical community when he received the Fields Medal at the International Mathematical Congress of 1966. The medal is given every four years to the outstanding mathematician under the age of 40. He was the first recipient of the medal for work in logic and helped to give the discipline an added boost in the judgment of the rest of the mathematical community.

The technique of forcing became a standard part of the repertoire of logicians working in the area of mathematics known as recursion theory. This field studies the complexity of mathematical subsets of the positive whole numbers and leads to a hierarchy of sets. Varieties of forcing have continued to be introduced in an effort to achieve more and more sophisticated models of the axioms for set theory. The term "Cohen reals" is used to refer to the numbers introduced by the stipulations of forcing conditions.

In a philosophical sense the issue of the truth of the continuum hypothesis has been a matter for much speculation in light of Cohen's result. If the continuum hypothesis is not implied by the standard axioms of set theory nor is its negation, then somehow the standard axioms of set theory leave open a rather fundamental issue about the relationship between the set of whole numbers and the set of the reals. Some philosophers of mathematics have urged the introduction of new axioms, especially those asserting the existence of extremely large infinite numbers, as a way of resolving the question. Others have suggested variants on the axiom of choice in which Zermelo was interested as more intuitive but capable of settling the truth of the continuum hypothesis. Still others have argued that there is no actual truth about statements of set theory, since the objects in question are so far removed from human intuition. The same kind of objection that had been raised to the axiom of constructibility was brought up with regard to the other proposed axioms as well. After the introduction by Cantor of the techniques that first led to the inclusion of infinite sets within the arsenal of mathematicians and not just philosophers, the demonstration by Cohen of the independence of the continuum hypothesis has taken the subject back into the domain for philosophers as well as mathematicians.

THOMAS DRUCKER

Further Reading

Bell, J.L. *Boolean-valued Models and Independence Proofs in Set Theory.* Oxford: Oxford University Press, 1977.

Cohen, Paul J. *Set Theory and the Continuum Hypothesis.* New York: W.S. Benjamin, 1966.

Gödel, Kurt. *The Consistency of the Axiom of Choice and of the Generalized Continuum Hypothesis with the Axioms of Set Theory.* Princeton: Princeton University Press, 1940.

Moore, Gregory H. *Zermelo's Axiom of Choice: Its Origins, Development, and Influence.* New York: Springer-Verlag, 1982.

The Rise and Fall of Catastrophe Theory

Overview

In the 1960s a French mathematician named René Thom (1923-) developed a mathematical tool known as catastrophe theory. Thom used his theory to study and make predictions of processes involving sudden changes. His ideas became popular with mathematicians and scientists in a variety of fields during the 1970s. However, catastrophe theory was sometimes applied to areas outside its scope, and for this reason it eventually became somewhat discredited.

Background

Scientists have found that there are two basic types of processes in nature: continuous and discontinuous. An example of a continuous process

René Thom, the founder of catastrophe theory, which has numerous applications in the exact and social sciences. *(Mathematisches Forschungsinstitut Oberwolfach. Reproduced by permission.)*

is the increase in temperature of a gas as it is heated. As one variable is changed at a constant rate (heat is added to the gas), a second variable also changes at a constant rate (the temperature of the gas increases). Because continuous processes are "smooth," they are relatively easy to predict. The branch of mathematics used to study continuous processes is called calculus and was developed by Isaac Newton (1642-1727) and Gottfried Leibniz (1646-1716) more than 300 years ago.

Discontinuous processes, on the other hand, are "abrupt" rather than smooth. An example of a discontinuous process involves an arched bridge to which more and more weight is added. At first, little effect is seen as the weight on the bridge is increased—the bridge begins to bend almost imperceptibly. At a certain point, however, enough weight is added to the bridge that it collapses. A steady change in one variable (the amount of weight on the bridge) results in almost no change in a second variable (the shape of the bridge distorts slightly) followed by a sudden change to a very different state (the bridge collapses).

A sudden change in a discontinuous process is called a *catastrophe*. In mathematics catastrophes can include sudden disasters, such as a bridge collapse or an earthquake, but they can also include much less dramatic events, such as the boiling of water. As room temperature water is slowly heated, it remains a liquid. Once it reaches its boiling point, however, the water suddenly begins to change state, from a liquid to a gas. In other words, a catastrophe has occurred. The values of the variables for which a catastrophe occurs are called the *catastrophe set*. For the boiling-water catastrophe, there is only one variable, that of temperature, and the catastrophe set consists of only one temperature, that of 100°C. Most discontinuous processes, however, involve more than one variable, and the catastrophe set may be quite large.

Because discontinuous processes involve sudden changes, they are usually much more difficult to predict than continuous processes. In the 1960s René Thom developed a way of studying discontinuous processes, which he called catastrophe theory. Thom became interested in catastrophes because he hoped to apply mathematics to the "inexact" science of biology. (Biology and sociology are said to be inexact sciences because they involve primarily discontinuous processes.) Thom presented his ideas in two books: *Structural Stability and Morphogenesis*, which was published in 1972, and *Catastrophe Theory in Biology*, which appeared in 1979.

As he was developing his theory, Thom collected data relating to the variables involved in sudden changes. When he then plotted these data on three-dimensional graphs, the result was a curved surface representing a catastrophe in mathematical form. Therefore, catastrophe theory allowed mathematicians to study not only numerical data from discontinuous processes but also visual data in the form of three-dimensional shapes. For this reason, catastrophe theory is considered to be a branch of geometry.

Thom showed that even though the number of discontinuous processes in nature is essentially infinite, the graphs of these processes could be categorized into a few basic shapes. For processes involving four variables, he discovered that there are seven basic types of catastrophes. They are named for the shapes formed when their variables are graphed: fold, cusp, swallowtail, butterfly, wave, hair, and fountain. To picture the graph of a fold catastrophe, for example, imagine taking a sheet of a paper and bending it into the shape of the letter C. The upper curve of the C would represent one stable state, and the lower half of the C would represent a second stable state. A catastrophe would be represented by a jump from the upper curve to the lower or vice versa. Thom also found that if a catastrophe depends on more than six variables, its graph becomes too complicated and results in no clear shapes that can be studied.

Impact

The ultimate goal of catastrophe theory was to produce a model of a discontinuous process that could then be used to make predictions. First, a scientist or mathematician would select variables related to the process being studied. For a chemical process, for example, these variables might be temperature and the concentration of reactants. Next, the scientist would collect as much data as possible about the effects of different combinations of temperature and concentration on the process. With the application of complex calculations and the aid of computer software, the scientist could transform the data into a three-dimensional graph, which could then be used as a model.

One scientist who used catastrophe theory to examine a discontinuous biological process—that of the fight-or-flight response in dogs—was E. Christopher Zeeman. Zoologist Konrad Lorenz (1903-1989) had studied aggressive behavior in dogs and believed that it was controlled by two variables: anger and fear. Zeeman drew upon Lorenz's work and defined a dog's mood as the combination of its anger, as indicated by the degree to which its teeth are bared, and its fear, as indicated by the degree to which its ears are flattened. He then proceeded to collect data on how a dog's mood affects its behavior. If a dog is angry, but not afraid, it will attack (the fight response). When the dog is afraid, but not angry, it will flee (the flight response). If the dog is neither afraid nor angry, it is unlikely to either flee or attack. What Zeeman was interested in, however, was what would happen (and how to predict what would happen) when a dog is angry and afraid at the same time.

Zeeman used the data he gathered to create a three-dimensional graph with anger plotted on one axis, fear plotted on a second, and the dog's behavior plotted on a third. The result was a curved surface. This surface matched one of the basic catastrophes described by Thom: a cusp catastrophe. The graph predicted that an angry dog that is slowly made afraid will continue to behave aggressively until its fear increases to a point that a catastrophe occurs. At this point, the dog suddenly flees—a flight catastrophe. Similarly, the opposite can also occur. A fearful dog that is slowly made angry will eventually attack—an attack catastrophe. Therefore, Zeeman predicted that the effects of frightening an angry dog would be different from those of angering a frightened dog.

Zeeman's use of catastrophe theory to study flight-or-fight responses showed that it could be used to approach certain problems in a new way. However, catastrophe theory and the models produced by it soon became somewhat of a fad. It was quickly embraced by mathematicians and scientists in diverse fields, and several national magazines, including *Newsweek* and *Scientific American,* published articles that explained it to the public in broad terms. One reason for the popularity of catastrophe theory was the belief that it could be applied to every branch of science. Some hoped that it would play the same role for inexact sciences as calculus had for the more exact sciences of physics and chemistry.

Despite the initial acceptance of the theory, it eventually became controversial. The number of variables involved in a discontinuous process must be small in order for catastrophe theory to model it with any accuracy. In the real world, however, especially in inexact sciences such as biology and sociology, these conditions rarely occur. One less than practical application of catastrophe theory involved its use to model the escalation of hostilities between nations. The variables used were threat and cost. It was argued that catastrophes—in this case, sudden attacks or surrenders—would occur when threat and cost were both high. Although such a model might be used to describe theoretical nations in very general terms, many more variables come into play when real people and real nations are involved. Therefore, such a model could not be used to make predictions of any practical value. Catastrophe theory was also applied with varying degrees of success and failure to social topics ranging from the stock market to prison riots to eating disorders.

Almost all biological and sociological systems are infinitely more complex than can be described adequately by catastrophe theory. In other words, they are essentially impossible to predict by this method. Therefore, catastrophe theory has turned out to be most useful in the exact sciences of physics, engineering, and chemistry even though Thom had originally intended it as a tool for studying the inexact sciences. One problem that catastrophe theory can be used to effectively study, for example, is whether light will reflect from or pass through moving water. Even in the arena of exact science, however, other mathematicians soon pointed out that many of the most useful ideas of catastrophe theory had already been developed under other names.

STACEY R. MURRAY

Further Reading

Ekeland, Ivar. *Mathematics and the Unexpected.* Chicago: The University of Chicago Press, 1988.

The MacTutor History of Mathematics Archive. University of St. Andrews, 1999. http://www-history.mcs.st-and.ac.uk/history/

Zeeman, E.C. "*Catastrophe Theory.*" *Scientific American* 234 4 (April 1976).

Fractal Theory and Benoit Mandelbrot

Overview

In 1975 Benoit B. Mandelbrot (1924-) wanted a word to describe the strange group of mathematical sets he was studying, and looked for inspiration in his son's Latin dictionary. The term he created was "fractal" to describe sets that modeled such diverse phenomena as cloud boundaries, stock market prices, plant growth, and even the distribution of matter in the universe. Mandelbrot's attempts to make the mathematical, scientific, and business communities, as well as the general public, aware of fractals have led some critics to see him as obsessed, both with fractals and his own place in history. While practical uses of fractals have been few, these unusual mathematical sets have left their mark on many areas, from financial analysis to Hollywood special effects.

Background

In the late nineteenth and early twentieth centuries a number of mathematicians described strange mathematical sets that seemed to defy logic. One such set, the Koch curve—named after Nils Fabian von Koch (1870-1924)—is constructed by taking a line segment and replacing the middle third of the line with the other two sides of an equilateral triangle. The result has four line segments, each one-third the length of the original line. The next step takes these four line segments and replaces the middle third of each with the other two sides of an equilateral triangle, and so on to infinity. It is a simple construction but produces unusual results.

The length of the created curve is 4/3 to the power of the number of steps made. So for the first step the length is 4/3, for the second it is 16/9, and so on. Therefore as the number of steps tends to infinity, the length of the Koch curve also tends to infinity, yet mathematically the curve has no area (in the same way that the original line segment has no area).

A further, perhaps more simple, example is the Cantor Set, named after Georg Cantor (1845-1918). Starting again with a line segment, the middle third is removed, creating two line segments. The middle third is then removed from these line segments, and so on to infinity. As the number of steps tends to infinity the overall length of the set tends to zero, as there have been an infinite number of subtractions. Yet the set also has an infinite number of points, for at each step more line segments are created.

Such constructions seem to defy common sense, and to a degree they do, for common sense is based on experiences in our 3-dimensional world. Mathematically, objects are generally analyzed in terms of 0, 1, 2, or 3 dimensions. For example, a cube or sphere is 3-dimensional, a circle or square is 2-dimensional, a line or curve is 1-dimensional, and a point is 0-dimensional.

Mandelbrot's key point was that objects such as the Cantor Set and Koch curve have fractional dimensions, which is why they seem so strange. The Koch curve has a fractal dimension of about 1.262, which is greater than the 1-dimensional line we started with, but smaller than a 2-dimensional square. The Cantor Set has a fractal dimension of 1.585. Just to confuse the issue, not all fractals have a fractional dimension; rather a fractal is defined as having a measured dimension greater than its topological (or, if you like, standard, common sense, or integer) dimension.

Fractals generally have fine structure, visible no matter how far you zoom in, and these details seem similar at all scales. Fractals are often too irregular to be described in traditional geometric terms, and can generally be constructed

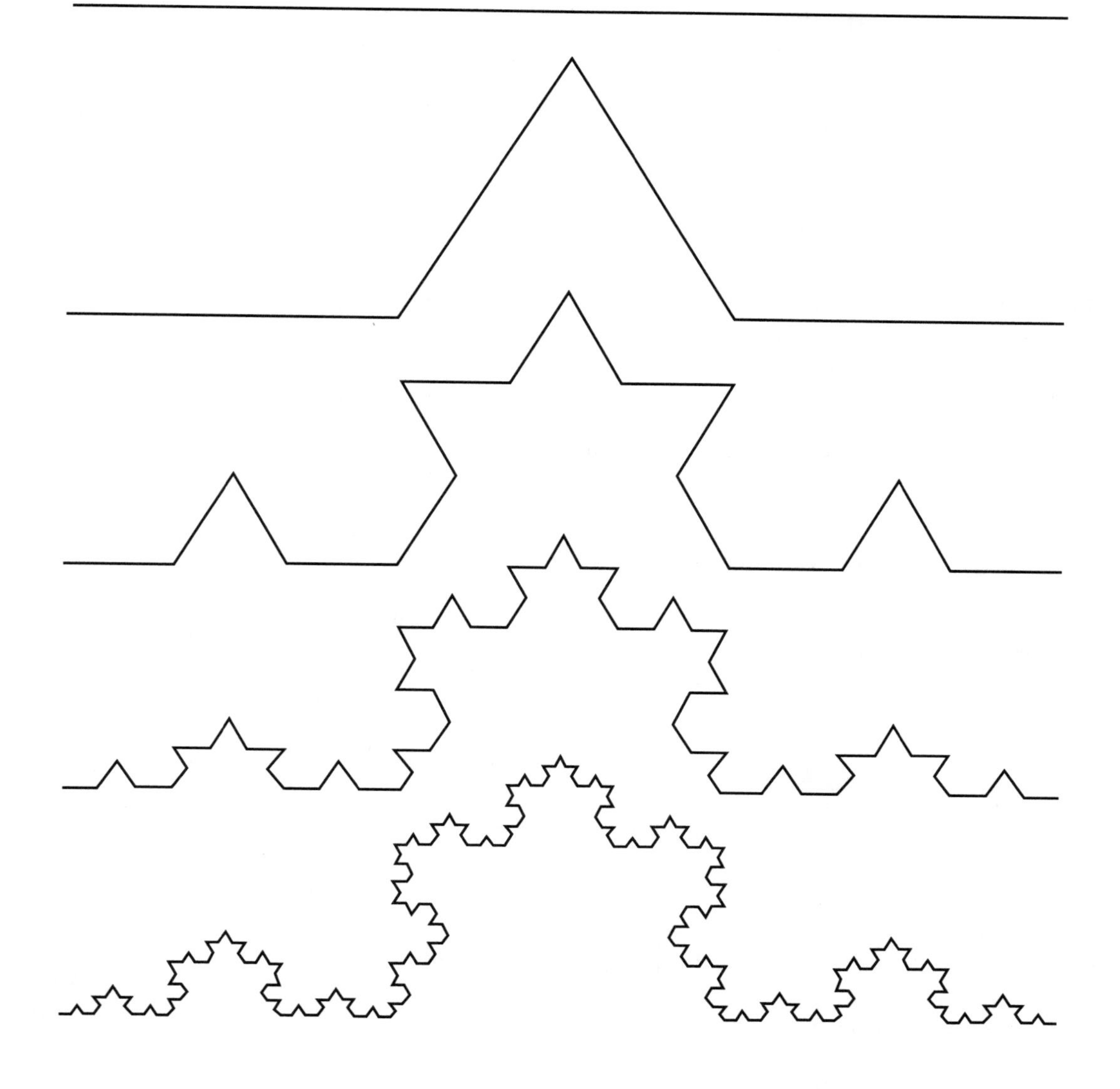

The Koch curve. *(Electronic Illustrators Group.)*

from simple methods. Unfortunately not all fractals have all of these characteristics; indeed they are hard to generalize, as some are curves, some are surfaces, others are disconnected points, and yet others are so oddly shaped there are no good terms for them.

Mandelbrot has often acknowledged those whose pioneering work led to his own theories on fractals, with a number of his books including short biographies of such individuals. He has a particular fondness for theorists on the edge, or even outside, of mathematics, especially those whose work was forgotten, ignored, or misunderstood. However, Mandelbrot admits that it is only in hindsight that such forerunners were found. While Koch and Cantor are obvious examples, and Felix Hausdorff's (1868-1942) work on measuring dimension has become an important foundation of fractal theory, it was only after-the-fact that many historical precedents were uncovered.

Impact

The road to fractals was a long one for Mandelbrot, starting with investigations into the noise in electronic transmission beginning in the

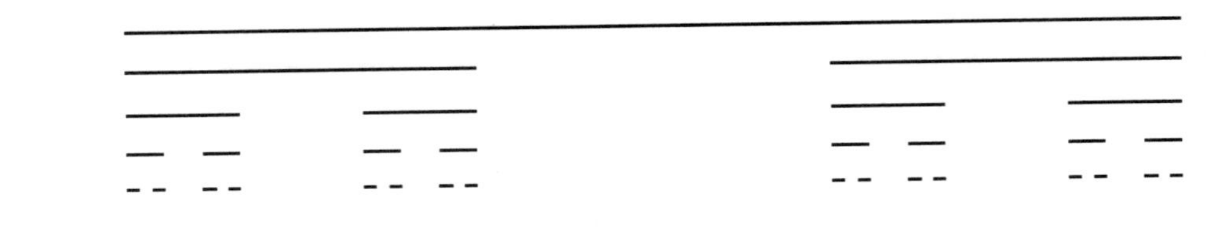

The Cantor set. *(Electronic Illustrators Group.)*

1950s. Mandelbrot noticed that the result always looked similar, no matter how long or short the time duration chosen to measure noise was made. He discovered that the construction of the Cantor Set was a good mathematical approximation of measuring noise, with each construction step being equivalent to a shorter and shorter measuring time.

Mandelbrot then turned to economics, with a 1963 article entitled "The Variation of Certain Speculative Prices." He argued that previous mathematical models did not adequately describe the complexity of the rise or fall of stock prices. Mandelbrot's theories attracted attention, and he was listed in an edition of *Who's Who of Economics*. However, his writing was hard to follow, his ideas complex, and ultimately they offered no power to predict real stock prices, so interest faded and the next edition of *Who's Who* dropped his name.

Moving on from finance, Mandelbrot turned to more theoretical and mathematical approaches for his coalescing ideas. In 1967 he published a paper entitled "How Long Is the Coast of Britain?" in which he showed that while "common sense" suggests you could, for instance, take an aerial photograph and measure the outline, in actuality the result obtained depends on the smallest resolution of measurement used. As the scale becomes finer, bays and peninsulas reveal new sub-bays and sub-peninsulas. The Koch curve is a reasonable approximation of a coastline, with each step in the construction being equivalent to using a smaller and smaller resolution of measurement. Mathematically any coast, or any river for that matter, is of infinite length if you use an infinitely small resolution of measurement. This helped explain why some countries give different lengths than others for international borders, as they may have used different resolutions of measurement.

In 1973-74 Mandelbrot gave a series of lectures at the Collège de France in which he refined his earlier works into a more coherent whole. In 1975 he coined the word "fractal," and gathered the lectures together as a book, *Les objets fractals: forme, hasard et dimension.* In 1977 an English version, *Fractals: Form, Chance, and Dimension,* appeared with further revisions and expansions. The book was not overly mathematical yet was still difficult to follow and did not generate much interest.

In 1982 Mandelbrot tried again, releasing *The Fractal Geometry of Nature*, which was a further reworking of his previous material; indeed portions were identical to earlier works. However, the 1982 book included something his previous writings had been lacking: colorful computer-generated pictures. Fractal landscapes, planetscapes, and unusual patterns showed the beauty and potential of fractals.

Quickly fractals entered the "popular" science press with articles often containing bright, colorful illustrations of fractal designs. Computer magazines offered code to display fractals on the home computer screen, and many programs and screen savers used fractal effects. Fractals have appeared on t-shirts, beach towels, calendars, posters, in music videos, and have been displayed as art. The most popular fractal image is named the Mandelbrot Set, once called the most complicated object in mathematics, which consists of "bubbles" with infinitely fine surface detail.

The high profile of fractals in the 1980s led to a rediscovery of Mandelbrot in fields that had overlooked his early work. Reprints of Mandelbrot's economic articles were updated to include new terms and diagrams. The language of many of Mandelbrot's early papers was difficult to read, as English was not his first language, and the re-editing of papers has made them far more accessible and readable.

The popular success of fractals was based on the appeal of colorful images, which has provided fuel for critics. Many pure mathematicians have found Mandelbrot's lack of solid proofs frustrating. Some opponents have seen Mandelbrot as obsessed with a field that is no more than a novelty, and obsessed with his place in history. The way Mandelbrot has popped in and out of many disciplines, leaving behind only partly formed ideas for others to sift through, has been a source of criticism for some, but a source of opportunity for others.

Practical applications for fractals are few, but at the same time surprising in their range. Fractals have contributed to computer image compression techniques. Realistic movie special effects, such as landscapes and planets, have been generated by fractals. However, it is in the mathematical modeling and graphical analysis of irregular natural phenomena that fractals flourish: snowflakes, coastlines, rivers, cloud edges, mountains, galactic dust, turbulent fluids, moon craters, and the list could go on and on.

Yet there are no true fractals in nature, with "natural fractals" like coastlines breaking down somewhere, often at the atomic level (although it should be remembered that there are no true straight lines in nature either). Just as a sphere is only an approximation of the shape of the Earth, and an ellipse is only an approximation for the orbit of planets, so a fractal model is only an approximation within a given range of scale. An elliptical model for an orbit allows for accurate predictions of the location of a planet, and a spherical earth can successfully model phenomena such as its magnetic field. However, a fractal model of stock prices, crater impacts, or broccoli growth cannot predict how prices will change, where the next meteor will strike, or the final shape of a broccoli floret—all they can do is offer an imaginary version, with no predictive power at all.

DAVID TULLOCH

Further Reading

Books

Devaney, R.L. and L. Keen, eds. *Chaos and Fractals: The Mathematics Behind the Computer Graphics.* Proceedings of Symposia in Applied Mathematics, Vol. 39, August 1988.

Falconer, Kenneth J. *Fractal Geometry: Mathematical Foundations and Applications.* John Wiley & Sons, 1990.

Mandelbrot, Benoit B. *Fractals: Form, Chance, and Dimension.* San Francisco, 1977.

Mandelbrot, Benoit B. *The Fractal Geometry of Nature.* New York, 1982.

Mandelbrot, Benoit B. *Fractals and Scaling in Finance.* New York, 1996.

Stephen Cook Advances Knowledge of NP-Complete Problems, Assisting Computer Scientists

Overview

In 1971 mathematician Stephen Cook (1939-) was able to show that a solution to a certain family of computational problems (computer science problems) could not be computed in a reasonable amount of time on even the fastest computers that can exist. He was also able to show that this family of problems was related, so that if a "fast" solution could be found for a single one of them, they could all be solved in the same manner. These problems, called "NP-Complete" problems, turn out to be very important because they deal with optimizing many sorts of activities, including travel itineraries, computer architecture, scheduling, electrical circuits, and more. As a result of the work done by Cook and others, we now have a better idea of problems that are hard to solve, those that can be solved easily (and quickly), and how to differentiate between them. This, in turn, is valuable to computer scientists and others who work to compute solutions to complex problems.

Background

An algorithm is a methodology that can be used to solve a problem. It is, in essence, a list of instructions that can be followed to arrive at a solution in a predictable number of steps. For example, someone consistently late for school could generate the following algorithm to solve this problem:

> 7:00: Alarm sounds; 7:01: Turn off alarm, get out of bed; 7:10: Choose clothes, get dressed, brush teeth; 7:15:

Eat breakfast, gather books, leave house;
7:30: Arrive at school

If followed precisely, this algorithm will guarantee a predictable outcome, as mathematical algorithms will help reach a solution to mathematical problems.

In general, complex problems require complex algorithms if they are to be solved in a methodical manner. However, it is not a simple matter to determine how long an algorithm must take to reach a solution, based only on the type of problem. Certain types of problems, called "P" problems, can be reliably solved in what is called "polynomial time." In other words, a relatively simple mathematical relationship can be developed, using polynomials, to find out how long it will take to solve a "P" problem. P problems are usually cast as decision problems with an answer that must be either yes or no. One example of a P problem would be, given a group of numbers, is there any number that is evenly divisible by three? This is a P problem because, no matter how many variables you have, the number of steps to find an answer is easily calculated. For example, with any number of variables, the largest number of steps to completely solve the problem is equal to the number of variables because the algorithm would simply be to divide every number by three to see if the answer is yes or no. The amount of time to calculate the answer to such problems is not difficult to determine. In addition, since one *can* determine a series of steps to find an answer and the length of time that the process will take, such problems are called "deterministic."

Another set of problems may not be solvable deterministically and are called, appropriately enough, "NP," for non-deterministic polynomial problems. Proposed solutions to these problems can be verified or rejected in polynomial time, but a precise solution, to the best of our knowledge, cannot be. In other words, if you make a guess at a solution, it is easy to determine if the guess is correct. On the other hand, it is difficult (perhaps impossible) to arrive at an answer using a methodical algorithm. A famous example of such problems is the "Traveling Salesman" problem.

The Traveling Salesman problem is an exercise in optimization. A traveling salesman has a number of stops to make that are scattered all over. He is supposed to find a route that will visit each stop only one time and that will cover the shortest distance possible. If there are only a few stops, the problem is not difficult to solve and can usually be solved by simply taking the distance between each set of stops and adding them up in all the various ways possible to find the shortest route. However, as the route grows larger, the problem becomes very difficult very quickly. The reason for this is that the number of options grows much faster than the number of stops. In fact, the number of possible itineraries grows as the factorial of the number of stops. A route with three stops will have six possible itineraries because three factorial (3x2x1, written as 3!) is equal to six. Four stops brings us to 24 itineraries, five will give 120, and six, 720 itineraries. Writing an algorithm that will give an answer to this problem is easy, but the time to solve the algorithm cannot be represented by a polynomial, and success in solving such problems is not guaranteed.

Now, consider trying to solve the traveling salesman problem with a computer. If your computer can test one complete itinerary in a microsecond, the 10 stops will take over 3.5 seconds to solve (10! equals 3,628,800). Fifteen stops will take slightly over fifteen days to solve, and 20 stops will take over 77,000 years to reach a solution by testing every single possible itinerary. More complex itineraries can't be solved in the life of the universe using any known algorithm.

What Cook was able to do was to show that some problems, called "NP-complete" problems, are similar to *all* NP problems and to each other. Therefore, if a single NP-complete problem can be solved in polynomial time, then all NP problems can be solved in polynomial time. This would mean that computers could be programmed to solve even very large problems (problems with a large number of variables) in a relatively short time.

This sounds, in many ways, like an abstract mathematical problem with very little bearing on the "real world." In some ways, this may be true, but there are a great many practical applications for Cook's work and for NP-complete problems in general.

Impact

One thing that was found as work on NP problems progressed was that it might not be necessary to test every single itinerary. For example, you wouldn't want to calculate the distance from a city to itself—even though this is one of the possibilities—because the answer is going to be zero and because it doesn't help solve the problem. So you can automatically throw out all route segments that start and end in the same place. In addition, for specific routes, you might choose to say that the salesman has to start and end in specific cities. Removing even just one or two terms from a factorial can drastically cut down on computation time, possibly making the problem solvable. This, in turn, helps to simplify the problem.

It was also found that it is not always even necessary to find a shortest route. For example, you might decide that the salesman simply has to be back in the office within a week. In such a case you need not find the single shortest route, but just one that will take a week or less. If there are many routes that can be traveled in a week, the problem will be solved more rapidly. From a practical standpoint, it might be preferable to send a salesman out, knowing that he might take five days to complete his route instead of four, but also knowing that it might take months or years to calculate the perfect route. Knowing in advance that a problem is NP, then, can help to set the strategy you will use to attack it—that is, whether to try to find the single best solution or simply one that is good enough, given everything else you know about the situation.

There are many problems that fall into the NP-complete category besides the classical traveling salesman problem. For example, finding the optimum route for internet packets to take when traveling from computer to computer might fall into this category and, as the global computer network becomes increasingly complex, so does the problem of routing e-mail, making internet phone calls, downloading information, and so forth. Another problem of this sort would be scheduling work for the different processors in parallel or multiprocessor computers. Developing production schedules for industry, plowing roads after a heavy snowfall, optimizing production of different products with different profit margins and popularity, and so forth are other problems that are NP or NP-complete.

In some of these cases, it is tempting to simply guess, knowing that a methodical approach could take many years while a lucky guess could solve the problem in the first try. There are two reasons why guessing is not a good strategy, though. First, if you are trying to find the single shortest path, even if you guess it on the first try, you still have to try every path to prove that it is the shortest. So, in this case, guessing saves no time at all.

In spite of this, guessing still looks attractive if you are simply trying find an acceptable route. To use a previous example, if you are merely trying to find a route that will get the salesman back in a week, you can stop looking as soon as a single such route is found, which could happen by sheer chance with the first guess, the tenth, or the hundredth. However, it might also take until the 77 trillionth guess, making you wait nearly 2.5 years. And that is the problem with guessing—you have no idea how long it will take to reach a solution. This lack of certainty makes guesswork a poor method for solving a problem of any sort, especially a real-world problem in which large sums of money may be at stake.

Put simply, Cook's research helped to clarify the boundaries between the problems we know can be solved in a reasonable amount of time and those that we know probably can't. By making this clear and by showing that solving a single NP-complete problem will erase those boundaries, Cook was able to also help us to determine which problems are worth trying to solve exactly and which aren't.

P. ANDREW KARAM

Further Reading

Devlin, Keith. *Mathematics: The New Golden Age.* New York: Columbia University Press, 1999.

Eric Weisstein's World of Mathematics. http://mathworld.wolfram.com

Efron's Development of the Bootstrap

Overview

Powerful computer-based data-analysis techniques referred to by statisticians as "bootstrap statistics" allow mathematicians, scientists, and scholars working with problems in statistics to determine, with great accuracy, the reliability of data. The techniques, invented in 1977 by Stanford University mathematician Bradley Efron, allow statisticians to analyze data, draw conclusions about data, and make predictions from smaller, less complete samples of data. Bootstrap techniques have found wide use in almost all fields of scholarship, including subjects as diverse as politics, economics, biology, and astrophysics.

Background

Almost all of the research and innovation in statistics during the last two decades of the twentieth century was a result of, or was deeply influenced by, the increasing availability and power of computers. In 1979, Efron's important and seminal article titled "Computers and Statistics: Thinking the Unthinkable" argued that statistical methods once thought absurd because of the large number of calculations required, would—given the growth of computing—soon be common mathematical tools. Subsequently, Efron's Bootstrap, a technique involving resampling an initial set of data thousands of times, did indeed become a standard tool of statistical analysis.

Descriptive statistics is a branch of mathematics concerned, in general, with determining quantities, means (averages), and other characteristics of a set of data. In contrast. inferential statistics is concerned with the reliability of descriptive statistics. Prior to the invention of bootstrap techniques, inferential statisticians largely relied on mathematical equations and techniques developed during the nineteenth century. Many of these techniques required a large number of calculations that were demanding in terms of time, and as a result, were often fraught with human error. At best, accurate statistical computations were often the result of a slow and tedious process. During the last quarter of the twentieth century, high-speed computing power made the development of bootstrap techniques possible. In contrast to the traditional limitations of statistical calculations, computers could quickly and accurately perform millions of calculations. In addition, computers could quickly incorporate and make accessible, large and complex databases.

Regardless of the subject (e.g., a poll on political preferences or the depiction of subatomic particle behavior in physics experiments), statistical sampling of data from small numbers is scrutinized in order to draw conclusions about the attributes of a larger group. Bootstrap statistics allow statisticians to specifically determine whether the selection of a particular sampling group influences the ultimate conclusions. In other words, statisticians can determine whether selecting another group of data would alter the outcome. In the alternative, bootstrap techniques allow statisticians the ability to determine exactly how small a data sample can be before the sample no longer is a valid representative of the larger group.

Older statistical methodologies determine the accuracy of a sample by comparing it with artificial samples projected from assumptions that data usually fall into one of several patterns of distribution (e.g., a bell-shaped curve). In contrast, computers running bootstrap statistical programs randomly select data elements from a set of data. If, for instance, astronomers made 10 measurements of radiation (e.g., gamma rays or x rays) emanated by a particular star, computers programmed to use bootstrap techniques would then review those 10 sets of data. Computers would then randomly select data from one of the 10 measurements and copy (i.e., the original data would remain a part of the larger set of all measurements) that data to a new data set for subsequent analysis. By continuing to randomly choose data from the original data set in this manner, a complete and new data set is eventually formed. Computers often repeat this procedure hundreds or thousands of times to form artificial data sets that might, for instance, contain multiple copies of the original data or, in another variation, be missing certain data present in the original measurements. With computer capability, statisticians can then look at how these artificial data sets vary from the original measurements to determine the reliability of conclusions derived from the original data.

Efron labeled his techniques "bootstrap" because of their ability allow measurements to "pull themselves up" (i.e., to provide information regarding their reliability) by their own bootstraps (i.e., data). For his important advancement of "the bootstrap," Efron, a 1983 MacArthur Prize Fellow, was elected to the National Academy of Sciences. In 1990, Efron was awarded the S. S. Wilks Medal, the most prestigious prize of the American Statistical Association.

After Efron advanced his bootstrap techniques, they received considerable scrutiny from the mathematics community. Although bootstrap techniques gave substantially the same results as conventional techniques when the application of a bell-curve was appropriate, the bootstrap proved more broadly useful when sets of data were not distributed according to the traditional bell-curve.

Bootstrap techniques allowed statisticians a way to determine the trustworthiness of data and of statistical measurements derived from that data. The techniques are comparable to other statistical measures such as standard deviation. With many measurements it is important to determine the mean (i.e., statistical average) and the standard deviation of that mean. Using these common statistical methods, predictions can be made that an event (datapoint) will about 68% of the time be found within one standard deviation and 95% of the time within two standard deviations from the mean. Statisticians also refer to such intervals as "confidence intervals"—a standard interval allowing for a certain error above and below an estimate. The bootstrap allows mathematicians a method to improve on the reliability of such error measurements. In some cases estimates of error could be improved by whole orders of magnitude.

Although the classifications are not agreed upon by all mathematicians, in general, there are four basic types of resampling techniques: Randomization exact tests (i.e., Permutation tests), Cross-validation, Jacknife, and Bootstrap. Bootstrap techniques often provide a more accurate statistical picture than traditional first-order asymptotic approximations (e.g., standard curves) but—using the power of computers—without the laborious computation or mathematical complexity. Bootstrap techniques are also important in the fitting of curves to data and to the elimination of "noise" in data.

Impact

Many statisticians consider the advancement of bootstrap methods to be among the most important innovations in mathematics during the last half of the twentieth century.

By the end of the twentieth century, bootstrap techniques were widely used to analyze both applied and theoretical problems. One attraction of bootstrap techniques was that, regardless of the underlying theory or application, the essential techniques remain essentially unchanged. Moreover, one of the major impacts of using bootstrap techniques was the ability of statisticians to extract more information from data than could be obtained without using the techniques. This increased statistical capability allowed statisticians and scientists to solve many problems previously thought too complicated to tackle.

In politics, bootstrap techniques led to explosive growth in polling. Polls alleging to portray how people would vote, for which candidate people would vote, or polls of voter's opinions regarding key issues came to be a dominant force on the political landscape in America and other technologically advanced nations. The use of bootstrap techniques allowed pollsters to quickly, inexpensively, and confidently make predictions from very small voter samples. The results of this increased statistical efficiency also provoked opposition from those who felt that use of such polls was potentially disruptive to elections and the process of representative government. There was a fear that in a close election the process of "calling an election" based on small voter samples might discourage people from voting and thus sway the election. Other critics decried political leadership based on polling because they felt that political leaders might too easily decide policy affecting millions based not on what was proper, but rather on what was merely popular with a few hundred people in a statistical database.

Bootstrap techniques are useful to biologists because they allow construction of evolutionary trees based on small samples of DNA data. In medicine, pharmacologists are able to evaluate the effectiveness and safety of new drugs from small clinical trials involving only a relatively few patients. For both government and business, bootstrap techniques are increasingly used to analyze and forecast economic data. For business, the impact of bootstrap techniques on marketing can not be overstated. Major purchasing and manufacturing decisions are now routinely based on small samples of consumer preferences. In archaeology and forensic science, bootstrap techniques allow for identification and characterization of remains. In the environment sciences the bootstrap allows widespread predictions of climate change based upon historically limited observational data. In geology, researchers using the bootstrap can better develop earthquake models based on preliminary seismic data that, in turn, can enhance earthquake warning systems.

Over the last decade of the twentieth century, the availability and accessibility of bootstrap techniques increasingly influenced the teaching of statistics. Traditionally, difficult statistical theories were often reduced to oversimplified and impractical teaching models designed to reduce the amount of calculations required of students. These models were often of little practical use. Resampling bootstrap techniques made "real" analysis possible for students and allowed mathematics teachers to use more "real-world" examples in the teaching of probability and statistics.

Late in the twentieth century, as understanding of statistics and probability became increasingly critical quantitative skills needed to cope with mountains of data made possible by computers and improved sensing technologies, bootstrap statistical techniques provided scholars with mechanisms to avoid data "overload" (i.e., too much data to handle through routine calculation). In addition, armed with more accessible statistical techniques, scholars in many fields were able to reexamine old data in which meaningful statistics were often found that were once lost in a fog of algebraic complexity.

K. LEE LERNER

Further Reading

Books

Barlow, R. *Statistics.* New York: Wiley, 1989.

Edgington, E.S. *Randomization Tests.* New York : M. Dekker, 1995.

Efron, B. and R.J. Tibshirani. *An Introduction to the Bootstrap.* New York: Chapman & Hall, 1993.

Maddala, G.S. and J. Jeong. *A Perspective on Application of Bootstrap Methods in Econometrics,* in Handbook of Statistics, *vol. 11.* Amsterdam: North-Holland, pp. 573-610, 1993.

Mammen, E. *When Does Bootstrap Work? Asymptotic Results and Simulations.* New York: Springer-Verlag, 1992.

Periodical Articles

Diaconis, P. and B. Efron. Computer-Intensive Methods in Statistics. *Scientific American* (May 1983): 116-130.

Singer, J. D. and Willett, J. B. "Improving the Teaching of Applied Statistics: Putting the Data Back Into Data Analysis." *The American Statistician* vol. 44, no. 3 (August 1990): 223-30.

Mathematicians Complete the Classification of All Finite Simple Groups

Overview

After 150 years of work capped with several decades of intense effort, mathematicians were able to demonstrate that they had classified all mathematical entities known as finite simple groups. This proof was the longest ever completed, consisting of more than 500 scientific papers that filled over 15,000 pages in mathematical journals. The proof, in final form, was over 5,000 pages in length. Mathematics' acceptance of this proof, which will likely never be reviewed in its entirety by a single person, marked an interesting change in the manner in which increasingly complex research is performed and accepted by the scientific community.

Background

In 1980 Ohio State University mathematician Ronald Solomon completed work on a mathematics problem that had begun in 1832 with a mathematics paper that was rejected by the French Academy of Sciences. The author of this paper, 20-year-old Evariste Galois (1811-1832), died in a duel just a few months later. His paper went unremarked for 12 years, until the great mathematician Joseph Liouville (1809-1882) personally presented it to the French Academy as a worthy achievement.

Galois's paper, which explored methods for solving the quintic equations (in which one of the terms is raised to the fifth power), introduced the mathematical concept of a group as a tool for attacking these equations. What Galois did not realize was that group theory would prove enormously useful in many areas of mathematics and physics, culminating in Solomon's paper.

Mathematically speaking, a group consists of five elements:

1) A set, usually designated G.

2) An operation * performed on group elements x and y such that $x * y$ is also a member of G.

3) This operation must be associative, i.e. $(x * y) * z = x * (y * z)$.

4) The group must have an "identity element."

5) Every member in the group must have an inverse.

There are an infinite number of groups. Most of these groups can have an infinite number of members and are quite complex, some with a daunting array of operations or complex operations. However, these groups are entirely composed of a class of groups called finite simple groups, which can be thought of as the building blocks of group theory. In a way, they are analogues to the prime numbers, which can be multiplied and added to form every other number.

Because of the fundamental nature of the finite simple groups, mathematicians made many efforts to understand them. This work proceeded sporadically until the 1950s, when efforts and progress picked up significantly. Around this time, it was realized that the finite simple groups fell into certain categories, with all of the members of each category sharing some fundamental properties with each other. The last of these groups, called the monster group, was described mathematically in 1980. This was followed by Solomon's mathematical proof stating that, with the description of the monster, all the finite simple groups had been discovered and described. When all the published papers in this area were compiled, it was found that more than 100 mathematicians had published over 500 papers with a total of 15,000 pages of text over more than a century to bring this problem to its conclusion.

Solomon's proof was a mathematical argument, designed to convince other mathematicians that a certain premise or statement was true. His specific proof was not long and was essentially intended to tie up the remaining loose ends in this long quest. The entire proof for this work will eventually be published in nearly a dozen volumes, reflecting the contributions of many mathematicians. But this raises the question of what a mathematical proof really is.

As mentioned above, a mathematical proof is an argument, couched in the language of mathematics, that is designed to show that something either is true or cannot be true. Mathematical proof dates back to the ancient Greeks. The first recorded proofs were generated in the sixth century B.C. when Thales (c. 624-547 B.C.) was able to use basic facts about the nature of length and area to prove the truth of common geometrical observations. Not much later, Pythagoras (c. 580-500 B.C.) developed a proof of the theorem that bears his name. Since that time, mathematical proof has evolved considerably in both length and complexity, but it retains its original character: a series of logical steps, based on mathematical concepts that are basic, that leads inexorably to the conclusion.

Over the course of centuries, mathematics has grown immeasurably more complex than in Pythagoras's time, and proofs have become correspondingly longer, more complex, and less accessible to the nonspecialist. The proof of the finite simple groups is one of the longest mathematical proofs ever generated, and it is doubtful that any single person could competently review or critique it. In addition, we entered the era of computer-generated proof in 1976, when a proof of the four-color map problem was generated, relying heavily on computer programming. In this case, the calculations required are too complex and lengthy for a person to perform, necessitating the use of computer power. This reliance on computers for many high-level problems in mathematics is increasingly common but makes many mathematicians uneasy. However, it is likely to continue and, however limited people may be with respect to performing calculations by hand, it should be some solace to remember that nobody has yet programmed human intuition or insight into a computer.

Impact

The proof that all finite simple groups had been accurately classified raises some interesting questions about the nature of mathematical proof, questions that are as much philosophical as mathematical in nature. These questions will be discussed following a short summary of the impact of group theory in mathematics and physics.

Group theory is one of the more fruitful areas of mathematics. Galois used it to help answer questions regarding quintic equations. Chemists and crystallographers use it to better describe and understand a variety of crystal structures. Group theory has also proved useful in particle physics and atomic physics. This widespread use of group theory utilizes all groups, finite and infinite, simple and complex. However, the finite simple groups are the fundamental units of which all groups are composed; understanding and classifying these fundamental units leads to a better understanding of all groups.

These aspects of group theory notwithstanding, this particular proof has implications that reach further. The most fundamental implication deals with the nature of mathematical proof itself. As originally intended, and as still assumed by most mathematicians, a mathematical proof is a series of arguments that, when reviewed by a knowledgeable mathematician, present a convincing argument that a particular statement is true. However, the proof for the classification of all finite simple groups is so long and complex—drawing on many mathematical specialties—that it is unlikely ever to be reviewed in its entirety by a single person. This means that the mathematical community is accepting as true a proposition that is unlikely ever to be verified in its entirety. Instead, mathematicians are trusting that a team of reviewers has the patience, attention, mathematical skills, and ability to coordinate and communicate their activities sufficiently well to say that the proof is accurate and complete. In turn, the mathematicians whose work has gone into this proof are placing a large degree of trust in each others' work, assuming that their colleagues are all competent and conscientious. In the case of the four-color map theorem, a further assumption is made that the computer running the program that helped prove this theorem was operating properly and the program was correctly written. These assumptions may never be fully tested or confirmed.

These assumptions seem warranted in both cases. However, by accepting these proofs, mathematicians have tacitly accepted that some areas of mathematics have grown beyond the abilities of any single person to fully comprehend. In addition, unlike proofs that came before, it may never be known absolutely that these theorems are entirely correct. There may always remain some degree of doubt about them because, in a team of mathematicians, there is always the chance that some error, small or large, has been missed because it fell between the work of two reviewers or two of the mathematicians working on the proof. In essence, for the first time, mathematicians were asked to accept a proof, not because unassailable logic showed that it must be correct but because the logic appeared correct and nobody was able to show otherwise.

This same trend towards complexity has been noted in other scientific disciplines, too. In particle physics, for example, teams of scientists sometimes a hundred strong collaborate to produce scientific research, the results of which rival these mathematical proofs in complexity. In multidisciplinary research it is not uncommon for work to be reviewed by a team of scientists, each of whom understands one or two aspects of the research, but none of whom can understand all the work in its entirety because it may span too many areas of science. And, as research becomes more arcane and research subjects smaller or more remote, these tendencies show every sign of continuing.

P. ANDREW KARAM

Further Reading

Davis, Philip and Reuben Hersh. *The Mathematical Experience.* Boston: Houghton-Mifflin, 1981.

Devlin, Keith. *Mathematics: The New Golden Age.* New York: Columbia University Press, 1999.

Eric Weisstein's World of Mathematics. WWW document. http://mathworld.wolfram.com

American Public Schools Begin Teaching New Math

Overview

Since the end of World War II, methods for teaching mathematics in the United States have changed in style repeatedly, often with controversial results. The first major change, coined New Math, was launched into American schools in the early 1960s. New Math stressed conceptual understanding of the principles of mathematics and de-emphasized technical computing skills. With New Math, out went the rote drill and practice of math facts and formulas. Instead, along with a whole new vocabulary of terms related to mathematical operations, students were taught abstract concepts involving operations on sets of numbers grouped by their characteristics and properties. The intended focus of New Math was to teach children basic mathematical truths that they would then be able to apply to specific problems in a rapidly specializing scientific and technical world. New Math, however, stirred controversy among educators and students alike and brought into focus sharp divisions in opinions regarding the goals and objectives of math education.

Background

The cultural and political climate of World War II through the 1950s sparked the need for a new way to teach mathematics. During World War II the demands of understanding and operating new technologies such as radar highlighted the weakness of American soldiers' math skills. The immigration of eminent mathematicians and scientists, who played major roles in the development of theories and technologies vital to the Allied victory, brought home the need to improve math and science education in the U.S. Throughout the Korean conflict and start of the arms race with the Soviet Union in the 1950s, military technologies continued to develop rapidly. During these years mathematics students were educated in groups according to ability, and they learned mathematical concepts

The use of new math in American public schools beginning in the 1960s has remained a major source of controversy among mathematics educators. *(Bettmann/CORBIS. Reproduced with permission.)*

through repetition, memorization (multiplication tables, for example), and timed drills. Two-thirds of students studied math only through the ninth grade.

In 1952 the National Science Foundation, in conjunction with a group of mathematicians at the University of Illinois and Yale University, began to develop a reformed method of teaching mathematics. By the mid-1950s a small number of schools were testing the group's recommended curriculum, which relied heavily on incorporating set theory and abstract mathematical laws. This curriculum served as the basis for the New Math movement.

The concepts and terminology of basic set theory were pillars of the New Math that introduced into American public school curricula during the 1960s. Instead of simply using standard arithmetic operations to solve relatively straightforward problems—where students were required to know which operations to perform and in what order to do them—an attempt was made to teach students to identify sets (groupings of numbers or objects) and elements of sets upon which operations were to be performed.

Prior to the introduction of New Math, for example, a typical student might be asked, given a sales price (e.g., $20) for a pair of widgets and a manufacturing cost of 3/5 the sales price (i.e., $12), to calculate the profit realized from the sale of the widgets. The same problem reworded according to the dictates of New Math might—in the extreme—assert:

> A set of widgets (designated as set *W*) was exchanged for a set of money (designated as set *M*). The cardinality (the number of elements in a set) of set *M* was equal to 20—with each element (i.e., currency) being equal to one (e.g., a monetary denomination of $1). If *x*'s are used to designate the elements of each set, then set *P* (representing manufacturing costs) has eight fewer *x*'s than set *M*. Represent set *P* as a subset of set *M* and determine the cardinality of the elements to determine the profit realized from the sale of widgets.

Although the construction of such densely complex problems was, in scope, far from actual problems presented to students, it is illustrative of many of the conceptual challenges and obstacles facing students studying New Math.

The Soviet Union's launch of *Sputnik 1* (the first satellite to successfully orbit the Earth) in October 1957 raised revision of math and science education in the U.S. to a strategic national crisis. Americans, seemingly behind the Soviets at every turn in the early space race, also perceived the

U.S. as behind in mathematics and scientific achievement. Throughout the country school districts rushed to boost their mathematics curricula with New Math. Although primary education received its share of the blame, special emphasis was placed on high schools, as perception grew during the Cold War that the U.S. was not producing enough mathematicians and scientists.

The introduction of New Math forced educators to abandon many of the long-cherished institutions of mathematics with regard to rote practice of operational skills. Out went computational worksheets and memorization of tables, facts, and formulas. In many cases, parents were not only incapable of helping with homework, they were incapable of understanding what they often argued was a seemingly unnecessary complication of what had once been straightforward problem solving. To make matters worse for the proponents of New Math, its introduction came at a time of radical and sweeping change in American society that ultimately resulted in changes in the public education system. Opponents of New Math, fueled by fears of falling test scores and a perceived educational gap between the United States and other industrial powers, laid the blame for many ills at the doorstep of New Math.

Impact

New Math, swiftly implemented in high schools, was widely introduced in kindergarten through grade eight by the early 1960s. Despite initial enthusiasm from researchers and endorsement from the National Council of Teachers of Mathematics (NCTM), the New Math curriculum proved difficult to maintain in real-life classroom situations. Teachers, whose classrooms were bulging with baby-boom students, struggled to introduce New Math concepts, often with little training. Parents were doubly confused. They could not understand the New Math concepts themselves, and they became concerned that many of the fundamental concepts they had learned (multiplication tables, for example) were foreign to their children. Although many school districts attempted to allay parents' concerns by conducting classes in understanding New Math, both parents and students continued to struggle.

As early as 1962 academic journals began to publish articles opposing New Math. Publication of a five-year study in 1967, showing American students lagging in math skills among other Western nations, dealt the New Math curriculum a serious blow. By the early 1970s opinion polls indicated that Americans, concerned that students were not learning basic skills, favored a "Back to Basics" approach in their children's' mathematics education, emphasizing computation, formulas, and mathematical laws. When an indictment of New Math entitled *Why Johnny Can't Add* was published in the mid-1970s, two new but similar approaches to teaching mathematics—the back to basics movement and the reform movement—found their way into American mathematics curricula.

Historians and educators often argue that Americans did not embrace the New Math of the 1960s for several reasons. Most notably, any decline or stagnation in standardized test scores, which were increasingly the index of success or failure for schools, was often blamed on confusion caused by New Math. Additionally, some fundamental concepts taught in New Math did not translate as expected into understanding New Math based technologies. Teaching arithmetic using numbers written in bases other than the often-used base ten, for example, was intended to ready Americans for the dawning computer age. The ability to write assembly language computer programs in binary (base two) code was presumed to be an essential future task. That Americans would buy their computer software rather than write it, and would deal with computer language in a point-and-click environment was, at best, an unforeseen development.

By the 1980s educators re-evaluated the New Math movement. Some educators argued that New Math concepts were beneficial and that any fault lay in the implementation process. After studying how teachers prepared to teach New Math, educators and schools of education adjusted the classroom-teacher education process to include more training in the use of manipulatives (i.e., objects such as blocks and figures that could be used to build models of mathematical concepts, or to allow physical representation of mathematical formula).

A "newer" New Math was reintroduced in some states in the 1990s, partially funded by the U.S. Congress, and again supported by the National Council of Teachers of Mathematics. Coined "Fuzzy Math" by its critics, it also shuns rote computation while embracing "experiential learning" where students figure out answers for themselves through group learning and an increased use of manipulative tools and computer simulations designed to stimulate conceptual thinking.

Regardless of its actual merit or intents, New Math was alternatively lampooned and despised by the general public. The major impact of the controversy surrounding New Math, however, was that it became the focal point for passionate social and political debate regarding pedagogy (teaching methods) in American schools. As mathematical concepts became increasingly distant, the influence of scholars over the merits of curricula became increasingly obtuse to a general public and political process comfortable only with "bottom-line" analysis of test scores. As the twentieth century drew to a close, mathematics education remained in a controversial quest for solutions that might close the documented gap in mathematics education between trailing American students and students in the rest of the industrialized world.

BRENDA WILMOTH LERNER

Further Reading

Barrow, John D. *Theories of Everything: The Quest for Ultimate Explanation.* Fawcett Columbine, 1991.

Boyer, C.B. *A History of Mathematics.* New York: John Wiley & Sons, Inc., 1968.

Kramer, Edna E. *The Nature and Growth of Modern Mathematics.* Princeton, NJ: Princeton University Press, 1981.

Miller. Jeffrey W. "Whatever Happened to New Math?" *American Heritage Magazine* (December, 1990).

Patterns of Chaos

Overview

For centuries, scientists ignored or avoided chaotic, or nonlinear, data. Real but messy results were often ascribed to experimental error or "noise." In the 1960s, starting with the work of Edward Lorenz (1917-), new approaches began to reveal the structures, dependencies, and patterns of nonlinear data. These data were found everywhere—in the price of cotton, the rise and fall of animal populations, and the shape of clouds and mountains, for example. Thanks largely to visual approaches that were supported by new, more powerful computers, general principles such as sensitivity to initial conditions and self-similarity were revealed. Chaos theory has given us a deeper understanding of nonlinear systems, both natural and artificial; pointed to solutions in communication, medicine, ecology, and other fields; it has even entered popular culture through images of fractals, the novel and film *Jurassic Park,* and popular science fiction stories.

Background

Linear equations, such as Newton's Laws, are well behaved. They provide exact answers and can be applied to practical, real-life problems. Much of engineering, for example, is based on linear equations. Nonlinear equations, on the other hand, contain infinite components and cannot be solved exactly; their results are unpredictable. For centuries, most scientists ignored and avoided nonlinear results, which they labeled useless "noise." Ignoring chaos, however, distorts our view of nature and blocks our understanding of many natural phenomena.

In the 1960s, a new way of looking at the world, called chaos theory, began to emerge. Mathematicians, biologists, physicists, and other scientists started to tackle nonlinear equations. The solution of these equations finally revealed order and structure within what looked like noise.

A milestone in chaos theory is the work of meteorologist Edward Lorenz. In 1960, Lorenz was trying to understand, model, and predict weather systems. He found that even when he was doing simple calculations, he could end up with messy and apparently random results—the hallmark of nonlinear equations. Rather than ignore these effects, Lorenz took a closer look and was able to abstract three key ideas: sensitivity to initial conditions, infinite variation, and strange attractors.

The idea of sensitivity to initial conditions came from a rounding error. When Lorenz tried to pick up where he had left off with a calculation, he found that the difference between 0.506 and 0.506127 led to wildly different weather patterns, despite the simplicity of the initial equations. Small effects could have big consequences. Infinite variation came from the discovery that, as results were tracked for certain equations, they stayed within a designated area, but they never repeated themselves. It was like having an infinite line in a box. The third idea, which came to be called strange attractors, indicated that there could be stable points that were unreachable. The typical attractors of a pendulum, for instance, are the regular consistent arc and the straight down, stopped position. Both states are energy minima that a pendulum will return to even if it is bumped and set, temporarily, into another motion. Strange attractors create motion that is just as stable, but totally irregular.

Four cultural conditions helped Lorenz establish chaos theory. First, the 1960s was a decade when authority and exact answers were questioned and challenged. Novelty and nonconformist thinking found expression in science, as well as in politics, music, and culture. Second, large investments were being made in research, much of it resulting from Cold War tensions. Pure research projects, without defined payoffs, were tolerated and, in some cases, even nurtured. For instance, Thomas Watson, CEO of IBM, personally advocated the support of "wild ducks" in his company's research division. These first two conditions provided a level of tolerance, if not acceptance, for thinkers who worked outside the mainstream. Third, as scientists in a variety of disciplines tried to solve particularly difficult problems, they all hit the same limits—a lack of tools to handle nonlinearity. Fourth, just as chaos gained a footing in the scientific world, computers became more powerful and widely available. Since chaos equations require an extreme amount of iterative (repetitive) calculation, computers offered a way through the mind-numbing computations of nonlinear systems. They also produced pictures that made complex patterns visible and accessible to intuition. This helped break down the preference for linear equations and more easily solved problems.

Of course, Lorenz was not the first to observe nonlinear systems. Jules Poincaré (1854-1912) found that the three-body problem (predicting positions of three bodies in space) was chaotic. Even knowing the gravities, velocities, and positions of each body failed to make such predictions practical because unmeasurable differences led to large effects. Experiments done in the 1920s put one of the first cracks in the linear wall. Balthasar van der Pol was looking at frequency changes driven by current changes in a vacuum tube. Because oscilloscopes were not yet available, he listened for a pattern of sounds and he consistently heard irregular noise just before frequencies locked in. Years later, these unusual results were brought to the attention of Stephen Smale (1930-). Van der Pol's work demonstrated the incorrectness of a conjecture Smale had put forth, that erratic behavior could not be stable. It set his thinking on a track that was, ultimately, to lead him to strange attractors.

In another development, Mitchell Feigenbaum (1945-) determined mathematical formulations of phase boundary conditions. By finding consistency across different scales, he showed that many systems changed phase—went from having one stable point to having two—and did so under consistent, predictable conditions. Feigenbaum's work had neither the rigor of a mathematical proof, nor the demonstrated connection with real-world phenomena that physicists demand. His discovery of the quantitative universality of the behaviors of a variety of different systems might have ended up as an academic exercise but for the work for Albert Libchaber (1934-). Libchaber did the simplest and most controlled experiment with turbulence ever. In an environment protected from vibration, at a temperature near absolute zero, he measured thermally-induced turbulence in a tiny cell filled with liquid helium. The complexity of the turbulence increased exactly as much and as frequently as Feigenbaum's equations predicted.

Libchaber's work helped to validate the work of others involved in the study of chaos, scientists who were looking at the rise and fall of animal populations, the scaling of earthquakes, and genetic variation. At the same time, the work of Benoit Mandelbrot (1924-) was helping to elucidate and popularize chaos both inside and outside the scientific community. Mandelbrot had come up with the idea of fractional dimensions—*fractals.* Fractals provided a way to express the self-similarity of nature across different scales. The quality of the shape of a coastline, for instance, does not vary whether it is seen from space, an airplane, three feet away, or under a microscope. Fractals help explain how you can have Lorenz's infinity in a box. They characterize Feigenbaum's phase changes. They also provide a means to produce appealing, evocative computer visualizations.

Impact

Chaos science has become an accepted discipline with academic departments, journals, degrees, and funding. The study of nonlinear phenomena has established common ground for many diverse scientists, particularly those who share a holistic outlook, to talk across their disciplines.

For all scientists, chaos forces a reanalysis of the meaning of the data when noise, complexity, and anomalous results appear. It demands a reexamination of edge conditions, where approximations may hide the real story. Chaos has revealed that the linear and most easily manipulated aspects of nature are actually the exceptions to a more complicated and interesting reality.

Virtually every area of science, including biology, astrophysics, thermodynamics, ecology, and chemistry has identified and studied chaotic systems. Even medicine and social sciences, such as economics, have used the tools of chaos to provide better understanding, to recognize the limits of linear approaches, and to suggest solutions to complex problems. Heart fibrillations, turbulence, chemical reactions, and the distribution of galaxy clusters are just some of the real-world phenomena that have been shown to be governed by nonlinearity.

Chaos has provided more reliable communications, guidance on vaccination programs, equations for artificial life, and simulations and techniques for encryption. It has led to improved weather predictions and new ways of compressing data.

One frontier of chaos involves public policy issues. Problems like health care, political realignments, and global sustainability have aspects—self-organization, nonlinearity, sensitivity to initial conditions, and decentralization—that seem analogous to natural systems that are known to be chaotic. The hope is that the theories and analyses developed in chaos science (often referred to as complexity in this context) may provide insight and direction in these areas. For instance, there is a growing realization that much of the detail work of economics research is based on an outdated and incorrect linear view of what is essentially a nonlinear phenomenon. Research resources are being reallocated accordingly.

Chaos theory has also changed the way science in general is done. Computers have provided a visual and dynamic means to rapidly explore and understand complex systems and are an important way for a scientist to develop intuition about abstract, complicated phenomena.

Chaos not only benefited from computers, but it became the first discipline to establish computers as valid experimental tools. The acceptance and recognition of computer simulations has since led to an understanding of artificial life, rational drug design, and ecological studies. These are considered to be real experimental contributions to their disciplines, not simply theoretical constructs. New devices, including jet aircraft, have been designed using computers. Simulations have become realistic enough, even at the video game level, to teach skills like flying and urban design. In fact, some popular games, like SimCity, even take advantage of cellular automata, which are chaos-driven artificial life forms.

Perhaps the most dramatic impact on popular culture by chaos has been its inclusion in Michael Crichton's *Jurassic Park*. This 1990 novel, and the subsequent Hollywood film, both provide popular explanations of chaos concepts and use chaos as a plot device. Fractal art has become a visible instance of chaos and can be found on T-shirts, ties, coffee mugs, and calendars. Fractal equations have become critical to computer animation for commercial movies and games, and have provided a means of data compression. And chaos continues to inspire science fiction writers and to provide rationales for their extrapolations.

PETER J. ANDREWS

Further Reading

Gleick, James. *Chaos: Making a New Science*. New York: Penguin, 1988.

Pickover, Clifford A. *The Loom of God: Mathematical Tapestries at the Edge of Time*. New York: Plenum Press, 1997.

Waldrop, M. Mitchell. *Complexity: The Emerging Science at the Edge of Order and Chaos*. New York: Touchstone Books, 1993.

The Proliferation of Popular Mathematics Books in the 1990s

Overview

The 1990s saw a great increase in the number of books on mathematical subjects addressed to the general public. The content of these books was, for the most part, readily understood by anyone with a basic high school background in mathematics. The following essay provides an overview of this book-publishing and public-interest phenomenon, including discussion of the types of general mathematical works published, focusing on the various formats used by their authors rather than providing an exhaustive bibliography, and brief comment on some specific representative titles. Finally, some reasons will be offered to explain why the 1990s witnessed such a bumper crop of mathematical books.

Background

One type of book that was very popular in the 1990s addresses the history of mathematics, either through biographies of mathematicians or the history of particular mathematical subjects. Biographies, both modern and ancient, give readers insight into the lives and works of mathematicians, and help readers understand the training, talent, dedication, and frustrations that are all part of being a research mathematician. Histories of particular areas of mathematics were also published in the 1990s. The story of the proof of Fermat's Last Theorem, for example, inspired several books addressed to a general audience. Others addressed a range of topics, usually tied together in some way by the author. This approach gave the author the opportunity to sift through various subjects in the history of mathematics and choose those that would make interesting reading for the non-mathematician.

A second format concentrates on a particular subject area, such as calculus, number theory, or statistics. Here the author attempts to make a difficult subject both readable and understandable. This requires some liberty with mathematical rigor, but makes the subject more accessible to the less mathematically inclined reader. A subset of this category includes many books about practical mathematics. In these, the author usually attempts to make readers more knowledgeable about mathematics. They appeal to an audience that wants to know more about everyday applications of mathematics.

In terms of sheer numbers, the most popular type of mathematics book in the 1990s was that which attempted to convey the beauty and excitement of mathematics. Some compared the elegance of mathematics to poetry; others made a case for mathematics as an art rather than a science; still others provided mathematical puzzles and curiosities for their readers' enjoyment. Whatever their forms, these books usually had one primary purpose: to inspire in readers an appreciation for the beauty and elegance of mathematics.

Impact

In histories of mathematics, biographies are often effective for telling a story. Four published in the 1990s are of particular interest. The first two chronicle the lives of twentieth-century mathematicians: *The Man Who Knew Infinity,* by Robert Kanigel, chronicles one of the most fascinating characters in history, the Indian mathematician Srinavasa Ayengar Ramanujan (1887-1920). It is a story of a true genius that ends in tragedy. Another excellent book is *The Man Who Loved Only Numbers,* by Paul Hoffman. This is a story about Paul Erdős (1913-1996), an eccentric number theorist who traveled the world throughout his lifetime with a singular purpose: to do mathematics. Neither of these biographies is dry reading—both were fascinating men who led incredibly eventful lives.

Two other 1990s biographies chronicle the lives of mathematicians from the past. *Archimedes: What Did He Do Beside Cry Eureka?,* by Sherman Stein, details the life of the Greek mathematician famous for, among other things, the theory of buoyancy and the law of the lever. In his book *Euler: Master of Us All,* William Dunham presents the life and works of the eighteenth-century mathematician who is generally regarded as the most prolific mathematician in history.

Authors in the 1990s also wrote on a wide variety of historical topics. While some address whole branches of mathematics, most focus either on one narrow subject or combine essays on related subjects. For instance, Eli Maor's book *e: The Story of a Number,* traces the history of *e*, the number that is the base for natural logarithms. Like π, *e* is extremely important in many areas of mathematics.

Other books combine separate historical topics into one work. *Five Golden Rules: Great Theorems of 20th-Century Mathematics and Why They Matter,* by John L. Casti, and two books by William Dunham, *Journey through Genius: The Great Theorems of Mathematics* and *The Mathematical Universe: An Alphabetical Journey Through the Great Proofs, Problems, and Personalities,* all give the reader an eclectic tour of the many facets of the mathematical world.

A single event of the 1990s spawned the greatest number of attempts to communicate mathematics to the general public. This event was the successful proof of Fermat's Last Theorem by Princeton mathematician Andrew Wiles (1953-). Wiles proved a theorem that had become the Holy Grail of mathematics and had eluded the greatest mathematical minds for over three centuries. In addition to many newspaper and magazine articles, several books were written to explain the importance of Wiles's accomplishment to the nonmathematician. Simon Singh's book, *Fermat's Enigma,* is probably the best known of these works, but other well written books such as *Notes on Fermat's Last Theorem* by Alf van der Poorten and *Fermat's Last Theorem: Unlocking the Secret of an Ancient Mathematical Problem,* by Amir D. Aczel, also helped spread the story of Wiles's triumph.

The second category of popular mathematical works included books on a variety of subjects, from basic arithmetic to calculus. Many of these say that they contain tools, tricks, and techniques to help the average person become more adept at mathematics. Some, such as David Berlinski's *A Tour of the Calculus,* present a very serious subject in a somewhat informal, even light-hearted, manner. These appeal to a less general audience than those in the first category, but have found their niche in a world that places more and more emphasis on mathematical ability.

Also falling within this category is a large number of books whose purpose is to teach something about mathematics while at the same time showing how mathematics is used in everyday life. Among the many books written for this purpose are a series of works by John Allen Paulos, including *Innumeracy: Mathematical Illiteracy and Its Consequences* (1990), *Beyond Numeracy: Ruminations of a Numbers Man* (1992), *A Mathematician Reads the Newspaper* (1996), and *Once Upon a Number* (1998). In these books, Paulos draws attention to the mathematics that we are confronted with daily. He also emphasizes the consequences we face if we do not understand the meaning of the statistics that have become part of out everyday experience.

The final category of popular mathematics is the largest in terms of numbers of books published in the 1990s. A search of any database containing published books yields a huge number of works devoted to the beauty and fun that mathematicians find in their subject. These books attempt to convey these feelings towards mathematics to the reader. Titles of books include words like "joy" and "beauty" along with the word "mathematics." Although this seems unusual, possibly even ridiculous, to many people, such books attempt to convey the essence of mathematics to the general public. *The Joy of Mathematics: Discovering Mathematics All Around You,* by Theoni Pappas, is just one example of such a work.

Another important contribution to the popularization of mathematics in the 1990s did not start out as a book. *Life by the Numbers* was a television series broadcast by PBS that explored the importance of mathematics in today's society. A companion book to this series by Keith J. Devlin, also entitled *Life by the Numbers,* explores in written form what the program addressed in video form. Such topics as the statistics of gambling, mathematics in nature, and mathematics in sports, created an extremely informative and entertaining look into the world of mathematical applications.

This is not an exhaustive list of popular works pertaining to mathematics published in the 1990s, and the 1990s were by no means the first time that such works had been published. It does seem, however, that many more books of this nature appeared then than in previous times. We might attribute this to an increasing awareness that mathematical knowledge is an important component of successful modern life. It may also be attributed to the new and creative ways authors have found to present mathematics to the public. Computer-generated graphics, videos, and other technologies have combined to make mathematics more accessible than ever before. Whatever the reasons, these new books have led to an increased awareness of, and appreciation for, the importance of mathematics in our society.

TODD TIMMONS

Further Reading

Casti, John L. *Five Golden Rules: Great Theories of Twentieth-Century Mathematics and Why They Matter.* New York: Wiley & Sons, 1996.

Devlin, Keith. *Life By the Numbers.* New York: Wiley, 1998.

Dunham, William. *Journey Through Genius: The Great Theorems of Mathematics.* New York: Wiley, 1990.

Dunham, William. *The Mathematical Universe: An Alphabetical Journey Through the Great Proofs, Problems, and Personalities.* New York: Wiley & Sons, 1994.

Pappas, Theoni. *The Joy of Mathematics: Discovering Mathematics All Around You.* San Carlos, CA: Wide World, 1989.

Paulos, John Allen. *Once Upon a Number: The Hidden Mathematical Logic of Stories.* New York: Basic Books, 1998.

Singh, Simon. *Fermat's Enigma: The Epic Quest to Solve the World's Greatest Mathematical Problem.* New York: Anchor Books, 1998.

The Contributions of Japanese Mathematicians since 1950

Overview

The latter part of the twentieth century has been called the golden age of mathematics. Much of the luster of this period comes from the contributions of European, Chinese, and American mathematicians. However, to dwell entirely on their contributions does a disservice to the mathematicians from Japan because this era has certainly been a golden age for Japanese mathematicians, too. In the latter decades of the twentieth century, Japanese mathematicians made significant contributions to several fields of mathematics and helped launch the field of computational mathematics. In so doing, they not only helped expand knowledge of traditional mathematics, but helped enlarge the very boundaries of their profession by taking advantage of increasingly powerful computer technology. This, in turn, had important effects on mathematics and many aspects of modern society.

Background

Japan has long been accused of being a nation that can innovate but not invent. Critics, pointing to Japan's culture of conformity and traditionalism, suggest that the Japanese lack true originality and creativity, excelling instead at exploiting discoveries made by other nations. While this stereotype carries with it some degree of truth, it must be noted that Japan has produced a great many innovative scientists who have won acclaim for a number of important discoveries. The field of mathematics is no exception.

Japanese mathematicians came into their own in the years following World War II. Kunihiko Kodaira (1915-1997) was awarded the Fields Medal in 1954 for his outstanding contributions in harmonic analysis, algebraic geometry, complex manifolds, and other areas of mathematics. In 1970 Heisuke Hironaka won the Fields Medal for his work on algebraic varieties, and Shigefumi Mori (1951-) earned the 1990 Fields Medal for his work in algebraic geometry. Other Japanese mathematicians who have won high honors for their original and important work in mathematics include Kiyoshi Ito, Kenkichi Iwasawa (1917-1998), Goro Shimura, Tosio Kato, and others.

Still other Japanese mathematicians made significant contributions to understanding one of the most important problems of the last 300 years: the proof of Fermat's last theorem. In this endeavor, Yutaka Taniyama and Goro Shimura helped lay the groundwork that led to Andrew Wiles's (1953-) 1994 proof of the validity of Fermat's last theorem. The Shimura-Taniyama conjecture, which describes the relationship between certain families of curves, proved key to solving this theorem.

In addition to pure mathematical research, Japanese mathematicians have helped develop computational techniques that use computer technology to solve a variety of problems. In this endeavor they are building on the many contributions of Japanese computer scientists and electronics corporations. Of particular importance are the exceptionally sophisticated computational techniques to determine airflow around vehicles developed by Kunio Kuwahara and Susumu Shirayama. These simulations have been extremely useful in designing cars that are safer and more fuel efficient. In another application of using computers to solve mathematical problems, Yasusi Kanada has performed very interesting and important work in the areas of cellular automata and has used supercomputers for a wide variety of mathematical research.

Impact

Andrew Wiles's proof of the solution to Fermat's Last Theorem stands as one of the most important mathematical works of the twentieth century. Because it had been unproven for three centuries, its proof won Wiles worldwide acclaim. It is entirely likely that this problem would not have been solved without the work of Shimura and Taniyama on elliptic functions.

These mathematicians showed that a class of curves called elliptic curves were related to another class of curves known as modular curves. When this association was made, other aspects of Fermat's Last Theorem fell into place. It took Wiles seven years of hard work to take these observations and turn them into a rigorous mathematical proof, but he did just that, presenting his proof to the world in 1993. Since that time, work on the Shimura-Taniyama conjecture has continued and seems to indicate that other, deeper aspects of mathematics may be explored using this relationship. If so, then the Shimura-Taniyama conjecture will prove to be much more important than suggested even by its use in proving Fermat's last theorem.

It is also interesting to note that Fermat's last theorem is one of the few mathematics problems to capture widespread attention in the media and among the general public. From this perspective, too, the work of Shimura and Taniyama, while not subject to the same acclaim as that of Wiles, helped to raise public awareness and appreciation of mathematics and mathematicians, even if for only a short time.

Other mathematicians also made singular and lasting contributions to the study of pure mathematics. The work of Shigefumi Mori seems to have been of particular importance because his outstanding mathematical creativity clearly disproves the preconception that Japanese scientists lack creativity. In fact, a fellow mathematician said of Mori's work that "the most profound and exciting development in algebraic geometry during the last decade or so was the minimal model program or Mori's program in connection with the classification problems of algebraic varieties of dimension three. Shigefumi Mori initiated the program with a decisively new and powerful technique, guided the general research direction with some good collaborators along the way, and finally finished up the program by himself overcoming the last difficulty." For this work, Mori was awarded the 1990 Fields Medal, mathematics' highest honor.

Impressive as these accomplishments are, other Japanese mathematicians also made significant contributions in the use of computers to solve mathematical problems. In 1988 Yasumasa Kanada used a computer to calculate the value of π to over 201 million decimal places. While such calculations may not seem terribly important or relevant to everyday life, they are not only an important test of a computer's power, but also help to demonstrate efficiency and effectiveness of new computing techniques. In effect, they may be used as a sort of computational (or programming) test drive.

Kanada also worked in the area of cellular automata (CA or "computer life"), small programs that interact within a computer under certain rules governing their behavior. Researchers, including Kanada, noticed that CA tend to build complexity when left to their own devices, constructing increasingly complex systems of interactions. In this, they seem to mimic what we see in the real world, where life has built a very complex biosphere from the relatively simple building blocks that existed in the early earth. One of Kanada's insights into CA mathematics has been to introduce noise—random variations—into his scenarios. This makes them more realistic, giving deeper insights into the phenomena under study. For example it might be possible in a computer program to stand a baseball bat on top of a baseball because, if the forces are aligned correctly and no outside influences are permitted into the program, the bat will balance indefinitely. However, we know that this does not happen in the real world, so such a model is inaccurate. Kanada's programming would include random fluctuations in such a problem, so that in this analogy the bat would tumble to the ground because, in real life, imperfections in the covering on the ball, small gusts of wind, or spare vibrations all occur and cause the bat to fall. By introducing similar random variations into systems of cellular automata, Kanada is able to make them more closely resemble real-world systems.

Finally, we must mention the contributions of Kunio Kuwahara and Susumu Shirayama as examples of similar work performed by many other Japanese scientists. Kuwahara and Shiryama have taken a leading role in the use of computers to solve mathematical equations governing complex physical phenomena. These equations cannot be solved analytically (that is, by plugging numbers into the equations). Their most impressive achievements have been in the area of fluid flow, specifically around automobiles, in which they examined the various factors that cause drag and reduce fuel efficiency. The basic equations that govern fluid flow, the Navier-Stokes equations, are well known, but are also very difficult (possibly impossible) to solve analytically. Kuwahara and Shiryama have translated mathematical algorithms into computer code that allow these complex equations to be solved numerically at a large number of locations in space. They then program the important parameters of the automobile to be tested (in effect, telling the computer what the car looks like) and then simulate releasing smoke into mathematical streamlines when the car is put into an electronic "wind tunnel." These models, then, are used to verify actual airflow characteristics around the proposed new car. By helping design fuel-efficient cars, much of this work helps to save money and natural resources as well.

P. ANDREW KARAM

Further Reading

Books

Devlin, Keith. *Mathematics: The New Golden Age*. New York: Columbia University Press, 1998.

Kaufmann, William and Larry Smarr. *Supercomputing and the Transformation of Science*. Scientific American Library, 1993.

Other

Eric Weisstein's World of Mathematics. (http://mathworld.wolfram.com)

The MacTutor History of Mathematics Archive. (http://www.vma.bme.hu/mathhist/).

Kepler's Sphere-Packing Conjecture Is Finally Proved

Overview

For nearly four centuries, the Kepler conjecture regarding the most efficient geometrical arrangement for stacked spheres remained one of the most complex and vexing problems in mathematics. Kepler's conjecture—a mathematical expression of commonplace packing techniques—states that the most efficient (i.e., the densest arrangement with the least unused or empty space) packing of spheres results from placing a layer of spheres as tightly as possible on top of an underlying layer. Although Kepler's conjecture seemed proved by everyday experience, a mathematical proof that no better packing arrangement existed seemingly eluded mathematicians until the last decade of the twentieth century.

Background

Although stacking problems are an intuitively ancient exercise for mankind, the formal incorporation of stacking theory dates to the sixteenth century when Sir Walter Raleigh (1554?-1618) challenged his assistant, English mathematician Thomas Harriot (1560-1621), to find a reliable and efficient manner to estimate the number of cannonballs in a munitions pile. Such an estimate was important both to the efficient movement of an army, especially aboard ships, and to the strategic estimate of an opponent's ability to wage war. Although Harriot was quickly able to come up with a useable calculation he sought a deeper mathematical proof that the intuitive way to pack spherical objects such as cannonballs was indeed the best method available.

Harriot sought the advice of German mathematician and astronomer Johannes Kepler (1571-1630) who was also interested in such stacking problems. Kepler was attempting to reconcile the orbits of planets in the then controversial heliocentric universe proposed by Polish mathematician Nicholas Copernicus (1473-1543) . In asserting that the known planets traveled in orbits around the Sun rather than about the Earth, Copernicus challenged the long-standing Ptolemaic concepts (named after Alexandrian astronomer and mathematician Ptolemy who published his theories in approximately A.D. 140) of a universe consisting of stacked crystalline spheres upon which the heavens moved. Although he eventually made courageous and fundamental discoveries regarding the elliptical orbits of planets, Kepler spent a lifetime attempting to relate the orbits of planets to the shape of perfect solids. Kepler was convinced that the correct model for the universe consisted of tightly packed stacked perfect solids located within spheres having a common center located at the Sun. Kepler's passion to find the most efficient way to arrange these spheres drove him to study other stacking and crystalline arrangements in ice or snowflakes.

Although Kepler could not mathematically deduce a more efficient way to pack spherical objects he was able to determine that the intuitive method used for centuries was indeed the most efficient known manner to pack. In essence, Kepler's problem came from his inability to construct a mathematical proof that there were no other arrangements that might prove more efficient.

After studying the problem, in 1611 Kepler asserted that the so-called "face centered" cubic lattice was the most efficient packing arrangement for spheres. This assertion regarding the problems related to sphere packing became known as the Kepler's conjecture.

Kepler's conjecture stated that the tightest pack of any spherical objects (i.e., balls) of the same radius was achieved by stacking layers of spheres one upon the other. Each layer being added, however, was shifted to allow the new layer of spheres to partially fit into, and thus partially fill, the holes between spheres located in the lower layer.

Although efficiency with regard to packing problems is a simple mathematical concept, it is an economically critically concept to agriculture and industry. Spheres (e.g., grapefruit) can be laid in layers to form a crystal lattice wherein the centers of the spheres are directly on top of each other. Such an arrangement, however, is not nearly as an efficient means of packing as method suggested by Kepler's conjecture. When the centers of the upper layer of spheres are shifted to a position over the gaps between spheres in the first layer, almost twice the number of spheres (e.g., fruit, balls, spherical mechanical parts) can be packed into the same space. Moreover, without such shifting, the wasted space in a lattice nearly equals that of the space occupied by the spheres. A shift in the centers of the spheres located in the initial (i.e., lowest) base layer, however, so that neighboring rows also have centers filling the gaps in adjoining rows (upon which additional layers are then laid using the gap filling method) provides increased efficiency. Using such a packing arrangement, nearly three-fourths of the space in a box is comprised of spheres and only slightly more than a quarter of the space remains unused. (i.e., an efficiency of about 74%).

Although most people are unaware of the mathematics behind the proof of Kepler's conjecture, they deal with the conjecture on a practical basis almost daily. Grocers, for example, use this method to stack fruit for shipping and display.

Efficiency in sphere packing is also critically important to physicists and chemists trying to understand and predict the behavior of atoms and molecules. Because nature seeks the lowest entropic state, there is a drive toward efficiency in packing. Understanding the mathematics related to sphere packing allows physicists and chemists to study areas such as metal structure and the dispersion of gas molecules.

German mathematician and physicist Carl Friedrich Gauss (1777-1855) offered a partial two dimensional solution to the Kepler conjecture, by proving that the most efficient arrangement of circular disks was one that allowed a disk to be surrounded by six others in a hexagonal arrangement. At the close of the nineteenth century, mathematician David Hilbert (1862-1943) included Kepler's conjecture regarding the sphere-stacking problem in his famous list of 23 problems to be solved during the twentieth century.

Impact

Although Kepler's conjecture regarding sphere-packing seemed true to everyday experience and remained mathematically undisputed, a formal mathematical proof remained elusive. Kepler's conjecture vexed mathematicians well into the twentieth century when two independent solutions were put forth by mathematician Wu-Yi Hsiang of the University of California and mathematician Thomas Hales (1958-) of the University of Michigan.

In 1993 Hsiang claimed a proof to Kepler's conjecture. Although well-constructed, Hsiang's proof eventually ran into problems because of the fundamental assumptions upon which it relied. Hales and his research student Samuel P. Ferguson subsequently announced another proof to Kepler's conjecture that avoided the problems found in Hsiang's proof. Although Hales himself refrains from actually claiming the Kepler conjecture as "proved," his proof led most, but certainly not all, mathematicians to consider the Kepler conjecture as probably proved. At the close of the twentieth century, the issue had not yet been finally settled.

Hales's work was based, in part, on the 1950's work of mathematician L. Fejes Tóth who proposed that the Kepler conjecture could be reduced to a finite but impossibly large calculation. With amazing foresight, Tóth proposed that computers might one day be used to construct a proof to the conjecture. In addition to the use of computers, Hales's work utilized the concept of hyperplanes to overcome the limitations of the attempting to maximize functions through the standard techniques used in calculus (e.g., taking derivatives) that are often too cumbersome for complex problems.

The proof of Kepler's conjecture did prove highly complex, requiring hundred of pages of analysis and equations with more than a hundred variables. Accordingly, proof that there was no more efficient arrangement than the standard "grocer's solution" proved difficult without the use of powerful computers. Similar to the 1976 solution put forth by Wolfgang Haken and Kenneth Appel (1932-) regarding proof of the Four Color Conjecture (i.e., English mathematician Francis Guthrie's conjecture, put forth in 1852, that only four colors were required when attempting to color maps in such a way that no neighboring areas would be forced to have the same color), Hales's proof wasn't purely theoretical because it relied, in part, on computer-based algorithms.

Of special significance during the information revolution was the way Hales announced and argued his proof of Kepler's conjecture. Although he subsequently published in established journals, Hales initially set out his proof on the Internet. This was a radical departure from the traditional methods of submitting proofs to the mathematical community for review. Prior to late twentieth-century advancements in information technology, scholars were limited in ways to present new ideas. Essentially, scholars could submit articles for selected peer review and possible publication in a scholarly journal or they could present their work at scholarly (e.g., mathematical) conferences. Similar scrutiny had revealed the flaws in a previously offered proof of Kepler's conjecture put forth by Hsiang.

Such computer-reliant proofs remained a source of controversy in late twentieth-century mathematics. Critics of such proofs asserted that potentially undiscovered hardware or programming errors (e.g., potential flaws with linear inequalities used in the linear programs) cast a shadow of doubt over such proofs not usually present in traditional pen and paper proofs. Pure mathematicians claimed that such computer-dependent proofs lacked elegance (i.e., simplicity) required in theorems. More strident critics of computer usage classified such solutions as brutish methodology that attempted to prove theorems by volume of calculation. Other scholars, however, lauded the use of computers and the concurrent programming advancements as careful, rigorous, and realistic treatment of complex problems.

The potential proof of Kepler's conjecture excited researchers looking for potential solutions to other "stacking" problems. In particular, Hales's potential proof of the Kepler conjecture caught the interest and scrutiny of mathematicians and engineers concerned with coding theory (i.e., the branch of mathematics concerned with Keplerian-type packing problems). Hope ran high that the methodology developed to prove Kepler's conjecture might provide potentially new and innovative ways to stack (compress) electronic computer data without corrupting or damaging the data. Researchers in this field also speculated that advances in packing theories could allow for greater memory capacity, increased encryption security, and more efficient transmission of data over the burgeoning Internet.

Proof of packing theory impacts areas as diverse as the search for oil (e.g., through petrological analysis of reservoir capacity) and the quest for formulating theories regarding nature. Physicists seek to use methodologies derived from the proof of Kepler's conjecture to study how holes or voids are displaced in space relative to the positions of the spheres and how holes or voids are displaced relative to each other.

K. LEE LERNER

Further Reading

Books

Gruber, P.M. and C.G. Lekkerkerker. *Geometry of Numbers.* 2nd ed. Amsterdam: North-Holland, 1987.

Periodical Articles

Hales, T.C. "The Status of the Kepler Conjecture." *The Mathematical Intelligencer* 16 (1994): 47-58.

The Intimate Relation between Mathematics and Physics

Overview

Since the 1960s physics has seen a rebirth of the use of advanced mathematics. Much of this revival occurred after the study of black holes was greatly expanded in the 1960s and 1970s by the English scientists Stephen Hawking (1942-) and Roger Penrose (1931-). In the 1980s physicists' 10-dimensional superstring theories received another mathematical boost from tools developed by Edward Witten (1951-) and others.

Background

Physics and mathematics have always enjoyed a close relationship, beginning in the Renaissance with Johannes Kepler's (1571-1630) 1609 discovery of the three laws of planetary orbits. In 1687

Isaac Newton (1642-1727) introduced the theory of gravity. James Clerk Maxwell (1831-1879) was able to unify the forces of electricity and magnetism in 1865 with the theory of electromagnetism. In the twentieth century mathematical theories from the fields of geometry were instrumental in constructing Albert Einstein's (1879-1955) theory of general relativity as well as in the later development of superstring theory. All of these theories have been predicated upon the prior development of mathematical techniques that had been invented for pure and applied purposes.

In the late seventeenth century Isaac Newton could not have developed the theory of gravity without calculus, a set of mathematical techniques he had developed for studying rates of change. (Calculus was also developed independently by the German mathematician and physicist Gottfried Leibniz (1646-1716.) The definition of gravity underwent another significant revision in 1916 when Albert Einstein showed that gravity could be interpreted as curvatures of space and time. But Einstein could never have developed his theory, now called general relativity, without the non-Euclidean geometry developed by the German mathematician Bernhard Riemann (1826-1866). Riemann's geometry system, developed in 1854, was able to handle descriptions of space where curves predominate and all lines must eventually meet. This was an entirely new way of describing space that the 2000-year-old Euclidean system could not handle.

In the twentieth century several scientists, including Niels Bohr (1885-1962) and Erwin Schrödinger (1887-1961), developed quantum mechanics, which describes the structure of atoms with great precision. Quantum mechanics deals with the microscopic world by treating particles as both particles and waves. Mathematics became increasingly important to physicists during the latter half of the twentieth century, usually because the physical objects under investigation were inaccessible to experimental physics. These objects are as large as black holes and as small as the tiny strings and branes of superstring theory.

Impact

Black Holes

Black holes were initially thought to be strange quirks of Einstein's general relativity theory. These objects are incredibly dense and have a gravity so strong that neither light nor matter can escape. Since they can never be observed directly, descriptions of their shape, size, temperature, and mass remain almost wholly mathematical. (Black hole masses can be calculated relatively easily by observing their effects on surrounding matter.)

In the 1960s Stephen Hawking and Roger Penrose collaborated to study the centers of black holes, regions known as *singularities,* where time and space become so warped and twisted that they cease to have meaning under normal physical laws. Penrose and Hawking showed that singularities were possible and that under certain conditions would have to be formed. After this discovery, black holes, whose study had formerly been a rather esoteric field, suddenly became a hot subject for many theoretical physicists.

Because black holes seemed to suck in everything and never release anything, they appeared to violate certain physical laws. Then in 1973 Stephen Hawking showed that black holes actually radiate a tiny amount of heat. Hawking proved this by combining mathematics and theories from quantum physics, general relativity, and the laws of thermodynamics—the first time these theories had ever been used simultaneously.

Penrose Tiling and Quasi-Crystals

Penrose also made significant contributions to the field of geometry. In 1974 he showed that nonrepeating patterns could be made out of just two repeating geometric figures. These shapes, called tiles, could be arranged on a flat surface without ever encountering a repeating pattern. In his honor these patterns are now called Penrose tiles. For a decade, Penrose tiles seemed to be an interesting mathematical proof without a real-world application. Then in 1984 a crystalline alloy of aluminum and manganese was discovered that arranged itself into Penrose-tiling patterns. These crystals are called quasi-crystals since they do not exhibit the repeating molecular structure of regular crystals.

The Shape of the Universe

The Big Bang theory is currently the best description of the universe's history. It suggests that the universe began as a small, compressed point of matter—possibly a singularity similar to those studied by Penrose and Hawking—that exploded violently outward to form everything we see today. Since we cannot observe the universe in its earliest days—no matter how far we look out with telescopes—describing our universe's birth is a problem for both mathematics and physics. Astrophysicists and cosmologists are also curious about the fate of the universe: Will it continue racing and expanding outward, will the expansion come to a stop, or will it collapse back upon itself? Most observations to date suggest that the universe will keep expanding; whether the expansion will slow to zero or continue forever, however, is still unknown.

A theory first proposed in 1980 by Alan Guth of the Massachusetts Institute of Technology, called inflation theory, predicts another outcome. Inflation theory states that the universe underwent an exponential expansion when it was about the size of a pea. This expansion will eventually come to a stop, making the universe "flat" and obeying Euclidean geometry. Most calculations of the universe's total mass do not support this view, however. Instead, astronomers' observations show that the universe does not have enough mass to halt the runaway expansion and will, therefore, continue expanding forever.

Inflation theory might work, however, if scientists could account for this apparent lack of matter. One theory proposes the existence of "dark matter," which is matter that does not glow with enough energy to be seen from earth. A large-enough amount of dark matter could halt the expansion. The other potential solution involves a mysterious form of energy often called the cosmological constant. It was a "fudge factor" introduced by Einstein to make the theory of general relativity account for observations regarding the unchanging size of the universe. He later retracted the cosmological constant, calling it the biggest mistake of his life, when Edwin Hubble (1889-1953) showed that the universe was indeed expanding. Today cosmologists have resurrected the cosmological constant in a different form, believing that it could be an undiscovered form of energy that will slow the universe's expansion, keeping it flat and in agreement with inflationary theory.

Cosmologists who accept that the universe does not contain enough mass or energy to halt expansion say that our universe is "negatively curved." Mathematicians have shown that many negatively curved spaces with hyperbolic geometry can fold up in ways that could still contain a finite universe. These shapes also give rise to some rather interesting conjectures, one of which is that you could travel in a straight line across the universe and eventually end up at your starting point. Another is that we could conceivably look out and see our own Milky Way galaxy at a young age after its light had traveled around the entire universe. Proving this theory would require very detailed observations of the skies. In 2000 NASA's Microwave Anisotropy Experiment satellite will begin to make some of these observations.

String Theory

Superstring theory had its beginnings in the 1970s. At that time, some physicists were dissatisfied with contemporary theories that treated subatomic particles as points. They encountered too many complexities when using a dimensionless point to describe fundamental particles (these were the same kinds of difficulties encountered by Penrose and Hawking in their study of singularities.) These physicists tried instead to come up with a method for treating the particles as loops, or strings. The vibrational patterns of these loops, in turn, determined their characteristics and how they formed larger particles, such as quarks, electrons, and protons. But the theory, simply called string theory back then, was not very successful.

In the 1980s the theory was reborn as supersymmetric 10-dimensional string theory, or superstring theory for short. Theoretical physicists found that using one-dimensional strings in a 10-dimensional universe showed promising results. The theory blossomed with investigations by Edward Witten and others. This period in the mid-1980s is now known as the first golden age of string theory, during which physicists were able to make giant leaps forward. Since then the development of string theory has slowed its rapid pace of advancement.

Unfortunately, superstring theory deals with particles so small that physicists have little hope of ever detecting them or even building a large-enough detector. A particle accelerator large enough for the task would need to be larger than the solar system. This difficulty leaves many scientists leery of superstring theory. If the theory remains unverifiable by experiment, how can it ever be proved? In addition, a theory that states that there are 10 dimensions, six of which are curled up so tightly that we cannot see them, rubs against today's four-dimensional viewpoint, which involves three dimensions in space and one of time. String theory supporters remain convinced of their theory's veracity because it fits so well with everything else we know. They believe that superstring theory is one of the few instances in which theory has been able to leap ahead of experiment. Also, since the mathematics involved are so advanced, many string theorists admit that they still do not understand all the implications of the mathematics they have discovered. As Witten has said, it's twenty-first-century mathematics that miraculously landed in the twentieth.

The Seiberg-Witten Equations

Transfers between mathematics and physics have not just been in one direction. At the end of 1994 physics made a huge contribution to the mathematical field of topology when Ed Witten and Nathan Seiberg found that some of the mathematics they had developed to study singularities was applicable to the study of manifolds, which are descriptions of how space and curves can be deformed and stretched without changing certain properties. The equations caused a revolution in the study of manifolds and gave mathematicians an entirely new classification scheme. This development, in turn, has been important for studying the effects of gravity on space and time.

PHILIP DOWNEY

Further Reading

Books

Greene, Brian. *The Elegant Universe.* New York: W.W. Norton, 1999.

Penrose, Roger. *The Emperor's New Mind.* New York: Oxford University Press, 1989.

Peterson, Ivars. *The Mathematical Tourist.* New York: W.H. Freeman, 1998.

Periodical Articles

Cowen, Ron. "Cosmologists in Flatland." *Science News* (February 28, 1998): 139-141.

Glanz, James. "Radiation Ripples from Big Bang Illuminate Geometry of Universe." *New York Times* (November 26, 1999): A1.

Horgan, John. "The Pied Piper of String Theory." *Scientific American* (November 1991): 42-47.

Horgan, John. "Quantum Consciousness." *Scientific American* (November 1989): 30-33.

Peterson, Ivars. "Circles in the Sky." *Science News* (February 21, 1998): 123-125.

Internet

Weburbia Press. http://www.weburbia.demon.co.uk/pg/contents.htm

"The Superstring Mystery." *The Internet Science Journal.* http://www.vub.ac.be/gst/sci-journal/Disciplines/QGD/notebooks/strings.htm

The Flowering of Differential Topology

Overview

A number of important advances in understanding the curvature of surfaces in three- and higher dimensional space have occurred in the decades following 1950. A method of cutting up surfaces, called surgery on manifolds, enabled the resolution of some long-standing conjectures about surfaces in higher dimensional spaces. In ordinary three-dimensional space, computer-assisted investigators discovered families of new minimal area surfaces. René Thom's catastrophe theory claimed to provide a means of explaining abrupt changes in the stable behaviors of complex systems, but met a varied reception among scientists and mathematicians.

Background

Topology is concerned with the behavior of geometrical forms as they are stretched or squeezed or twisted. To a topologist a billiard ball and a soup bowl are related because they can be gradually transformed into each other without separating any points that were originally very close to each other. In the same sense, a donut and a teacup are topologically related to each other but not related to a billiard ball since they both involve an opening. One of the simplest objects studied by topologists is the Möbius strip, which is obtained when a long strip of paper is given a half twist and its ends pasted together. This object has only one surface and one edge, as can be seen by coloring the middle of the strip or the edge with a crayon. But topologists are also interested in the properties of objects that cannot be visualized in any usual sense.

As in other areas of mathematics, topologists try to be as general as possible in drawing conclusions. Often they do not restrict themselves to the two- or three-dimensional space of experience, but ask about the characteristics of objects in four, or five, or 500 dimensions. These studies are not necessarily sheer flights of fancy. An equation in five variables defines a surface or "hypersurface" in five-dimensional space. Topologists will want to know whether the surface is closed or infinite in extent, and about how curved different parts of the surface might be.

Differential topology is the study of the curvature of generalized surfaces, or, as topologists call them, manifolds. Measuring the curvature of a surface in a space of more than three dimensions is difficult and topologists often deal with the problem by taking what might be called an "ant's eye view." Suppose that a mathematically inclined ant is walking over the surface of a very large mound. Since he is small, he is unable to tell if the mound is flat, or has a bit of curvature to it. If he finds a bit of string, however, he can tuck one end into the ground and trace out a circle, and then measure, perhaps by counting paces, both the apparent circumference of the circle and its diameter. If the ratio of circumference to diameter equals the number π, that is 3.14159 . . . , he can conclude that the surface is flat. If it is larger, then the surface is said to have a negative curvature. If it is smaller, then it has a positive curvature. In either case he can use the difference between the ratio and π as a measure of the curvature.

One of the means by which topologists classify manifolds is by their connectedness. If one draws a circle on the surface of a sphere, one can deform it gradually so that it gets smaller and smaller and shrinks to a point. Because this is true for any circle that can be drawn on a sphere, or those other three-dimensional objects that a sphere can be transformed into, we say that the sphere is simply connected. This property is not shared by the torus (donut shape), because a circle drawn around the hole can never be shrunk to a point.

One of the founders of topology, the mathematician Jules-Henri Poincaré (1854-1912), asked whether the same shrinking circle argument could be used for a manifold that was the three-dimensional surface of a four-dimensional sphere; that is, could the so-called "three-spheres" and the topologically related objects be distinguished from all other three-dimensional manifolds by this shrinking circles test. This question has remained unanswered, mainly because topologists have not found a completely effective way of classifying three-dimensional manifolds. In 1961 the American mathematician Stephen Smale (1931-) developed a method for classifying higher dimensional manifolds by breaking them up into small pieces that could be moved around. Smale was able to prove that the circle test would hold true for manifolds of five or higher dimensions, but was unable to find a proof for the three- or four-dimensional case. The study of higher dimensional manifolds has provided a number of surprises. In 1959 American mathematician John Milnor (1931-) showed that a seven-dimensional sphere could be rearranged to form a smooth manifold in 28 different ways.

Another area of interest in differential topology is the existence of so called minimal surfaces. If one forms a closed curve out of wire and then dips it into a soap solution, the film that is formed when the wire is removed will be one of minimum area for the curve selected. For such a minimum area surface, it can be shown that the average of the curvature around any point will be zero. The property of average zero curvature can then be used to define a minimal surface of infinite extent. For many years topologists knew of only three surfaces of minimal curvature in three-dimensional space that did not intersect themselves, technically known as embedded minimal surfaces. These were the infinite plane, which has zero curvature in all directions at each point, the helicoid, a sort of smoothed spiral staircase, and the catenoid, a sort of hourglass, all known in the nineteenth century. The first two surfaces were topologically related to a spherical surface with a circular opening cut in it, and the last to a spherical surface with two openings. In the 1980s a Brazilian graduate student named Celso Costa devised a set of equations that he could prove represented an infinite minimal surface, but could not determine whether it intersected itself. It was left to two American mathematicians, David Hoffman and Bill Meeks, and a graduate student in computer science, James Hoffman, at the University of Massachusetts to determine that it was in fact a non-intersecting and thus minimal surface. To truly understand the nature of these surfaces, these researchers relied heavily of new computer graphics programs. Not only were they ably to devise a proof Costa's surface did not intersect itself, but they were also able to demonstrate the existence of an infinite number of such embedded surfaces.

The work of topologists seldom attracts attention from the general public. This was not so in the case of catastrophe theory, developed by French mathematician René Thom (1923-) and presented to the public in a book entitled *Structural Stability and Morphogenesis,* first published in French in 1972. Thom's theory deals with the type of system frequently encountered in the sciences in which the state of a system is described by two kinds of variables, called observables and parameters. The parameters may be quantities that could be externally controlled, or that would vary slowly from one instance to the other, while the observables are measurable properties of the system. For a given set of values of the parameters, the relatively stable states would correspond to minimums, or spots of greatest positive curvature in a manifold. In Thom's terminology, a catastrophe occurs when a stable condition becomes unstable when the parameters change by a small amount. Thom was able to argue that, with very limited exceptions, for systems with six or fewer parameters, there were only seven types of catastrophe possible—that is, seven different types of manifold that would allow catastrophe's to occur.

Impact

Thom's book met an initially highly positive response from reviewers. A reviewer for the (London) *Times* compared Thom's book to Sir Isaac Newton's *Mathematical Principles of Natural Philosophy.* In 1974, at the International Congress of Mathematicians held in Vancouver Canada, English mathematician Christopher Zeeman gave a major lecture on the topic, suggesting that catastrophe theory would find its main applications in the behavioral and social sciences. When the English translation of Thom's book was published in 1975, enthusiastic reviews appeared in a number of scientific periodicals, as well as in the *New York Times* and the widely read *Newsweek.*

There were negative responses to catastrophe theory as well. One mathematics professor at Rutgers University spent two years giving seminars at different meetings, claiming not that catastrophe theory was mistaken, but that the clams made for it in areas far from mathematics would eventually not be born out. Understanding was not helped by the fact that the term "catastrophe" has a less dramatic meaning in French than the same word in English, which is synonymous with calamity. Enthusiasm for catastrophe theory has died down somewhat, while the basic mathematical soundness of Thom's ideas has not been questioned.

The desire for a consistent theory of change has been a recurrent theme in philosophy. In his *Physics,* the Greek philosopher Aristotle (384-322 BC) attempted to provide a uniform explanation for change in the physical world, in living things and in societies, and to do so without mathematics. Thom's work, to some interpreters, represents a new, and mathematically informed, attempt towards such a synthesis.

DONALD R. FRANCESCHETTI

Further Reading

Casti, John L. *Reality Rules: Picturing the World in Mathematics.* Vol. 1. New York: Wiley, 1992.

Casti, John L. *Searching for Certainty.* New York: William Morrow, 1990.

Eckland, Ivar. *Mathematics and the Unexpected.* Chicago: University of Chicago Press, 1988.

Peterson, Ivars. *The Mathematical Tourist.* New York: Freeman, 1998.

Thom, René. *Structural Stability and Morphogenesis.* Reading, MA: Benjamin, 1975.

Advances in Harmonic Analysis

Overview

Most everyone watches television or listens to the radio, and many people use the Internet to obtain graphic images, sound, and video. How can this information be transmitted electronically without requiring an immense amount of time and space? The FBI has accumulated approximately 200 million fingerprint files. How can a current suspect's fingerprints be compared with the contents of these files? The answer is *harmonic analysis,* which allows us to compress these images and sounds into their main components before transmission, and reconstruct them on the receiving end. Harmonic analysis grew out of a study of the way a string vibrates, and continues to reinvent itself. Harmonic analysis bridges the gap between mathematical theory and engineering practice.

Background

In *Recherches sur les cordes vibrantes* (1747), the French mathematician Jean d'Alembert (1717-1783) described his study of the shape of a vibrating string fixed at both ends. He discovered that the height of a point on the string (measured from its beginning taut state) is a function of two variables: its horizontal position and the time since the string was plucked. This height satisfies a partial differential equation known as the wave equation; D'Alembert solved a special case of this equation. In 1750 he derived a solution that involved the sine and cosine functions. A few years later in 1753, Daniel Bernoulli (1700-1782) investigated vibrating strings by using the idea of tones, or frequencies, reasoning that the movement of the vibrating string could be represented by a sum of sine and cosine functions with varying frequencies.

Mathematicians debated the merits of these solutions into the early 1800s. D'Alembert's and Bernoulli's ideas were revisited by Jean-Baptiste-Joseph Fourier (1768-1830), who used infinite sums of sine and cosine functions of varying amplitudes and frequencies to find the temperature distribution on a rectangular plate, known as the heat equation. In his *Théorie analytique de la chaleur* (Analytic theory of heat) Fourier discussed how certain functions could be approximated using the sums of sine and cosine functions. For example, the graph of

$$\cos(x) - 1/3\cos(3x) + 1/5\cos(5x) - 1/7\cos(7x) + \ldots$$

approximates a square wave, an alternating series of straight line segments above and below the x axis. In a very real sense, this breaks the function into its component frequencies. This is the heart of *Fourier analysis,* which grew into the broader harmonic analysis.

Anything that can be modeled with waves, such as earthquakes, tides, electromagnetic waves, satellite communications, and medical imaging can be studied using harmonic analysis. When Fourier analysis is used in these fields it is often called signal processing, because the function is a signal emitted from some source. Harmonic analysis can also analyze chemical compounds by examining their electromagnetic spectra—which are the frequencies of light emitted or absorbed by the compound. A light wave analyzed by Fourier transform reveals its component frequencies, the sine and cosine functions that make up the wave. Sines and cosines with large amplitudes (large coefficients) allow the wave (and the substance) to be identified.

During the remainder of the nineteenth century, the types of functions (not all of which appear to be wavelike) that could be represented by these methods were expanded, and many of the questions concerning how well a series of sines and cosines approximate a function were answered. From these theoretical concerns, general harmonic analysis was born.

Signal processing, while important during World War II, expanded significantly after the war, when the availability of computers opened many applications for harmonic analysis. Although computational speed and power was limited at first, analog signals (continuous functions) were eventually replaced with digital signals (discrete functions). The input and output functions became a sequence of numbers, and the *discrete Fourier transform* was adopted. The process of computing a Fourier transform became mathematically equivalent to multiplying a vector of n data points by an $n \times n$ matrix. This process involves n^2 multiplications. In analyzing a signal, this is repeated many thousands of times. For large values of n this computation becomes time consuming.

All of this changed in 1965. In "An Algorithm for the Machine Calculation of Complex Fourier Series," James W. Cooley and John W. Tukey laid out the plan that sped up the computation of Fourier transforms. This algorithm became known as the *fast Fourier transform* (FFT), which made many innovations in technology possible by significantly increasing the speed with which Fourier transforms were calculated. The FFT reduced the n^2 multiplications previously required to $1/2\ n \log_2 n$. So if $n = 2^{12} = 4096$, the $2^{24} = 16{,}777{,}216$ multiplications are reduced to $6 \times 2^{12} = 24{,}576$; this is almost 700 times faster.

One of the difficulties with Fourier analysis, however, is that when adding sine and cosine functions it is difficult to approximate discontinuous functions, which have sudden changes in function values. If the function contains information about an image, for example, this would occur when you have an edge in the image. Similarly, sines and cosines will not recognize erratic heartbeats on an electrocardiogram. In simple terms, sines and cosines can give you an accurate model of the big picture, but not of what happens in a small window. Modified Fourier analysis has been used to solve this problem, but *wavelet analysis* is better suited to the task.

In wavelet analysis, functions that are defined on small intervals are used to decompose and reconstruct the signal. These small functions are designed to make computation with them very efficient. Since they are defined on a small region, they can approximate functions with sudden changes. A main function (called a mother wavelet) is scaled to get either a more global or a more local picture of the function. This scaling is accomplished by "stretching" and "shrinking" the graph of $y = f(x)$, using $f(ax)$ where a is a constant either between 0 and 1 (for stretching) or greater than 1 (for shrinking).

The idea of wavelet analysis can be traced back to the 1910 work of Alfred Haar (1885-1933) who studied the types of functions used to approximate other functions, much like the sine and cosine functions are used in the Fourier transform. During the 1930s researchers also began to investigate the idea of scaling the functions used to approximate other functions. In the early 1980s, the idea for wavelets grew out of a study of seismic data. Geophysicist Jean Morlet used the idea and shared it with physicist Alex Grossman, who saw the merit in the idea. Mathematician Yves Meyer began developing the mathematical theory of these wavelets, and Ingrid Daubechies popularized the theory of wavelet analysis. Finally, in 1985 Stephane Mallat realized that wavelets could be streamlined to improve computation speed. This was the breakthrough that unleashed wavelet analysis applications, much as the FFT provided applications for Fourier analysis. The history of wavelet analysis has yet to be completed, as more important ideas and applications continue to be developed.

Impact

It is hard to overestimate the impact of Fourier analysis, the FFT, and wavelet analysis. Fourier analysis has found many applications in modern technology. In medicine it is used in electrocardiograms and magnetic resonance imaging (MRI); in mass communications it is used to compress and reproduce signals and to filter noise from signals. It's fundamental to digital audio (compact disks), video (HDTV) and satellite television. Fourier analysis is also the basis of the JPEG image compression standard. In sciences such as geology and chemistry, Fourier analysis is used to identify substances by looking at their electromagnetic spectra (wavelengths), and is also the basis of mass spectroscopy. These applications can be done efficiently because the FFT has shortened computation time.

Applications for wavelet analysis are becoming more prevalent and are taking the place of Fourier analysis in some areas. A new image compression standard, being developed to take the place of the JPEG, will use wavelet transforms. New digital filtering techniques based on wavelet analysis are being used to restore old recordings. The HDTV standard is based on Fourier transforms, but with the emergence of wavelet theory this could change; companies are at work on the problem right now.

The new theory has already taken over at the FBI, which has about 200 million (2×10^8) sets of fingerprints on cards at their headquarters in Washington, D.C. When suspects are fingerprinted, their prints are sent by mail to the fingerprint file to be compared. The turnaround time is quite long, which is a problem for the justice system. A digitized fingerprint card with 10 prints would require around 10 megabytes (10^7 bytes) of data. This could easily be sent electronically to the FBI. Unfortunately, with 200 million cards, digitized storage would require 2×10^{15} bytes, or about 200,000 ten-gigabyte (10^9) hard drives. This is too much to store, let alone search electronically. Wavelet analysis, however, can compress the images; a rate of about 15:1 would require about 13,000 ten-gigabyte hard drives. If the database is restricted only to current suspects, it could be reduced to about 1,300 ten-gigabyte hard drives, a manageable number in today's terms. The FBI is now adopting just such a system. Undoubtedly, wavelet analysis will continue to find its way into other applications as new techniques are designed to take the place of the Fourier methods.

GARY S. STOUDT

Further Reading

Books

Hubbard, Barbara. *The World According to Wavelets: The Story of a Mathematical Technique in the Making*. Burke-Natick, MA: A. K. Peters, 1998.

Advances in Algebraic Topology since 1950

Overview

The last half of the twentieth century saw numerous advances in the field of topology, especially in the study of manifolds, or surfaces, as well as in the theory of knots. These advances have had a significant impact on mathematics as a whole. In addition, the mathematical tools developed in these areas have proven useful in several branches of physics and may help make exploration of the Solar System more efficient, although at the expense of increased travel time.

Background

Topology is a branch of mathematics that resembles geometry in that it is concerned with the properties of shapes and surfaces. However, unlike geometry, topology is interested in examining only those properties that do not change with folding, stretching, and other manipulations that do not penetrate the surface. From a topological standpoint, a sausage, a sphere, and a cube are identical because they are all closed surfaces with no holes or penetrations. Similarly, a coffee cup and a donut are topologically identical because they each have exactly one hole. In both of these examples, it is not difficult to see how, by stretching, folding, and so forth, one shape can be turned into another.

One of the most important branches of topology is the study of manifolds, or surfaces. The simplest manifold is a line, which is a surface with only one dimension. A sheet of paper is a manifold in two dimensions and, because a sheet of paper can be formed into a sphere, the surface of a sphere is also considered a two-dimensional manifold. Manifolds can exist in higher dimensions, and the general form is to refer to an *n*-manifold in which *n* refers to the dimensionality of the manifold.

There are many other branches of topology besides manifolds. Algebraic topology is the subdiscipline that uses algebraic methods to study manifolds. For example, it is intuitively obvious that a closed curve that does not cross itself forms a shape with an inside and an outside. Algebraic topology techniques can be used to prove this. Other tools that topologists use to classify manifolds include cordobism and homotropy. Homotropy tells us how one map may be transformed into another, similar map. An example of this would be to take an inflatable sphere with a map of the world on it. Deflating the sphere changes the way the map looks, as do stretching, folding, and wrinkling the sphere. Yet the map is fundamentally the same. Using homotropy theory, French mathematician René Thom (1923-) developed cordobism to classify surfaces such as the two-dimensional manifold represented by the sphere. Another tool that can be used in investigating manifolds is spectral sequences, developed by another French mathematician, Jean-Pierre Serre (1926-). A spectral sequence is a sequence of numbers with certain properties that make it valuable in analyzing manifolds.

Another related area is that of knot theory. From a topological perspective, a knot is a line that does not intersect itself. Knots are contained in three-dimensional space, but they can be projected onto a plane where they are classified in terms of the number of times that the line crosses itself. Please note that crossings are not the same as intersections. In a crossing, one segment of the line *passes over the top* of another line segment. In an intersection, the line segment is *joined* by another line segment. It is like the difference between a highway overpass and a street intersection.

In the preceding paragraphs, a number of related topics in topology have been described briefly. In the rest of this essay, the importance of the recent work in these areas and the manner in which they interact will be explored in more detail.

Impact

One of the first problems in algebraic topology, the Poincaré conjecture, was posed in 1904 by the great French mathematician Henri Poincaré (1854-1912). Poincaré noted that it is possible to draw a circle on a sphere and to successively shrink it to the size of a dot. The sphere is the only closed two-dimensional surface that we know of for which this must always be true. For example, a circle drawn on a torus can be drawn around the circumference, and the smallest it can be is the diameter of the inner hole in the torus. Poincaré conjectured that the sphere and higher-dimensional versions of the sphere were the only shapes for which this was true.

The Poincaré conjecture went totally unproved until 1960, when American mathematician Stephen Smale (1930-) showed it to be true for any dimensions greater than seven. In other words, a five-dimensional sphere surface (which would exist in six dimensions) has the property of being the only closed five-dimensional surface for which a circle (or loop, as topologists would say) can be drawn where the loop can be shrunk to a point while remaining attached to the surface of the sphere at all times and at all points along the loop. John Stallings showed it to be true for six-dimensional surfaces, and Christopher Zeeman (1925-) proved the five-dimensional case in the 1960s. In 1982 Michael Freedman (1951-) showed that the Poincaré conjecture was also true for four-dimensional sphere surfaces, but nobody has yet proved the Poincaré conjecture for the three-dimensional case. This makes the Poincaré conjecture the most famous unsolved problem in topology, the equivalent of Fermat's last theorem for topologists.

Knot theory became important in the late nineteenth century when Lord Kelvin (1824-1907) speculated that atoms were actually small knots in space. He felt that by studying the properties of knots, we could learn more about atoms. He was wrong in this speculation, but knot theory has become an interesting branch of topology with some applications outside of mathematics. Recently, knot theory has had somewhat of a resurgence in this area, although in a manner Kelvin could never have imagined. Superstring theory, a physics theory that suggests subatomic particles can be treated as tiny loops of string vibrating in ten-dimensional space, uses many of the tools developed in knot theory. While this theory has yet to be proven, it has gained a lot of attention in recent years because it seems to describe some areas in physics quite nicely.

With the exception of knot theory, the other branches of topology mentioned above all deal directly with manifolds, their classification, and their properties. Although these concepts seem almost disconnected from everyday utility, they do have relevance to other fields of scientific inquiry. For example, the universe itself is likely a manifold, although we are still uncertain of its actual shape. A more immediate use of manifold theory lies in the mathematical properties of multi-dimensional surfaces. Specifically, if you take a sufficiently small portion of any surface, it will look flat, like the space in which Euclidean geometry occurs. The surface of Earth, for example, is curved since Earth is a sphere. However, our living room floor can be treated as being flat because it is such a small part of Earth's surface that it appears flat. Treating it in this way makes many physics calculations much easier to perform because we know how to do physics, including taking derivatives and integrals, on a flat, Euclidean surface. Other manifolds can be treated the same way, regardless of their dimensionality. This means that, in local areas of even high-dimensional objects, we can still use our "standard" mathematics, in spite of the fact that this mathematics was developed for use on flat surfaces in Euclidean space. In turn, this makes working with high dimensions, such as the ten-dimensional space necessary for physics' string theory, possible.

Another application of manifolds may lie in space travel. Strictly speaking, a manifold is *any* surface, whether "real" or a mathematical construct. Consider, for example, a surface constructed by measuring the force of gravity at different distances from a planet. There will be places where the gravitational force will be the same, what is called an *equipotential surface*. In the case of Earth, some equipotential surfaces will link up with those of the Moon because the gravitational force felt by a spacecraft will be, say, one meter per second (3 feet per second) at any place on that surface. It is not important that in some locations, the gravitational attraction is towards the Moon and not towards Earth, because we are only measuring the pull of gravity. This equipotential surface is a manifold. A spacecraft on this manifold can change orbits very easily, and by staying on this manifold, it can easily move from Earth orbit to a lunar orbit. If the manifold is chosen correctly, it might also intersect a similar surface from the Sun, giving the spacecraft even more possibilities for changing orbits and traveling around the Solar System. In addition, by examining how these manifolds lie in space, a spacecraft may be able to follow one manifold to a location very near another manifold that extends to the inner or outer Solar System. At that point, with a small expenditure of energy, the craft could then put itself on a manifold that would take it to Jupiter, say, much more efficiently than by simply blasting directly. The trade-off, of course, is that traveling around the Solar System in this manner can take a lot longer than even chemical rockets.

P. ANDREW KARAM

Further Reading

Books

Devlin, Keith. *Mathematics: The Science of Patterns*. Scientific American Library, 1994.

Devlin, Keith. *Mathematics: The New Golden Age*. New York: Columbia University Press, 1999.

Other

"Topology." *Eric Weisstein's World of Mathematics*. http://mathworld.wolfram.com/topics/topology.html

"Topology Enters Mathematics." *The MacTutor History of Mathematics Archive*. http://www. vma.bme.hu/mathhist/histopics/topology_in_mathematics.html

Applications of Number Theory in Cryptography

Overview

Cryptography is a division of applied mathematics concerned with developing schemes and formulas to enhance the privacy of communications through the use of codes. Cryptography allows its users, whether governments, military, businesses, or individuals, to maintain privacy and confidentiality in their communications. The goal of every cryptographic scheme is to be "crack proof" (i.e, only able to be decoded and understood by authorized recipients). Cryptography is also a means to ensure the integrity and preservation of data from tampering. Modern cryptographic systems rely on functions associated with advanced mathematics, including a specialized branch of mathematics termed number theory that explores the properties of numbers and the relationships between numbers.

Background

Attempts to preserve the privacy of communications is an age-old quest. From the use of hidden text, disappearing inks, and code pads has evolved the modern science of cryptography. The word cryptography originally derives from the Greek, *kryptos* (to hide). In essence, cryptography is the study of procedures that allow messages or information to be encoded (obscured) in such a way that it is extremely difficult to read or understand the information without having a specific key (i.e., procedures to decode).

Encryption systems can involve the simplistic replacement of letters with numbers, or they can involve the use of highly secure "one-time pads" (also known as Vernam ciphers). Because one-time pads are based upon codes and keys that can only be used once, they offer the only "crack proof" method of cryptography known. The vast number of codes and keys required, however, makes one-time pads impractical for general use.

Many wars and diplomatic negotiations have turned on the ability of one combatant or country to read the supposedly secret messages of its enemies. In World War II, for example, the Allied Forces gained important strategic and tactical advantages from being able to intercept and read the secret messages of Nazi Germany that had been encoded with a cipher machine called Enigma. In addition, the United States gained a decided advantage over Japanese forces through the development of operation MAGIC, which cracked the codes used by Japan to protect its communications.

In step with the growth of computing technologies and the decline of paper and pen record keeping, the importance of cryptography rose during the later half of the twentieth century. Increasing amounts of data began to have permanent storage only in computer memory. Although the technological revolution and rise of the Internet presented unique security challenges, there were also challenges to the basic security of mounting levels of information stored and transmitted only in electronic form. This increasing reliance on electronic communication and data storage increased demand for advancements in cryptologic science. The use of cryptography broadened from its core diplomatic and military users to become of routine use by companies and individuals seeking privacy in their communications. Governments, companies and individuals, required more secure—and easier to use—cryptologic systems to secure their databases and email.

In addition to improvements made to cryptologic systems based on information made public from classified government research programs, international scientific research organizations devoted exclusively to the advancement of cryptography (e.g., the International Association for Cryptologic Research, or IACR), began to apply mathematical number theory to enhance privacy, confidentiality, and the security of data. Applications of number theory were used to develop increasingly involved algorithms (step-by-step procedures for solving a mathematical problem). In addition, as commercial and personal use of the Internet grew, it became increasingly important not only to keep information secret, but also to be able to verify the identity of message sender. Cryptographic use of certain types of algorithms called "keys" allow information to be restricted to a specific and limited audience, whose individual identities can be authenticated.

In some cryptologic systems, encryption is accomplished by choosing certain prime numbers and then products of those prime numbers as the basis for further mathematical operations. In addition to developing such mathematical keys, the data itself is divided into blocks of specific and limited length so that the information that can be obtained even from the form of the message is limited. Decryption is usually accomplished by following an elaborate reconstruction process that itself involves unique mathematical operations. In other cases, decryption is accomplished by performing the inverse mathematical operations performed during encryption.

Although it may have been developed earlier by government intelligence agencies, in August 1977 Ronald Rivest, Adi Shamir, and Leonard Adleman published an algorithm destined to become a major advancement in cryptology. The RSA algorithm underlying the system derives its security from the difficulty in factoring very large composite numbers. By the end of the twentieth century the RSA algorithm became the most commonly used encryption and authentication algorithm in the world. The RSA algorithm was used in the development of Internet web browsers, spreadsheets, data analysis, email, and word processing programs.

More than simply publishing a mathematical algorithm, however, Rivest, Shamir, and Adleman developed the first public key cryptologic system widely available to commercial and private users. The most important of the modern cryptographic systems to be based on the RSA algorithm (and its modifications and derivations) are termed "public key" systems. These systems are considered to be among the most secure of cryptographic techniques. Encoding and decoding is accomplished using two keys—mathematical procedures to lock (code) and unlock (decode) messages. In such "two-key" cryptologic systems, those wishing to use the public key system distribute the "public" key to those intended to have the capability to encode messages. The sender uses the recipient's public key to encode the message, but the message can only be decoded with the recipient's private key. This assures that only the holder of the private key can decode an encoded message. Beginning in 1991 the public key method was used to enhance Internet security through a freely distributed package called Pretty Good Privacy (PGP).

Impact

Applications of number theory allow the development of mathematical algorithms that can make information (data) unintelligible to everyone except for intended users. In addition, mathematical algorithms can provide real physical security to data—allowing only authorized users to delete or update data. One of the problems in developing tools to crack encryption codes involves finding ways to factor very large numbers. Advances in applications of number theory, along with significant improvements in the power of computers, have made factoring large numbers less daunting.

In general, the larger the key size used in PGP-based RSA public-key cryptology systems, the longer it will take computers to factor the composite numbers used in the keys. Accordingly, RSA cryptology systems derive their reliability from the fact that there are an infinite number of prime numbers— and from the difficulties encountered in factoring large composite numbers composed of prime numbers.

Specialized mathematical derivations of number theory such as theory and equations dealing with elliptical curves are also making an increasing impact on cryptology. Although, in general, larger keys provide increasing security, applications of number theory and elliptical curves to cryptological algorithms allow the use of easier-to-use smaller keys without any loss of security.

Another ramification related to applications of number theory is the development of "non-reputable" transactions. Non-reputable means that parties cannot later deny involvement in authorizing certain transactions (e.g., entering into a contract or agreement). Many cryptologists and communication specialists assert that a global electronic economy is dependent on the development of verifiable and non-reputable transactions that carry the legal weight of paper contracts. Legal courts around the world are increasingly being faced with cases based on disputes regarding electronic communications.

Advancements in number theory have been equally applied, however, in an attempt to crack important cryptologic systems. In RSA composite number-based, two-key cryptologic systems, there are public keys and private keys. Trying to crack the codes (the encryption procedures) requires use of advanced number theories that allow, for instance, an unauthorized user to determine the product of the prime numbers used to start the encryption process. Factoring this product is a difficult and tedious procedure to determine the underlying prime numbers. An unsophisticated approach might be simply to attempt to try all prime numbers. The time to accomplish this task, however, can defeat all but the most determined of unauthorized users. Other more exotic attempts involve algorithms termed quadratic sieves, a method of factoring integers developed by Carl Pomerance that is used to attack smaller numbers, and field sieves algorithms, which are used in attempts to determine larger integers.

Within the last two decades of the twentieth century, advances in number theory allowed factoring of large numbers that by hand might take billions of years to procedures that with the use of advanced computing might be accomplished in a matter of months. Further advances in number theory may lead to the discovery of a polynomial time factoring algorithm that can accomplish in hours what now takes months or years of computer time.

Advances in factoring techniques and the expanding availability of computing hardware (both in terms of speed and low cost) make the security of the algorithms underlying cryptologic systems increasingly vulnerable. These threats to the security of cryptologic systems are, in some regard, offset by continuing advances in the design of powerful computers that have the ability to generate larger keys by multiplying very large primes. Despite the advances in number theory, it remains easier to generate larger composite numbers than it is to factor those numbers.

The National Institute of Standards and Technology (NIST) oversees the development of many cryptography standards. One such standard, developed by commercial entities and the United States National Security Agency (NSA) in the 1970s, was termed the Data Encryption Standard (DES). In anticipation of increasing security needs, late in the twentieth century NIST began to work toward the implementation of the Advanced Encryption Standard (AES) to replace DES. Although there were efforts to liberalize trade policies, at the end of the twentieth century security algorithms used in PGP-type programs were classified as munitions by the United States government. As such, they remained subject to severe export control and restrictions that inhibited their widespread distribution and use.

K. LEE LERNER

Further Reading

Beckett, B. *Introduction to Cryptology.* Blackwell Scientific, 1988.

Burn, R.P. *A Pathway into Number Theory.* Second ed. Cambridge University Press, 1997.

Menezes, A., P. van Oorschot, and S. Vanstone. *Handbook of Applied Cryptography.* CRC Press, 1997.

Rosen, K. H. *Elementary Number Theory and Its Applications.* Addision-Wesley, 1986.

Seberry, J., and J. Pieprzyk. *Cryptography: An Introduction to Computer Security.* Prentice-Hall, 1989.

Lie Algebra Is Used to Help Solve Hilbert's Fifth Problem

Overview

Advances in the theory of Lie algebra (also called Lie groups), developed by Norwegian mathematician Sophus Lie (1842-1899), have long enriched mathematics, particularly in the area of group theory. In 1952 Lie algebra was used to help solve one of the most famous problems in mathematics: Hilbert's fifth problem, posed in the year 1900 by David Hilbert (1862-1943). In addition to its use in solving this problem, Lie algebra has been used to gain a better understanding of the properties of many-dimensional surfaces in general, helping to advance the mathematical discipline of topology as well.

Background

One of the keynote speakers at the 1900 International Congress of Mathematicians, David Hilbert spoke about the future of mathematics. In his address, the renowned mathematician posed 23 major unsolved problems in mathematics to the Congress, noting that their solutions would help to push mathematics forward. One of these problems, the fifth, was to challenge mathematicians for over half of the twentieth century.

Hilbert's fifth problem was phrased as such: "Can the assumption of differentiability for functions defining a continuous transformation group be avoided?" Another way to state this question is to ask if any locally Euclidean topological group can be given the structure of an analytic manifold to become a Lie group. These individual terms can be confusing and will be explained further before going on.

A topological group is any group of numbers that are points in topological space for which group operations are continuous. An example of topological groups is the set of real numbers where addition or subtraction are group products. Another is the set of rigid motions that a group of points can take in Euclidean space (that is, space that follows the geometry described by Euclid [c. 330-260 B.C.]) when one point remains fixed. In this case, if one visualizes a rotating block of wood as representing a group of points, these points would form a topological group because they stay together as the block rotates in space.

A manifold is another concept entirely. Put most simply, a manifold is a surface with a given number of dimensions. For example, the surface of a sphere is a manifold with two dimensions because there are only two directions an object on that surface can move. The fact that the surface encloses a three-dimensional structure is irrelevant in this case. An analytic manifold can be mathematically described. One interesting property of manifolds is that, in Euclidean space, small patches appear to be flat and can be treated as such. For example, even though Earth is a sphere, we treat a floor as flat because, for all practical purposes, it looks that way to us at our scale. This makes it possible to use "ordinary" mathematics to describe small sections of the manifold and any shapes or curves that might be drawn on them.

The final term to describe is Lie algebra or Lie group. In general, a group is a set (either finite or infinite) of items (called operands) that can be combined via some sort of mathematical operation to form defined products. A Lie group is a special type of group in which the underlying space is an analytic manifold and, on that manifold, group operations are analytic (that is, they are described by equations). At this point, too, it must be noted that a Lie group is not, in the strictest sense, a group at all. Rather, it is a series of operations that, together, comprise a system for finding solutions to problems. In other words, the more correct term for a Lie group is Lie algebra, and these terms are often used interchangeably.

This brings us back to Hilbert's fifth problem, phrased as "Can any locally Euclidean topological group can be given the structure of an analytic manifold to become a Lie group?" With the definitions mentioned above, we see that this is asking if there is a general way to take any group of numbers (which will define a surface or a shape in space) in Euclidean space, place them on a smooth surface, and make them into a Lie group, so that the surfaces and the underlying space are still analytic.

Finally, we need to introduce the concept of "compactness" for groups. A group is "compact" if all of the members are adjacent when plotted in space or on a surface. For example, a ball would be compact because all of the points that make up the ball lie together in space. On the other hand, the stars in the sky are not compact as we see them because they are sprinkled more or less randomly across the sky.

Hilbert's fifth problem was solved in part in 1929 by John von Neumann (1903-1957), when he demonstrated that functions could be integrated on general compact groups. However, this solution was only partial. Three years later, Alfred Haar added to von Neumann's proof, and in 1934 Soviet mathematician Lev Semyonovich Pontryagin proved Hilbert's conjecture for a different set of groups, the Abelian groups. Finally, this problem was completely solved by Gleason, Montgomery, and Zippin in 1952. They showed that any locally compact topological group is a limit of Lie groups, making this finding even more important.

Impact

In general, by posing his problems and daring the world's mathematicians to solve them, Hilbert contributed to many advances in mathematics in each of the areas upon which these problems touched. The field of mathematics is richer because of this, and those fields that use mathematics have more mathematical tools they can bring to bear on their problems. Thus, in a sense, all of mathematics, physics, computer science, and many other fields have benefited from the posing of and solutions to Hilbert's problems.

In this case, the solution to Hilbert's fifth problem has led to a greater understanding of the mathematical properties of surfaces. Such surfaces include the shape of machine parts, the shape of planets and black holes, or the shape of the universe in its entirety.

However, the impact of this line of inquiry is not limited to solving this specific problem. The application of Lie group theory to solving

Hilbert's fifth problem is just one of the impacts of this field on the whole of mathematics. In particular, Lie examined "contact transformations," which have since become important. A contact transformation is a way of describing the mathematical properties of a very small part of a surface by determining the equation of a line or vector tangent to the surface (i.e. touching the surface at only a single point). This is analogous to drawing a line tangent to a curve to determine the slope of the curve at that point or, in three dimensions, constructing a plane that touches a sphere at only one location to better determine the properties of the sphere. In the case of Lie algebra, a surface of any number of dimensions (called an n-dimensional surface) can be described in terms of a field of such vectors, in the same way that a curve can be described in terms of all of the lines that are tangent to it. What Lie did was to generalize this concept and to provide a way to work with surfaces of any number of dimensions. This, in turn, has an application when mathematically describing multi-dimensional problems, such as the shape of the universe, super-string theory (which assumes ten dimensions in space), and some problems with a large number of variables.

P. ANDREW KARAM

Further Reading

Arfken, G. *Mathematical Methods for Physicists.* Academic Press, 1985.

Browder, Felix. *Mathematical Developments Arising from Hilbert Problems.* American Mathematical Society, 1976.

Lipkin, H .J. *Lie Groups for Pedestrians.* North-Holland Press, 1966.

The Resurrection of Infinitesimals: Abraham Robinson and Nonstandard Analysis

Overview

For centuries prior to 1800, *infinitesimals*—infinitely small numbers—were an indispensable tool in the calculus practiced by the great mathematicians of the age. Between the mid-1800s and the mid-1900s, however, infinitesimals were excluded from calculus because they could not be rigorously established. This changed in 1960, when Abraham Robinson resurrected their use with his creation of nonstandard analysis. Since that time, nonstandard analysis has had an important effect on several areas of mathematics as well as on mathematical physics and economics.

Background

An infinitesimal is an infinitely small number. More precisely, it is a nonzero number smaller in absolute value than any positive real number. But merely *defining* a mathematical entity does not guarantee its *existence.* For example, we can define an obtuse-angled triangle as a triangle all of whose angles are greater than 90 degrees. But such triangles do not exist (in Euclidean geometry)! In the case of infinitesimals, there are no *real* infinitesimals, since given any positive real number a, $a/2$ is a smaller positive real. To accommodate infinitesimals we must extend the real numbers.

This idea of extending a mathematical system in order to obtain a desired property not already present is common and important in mathematics. For example, while the positive integers are prehistoric, the other number systems, such as the integers, rational numbers, real numbers, and complex numbers, arose over the centuries as human constructs. Thus, the integers were introduced so as to make sense of numbers such as -1, the real numbers to give meaning to numbers like $\sqrt{2}$, and the complex numbers to accommodate such numbers as $\sqrt{-1}$. In each case, these numbers were introduced because they turned out to be *useful.*

This was also the case for infinitesimals—or *differentials,* as Gottfried Wilhelm Leibniz (1646-1716) called them. They were indispensable in the calculus of the seventeenth, eighteenth, and early nineteenth centuries. For example, to find the slope of the tangent line (later called the *derivative*) to the parabola, $y = x^2$ at the point (x, x^2), seventeenth-century mathematicians would argue as follows:

Let e be an infinitesimal. Then

$(x + e, [x + e]^2)$

is a point on the parabola infinitesimally close to (x, x^2), hence the tangent line to the parabola can be identified with the line joining these two points. Its slope is

$[(x + e)^2 - x^2] / [(x + e) - x] =$
$(2xe + e^2) / e = 2x + e$

by canceling e. Finally, $2x + e$ can be identified with $2x$ since e is infinitesimally small compared to $2x$ and can therefore be deleted.

Of course the answer is correct since the derivative of x^2 is indeed $2x$. But what about the method used to obtain it? It was severely criticized even in the seventeenth century. Canceling e, the critics argued, meant regarding it as not zero; but deleting e implied treating it as zero. This is inadmissible, they rightly claimed. In a trenchant critique of infinitesimal methods, the philosopher George Berkeley (1685-1753) called such e's "the ghosts of departed quantities," arguing that "by virtue of a twofold mistake one arrived, though not at a science, yet at the truth."

Most mathematicians, however, were unperturbed by such objections. Although they recognized that their methods were logically questionable, these methods yielded correct results. Leibniz, for example, said of his differentials (infinitesimals) that "it will be sufficient to simply make use of them as a tool that has advantages for the purpose of calculation, just as the algebraists retain imaginary roots with great profit." In Leibniz's time complex numbers (imaginary roots) had no greater logical legitimacy than infinitesimals. These are two important examples of a common phenomenon in mathematics, namely the use of objects *before* their existence is rigorously established.

The nineteenth century ushered in a critical spirit in calculus, in which fundamental concepts were reexamined and put on a logical basis. This resulted in the replacement of the logically problematic infinitesimals with limits, defined rigorously by Karl Weierstrass (1815-1897) in the mid-nineteenth century in terms of the Greek letters epsilon, ϵ, and delta, δ.

About a century after Weierstrass had banished infinitesimals "for good"—so we all thought until 1960—they were brought back to life as rigorously defined mathematical objects in the nonstandard analysis conceived by the mathematical logician Abraham Robinson (1918-1974). His idea was to provide a rigorous development of calculus based on infinitesimals rather than on limits.

While standard analysis—the calculus we inherited from Weierstrass (and others)—is based on the real numbers R, nonstandard analysis is grounded in an extension of the real numbers called "hyperreal" numbers, R^*. The hyperreal numbers contain infinitesimals, where (by definition) $e \in R^*$ is infinitesimal if $e \neq 0$ and $-a < e < a$ for all positive $a \in R$. They also contain infinite numbers, since if e is an infinitesimal, $1/e$ is an infinite (hyperreal) number.

Robinson was inspired to create nonstandard analysis in part by his work in the newly emerging subfield of mathematical logic called *model theory*. Here is how he put it:

> *In the fall of 1960 it occurred to me that the concepts and methods of contemporary Mathematical Logic are capable of providing a suitable framework for the development of the Differential and Integral Calculus by means of infinitely small and infinitely large numbers.*

It is ironic that infinitesimals were excluded from calculus in the nineteenth century because they proved to be logically unsatisfactory, and they were rendered mathematically respectable in the twentieth century thanks to logic. Robinson was very gratified that it was mathematical logic that made nonstandard analysis possible. The great mathematical logician Kurt Gödel (1906-1978) valued Robinson's work because it made logic and mathematics come together in such a fundamental way. The contemporary mathematician Simon Kochen echoed this: "Robinson, via model theory, wedded logic to the mainstream of mathematics."

Robinson was also guided in his work in nonstandard analysis by a sense of history. He saw it in the tradition of the great analysts Leibniz, Leonhard Euler (1707-1783), and Augustin-Louis Cauchy (1789-1857). In fact, he argued that "Leibniz's ideas can be fully vindicated" by his own rigorous theory of infinitesimals.

Leibniz, as we mentioned, tried to justify his work with infinitesimals on pragmatic grounds—that it yielded correct results. He also attempted to rationalize his handling of infinitesimals with a rather vague *principle of continuity*—that (in our language) properties of the reals also hold for the hyperreals. But as mathematicians of the seventeenth century realized, not all properties of the former carry over to the latter. For example, the *Archimedean property*, which says that given real numbers a and b, with

b positive, there exists an integer n such that $nb > a$, does not hold in R^*. For if $a = 1$ and $b = e$, a positive infinitesimal, then $e < 1/n$ for every positive integer n, by definition of an infinitesimal, so that $ne < 1$ for all n. Robinson observed that

> What was lacking at the time [of Leibniz] was a formal language which would make it possible to give a precise expression of, and delimitation to, the laws which were supposed to apply equally to the finite numbers and to the extended system including infinitely small and infinitely large numbers.

It is the working out of this program for which Robinson is responsible. More specifically, he 1) provided a rigorous construction of the system of hyperreal numbers R^* in which he was able to *prove* the existence of infinitesimals; and 2) formulated a transfer principle that gave formal expression to Leibniz's principle of continuity and thus rendered precise those properties that are transferable from the reals to the hyperreals. He accomplished both tasks with the aid of mathematical logic.

Impact

What has nonstandard analysis accomplished? First, it has supplied a rigorous presentation of calculus based on infinitesimals that, some have argued, is much preferable to the standard treatment via limits. Second, the methods of nonstandard analysis have been introduced into important branches of mathematics aside from calculus, such as topology, differential geometry, measure theory, complex analysis, and Lie group theory. They have also been applied in functional analysis, differential equations, probability, areas of mathematical physics, and economics. These inroads of the subject, in such a short time-span, are indeed most impressive.

ISRAEL KLEINER

Further Reading

Books

Boyer, Carl. *The History of the Calculus and its Conceptual Development*. New York: Dover, 1949.

Dauben, Joseph. *Abraham Robinson: The Creation of Nonstandard Analysis; a Personal and Mathematical Odyssey.* Princeton, NJ: Princeton University Press, 1995.

Edwards, Charles. *The Historical Development of the Calculus*. New York: Springer-Verlag, 1979.

Keisler, Jerome. *Elementary Calculus: An Infinitesimal Approach*. 2nd ed. Boston: Prindle, Weber & Schmidt, 1986.

Keisler, Jerome. *Foundations of Infinitesimal Calculus*. Boston: Prindle, Weber & Schmidt, 1976.

Robinson, Abraham. *Non-Standard Analysis*. Amsterdam: North-Holland Pub. Co., 1966.

Periodical Articles

Davis, Martin and Hersh, Reuben. "Nonstandard analysis." *Scientific American* 226 (June 1972): 78-86.

Harnik, Victor. "Infinitesimals from Leibniz to Robinson: Time to Bring Them Back to School." *Mathematical Intelligencer* 8, 2 (1986): 41-47, 63.

Lakatos, Imre. "Cauchy and the Continuum: The Significance of Non-Standard Analysis for the History and Philosophy of Mathematics." *Mathematical Intelligencer* 1 (1978): 151-61.

Laugwitz, Detlef. "On the Historical Development of Infinitesimal Mathematics," I, II. *American Mathematical Monthly* 104 (1997): 447-455, 660-69.

Politics Impinges upon Mathematics

Overview

In the second half of the twentieth century, the many profound changes in society made it inevitable that intellectuals, including mathematicians, would get caught up in political conflict. Some actively bore the standard of their personal beliefs. Others were simply targets of other people's crusades. Political conflicts, sometimes virulent enough to destroy academic careers, continue to be a feature of university life.

Background

While mathematicians often spend their careers working on purely abstract concepts, they do not work in a vacuum. Most are employed either by universities or governments, both of which are extremely political institutions. All function within larger societies that vary in their tolerance of those with different backgrounds or ideas.

The ideal of academic life is the freedom to speak the truth as one sees it, and to exchange ideas without threat of retaliation. However, like most ideals, this one is not always realized in practice. Universities operate under governmental regimes that vary in their tolerance for free speech. Even absent governmental control, the academy is not completely unconstrained. Faculty members depend upon grant-making organizations for their financial support. Their careers and credentials are dependent upon the university organization and their place in its hierarchy. Such institutions are made up of human beings, with their various likes, dislikes, and prejudices. Often these are framed in political terms.

In government research facilities around the world, national security arguments may be brought to bear to discriminate against someone who has unpopular ideas or who belongs to an unpopular group. Many mathematicians work in particularly sensitive areas, for example cryptography and ballistics, where such arguments are easy to make. Governments also control a significant percentage of the grants and other funds that flow into universities.

Some mathematicians refused on principle to go along with the "witch hunt" atmosphere of the McCarthy era in the United States in the 1950s, during which many academics were forced to sign declarations that they were not members of the Communist Party. Others became involved in anti-war movements and fell afoul of their government or university administration. Some belonged to ethnic, religious, or other groups that were discriminated against. As a result, many left their universities, their academic discipline, or their country.

At the height of the Cold War in the 1950s, Congress became consumed with rooting out Communist influences in American life. Under the influence of Senator Joseph McCarthy, the Senate Permanent Investigating Subcommittee and the House Un-American Activities Committee were established. During this period many writers, artists, and performers as well as academics were "blacklisted" for their political views and pressured to inform upon their colleagues. Officials of many universities bowed to government pressure and required their faculty to sign "loyalty oaths."

At the University of California 31 professors were driven off the faculty for refusing on principle to sign the loyalty oath stating that they were not members of the Communist Party. None had been accused of being a Communist, and all had been found by the tenure committee to be competent teachers and scholars. Among the mathematicians who opposed the University of California loyalty oath was Julia Robinson (1919-1985). She was holding the rank of part-time lecturer, the lowest position in the academic hierarchy, when in 1976 she became the first female mathematician elected to the National Academy of Sciences and was finally offered a professorship. Mathematicians Hans Lewy and Pauline Sperry (1885-1967) were fired from Berkeley for refusing to sign the oath, then reinstated years later by court action.

When called before congressional investigating committees, many began pleading their Fifth Amendment rights, not only to avoid self-incrimination but also to head off questioning about their colleagues. "Taking the Fifth" was soon interpreted in itself as evidence that the subject of the investigation was a Communist. In 1952 the mathematician Simon Heimlich, as well as the eminent classicist M.I. Finley, were fired from Rutgers University for pleading the Fifth. The Board of Trustees was undeterred by the results of a faculty investigation that had cleared the two professors.

The brilliant and eccentric Jewish mathematician Paul Erdös (1913-1996) fled his native Hungary for the United States in 1938, thereby escaping the Holocaust. However, in 1954 he was refused a U.S. re-entry visa and could not return to the country for nine years. Mathematicians Chandler Davis, Lee Lorch, Louis Weisner, and Dirk Struik were among those who found themselves blacklisted across the United States and had to leave the country to find employment.

By the mid-1960s a new era had begun. Mathematicians were among the many professors who were active in protests against the war in Vietnam and military buildup in general, especially nuclear weapons. Mathematician William Davidon founded a nationwide movement called Resist, and was on the national board of the Committee for Sane Nuclear Policy. He also worked to establish peace education projects in the local community. Professors were no longer generally punished by their universities for dissident views, and Davidon was particularly fortunate in that he taught at Haverford College, a Quaker institution that supported his political activities.

Still, it was not only the professor's own university he or she had to worry about. Standing in the academic community is also dependent on professional societies and organizations. Steven Smale (1930-), a mathematician at Berkeley, won the Fields Medal, the mathematical equivalent of the Nobel Prize, in 1966 for work he did five years earlier. However, he was not elected to the National Academy of Sciences until 1970. Many including Smale himself attribute the delay to his active opposition to the Vietnam War.

On many university campuses symbols of the military such as ROTC programs were vociferously opposed during the Vietnam era. One such symbol was the Mathematics Research Center (MRC) at the University of Wisconsin at Madison. The MRC had been established in the mid-1950s as a "think tank," where the Army could have access to the expertise of university mathematicians.

In August 1970 a small group of students loaded up a stolen van with explosives and detonated it in the loading dock of the building that housed the MRC. The ensuing blast destroyed several floors of the building, mostly laboratories that had nothing to do with the Army facility. Until the World Trade Center bombing two decades later, it was the most destructive act of sabotage in U.S. history. A physics post-doctoral student in one of the laboratories was killed. The leader of the group, Karl Armstrong, pleaded guilty to second-degree murder.

Conflicts between politics and academia are by no means confined to the United States. The English mathematician Alan Turing (1912-1954) was a key figure in the early development of computers, and instrumental in British code-breaking operations during World War II. In 1952 he was put on trial for homosexuality, then considered both criminal and scandalous as well as a security risk with regard to his military work. He committed suicide in 1954, using the cyanide he had in his home for chemical experiments.

Algeria, which was once a French territory, won its independence in 1962 after seven years of bloody battles. The protracted fighting has been called "France's Vietnam" in recognition of the civil unrest it caused in France. Some French intellectuals opposed the colonial regime in Algeria and thought it was useless to continue a war that left a quarter of a million people dead. Army officers and French settlers in Algeria, on the other hand, objected to giving up the country. The rebellion resulted in the return to power of the World War II hero Charles de Gaulle, and a new French constitution establishing a strong presidency.

In Russia and previously in the U.S.S.R., anti-Semitism has contributed to the mass emigration of Jewish mathematicians and scientists. The long-time head of the major mathematical institute in Moscow, Ivan Vinogradov (1891-1983), maintained a policy of trying to push Jewish mathematicians out of the field. Between anti-Semitism affecting the Jewish mathematicians and the difficult economic conditions affecting Russia as a whole, it has been estimated that as many as 70-80% of all mathematicians in Russia have emigrated, most to Western Europe, the United States, Canada, Australia, or Israel. Chinese mathematicians have also fled periodic episodes of persecution directed at intellectuals.

Impact

Political conflicts continue to trouble the academic world. In the United States many involve the degree to which, and the means by which, historical discrimination against women and minorities should be redressed. Unfortunately, incidents demonstrating that prejudice still exists are all too easy to find. Some argue that discrimination is in fact built in to the very structures of our society, and special efforts such as hiring preferences and multicultural awareness exercises are necessary to combat it. Others maintain that hiring preferences simply replace one brand of discrimination with another. Some make efforts to change or expunge language or behavior that particular groups find offensive, while others object to this "political correctness" on free speech grounds.

Issues of equity are particularly critical in mathematics because university mathematics courses such as calculus often serve as critical filters, allowing or denying admission to high-paying occupations including engineering and medicine. Since college-level calculus requires years of preparation at the elementary and secondary level, mathematics teachers are being challenged to find the best ways to work with a diverse population of children. Debates rage over whether or not to group children by ability, and whether equality of opportunity or equality of outcome is the more appropriate goal. It is not likely that politics will disappear from the field of mathematics in the foreseeable future.

SHERRI CHASIN CALVO

Further Reading

Alker, Hayward R. *Mathematics and Politics.* New York: Macmillan, 1965.

Bates, Tom. *Rads: The 1970 Bombing of the Army Math Center at the University of Wisconsin and its Aftermath.* New York: HarperCollins, 1992.

Fried, Albert. *McCarthyism: The Great American Red Scare.* New York: Oxford University Press, 1996.

Frieman, Grigori. *It Seems I Am a Jew: A Samizdat Essay.* Carbondale, IL: Southern Illinois University Press, 1980.

Hentoff, Nat. *Free Speech for Me but Not for Thee: How the American Right and Left Relentlessly Censor Each Other.* New York: HarperCollins, 1992.

Porter, Theodore M. *Trust in Numbers.* Princeton: Princeton University Press, 1995.

Reid, Constance. *Julia: A Life in Mathematics.* Washington DC: Mathematical Association of America, 1996.

Restivo, Sal. *Mathematics in Society and History.* Dordrecht: Kluwer Academic Pubolishers, 1992.

Schechter, Bruce. *My Brain Is Open: The Mathematical Journeys of Paul Erdos.* New York: Simon and Schuster, 1998.

Schrecker. E.W. *No Ivory Tower.* Oxford: Oxford University Press, 1986.

Biographical Sketches

Sir Michael Francis Atiyah

1929-

English Mathematician

Sir Michael Atiyah is best known as one of the mathematicians responsible for K-theory, a technique in topology that made possible the solution of numerous problems. (Friedrich Hirzenbruch and Alexander Grothendieck (1928-) also made significant contributions to the creation of K-theory.) He also helped create a significant theorem relating to the number of possible solutions for elliptic differential equations, called the index theorem. This discovery in turn led to a number of other advances, including a fixed-point theorem, which Atiyah helped define. These discoveries earned him the Fields Medal in 1966.

The son of a Lebanese father and a Scottish mother, Atiyah received his grade-school education in Egypt at Cairo's Victoria College and in England at the Manchester Grammar School. He later performed military service, then enrolled at Trinity College, Cambridge.

After earning his B.A., Atiyah immediately commenced work on his doctorate. He became a fellow of Trinity College in 1954 and the following year went to the Institute of Advanced Study in Princeton, New Jersey, as a Commonwealth Fellow. Upon his return to Cambridge, Atiyah became a lecturer in 1957, then a Fellow of Pembroke College in 1958. In 1961 he took a readership at Oxford and became a Fellow of St. Catherine's College.

During the 1950s and 1960s, Atiyah undertook some of his most important work in the field of cohomology, an area of topological theory that uses groups or vector bundles to study the properties of topological spaces. Later, while working with Isadore Manuel Singer, Atiyah developed the index theorem. This work, along with K-theory, earned him the Fields Medal, awarded at the International Congress in Moscow in 1966.

Michael Atiyah. *(Mathematisches Forschungsinstitut Oberwolfach. Reproduced by permission.)*

The following year Atiyah published *K-Theory,* in which he explained both the new concept and the index theorem. The latter gained new definition as Atiyah and Raoul Bott (1923-) developed the fixed-point theorem. In time, these ideas would be applied to quantum theory, and theoretical physicists made extensive use of them. This was particularly so when Simon Donaldson (1957-), one of Atiyah's students, applied the index theorem to four-dimensional geometry. Atiyah's ideas also made possible advances in the theories of superspace and supergravity, which attempt to unify relativity and quantum theory; they have also been applied to the string theory of fundamental particles.

In 1962 Atiyah was named a Fellow of the Royal Society, and in 1968 received its Royal Medal. He was knighted in 1983, and in 1988 received the Royal Society's Copley Medal. He has received numerous other awards as well, including election as a foreign member of national academies in a variety of countries, and honorary degrees from more than a dozen universities.

Atiyah remained at Oxford until 1990, when he was named Master of Trinity College, as well as first director of the Isaac Newton Institute for Mathematical Sciences in Cambridge. From 1990 to 1995 he served as President of the Royal Society. His work at Trinity College has focused on the application of mathematics to physics, and specifically on the linkage between geometry and particle physics.

JUDSON KNIGHT

Paul Joseph Cohen
1934-

American Mathematician

Paul Joseph Cohen is a leader in the advancement of mathematics. In addition to being given the National Medal of Science, he was awarded the Fields Medal—mathematics' highest honor—for his accomplishments with set theory. He solved the first of 23 problems in set theory proposed by David Hilbert (1862-1943) to the Second International Congress of Mathematics in 1900.

Having been intrigued with mathematical studies at an early age, Cohen chose to pursue the field while studying at Brooklyn College between 1950 and 1953. Cohen earned his master's degree in 1954 and his doctorate degree in 1958, both from the University of Chicago. He began teaching mathematics one year prior to receiving his Ph.D. This position was at the University of Rochester in 1957. The following year he began teaching at the Massachusetts Institute of Technology, before spending two years working at the Institute for Advanced Study at Princeton. In 1964 he became professor of mathematics at Stanford University, a position he has held since that time.

Two years following his professorship at Stanford, Cohen received the Fields Medal based on his solution to the continuum hypothesis in set theory. Originally created in 1877 by Georg Cantor (1845-1918), who discovered that there were multiple levels of infinity in numbers, the problem was highlighted by Hilbert in his historic 1900 challenge to mathematicians. Cohen proved that the continuum hypothesis was independent of standard set theory, and that the continuum hypothesis could not be regarded as an axiom of set theory. He arrived at his solution by introducing a technique called "forcing" that revealed his theory of independence. Cohen published his findings in *Set Theory and the Continuum Hypothesis* in 1966. The further use of his technique of forcing has continued to cause revelations in mathematics today.

Cohen's other honors include being awarded the Bocher Memorial Prize from the American Mathematical Society in 1964. This award is given every five years to an outstanding research analyst who has been recognized in any North American journal. Cohen also received the National Medal of Science in 1967. He is a member of the National Academy of Sciences and continues his analytical research in set theory as well as other mathematical areas including harmonic analysis and differential equations.

BROOKE COATES

Stephen Arthur Cook
1939-

American Mathematician

Stephen Cook, an American mathematician who teaches at the University of Toronto in Canada, is a specialist in computational complexity. In 1971 he advanced the theory of NP completeness, which addresses the solvability of certain problems.

Cook was born in Buffalo, New York, on December 14, 1939. His family later moved to Clarence, New York, where he met Wilson Greatbach (1919-), inventor of the implantable pacemaker. Greatbach became the young man's mentor, and influenced him toward a career in electrical engineering. Cook duly enrolled in the science engineering program at the University of Michigan in 1957, but after taking a course in computer programming, he switched his major to mathematics.

After graduating from Michigan with a B.S. in mathematics in 1961, Cook went on to Harvard University, where he became interested in propositional calculus and complexity theory. He earned his M.S. from Harvard in 1962, and his Ph.D. in 1966. He then took a position as assistant professor of mathematics and computer science at the University of California, Berkeley, where he remained until 1970, when he joined the faculty of the University of Toronto.

In 1971 Cook presented a paper entitled "The Complexity of Theorem Proving Procedures," which introduced the theory of NP completeness. Computation complexity classifies problems by difficulty: *P* for problems that can be solved in polynomial time; *NP* for problems that can be solved in nondeterministic polynomial time (that is, a problem for which a solution can be guessed and tested in an undefined period of time); and so on. NP completeness addresses the distinction between *P* and *NP.*

Cook was promoted to professor in 1975, and in 1977 received a Steacie Fellowship. In 1982 he won the Alan M. Turing Award for his work on NP completeness, and received a Killam Research Fellowship. Promoted to university professor in 1985, he received Computer Science Teaching awards in 1989 and 1995. He is a fellow of the Royal Society of Canada, and was elected to membership in the (U.S.) National Academy of Sciences and the American Academy of Arts and Sciences.

Married in 1968, Cook has two children. He continues to pursue research in computational complexity, as well as programming language semantics, parallel computation, and the relationship between logic and complexity theory.

JUDSON KNIGHT

Simon Kirwan Donaldson

1957-

English Mathematician

While still a graduate student, English mathematician Simon Donaldson published advances in topology (a division of geometry concerning the mathematical properties of deformed space) heralded by scholars as an important contribution to the understanding of four-dimensional "exotic" space. Besides sparking worldwide attention, Donaldson's work earned a 1986 Fields Medal for advancing understanding of four-dimensional space and, in particular, the topology of four-dimensional manifolds. Donaldson's work established that four-dimensional space has properties that are different from all other dimensions.

Donaldson used ideas taken from theoretical physics (for example, Yang-Mills equations and gauge theories) to develop classifications for four-dimensional space. Using elegant methodologies involving nonlinear partial differential equations and algebraic geometry, Donaldson's discoveries are regarded by many mathematicians as a seminal new synthesis between pure and applied mathematics—that is, between advancements of mathematical theory and mathematics used to solve practical problems in science, engineering, and economics. In essence, Donaldson utilized mathematical concepts used by physicists to solve purely mathematical topological problems.

Donaldson undertook his undergraduate education at Cambridge and completed his doctoral work at Oxford University. After earning his doctorate, Donaldson did further postdoctoral work at Oxford, completed an appointment at the Institute for Advanced Study at Princeton, then returned to Oxford as a professor of mathematics. In 1986 Donaldson was elected a Fellow of the Royal Society. In 1997 Donaldson took on an additional academic appointment at Stanford University.

Donaldson's work was based, in part, on the work of French mathematician Jules Henri Poincaré's (1854-1912) system of classifying manifolds that contain local coordinate systems that are related to each other by coordinate transformations belonging to a specified class. The concept of manifolds was first advanced by German mathematician Georg Friedrich Bernhard Riemann (1826-1866) to describe relationships in topological deformed space, where relative position and general shapes are defined rather than absolute distances or angles. Grounded in this mathematical heritage, Donaldson's work then proceeded to reverse the normal flow of ideas from pure math to applied math by utilizing physics applications—for example, the mathematics used to describe the interactions among subatomic particles—to fuel the engine of theoretical advancement.

Although a complete understanding of Donaldson's work requires a familiarity with mathematics beyond calculus (particularly a graduate level understanding of differential topology, geometry, and differential equations), Donaldson's work has been widely recognized for its significant advancement of low-dimensional topological geometry through the application of mathematical concepts used by particle physicists to describe short-lived subatomic particle-like wave packets called instantons.

Donaldson's important discoveries regarding the unique applications of instantons were substantially based on equations used in quantum mechanics—for example, equations published by physicists Chen Ning Yang (1922-) and Robert L. Mills—that were originally derived from the fundamental electromagnetic equations put forth by Scottish physicist James Clerk Maxwell (1831-1879). Although the Yang-Mills equations utilized solutions termed instantons to calculate interactions among nuclear particles, Donaldson insightfully used the assumptions about the concepts of dimensional space underlying instantons to help analyze four-dimensional space. Subsequently, easier and more direct methods also derived from theoretical physics (such as monopoles and string theory) have been found that confirm Donaldson's conclusions and contributions.

Donaldson has also made significant contributions to the differential geometry of holomorphic vector bundles, applications of gauge theory. Donaldson's publications include studies on symplectic topology, the application of gauge theory to four-dimensional topology, complex algebraic surfaces and stable vector bundles, the intersection forms of four-manifolds, infinite determinants, and a co-authored work with P. B. Kronheimer titled *The Geometry Of Four-Manifolds* (1990).

ADRIENNE WILMOTH LERNER

Paul Erdös
1913-1996
Hungarian Mathematician

Paul Erdös was widely assumed by his colleagues to be one of the most brilliant, and perhaps most eccentric, mathematicians of the twentieth century. Erdös established the field of discrete mathematics, which set the stage for the emergence of computer science. Although Erdös studied varied fields of mathematics, it was number theory that occupied him most of his life. He was widely known for posing simply stated mathematical problems that required complex solutions involving the relationships between numbers—then simplifying the complex solutions his colleagues had labored to deliver.

Erdös was born in Budapest, Hungary, to a mother and father who were both mathematics teachers. His genius presented itself early when,

Paul Erdos. *(Mathematisches Forschungsinstitut Oberwolfach. Reproduced by permission.)*

at three years of age, Erdös multiplied three-digit numbers in his head, and found negative numbers for himself when he subtracted 250 degrees from 100 degrees—arriving at the solution of 150 degrees below zero. Erdös briefly attended public school as a child, though left due to his mother's fear that the schools were germ-laden (Erdös's two sisters had died of scarlet fever). She preferred that her son learn at home with the help of a governess.

When Erdös reached age 17, his mother allowed him to attend the University of Budapest, where he earned a doctorate degree in mathematics within four years. Erdös completed post-doctoral study in Manchester, England, in the late 1930s. Afterwards, fleeing the persecution of Jews in Hungary, Erdös moved to the United States. Erdös did not affiliate solely with any one university or institution. Instead, he began a life of travel, owning few personal possessions and focusing his efforts on the research and exchange of his mathematical ideas.

By the age 20, Erdös made his first major contribution to number theory, a simpler and more elegant proof of a theorem regarding prime numbers proved previously by Russian mathematician Pafnuty Chebyshev (1821-1894). The theorem states that for every prime number greater than one, there always exists at least one more prime number between it and its double.

In 1949 Erdös and Norwegian mathematician Atle Selberg (1917-) found an elementary proof of the Prime Number Theorem, which involves the patterns of prime number distribution. Though a joint effort, Selberg published his research first and won the Fields Medal in 1950 for his contribution to this milestone in mathematics. Characteristically indifferent to the politics of academics, Erdös remained unaffected by the oversight. Throughout his life he donated most of the funds he earned from major prizes—including the Cole Prize of the American Mathematical Society in 1961 and the Wolf Prize in 1983—to colleagues or institutions of mathematics. In fact, Erdös often created his own awards, paying as much as $1,000 to students or colleagues who were able to solve problems set up by Erdös.

Erdös was one of the most prolific mathematicians in history, authoring more than 1,000 articles. Fellow mathematicians who co-authored works with Erdös were so numerous that the concept of the "Erdös Number" became popular conversation among the world mathematics community. Erdös numbers describe the number of collaborative efforts with Erdös himself, or those who have a published Erdös collaboration in common. In fact, Erdös's collaborations were so numerous that articles stemming from research partnered before his death continue to be published today.

To foster these partnerships and gain inspiration, Erdös spent little time in one place. A lifelong bachelor, he shuffled from one professional conference to another at universities and research institutions throughout the world. Erdös accepted the hospitality of his colleagues who housed him, fed him, and performed many of the everyday tasks of living, such as filing his taxes and arranging for dental appointments and travel. In return, Erdös would often announce to his host, "My brain is open," and thereby begin to shower his host with the latest mathematical topics and challenges. Then, with his half-packed suitcase, Erdös departed for the next engagement. During one such meeting, a mathematics conference in Warsaw, Poland, Erdös suffered a heart attack while working on an equation, and died at the age of 83.

BRENDA WILMOTH LERNER

Gerd Faltings

1954-

German Mathematician

Mathematician Gerd Faltings was born in Gelsenkirchen-Buer, Germany. Faltings undertook his graduate work at the University of Münster and completed his doctoral work in 1978. He then went on to do postdoctoral research at Harvard, taught at the University of Wuppertal, and eventually accepted a professorial appointment at Princeton University. In 1996 Faltings was invited to study at the Max Planck Institute for Mathematics in Bonn.

Faltings's proof of Mordell's conjecture earned him a 1986 Fields Medal and provided a major stepping stone toward the elusive proof of Fermat's Last Theorem, advanced in 1995 by fellow Princeton mathematics professor Andrew Wiles (1953-).

Fermat's Last Theorem is a variation on the famous Pythagorean theorem stating that for right triangles the square of the hypotenuse is equal to the sum of the squares of the sides ($x^2 + y^2 = z^2$). Both are based on Diophantine equations of the form $x^n + y^n = z^n$ where x, y, z and n are integers—equations first explored by the Greek mathematician Diophantus (circa 250 A.D.). Based on the observation that it was impossible for a cube to be the sum of two cubes and a fourth power to be the sum of two fourth powers, in 1637 French mathematician Pierre de Fermat (1601-1665) offered the theorem that, in general, it was impossible for any number that is a power greater than the second to be the sum of two like powers (that is, there were no non-zero solutions for equation $x^n + y^n = z^n$ where n is greater than 2). In his notes, Fermat claimed to have found proof for his theorem, but none was discovered prior to Wiles's proof. The problem vexed mathematicians for centuries.

Faltings's contribution to proof of Fermat's Last Theorem came in 1983 when Faltings's proved Louis Mordell's 1922 conjecture regarding systems of polynomial equations that define curves as having a finite number of solutions (that is, a finite number of prime integers x, y, z where $x^n + y^n = z^n$). Mordell's conjecture stipulated that within three-dimensional space, two-dimensional surfaces are grouped according to their genus (the number of holes in the surface). A ring, for example, has one hole and therefore its genus is considered one. If the surface of solutions for equations contained two or more holes, then the underlying equations must have a finite number of integer solutions.

By proving Mordell's conjecture, Faltings showed (using methodology based in arithmetic algebraic geometry) that the Diophantine equations $x^n + y^n = z^n$ underlying Fermat's Last Theorem could only contain a finite number of integer solutions for each exponent of Fermat's Last Theorem (that is, for $n > 2$).

Faltings's proof stopped short of proof that the finite number was, in accordance with Fermat's Last Theorem, zero—meaning that there were no solutions. Although proof of Fermat's theorem was not Faltings's goal (a proof of Fermat's Last Theorem would have required a proof encompassing all exponents $n > 2$), his proof was a major step along the road to a solution.

Faltings's work was of such siginificance that when Wiles sought verification of his proof of Fermat's Last Theorem he turned to Faltings for critical review. Among other works, Faltings published commentary and analysis of Wiles's proof for the American Mathematical Society.

The proof of Fermat's Last Theorem captured popular attention and propelled mathematicians, including Faltings, from scholarly renown to wider celebrity.

Much of Faltings's work has dealt with the arithmetic of elliptic curves, the determination of algebraic structure, and geometric combination of the solution. Faltings's well-regarded work ranges over, and mixes, algebra, arithmetic, and analysis. His published works include studies of the Shafarevich and Tate conjecture (the proof of which is often credited to Faltings), Arakelov theory, degeneration of Abelian varieties, vector bundles on curves, and the arithmetic Riemann-Roch theorem.

ADRIENNE WILMOTH LERNER

Alexander Grothendieck

1928-

French Mathematician

Alexander Grothendieck is regarded by many as one of the preeminent mathematicians of the twentieth century. He is credited with establishing a new school of algebraic geometry, and his work garnered a Fields Medal in 1966 for advancement of K-theory (Grothendieck groups and rings).

Born in Berlin, Grothendieck emigrated to France 1941. He earned his doctorate at the University of Nancy in 1953, and thereafter served in several academic posts around the world, including Harvard, before returning to France. Although he concentrated his early efforts on advances in functional analysis, during his international travels he shifted the emphasis of his work and subsequently made substantial contributions to topology and algebraic geometry.

In 1959 Grothendieck accepted an appointment at the French Institute des Hautes Etudes Scientifiques (Institute for Advanced Scientific Studies). Concerned, however, over military funding of the Institute, Grothendieck eventually resigned his post in 1969. An ardent pacifist and environmentalist, Grothendieck returned to his undergraduate institution, the University of Montpellier, where he actively promoted military disarmament and farming without the use of pesticides. Although he remained a diligent teacher, by the 1980s Grothendieck had so withdrawn himself from the international mathematics community that he made few public appearances. In 1988 he rejected the Swedish Academy's Crafoord Prize (along with its monetary award) because of what Grothendieck publicly characterized as a growing dishonesty and politicization of science and mathematics.

Nearing retirement in the late 1980s, Grothendieck's published works branched into the philosophy of science and mathematics. Though his memoir, titled *Récoltes et semailles* (Harvest and Sowing), deals with a great many nonmathematical topics, Grothendieck wrote that "mathematical activity involves essentially three things: studying numbers, studying shapes and measuring distances." Grothendieck contended that all mathematical reasoning and divisions of study (for example, number theory, calculus, probability, topology, or algebraic geometry) branched from one or a combination of these methodologies.

Grothendieck's abstract and highly scholarly work built largely upon the work of French mathematicians André Weil (1906-1998), Jean-Pierre Serre (1926-), and Russian-born mathematician Oscar Zariski (1899-1986). Together, these mathematicians laid the foundation for modern algebraic geometry. Because algebraic geometry borrows from both algebra and geometry, it has found practical application in both areas. The geometry of sets (elliptic curves, for example) can be studied with algebraic equations; it also enabled English mathematician Andrew John Wiles (1953-) to formulate a proof of Fermat's Last Theorem. Other applications include solutions for conics and curves, commutative ring theory, and number theory (especially for the Diophantine type problems, including Fermat's Last Theorem).

The depth of Grothendieck's work remains largely inaccessible to all but the most learned and nimble mathematical minds. More accessible is his theory of schemes (that provided a base upon which certain Weil conjectures regarding number theory were solved) and his work in mathematical logic. Grothendieck placed a special emphasis on defining geometric objects in accordance with their underlying functions. In addition, he is credited with providing the algebraic definitions relevant to grouping of curves. Grothendieck's work ranged over a veritable landscape of modern and postmodern mathematics, borrowing from and making substantial contributions to various topics, including topological tensor products and nuclear spaces, sheaf cohomology as derived functions, number theory, and complex analysis. Grothendieck also won the attention of the mathematical community with his highly regarded major work on homological algebra, now commonly referred to as the Tohoku Paper.

Grothendieck's publications also include his celebrated 1960 work *Eléments de géométrie algébrique* (Elementary Algebraic Geometry).

K. LEE LERNER

Thomas C. Hales
1958-
American Mathematician

Thomas C. Hales, a professor of mathematics at the University of Michigan, suddenly burst onto headlines around the world in the summer of 1998. The occasion was his proof of Kepler's sphere-packing conjecture, which had bedeviled mathematicians for some 400 years, and for which Hales offered a solution in more than 250 pages of proofs. A year later, he solved another nettlesome problem similar to the sphere-packing conjecture: the hexagonal honeycomb conjecture.

Hales was born in 1958, and became a professor of mathematics at the University of Michigan during the 1990s. He began working on the Kepler sphere-packing conjecture in 1988, 10 years before he offered his proof. Indeed, mathematicians had been studying the problem for four centuries.

The Kepler sphere-packing conjecture began with Sir Walter Raleigh (1554-1618), who asked his friend Thomas Harriot (1560-1621), a mathematician, for a simple formula to determine the number of cannonballs in a pile on a ship's deck. Harriot provided the formula, but realized that he was not certain of the most efficient way to stack cannonballs. He then turned to German astronomer Johannes Kepler (1571-1630).

Kepler considered the problem, and in 1611 stated that the most efficient way to pack equal-sized spheres would be in what is formally termed a face-centered cubic lattice. This method is commonly used, for example, in the packing of oranges in a crate, where packers use staggered layers so that the oranges in each layer sit in the hollows made by the oranges in the layer below them.

Of course grocers and fruit-packers around the world, though they may never have heard of a face-centered cubic lattice, know that this is the most efficient way to stack oranges. But a theoretical possiblilty remained that there was a more efficient way to pack spheres—placing them, for instance, in identical layers, one on top of the other—and neither Kepler nor any mathematician before Hales proved that the face-centered cubic lattice was indeed the most efficient method.

In the early 1800s German mathematician Karl Friedrich Gauss (1777-1855) made the first major step toward solving the problem by proving that out of all possible lattice packings—in which the centers of spheres are arranged in a lattice, or a three-dimensional grid—the face-centered cubic lattice was indeed the most efficient. Useful as Gauss's proof was, however, it did not rule out the possibility that some nonlattice arrangement might be more efficient.

Another 140 years or so passed before Laszlo Toth, a Hungarian mathematician, provided another piece of the puzzle in 1953. The only way to approach the problem, he indicated, would be with a lengthy calculation involving all possible specific cases.

Finally, in 1991, Wu-Yi Hsiang, a mathematician at the University of California, Berkeley, announced that he had proven the sphere-packing conjecture. However, a number of mathematicians found flaws in the proof. Among them was Hales, who in 1994 published a critique of Hsiang's proof in the *Mathematical Intelligencer.* Hales called for Hsiang to withdraw his claim, but Hsiang reportedly continued to stand by his proof up to the time that Hales presented his own.

As for Hales, he approached the problem along the lines suggested by Toth, testing each possibility, and in 1994 developed a five-step strategy for solving the problem. He ran a computer program that found 5,094 possible packings, and reduced these to about 50 that seemed to challenge Kepler's conjecture. These he gradually whittled away, partly with the help of graduate student Samuel Ferguson, whose Ph.D. thesis (written under Hales's direction) concerned the single-most challenging of all the arrangements. This was called a pentahedral prism, consisting of 12 spheres surrounding a 13th central sphere.

Hales's complete proof, presented in August 1998, ran to more than 250 pages. He posted it on the World Wide Web, and by making it easy for the world's mathematicians to view his proof, he was able to test it against their criticisms much more quickly. As time passed, Hales's proof seemed successful. As for its practical applications, Hales's proof may point the way to more efficient means of packing, storing, and encrypting electronic data.

In June 1999 Hales announced that he had also solved another long-standing problem, the hexagonal honeycomb conjecture. This holds that hexagons are the most efficient (i.e., least-perimeter) way to enclose infinitely many unit areas in a plane. This could explain why hexagons occur so often in nature, including, for example, the honeycomb in a beehive. As with Kepler's conjecture, this one had been asserted by a great theorist—Hermann Weyl (1885-1955)—but remained unproven until Hales again posted the proof at his Web site.

JUDSON KNIGHT

Donald E. Knuth

1938-

American Mathematician and Computer Scientist

Donald E. Knuth's *The Art of Computer Programming,* published in 1968, became the authoritative textbook in the subject during the 1970s. But just as computers have evolved, so has Knuth's monumental text: the fourth of a planned seven-volume series was published in 1995. As a mathematician and computer scientist, Knuth has established a number of key ideas in both fields, and has developed a pair of computer languages, TEX and METAFONT.

Born on January 10, 1938, in Milwaukee, Wisconsin, Knuth was the son of Ervin and Louise Bohning Knuth. His father was the first college graduate in the history of the family, and worked first as a grade-school teacher before taking a position as a bookkeeping teacher at a private Lutheran high school. From his father, Knuth learned an appreciation not only for learning in general, but for mathematics—and music—in particular.

As a junior-high and high-school student, Knuth showed prodigious talents. No doubt drawn by the almost mathematical precision of grammar, he was that rare child who loved to diagram sentences, and often practiced diagramming ones he saw outside of textbooks. When Ziegler's Candies ran a contest to see who could find the most word combinations in the phrase "Ziegler's Giant Bar," Knuth feigned sickness for two weeks so that he could stay home and pore over a dictionary for words. He came up with a list of 4,500 words, even without using the apostrophe—far more than the Ziegler's master list of 2,500 words. As a result, his school won a television set and enough Ziegler candy bars for the entire student body.

Graduating with the highest grade-point average in the history of his high school, Knuth enrolled at Case Institute of Technology, planning to study physics. But while at Case, he had his first encounter with a computer, an IBM 650 whose manual he studied thoroughly. From this examination Knuth concluded that he, too, could write programs, and soon he began to do just that. In 1958 he developed a program to assist his school's basketball coach in analyzing players; not only did this help the team win a league championship, according to the coach, but *Newsweek* ran an article about Knuth.

Having proved himself every bit as talented in college as he had been in high school, Knuth graduated in 1960 with a Bachelor of Science degree—and the Case faculty made the unprecedented move of simultaneously awarding him a Master's degree in mathematics. He then moved to California and enrolled at the California Institute of Technology, or Cal Tech, from which he earned his doctorate in mathematics in 1963.

Knuth became an assistant professor of mathematics at Cal Tech, but he remained fascinated with computers, and in particular compilers—programs that convert programs written in human language to codes that can be understood by computers. In 1962 publisher Addison-Wesley approached Knuth, still in graduate school at the time, about writing a book on the subject of compilers. The result was *The Art of Computer Programming,* the first three volumes of which appeared in 1968, 1969, and 1973.

Among the ideas pioneered by Knuth was "lookahead," a concept whereby a compiler looks ahead by a few words to decide on the grammatical context of prior words. His idea of inherited attributes, an expansion on the "attribute grammar" of Backus and Naur, provided a basis for

Donald Knuth (left). *(UPI/Corbis-Bettmann. Reproduced by permission.)*

the object-oriented techniques that dominated computer programming during the 1990s.

Knuth's interest in typography led to the creation of his computer languages TEX, a typesetting program, and METAFONT, which shapes letters. At one point he wrote an entire paper on the subject of "The Letter S"—its shape throughout history, and the mathematical properties of that shape.

Knuth became a professor at Stanford University in 1968, and remains with Stanford as a professor emeritus. He has received numerous awards, including the Alan M. Turing Award (1974), the National Medal of Science (presented by President Jimmy Carter in 1979), the Steele Prize (1978), a Guggenheim fellowship, and honorary degrees from universities around the world. Knuth married Jill Carter in 1961, and the couple have two children, John and Jennifer.

JUDSON KNIGHT

Kunihiko Kodaira
1915-1997
Japanese Mathematician

In a long career on two continents, Kunihiko Kodaira conducted ground-breaking research in the areas of algebraic varieties, harmonic integrals, and complex manifolds. He also became the first Japanese recipient of the Fields Medal in 1954, and he authored a number of textbooks that served to increase understanding of mathematics among Japanese students.

Born on March 16, 1915, in Tokyo, Kodaira was the son of Gonichi and Ichi Kanai Kodaira. He lived in Japan's capital during his early years, including his education at the University of Tokyo, which began when he was 20. There he studied a variety of mathematical fields and published his first paper (written in German) a year before he graduated in 1938. Among his early influences were John von Neumann (1903-1957), André Weil (1906-), Hermann Weyl (1885-1955), M. H. Stone, and W. V. D. Hodge. Also significant was the influence of the book *Algebraic Surfaces* by the Italian geometer Oscar Zariski, which sparked Kodaira's interest in algebraic geometry.

Kodaira continued his university education, earning a second degree in theoretical physics in 1941. By this point, of course, Japan was at war with many of the world's powers, and this left Japanese mathematicians isolated from the research taking place in the university centers of Britain, the United States, and other Allied nations. Yet he managed to investigate a number of challenging problems, and during the war years formed many of the ideas that would undergird his doctoral thesis. In 1943 Kodaira married Sei Iyanaga, and they later had four children.

He received an appointment as associate professor of mathematics at the University of Tokyo in 1944, and continued working on his thesis, which concerned the relation of harmonic fields to Riemann manifolds. A harmonic entity in mathematics is something whose interior can be described simply by examining only the boundary of its surface, and Weyl had been noted for his research in Riemann surfaces, a realm of topology associated with the non-Euclidean geometry of G. F. B. Riemann (1826-1866). Now Kodaira sought to establish the concept of a Riemann manifold, a type of topological space.

Just after Kodaira received his doctorate in 1949, Weyl read his dissertation in an international mathematics journal, and was so impressed that he invited the young Japanese mathematician to join him at the Institute for Advanced Studies (IAS) in Princeton, New Jersey. Kodaira accepted the invitation, and moved his family to the United States, where they would remain for the next 18 years.

During his time at Princeton, Kodaira had an opportunity to work with some of the world's leading mathematicians, including W. L. Chow, F. E. P. Hirzenbruch, and D. C. Spencer. In the course of his research there, he discovered that by use of harmonic integrals, he could more completely define Riemann manifolds. Complex manifolds—that is, manifolds involving complex numbers—are essential to the study of modern calculus, but at that time the properties of many such manifolds were undefined; hence Kodaira's work was highly significant.

Kodaira also investigated Köhlerian manifolds, which he sought to prove were analytic in nature, just like Riemann manifolds—i.e., that they could be solved or defined by calculus. This led him into investigation of a Köhlerian subset, Hodge manifolds. Using what he called "the vanishing theorem," an idea he had developed, Kodaira was able to prove that Hodge manifolds were analytic. He then proved, by using another theorem he had conceived, the "embedding theorem," that this in turn meant that all Köhlerian manifolds were analytic as well.

Needless to say, this was all highly specialized work at the furthest cutting edge of contemporary mathematics, and given the great differences between various types of manifolds, Kodaira's contributions to their classification was a significant one. For this he received the Fields Medal in Amsterdam in 1954, with his mentor Weyl pinning the medal on him. His native country later awarded him two of its highest honors, the Japan Academy Prize and the Cultural Medal.

Kodaira spent a year at Harvard in 1961, then received an appointment to Johns Hopkins University. After three years at Johns Hopkins, he accepted the chair of mathematics at Stanford University, but in 1967 he returned to the University of Tokyo. There he began writing a number of works, including a series of textbooks commissioned by the Japanese government. He died in Kofu, Japan, on July 26, 1997.

JUDSON KNIGHT

Hendrik W. Lenstra, Jr.

1949-

Dutch Mathematician

In 1981 Hendrik W. Lenstra, Jr. demonstrated that a certain class of integer programming problems does not exhibit a pattern of exponential growth. This greatly assisted mathematicians in calculating the solution time needed for such problems. Four years later, in 1985, he developed a means of factoring based on so-called elliptic curves, the first great improvement in this area since the time of Carl Friedrich Gauss (1777-1855) more than 150 years before.

Lenstra was born on April 16, 1949, in the Netherlands. He studied at the University of Amsterdam, where he obtained his Ph.D. in mathematics in 1977. In 1978 he became a professor of mathematics at the University of Amsterdam. The first of Lenstra's two most notable contributions to mathematics came in 1981, when he showed that a pattern of exponential growth does not occur with a certain class of integer programming problems. Previously mathematicians had supposed that as the complexity or size of a problem grew, likewise the solution time would grow exponentially.

In 1985 Lenstra devised a method of factoring based on what are sometimes called elliptic curves. In many cases, the behavior of these curves makes it possible to factor large numbers that have resisted all other methods. Up to this time, all theories of factoring had been based on the properties of certain quadratic equations introduced by Carl Friedrich Gauss in the late eighteenth century.

During the 1990s Lenstra divided his time between fall semesters at the University of California, Berkeley, where he taught algebraic number theory and algorithms, and spring semesters at the Universiteit Leiden in the Netherlands.

His research focuses on algorithmic number theories that interface with the computer sciences and algebraic number theories. Among his students, Lenstra is known for his sense of humor, which he has demonstrated, for instance, by collaborating on a limerick about Fermat's famous last theorem.

Awards received by Lenstra include the Fulkerson Prize of the American Mathematics Society (AMS) and Parisienne Society in 1985, and the Royal Dutch Academic Service Prize. He is a member of the AMS, and of the Dutch Mathematical Society.

JUDSON KNIGHT

Benoit B. Mandelbrot

1924-

Polish-born French Mathematician

Polish-born French mathematician Benoit Mandelbrot is widely acclaimed as one of the founding fathers of fractal geometry. Educated in France, Mandelbrot shunned the prevailing French emphasis on pure mathematics to take and early and strong interest in applied mathematics.

Due to the Second World War, Mandelbrot's education was, at times, sporadic. In many academic areas, including some mathematical subjects, he was self-taught. This reliance on his own ability to investigate and prove mathematical concepts lead him to construct geometrical proofs. Mandelbrot often credits these formative years as a source of stimulus and necessity that helped him develop his capacity for geometric thought and his geometrical approach to mathematics.

Mandelbrot eventually studied at the Ecole Polytechnique. After obtaining his doctorate, he took on academic posts at the California Institute of Technology and at Princeton's Institute for Advanced Study. In 1955, for a brief interval, Mandelbrot returned to France to accept a post at the National Center for Scientific Research (Centre National de la Récherche Scientific) before returning to the United States to accept a position with IBM laboratories. Mandelbrot has also held teaching and research posts in mathematics, engineering, physiology, and economics at various institutions, including Yale, the Einstein College of Medicine, and the French Ecole Polytechnique.

Using the emerging tools of computer graphics, Mandelbrot developed many of the well known concepts associated with fractal geometry. In 1963 Mandelbrot published what was termed the fractal concept. Fractals are geometric shapes that maintain their similar properties and relationships at all levels of magnification.

In contrast to ordinary geometry, a fractal geometric object may have fractional dimensions or use infinities to represent dimensions. Ordinary geometric objects have integer representations of their dimensions. Planes are two dimensional, and lines are one dimensional objects. Fractal objects exhibit a property described as self-similarity. The degree of self-similarity between fractals may vary, but, in general, within a self-similar object the component parts resemble the whole regardless of scale (this property is also termed scaling symmetry). In practical terms, this means that if any fractal component of an object is magnified it resembles the structure as a whole.

This fractal scaling is not the same type of scaling found in the objects familiar to classical geometry involving translational symmetry (that is, objects with translational, rotational, or reflective symmetry.) Natural fractals can be seemingly chaotic or random, yet, when this is the case, they retain the overall structure in only a statistical sense. What always remains invariant with fractals is their factual dimension.

Mandelbrot's work in fractal geometry created a mathematical school with broad scope and application. Fractals seemed to be everywhere—a universality in nature. Mandelbrot's vision was, however, inconsistent and at odds with the mathematical descriptions of natural events routinely used by physicists. In essence, while physicists tried to smooth data to explain seemingly chaotic phenomena such as turbulence, Mandelbrot saw the profound differences in behavior that could be characterized using the methodologies of factual geometry.

Fractal concepts are now used by astrophysicists to construct computer simulations depicting the collapse of systems in a gravitational field. As such, they may help formulate an understanding of the dynamics involved in the highly complex formation of cosmic structures (for example, galaxies, galactic clusters, and planetary systems).

During his career, Mandelbrot moved fractal concept from the realm of geometry into a vast array of scientific disciplines. Fractals were used to describe, unite, and relate seemingly far-flung and separate phenomena. The compartmentalized fractal concept was even used to describe cellular processes and behaviors.

In addition to his mathematical work, Mandelbrot worked on the development of computer graphics programs that could be used to represent his concepts. Fractal concepts have found widespread use in computer animation.

Mandelbrot's long and productive career has garnered significant honors, including the 1993 Wolf Prize for Physics. His published works include the 1982 book *The Factual Geometry Of Nature.*

ADRIENNE WILMOTH LERNER

John Willard Milnor

1931-

American Mathematician

In 1962 mathematician John Willard Milnor was awarded a Fields Medal for his work in differential topology, including his proof that a seven dimensional sphere could have 28 differential structures. In addition to the prestigious Fields Medal, Milnor has received many other honors, including the National Medal of Science in 1967.

In 1989 Milnor received the Wolf prize, an international prize intended to promote science and art for the benefit of mankind. The Wolf Foundation praised Milnor "for ingenious and highly original discoveries in geometry, which have opened important new vistas in topology from the algebraic, combinatorial, and differentiable viewpoint."

Milnor was born in New Jersey and undertook his undergraduate and graduate studies at Princeton, completing his doctoral thesis on the "Isotopy of Links" in 1954. Milnor displayed such exceptional brilliance as a student that he was offered an appointment to the Princeton faculty prior to the actual completion of his studies. After holding several academic positions, including a faculty appointment at UCLA, in 1970 Milnor was invited to join the Institute for Advanced Study at Princeton. In 1989 he took over as director of the Institute for Mathematical Sciences at the State University of New York in Stony Brook.

Milnor has served as vice-president of the American Mathematical Society (1975-76), editor of the *Annals of Mathematics,* and as an American Mathematical Society Colloquium Lecturer in 1968. He was also elected as a member of the National Academy of Science and the American Academy of Arts and Science.

Milnor is often credited with returning a geometrical emphasis to topological mathematics. Prior to Milnor there had been an increasing move toward using algebraic approaches to define and solve problems in topology. Milnor's 1956 publication on multiple differential structures for seven-dimensional spheres was considered a milestone in the return to geometric approaches in differential topology. Differential topology encompasses the study of differential manifolds, differentiable maps, and vector fields, among other topics, and utilizes qualitative concepts originally designed for differential geometry and differential calculus. Basic concepts of differential topology rely on the notion of a jet, embedded manifolds and intrinsic manifolds, manifold boundaries, and isomorphisms of differentiable structures. Problems and methodologies concern differential topology, differentiable manifolds, morphisms, cobordant manifolds, and what are termed surgery techniques.

Milnor's work ranges over—but is certainly not limited to—contributions in algebraic K-theory, differential geometry, algebraic topology, total curvature, duality theorems, differentiable structures, Betti numbers, polylogarithms, n-person games, and directional entropies. In 1961 Milnor provided important clarifications of the limitations of certain theories regarding manifolds. Milnor's work in the late 1990s included research in dynamics, especially holomorphic dynamics, and the relationship of low-dimensional dynamics to systems theory, properties of rational maps, periodic orbits, and Mandelbrot sets.

Regarding his approach to mathematics, Milnor is quoted as saying: "If I can give an abstract proof of something, I'm reasonably happy. But if I can get a concrete, computational proof and actually produce numbers I'm much happier. I'm rather an addict of doing things on the computer, because that gives you an explicit criterion of what's going on. I have a visual way of thinking, and I'm happy if I can see a picture of what I'm working with."

Milnor's advancements have allowed for refined conceptualization of curved space-time demanded by general relativity theory. Milnor has also advanced what is regarded as Milnor's theorium regarding the curvature of a knot.

K. LEE LERNER

John Forbes Nash

1928-

American Mathematician

Mathematician John Forbes Nash was born in Bluefield, West Virginia. He undertook

John Milnor (left) receiving the Medal of Science from President Lyndon Johnson in 1966. *(The Library of Congress. Reproduced by permission.)*

his undergraduate work at Carnegie Mellon University, and his graduate work at Princeton. At age 22 Nash earned his doctorate with papers and a thesis on non-cooperative games.

Nash's thesis work established the theory of equilibria in non-cooperative games, earning him a share of the 1994 Nobel Prize for Economics nearly half a century later. Nash received the Nobel Prize with Hungarian-American economist John C. Harsanyi and German mathematician Reinhard Selten. The Nobel Prize for Economics is designated to recognize substantial and important contributions to political science, psychology, and sociology.

The brilliance of Nash's work was obscured by a three-decade-long battle with paranoid schizophrenia. Despite frequent hospitalizations and the sometimes debilitating symptoms of his illness, Nash continued his work whenever able. However, though Nash strove to continue the excellence of his research, his personal and professional life suffered from his illness. He had a short faculty tenure at MIT until his illness forced him to resign his post. During the height of Cold War political paranoia in McCarthy-era America, Nash lost his security clearance and an opportunity to work for several strategic think-tanks that employed eminent mathematicians outside of academia.

During the late 1950s and early 1960s, Nash eventually faced destitution and homelessness until his ex-wife and friends within the mathematical community brought him back to Princeton, where Nash worked in obscurity while he maintained tenuous and informal links to the scholarly community between hospitalizations for his illness. By 1970 Nash had became an enigmatic figure on the Princeton campus. Nash would leave mathematical notations and writings scrawled on empty classroom blackboards in the Princeton mathematics building.

The significance of Nash's work was not limited to mathematics. Non-cooperative game theory is viewed by many as one of the most important developments in twentieth-century economic and social science. Game theory has found application in the development of strategy for modern warfare, governmental science (including political campaigns and lobbying efforts), and business strategy.

In essence, game theory is a division of mathematics that sets out theories regarding the behavior of rival competitors operating with a mixture of interests and goals. Nash provided a solution—known now as the Nash equilibrium—that allowed prediction of the dynamic interchange of needs, wants, and threats among competitors. Most importantly, game theory offers important insights into how people and entities (governments and corporations) arrange affairs and make strategic choices.

Critics of game theory, however, contend that its influence is still debatable and that one of the great flaws of the theory is its inability to make predictions in addition to offering explanations of human behavior (for example, how players react). Some critics contend that game theory, rather than being broadly applicable, is relevant to only a few selected problems that have a limited number of competitors. Other critics contend that game theory is incompatible with the idea of free economic competition because the theory assumes an inability of players to influence eventual outcomes.

Supporters of game theory maintain that it is a useful tool for both simple and complex problems, and cite the successful incorporation of game theory in the lucrative action of telecommunication frequencies in the 1990s.

Despite the recognition accorded as a Nobel laureate, many scholars contend that social and academic prejudices regarding Nash's illness prevented him from receiving greater acclaim for work that influenced two generations of scholars in the formulation of mathematical models concerning human conflict and cooperation. Game theory, supporters contend, offers the best opportunity to study economic behavior ranging from price wars to illegal collusion.

Prior to receiving the Nobel Prize, Nash made a sudden recovery from his illness. In addition to his work on game theory, he made significant contributions to the study of real algebraic varieties, differential equations, and geometry.

ADRIENNE WILMOTH LERNER

Sir Roger Penrose
1931-
English Mathematician and Physicist

Sir Roger Penrose, one of the leading mathematicians of the later half of the twentieth century, is perhaps best known for his collaborative work with British physicist Stephen Hawking (1942-) regarding the calculation and prediction of the fundamental properties associated with black holes. Penrose served on Hawking's Ph.D. thesis committee, authored books with Hawking, and shared the 1988 Wolf Prize for Physics with Hawking for their contributions to theoretical understanding of black holes, singularities, and cosmology. In 1994 Penrose was knighted for his contributions to mathematics and science.

Penrose was born in Colchester, England. He undertook undergraduate studies at University College School, London, and doctoral work in algebraic geometry at Cambridge. Penrose earned a Ph.D. in 1957 and subsequently held a number of academic posts, including a professorship at Oxford. In addition to many honors and awards for a distinguished scholarly career, Penrose, was elected to the Royal Society of London in 1972 and was elected a Foreign Associate of the United States National Academy of Sciences in 1998.

John Forbes Nash. *(Robert P. Matthews; courtesy of John Nash. Reproduced by permission.)*

Penrose has made substantial contributions to theories of the very vast and the very small. His work has provided deeper insight into relativity theory, especially general relativity theory, applications of mathematics to physics and cosmology (the study of the origins and nature of the Universe), and to the subatomic world of quantum theory. Penrose and Hawking's efforts toward advancing the theory of general relativity, put forth by Albert Einstein (1879-1955) in the first half of the twentieth century, established the necessity for cosmological singularities (black holes). This insight was a significant contribution toward understanding of the life cycles and deaths of stars, as well as toward models concurring the origin and fate of the Universe.

Working with Hawking, Penrose provided mathematical proof that the matter within a black hole must eventually collapse under gravity into a singularity, a geometric point in space with no size (and hence no volume) where the mass of the black hole is compressed to infinite density. Penrose developed what are know known as Penrose Diagrams that map the space-time and the gravitational environment surrounding a black hole.

Penrose has also contributed to the advancement of theories of consciousness, quantum cosmology, the development of artificial intelligence, and other topics that explore the nature of physics, knowledge, and reality.

Seemingly far removed from his work in mathematical physics, Penrose's personal interest in tessellation type puzzles lead to important practical discoveries regarding the nature of quasicrystals (crystals that form in a quasi-periodic fashion), subsequently used in industry to provide cookware with protective non-stick coatings. In 1958 Penrose and his father advanced the abstract concept of strange loops with the hypothetical Penrose square stairway, wherein travel in either direction resulted in no net loss or gain in elevation, and another impossible figure known as the tribar. Penrose also developed the concept of what is now known of Penrose tiling (the complete covering without gaps or overlaps of a two-dimensional planar surface with two sizes of rhomboid or pentagon shaped tiles).

The author of a substantial volume of scholarly work, Penrose is also well known for his books designed for the lay audience. In 1989 Penrose's book *The Emperor's New Mind* moved onto the bestseller lists and earned Penrose wide acclaim and fame outside the scholarly community. Penrose has also authored other well-received books, including *Shadows of the Mind* (1994) and *The Large, the Small and the Human Mind* (1997). In 1996 Penrose worked with Hawking to produce another well-regarded book titled *The Nature of Space and Time.*

ADRIENNE WILMOTH LERNER

Daniel Grey Quillen
1940-
American Mathematician

In 1977 Daniel Quillen and Russian mathematician A. A. Suslin independently discovered two quite distinct proofs of a 20-year-old theory concerning the structure of generalized vector spaces. These proofs established that all abstract spaces of certain common types were constructed in direct analogy with two- and three-dimensional Euclidean space. In the following year, 1978, Quillen received the Fields Medal in recognition of this and other outstanding contributions to mathematical study.

The son of a chemical engineer who went on to become a physics teacher, Daniel Quillen was born in Orange, New Jersey, on June 27, 1940. He attended grade school at the private Newark Academy, then entered Harvard University. Then, after he received his B.A. in 1961, he continued his work at Harvard under the supervision of Raoul Bott. In 1964 Quillen earned his Ph.D. with a thesis on partial differential equations, *Formal Properties of Over-Determined Systems of Linear Partial Differential Equations.*

By this point the 24-year-old Quillen and his wife Jean, a violinist, had two children; eventually they would have five. Quillen went on to join the faculty of the Massachusetts Institute of Technology (MIT), but many of his most pivotal experiences during the 1960s and 1970s took place while serving as a visiting scholar at other universities. In the 1968-69 academic year, Quillen was a Sloan Fellow in Paris, where he came under the influence of Alexander Grothendieck (1928-). During the following year, as a visiting member of the Institute for Advanced Study at Princeton, he met Michael Atiyah (1929-), who would have an even more profound effect on Quillen's work. Quillen also spent a year in France as a Guggenheim fellow from 1973 to 1974.

Quillen's principal interests lay in the homology of simplicial objects. A simplicial object, or simplex, is one that has n dimensions determined by $n + 1$ points, in a space equal to or greater than n; thus a triangle and its interior, determined by three points, is a two-dimensional simplex. Homology is an area of topology concerned with the partitioning of space into geometric components such as points, lines, and triangles, and with the interrelationships of these components, particularly as this relates to group theory.

Quillen had begun in the 1960s by identifying the means of defining the homology of simplicial objects over a variety of categories such as sets; then he turned to developing a conjecture in homotopy theory posited by Frank Adams. Using techniques from algebraic geometry and from the modular representation theory of groups, Quillen and one of his students were able to prove Adams's conjecture. Quillen later applied these concepts to the study of K-theory, which Atiyah had developed as a means of dealing with cohomological questions. He expanded K-theory, establishing a higher algebraic version of it in 1972, and this made possible a variety of discoveries which in turn led to his receiving the Fields Medal six years later.

Quillen has often been lauded by other mathematicians for his fresh approach to challenging questions. He is noted for his rigor and the breadth of his thinking, and in his personal life is known as a quiet family man. During the 1990s Quillen worked at Oxford in England.

JUDSON KNIGHT

Julia Bowman Robinson
1919-1985
American Mathematician

Julia Robinson was a prominent mathematician who devoted her career to applying number theory methodology to the resolution of mathematical logic problems. The Julia Robinson Hypothesis led to the solution of Hilbert's Tenth Problem, which mathematicians had pondered for decades and feared was unsolvable. She also achieved scientific leadership positions previously not held by female mathematicians and used her influence to seek equal opportunities for all scholars. Considered a mentor and exemplary figure, Robinson inspired mathematicians of both genders to approach mathematical puzzles with ingenuity and resourcefulness.

Robinson was born in St. Louis, Missouri, on December 8, 1919, to Ralph Bowers and Helen (Hall) Bowman. When she was two-years-old, her mother died. Robinson lived first in her grandmother's home in Phoenix, Arizona, and then in San Diego, California, when her father remarried. Suffering scarlet fever when she was nine, Robinson was quarantined. Within a year, she contracted rheumatic fever and was confined to bed. During this time, her tutor inspired her interest in mathematics. Initially, she began taking classes at San Diego State College and planned to qualify as a public school teacher, specializing in mathematics.

As she became interested in researching more abstract mathematical concepts, Robinson decided to study at the University of California at Berkeley, from which she graduated in 1940. Discouraged by employers' lack of respect for her intellectual abilities, Robinson returned to Berkeley for graduate school. Her assistant professor Raphael M. Robinson taught her number theory. She married him in 1941, but could not continue her employment in Berkeley's mathematics department because of nepotism rules. Instead, she joined the Berkeley Statistical Laboratory as an assistant to Jerzy Neyman, who was working on classified World War II projects. After the war, Robinson focused on obtaining a doctorate. Directed by Alfred Tarski, a logician, she wrote a thesis on number theory that explored how integers could be defined by the addition and multiplication of rational numbers. She graduated in 1948.

Along with her colleagues, Robinson then concentrated on solving Hilbert's Tenth Problem, so named because it was the tenth query posed on a list compiled by David Hilbert, who had wondered if a method could be devised to solve a Diophantine equation (a polynomial equation with several variables) using integers. Robinson wrote academic papers addressing various mathematical questions but essays on Hilbert's Tenth Problem comprised the majority of her professional publications. In 1961 she coauthored a paper with Martin Davis and Hilary Putnam. The two, who had contacted her after she spoke at the 1950 International Congress of Mathematics, sent her a theorem that they had been contemplating. The trio's paper stated that no algorithm existed to determine if an exponential Diophantine equation could produce a solution with natural numbers. Their ideas ultimately helped Yuri Matijasevic prove the absence of the desired method in 1970. Robinson later initiated working with Matijasevic and traveled to the Soviet Union to further explore his findings.

Robinson also applied her mathematical abilities to theoretical analysis for the RAND corporation and to hydrodynamics projects conducted by the Office of Naval Research. Appointed a full professor at Berkeley in 1976, she taught classes there, despite health problems exacerbated by scar tissue from her childhood ailments. She was the first female mathematician elected to the National Academy of Sciences, being honored in 1975. Five years later, she was the second woman to present an American Mathematical Society Colloquium Lecture. She also delivered the 1982 Emmy Noether Lecture for the Association for Women in Mathematics. In 1982 Robinson became the first woman selected as president of the American Mathematical Society, where since 1978 she had been one of that group's first female officers. She also received an honorary degree from Smith College and the $60,000 MacArthur Foundation Prize Fellowship in acknowledgement of her accomplishments in mathematics. Suffering from leukemia, Robinson died on July 30, 1985, in Oakland, California.

ELIZABETH D. SCHAFER

Jean-Pierre Serre

1926-

French Mathematician

As do many mathematicians working at the frontiers of the discipline, Jean-Pierre Serre specializes in topology, the study of geometric figures whose properties are unaffected by physical manipulation. Despite the fact that he has grappled with subjects far beyond the understanding of all but a few highly trained specialists, Serre has proven his ability to write about mathematics in a lucid, easily understood style. He received the Fields Medal in 1954.

Serre was born in Bages, France, on September 15, 1926, to Jean and Adèle Diet Serre. Both parents were pharmacists and they brought their son up with a fascination for chemistry. From an early age, however, Serre took an interest in calculus and began poring over his mother's books in advanced mathematics. When he was just 15, he taught himself the fundamentals of derivatives, integers, series, and other aspects of calculus that most people only absorb—if indeed they ever do—under the guidance of an instructor. As a boarding student at the Lycée de Nîmes in high school, he kept himself from being bullied by the older students by helping them with their math homework-even though they were taking more advanced classes than he.

In 1944, the 19-year-old Serre won the Concours Général, a national mathematics competition, and in the following year he entered the highly prestigious Ecole Normale Supérieure in Paris. Up to this point he had planned to become a high school mathematics teacher, but, as he later recalled, his acceptance to the school, which had required him to pass extremely competitive examinations, gave him the confidence to believe that he could make a living as a research mathematician. In 1948 he married Josiane Heulot, a chemist, with whom he later had a daughter.

In the late 1940s and early 1950s Serre did some of his most important work in topology, the subject in which he earned his doctorate from the Sorbonne in 1951 with a dissertation on homotopy groups. He spent two years on the faculty of the University of Nancy before becoming chairman of the department of algebra and geometry at the Collège de France.

During this time, Serre investigated spectral sequences, an algebraic construction formulated by the French mathematician Jean Leray. He used them to study the relations of homology groups—that is, groups of geometric components such as points, lines, or triangles. This work suggested numerous connections between homology groups and homotopy groups (classes of functions that are equivalent under a continuous deformation), and gave him insights concerning the homotopy groups of spheres.

His work on spectral sequences as applied to sheaves (a group of planes passing through a common point) earned Serre the Fields Medal, which is awarded every four years to an outstanding mathematician under the age of 40, when he was 28 years old. Serre had already stimulated thinking among mathematicians concerning the loop space method in algebraic topology, and as early as 1952, in a lecture at Princeton University, he was discussing the extension of homotopy groups in a field he termed C-theory.

After receiving the Fields Medal, Serre went on to investigate new mathematical territories, including complex variables, cohomology, algebraic geometry, and number fields. Beginning in the early 1950s, he published a major work every few years and earned widespread recognition as a mathematician and writer. The Royal Society of London made him an honorary fellow, and he became a member of the French, Dutch, Swedish, and American academies of science. He was awarded the Medaille d'Or of the Centre Nationale de la Recherche Scientifique and received the Balzan Prize.

Serre retired from the Collège de France in 1994, assuming the position of honorary professor. In 1995 he received the Leroy P. Steele Prize for Mathematical Exposition from the American Mathematical Society. This award came in recognition forhis writing, specifically the book *A Course in Arithmetic,* first published in 1970. Nevertheless, the award's citation read, "Any one of Serre's numerous other books might have served as the basis of this award. Each of his books is beautifully written, with a great deal of original material by the author, and everything smoothly polished."

JUDSON KNIGHT

Stephen Smale

1930-

American Mathematician

The work of Stephen Smale has focused on a number of topics, including topology and dynamical systems. During the 1970s, he turned his attention to the application of mathematical theory to economics, and, in later years, he explored questions in computer science and their application to principles of mathematics. Recipient of the Fields Medal in 1966, Smale has also been honored with the Veblen and Chauvenet prizes.

Smale was born on July 15, 1930, in Flint, Michigan, to Lawrence and Helen Smale. He later described his father as an "armchair revolutionary," who had been expelled from Michigan's Albion University for publishing a radical newsletter. Following his expulsion, Lawrence Smale had gone to work as an assistant in the ceramics laboratory at the AC Spark Plug factory in Flint, the home of General Motors.

When Smale was five years old, his family moved to a small farm outside of Flint, and soon afterward he began his education in a one-room schoolhouse. By the time he was in high school, Smale's interests included chemistry—the field in which he intended to work—and politics. Like his father, he was a student radical, but, as he later recalled, his attempts to organize a protest against the omission of evolution from the biology curriculum failed to stir much support.

By 1948, when he enrolled at the University of Michigan, Smale had switched from chemistry to physics as his intended major. When he began doing poorly in physics, however, he gradually shifted his focus to mathematics, a subject in which he had always excelled. Still, physics was the subject in which he earned his B.S. degree in 1952, and he went on to earn an M.S. in physics the following year. His political radicalism increased during his college years; Smale joined the American Communist Party and protested U.S. involvement in the Korean War. Later, he admitted that his greatest motivation for staying in college during this period was to avoid the draft.

In 1955 Smale married Clara Davis, a classmate at the University of Michigan. The couple later had two children: Laura, who became a biological psychologist, and Nat, who became a mathematician. Smale earned his Ph.D. in mathematics in 1956 and for a period of many years focused on the area of topology, work which earned him the Fields Medal 10 years later. In the meantime, however, Smale had turned from topology to dynamical systems. (The term "dynamical systems" refers to methods for mathematically describing changes that take place over time in some real or abstract system.) Smale's shift in attention to dynamical systems coincided

Stephen Smale. *(AP/Wide World Photos. Reproduced by permission.)*

with his acceptance of an appointment as professor of mathematics at Columbia University.

Ever restless, however, Smale experimented with a number of mathematical fields, including the calculus of variations and infinite dimensional manifolds, during the early 1960s. In 1964 he took a position as professor of mathematics at the University of California at Berkeley, and there he began concentrating once again on dynamical systems. At one of the nation's leading centers for the student protest movement of the 1960s, Smale worked with student activist Jerry Rubin and others, organizing and leading protests against the Vietnam War.

As the 1960s became the 1970s, however, Smale began to change both his mathematical interests and his politics. The awarding of the Fields Medal in 1966 had taken place in Moscow and Smale had several other opportunities to visit the Soviet Union, experiences that led to his becoming highly disillusioned with Communism. Eventually, he adopted a stance that he has described as radical in some regards, conservative in others—but always anti-military. As for his professional focus, he turned to mathematical applications in economics, the result of several conversations with Nobel economics laureate Gerard Debreu.

With the explosion of knowledge that took place in the field of computer science during the

1980s, Smale has increasingly turned his attention to that area. He has observed that if the mathematics used in computer science could be brought into closer relationship with mainstream mathematics, revolutionary changes within the discipline of mathematics itself could result.

JUDSON KNIGHT

René Frédéric Thom

1923-

French Mathematician

Best known for his formulation of catastrophe theory, which has wide-ranging applications in both the natural sciences and the social sciences, René Thom is a topologist and mathematical philosopher of distinction. Not only was he the recipient of the Fields Medal in 1958, he has also been named a Knight of the Legion of Honor in France, and has received numerous other awards.

Born on September 2, 1923, in Montbéliard, France, Thom was the son of Gustav (a pharmacist) and Louise Ramel Thom. After earning an academic scholarship at the primary school in his hometown, Thom went on Collège Cuvier, also in Montbéliard, and earned his bachelor's degree in elementary mathematics from the University of Besançon in 1940. Soon afterward, however, the German army invaded France and Thom's parents encouraged him and his brother to flee. They made their way to the southern part of the country, and from there to Switzerland, where they were received with such a warm welcome that Thom fondly remembered it years later.

They helped with the harvest in the town of Romont, but soon Thom returned to France, where he earned another degree in philosophy. He migrated to Paris, where he attended the Lycée Saint-Louis before applying for admission to the highly prestigious Ecole Normale Supérieure in 1942. He failed to gain acceptance, however, but applied again the following year and was admitted.

Despite the fact that he was studying under conditions of wartime occupation, Thom's experience at the Ecole was a fruitful one. During this period, he was exposed to the approach of the Bourbaki group and to the work of Henri Cartan, which would have a great influence on Thom. His last year at the Ecole Normale Supérieure, after the liberation of Paris by Allied forces, was a particularly happy one in Thom's memory. In 1946 he moved to Strasbourg to work with Cartan, who served as advisor on his thesis, *Fibre Spaces in Spheres and Steenrod Squares.* Thom earned his doctorate in 1951, and taught at Grenoble from 1953 to 1954 before going to Strasbourg, where he remained until 1963.

Thom's thesis contained the basic ideas of cobordism, a classification scheme for manifolds—multidimensional topological surfaces—based on homotopy, a continuous deformation of one function into another. It was cobordism, presented by Thom in fully developed form in 1954, which earned him the Fields Medal in 1958. Many years later, Thom revealed that after receiving the coveted award, he had doubts about his own ability to continue generating meaningful results as a mathematician, and therefore turned his attention to singularities of differentiable maps, a topic he viewed as "more flexible and more concrete."

Appointed professor in 1957, Thom left Grenoble in 1963 and began teaching at the Institut des Hautes Etudes Scientifiques (IHES) in Bures-sur-Yvette. Without teaching responsibilities, he was free to concentrate on research, and prepared an article criticizing the movement to remove geometry from the general mathematics curriculum. In the years that followed, he moved increasingly into applied mathematics, turning his attention from optical problems in physics to the biological discipline of embryology.

The latter move was to prove pivotal, because from it arose Thom's famous catastrophe theory. Thom found that biological forms are subject to sudden changes, which he called catastrophes, findings he published in *Structural Stability and Morphogenesis* (1972). Catastrophe theory, grouped with geometry because its results are typically shown as curves and surfaces, attempts to explain predictable discontinuities in the output of systems subject to continuous inputs. Such predictions are beyond the reach of calculus, making Thom's theory invaluable to a wide array of disciplines outside of mathematics.

Catastrophe theory has been subject to criticism due to the fact that, as Thom himself has observed, "the theory did not permit quantitative prediction." It has nonetheless proven applicable in fields as diverse as hydrodynamics, linguistics, and industrial relations. Since the early 1970s, Thom has concerned himself primarily with matters of mathematical philosophy, a role that evolved in part from the need to defend and develop catastrophe theory.

JUDSON KNIGHT

Andrew Wiles

1953-

British Mathematician

Andrew Wiles, born in 1953 in Cambridge, England, is perhaps the most celebrated living mathematician. He became an instant celebrity when he announced in 1993 that he had proven Fermat's Last Theorem. The proof of this theorem, one of the most famous unsolved problems in the history of mathematics, had eluded mathematicians for over three centuries.

HOW DOES A MATHEMATICIAN WORK?

Most people assume that a mathematician works by plugging numbers into formulas. But to create new mathematical knowledge, mathematicians must, of course, be creative. Creativity involves trying all sorts of ideas before finding one that makes sense. (Picture a novelist beginning a new work, with piles of crumpled-up paper on the floor, each representing a failed attempt to find the perfect opening line.) Andrew Wiles has compared his experience in creating mathematics to exploring a dark mansion:

> *You enter the first room of the mansion and it's completely dark. You stumble around bumping into furniture, but gradually you learn where each piece of furniture is. Finally, after six months or so, you find the light switch, you turn it on, and suddenly it's all illuminated. You can see exactly where you were. Then you move into the next room and spend another six months in the dark. So each of these breakthroughs, while sometimes they're momentary, sometimes over a period of a day or two, they are the culmination of—and couldn't exist without—the many months of stumbling around in the dark that proceed them.*

Fermat's last theorem states that the equation $x^n + y^n = z^n$ has no whole number solutions for any values of equation n greater than 2. In other words, equation $x^2 + y^2 = z^2$ has whole number solutions (such as $x=3$, $y=4$, $z=5$), but $x^3 + y^3 = z^3$, $x^4 + y^4 = z^4$, etc., have no whole number solutions. The French mathematician Pierre Fermat (1601-1665) wrote in the margin of a book that he had discovered a remarkable proof for this theorem, but the margin was to small to contain it. For the next 350 years, many of the best mathematicians in the world attempted to find a solution to this problem.

Andrew Wiles first encountered the theorem that would be so important in his life when he was only ten years old. As a boy who had already taken an interest in mathematics, he could easily understand Fermat's Last Theorem. Later, as a teenager contemplating mathematics as a career, Wiles actually spent some time trying to prove it.

Wiles did become a mathematician, but his dream to prove Fermat's Last Theorem was pushed into the background while he attended to more immediate and attainable goals in mathematics. Wiles attended Oxford University, where he received a B.A. in 1974. He then received a Ph.D. from Cambridge University in 1980. In 1981 he came to the United States, where he spent a year at the Institute for Advanced Study at Princeton. In 1982 he was appointed professor of mathematics at Princeton. During this time Wiles was establishing himself as a young mathematician with a promising future. Although his research had seemingly led him away from Fermat's Last Theorem, it was actually preparing him for the great work that lay ahead.

Andrew Wiles's interest in the famous problem was rekindled when he heard that the American mathematician Ken Ribet had shown that there was a link between Fermat's Last Theorem and the Taniyama-Shimura conjecture, which had been proposed by two Japanese mathematicians, Yutaka Taniyama and Goro Shimura, in the 1950s. Ribet showed that if this conjecture were proven to be true, then Fermat's Last Theorem must also be true. Suddenly, Wiles had a legitimate reason to turn his interests to the problem that had captured his imagination as a youth. Interestingly, the Taniyama-Shimura conjecture involved the theory of elliptic curves, Wiles's research specialty since his days as a graduate student at Cambridge. The stage was set for one of the most unusual quests in the history of mathematics.

What made Wiles's eventual success so unusual was that he worked on the problem in almost total isolation for seven years. He was concerned that any announcement that he was working on Fermat's last theorem would result in much unwanted publicity. He preferred to work alone, usually at home, not even sharing with his colleagues at Princeton the true nature of his ongoing research.

Finally, in the summer of 1993 Wiles was ready to share his completed proof with the world. He chose to unveil his work at a conference at the Isaac Newton Institute in Cambridge. In a dramatic series of lectures, Wiles stunned the mathematical world with the solution to the most famous problem in the history of mathematics. The quest for a proof of Fermat's Last Theorem appeared to be over. All that remained was for the proof to pass the scrutiny of Wiles's peers.

Unfortunately, a section of Wiles's work did not stand up to this intense scrutiny. An error was found in one part of the long and difficult proof. At first Wiles was confident that he would be able to fix the error and save the proof, but as time went on the "correction" became more and more difficult. Wiles even asked for help from one of his former students, Richard Taylor. After a year of work, Wiles and Taylor appeared to be no closer to correcting the error in the proof. Remarkably, just as Wiles was about to give up, a sudden inspiration helped him to find the new approach that was needed to finish the proof.

The proof of Fermat's Last Theorem brought with it many honors for Andrew Wiles. He was elected a Fellow of the Royal Society of London and a foreign member of the National Academy of Sciences in the United States. In addition, he has received many other prizes and awards from all over the world. Today, Wiles continues to work as a research mathematician. Although he admits that solving Fermat's problem was a once-in-a-lifetime opportunity, he continues to search for answers to other difficult mathematical riddles.

TODD TIMMONS

Biographical Mentions

Leonard M. Adleman
1947-

American mathematician and computer scientist known for his work in computational complexity, number theory, molecular biology, immunology, and computer viruses. While at the Massachusetts Institute of Technology he helped invent the RSA Public-key Cryptosystem. He received his BA in mathematics (1968) and his Ph.D. in computer science (1976) from the University of California at Berkeley. He later became Professor of Computer Science at the University of Southern California.

Kenneth I. Appel
1932-

American mathematician known for his work in topology, the branch of mathematics that explores certain properties of geometric figures. In 1976 he solved the four-color problem, originally posed in 1850, that has enabled maps to be drawn and color-oriented so that no two adjacent regions are of the same color. He received his B.S. from Queens College (1953) and his Ph.D. in mathematics from the University of Michigan (1956). He later taught at the University of New Hampshire.

Kenneth J. Arrow
1921-

American economist and 1972 Nobel Prize recipient with John Hicks for his groundbreaking work in general equilibrium theory that examines the relationship between the processes of production, distribution, and consumption in the economy. Using modern mathematical methods, he has expanded his theoretical and mathematical models to social concerns such as medical care, education, and racial discrimination. He received his masters degree in mathematics (1941) and Ph.D. in economics (1951) from Columbia University. He later taught at Harvard University.

Richard Askey
1933-

American mathematician known for his work in classical mathematics, especially in the area of special functions. His work attracted attention in 1984 for a paper that provided a link to the Bieberbach conjecture, a problem having to do with functions of a complete variable. He earned his master's degree in 1956 from Harvard University and his Ph.D. in 1961 from Princeton University. He taught at the University of Chicago before joining the University of Wisconsin.

Allen Baker
1939-

British mathematician awarded the Fields Prize in 1970 for his work on Diophantine equations. These equations have been debated for nearly 1,000 years, and his work contributed to the understanding of them. Baker received his doctorate and was elected a Fellow of Trinity College, Cambridge, in 1964. He became a member of the Institute for Advanced Studies at Princeton in 1970 before returning to teach at Trinity College.

Ruth Aaronson Bari
1917-

Polish-American mathematician considered the world expert on chromatic polynomials. She attended an all-girl's high school where it was thought that mathematics was not important. Tutoring herself, she went on to receive her undergraduate degree in 1939 from Brooklyn College and her master's from John Hopkins University in 1943. She worked for Bell Telephone Laboratories and in 1966 received her Ph.D. from George Washington University. She later taught at George Washington University.

Alexandra Bellow
1943-

Romanian-American mathematician known for her work in Ergodic theory, harmonic analysis, and number theory. She graduated from the University of Bucharest in 1957 before traveling to the United States with permission of the then-Communist regime. While at Yale, she studied with Shizuo Kakutani, a founder of Ergodic theory, and received her Ph.D. in 1959. She later taught at Northwestern University.

Dorothy Lewis Bernstein
1914-

American applied mathematician whose research focused on the Laplace transform, a mathematical function named after the French mathematician Pierre Simon Laplace. Bernstein was a pioneer in incorporating applied mathematics and computer science into the undergraduate mathematics curriculum in the United States. Bernstein spent the majority of her career at Goucher College in Maryland. She was also closely associated with the Mathematical Association of America.

Lipman Bers
1914-1993

Russian-American mathematician known for his work on Teichmuller theory, pseudoanalytic functions, quasiconformal mappings, and Kleinnian groups. He studied in Russia and Switzerland and received his Ph.D. from Charles University in Prague, Czechoslovakia, before escaping World War II by coming to the United States in 1940. He conducted research for Brown University on the two-dimensional subsonic fluid flow for the United States war effort. He held teaching posts at Princeton, Courant Institute, and Columbia University, where he remained until he retired.

David Harold Blackwell
1919-

American mathematician who is the only African-American mathematician to be elected to the National Academy of Sciences. Born in Centralia, Illinois, Blackwell earned his bachelor's degree (1938), master's (1939), and Ph.D. (1941) from the University of Illinois at Urbana-Champaign. He holds numerous honorary doctorates from other institutions.

Lenore Blum
1943-

American mathematician who is one of the most accomplished women in the field. She has been president of the Association for Women in Mathematics and chairperson of the American Association for the Advancement of Science. She earned her Ph.D. in 1968 from the Massachusetts Institute of Technology and is the author of *Complexity and the Real Computation.*

Enrico Bombieri
1946-

Italian-born mathematician who won the Fields Medal in 1974 for his development of Bombieri's mean value theorem. Bombieri has lived in the United States for many years and is a member of the National Academy of Sciences.

Richard Ewen Borcherds
1959-

South African mathematician who proved the "Monstrous Moonshine" conjectures, so-called because they were so bizarre. Borcherds introduced new techniques into mathematics, including vertex algebras, for which he was awarded a Fields Medal in 1998, and a new class of algebras, now called Borcherds algebras. His tools are used for studying the representations of the monster simple group, a group with more than 8 x 10^{54} elements, and uncovering the mysterious connections between the Monster and other areas of mathematics.

Raoul Bott
1923-

Hungarian mathematician who helped to develop techniques for analyzing geometric figures representing closed paths (loops) in multidimensional space. While techniques for determining the lengths of such paths existed, they all required approximating the path in a "piecewise" manner, similar to approximating the circumference of a circle by constructing many-sided polygons. Bott helped to develop more exact methods of doing this task in any number of spatial dimensions.

Jean Bourgain
1954-

Belgian-born mathematician who was awarded a Fields Medal for his innovative work in several areas of mathematical analysis. Bourgain's much-lauded work includes solutions for long-standing problems in Banach space theory and harmonic analysis. His work has improved the understanding of convexity in high dimensions, ergodic theory, the circle maximal function, and nonlinear partial differential equations, which are used in theoretical (mathematical) physics. In 1994 Bourgain was appointed to the Institute for Advanced Study at Princeton.

Louis de Branges

French-American mathematician who in 1984 proved the Bieberbach conjecture. This conjecture, posed by German mathematician Ludwig Bieberbach in 1916, dealt with properties of maps, specifically with mapping the surface of one shape onto another. One example of this procedure is the mapping of Earth's spherical surface onto a flat map. De Branges was able to show that transferring maps from generalized mathematical surfaces onto others was topologically possible provided that certain mathematical conditions were met.

Hans-Joachim Bremermann
1926-1996

German-American mathematician who made important contributions to the theory of locally convex spaces, which are spaces having local areas that are convex and are mathematically described with rather complex equations. Bremermann was able to take his contributions to this branch of mathematics and apply them to areas of physics, electrical theory, and electrical circuit design, emphasizing the importance of the boundary values of analytic functions.

Marjorie Lee Browne
1914-1979

African-American mathematician renowned for her work in topology. Browne, one of the first African-American women to receive a Ph.D. in mathematics, won several awards for her ground-breaking work in topology, the study of surfaces. She also served for nearly 20 years as Chair of the North Carolina College Department of Mathematics and was respected for both her teaching and sponsorship of aspiring mathematicians. Browne was awarded the W. W. Rankin Memorial Award for Excellence in Mathematics Education in 1975.

Albert P. Calderón
1920-1998

Argentinean engineer and mathematician who made fundamental contributions to mathematical analysis. Along with Antoni Zygmund, Calderón developed the Calderón-Zygmund theory of singular integral operators. Calderón's work provided solutions for elliptic differential operators and, subsequently, pseudodifferential operators. He perfected techniques that became important tools of harmonic, nonlinear, and complex analysis. He also published important results regarding the Cauchy problem involving partial differential equations, the Atiyah-Singer index theorem, and the propagation of singularities in nonlinear equations.

Lennart Axel Edvard Carleson

Swedish mathematician who showed in 1966 that, if a function is squarewise integrable, then its Fourier series is convergent from nearly every point. This announcement caused a stir because, earlier, Kolmogorov had shown that functions existed that could be integrated but whose Fourier series were divergent. While seemingly abstract and of little common interest, methods of Fourier analysis are important, for example, in reducing scientific data, computer enhancement of "grainy" photos, and looking for signs of intelligent life using interstellar radio signals.

Sun-Yung Alice Chang
1948-

Chinese-born American mathematician who was awarded the 1995 Ruth Lyttle Satter Prize in Mathematics (a prize awarded every two years to a woman who has made outstanding contributions to mathematics research) for her contributions to the understanding of Riemannian manifolds, nonlinear partial differential equations, and isospectral geometry. Chang has served as an editor for leading mathematical journals and as vice-president of the American Mathematical Society.

Fan R. K. Chung
1949-

Taiwanese-American mathematician who earned her Ph.D. in mathematics from the University of Pennsylvania in 1974 and worked for 20 years at Bell Labs in the area of combinatorics. She is well respected for her teaching, research, and support of women in mathematics and the sciences. Her research interests include spectral graph theory, discrete geometry, algorithms, and communication networks.

Alain Connes
1947-

French mathematician who was awarded a Fields Medal in 1982 for his contributions to theories on operator algebras (including the classification and structure theorem of type III factors), C*-algebras, and the classifications of injective factors and automorphisms of the hyperfinite factor. Connes's work in operator algebras has applications to differential geometry, non-commutation integration theory, and noncommutative geometry. His work also has important applications to theoretical physics, especially in the development of mathematical models for quantum mechanical systems.

Gertrude Mary Cox
1900-1978

American mathematician who was the first woman elected into the International Statistical Institute. In 1956 she was elected President of the American Statistical Association and in 1975 was elected into the National Academy of Sciences. Born in Dayton, Iowa, Cox took an unusual path for women in those times: she earned a master's degree in statistics at Iowa State College. She continued her graduate work in California at the University of California at Berkeley and returned to her alma mater, where she soon was appointed assistant professor of statistics. Cox was later appointed Director of Statistics at the Research Triangle Institute in Durham, North Carolina, where she remained until her retirement in 1964.

Ingrid Daubechies
1954-

Belgian-American mathematician who received her Ph.D. in mathematics from the Free University in Belgium. She held a research position at AT&T Bell Laboratories and later conducted research at Princeton University, where she was first woman professor in the department of mathematics. In 1997 she was awarded the Ruth Lytlle Satter Prize by the American Mathematical Society for her book *Ten Lectures on Wavelets*. Her research interests are mathematical aspects of time-frequency analysis and wavelets.

Pierre R. Deligne
1944-

Belgian mathematician who received his training and Ph.D. in both Belgium and France. He later served on the mathematics faculty at Princeton University at the Institute for Advanced Study. He is the recipient of the Fields Medal in 1978 for his work in algebraic geometry. His work has provided important insight into the relationship between algebraic geometry and algebraic number theory. He also developed weight theory, a division of mathematics concerned with differential equations.

Vladimir Gershonovich Drinfeld
1954-

Ukranian-born mathematician who introduced quantum groups. A brilliant and prolific mathematician, Drinfeld proved an important example of the Langlands conjecture, introducing many deep new ideas. In 1985 Drinfeld (and, independently, the Japanese mathematician Jimbo) introduced the idea of quantum groups. Quantum groups have since proved enormously important in both mathematics and physics. Drinfeld was awarded a Fields Medal in 1990 for his groundbreaking work.

Bradley Efron

American mathematician known for his innovative work in statistical methods that substitute computer power for mathematical formulas. He applies this to statistical analysis of health care evaluations in order to make analysis more realistic and applicable. He was the Max H. Stein Professor of Humanities and Sciences in the Department of Statistics and Department of Health Research and Policy at Stanford University.

Samuel Eilenberg
1913-1988

American mathematician and legendary collector of south Asian art. He received his Ph.D. from Warsaw University in Poland in 1939 and later went to Columbia University in 1947. His work created a new discipline of mathematics, algebraic topology, using algebra to describe how certain properties of multidimensional forms remain unchanged even when the forms are bent, twisted, or stretched.

Etta Zuber Falconer
1933-

African-American mathematician whose work in group theory has won acclaim. In addition to her mathematical accomplishments, Falconer was awarded the Louise Hays Award for her outstanding achievements in mathematics education, particularly her efforts to improve the mathematics education of African-Americans at all levels. Falconer served as Chair of the Spelman College Department of Mathematics from 1971 to 1985 and continues to remain active in both mathematics and mathematics education.

Charles Louis Fefferman
1949-

American mathematician and child prodigy who became the youngest full professor ever appointed in the United States. Fefferman earned a Fields medal for his revolutionary study of multidimensional complex analysis, in which he found correct generalizations of low-dimensional results. He discovered important applications for Stefan Bergman's ideas on biholomorphic mappings. Fefferman has also made significant contributions to the study of partial differential equations, Fourier analysis, and Hardy spaces.

Mitchell J. Feigenbaum
1945-

American mathematical physicist who made pioneering advances in the study of chaotic systems. Feigenbaum found that systems show consistent patterns as they verge towards chaos. These patterns are now known as Feigenbaum numbers. Chaos theory attempts to deal with complex natural events such as weather patterns, fluid dynamics, and the movement of quantum particles.

Michael H. Freedman
1951-

American mathematician who received his Ph.D. from Princeton University in 1973. He is best known for his work in solving the longstanding Poincaré conjecture in four dimensions, for which he received the Fields Medal, the highest honor in mathematics. He has been the recipient of numerous awards including Sloan and Guggenheim fellowships, a MacArthur Fellowship, and the National Medal of Science. He was an elected member of National Academy of Science.

Andrew Mattei Gleason
1921-

American mathematician who in 1952 joined colleagues Deane Montgomery and Leo Zippin in developing a complete solution to Hilbert's Fifth Problem, which considered properties of analytic manifolds. Specifically, Gleason was able to show that any locally compact topological group is a limit of Lie groups, serving to emphasize the importance of Lie groups in the theory of continuous groups. This proved to be the definitive solution to the fifth of 23 problems posed by David Hilbert to the International Mathematical Congress 52 years earlier.

Timothy W. Gowers
1963-

American mathematician who received his Ph.D. at Cambridge University in 1990. He teaches at Trinity College at Cambridge. He received the Fields Medal—often referred to as the Nobel Prize for mathematics—in 1998 for his contribution to a branch of mathematics known as functional analysis, making extensive use of combinatorial theory. He has been successful in combining these disparate fields together. His work solved one of the most famous problems in functional analysis, the homogeneous space problem.

Evelyn Boyd Granville
1924-

American mathematician who was one of the first African-American women to earn a mathematics doctorate. Granville graduated from Yale University, specializing in functional analysis. She taught at several universities and conducted mathematical assignments for various governmental agencies. By 1956 Granville began working for private companies, writing computer programs for aerospace applications such as the determination of spacecrafts' trajectories and orbits. She also lectured at California State University, Los Angeles, and developed curriculum guides for elementary school mathematics teachers.

Wolfgang Haken

German-American mathematician who in 1976 submitted the first mathematical proof that depended on computer assistance. This proof, that the four-color map conjecture was true, was a watershed event in mathematics, ushering in an era of proofs increasingly dependent on computers for their success. The four-color map conjecture, which had defied proof for over a century, stated that it was impossible to construct a map that needed more than four colors to make all adjacent regions a different color.

Chandra Harish
1923-1983

Indian-American mathematician who studied theoretical physics and received his Ph.D. from Harvard University in 1947. While there, he attended a lecture by Nobel laureate Wolfgang Pauli, during which he pointed out a mistake by Pauli; the two became life-long friends. At Columbia University (1950-63) he collaborated with many well known mathematicians and worked on theoretical geometry and representations of semi-simple Lie groups. He received many awards, including election to the National Academy of Science.

Louise Schmir Hay
1935-1989

Polish-American mathematician who developed an international reputation for her work in mathematical logic and theoretical computer science. As Chair of the Department of Mathematics at the University of Illinois at Chicago, Hay was the first woman to have such a position at a major American research university. Hay was also a founding member of the Association of Women in Mathematics, which after her death established the Louise Hay Award for Contributions of Women in Mathematics in her honor.

Heisuke Hironaka
1931-

Japanese-born mathematician who was awarded the Fields Medal in 1970 for his insight into algebraic varieties. Hironaka is credited with generalizing the work of Oscar Zariski, who had proved for three or fewer dimensions a theorem concerning the resolution of singularities, by proving results for any dimension. A Harvard University professor *emeritus,* Hironaka has also served as executive director of the Japanese Association for Mathematical Sciences and is credited with providing the inspiration for the International Math Olympics, a competition for schoolchildren that encourages mathematical accuracy and speed.

Lars Hormander
1931-

Swedish mathematician whose work on partial differential equations—fiendishly difficult to solve—was rewarded with the Fields Medal, the highest honor in mathematics, in 1962. Hormander followed this achievement with his monumental work, *The Analysis of Linear Partial Differential Operators,* a four-volume text that expanded on his Fields Medal-winning work. After a four-year appointment at the Institute for Advanced Study, Hormander returned to Sweden as Chair of Mathematics at the University of Lund.

Kenkichi Iwasawa
1917-1998

Japanese-American mathematician who received his Ph.D. in 1945 from Tokyo University. He was a member of the Institute of Advanced Studies at Princeton (1951-52) and taught at the Massachusetts Institute of Technology and Princeton University. He is known for his Iwasawa theory, a work on the theory of algebraic number fields, which won the American Mathematical Society's Cole award in 1962. He was a member of the American Mathematical Society and the Mathematical Society of Japan and a fellow of the American Academy of Arts and Sciences.

Vaughan Fredrick Randall Jones
1952-

New Zealand-American mathematician who received his Ph.D. in Switzerland in 1979. He taught mathematics at many prestigious universities, including the University of California at Los Angeles, the University of Pennsylvania, and the University of California at Berkeley. He received the Fields Medal in 1990 for his work on the index theorem for von Neumann algebra as well as the discovery of a new polynomial invariant for knots.

Mark Kac
1914-1984

Polish-born American mathematician who was a pioneer in the development of important concepts in probability theory and statistical mechanics. Kac's work is best known for its wide application to statistical physics and its attempt to explain the relationships between space dimensionality (i.e., dimensions of space-time) and physical phenomena. The Feynman-Kac path integral (named after Richard Feynman, the physicist and Nobel laureate, and Kac) remains an important method for quantization. Kac's work also included renowned publications dealing with statistical probability and number theory.

Yasumasa Kanada

Japanese mathematician who pioneered new computer methods for calculating the value of π to unprecedented levels of precision. Prior to Kanada's work,π had been calculated to 2,037 places using computers while, calculating by hand, solving π to 707 places had required 15 years of effort. In 1988 Kanada calculated the value of π to over 200 million decimal places in six hours of "supercomputer" time. Since then, using other supercomputers, π has been calculated to over 1 billion decimal places.

Tosio Kato
1917-

Japanese-American mathematician whose work on the functional analysis of differential equations proved to be of great importance. Kato also wrote a significant work on perturbation theory, describing how small perturbations in the initial conditions of a mathematical problem can affect the ultimate solution—important in many areas of analysis. Finally, Kato contributed to the understanding of the application of functional analysis to the mathematics of differential equations.

Linda Keen
1940-

American mathematician who has improved analytical problem resolution methods. Keen earned her doctorate at the Courant Institute of Mathematical Sciences. Focusing on dynamic systems and complex analysis, she has been a professor at Lehman College since 1968. Keen served as president of the Association for Women in Mathematics, vice president of the American Mathematical Society, and associate editor of the *Journal of Geometric Analysis*. She has addressed numerous mathematical meetings, including the presentation of the 1993 Emmy Noether Lecture.

Maxim Kontsevich
1964-

Russian mathematician who received his Ph.D. from the University of Bonn, Germany, in 1992. He has established a reputation in pure mathematics and theoretical physics and in 1998 was awarded the Fields Medal for work in string theory, quantum field theory, and knot theory. Mathematicians are looking for ways of classifying knots, as there are scientific applications in cosmology, statistical mechanics, and genetics. Kontsevich taught at the Institut des Hautes Etudes Scientifiques in France.

Krystyna Kuperberg
1944-

Polish mathematician who has resolved topological puzzles. Kuperberg graduated with mathematics degrees from universities in Warsaw before moving to the United States and earning a topology doctorate from Rice University. A professor at Auburn University, Kuperberg solved a bihomogeneity problem posed by B. Knaster in 1921. She next developed a counterexample to Herbert Seifert's conjecture regarding three-dimensional shapes that benefitted dynamic systems theory. Kuperberg has received the Alfred Jurzykowski Award and has presented the 1999 Emmy Noether Lecture.

Thomas E. Kurtz
1928-

American mathematician who received his Ph.D. from Princeton University in 1956. He was cofounder with John Kemeny of the computer programming language BASIC (beginners all-purpose symbolic instruction code), the first common programming language of microcomputers. BASIC uses English words and decimal notation rather than binary numbers. He served on the President's Science Advisory Committee in 1965-66 and later became vice-chairman and director of True Basics, Inc., the company he founded with Kemeny.

Robert Phelan Langlands
1936-

Canadian-American mathematician who formulated the Langlands program. Langlands developed a series of conjectures and problems that related seemingly disparate areas of mathematics, including algebra, algebraic geometry, and number theory. The known results coming from his conjectures have been very important in linking number theory and representation theory. Many of his problems and conjectures are still unsolved. Langlands has received numerous honors and awards, including the Wolf Prize, which he shared with Andrew Wiles in 1995.

Edna Kramer Lassar
1902-1984

American mathematician who received her Ph.D. in mathematics from Columbia University in 1930. She was the first woman math instructor at New Jersey State Teachers College. During the 1930s she became acting department chairman of mathematics with the New York City School System and taught curriculum methods at Brooklyn College. During 1943-45 she worked at Columbia University's Division of War Research. She is well known for her book *The Nature and Growth of Modern Mathematics*, published in 1970, the culmination of 14 years of work.

Chia-Chiao Lin
1916-

Chinese-American mathematician whose work on density waves helped to explain the structure of spiral galaxies. Lin was able to show mathematically that the spiral arms seen in many galaxies are actually the result of basic physical processes that occur during the formation of a galaxy. Many of his predictions, which now form the basis for much of galactic dynamics theory, have been borne out by detailed Hubble Space Telescope observations.

Pierre-Louis Lions
1956-

French mathematician and Fields Medal winner who contributed critical insights into theories on nonlinear partial differential equations involving viscosity solutions. Lions also contributed fundamental insights into variational problems, the Boltzmann equation, and other kinetic equations that describe interactions between colliding particles. Lions's work has allowed for the improved modeling of applied problems in statistical mechanics, including atmospheric and oceanic studies.

Bernard Malgrange
1928-

French mathematician who made important contributions to the application of differential analysis to areas of topology. Topology is the study of surfaces; differential analysis of topology is a powerful tool to describe some of the mathematical properties of these surfaces. Malgrange's book, *Ideals of Differentiable Functions,* looks at this topic in some detail.

Vivienne Lucille Malone-Mayes
1932-1995

Malone-Mayes, a pioneering African-American woman in mathematics, became the fifth African-American woman to receive a Ph.D. in mathematics when she earned her degree from the University of Texas in 1966. She was also the second African-American and first African-American woman to earn this degree from the University of Texas. Also in 1966 Malone-Mayes became the first African-American faculty member at Baylor University. Malone-Mayes was active in mathematical, community, and religious organizations. Her mathematical interests were in summability theory and mathematics education.

Gregori Aleksandrovic Margulis
1946-

Russian-born mathematician who received a 1978 Fields Medal for his work in combinatorics, ergodic theory, differential geometry, dynamical systems, and Lie groups. Lie groups, a branch of group theory, have algebraic structure and are subsets of space with discrete geometries. Some of these geometries resemble Euclidean space and thus can be studied by analytical techniques (e.g., differential equations). Margulis's work made Lie groups useful in applications of mathematics to science. In 1991 Margulis assumed a professorial chair at Yale University.

Margaret Dusa McDuff
1945-

English mathematician who received her Ph.D. from Cambridge University in 1971. After finishing her doctorate, she held a two-year Science Research Council Fellowship at Cambridge University, a fellowship at the Institute of Advanced Study and the Massachusetts Institute of Technology. She has held faculty positions at the University of York, University of Warwick, and the State University of New York at Stonybrook. She is known for her work in von Neumann algebras and global symplectic geometry. She is active in the Women in Science and Engineering (WISE) program, which offers support to undergraduate women in the sciences.

Curtis T. McMullen
1958-

American mathematician who received his Ph.D. from Harvard University in 1985. He has held teaching positions at various universities, including the University of California at Berkeley and Harvard University. He was awarded the Fields Medal in 1998 for his work in the fields of geometry and complex dynamics, a branch of the theory of dynamic systems, also known as chaos theory. In 1998 he was elected to the American Academy of Arts and Sciences.

Yoichi Miyaoka

Japanese mathematician who in 1988 developed the outline of a proof of Fermat's last theorem, a problem that had bedeviled mathematicians for over 300 years. This theorem, inspired by the Pythagorean theorem for determining the length of the sides of a right triangle ($a^2+b^2=c^2$), states that there is no solution for similar problems with exponents greater than 2. Proof of this theorem was not, however, provided by Fermat. Unfortunately, Miyaoka's outline was later shown to have some flaws.

Edwin Evariste Moise
1918-1998

Moise was born in 1918 in New Orleans, Louisiana. Moise served in the Navy during World War II and then earned a Ph.D. in topology at the University of Texas under R. L. Moore. He is best known for his work in mathematics education, writing an influential high school geometry text for the School Mathematics Study Group in 1964 and the accompanying advanced text for training teachers. These texts were part of the reform of secondary school mathematics in the United States during the "space race" of late 1950s and early 1960s.

Cathleen Synge Morawetz
1923-

Canadian-American mathematician who received her Ph.D. from New York University in 1951. The first woman in the United States to head a mathematical institute, she was appointed director of the Courant Institute of Mathematical Sciences in 1982. Her research has focused on the wave equation, attempting to answer the classical problem of whether light should be treated as waves or streams of particles. She has been the recipient of many awards, including from the National Academy of Sciences and from the National Organization of Women for successfully combining career and family.

Shigefumi Mori
1951-

Japanese mathematician who received his Ph.D. from Kyoto University in 1978. In 1990 he joined the faculty of the Research Institute for Mathematical Sciences at Kyoto University. He was a recipient of the 1990 Fields Medal for his work in algebraic geometry. In 1979 he proved Hartshorne's conjecture, an unsolved problem in algebraic geometry. This work focused on developing new techniques for the classification of different algebraic varieties in dimension three.

Michael Morrison
1954-

American computer programmer who is the developer of Java, an advanced computer programming language. Java is able to run, unaltered, on just about any computer in operation today. Morrison obtained his B.S. in electrical engineering from Tennessee Technological University.

David Bryant Mumford
1937-

English mathematician whose major contributions have been in the areas of algebraic surface theory and other branches of geometry. Mumford was awarded the 1974 Fields Medal for his contributions in mathematics, largely in these areas. In particular, Mumford has helped achieve a deeper understanding of moduli, parameters that help to describe various properties of geometric objects. Mumford helped to prove that these moduli exist and to describe their mathematical structure.

Evelyn M. Nelson
1943-1987

Canadian mathematician whose work in algebraic problems in theoretical computer science won her much acclaim. Nelson began working in algebra theory in the late 1970s, seeing applications to some problems in computer science. Her work in this area was deemed important, and in 1982 she was asked to chair the McMaster University Computer Science Unit. Nelson died of cancer at age 44 and is memorialized by the Kreiger-Nelson Prize Lectureship for Distinguished Research by Women in Mathematics, awarded by the Canadian Mathematical Society.

Hanna Neumann
1914-1971

German mathematician who developed theories concerning pure mathematics. Neumann earned an Oxford University doctorate, focusing on group theory problems. She lectured at Hull University and conducted research at the Courant Institute of Mathematical Sciences. In 1963 she became head of the Department of Pure Mathematics at Australian National University. Neumann was founding vice president of the Australian Association of Mathematics Teachers. She devised abstract mathematical puzzles that required comprehension of mathematical laws and logic to achieve solutions.

Sergei Petrovich Novikov
1938-

Russian mathematician who was educated at Moscow University and received his Ph.D. in 1964. He held many prestigious positions at the Steklov Institute of Mathematics, Moscow University, and the Academy of Sciences in the USSR. Beginning in 1966 he became affiliated with the University of Maryland. He received a Fields Medal in 1971 for his work in mathematical systems, dynamical systems, and the Novikov conjecture. He is a recipient of many other international awards, including from the Academy of Sciences in the USSR and the National Academy of Science in the United States.

Olga Oleinik
1925-

Russian mathematician who has developed applied mathematics techniques. Oleinik earned a doctorate from the Institute of Mathematics at Moscow State University, where she has since taught. In 1973 she became head of the Department of Differential Equations. Oleinik also specializes in algebraic geometry and mathematical physics. She applies mathematics to heat distribution and elasticity problems. A prolific author, Oleinik has received numerous prizes recognizing her mathematical achievements. She has also presented the 1996 Association for Women in Mathematics Emmy Noether Lecture.

Vera Pless
1931-

American mathematician who is an authority on coding theory. Pless earned a Northwestern University doctorate, specializing in algebra. At the Air Force Cambridge Research Laboratory, Pless joined mathematicians who pioneered the exploration of error-correcting codes. She next conducted research at the Massachusetts Institute of Technology before becoming a professor in the University of Illinois at Chicago's Department of Mathematics, Statistics, and Computer Science in

1975. Her writings include the book *Introduction to the Theory of Error-Correcting Codes.*

Abraham Robinson
1918-1974

German-American mathematician who received his Ph.D. from London University in 1949. In 1933 his family immigrated to Palestine, where he studied mathematics with respected teachers such as Fraenkel and Levitzki. He studied at the Sorbonne in 1939 but was forced to flee when the Germans invaded France. He worked on aerodynamics during the remainder of World War II. He is known for his pioneering work in model theory, the metamathematics of algebraic systems, and non-standard analysis.

Mary Ellen Rudin
1924-

American mathematician who received her Ph.D. from the University of Texas in 1949. She held teaching positions at Duke University, the University of Rochester, and the University of Wisconsin, where she served until her retirement. Her research interests centered upon set-theoretic topology, with an emphasis on the construction of counter examples, authoring nearly 70 papers on the subject. In addition to raising a family of four children, she was the recipient of numerous professional awards and a member of many mathematical associations.

Alice T. Schafer
1915-

American mathematician who has encouraged other women to become mathematicians also. Earning a doctorate at the University of Chicago, Schafer specialized in abstract algebra and differential geometry. She taught at several universities and was the mathematics department head at Wellesley College. Schafer helped establish the Association for Women in Mathematics, serving as its second president. The organization created the Schafer Prize, which annually recognizes outstanding undergraduate female mathematics students. Schafer has received several awards for improving professional conditions for women mathematicians internationally.

Laurent Schwartz
1915-

French mathematician who received his Ph.D. in mathematics in 1943 from University of Strasbourg. He taught at Grenoble, Nancy, and the Ecole Polytechnique in Paris. He is recognized for his theory of distributions in differential and integral calculus. In 1950 he received the Fields Medal for work in the theory of distributions. He has received many professional awards and is a member of the Paris Academy of Sciences.

Igor Rostislavich Shaferevich

Russian mathematician best known for his views on the relationship between religion and mathematics. An expert on algebraic geometry, Shaferevich created a stir when he advocated a variety of Neoplatonism. He argued that the independent discovery of many mathematical theories suggested their actual existence on another plane. He also felt, as did the ancient Greeks, that mathematics could serve as a bridge between religion and the physical world.

Saharon Shelah

Israeli mathematician whose work in mathematical logic and the mathematics of semantics has contributed greatly to the fields of logic and metalogic (logical systems that encompass other, lower-level logical systems). Shelah's most important work deals with advanced concepts in semantics and the science of concepts.

A.A. Suslin

Russian mathematician who posed the "Suslin Problem" in topology. The Suslin Problem refers to separable, connected linear spaces and setting limits on the properties of such spaces. The problem was shown to be unsolvable within the framework set by the axioms (basic rules) of set theory. This result, in turn, has had interesting implications for both set theory and topology.

Michio Suzuki

Japanese-Canadian mathematician whose work on sporadic simple groups helped to advance this part of group theory significantly. Simple groups are those with no normal subgroups and are the building blocks for all groups. Sporadic simple groups, in contrast, have other properties that make them unique. Mathematicians do not yet know if sporadic simple groups are actually part of an as-yet unknown family of groups of if they are truly "exceptional" groups. Suzuki's contribution lay in helping to describe fully some of these sporadic simple groups, known as the Suzuki Groups.

Richard Tapia
1946-

American mathematician who received his Ph.D. from the University of California at Los Angeles in 1967. In 1990 he was named one of the 20 most influential leaders in minority mathematics education by the National Research Council. His research interests have focused on improving interior point methods for linear programming and designing quadratic methods for nonlinear programming. He is involved in outreach programs aimed at encouraging minority students in science, engineering, and mathematics.

John Griggs Thompson
1932-

American mathematician who received his Ph.D. from the University of Chicago in 1959. He taught at Harvard University, the University of Chicago, and Cambridge University. He was awarded the Fields Medal in 1970 for his work in finite group theory and non-Abelian finite simple groups. His later research interests centered on coding theory and Galois groups. He was elected to the National Academy of Sciences in the United States and the Royal Society of London.

William Paul Thurston
1946-

American mathematician who received his Ph.D. in 1972 from the University of California at Berkeley. After completing his Ph.D. he taught at the Massachusetts Institute of Technology and Princeton University. In 1982 he was a recipient of the Fields Medal for his work that revolutionized the study of topology in 2 and 3 dimensions and brought a new dialogue between analysis, topology, and geometry.

Karen Uhlenbeck
1942-

American mathematician who has developed unique analytical methodologies. Uhlenbeck received a doctorate from Brandeis University and taught at several colleges before accepting a mathematics chair at the University of Texas. A partial differential equations expert, she mathematically describes shapes of abstract spaces. Mathematicians have applied her techniques to solve complex geometric problems. Uhlenbeck is interested in diverse mathematical theories, seeking interrelationships among approaches. She has served as vice president of the American Mathematical Society and has lectured internationally.

Argelia Velez-Rodriguez
1936-

Cuban-American mathematician who received her Ph.D. from the University of Havana in 1960. She was the first black woman to receive a doctorate there. Her research interests are differential equations and astronomy. After immigrating to the United States during the Cuban Revolution, she joined the mathematics faculty at Bishop College in Dallas Texas. In 1979 she worked with the Minority Institutions Science Improvement program and since 1980 has been a program director for the U.S. Department of Education.

Mary Catherine Bishop Weiss
1930-1966

American mathematician who contributed to the advancement of geometric methodology. Earning a doctorate from the University of Chicago, Weiss focused on trigonometric series. She taught at DePaul University, Washington University, and Stanford University. Weiss proved assumptions her colleagues had developed regarding harmonic analysis. She then concentrated on Hardy spaces in higher dimensions. A National Science Foundation postdoctoral fellowship enabled Weiss to conduct research for a year at Cambridge University. She taught briefly at the University of Illinois before her death.

Hugh C. Williams

American mathematician who used computers to help find solutions to difficult classical problems in mathematics. Williams's first such solution, a number containing over 200,000 digits, solved Archimedes's 2,000-year-old "cattle of the sun" problem. This solution was completed in 1965. In another notable achievement, completed in 1985, Williams and his collaborators determined that the number consisting of a string of 1,031 consecutive ones is a prime number. This accomplishment showed the power of computers to help resolve lengthy mathematical problems.

Edward Witten
1952-

American physicist and mathematician who is the leading theorist behind superstring theory. Superstring theory is an attempt at a "theory of everything" that can meld quantum mechanics and Albert Einstein's general relativity theory. Witten and others have shown how tiny particles called strings can form all of the elementary particles we see today, such as protons, electrons, and quarks, and account for gravity. Witten also developed the Seiberg-Witten equations with Nathan Seiberg; these have revolutionized the study of four-dimensional objects. He is a winner of the Fields Medal, the Nobel Prize of mathematics.

Shing-Tung Yau
1948-

Chinese mathematician who was awarded a Fields Medal in 1982 for advancing the study of partial differential equations, the positive mass conjecture of general relativity, the Calabi conjecture, and Monge-Ampère equations. Yau, along with others, has studied the stability and construction of minimal surfaces with the goal of developing methods to analyze their behavior in the space-time of general relativity. Yau's work has provided insight into the formation of black holes.

Jean-Christophe Yoccoz
1957-

French mathematician who received his Ph.D. in 1977 from Ecole Polytechnique. In 1994 he was awarded the Fields Medal for his work on dynamical systems, an area developed by Henri Poincaré to study the stability of the solar system. These techniques are applicable to problems in biology, chemistry, mechanics, and ecology—areas where stability is needed. "Yoccoz puzzles" also produce aesthetically pleasing objects, such as the Julia and Mandelbrot fractal sets. He later taught at the University of Paris.

Lai-Sang Young
1952-

Chinese-American mathematician whose work focuses on "strange attractors." Young was awarded the Ruth Lyttle Satter Prize for her role in "the investigation of the statistical properties of dynamical systems and [having] developed important and difficult techniques which have done much to clarify the subject." Born in Hong Kong, Young has studied and worked at a number of prestigious institutions, including the University of Michigan, the Institute of Advanced Study at Princeton, and Berkeley.

Erik Christopher Zeeman
1925-

British mathematician who received his Ph.D. from Cambridge University in 1949, after serving in the Royal Air Force. He is best known for his work on catastrophe theory, and he has publicized the applications of catastrophe theory in the biological, behavioral, and physical sciences. In 1978 he gave the Christmas Lectures at the Royal Institution, which ushered in the popular Mathematics Master classes for 13-year-old students. He received many honors and awards in England and abroad.

Efim I. Zelmanov
1955-

Russian mathematician who solved the Restricted Burnside Problem, one of the most difficult and long-lasting problems in the branch of mathematics called group theory. Author of more than 60 research papers in mathematics, in 1994 Zelmanov was awarded a Fields Medal (the highest honor in mathematics) for his solution to the Restricted Burnside Problem. He proved that all groups of a certain type had to be finite. He has also proved many other deep results in algebra.

Bibliography of Primary Sources

Books

Cohen, Paul. *Set Theory and the Continuum Hypothesis* (1966). In this work Cohen discussed his proof that the continuum hypothesis was independent of standard set theory, and that the continuum hypothesis could not be regarded as an axiom of set theory. He arrived at his solution by introducing a technique called "forcing" that revealed his theory of independence.

Knuth, Donald E. *The Art of Computer Programming* (1968). This work became the authoritative textbook in the subject during the 1970s. But just as computers have evolved, so has Knuth's monumental text: the fourth of a planned seven-volume series was published in 1995.

Mandelbrot, Benoit. *The Fractal Geometry of Nature* (1982). After several unsuccessful attempts to publicize his work on fractals, Mandelbrot tried again with this 1982 book; indeed, portions were identical to items in his earlier works. However, the 1982 book included something his previous writings had been lacking: colorful computer-generated pictures. Fractal landscapes, planetscapes, and unusual patterns showed the beauty and potential of fractals. Quickly fractals entered the "popular" science press with articles often containing bright, colorful illustrations of fractal designs.

Penrose, Roger. *The Emperor's New Mind: Concerning Computers, Minds, and the Laws of Physics* (1989). Penrose, an acclaimed mathematician who did pioneering work on black holes with Stephen Hawking, here offered his views on finding a Grand Unified Theory, or "theory of everything."

Robinson, Abraham. *Non-Standard Analysis* (1966). In this work Robinson summarized his groundbreaking work in conceiving non-standard analysis, which he used to provide a rigorous development of calculus based on infinitesimals rather than on limits.

Thom, René. *Structural Stability and Morphogenesis* (1972). Thom developed a way of studying discontinuous processes, which he called catastrophe theory, and he presented his ideas in this 1972 work.

Periodical Articles

Donaldson, Simon. "An Application of Gauge Theory to Four-Dimensional Topology" (1983). Donaldson won a Fields Medal for his work in four-dimensional topology, the subject of this 1983 article.

Efron, Bradley. "Bootstrap Methods: Another Look at the Jackknife" (1979). Efron here discussed his acclaimed "bootstrap method" of statistics, which relies on the use of high-speed computers.

Milnor, John. "On the Geometry of the Kepler Problem" (1983). Known for his groundbreaking work in differential topology, this article featured Milnor's work on the Kepler problem.

Wiles, Andrew. "Modular Elliptic Curves and Fermat's Last Theorem" (1995). Fermat's Last Theorem was the most famous unsolved problem in mathematics until Andrew Wiles found a solution in the 1990s, an achievement that instantly made him the best-known living mathematician in the world.

Medicine

Chronology

1951 The introduction of the first successful oral contraceptive, based on discoveries by American biologist Gregory Pincus, sparks a social revolution with its ability to divorce the sex act from the consequences of impregnation.

1952 Douglas Bevis of Great Britain is the first physician to use amniocentesis for a fetal diagnosis.

1954 American virologist Jonas Salk produces the first successful anti-polio vaccine.

1954 American surgeon Joseph Murray conducts the first successful organ transplant when he transfers the kidney of one twin to another.

1958 The first use is made of closed-chest cardiac massage combined with mouth-to-mouth respiration for cardiac resuscitation, leading to the widespread adoption of cardiopulmonary resuscitation (CPR) methods.

1967 Christiaan Barnard, a South African surgeon, conducts the first successful human heart transplant; recipient Louis Washkansky lives for 18 days.

1970 English physicians Patrick Steptoe and R. G. Edwards accomplish in-vitro fertilization of human ova; seven years later, the first test-tube baby is born, from an egg implanted by Steptoe and Edwards in the mother's uterus after fertilization.

1971 A hospital in Nimbledon, England, becomes the first facility to install a computerized axial tomography (CAT) scanner, based on technology developed by electrical engineer Godfrey Newbold Hounsfield.

1977 Two homosexual men in New York City, diagnosed as suffering from Kaposi's sarcoma, are the first reported cases of AIDS (acquired immune deficiency syndrome).

1978 Smallpox is eliminated worldwide.

1985 Lasers are used for the first time in artery surgery.

Overview: Medicine 1950-present

During the twentieth century, medical theory and practice underwent more profound changes than in all of the years since the time of Hippocrates (460?-377? B.C.). Since World War II, changes in science and society have transformed the theoretical, institutional, educational, economic, and ethical foundations of medicine. For example, research on the growth of stem cells has sparked both hope for revolutionary medical applications and profound ethical challenges. In 1999 the editors of the journal *Science* selected stem cell research as the "Breakthrough of the Year." Embryonic stem cells could be used to produce specific types of cells and tissues. Eventually, stem cells might be used to build new body parts to replace failing human organs.

Nineteenth-century scientists and physicians introduced the modern germ theory of disease and made it possible to identify the cause and means of transmission of many infectious diseases. Since 1950 many of the most feared infectious epidemic diseases have been brought under control by means of preventive vaccines, antibiotics, public health measures, and improvements in sanitation. Preventive immunizations for infectious epidemic diseases began with inoculation and vaccination against smallpox. The global eradication of smallpox in the 1970s is one of the greatest achievements of twentieth-century public health medicine and a model for international cooperation. Public health authorities hoped that the lessons learned in the smallpox campaign would lead to global immunization programs for controlling or eradicating poliomyelitis, measles, whooping cough, diphtheria, and tuberculosis. The Salk and Sabin vaccines for poliomyelitis essentially ended the threat of this disease in the wealthy, industrialized nations. Nevertheless, debates about the safety and efficacy of preventive vaccines continue.

Advances in the science of virology, bacteriology, immunology, and molecular biology have led to new approaches to the construction of vaccines. At least 15 new or improved vaccines were developed between 1980 and 2000. Experimental vaccines developed in the 1990s may offer enhanced protection against influenza, pneumococcal pneumonia, pertussis, rubella, rabies, bacterial meningitis, hepatitis B, and adenovirus-associated respiratory disease. New vaccine technologies also offer hope of ameliorating the effects of autoimmune disorders and allergies. Of course, in much of the world old specters, such as tuberculosis, malaria, measles, poverty, and malnutrition, remain major threats to life. Historians have called malaria the most devastating disease in history. Malaria and other "tropical diseases" are still major public health threat in many parts of the world.

In the twentieth century, the chronic diseases, especially those that seem to be related to diet and obesity, have eclipsed the threat of infectious disease. By the second half of the twentieth century, heart disease, cancer, and stroke had replaced the infectious, epidemic diseases as the leading causes of death in the United States. Epidemiologists and health policy experts in the 1960s predicted that, at least in the wealthy, industrialized nations, chronic, degenerative diseases would replace infectious diseases as the only significant public health problems. Newly emerging diseases, such as Ebola fever and newly discovered forms of pathogens, such as prions, have challenged that concept. Prions have been identified as the cause of a virulent new form of Creutzfeldt-Jakob Disease that is related to "Mad Cow Disease."

Moreover, the global epidemic of AIDS continues to escalate. AIDS first appeared as a diagnostic entity in 1981 when the Centers for Disease Control (CDC) began to report strange clusters of rare illnesses associated with a severely compromised condition of the immune system in previously healthy homosexual men in New York and California. By 1984, when the human immunodeficiency virus (HIV), the retrovirus that causes AIDS, was discovered, more than one million Americans were infected with HIV. The World Health Organization reported that five to 10 million people were infected with HIV by the 1990s. Advances in anti-viral therapy have extended the life expectancy of patients with HIV/AIDS, but such drugs cause adverse side effects and are extremely expensive.

In the first half of the twentieth century Paul Ehrlich's (1854-1915) discovery of Salvarsan for the treatment of syphilis and Gerhard Domagk's (1895-1964) discovery of sulfanilamide stimulated the search for more "magic bullets" to fight bacterial infections. By the end of World War II, the problem of drug-resistant strains of bacteria was already compromising the effectiveness of the "sulfa" drugs. The same problem would plague all the antibiotics that were discovered in the second half of the century. Although Alexander Fleming (1881-1955) discovered penicillin in 1928, it remained a laboratory curiosity until research and development during World War II led to the isolation and large-scale production of the drug that has been called the greatest therapeutic advance of all time.

Immunology is a relatively young field, but its twentieth-century evolution has been so dynamic that it has become one of the fundamental disciplines of modern medicine and biology. Throughout the 1970s and 1980s immunologists were awarded Nobel Prizes for achievements of remarkable theoretical and practical significance in understanding organ transplant rejection and autoimmune diseases. One of the most important methodological breakthroughs in medical science in the 1970s was the production of monoclonal antibodies. As cardiovascular disease and cancer replaced infectious diseases as the leading causes of morbidity and mortality in the wealthy, industrialized nations, immunology seemed to offer the answer to the riddle of health and disease just as microbiology had provided answers to the diseases that had posed the greatest threats in the nineteenth century. Research linking immunology and molecular biology has created a new generation of weapons in the battle against cancer, autoimmune disorders, organ rejection, and infectious diseases.

Immunobiology and neuroendocrinology are providing insights into AIDS, cancer, rheumatoid arthritis, and mind-body medicine. After centuries of confusion about the basic causes of cancer, scientists are gaining insights into the factors that cause the loss of control of cell multiplication that is the fundamental characteristic of cancer cells. A cure for cancer remains elusive, but many promising experimental approaches to diagnosis and treatment, based on developments in immunology and molecular biology, were undergoing clinical trials by the end of the 1990s.

Among the most remarkable changes in medical practice over the course of the last 200 years involves the use of technological aids for the diagnosis of disease. With the development of sophisticated new instruments and ways of looking at the human body, specialization became a fundamental aspect of the medical profession. Beyond their obvious role in transforming the art of diagnosis, medical instruments have profoundly affected the relationship between patient and physician, the education and practical training of physicians, the demarcation between areas of medical specialization, the locus of medical practice, and even the financial base of medical care and treatment.

With medical costs skyrocketing, critics of our very unsystematic "health care system" have called for increased attention to preventive medicine and health education. Progress in surgery made possible by advances in technology, such as the heart-lung machine, diagnostic screening devices such as magnetic resonance imaging (MRI), critical care units, and intensive care units are among the factors that increase the costs of modern medicine. Organ transplant operations and the development of artificial organs have saved lives, but are among the most expensive interventions of modern medicine.

Over half of the deaths in the United States are caused by cardiovascular diseases. Many of these deaths could be prevented by aggressive management and surgical procedures, including heart transplant operations. The shortage of donor hearts led to hope that a totally implantable mechanical device could overcome the shortage of donor hearts and avoid the problem of immunological rejection. Early attempts to implant a permanent artificial heart were criticized as premature human experiments. In addition to human heart transplants, heart assist devices, and artificial hearts, animal tissues and organs, or combinations of living cells and artificial materials might eventually be used to assist or replace ailing hearts. Scientists are attempting to grow heart muscle tissue, heart valves, and blood vessels in the laboratory, or modify animal organs so human recipients will not reject them. Other scientists believe that much of the social and individual burden of heart disease could be prevented through changes in diet, exercise, and the elimination of smoking.

Because of changing patterns of mortality and morbidity, the wealthy industrialized nations are increasingly faced with concerns about the medical needs of aging populations. In 1975 the National Conference on Preventive Medicine warned that therapeutic medicine might have reached a point of diminishing returns. Doctors are increasingly called upon to treat the preventable diseases of affluence, such as obesity and arteriosclerosis, with every technological weapon at their disposal. When misused or misapplied, however, the results can be as deadly as the diseases they were designed to treat. Indeed, a study of medical errors released in November 1999 by the National Academy of Sciences and the Institute of Medicine (entitled "To Err is Human") revealed that between 44,000 and 98,000 Americans die each year as a result of medical mistakes. Even the lower range of the estimate is greater than the toll taken by automobile accidents, cancer, or AIDS.

Understanding the complex relationships that link health, disease, demography, geography, ecology, and economics, and the differences in patterns of disease found in wealthy nations and developing nations, remains a major challenge. New approaches to understanding, diagnosis, preventing, and treating disease, and the emergence of new diseases in the late twentieth century demonstrate the need for a global and historic perspective in medicine and the biomedical sciences.

LOIS N. MAGNER

The Invention of the Heart-Lung Machine Launches the Era of Open-Heart Surgery

Overview

One of the most important advances in cardiac medicine in the twentieth century was the invention of the heart-lung machine. Scientists knew that delicate heart surgery was impossible without a console to take over the function of the human heart and lungs, but few deemed it possible. After decades of trial and error, John "Jack" H. Gibbon successfully tested the first human heart-lung machine in 1953, fueling the visions of his peers and ushering in a new era of open-heart surgery. Today, the heart-lung machine is an indispensable device that has extended the bounds of operative treatment beyond the most imaginative dreams.

Background

From the earliest days of medicine until the mid-1900s, tampering with the heart was considered taboo. For centuries, people regarded the human heart as the seat of our soul, our spirit, and our emotions, and as such the organ was off-limits to surgeons and doctors. Evidence of this belief is cemented in the writings of some of the earliest French surgeons. In 1648 Riolanus (a.k.a. Jean Riolan) described the heart as the "noblest organ in the body and the source of a life-giving substance which supplied the rest of the body with nourishment."

From the Middle Ages through the Renaissance, physicians observed heart functions and failure but did more to impede medical progress than to further it. With the exception of a few bold scientists, the heart remained surgically untouchable for two and a half more centuries.

Those who dared to suggest that heart surgery was, indeed, possible, were looked upon with doubt and disbelief by their peers. Heart wounds could be sutured and the cavity surrounding the heart could be drained, but little beyond these procedures found acceptance.

By the turn of the twentieth century, medical science was booming, and yet cardiac surgery remained virtually non-existent. Scientists had pioneered major advances in anesthesia, antibiotics, and blood transfusions, but successful surgery of the heart and chest was still decades away. Doctors carrying out experimental work in the early 1900s failed more often then they succeeded, and open-heart surgery remained an evasive pursuit.

In the early 1930s a young medical resident, touched by the death of a patient suffering from a pulmonary embolism, gave birth to an idea that he pursued vigorously for the next twenty years. John "Jack" H. Gibbon (1903-1973) proposed building a machine that would perform the function of the heart and lungs during surgery. Little did Gibbon know that his novel idea would one day revolutionize surgery of the heart.

While Gibbon struggled with various prototypes and experiments, World War II broke out. The need for improvements in cardiac surgery escalated with a fervor never seen before. Soldiers with shell fragments and bullets lodged inside their hearts begged for help as military physicians pondered how to save them. To do nothing was dangerous, but to remove the foreign objects was almost surely fatal.

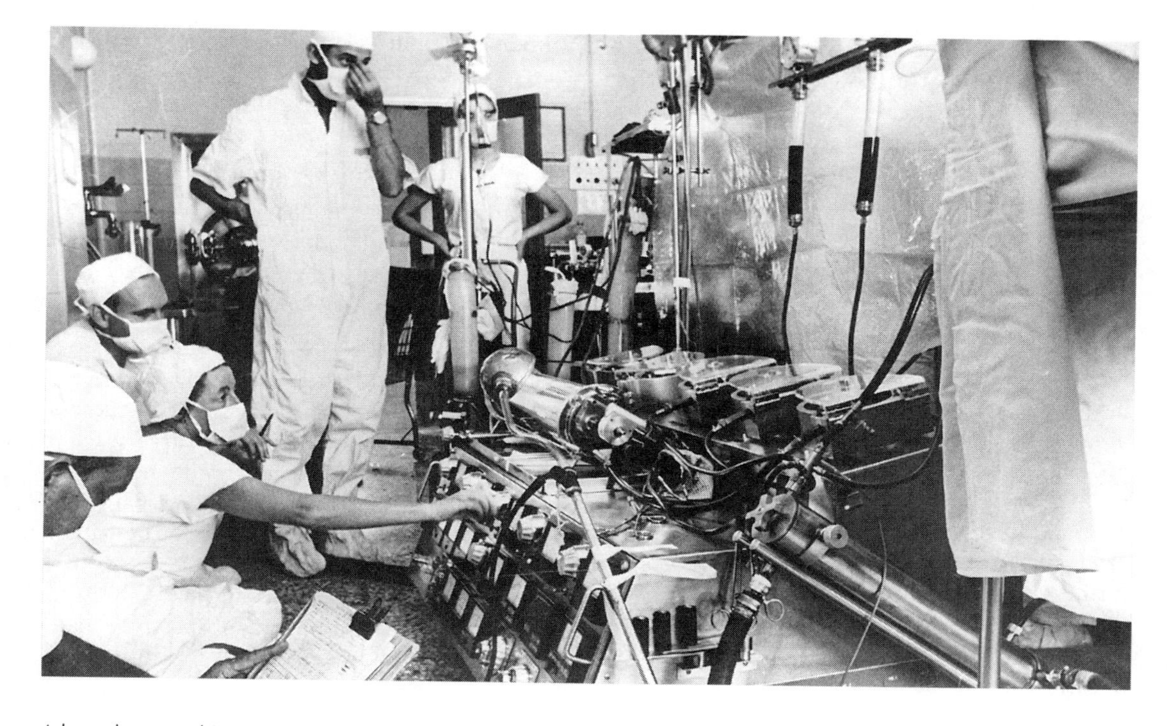

A heart-lung machine, 1968. *(Hulton-Deutsch Collection/CORBIS. Reproduced with permission.)*

One of the first surgeons to cut into a beating heart was Dr. Dwight Harken, a young U.S. Army surgeon stationed along the European front. Harken conducted three different animal studies to see if he could slice open a beating heart and remove the shrapnel with his own finger. His first experiment was a complete failure; all 14 animal subjects died. The second time, 7 died. But by the third test, only 2 of the 14 died. Harken was ready to try the technique on humans. He eventually removed 134 missiles from the chest, including 13 in the heart chambers, without losing a single patient. His success proved that human heart surgery was possible.

After Harken's incredible accomplishment, advances came one right after another. In 1948 doctors performed a bold procedure to widen a narrowed mitral heart valve, following the same finger-insertion procedure used by Harken. Early results were devastating and most-often fatal. Gradually, surgeons refined Harken's technique, and the mitral valve procedure became quite safe.

Despite the tremendous leap in cardiac surgery, the advancement made little difference to patients suffering from more serious heart problems, such as congenital heart disease. Surgeons could not open the human heart without fear of patients bleeding to death. While it was possible to temporarily halt a patient's circulation, doing so gave doctors only four short minutes to work magic before brain damage from oxygen deprivation would occur.

Then, a young Canadian surgeon suggested a highly bizarre, but quite plausible, solution. Taking a cue from hibernating animals, which survived the bitterly cold Canadian winters, Dr. Bill Bigelow theorized that cold temperatures might be the key to open heart surgery. Bigelow discovered that by cooling animals considerably, open-heart surgery could be performed for long periods without death. At hypothermic levels, the tissues of the body and brain required less oxygen and, therefore, could survive much longer.

Bigelow's discovery allowed physicians to embark on an exciting new level of cardiac surgery. Daring to push science to the limit, two University of Minnesota doctors attempted the first open-heart surgery in 1952. Drs. Walton Lillehei (1918-) and John Lewis operated on a five-year-old girl who had been born with a hole in her heart. They cooled her body to 81 degrees Fahrenheit (27 degrees Celsius) and quickly repaired her injured heart. The child was immersed in warm bath water to raise her body temperature back to normal. The operation was a success.

While 10 minutes gave physicians enough time to treat small heart defects, it was insufficient for more complex problems. Doctors needed time to first visualize a major defect, and even more to repair it. Cardiac surgeons had hit a roadblock.

Gibbon had what they needed. After 23 years of trial and error, the Philadelphia native finally fulfilled the dreams of surgeons across the country. Gibbon built a machine that was successfully capable of assuming the functions of the human heart and lungs. The innovative and indispensable device was the result of more than two decades of hard labor and often discouraging results.

Gibbon's early experiments were cumbersome and dangerous; the machine often leaked blood, and hemorrhaging was a common form of failure. One of his first successes came when his heart-lung console proved capable to support the vital heart and lung functions in small animals during occlusion of their pulmonary artery. The results were very encouraging and further tests were planned. But Gibbon, now the director of surgical research at the Jefferson Medical College, lacked the funding and support to carry out such plans.

In a stroke of incredible luck, one of his medical students introduced him to Thomas Watson (1874-1956), the visionary chairman of International Business Machines (IBM). Watson agreed to underwrite the cost to develop a heart-lung device for human adults.

With money in his pocket, the aggressive Gibbon refused to let earlier complications discourage him. By January 1952 he was ready to test his machine on human patients. Gibbon announced his results at the May 1952 American Association for Thoracic Surgery: "I believe we are approaching the time when extracorporeal blood circuits of the heart-lung type can be safely employed in the treatment of human patients."

Gibbon had perfect timing. By the time of his 1952 speech, surgeons in a number of medical centers were constructing similar machines; three of them had used the machines in human cases. Gibbon believed his thoroughly tested console would safely permit precise, unhurried operations inside the human heart, and he vigorously sought to prove his point. Gibbon made certain his tests were performed on patients who would not have survived their ailments without surgical intervention.

Gibbon's first opportunities to use his machine on humans failed tragically. His first attempt was in March 1952 on a desperately ill 42-year-old man who had a possible clot or tumor in the right atrium. The outlook was bleak from the very beginning of the operation. The diagnosis was incorrect: neither a clot nor tumor was found. The patient's right ventricle was swollen to about twice its normal size. Doctors concluded nothing could be done and the procedure was terminated.

A year later Gibbon conducted the first trial with total cardiopulmonary bypass. The patient this time was a 15-month-old girl whom doctors believed had a heart defect. No such defect was found and the child's enlarged heart began to weaken from the moment it was exposed.

Three months after the tragic experience with the infant, Gibbon finally achieved success. The first successful open-heart surgery using cardiopulmonary bypass with the heart-lung machine took place on May 6, 1953, on a young woman with an atrial septal defect. The patient survived. News of Gibbon's success spread rapidly across the states. His dream had been realized and his accomplishment instantly made him a national and international celebrity.

Impact

A new era of heart surgery had begun. IBM began revising the original heart-lung machine, and other companies quickly followed suit. In 1955 John Kirklin at the Mayo Clinic reported success with a slightly modified Gibbon apparatus.

In the early 1960s scientists discovered that combining Gibbon's technique with the hypothermic cooling method originally proposed by Bigelow brought even better results. The method became known as extracorporeal cooling. The machine cooled the blood as it circulated through it, and after the operation, warmed it back to the correct temperature. With the heart dry and motionless, surgeons could operate more efficiently on the coronary arteries.

Gibbon and the earlier pioneers opened the door for cardiac surgery's exciting climax: organ transplantation. In 1967 South African surgeon Christiaan Barnard (1922-) transplanted the first human heart. While Barnard's surgical triumph was short-lived, Dr. Norman Shumway was not discouraged. By the mid-1970s Shumway had devised a way of spotting rejection attacks and transformed the picture for heart transplant patients. In little over a half of century, the heart had been transformed from a forbidden organ to one that could remarkably be repaired, or even replaced.

By January 2000 261 medical institutions in the United States operated an organ transplant program. From 1988 to 1999 doctors performed more than 20,000 heart transplant surgeries in the United States.

KELLI A. MILLER

Further Reading

Books

Austin, Harner. *Heart and Lung Machine and Related Technologies of Open Heart Surgery.* Phoenix Medical Communication, 1986.

Romaine-Davis, Ada. *John Gibbon and His Heart-Lung Machine.* Philadelphia: University of Pennsylvania Press, 1992.

Shumacker, Harris B. *A Dream of the Heart : The Life of John H. Gibbon, Jr., Father of the Heart-Lung Machine.* Fithian Press, 1999.

Internet Sites

The American Heart Association. http://www.americanheart.org

The Development of Organ Transplantation

Overview

By the mid-twentieth century surgeons began successfully transplanting human organs in order to save the lives of patients whose organs were failing from disease. These procedures were at first sensational, sparking debate among the medical community and the general public. In order for a transplantation to take place, a donor was required. With many organ transplantations, the donor was deceased at the time of donation. This startling development in medical history was, in fact, less sensational than the product of years of careful research. The field of human organ transplantation required many of the fields of medicine—surgery, histology, and immunology, for example—to unite in its cause. The research and performance of human organ transplants throughout the second half of the century resulted in more Nobel Prize awards than any other medical field in history. Within 50 years, kidney, liver, and heart transplants moved from experimentation to mainstream medical treatment for patients with failing organs and few other options to regain health.

Background

The transplant barrier was broken in 1954, with the first successful human kidney transplant. American surgeon Joseph E. Murray (1919-) led a team of surgeons performing the transplant procedure at a hospital in Boston, Massachusetts. The patient received a transplanted kidney from his living twin brother, and the kidney functioned normally for eight years. Murray attempted the procedure after years of studying how the body accepts or rejects donor tissue. Murray noticed that skin grafts applied to burn patients as a temporary measure were slowly rejected by the body, while skin grafts among identical twins were successful. By 1962, as the first generation of drugs were available to suppress the body's immune response and therefore counteract the rejection of the new organ, the first successful kidney transplant from a cadaver (deceased) donor was performed, again in Boston. The kidney functioned for 21 months. In 1990 Murray was awarded the Nobel Prize in medicine for his pioneering work in transplantation.

Although a kidney from a related donor is less likely to be rejected from the body, cadavers became the most common source for donor kidneys, due to greater availability and eliminating the risk to living donors. By the end of the century kidneys were the most commonly transplanted organ, and kidney transplantation became mainstream treatment for end-stage renal (kidney) disease.

American surgeon Thomas Starzl (1926-) performed the first human liver transplant at the University of Colorado in 1963. The procedure garnered much initial enthusiasm among the medical community as a possible treatment for end-stage liver disease, at that time almost always fatal, as no supplementary technology was available (such as in kidney dialysis) to help perform the work of the liver. This enthusiasm was diminished when the first seven patients to undergo liver transplant, at three different medical centers, all died within one month of the transplant. Liver transplants were temporarily suspended in order for scientists to explore the complex and serious post-operative complications the liver transplant patients suffered, notably organ rejection, infection, and pulmonary emboli (blood clots or air trapped in the lungs). Starzl's research focused on the body's immune response, which rejected the new liver. Important improvements in surgical technique were also learned from ongoing kidney transplantation. In 1967 Starzl and his team attempted another liver transplant, aided by the introduction of more potent anti-rejection drugs. The liver functioned successfully for 13 months. By the late 1990s approximately one hundred transplantation centers across the United States had performed liver transplants, with Starzl training physicians in many of these centers. The demand for liver transplants grew until, by the end of the century, nearly 30% of patients eligible for liver transplants died while waiting for a donor organ.

Transplantation captured world-wide attention in 1967, when South African surgeon Christiaan Barnard (1922-) performed the world's first human heart transplant. Barnard and his team of physicians and nurses at Groote Schuur Hospital in Cape Town, South Africa, removed the heart of a 55-year old man and replaced it with the healthy heart of a 25-year old woman who had died earlier of injuries sustained in an automobile accident. The patient survived 18 days after the transplant, dying from pneumonia as a result of an immune system suppressed to prevent rejection of the donor heart. Barnard performed his second heart transplant in 1968. The patient achieved notoriety as a symbol of hope for victims of heart disease and spurred the transplantation process, surviving 563 days after the operation. Also in 1968 American surgeon Norman Schumway performed the first successful heart transplant in the United States. Both surgeons continued to develop and refine surgical techniques for the burgeoning field of heart transplantation. It was not until the early 1980s, however, with the advent of cyclosporin and other next-generation anti-rejection drugs, that the heart transplant procedure became widely accepted. By the 1990s heart transplantation evolved from an experimental operation to an established treatment for advanced heart disease, with over two thousand performed yearly in the United States.

Impact

Transplants are considered when a major organ of the body is failing and does not respond to all other therapies, but otherwise the health of the patient is good. Patients receiving successful transplants are often able to resume their daily lives with no dependence on complicated medical machinery, such as a kidney dialysis machine or a heart pump assistive device. Although transplant recipients must adhere to strict regimens of medications and frequent examinations, increased survival rates at the turn of the century enabled over 75% of successful transplant recipients to return to a daily work schedule and to recreational activities enjoyed prior to becoming ill.

As organ transplant procedures increased and became standard treatment for otherwise fatal illnesses, both the medical community and the public at large considered ethical issues brought forth by organ donation. The National Transplantation Act, passed by the U.S. Congress in 1984, mandated a centralized system for sharing available organs along with a scientific register to collect and report transplant data. The act also made illegal the sale or purchase of organs.

A national system was established to match donors and recipients; it is managed by the United Network for Organ Sharing (UNOS). UNOS members work with all transplant centers in the United States to ensure that the limited supply of organs is distributed fairly to patients in need regardless of age, sex, race, lifestyle, or financial or social status. Through the UNOS Organ Center, organ donors are matched to waiting recipients every day of the year, around the clock. Organ sharing is based upon scientific criteria including the recipient's acuity (urgency state) of the disease process, compatibility of body size and blood chemistries, as well as length of time on the waiting list. At the close of the century, new laws were under consideration designed to remove any geographical bias in organ allocation. The Scientific Registry maintained by UNOS contains data on every solid organ transplant since 1987 and is one of the most comprehensive data analysis systems targeting a single therapy in the world. Patient confidentiality is maintained with a number system, and scientists are able to quickly exchange information vital to the progress of transplantation.

With increases in the number of transplants performed, UNOS and public health organizations attempted to raise public awareness of the importance of organ donation. States included organ donor status on citizens' driver's licenses, and a universal donor card was widely publicized. By the 1990s most states had passed legislation requiring medical personnel to approach all potential donor patients, or if the patient is unable, their families, for a donation decision. The criteria for brain death was clarified in 1981 by a presidential commission on medical ethics to allay public concern regarding time of death and organ recovery. All major religions in the

United States voiced opinions encouraging personal choice and organ donation.

In spite of these efforts, the demand for donated organs has far outnumbered the supply. By the turn of the century, over 67,000 Americans were on the UNOS national patient waiting list. At the same time, transplantation procedures were quickly growing. Organs and tissues were needed for additional types of transplants added to the medical arsenal against disease. Lung, pancreas, bone marrow, small intestine, cornea—all were considered an acceptable part of medical treatment. To potentially ease the shortage, some scientists experimented with xenotransplantation, or transplanting an organ of another species into a human. A celebrated xenotransplantation case was that of "Baby Faye," into whom a baboon heart was transplanted in 1984 at the Loma Linda Medical Center in California. Baby Faye's baboon heart functioned for 20 days. Xenotransplantation remains experimental, as do artificial mechanical organs—scientists continue to study both of these measures as potential "bridges" to serve a critically ill patient until a donor organ can be located.

BRENDA WILMOTH LERNER

Further Reading

Caplan, Arthur L., and Daniel H. Coelho, eds. *The Ethics of Organ Transplants: The Current Debate*. Prometheus Books, 1999.

Hakim, Nadey, ed. *Introduction to Organ Transplantation*. London: Imperial College Press, 1996.

Pensak, Robert and Dwight Williams. *Raising Lazarus*. New York: G. B. Putman & Sons, 1994.

Starzl, Thomas E. *The Puzzle People: Memoirs of a Transplant Surgeon*. Pittsburgh, PA: Pittsburgh University Press, 1992.

Advances in Diagnosis and Treatment of Diseases of the Eye

Overview

Major innovations in diagnosis and treatment of diseases of the eye occurred in the last half of the twentieth century. From the development of the contact lens to precise lasers used in surgery of the eye, technology played a vital role in the correction of visual difficulties. Recent research has advanced scientific and clinical knowledge of how the eye functions, providing a basis for future sight-saving treatments.

Background

Although progress in understanding and treating diseases of the eye was significant from 1950 until the end of the century, the need for eye care increased dramatically. In the United States alone, one third of the population is estimated to need corrective lenses in order to see properly. The economic impact of visual disabilities climbed to over 35 billion dollars in 1995 in direct medical costs and indirect costs to society through lost productivity.

The National Eye Institute, created by Congress in 1968, joined with similar organizations world-wide to encourage innovations in medical technology and therapies for the treatment of eye disease. The developed world reaped most of the benefit from recent advances, as new drug therapies and surgical techniques quickly spread, fueled by consumer interest for better eyesight without eyeglasses. While patients in North America and Europe queued for voluntary high-tech surgeries, preventable blindness became a disproportionate problem in developing countries. The World Health Organization identified some diseases of the eye, such as glaucoma, as major targets for reduction in the twentieth-first century.

Impact

Prior to 1950, spectacles (eyeglasses) and the magnifying glass were the most common forms of treatment for visual problems, and impaired eyesight was considered an inevitable consequence of aging. Lenses for eyeglasses were often thick and cumbersome to wear, and their design allowed for peripheral vision distortion. Modern contact lenses, invented in 1948 by Kevin Tuohy, provided an alternative to eyeglasses. The lenses were made of hard plastic, were larger than the iris, and rested on the cornea (the transparent, curved surface of the eye) supported by a layer of tears. Contact lenses are used to treat myopia (near-sightedness), astigmatism (distorted vision due to slight irregularities in the shape of the cornea), and other refractory visual defects. Contact lenses also reduce peripheral vision distortion since the lenses move with the eye. Many considered contact lenses more cosmetically appealing than glasses, and by the mid-1960s, millions of Americans wore them—two thirds of whom were women. Corneal irritation was the main complication associated with contact lenses. During the 1970s and 1980s, improvements in the composition of the lenses reduced discomfort and irritation. Other improvements included the soft contact lenses made from water-absorbing plastics, extended-wear lenses, and gas-permeable hard lenses which allowed more oxygen to reach the eye. Tinted lenses to enhance or change eye color added to the appeal of contact lenses.

Beginning in the 1960s, modern refractive surgeries were performed on the eye to create better vision and decrease the need for wearing glasses. Russian surgeons corrected myopia by making multiple incisions on the anterior (front) surface of the cornea in a technique called Radial Keratotomy (RK). The results of RK demonstrated dramatic improvement of vision, and interest in RK quickly spread to the United States. A study sanctioned by the National Eye Institute demonstrated the initial effectiveness of RK, but also found a disturbing number of patients left with fluctuating daily vision after RK surgery. The procedure's popularity waned until the 1980s, when newly developed technologies such as microscopic guided lasers emerged. In PRK (photorefractive keratotomy), precise corneal incisions are made by laser energy which flattens the corneal surface, correcting myopia while leaving daily visual acuity stable. PRK has virtually replaced RK in the United States as the safest and most effective surgery of its kind for the correction of near-sightedness. In the 1980s and 1990s, thousands of PRK operations were performed across North America and Europe, and many PRK recipients experienced enough visual improvement to abandon their eyeglasses.

By the end of the century, the new surgical procedure Laser in Situ Keratomileusis, or LASIK, received limited approval and grew rapidly due to customer demand. LASIK involves creating a hinged flap in the cornea, and using a laser to remove minute tissue from the inside of the corneal stroma (body). The surface flap is then reattached, and the patient in most instances experiences a swift improvement in vision with less discomfort than other refractive surgeries. LASIK is used to correct myopia, hyperopia (far-sightedness), and astigmatism. LASIK surgeries were performed by the thousands by the late 1990s, just in time for the baby-boomer population to experience age-related refractory eye changes. LASIK and PRK surgeries have made it possible for thousands of people around the world to see clearly without depending on glasses or contact lenses.

Lasers were also used to treat glaucoma, a disease in which the fluid pressure inside the eye elevates, damaging the optic nerve and leading to vision loss. When medications fail to control the pressure, laser surgery is considered. By focusing many tiny laser burns to the filtering angle area of the eye, an area is created which improves fluid drainage, and pressure is reduced. Another laser surgery for glaucoma focuses on creating a small opening in the peripheral opening of the iris. This helps keep fluid from building up behind the iris, and preserves the ability of the iris to filter light. Before the advent of laser-guided microsurgery in the 1980s, those affected with glaucoma had few options to preserve sight when conventional medicines failed them.

Beginning in 1950, surgeons implanted artificial intraocular lenses in patients whose vision was clouded by cataracts. A cataract is an opacity of the normally clear lens of the eye that restricts vision. World-wide, cataracts are the cause of half of all blindness, and are the result of advancing age, trauma, genetic predisposition, or metabolic disorders. By the 1970s, the overwhelming success of new, flexible, light-weight plastic lenses created a revolution in cataract treatment in the developed world. Surgeons were able to remove the clouded lens, and select from among several therapeutic artificial lenses the one which would best minimize the patient's need for glasses after surgery. By the end of the twentieth century, an artificial lens was nearly always implanted in the eye at the time of cataract surgery, and the United States had spent more than three billion dollars on this successful procedure through the Medicare program alone.

Complications from diabetes often lead to eye disease, especially those of the retina (the inside lining of the back surface of the eye where light energy is converted into electrical impulses). Diabetic retinopathy involves tiny hemorrhages and deposits in the retina, as well as abnormal retinal blood vessel growth, all of which

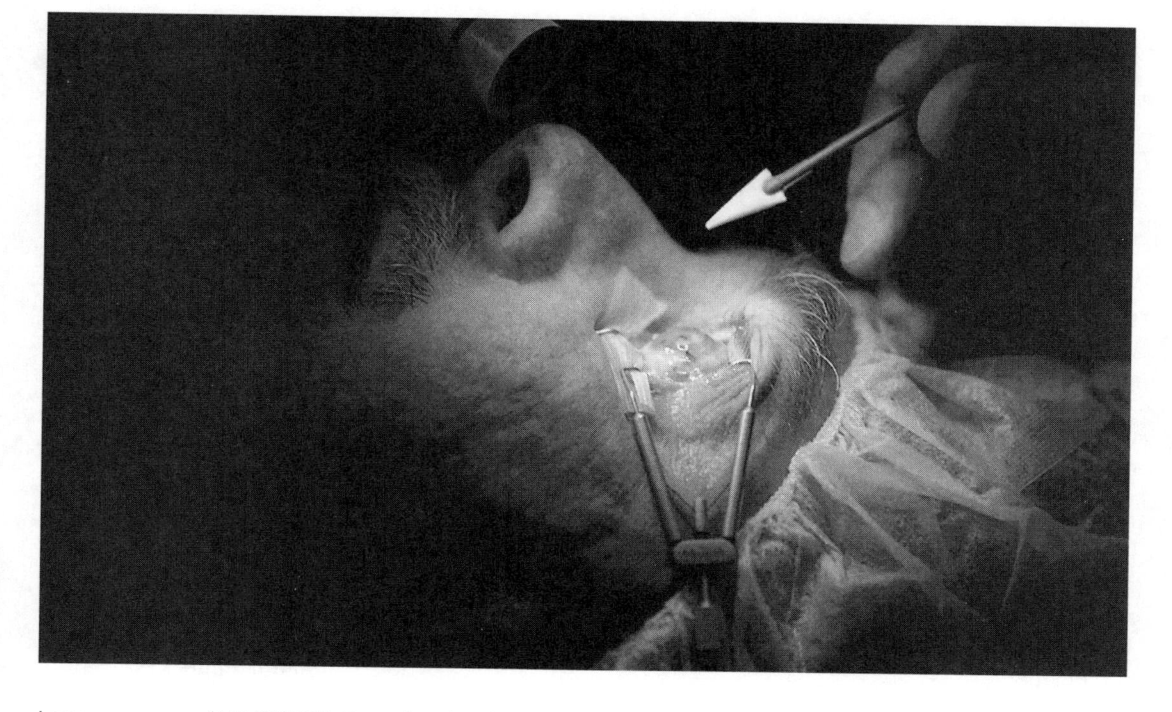

Laser eye surgery. *(AFP/CORBIS. Reproduced with permission.)*

may lead to blindness. Surgery known as laser photocoagulation has proved effective in maintaining and sometimes partially restoring vision.

Gene replacement therapy may one day prevent vision loss by delivering healthy genes to patients with retinal degeneration. Researchers conducted studies in rodents in the 1990s to find an appropriate vector, or delivery system, capable of introducing therapeutic genetic information to retinal cells while eliminating the harmful genetic information. In gene therapy experiments, certain viruses have successfully vectored therapeutic genetic material, resulting in delayed vision loss for long periods of time. Human gene therapy trials for retinal degeneration were anticipated by researchers as the twentieth century came to an end.

The computer chip also offered hope to patients with end-stage retinal degeneration. Designed for implantation on the surface of the retina, the chip mimics photoreceptor cell function. A camera mounted on a pair of glasses transmits visual information to the chip, which in turn transmits images to the brain via the optic nerve. Although prototypes of these high-tech visual aids were manufactured and initial testing in animals began, further work is needed before humans test the retinal chip. Researchers, however, remain optimistic about the retinal chip's future to restore functional vision for those with profound vision loss as a result of retinal degeneration.

Not all of the latest vision research and treatment involves high-tech processes. Vitamin supplements were found to benefit patients suffering vision loss from retinitis pigmentosa (RP), an inherited disease which causes the progressive destruction of specialized thin, light-absorbing cells lining the back of the retina. RP typically begins with night blindness and proceeds over time to continued loss of peripheral vision. The majority of people with RP are legally blind by age 40. Studies conducted in the 1990s showed that supplements of vitamin A palmitate in certain dosages slowed the progression RP. Additionally, vitamin E supplements were found to increase the rate of degeneration. While not touted as a cure, patients with RP have enjoyed added years of functional vision due to vitamin A supplementation therapy.

Concern for protecting eye health made its way into popular culture as the twentieth century ended. Many complained of eyestrain as demands of the workplace included long hours in front of a computer screen. Studies to determine potential adverse effects were under way, and colored filters for monitors became popular items. Computer manufacturers and software companies responded by creating large, adjustable monitor screens and programs, including choices of colors intended to soothe the eyes. Sales of sunglasses touting high filtration rates of the harmful rays of the Sun soared, and were marketed for babies as well as children and adults. Safety glasses once confined to industry became fixtures in many households and were used during everyday chores such as mowing the lawn. Athletes appeared on television sporting new high-tech protective eye wear.

BRENDA WILMOTH LERNER

Further Reading

Books

Armstrong, France and James J. Sales. *Beyond Glasses: The Consumer's Guide to Laser Vision Correction.* U.C. Books, 1998.

Casual, Gary H., Michael D. Billing, and Harry G. Random. *The Eye Book: A Complete Guide to Eye Disorders and Health.* Johns Hopkins University Press, 1998.

Lehmann, O.J., D.H. Verity, and A.G.A. Combos, eds. *Clinical Optics and Refraction.* Butterworth-Heinemann Medical Books, 1998.

Mock, Lyle G. and Marie Mock. *Macular Degeneration: The Complete Guide to Saving and Maximizing Your Sight.* Ballantine Books, 1998.

The Development of Polio Vaccines

Overview

Poliomyelitis, also known as infantile paralysis, is an acute viral infection that can invade the nervous system and cause paralysis. Where the disease is common, most infections probably go unnoticed or result in mild symptoms, such as fever, sore throat, headache, vomiting, and stiffness of the neck and back. Before the introduction of preventive vaccines, poliomyelitis was one of the most feared childhood diseases. During some epidemic years, over 10,000 paralytic cases occurred in the United States alone. Such epidemics formed the basis for the image of polio as the great crippler of children and exerted a profound influence on the direction of medical research. The history of poliomyelitis demonstrates the value of immunization. In 1981 about twenty years after the use of the polio vaccines became widespread, the number of recorded cases in the United States reached a record low of only six cases. By the end of the twentieth century, the global eradication of polio was considered a practical goal. In the wealthy, industrialized nations, diseases such as diphtheria, measles, mumps, pertussis, and rubella have been virtually eliminated or radically reduced. Nevertheless, fewer than half of all American children under the age of two are properly immunized.

Background

Although therapy for many diseases has improved, from the public health point of view vaccines are the most powerful and appropriate tools for preventing epidemic diseases. Whenever a large portion of a population has been vaccinated, the community achieves a form of protection against epidemics known as "herd immunity" because the number of susceptible people is significantly reduced. Advances in biotechnology have made possible the design of safer and more effective vaccines. Edward Jenner (1749-1823) introduced the term vaccination in the late eighteenth century to distinguish his method of inducing immunity to smallpox from older, more dangerous methods. Vaccines are made in a variety of ways, depending in part on the nature of the organism and the disease it causes. Weakened microbes, killed microbes, animal viruses that are not virulent in humans, and toxins are the most common components of the vaccines that have been in use throughout the twentieth century. Weakened live-virus vaccines have been used against rabies, mumps, measles and rubella, and poliomyelitis. Similar techniques are being used to create new vaccines against rotaviruses, respiratory syncytial virus, parainfluenza, and influenza viruses. Killed-virus vaccines are being used against poliomyelitis, whooping cough, and influenza.

In the battle against poliomyelitis, the Salk and Sabin polio vaccines have both been very successful, but they also illustrate the advantages and disadvantages of killed and live attenuated vaccines. Killed vaccines are generally easier to develop, but live vaccines induce longer-lasting protection. Albert Sabin (1906-1993) carried out pioneering research on poliomyelitis and developed a live attenuated vaccine for the prevention of the disease. Contrary to the prevailing views about the means of transmission of the polio virus, Sabin proved that the virus was spread by the fecal-oral route rather than the nasal route and that the virus multiplied in the human intestinal tract. By 1936, Sabin and his associates had been able to isolate and propagate the poliomyelitis virus in laboratory cultures of human embryonic nervous tissues. Sabin believed that long-lasting immunity could best be established with a live attenuated vaccine because antibodies to the virus were found in survivors of the disease many years after infection. The live vaccine could be administered by mouth, but the virus might spread from those who had been vaccinated to others who had not. During a period in which unvaccinated people might easily encounter the wild virus, the benefits of using the attenuated virus might outweigh the dangers. When the disease is rare, however, the transmission of even an attenuated form of the polio virus to non-immune people could be dangerous. Eventually, American epidemiologists found that the use of the live vaccine was associated with a small, but real, risk of paralytic polio.

Before Sabin perfected his vaccine, Jonas Salk (1914-1995) had developed a killed-virus vaccine that had to be administered by injection. A nationwide trial of the Salk vaccine in 1954 was successful. As a result of the Salk vaccine program, the incidence of paralytic polio in the United States decreased dramatically by 1961. The National Foundation for Infantile Paralysis, which had supported Salk's research, was committed to the Salk vaccine and unwilling to sponsor any other vaccines. The World Health Organization (WHO), however, supported tests of the oral polio vaccine in Mexico, Czechoslovakia, and the Soviet Union. In 1985 WHO began an effort to eradicate polio worldwide by the year 2000, but there is little likelihood that the goal will be achieved before 2040.

Impact

Advances in the sciences of virology, bacteriology, immunology, and molecular biology have led to new approaches in the construction of vaccines. Public health policies and procedures have been transformed by the development of vaccines against more than 20 infectious diseases. At least 15 new or improved vaccines were developed between 1980 and 2000. Experimental vaccines may offer enhanced protection against influenza, pneumococcal pneumonia, pertussis, rubella, rabies, bacterial meningitis, hepatitis B, and adenovirus-associated respiratory disease. New vaccine technologies also offer hope of ameliorating the effects of autoimmune disorders and allergies.

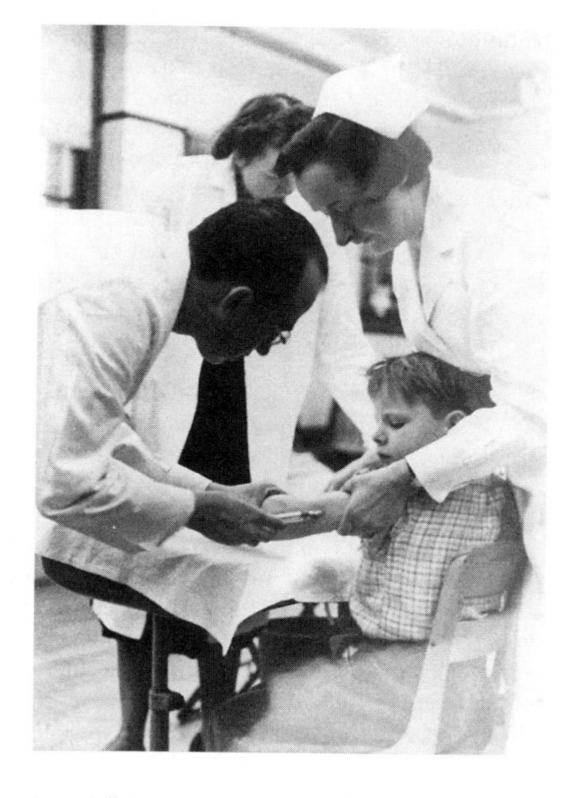

Jonas Salk immunizing a child with the polio vaccine. *(The Library of Congress. Reproduced with permission.)*

Microorganisms contain proteins called "antigens" that stimulate the host's immune response, resulting in the synthesis of proteins called "antibodies" that bind to the microbes and help destroy them. In addition, "memory cells" are produced and remain in the blood stream, ready to mount a quick response against subsequent infections by the same microbe. Vaccines have been made by using a portion of the disease-causing organism that contains the antigens that trigger the immune response. Such cell-free vaccines are generally safe but they may not induce long-lasting immunity. For microbes that produce toxins, vaccines can be prepared by changing the dangerous toxin into a harmless "toxoid." Improved vaccines against whooping cough (pertussis), for example, are made up of inactivated pertussis toxin. Based on Jenner's use of cowpox against smallpox, scientists have adopted a similar approach in the battle against rotaviruses, which are the leading cause of infant diarrhea, and parainfluenza viruses, which cause severe respiratory tract infections in children. Recently, biotechnology and genetic engineering techniques have been used to design "subunit vaccines," which are based on isolation of the genes that code for selected subunits of the genome of the infectious agent. The selected genetic material can then be produced in large quantities by growing it in bacteria or yeast host cells. Subunit vaccines are now available for meningitis, pneumonia, typhoid, and hepatitis B. Subunit vaccines for respiratory syncytial virus and parainfluenza virus infections are currently under development.

Some bacteria, such as those that cause pneumonia and meningitis, have an outer coat that protects them from the human immune response. When the outer coat of the pathogen is linked to proteins or toxins from another organism, the result is a combined, or conjugate, vaccine. In 1986 the first conjugate vaccine was licensed to protect against *Haemophilus influenzae* type b (Hib), the major cause of bacterial meningitis in babies and young children. By the end of the twentieth century, the widespread use of improved versions of this vaccine had virtually eliminated Hib meningitis in the United States.

Scientists have used genetic engineering technology to isolate specific genes and insert them into the DNA of certain microbes or mammalian cells grown in the laboratory. Such cells become "factories" for the mass production of the antigen that is the selected gene product. The antigen can then be separated from other material by the use of a monoclonal antibody that recognizes the antigen. A vaccine for the hepatitis B virus has been created through this approach.

Scientists can also insert genes for desired antigens into the DNA of related, but harmless, viruses such as the vaccinia virus or selected strains of bacteria. This approach is being used to develop novel vaccines for the viruses that cause hepatitis B, influenza, rabies, and AIDS. Another approach undergoing testing is the weakening of a dangerous microbe, such as the cholera bacillus or the herpes virus, by removing specific genes. The engineered microbes can produce immunity but not disease. One experimental DNA vaccine for AIDS involves isolating selected genes from the virus and injecting them into individuals. If these genes can enter host cells and cause the synthesis of viral proteins, these foreign proteins might elicit an immune response, which would protect the host against subsequent infection by the microbe. Some scientists believe that the creation of wholly synthetic vaccines made by isolating the gene that encodes for an appropriate antigen will be possible. Selected amino acid sequences of the resulting antigen (a protein molecule) could be synthesized and used as vaccines for malaria and diarrheal diseases.

Malaria parasites and the mosquitoes that serve as their vectors (means of transmission) are very ancient. The disease infects about 500 million people each year and claims about 2.7 million victims. Although the disease has resisted efforts to control it, during the second half of the twentieth century researchers acquired a great deal of information about various strains of malaria parasites and their characteristic antigens. Much of the malaria genome has been sequenced and chromosome mapping of the genome is nearly complete.

Progress towards anti-malaria vaccines has been slow—for both scientific and geopolitical reasons—but during the 1990s several promising experimental vaccines were being tested. New malaria vaccines are based on recombinant technology—a protein from the malaria parasite fused to hepatitis B surface antigen (HBsAg)—and highly effective adjuvants. (Adjuvants are substances added to vaccines to enhance the immune response.) Although experimental vaccines have not undergone rigorous field trials, research on these vaccines has provided insights into the immunological requirements for further improvements. Among the many obstacles to the development of a broadly effective malaria vaccine is the highly variable nature of the parasite's surface proteins. A vaccine that does well in limited clinical trials might not be effective against all variants of the parasite, and the duration of immunity is also problematic. The problems caused by variability in the malaria parasite are much more complex than those involved in working with the three major types of polio viruses.

Immunization is widely acknowledged as the most cost effective of all public health interventions. WHO estimates that vaccines for common infectious diseases, costing only pennies per dose, could save millions of lives and billions of dollars each year. According to WHO, however, 20% of the world's children, living mainly in the poorest countries, remain unprotected against polio, measles, diphtheria, pertussis, rubella, and other infectious diseases. Every year about 8 million children worldwide die from diseases that are preventable through the use of vaccines.

LOIS N. MAGNER

Further Reading

Carter, Richard. *Breakthrough: the Saga of Jonas Salk*. New York: Trident Press, 1966.

Dowling, Harry F. *Fighting Infection: Conquests of the Twentieth Century*. Cambridge, MA: Harvard University Press.

Klein, Aaron E. *Trial by Fury: The Polio Vaccine Controversy*. New York: Scribner, 1972.

McKeown, Thomas. *The Role of Medicine: Dream, Mirage, or Nemesis?* Princeton, NJ: Princeton University Press, 1979.

Paul, John R. *A History of Poliomyelitis*. New Haven, Yale University Press, 1971.

Plotkin, S., and S. Mortimer. *Vaccines*. New York: Saunders, 1995.

Rogers, Naomi. *Dirt and Disease: Polio before FDR*. New Brunswick, NJ: Rutgers University Press, 1992.

Silverstein, Arthur M. *A History of Immunology*. New York: Academic Press, 1989.

Smith, Jane S. *Patenting the Sun: Polio and the Salk Vaccine*. New York: W. Morrow, 1990.

Modern Advances in Surgery and in Medical Technology

Overview

The science of surgical care has advanced further in the last 50 years than it has in all preceding years combined. Complicated procedures such as natural and artificial organ transplants, xenotransplants (organs transplanted from non-human animals), neurosurgery (brain surgery), coronary artery bypass surgery, laparoscopic surgery, and "laser" surgery were rare, if not completely unknown 50 years ago, but these procedures are becoming more commonplace today. As with many aspects of our lives, computers have also extended the practice of medicine into previously unknown territory. Indeed, through the use of the Internet, "telemedicine" has become not just a possibility but a probability: soon, surgeons will be able to operate on patients remotely via live "webcasts."

What is more, surgeries generally have become far less invasive, thus requiring in many cases little, if any, hospital stay. As a result, the overall cost of many of these procedures has decreased dramatically in terms of both the financial costs to patients (and their insurance companies) as well as recovery costs to patients in terms of lost wages and physical and emotional strain. In short, advances in surgery and medical technology have allowed many more people to live healthier and longer lives than at any preceding time in history. Many diseases like cancer, for example, which used to nearly always be fatal, are now often eradicated entirely in patients due to the technological advances made in surgery over the past 50 years.

Background

Perhaps the most pervasive technological advance today in medicine is laparoscopic surgery. First introduced in the early 1970s, laparoscopic surgery is a technique whereby a surgeon makes four tiny "pinhole" incisions in a patient's body and then inserts a miniature camera, light, and the required surgical instruments for performing the procedure. Today's advanced digital technology allows magnification of the laparoscopic surgery site up to 20 times its actual size, thereby permitting surgeons to see anatomical structures in exquisite detail. Furthermore, new three-dimensional imaging technology allows the surgeon to view internal organs stereoscopically rather than in two dimensions only, as would be required if viewed on a regular monitor. Intraoperative ultrasound allows "real time" scans of the surgery site as the operation proceeds, providing additional valuable information. Because laparoscopic surgery is minimally invasive, patients recover much more quickly.

Laser surgery is also growing in popularity and application. As its name suggests, surgeons utilize a laser to perform various procedures, including during laparoscopic procedures. For example, lasers currently are used to excise cancerous tissue from the larynx, reshape the cornea of an eye to allow a patient to see better, and even to "resurface" the skin of a patient's face by burning off old layers skin so that new skin can grow. The growing popularity of lasers as surgical devices is due mainly to their ability to precisely destroy unwanted or abnormal tissue without bleeding.

Another well-known example of advancing surgical techniques involves combating cardiovascular disease. Because of lifestyle habits or genetic predisposition in many people, fatty acids (plaque) sometimes build up in the arterial walls of a person's heart. As more plaque builds up, less blood is able to flow through the artery to the heart. Ultimately, the plaque build-up may completely block the artery, preventing any blood from flowing through it. The result is cardiac arrest, which can be fatal. Surgeons have developed a technique known as angioplasty to combat the onset of cardiovascular disease. Using a technique similar to laparoscopy, a surgeon inserts a thin tube into the patient, working it up the artery to where the blockage resides. At the end of the tube is a small, balloon-like device that inflates, pressing the plaque against the arterial walls so that blood flow through the artery can be increased.

Surgeons have also developed another, more popular, procedure for dealing with coronary artery disease: the coronary bypass graft operation. By taking a portion of an artery from elsewhere in the patient's body—usually the internal mammary artery from inside the chest cavity—the new artery is grafted around the blockage of the old artery to allow blood to flow around the blockage via the new arterial route. Despite the fact that this procedure requires open-heart surgery, it is performed more than 300,000 times per year in the United States alone.

Like bypass grafting operations, transplantation procedures involve the replacement of organs and tissue from one location in the body to another location. Unlike bypass grafting operations, however, organ and tissue transplants often come from the bodies of another person, and sometimes even animals (known as xenotransplantation). Kidneys, hearts, lungs, and recently even hands have been transplanted successfully. According to the United Network for Organ sharing, there currently are more than 61,000 persons waiting for an organ transplant in the United States alone. Every 16 minutes, a new person is added to the waiting list. Though these procedures have become more commonplace with higher rates of success, unfortunately there simply are not enough organs available to meet the demand.

Just as computers pervade our everyday lives, so also do they pervade operating rooms. Currently, computer-aided surgery is being studied for various applications including simulating surgery in three-dimensional "virtual" environments. By using data from hundreds of x rays, or from magnetic resonance images, a composite picture of a patient's internal organs and skeletal frame can be produced with incredible detail. Such computer-designed images allow surgeons to practice complex procedures before they actually perform them. By exploring these three-dimensional environments for problematic areas, surgeons no longer are required to perform exploratory surgery in order to determine where a problem area lies; they simply may move about in the computer-reproduced virtual body to find the problem. When working on an organ as complex as the brain, computer-aided surgery helps to minimize the risk of needlessly or inadvertently damaging other areas.

Impact

According to American Hospital Association statistics, hospital outpatient surgeries, which do not require overnight hospital stays, have increased by more than 2 million, to 14.7 million, between 1993 and 1997; and inpatient surgeries, which require a stay of more than a day, have decreased by nearly 700,000 to 9.5 million during the same time period. As these statistics indicate, more people than ever before are "going under the knife" as the prospect of surgery becomes less daunting and more readily available.

Not surprisingly, a large and rapidly growing proportion of these surgeries consist of cosmetic, or "plastic," surgeries. Over the past two years alone, the number of plastic surgeries has increased by 50%, with more than a million performed in 1999 alone. According to the American Society of Plastic Surgeons, the most popular procedure in 1999 was liposuction—the removal of fat from the body by means of a suction device, followed by breast augmentation, eyelid surgery, facelift, and chemical peel—the removal of the top layer of skin from the face by chemical means. These surgeries are performed not so much to improve or maintain the health of the patient but to "aesthetically enhance" his or her physical features.

With many of these surgeries, however, there has come unforeseen risks. Breast augmentation surgery, for one, has sparked multi-billion-dollar class action law suits against the manufacturers of the breast implants, the silicon prostheses inserted during the surgery. Early studies suggested that silicon may leak into patients? bodies, causing significant physical damage, although more recent studies have indicated no convincing evidence that breast implants can be harmful. Other cosmetic surgical procedures such as liposuction, however, may result in unforeseen physical damage. Thus, although intended to improve one's physical image, plastic surgery sometimes can gravely endanger one's health.

Another problem, albeit more philosophical in nature, that has developed with the advancement of surgical techniques concerns the very definition of life. Fetuses can now be delivered as early as the beginning of the third-trimester of pregnancy and survive. As prenatal and postnatal care improves, this "viability" line may move back to even earlier times in a woman's pregnancy. Such viability issues, of course, have played crucial roles in abortion-rights debates over the past 25 years, most famously in the United States Supreme Court case of *Roe v. Wade*.

Similar to the problem of determining when life begins, modern medicine also has muddled the definition of death. Historically, when a person ceased to breath on her own, and/or her heart stopped beating, that person was considered dead. With surgical advancements, we no longer look to the heart and lungs to determine death, but the electrical activity of the brain. Indeed, human bodies can be kept functioning for years on cardio-respiratory machines—machines that pump blood for the heart and breathe for the lungs—long after the brain has died. Yet, at just what brain-activity level one can be considered dead is the focus of much debate. In the celebrated case of *Gannon v. Albany Memorial Hospital,* for example, an 86-year-old woman named Ms. Coon was hospitalized after a massive stroke. Soon thereafter she fell into a "persistent vegetative state," which has been described not as a coma but as a state of "wakefulness without awareness." Ms. Coon's sister, believing that Ms. Coon would never wish to be kept in such a state, and believing her to be essentially dead, successfully petitioned a court to allow her to take Ms. Coon off life-support so that she could "die." Luckily for Ms. Coon, prior to the withdrawal of her feeding tube, she woke-up! So, just when a person may be considered alive or dead is no longer a simple matter given advances in medical technology.

Finally, perhaps the most wide-ranging impact that advances in surgery in particular, and medicine generally, have had on society over the past 50 years is to increase the average life-span of human beings. In 1950 the average life-span for a man born that year in the United States was 66, and for a woman, 71; today a man born in the United States can expect to live, on average, to the age of 73, and a woman to the age of 79. The rise is even more dramatic in poor, third-world countries, although in places such as Africa factors ranging from war to famine to newly emerging diseases continue to hamper life-expectancy rates.

As medical technology improves, life expectancies will continue to grow. Ironically, though modern medical technology has allowed many people to enjoy healthier and longer lives—a seeming societal benefit—advancing medical technology has simultaneously created an enormous societal burden. It is no accident that the fastest growing segment of the U.S. population today consists of those aged 80 and over. Inevitably, as people grow older and live longer as a result of better medical care, their continued care can become quite expensive for the private insurance companies and public organizations, such as Medicare, that pay for it.

Likewise, because people over the age of 65 are entitled to social security benefits as well as funds from private pensions, as their ranks grow in proportion to the rest of the population, they will draw out more funds than can be replenished by the younger, working generations. This situation has caused many economists and politicians to worry that government programs like social security will become bankrupt in the near future, thereby leaving nothing for the younger generations.

Undoubtedly, as surgical techniques advance, more people will have the opportunity to enjoy the benefits of long and healthy lives. Though this is a noble endeavor, the impact that modern medical technology has had on society nevertheless can manifest itself in unusual, and sometimes undesired, ways.

MARK H. ALLENBAUGH

Further Reading

Chotkowski, L.A. *What's New in Medicine: More Than 250 of the Biggest Health Stories of the Decade.* Santa Fe, NM: Health Press, 1991.

Fuller, Joanna Ruth. *Surgical Technology: Principles and Practice.* W.B. Saunders Co., 1993.

Lewinwand, Gerald. *Transplants: Today's Medical Miracles.* New York: Franklin Watts, 1992.

Sherrow, Victoria. *Bioethics and High-Tech Medicine (Inside Government).* Twenty First Century Books, 1996.

Yount, Lisa. *Medical Technology.* New York: Facts on File, Inc., 1998.

Emerging Diseases since 1950

Overview

Emerging diseases, which come in a variety of forms, began to appear just when medical science was confident that scourges like smallpox and polio had been conquered. Emerging diseases include the Marburg virus, which was first recognized in 1967, and its close cousin, the Ebola virus. Both are deadly hemorrhagic diseases for which there is no cure. Equally lethal, the AIDS pandemic and infections from bacteria such as *E. coli* have sickened and killed many. Moreover, a number of diseases thought to have been under control, such as malaria and cholera, have suddenly increased in their rate of incidence. At the same time, because of increasing resistance to drugs, old conditions are re-emerging. For example, tuberculosis, thought to be under such control that tuberculosis hospitals were phased out, has re-emerged, killing more that two million people each year.

Background

Diseases and plagues have changed the course of history. Historians have theorized that malaria was more responsible for the fall of the Roman Empire than the barbarians. The Black Death, or bubonic plague, ended the institution of feudalism and caused such a general paralysis in the 1300s that countries such as England and France stopped their fighting and called a truce. The Spanish and other Europeans brought diseases to the New World that decimated whole populations of natives.

The influence of pandemics involving diseases such as cholera, typhus, and the Black Death was profound, but attitude and ignorance were also negative forces. This situation began to change in the middle of the nineteenth century with not only the discovery of bacteria and inoculation but with a more humane attitude toward life and health habits. Nevertheless, there were still problems because in a major epidemic, such as the one involving influenza in 1918, physicians were still powerless. In 1941 antibiotic drugs, for the first time, allowed an attack on bacterial organisms once they had invaded the body. The public began to think of medical science as a god able to take care of all ills and problems. At the beginning of 1950, the public attitude in the Western world was that science would cure anything. Plagues would be conquered; there would be no more disease.

Impact

Optimism about science continued to build in the 1960s and 1970s. For example, the last case of smallpox, registered in Somalia on October 26, 1977, touted the eradication of the disease. From remote areas of the past, however, began to emerge silent threats—quietly at first because they affected only a few people but then more loudly. With people travelling now more frequently to developing countries and with increased trade, conditions that were once isolated are now being seen in more locations. Few things have piqued the public's imagination more than scenes of zombie-like people dying with the god of medical science unable to intervene.

In 1967 workers in laboratories in Marburg and Frankfurt, Germany, and Belgrade, Yugoslavia, became ill after being exposed to African green monkeys imported for research. While the exact origin of the monkeys is unknown, they appear to have come from Uganda, Kenya, or Zimbabwe. How the monkeys originally got Marburg fever is also unknown, but humans contract it from the animals and then pass it on to other people. The disease is characterized by fever, chills, headaches, and myalgia, or muscle pain. Around the fifth day, the infected person may have nausea, vomiting, or diarrhea, which becomes more severe, leading to massive hemorrhaging and multi-organ dysfunction. The disease is almost always fatal. Many of the initial symptoms, unfortunately, are similar to other diseases, such as malaria and typhoid fever. Laboratory tests, such as ELISA (enzyme-linked immunoabsorbent assay) and other procedures, may be used to confirm a diagnosis in about three days. A specific treatment for the disease is not known. Hospitals treat all suspected cases with barrier nursing techniques, which include the use of protective gowns, gloves, masks, and strict isolation of the patient.

Ebola hemorrhagic fever caused a media frenzy in 1976 when it emerged. The disease, named after a river in the Democratic Republic of the Congo (formerly Zaire) is caused by an RNA virus. Since then, there have been a few outbreaks involving identified subtypes: Ebola-Zaire, Ebola-

Sudan, and Ebola-Ivory Coast. The fourth, Ebola-Reston, appeared in Reston, Virginia, among cymolgous monkeys (from the Philippines) in a research facility. Several workers were infected with the virus but did not become ill.

These viral hemorrhagic fevers (VHFs) are frightening because their origins are unknown and they receive a lot of media attention. These viruses share several common features. VHFs are RNA viruses, which means that they are composed of simple pieces of RNA covered with a tough lipid coat. They have a number of insect hosts; while humans are not the native host, once humans acquire the virus they may pass it on to others. (Contaminated needles or contact with any body fluid may spread the disease.) The emergence of these viruses cannot usually be predicted or controlled. Because more and more people are travelling, these diseases have appeared in places where they have never been seen before. In the first, or index case, the virus is spread through direct contact with blood or body secretions. The symptoms are similar to other diseases, especially at the beginning. Scientists do not know why some people live while most die. Since there is no cure, medical personnel treat the symptoms with transfusions of electrolytes or, sometimes, blood.

Other VHFs have emerged. The Machupo virus, for example, caused hundreds of deaths in Bolivia. In two towns in Nigeria—Lassa and Jos—both natives and foreigners died from the Lassa virus. In fact, this virus was so lethal that researchers in Yale University laboratories stopped work on it.

While many of the emerging diseases have not been traced, hantavirus pulmonary syndrome (HPS) has been found to be carried by rodents, especially the deer mouse. The disease, which appeared in the American Southwest, was traced to exposure to rodent droppings; after one to six weeks, flu-like symptoms with coughing appear. Hospitalization is required but, at present, there is no cure.

Insect vectors may also carry diseases. Lyme disease, for example, is caused by a bacterium carried in the bite of the deer tick. The disease is named after the placed in which it was identified, Lyme, Connecticut. A tell-tale red bull's eye indicates an infected area. If left untreated, the disease may cause serious arthritis-like symptoms. A vaccine for Lyme disease has been developed and approved by the Food and Drug Administration.

The word "encephalitis" means "inflammation of the brain." Many cases of the condition are caused by viruses transmitted by mosquitoes. St. Louis encephalitis (SLE) is found in areas such as Florida where mosquitoes are present. Another example of encephalitis involved a West Nile-like virus in the New York area. Ticks have been found to carry this condition in Asia and Africa but the transmission in New York was probably from mosquitoes feeding on birds infected with the virus. People who live in an area at the time of an outbreak are advised to stay indoors during the early evenings, cover their arms and legs, and apply insect repellent.

Cyclospora, a tiny, one-celled parasite, causes a disease first reported in 1979. Cases began to increase in the mid-1980s. One scare in the 1990s that was related to contaminated raspberries was called a "yuppie" condition because it was traced to luxury hotels that served gourmet dishes. Cyclospora infects the small intestine and causes diarrhea, low-grade fever, and fatigue.

Emerging diseases may be caused by new types of agents, such as prions, which are abnormal proteins that are less soluble and more resistant to enzymes than normal proteins. Prions, although still not completely understood, are known to cause bovine spongiform encephalopathy (BSE), a fatal neurological disorder in cattle sometimes called mad cow disease. BSE has been traced to scrapie, a similar disease found in sheep. Cattle who were fed contaminated sheep meat and bone developed BSE. A panic occurred in September 1997, when 168,000 cases of BSE were found in British cattle. Although the infected cattle were subsequently slaughtered, people worried that many cattle could still harbor the disease. They feared that eating meat from BSE-infected cattle could lead to Creutzfeld-Jakob disease, a rare but always deadly neurological condition. There is no proof that this could happen, but scientists are studying the link, if any, between the two diseases.

Vaccines do not exist for the vast majority of diseases. Scientists hope that when full sequences of viral DNA are deciphered, they be will able to develop prevention and treatment methods. In 1995 scientists completed the first genetic code for a bacterium, *Hemophilus influenzae,* a major cause of ear infections. By sequencing the genome and comparing it to others, researchers may locate what makes some strains harmless and others virulent. Likewise, this research could lead to the development of antibiotics. Researchers, however, believe sequencing activities have the greatest potential for victory against emerging diseases.

As the twenty-first century arrives, scientists are realizing that our knowledge of disease is never static and that emerging and changing forms of bacteria and viruses are always with us. Add to this equation new diseases such as AIDS and old plagues such as tuberculosis and it is evident that science still has much work to do.

EVELYN B. KELLY

Further Reading

Dubos, Rene. *The White Plague: Tuberculosis, Man, and Society.* Boston: Little Brown, 1952.

Garrett, Laurie. *The Coming Plague: Newly Emerging Diseases in a World Out of Balance.* New York: Viking Penguin, 1995.

Kolata, Gina. *Flu: The Story of the Great Influenza Pandemic of 1918 and the Search for the Virus That Caused It.* New York: Farrar, Straus, and Giroux, 1999.

Olshaker, Mark. *Virus Hunter: Thirty Years of Battling Hot Viruses Around the World.* New York: Doubleday, 1998.

Roueche, Berton. *The Medical Detectives.* New York: St. Martins, 1991.

Infant Mortality

Overview

Infant mortality is defined as the death of an infant between birth and one year of age. Sociologists often look at a nation's infant mortality rate to determine that particular nation's general state of health. International statistics show that the world's industrialized nations have a lower infant mortality rate than that of poorer nations. Infant death may result from a number of causes—stemming from congenital and environmental factors as well as poor diet and medical care—but premature birth is among the most common.

Background

A careful look at individual nations with lower rates often reveals a distinct difference among that nation's ethnic groups; wealthy white babies have a much lower mortality rate than do poor blacks, for instance. Poor mothers have less access to health care and thus have more premature babies, who are then at a higher risk of disease and death. It is every country's goal to improve its infant mortality rates so that every baby born has an equal chance of survival.

Among the countries of the world, the less industrialized nations have high infant mortality rates of about 100 per 1000 births. Western countries that provide good health care programs have lower rates, between 4 to 15 per 1000 births. Japan leads the world with the lowest infant mortality rate of 4 per 1000; Canada and Italy follow with 6 per 1000. In the United States, the rate is about 8.4 per 1000. Poor nations in Asia and Africa have appallingly high rates. For example, Afghanistan's is 147 per 1000, and Sierra Leone's is 133 per 1000. An interesting exception is the impoverished country of Cuba, which has a good infant mortality rate of 9.4 per 1000.

Within Western nations, the rates of infant mortality vary considerably. Different factors explain these rates: economic status and education are among the most important. Women who receive little or no prenatal care are at greater risk of complications during their pregnancies—both to themselves and to their babies. These women have a higher maternal death rate, and their babies have a higher infant mortality rate than mothers who receive adequate medical care during their pregnancies. If the baby does survive, there is more risk of its being premature and/or seriously sick.

A troubling gap exists between black and white infant mortality rates. Even black women who receive prenatal medical care have more premature babies and more low birth weight babies. The black infant mortality rate has been consistently higher than that of whites. Rates for both ethnic groups have declined since 1950, but the rate of white babies has declined almost twice as quickly as that for black babies. In 1950 the infant mortality rate was about 30 per 1000 for white babies, and 50 per 1000 for black babies. Though the figures for both blacks and whites declined dramatically to 15.7 for blacks and 7.4 for whites by 1991, the gap between the two groups had actually widened since 1950 and is not expected to decrease before the year 2010.

A major reason for the drop in infant mortality in recent years is better prenatal care. One part of this care is the use of fetal monitors during pregnancy and labor. Ultrasound helps the physician measure the size and growth of the fetus, and to determine if a problem exists, such as placenta previa, which might necessitate delivery by cesarean section. Monitors also can predict how the fetus will react during labor and if it can survive the stress of a natural delivery. During labor itself, an electrode can be attached to the baby's scalp; if the infant appears to be in distress, the doctor may decide to perform an immediate cesarean section.

Impact

The causes of infant mortality are several, such as certain congenital problems, premature babies and their low birth weights, and sudden infant death syndrome (SIDS). Researchers discovered that babies who sleep on their backs are much less likely to die from SIDS, with the result that the rate of SIDS has dropped by almost 40 percent by 1996. Of all the causes, however, premature births remain the major cause of infant mortality.

The number of premature babies has increased with the higher number of multiple births, which are caused by popular fertility treatments. If a baby is born before 37 weeks of gestation, it is known as a premature baby or a "preemie." These babies have not reached their full birth weight and can weigh as little as 800 grams or about two pounds. A female fetus matures more quickly than a male, so if a premature baby is a girl, she has a better chance of survival than a boy. All preemies have serious problems because their tiny organs are not fully developed. Premature infants need the special care provided by a neonatal intensive care unit where the baby is kept warm, fed, and protected in the proper environment.

An important part of the neonatal intensive care unit (NICU) is the incubator, a special bed or chamber that is kept at a constant temperature of 31° to 32° C (88° to 90° F). This is vital for premature babies because they lack sufficient body fat to keep warm. Incubators have been used since the latter part of the nineteenth century and helped premature babies survive. Today, neonatal intensive care units use both incubators and radiant warmers. Incubators are made of clear plastic and completely enclose the newborn infant, maintaining a certain temperature and level of humidity. A radiant warmer also keeps the baby warm but is completely open on top so that specialists can have easy access to the baby.

Despite the efforts of the neonatal specialists, the tiny baby still faces many health problems. The infant can develop jaundice and have serious respiratory problems. Jaundice causes the skin and eyes to be yellow; it is a common and easily treatable condition resulting from high levels of bilirubin. The premature baby's liver is not yet able to process the bilirubin, so it accumulates in the blood. If the levels are too high, the baby can suffer brain damage. Therefore, the neonatal staff closely monitors a premature baby's bilirubin levels and places the baby under special lighting to help eliminate the bilirubin. In an extreme case of jaundice, the baby may need a blood transfusion.

The most serious problem of the premature infant is breathing. The baby's lungs require a chemical, called surfactant, to function properly. If a premature infant, however, does not have sufficient surfactant in its lungs, the baby develops respiratory distress syndrome. Doctors try to measure the baby's level of surfactant even before it is born.

A test can be performed on the expectant mother if the doctor suspects a premature delivery. This test, an amniocentesis, will determine the level of surfactant in the lungs of the fetus. In an amniocentesis, a needle is injected through the mother's abdomen into the uterus; a sample of the amniotic fluid is taken and checked for surfactant. If the level is low, the doctor will inject the mother with a corticosteroid, which works in about 24 hours.

If the premature baby with respiratory distress syndrome is about to be born, however, there is no time for the injection. After the birth, the infant is placed in the neonatal intensive care unit, and a mechanical ventilator helps the baby breathe. The ventilator, or inhaler, administers the artificial surfactant and helps the baby's tiny lungs function more normally. The nurses continue to monitor the baby's levels of oxygen and carbon dioxide.

Before the NICU was developed in the 1960s, premature babies were treated with an excessive amount of oxygen to help their underdeveloped lungs. The result is now thought to have caused blindness by making the baby's retina become loosened. Medical researchers agree, however, that oxygen levels only partially explain the damage to the eyes of a premature infant. Babies in the NICU are given regular eye exams to monitor the blood vessels in their eyes. Doctors have turned to laser surgery on preemies in an effort to prevent detached retinas.

Until the late 1960s, hyaline membrane disease was another respiratory problem that was a major cause of death in premature babies. Doctors found that using artificial surfactant helped the infant's tiny lungs function more properly. The last two decades of the twentieth century saw the life expectancy of premature babies increase dramatically, due mainly to the use of surfactant and the NICU. Nevertheless, respiratory distress syndrome remains the major problem for babies born before the 28th week or weighing three pounds or less.

Various factors are responsible for premature delivery. The mother's lifestyle, for example, can definitely affect when her baby is born; smoking, drinking, and a poor diet can all cause her to go into labor early. In addition, there are also physical reasons for premature delivery, such as a mother's illness, infection, or a uterine abnormality. Access to prenatal medical care is also important.

Medical research and the development of the neonatal intensive care unit have allowed more premature infants to survive, but doctors have not yet discovered conclusively how to prevent babies from being born too small or too early. Many babies, about one-tenth of all babies born in the United States, have either a low birth weight or are pre-term. Researchers do know that smoking among pregnant women causes 20 percent of low birth weight babies, a factor that could be prevented. However, they have not yet determined how to prevent the other 80 percent of small babies being born or the specific causes of pre-term births.

The disturbing fact remains that African-American infants are twice as likely to be either pre-term or of low birth weight than either Asian-American or white babies. Until society addresses this discrepancy, the United States will continue to rank far behind many other countries in infant mortality.

M. E. ELGHOBASHI

Further Reading

Books

Clayman, Charles B., ed. *The American Medical Association Family Medical Guide.* 3rd ed. New York: Random House, 1994.

Eitzen, D. Stanley and Maxine Baca Zinn. *Social Problems.* Boston: Allyn and Bacon, 1994.

Oakley, Ann. *The Captured Womb: A History of the Medical Care of Pregnant Women.* New York: Blackwell, 1984.

Shapiro, Sam, Edward Schlesinger, and Robert Nesbitt, Jr. *Infant, Perinatal, Maternal, and Childhood Mortality in the United States.* Cambridge: Harvard University Press, 1968.

Periodical Articles

Singh, Gopal K., and Stella M. Yu. "Infant Mortality in the United States: Trends, Differentials, and Projections 1950 through 2010." *The American Journal of Public Health* 85, No. 7 (July 1995): 957-64.

The AIDS Pandemic

Overview

Acquired Immune Deficiency Syndrome (AIDS) was first identified in 1981. This complex disease is characterized by the breakdown of the body's immunologic defense system, which results in vulnerability to many normally harmless microorganisms. The causal agent, a human retrovirus now known as the human immunodeficiency virus (HIV), was discovered in 1984. Despite the growing toll of the global AIDS crisis and the absence of a preventive vaccine as well as the lack of safe and affordable therapeutic drugs, by 1995 epidemiologists in the United States were warning policy makers and the public about a growing complacency towards AIDS.

Background

AIDS first appeared as a diagnostic entity in 1981, when the Centers for Disease Control (CDC) began to report strange clusters of rare illnesses, such as *Pneumocystis carinii* pneumonia (PCP) and Kaposi's sarcoma, in previously healthy homosexual men in New York and California. These diseases were associated with a severely compromised condition of the immune system. Within five years of the first re-

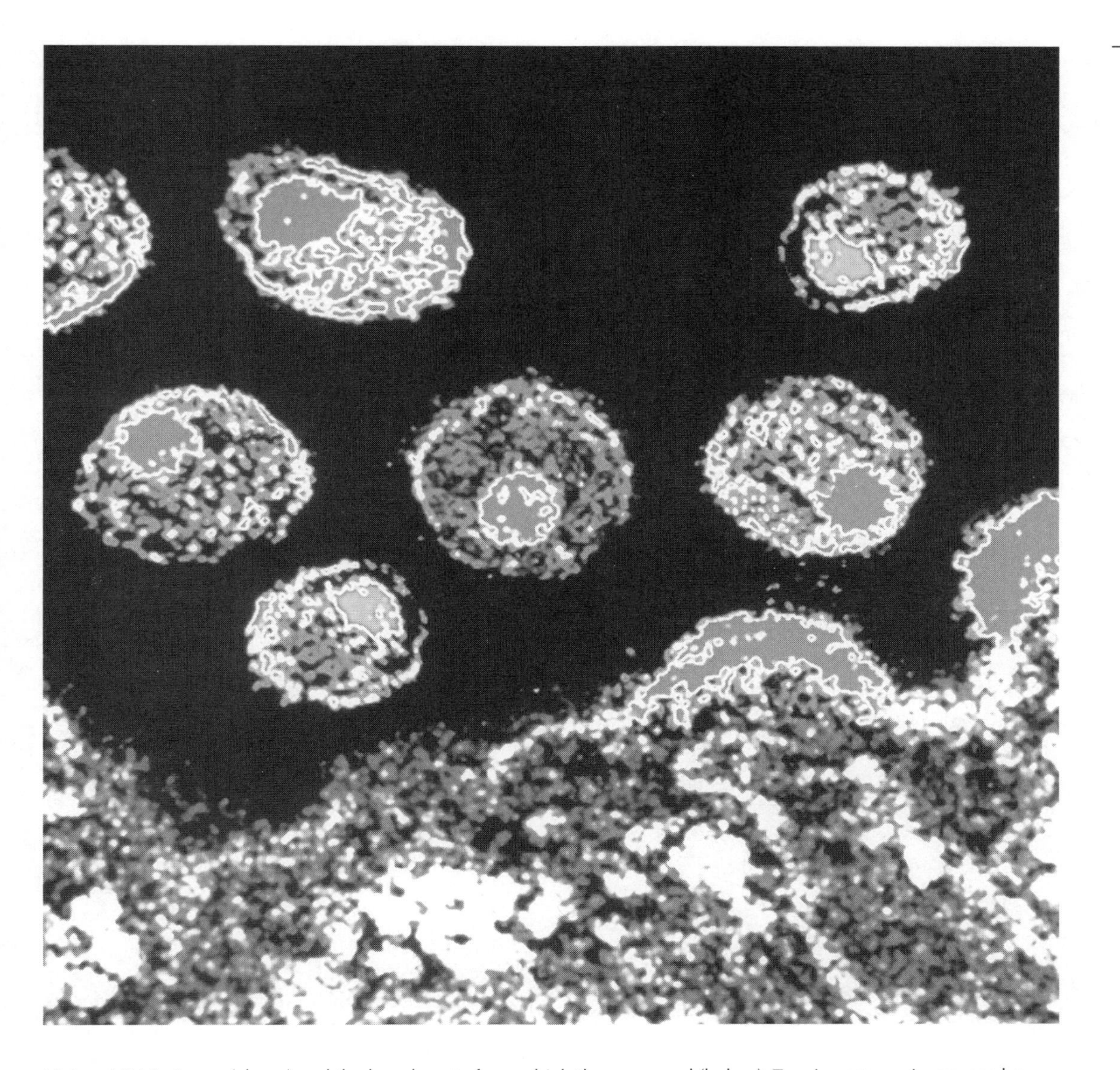

Mature HIV-1 viruses (above) and the lymphocyte from which they emerged (below). Two immature viruses can be seen budding on the surface of the lymphocyte (right of center). *(Scoot Camazinr, National Audubon Society/Photo Researchers. Reproduced with permission.)*

ports, the U.S. Public Health Service estimated that more than one million Americans were infected with HIV. By the end of 1994, the CDC had gathered reports of over 400,000 cases of AIDS, including 170,870 AIDS-related deaths. In the United States, AIDS quickly became one of the leading causes of death in adults aged 26 to 44. The disease also increased rapidly in Africa, South America, Western Europe, and Asia. Studies published by the World Health Organization (WHO) indicated that within ten years of the discovery of AIDS, five to ten million people were infected with HIV. In contrast to the pattern in the United States, in Africa AIDS was first recognized in sexually active heterosexuals, and AIDS cases in Africa have occurred at least as frequently in women as in men. Overall, the worldwide distribution of HIV/AIDS between men and women is approximately 1 to 1.

The human immunodeficiency virus (HIV), the retrovirus that causes AIDS, was discovered in 1984. Robert Gallo (1937-), at the National Institutes of Health, and Luc Montagnier (1932-), at the Pasteur Institute, are generally regarded as the discoverers of the virus. A test to identify AIDS antibodies, and thus screen for the disease, was developed at about the same time. Studies of previously stored blood samples suggest that the virus probably entered the U.S. population sometime in the late 1970s.

Retroviruses have genes composed of ribonucleic acid (RNA) molecules, rather than deoxyribonucleic acid (DNA). Although all viruses only can reproduce inside cells, retroviruses must use an enzyme called reverse transcriptase (RT) to convert their RNA into DNA. Because HIV RT makes many mistakes while making DNA copies from HIV RNA, many variants of the virus develop in an infected individual. A viral enzyme called HIV integrase splices the viral DNA into the host cell's DNA. When the viral DNA is incorporated into the cell's genes it is known as a "provirus." HIV belongs to a subgroup of retroviruses known as lentiviruses, or "slow" viruses. The interval between the initial infection and the onset of serious symptoms is generally quite long and variable. Eventually, immature viral particles form and acquire an envelope. In order to create infectious viral particles, an enzyme called protease has to cut long chains of proteins in the immature viral core into smaller pieces. HIV replicates so rapidly that several billion new virus particles can be produced every day. Some immune cells become infected but are not destroyed by the virus. Such cells can carry HIV to various organs, including the lungs and brain. Debilitating weight loss, diarrhea, neurologic conditions, and cancers, such as Kaposi's sarcoma and lymphomas, characterize the terminal stage of AIDS.

The time interval between infection with HIV and the development of AIDS-related symptoms is quite variable, sometimes as short as two years but usually about 10 to 12 years. Although the evidence that HIV is the etiological (causal) agent for AIDS appears to be very strong, some critics have proposed alternative explanations for the AIDS epidemic. One of the most vocal and prominent of these critics is Peter Duesberg, professor of molecular and cell biology at the University of California at Berkeley and author of *Inventing the AIDS Virus.* Nevertheless, increasingly sensitive testing methods, including the polymerase chain reaction (PCR), have strengthened the link between HIV infection and the development of AIDS. By the second decade of the AIDS crisis more than 1,000 journal articles and books on AIDS and HIV were being published every month. The National Library of Medicine dedicated AIDSLINE and other medical databases to writings on the disease and published a monthly *AIDS Bibliography.*

Impact

By 1999, the number of people worldwide living with HIV/AIDS was estimated at 33 million; almost 70% of these people lived in Sub-Saharan Africa. Through 1997, cumulative HIV/AIDS-associated deaths worldwide numbered approximately 12 million. By the end of 1998, the estimate for cumulative HIV/AIDS-associated deaths worldwide was about 13.9 million: 10.7 million adults and 3.2 million children. Africa has been particularly burdened by AIDS. In some African countries, the rate of HIV infection among adults aged 15 to 49 exceeds 20%.

When AIDS was first recognized in 1981, patients generally died within a year or two of the diagnosis. By 1995, studies of AIDS and HIV helped scientists develop more effective ways of combining anti-viral drugs and provided the basis for the development of new anti-HIV drugs that have prolonged the life span and enhanced the health of HIV-positive individuals. Basic research concerning the mechanism by which the virus attacks the immune system indicated that suppressing HIV replication could delay or prevent the progression of the disease.

The drugs originally used to treat HIV infection were nucleoside analogue reverse transcriptase (RT) inhibitors, which act by interfering with the action of HIV reverse transcriptase. The first of these drugs, zidovudine (AZT), was originally developed in 1964 as a possible cancer treatment. In the early 1980s AZT was found capable of suppressing HIV replication in the test tube; this discovery led to clinical trials of AZT and other nucleoside RT inhibitors, including zalcitabine (ddC), didanosine (ddI), stavudine (D4T), and lamivudine (3TC). HIV, however, mutates rapidly and quickly develops resistance to anti-HIV drugs. Researchers discovered that using combinations of drugs helped overcome this problem.

Although the drugs used to treat AIDS are known to have serious side effects and are very expensive, most authorities believe that their benefits outweigh the dangers. Critics, however, continue to raise doubts about the cause of AIDS as well as the safety of the drug regimens used to treat the disease. For example, South Africa's President Thabo Mbeki asserted in a speech before the South African parliament in October 1999 that AZT was unsafe. (South Africa has one of the highest incidences of AIDS cases; about 8% of the population, some 3.6 million people, are estimated to be HIV positive.)

AZT is one of the oldest and best known of the drugs used in the treatment of AIDS and has also been given to infected women to prevent the transmission of HIV to their babies during birth. By 1999, four protease inhibitors—saquinavir, ritonavir, indinavir, and nelfinavir—had been approved for marketing in the United States. Combinations of protease inhibitors and other categories of drugs can dramatically reduce the levels of HIV in the blood, although the virus appears to persist in certain cells. Other insights into the mechanisms of HIV replication suggest new targets for anti-HIV drugs as well as new ways to fight other diseases, such as hepatitis, influenza, and cancer.

Unfortunately, many patients cannot tolerate the available drug regimens. Moreover, most of the 30 million people worldwide who are infected with HIV cannot afford the drug regimens currently showing such promise in the United States. Given the expense and complexity of anti-HIV therapies, the development of a safe and effective vaccine for AIDS remains a public health priority throughout the world. At the end of the twentieth century, researchers continue to develop and test novel experimental HIV vaccines as well as vaccine adjuvants (compounds that enhance the immune response to vaccines).

Many social critics have argued that AIDS has exerted a profound influence on politics, science, art, the health care system, and the legal system as well as on biomedical research and resources. AIDS activists have succeeded in getting more research funding dedicated to AIDS research; they have changed the nature of clinical trials for AIDS drugs, asserted a right to privacy, called for nondiscrimination in the workplace and health care system, and renewed the debate about "pure research" vs. "mission-oriented research." Their successes have led those concerned with other diseases to adopt similar tactics. AIDS activists have demanded a more rapid review and approval process for new AIDS drugs, but the problem of evaluating quack remedies has become even more acute under current circumstances. While many critics argue that AIDS has had an adverse impact on American research priorities, the AIDS crisis is distinct because it is enmeshed in many complex social problems, including substance abuse, prostitution, homelessness, poverty, privacy issues, the legal rights of infected individuals, resource allocation, and the deteriorating infrastructure of our public health system.

Further studies of HIV have suggested that the virus had probably been incubating as a silent epidemic in areas of the world where the deaths of children and young adults from fevers and diarrheal diseases were not at all uncommon. AIDS has made it clear that the most powerful antibiotics are ultimately powerless against the onslaught of microbial agents if the body's own immunological defenses cannot participate in the battle. The AIDS pandemic may, therefore, stimulate renewed interest in public health medicine and the need for preventive immunizations. The fears generated by AIDS as well as the recognition of the general threat posed by newly emerging viruses may also reverse the tendency of wealthy nations to assume that infectious diseases have been conquered and that previously obscure third world diseases are inconsequential.

LOIS N. MAGNER

Further Reading

Books

Bellenir, Karen, ed. *AIDS Sourcebook.* Detroit, MI: Omnigraphics, 1999.

Cohen, P.T., et al., eds. *The AIDS Knowledge Base: A Textbook on HIV Disease from the University of California, San Francisco and the San Francisco General Hospital.* Boston, MA: Little, Brown, 1994.

Duesberg, Peter. *Inventing the AIDS Virus.* Lanham, MD: Regnery Publishing Inc., 1996.

Epstein, Steven. *Impure Science: AIDS, Activism, and the Politics of Knowledge.* Berkeley, CA: University of California Press, 1996.

Grmek, Mirko D. *History of AIDS: Emergence and Origin of a Modern Pandemic.* Princeton, NJ: Princeton University Press, 1990.

Hannaway, Caroline, et al., eds. *AIDS and the Public Debate: Historical and Contemporary Perspectives.* Amsterdam: IOS Press, 1995.

Kiple, Kenneth F., ed. *The Cambridge World History of Human Disease.* New York: Cambridge University Press, 1993.

Mann, Jonathan, et al., eds. *A Global Report: AIDS in the World.* Cambridge, MA: Harvard University Press, 1992.

Schoub, B.D. *AIDS & HIV in Perspective: A Guide to Understanding the Virus and Its Consequences.* New York: Cambridge University Press, 1999.

Stine, Gerald J. *Acquired Immune Deficiency Syndrome: Biological, Medical, Social, and Legal Issues.* Upper Saddle River, NJ: Prentice Hall, 1998.

Other

NIAID Home Page. "HIV/AIDS Research Agenda and Fact Sheets." http://www.niaid.nih.gov.

Medicine and Women: 1950-present

Overview

Since the 1950s, women's preventative medicine and health care have greatly improved in the developed world, particularly through the availability of new diagnostic techniques. The use of Pap tests and mammography has become standard methods for the early detection of potentially life-threatening diseases in women. In addition, the increasing participation of women in clinical studies and the greater number of women in the medical field have served to promote women's health, directing attention to medical ailments and issues specific to women.

Background

The Pap test, developed by G. N. Papanicolau in 1943, would prove one of the most important diagnostic techniques for women. This test was designed to enable small tissue samples to be scraped from the cervix for analysis. Any abnormal cells that indicate the possibility of cervical cancer or other uterine problems alert the doctor and patient to the need for careful monitoring and perhaps intervention.

Another major breakthrough for screening was the development of mammography. Albert Salomon studied breast cancer with x-rays in 1913 in Berlin, and in the 1930s Jacob Gershon-Cohen extended these studies at Jefferson Medical College in Philadelphia. These techniques were further developed in studies at the UCLA School of Medicine during the 1940s. In 1949 Raul Leborgne, working in Uruguay, introduced the idea of compressing the breast in order to enhance the image obtained from the x-rays. In 1951 Charles Gros developed the first radiological unit specifically for mammograms. The Compagnie Generale de Radiographie introduced Gros's Senograph machine to the meeting of the Radiological Society of North America in Chicago in 1967, and since then they have been marketed widely round the world.

Medical experimentation prior to 1950 typically excluded female participants. Such testing, during which drugs and other therapies are tried on samples of animals and humans to test their effectiveness and safety, was made complicated by ethical issues of informed consent and willing participation. Some medical researchers did not explain the research to the people they wanted to carry it out on. Some of the subjects were so young, or old, or ill, or mentally incapacitated that they would not have been able to give their consent based on full understanding.

These practices took place all over the world, including Europe and the United States. They reached their worst levels of abuse in the experiments carried out by Nazi doctors on Jewish and other concentration camp prisoners. These experiments were uncovered and many of their perpetrators were punished in the Nuremberg Trials after the Second World War. The Nuremberg Code was subsequently drawn up to establish ethical principles to prevent such abuses from happening again. This code was followed by other regulatory statements in the United States and elsewhere during the 1950s and 1960s. But dubious experimental practices still continued, and many important medical discoveries were made at the expense of mentally incapacitated children or prison inmates.

Impact

More than five decades after its development, the Pap test is now offered routinely to most women beginning in their late teen years, especially in countries that can afford it. This screening program has reduced the numbers of deaths from these conditions. In addition, mammogram technology has also improved. Better x-ray film has reduced the amount of radiation needed to "see through" the skin to the breast tissue to detect abnormalities. In the United States the major medical insurance companies and cancer prevention organizations have long supported universal mammogram screening for women over the age of 50 in order to reduce the numbers of deaths from this cancer.

There has been some debate about whether this screening program detects enough early cancers that can be cured to justify its costs. Some critics even suggest that the mammograms can themselves cause cancer because they use radiation. By the end of the century, ultrasound—which does not rely on radiation—was being used more widely for mammography. One unresolved problem with mammography is that it still requires health professionals to read or interpret the results, and the human eye doesn't always see or understand the information available on the x-ray, leading occasionally to failure to diagnose the disease in time for it to be effectively treated. The American Board of Radiology has only had a specific examination section on reading mammograms since 1990.

Over the last 50 years women have also insisted on participating more fully in clinical trials. It had long been thought important to protect women from becoming involved in such clinical trials, especially when they were still of child-bearing age or actually pregnant. This was to protect them and their children from unknown and unanticipated side effects of the drugs or other procedures that they might be subjected to. However, one consequence of this protectiveness was that medicines were developed based on experiments on men only. Even where the illnesses are the same, men and women can often react differently.

Yet, the drugs and dosages prescribed, as well as other aspects of new treatments, were developed and tailored according to the needs of men rather than both men and women. Even if women are not pregnant or breast-feeding, their body chemistry is different from that of men, especially in the field of hormones and menstruation. Many researchers found it too complicated to allow for the fluctuations of women's body chemistry and simpler to work with males. Similarly, women age differently than men and the role of menopause in their health was not properly considered in much of this medical research.

In the 1970s there began to be a reaction to this protectiveness, which was often an excuse for not addressing the separate issues involving women, and some women argued that men's health was receiving more research attention than women's. The issue came to the fore with the emergence of AIDS in the 1980s. Researchers have since attempted to find a cure or prevention for this disease, and many sufferers have been desperate to be involved in clinical trials that might lengthen their lives.

Women too suffer from AIDS, and they did not appreciate being excluded from these experiments, despite the risks. Many of the same problems of exclusion from clinical trials were felt by ethnic minorities in the United States and other countries. Together, with the women who were concerned about access to clinical trials, they fought some important legal cases to ensure their inclusion. In addition to legal victories, in 1991 the first woman director of the National Institutes of Health in the United States initiated a major long-term study of women's health, including research on the effects of smoking, exercise, hormone therapy, diet, and other factors on the incidence of osteoporosis (brittle bones), heart diseases, cancers, and other illnesses.

Another major area of change in women's health and medicine over the past 50 years has been the growing presence of women in the workforce. Women have entered into all professions, including medicine and health care. Previously relegated to auxiliary roles as nurses and paraprofessionals, women have since gained full entry to medical schools in the United States and Britain, where enrollments are commonly now at least half female.

As well as influencing the work patterns of medical doctors at the end of the twentieth century, the presence of women as physicians is changing the practice of medicine. There has been research to show that women behave differently as doctors than their male colleagues, and have different priorities in treating their patients. Certainly within medical schools in America and Europe there is a growing realization that the way doctors are examined for competence is changing because of the high numbers of women students. These changes are paralleled in the world of medical and scientific research, which are also seeing more women workers, though many still feel that there is a "glass ceiling" limiting their opportunities for promotion.

With the development of hormone replacement therapies many Western women are enjoying longer working lives, with some even beginning to work only after their children are old enough to work. The availability of more effective forms of contraception, even in less developed countries, has enabled women to exercise more control over the timing and frequency of their pregnancies.

Certainly, old paternalistic attitudes among doctors are less readily accepted in countries where patient charters and patient education have reached new levels of awareness and understanding through media exposure. The challenge of the next millennium is to ensure that these advances in women's health are shared equitably around the world, not least because healthy women produce healthy children.

SUE RABBITT ROFF

Further Reading

Laurence, Leslie, and Beth Weinhouse. *Outrageous Practice: The Alarming Truth About How Medicine Mistreats Women.* New York: Fawcett, 1994.

Litt, Iris. *Taking Our Pulse: The Health of American Women.* Stanford: Stanford University Press, 1997.

Northrup, Christine. *Women's Bodies, Women's Wisdom: Creating Physical and Emotional Health and Healing.* New York: Bantam Doubleday Dell, 1998.

Reichman, Judith. *I'm Too Young to Get Old: Health Care for Women After 40.* New York: Random House, 1998.

Seaman, Barbara, and Gary Null, eds. *For Women Only: Your Guide to Health Empowerment.* New York: Seven Stories Press, 1999.

Development of Prenatal Diagnostic and Surgical Techniques

Overview

Since 1950, the development of prenatal diagnostic and surgical techniques has transformed the field of obstetrics. Parents are now able to access detailed information about their children before birth. Birth defects are now not only discernable, but sometimes correctable before birth, resulting in a lower infant mortality rate and a healthier population of children. There is some concern, however, that this technology could be misused.

Background

Since the early 1970s, prenatal screening has become a routine part of a pregnant woman's medical care. Certain tests, including ultrasound and amniocentesis, not only diagnose a disorder in the unborn baby, but also give the mother the opportunity to have an abortion if she learns, for example, that her baby has Down syndrome and will thus be mentally impaired. Medical researchers, however, were not content merely to detect birth defects in the fetus; they then looked for ways to help the unborn baby. This search led to the first successful prenatal surgery at the University of California at San Francisco. Today, there are several centers in the United States and other countries that specialize in fetal surgery to correct birth defects that were discovered in prenatal testing.

Prenatal screening and testing are done at different stages in a woman's pregnancy. During the first trimester, the pregnant woman has a blood test to measure certain substances in her blood. Specific amounts of these substances could indicate if the fetus has Down syndrome. Between the 14th and 22nd week, during the second trimester, another blood test can tell if the fetus has neural tube defects.

Even if the results from the blood tests are positive, the fetus may not necessarily have a problem. Additional testing is done to ascertain whether or not the unborn baby has Down syndrome, spina bifida (a neural tube defect), or another problems. Although tests such as ultrasound and amniocentesis cannot diagnose every problem, they are nevertheless an important part of prenatal care.

The original concept of ultrasound dates back to the time when the Titanic sunk, and a British scientist developed a method to search for icebergs. His technique was then used during World War I to detect submarines. Also called sonar (for sound navigation and ranging), ultrasound is still used as a navigational tool by ships all around the world. Adapted for medical use after World War II, ultrasound became an important part of prenatal care by 1970. Ultrasound uses high frequency sound waves that are above the range of human hearing; tissues absorb some of the waves, and other waves are reflected back to form an image on a special type of television screen.

Impact

Because doctors learn so much from the reflected image of the fetus, ultrasound is now the most common prenatal test. More than 80 percent of pregnant women in the United States receive an ultrasound test between the 10th to 13th week of pregnancy. Doctors use ultrasound not only to diagnose a pregnancy, but also to determine the fetus's age and sex, observe a multiple pregnancy, or to detect certain abnormalities, such as an ectopic pregnancy. (Also called tubal pregnancies, ectopic pregnancies are serious conditions that can result in severe cramping and internal bleeding; they occur when the fertilized egg develops outside the uterus, often in a fallopian tube.) Ultrasound is also an invaluable tool during fetal surgery to guide the physician as he or she operates on the tiny fetus.

Another important prenatal test is amniocentesis, available to women whose baby may be

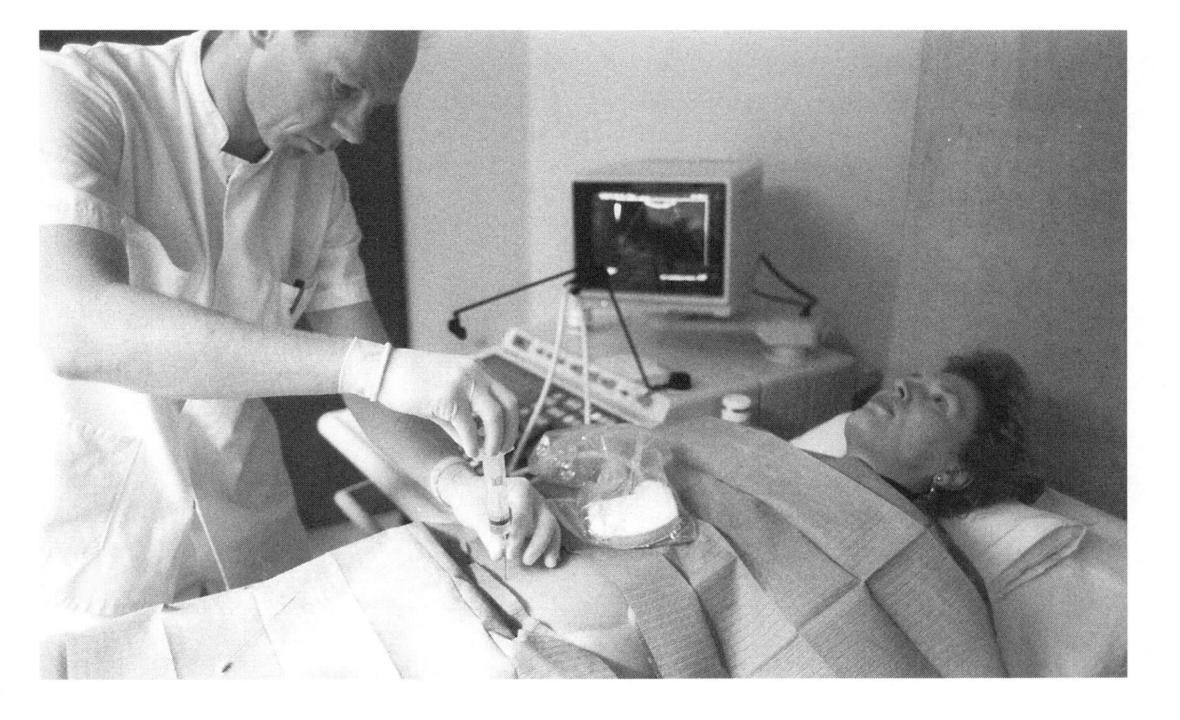

Amniocentesis, first developed in 1952, remains a crucial prenatal diagnostic technique. *(OwenFranken/CORBIS. Reproduced with permission.)*

at a high risk of a chromosomal abnormality. The early pioneer of amniocentesis was Douglas Bevis, whose work was done in the early 1950s. Unlike ultrasound, amniocentesis can determine if the fetus has either spina bifida or Down syndrome. To perform amniocentesis, the doctor uses ultrasound to guide a needle into the abdomen of the pregnant woman between the 16th and 18th week of her pregnancy. A sample of the fluid that surrounds the fetus (the amniotic fluid) is taken. Tests are then performed on the cells within the fluid to determine if the fetus has a chromosomal disorder.

Two other prenatal tests are cordocentesis and CVS. Cordocentesis is performed after 20 weeks of pregnancy; blood is taken from the fetus by way of the umbilical cord. CVS, or chorionic villus sampling, was developed by Dr. Evans at Wayne State University and is performed between the 10th and 12th weeks; this procedure uses a catheter to remove only a small sample of the chorionic villi from the placenta. Both cordocentesis and CVS help determine if the unborn baby has a genetic disorder.

Prenatal testing is not risk-free, however. Amniocentesis, cordocentesis, and CVS carry a miscarriage rate between 1 and 3 percent. Additionally, CVS can even cause certain birth defects. Therefore, the mother must be informed of the inherent risks and sign a consent form allowing the doctor to proceed with these prenatal tests.

With the advances in prenatal testing, doctors and their patients could literally see if the unborn fetus had any problems. But until about 1990, the mother and her doctor had few options if a test, for example, revealed that the unborn baby had a debilitating condition, such as spina bifida, or a life-threatening defect as a certain type of hernia or urinary obstruction. Today, the field of fetal surgery has advanced, and treatment centers around the world offer new hope to parents and their unborn child.

Prenatal or fetal surgery had its beginnings in the 1960s in New Zealand. There, Dr. A. William Liley (1929-1983) developed a technique for prenatal blood transfusions for a fetus with a blood disease. In 1963 Liley performed his first successful transfusion on a 32-week-old fetus. The parents were so grateful that they named their son after Liley. Although this type of transfusion is no longer necessary, Liley is still considered the "Father of Fetal Surgery."

There was little progress in the field of fetal surgery for the next 20 years. By 1980, at the University of California at San Francisco and at Wayne State University in Michigan, medical researchers were working first on the fetuses of lambs and monkeys. The doctors at both universities shared their ideas and data, and were soon ready to apply their findings to human fetuses. The first open fetal surgery was performed at UCSF in April 1981, when doctors placed a catheter in a fetus with a urinary obstruction. In 1987 Dr. Mark Evans of WSU was also successful in treating a 14-week-old fetus for obstructive uropathy. In June 1989 the UCSF team performed the first successful open fetal surgery for a diaphragmatic hernia.

There are two basic approaches to fetal surgery: open or in utero. In open fetal surgery, the doctor opens the mother's uterus and operates directly on the fetus, replacing it back in the uterus when the operation is completed; the pregnancy then continues normally. In utero surgery means that the doctor operates through the uterine wall with fiber-optic instruments. Physicians choose the open approach for some conditions; however, most now prefer in utero or endoscopic surgery. The term "fetoscopy" is used now to describe fetal surgery that uses fiber-optic instruments and is thought to be a safer and easier approach than open fetal surgery. Dr. Ruben Quintero, Director of Fetal Endoscopy at Wayne State University, has vigorously pursued the use of endoscopes—light scopes and viewers—not only to diagnose, but also to treat birth defects.

At Vanderbilt University Medical Center in Tennessee, open fetal surgery is the method of choice for treating fetuses with spina bifida. An opening in the spinal cord, spina bifida leads to severe mental and physical disabilities; some babies also develop hydrocephalus or water on the brain. At Vanderbilt, doctors have successfully treated several unborn babies diagnosed with spina bifida. In 1997 Dr. Noel Tulipan and Dr. Joseph Bruner perfected the open fetal surgical approach to close the lesion on a fetus's spinal cord. They now believe that the open surgery method offers a better outcome than endoscopic surgery for the baby with spina bifida. Tulipan and Bruner performed the first fetal surgery for hydrocephalus in 1999. They placed a shunt in utero, and a healthy baby was later born.

Fetal surgery does, however, have some drawbacks. There are risks for both mother and baby. For the mother, fetal surgery could lead to infections, diabetes, or blood loss. The most common fear is the possibility of the mother's going into early labor and delivery. A premature infant is then at risk for blood transfusions, brain and lung damage, and even death.

Fetal surgery is also expensive, costing tens of thousands of dollars and requiring a cesarean operation later to deliver the baby. Also, there are fewer than 20 centers in only five or six countries that perform fetal surgery.

Doctors and researchers have made wondrous strides in recent decades in the field of fetal medicine, but their progress has also raised troubling moral and ethical questions. After an ultrasound in China and India, where male children are so highly valued, parents often choose to abort a female fetus. In the United States, there are physicians who support abortion and those who are pro-life. Will these doctors offer the same prenatal testing to their patients, even if such a test means a parent may opt for an abortion of a defective fetus? One wonders if the future of mankind will be a "master race" of physically perfect human beings. Will genetic testing mean that we will no longer tolerate a newborn who has a disease or deformity? Will the expense of fetal surgery mean that only the wealthy will have perfect babies? Doctors and parents alike must struggle with these issues as the world enters the twenty-first century.

M. E. ELGHOBASHI

Further Reading

Casper, Monica J. *The Making of the Unborn Patient: A Social Anatomy of Fetal Surgery.* New Brunswick, NJ: Rutgers University Press, 1998.

Harrison, Michael R. *Atlas of Fetal Surgery.* New York: Chapman & Hall, 1996.

Manning, Frank A. *Fetal Medicine: Principles and Practice.* Norwalk, CT: Appleton & Lange, 1995.

Nightingale, Elena O. *Before Birth: Prenatal Testing for Genetic Disease.* Cambridge: Harvard University Press, 1990.

Oakley, Ann. *The Captured Womb: A History of the Medical Care of Pregnant Women.* New York: Blackwell, 1984.

New Frontiers in Dentistry

Overview

After World War II efforts were concentrated in the public health and preventative aspects of dentistry. Communities began to fluoridate their water supplies. Regular dental examinations were encouraged, and dental hygiene efforts were expanded. Over the next half-century, technological improvements had major effects both on preventative services and reconstructive techniques.

Background

In the early twentieth century, most people went to dentists only when they had a toothache. As a result, one out of five World War II military recruits in the United States failed to meet the requirement that they have at least twelve teeth: three pairs of matching incisors and three pairs of chewing teeth. Dental problems were the most common reason for rejecting potential soldiers. Finally, the dental standards had to be eliminated altogether in order to fill the ranks.

After the war, improving the dental health of the population became a priority both in the U.S. and in Europe. Dental schools began devoting time in the curriculum to public health. The American Board of Dental Public Health was established in 1950, and the U.S. became a leader in this field. Scientific and technical advances were fostered by the establishment of the National Institute of Dental Research near Washington, D.C., in 1948.

One weapon in the war against dental caries (tooth decay) was fluoride. In the first decades of the twentieth century, people whose water supply had a relatively high concentration of fluoride ions were observed to have brown mottling on their teeth but rarely experienced dental caries.

The new field of dental epidemiology began in the 1940s with large controlled studies of fluoride in the water supply. These studies were conducted in Michigan, Illinois, New York, and Ontario. To provide a basis for comparison, Dr. H. Trendley Dean of the Public Health Service developed a measure called the DMF Index, taking into account decayed, missing, and filled teeth. Gradually, municipal water districts all over the U.S. began adding fluoride to their systems at about 1 milligram per liter, a level that does not discolor the enamel but reduces dental caries by approximately 65%.

Dentists were not alone in working to prevent tooth decay. At the turn of the century, Dr. Alfred Civilion Fones (1869-1938) had trained an assistant to thoroughly clean the teeth of children he saw in his practice. Later, he coined the term "dental hygienist" and established a training clinic. The first dental hygienist licensing law was put into place in Connecticut in1917. Many of the women who graduated from the early training programs were employed by school systems, and the children they served exhibited drastic reductions in cavities.

Impact

As dentistry became part of routine health care after World War II, the organization of dental practice changed. Many dentists began to specialize in a particular field. *Orthodontists* handle misalignments of the teeth and jaws. *Periodontists* treat gum problems. *Pedodontics* is the branch of dentistry concerned with children's teeth. *Oral surgeons* handle a wide range of procedures involving the teeth and jaws, including repair of birth defects and fractures.

The one-dentist office gave way in large part to the group practice, two or more dentists working together and sharing office facilities. Group practices have allowed extended hours to accommodate patients' busy schedules without making unreasonable demands upon the dentists' time. Often a group includes one or more specialists, allowing patients easy access to these professionals. Dental assistants work as an integral part of the team; "four-handed dentistry" makes coping with equipment and dental materials much more efficient.

The postwar period also saw the beginning of full participation in the dental profession by African Americans. After the 1880s almost all African American dentists had been trained either at Howard University in Washington, D.C., or Meharry Medical College in Nashville, both educational institutions founded to serve black students. The 1954 desegregation decision by the United States Supreme Court mandated access to all schools, and slowly the doors opened. In 1962 the American Dental Association, having established bylaws prohibiting racial discrimination, threatened to bar state chapters that refused to do likewise.

From the patient's point of view, another important change was the advent of third-party payer systems. Dental insurance plans are offered both on their own and as part of some medical insurance packages. During the 1960s and 1970s collective bargaining resulted in dental benefits being offered to the members of many unions. They can also be obtained through corporate and other group insurance programs, as well as on an individual basis. Most encourage preventative care by covering a variety of routine services such as examinations every six to twelve months, x-rays, and cleanings. In Europe several countries have instituted government-sponsored dental insurance.

Science and technology have provided important new tools for improving dental health. Once the effectiveness of fluoridated water was demonstrated, communities quickly responded. By 1962 more than 2,000 municipalities in the United States had added fluoride compounds to their water supplies. Fluoride was also added to toothpaste formulations during the same period. Today about 135 million people live in regions with fluoridated water, and others consume foods or beverages processed there. Dental caries in schoolchildren declined by about one third in the 1970s, and another third in the 1980s.

However, fluoridation is still not universal. In some areas vociferous opposition has prevented it. Concerns focus on such issues as whether excess fluoride tends to make bones more brittle, especially in the elderly. Fluoride supplements may be recommended for children in unfluoridated-water communities, in addition to those children who drink well water or bottled water.

In 1959 Dr. Frank J. Orlond of the University of Chicago uncovered the organism principally responsible for tooth decay, the bacteria *streptococcus mutans*. On a susceptible tooth, especially one owned by someone partial to sugary and starchy foods, these bacteria flourish. The acid they produce demineralizes the tooth surface. Understanding the decay process is crucial to finding more effective ways to prevent it. An anti-caries vaccine is a possibility for the future, as is genetically engineering a version of the bacteria that produces less acid. Meanwhile, plastic coatings, or sealants, are commonly painted on children's permanent molars as they come in, protecting their irregular pits and crevices for about five years.

Biomedical engineering has contributed knowledge of the mechanics of chewing. Miniature sensors can be fastened to the tooth surface to study jaw alignment, movement, and pressure. Others allow precise measurement of the degree of demineralization of the surface. Periodontists can employ tiny sampling devices to analyze fluid in the crevices between the teeth and gums. Gum disease is caused by several types of bacteria that, unlike *streptococcus mutans*, are not generally found in the healthy mouth. By detecting these early, the dentist can treat the patient with antibiotics before permanent damage has been done. Lasers are now being used in periodontal treatment as well. In 1998 a "hard-tissue" laser was introduced, suitable for working on both gums and teeth.

Digital x rays, developed in the 1990s, are slowly finding their way into dentists' offices. They provide the ability to detect smaller changes, while exposing the patient to 75% less radiation. A matchbook-sized, plastic-covered sensor goes into the patient's mouth instead of the familiar "bitewing" film holder. With no need for film developing, the image can be displayed immediately on a screen and easily manipulated.

Despite the best efforts of preventative dentistry, repair and restoration are still sometimes necessary. In 1957 the S.S. White Company introduced high-speed dental drills, driven at 300,000 revolutions per minute by compressed air. The reduced vibration of the high-speed drills made them more comfortable for the patient, while at the same time providing increased control for the dentist. Today's dental drills are of the same basic design as the 1957 model. However, the handpieces can now be steam-sterilized, and they include fiber optics to cast light directly upon the working area.

Kinetic cavity preparation, in which pressurized microscopic aluminum oxide particles blast away the decay, is an alternative to drilling developed in the 1990s. It's faster and less painful, but the equipment costs about $18,000, so it will arrive in dentists' offices very gradually. It is also unsuitable for large cavities.

If the tooth is too damaged for simple cavity repair, more drastic measures are called for. In the 1960s crowns made of porcelain bonded onto metal were introduced. Earlier crowns were made of gold with an acrylic veneer, and the veneer would eventually wear away. With the porcelain and metal crowns, natural-looking, permanent restorations could be constructed. During the 1980s dental implants began to appear on the scene as an alternate to bridgework and removable dentures. Implants involve surgery to attach dentures directly and permanently to the jawbone. About 100,000 patients underwent the procedure by the early 1990s.

Dental plastics that harden when exposed to light or special chemicals were developed in the 1960s and 1970s. They are useful for cosmetic dentistry as well as restorations. *Bonding* procedures involve etching the tooth surface with a mild acid to create tiny irregularities onto which the liquid plastic will cling. The plastic can then be built up as desired, for example to hide an unsightly tooth, repair fractures, eliminate overly large spaces between teeth, or attach orthodontic brackets without metal bands. The material itself is generally a composite, with microscopic particles such as silicates for strength. Still, it lasts only about five years and can't stand up to very hard foods. Porcelain veneers, used when bonding can't produce the desired cosmetic result, last about twice as long.

SHERRI CHASIN CALVO

Further Reading

Books

Glenner, Richard A. *The Dental Office: A Pictorial History*. Missoula, MT.: Pictorial Histories, 1984.

Hoffmann-Axthelm, Walter. *History of Dentistry*. Chicago: Quintessence, 1981.

Ring, Malvin E. *Dentistry: An Illustrated History*. New York: Harry N. Abrams, Inc., 1985.

Wynbrandt, James. *The Excruciating History of Dentistry: Toothsome Tales and Oral Oddities from Babylon to Braces*. New York: St. Martin's Press, 1998.

Periodical Articles

Centers for Disease Control. "Fluoridation of Community Water Systems." *Journal of the American Medical Association* 267 24: 3264.

The Invention of the Artificial Heart

Overview

During the second half of the twentieth century coronary heart disease became the leading cause of death in wealthy, industrialized nations. Moreover, more than half of the deaths in the United States were caused by cardiovascular diseases. Many of these deaths could have been prevented by aggressive management and surgical procedures, including heart transplant operations. The shortage of donor hearts, however, led to hope that a totally implantable mechanical device could overcome the shortage and avoid the problem of immunological rejection, but early attempts to implant permanent artificial hearts were criticized as premature human experiments. Indeed, the controversies raised by experimental implantations in the 1960s may have inhibited the development of a permanent heart replacement. The poor quality of life provided by artificial hearts instead led to efforts to develop a new generation of left ventricular assist devices.

Background

The human heart is a remarkable organ—little bigger than a fist—that beats over 100,000 times a day without rest. In an average adult, the heart pumps more than 4,300 gallons (16,000 liters) of blood a day through nearly 100,000 miles (161,000 km) of blood vessels. Picturing the heart as just a pump, Michael E. DeBakey, a pioneer of heart surgery, who has been called the "Texas Tornado," predicted that a mechanical device could duplicate its main function. Artificial hearts actually date back to 1957, when Willem Kolff, inventor of the artificial kidney, and Tetsuzo Akutsu implanted an experimental heart into animals. Kolff's model heart kept a dog alive for 36 hours. Heart specialists and scientists have pursued four general approaches to heart replacement: artificial hearts, transplantation of donor hearts, assist devices that replace just part of the natural heart, and replacement hearts developed by tissue-engineering techniques in the laboratory, or hearts grown in genetically altered animals.

The ideal artificial heart would function essentially maintenance free for many years within the hot, humid, corrosive internal environment of the body. The design of a successful artificial heart would have to overcome the difficulties that have been revealed since the first such devices were tested in the 1960s: damage to the blood caused by contact with artificial materials, rejection of the replacement heart by the body's immune system, difficulties in delivering ade-

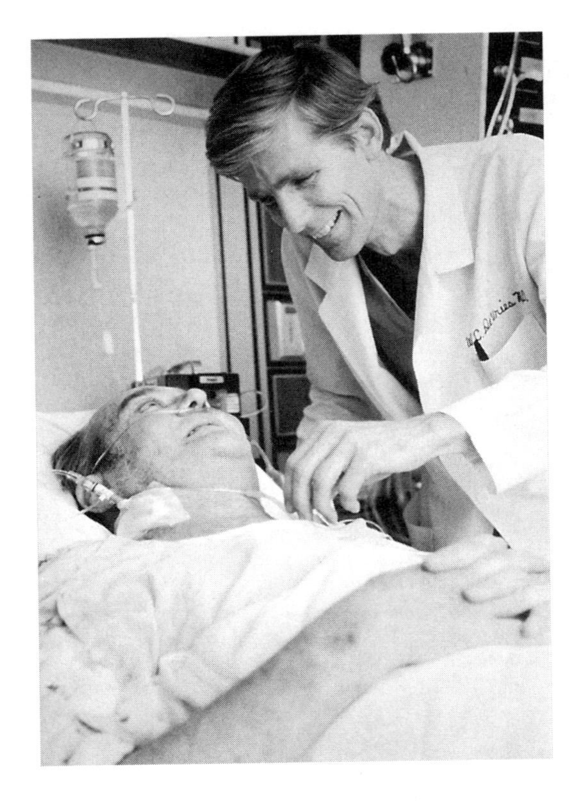

Dr. William DeVries shown with artificial heart patient Barney Clark after the historic 1982 operation. *(Corbis # U2100321. Reproduced with permission.)*

quate power to the pump without connections through the skin, miniaturizing the pumps enough for use in children and small adults, and adjusting blood flow in response to physiological stress. Although the artificial hearts in development during the 1990s may solve many of these problems, these devices will probably not become practical or routine for many years. Indeed, the history of the artificial heart is a history of controversial cases.

Impact

Studies conducted by the Institute of Medicine in the 1990s estimated that 10,000 to 20,000 Americans per year could be candidates for a total artificial heart and another 25,000 to 50,000 might need a left ventricular assist device. Heart failure affects about 5 million Americans per year; in addition, mortality from heart failure increased threefold between 1974 and 1994. Various forms of artificial pumps have provided temporary "bridges," keeping patients alive while awaiting a transplant, but the number of donated hearts is only about 2,000 per year. For many patients, assist pumps, also known as left ventricular assist devices (LVAD), may be more practical than replacing the entire heart. DeBakey began working on an artificial heart and related devices in 1960. He invented a simple blood pump, the LVAD, that could assist the heart while a patient waited for a transplant. In 1966 DeBakey performed the first human implantation of an LVAD.

One of the most dramatic events in twentieth-century surgery occurred in 1967, when Christiaan Barnard (1922-), a South African surgeon, performed the first human heart transplant. (In many cases, heart disease may be so severe that a patient may not survive the wait for a donor heart.) Attempts to use animal organs, such as Leonard Bailey's 1984 transplantation of a baboon's heart into a newborn, who was identified as Baby Fae, ended in failure. Therefore, the shortage of donor organs provided a great impetus to the development of an artificial heart.

On April 4, 1969, Denton A. Cooley performed the first human implantation of a total artificial heart when he used a device developed by Domingo Liotta to sustain the life of Haskell Karp. Karp was a 47-year-old patient who was in cardiac failure after surgery for a left ventricular aneurysm. Karp lived with the artificial heart in his chest for 65 hours but died shortly after receiving a heart transplant. DeBakey claimed that the heart Cooley used was identical to one under development in his laboratory and that Cooley had used it without permission. Because the device had been used with only limited success in calves, DeBakey considered human implantation premature and unwise. Although Cooley had obtained consent for the operation from the patient, he had not sought permission from the hospital review board or from federal agencies. He and Liotta thought that permission would not have been granted and that they would have lost a perfect opportunity to perform the experiment. The working relationship between Cooley and DeBakey was destroyed by the controversy surrounded the Karp operation.

Then Karp's widow brought a wrongful death suit against Cooley. She claimed that she and her husband had not been fully informed of the risks of the experimental procedure. The judge dismissed the case, ruling that the patient had given informed consent and that the hospital and surgeons had thoroughly informed the patient of the risks of the procedure and the low probability of complete recovery or survival. The decision in this case is regarded as a landmark in the development and implementation of medical technology.

In 1981 Cooley performed another controversial operation, the implantation of a total artificial heart developed by Tetsuzo Akutsu. The 36-year-old patient was sustained on the artificial heart for 55 hours until a donor heart was available for transplantation. Robert Jarvik, a physician and biomedical engineer, approached DeBakey about testing a similar device, known as the Jarvik-7, but DeBakey refused because he did not think that the device was ready for human use. One year later, William DeVries, in cooperation with Jarvik, implanted the Jarvik-7 heart into the chest of Barney Clark, a 61-year-old Seattle dentist dying of heart failure.

In contrast to the Karp case, in which the artificial heart was implanted as a bridge to transplant, DeVries and Jarvik intended to use their artificial heart as a permanent replacement for the diseased heart. Clark, who survived for 112 days on the artificial heart, was honored by members of the implant team as a "true pioneer" who understood that he was participating in an experiment that was unlikely to save his life but one that would provide information to help biomedical science and other patients.

Five similar implants were performed through 1985. The longest survivor was William Schroeder, who was supported by the Jarvik-7 for 620 days. The spectacle of the poor quality of life and painful complications endured by patients such as Clark and Schroeder created a significant public backlash against the artificial heart. Moreover, many doctors, scientists, ethicists, and policy makers concluded that use of the artificial heart was premature and that it would be well into the next century before a new generation of artificial hearts significantly improved patients' lives. The problems associated with implantable artificial hearts eventually led to a general consensus that an assist device would be more practical and beneficial to patients. The original purpose of LVADs was to keep people with terminal heart failure alive until a donor heart became available. In this manner, the Jarvik-7 was later used in hundreds of patients as a bridge to transplantation.

By the early 1990s, sophisticated LVADs were being used routinely in hospitals all over the world. Many of the early devices, however, were too large for use in children and small adults. Researchers thus focused on the development of a small, but still powerful LVAD. DeBakey and others had to carry out some of their experiments and clinical trials in Europe because governmental rules regarding clinical trials were more stringent in the United States. Innovative solutions to the problem of creating a better pump grew out of a collaboration between DeBakey and National Aeronautics and Space Administration (NASA) scientists. (This collaboration came about following an operation that DeBakey had performed on David Saussier, a NASA engineer.

The DeBakey Ventricular Assist Device (VAD), a miniaturized pump approximately one-tenth the size of the older devices, caused less damage to blood cells, required less than eight watts of power, and could be recharged through the skin. Many other experimental devices were also undergoing testing by 1998, when the 90-year-old DeBakey went to Germany to supervise the first human trials of his VAD personally. Six patients in all, the first, who was in critical condition at the time of the operation, died six weeks later. The second had his device removed because of the formation of a blood clot in the mechanism, but two other patients were able to leave the hospital with the device still in place.

The new generation of LVADs offers hope to many patients because of an unexpected phenomenon reported by several heart transplant centers. Some of the patients using LVADS while waiting for a donor heart were actually recovering. Apparently, the complete rest to the left ventricle provided by the LVAD reversed heart failure to a significant extent and enlarged heart cells were returning to normal size. Therefore, LVADs might also be used as a "bridge to recovery."

In addition to human heart transplants and mechanical hearts, some scientists think that animal tissues and organs or combinations of living cells with artificial materials will eventually be used to assist or replace ailing hearts. Scientists are now trying to grow heart muscle tissue, heart valves, and blood vessels in the laboratory; this approach is known as tissue engineering. Because an entire heart rarely fails, helping many patients with tissue-engineered heart muscle may be possible. In addition, in the field called xenotransplantation, scientists are already looking at ways to change animal organs so that they will not be rejected by human recipients. Opposition from animal activists and the threat of previously unrecognized viruses has made primates less desirable as sources of organs, but transgenic pigs may eventually provide organs for humans. Other scientists, however, believe that much of the social and individual burden of heart disease could be prevented through exercise, changes in diet, and the elimination of smoking.

LOIS N. MAGNER

Further Reading

Books

Ad Hoc Task Force on Cardiac Replacement. *Cardiac Replacement: Medical, Ethical, Psychological and Economic Implications.* Washington, DC: U.S. Government Printing Office, 1969.

Conrad, Peter, and Rochelle Kern, eds. *The Sociology of Health & Illness: Critical Perspectives.* 4th ed. New York: St. Martin's Press, 1994.

Hogness, John R., ed. *Artificial Heart: Prototypes, Policies, and Patients.* Washington, DC: National Academy Press, 1991.

Kolff, Willem. *Artificial Organs.* New York: Wiley, 1976.

Lubeck, D., and J.P. Bunker. *The Artificial Heart: Costs, Risks and Benefits.* Washington, DC: Office of Technology Assessment, 1982.

Reiser, Stanley Joel, and Michael Anbar, eds. *The Machine at the Bedside: Strategies for Using Technology in Patient Care.* New York: Cambridge University Press, 1984.

Shaw, Margery W., ed. *After Barney Clark: Reflections on the Utah Artificial Heart Program.* Austin, TX: University of Texas Press, 1984.

Periodicals

Jarvik, Robert. "The Total Artificial Heart." *Scientific American 244* (1981): 74-80.

Stover, Dawn. "Artificial Heart." *Popular Science 254* (February 1999): 11-17.

Issues and Developments in Birth Control since 1950

Overview

The conception of the birth control pill in 1956 was the beginning of a social and ethical reproductive revolution as well as the beginning of controversial debates that still continue. While groups throughout history had supported birth control and abortion, it was not until John Rock (1890-1984), Gregory Pincus (1903-1967), and others developed the pill that the era began. In 1961 Jack Lippes devised a flexible plastic intrauterine device (IUD) that ushered in an era of effectiveness as well as some controversy. A defective IUD called the Dalkon Shield clouded the IUD with concerns for safety. Implants were developed around 1990 and are still being researched. The advent of birth control coincided with a period in history of social unrest and the origin of many movements against establishment ideas. The rise of the feminist movement and of equal rights for women was supported by those who sought freedom from being controlled by the reproductive cycle. The whole question of reproductive freedom assisted by legal decisions like *Roe v. Wade* in 1973 made the issue of abortion or choosing to end pregnancy a hotly contested issue and one that has not gone away. With the advent of RU-486, the French "abortion pill," more fuel was added.

Background

In past centuries even educated women knew very little about the human body and even less about reproduction. Midwives who had served throughout history lost their jobs as male-dominated specialists became obstetricians. In the early 1800s groups of drug- and device-peddlers profited from women's desires for family control. Campaigns to curb the practices, along with the Victorian ideas that these things were wrong and should not be addressed, succeeded in passing groups of laws that criminalized abortion and even contraception information. Anthony Comstock formed a Committee for the Suppression of Vice and fought to get laws passed in 1873 that prohibited making, transporting, and disseminating "obscene, lewd, and lascivious matter." The laws affected all devices and information for preventing conception.

Nonetheless, women—at least in the Western world—did make progress in terms of gaining access to information about contraception. In Great Britain Marie Stopes (1888-1958) opened the first birth control clinic in 1921 to teach working class women about how to prevent unwanted pregnancies. American nurse Margaret Sanger (1883-1966), concerned with how low-income women of New York were devastated by self-induced abortions and unwanted pregnancies, opened up a store-front clinic in 1916. Sanger collided head-on with the groups supporting the Comstock laws but persisted in spite of arrests and threats. Her crusade that began in 1916 gave birth to the group Planned Parenthood and would be instrumental in encouraging the development of the birth control pill, as well as the legal decisions relating to abortions. Sanger in fact coined the term "birth control."

In October 1950 a wealthy 71-year-old widow named Katherine McCormick, heir of the fortune of Cyrus McCormick (1809-1884), inventor of the reaper, wrote a letter to Margaret Sanger. Sanger for all these years had been fighting for the right to disseminate information and had had some success in fighting the Comstock laws. McCormick told Sanger of her interest in contraceptive research and willingness to support the venture financially. Dr. Gregory Pincus, director of the Worcester Foundation of Experimental Biology, was studying the early development of mammalian eggs. The two women by chance met Pincus and revealed their vision of a method of birth control, a contraceptive that could be "swallowed like an aspirin."

The search for such a pill would follow a dramatic path. Russell Marker (1902-), a chemist, became very interested in the chemical composition of steroids as a possibility for treating arthritis, and he happened to find a corticosteroid called progesterone in the urine of pregnant women. Marker hit upon the idea of finding a plant with the steroid, and after a world-wide search he found the cabeza de nigra, a yucca-like yam in Mexico. Discouraged about setting up production in Mexico, Marker in 1949 destroyed all his papers and disappeared in Mexico.

Back in Massachusetts, Gregory Pincus accepted the challenge of finding that perfect pill. He met John Rock, a scientist working on the opposite problem of fertility. The two teamed up to search for Marker, whose steroid they had heard of. After several months, they found him and persuaded him to let them synthesize the molecule. After several other intriguing battles, they succeeded in making the pill. When they realized there would be a struggle for trials in the U.S., they got permission to clinically test in Puerto Rico and Haiti. The pill proved very effective. The most difficult trials were the religious and political struggles that would follow.

Oral hormones are synthesized compounds of the natural hormones estrogen and progesterone. Early in the menstrual cycle, the growth lining or endometrium of the uterus is controlled by estrogen. Progesterone, made in great quantities in the second half of the cycle, changes the lining of the uterus and causes mucus in the lower part of the reproductive tract to resist sperm. Both of these hormones act on the pituitary glands at the base of the brain. This gland then secretes hormones called gonadatrophins that regulate the estrogen and progesterone produced by the ovaries. Most oral contraceptives have both estrogen and progesterone. When one takes the pill, the presence of the hormones mimics a normal pregnancy and prevents the release of eggs from the ovaries. Different pharmaceutical companies produce pills with varying formulations. A few pills have only a type of progesterone that causes changes in the mucus lining that allows sperm to swim toward the uterus. Several years after the inception, the oral contraceptive was approved by the FDA in 1960.

Impact

The revolution brought by the pill not only enabled women to plan a family but to plan a life and career. The birth control pill was one of the wonders of scientific medicine of the 1960s and 1970s. However, as the years passed, some studies linked use with certain forms of cancer and other problems. Another advance in birth control is the injectable hormone depot mednoxyprogesterone acetate (DMPA). DMPA keeps the ovaries from releasing eggs to implant. A shot is needed every 12 weeks. The contraceptive called Depo-Provera is registered in some countries, but the U.S. FDA rejected it in 1974, 1978, and 1984. The side effects and idea of a shot has not made it very popular; however, it became available in Planned Parenthood clinics.

Intrauterine devices have been used in different forms throughout history to prevent pregnancy. They were invented in the nineteenth century but only became widespread in the late 1950s. Jack Lippes designed a flexible plastic device that was called the Lippes loop. IUDs, available only by prescription, come in varieties of shapes and are fitted into the uterus with a string that remains through the opening in the cervix. The string functions to check the position of the IUD as well as to aid in the removal of the device. IUDs work primarily by preventing fertilization of the egg by affecting the way the egg or sperm move. In 1970 Chilean physician Jaime Zipper added copper to the plastic device, which caused less bleeding and was more effective. However, IUDs were under great scrutiny. The Dalkon Shield was an IUD that sparked controversy because physical problems and infections became prevalent. Questions about the Dalkon Shield reflected on even those devices that were safe. Lawsuits that eventually caused the company A.H. Robbins to go bankrupt also gave all IUDs a bad name. By 1998 only four companies were manufacturing IUDs, and only two types were available in the U.S. One type contains copper and can be left in place for ten years. Another continuously releases a small amount of the hormone progestin and must be replaced every year.

A happening of the late twentieth century was the rise of litigation and prosecution for defective products. When women began to sue A.H. Robbins, the cost of litigation overwhelmed the IUD market with spillover lawsuits. For example, although the Searle CU-7 IUD was never determined defective, it drew more than 2,000 lawsuits.

In the 1990s a device called the Norplant implant began to attract attention. With this device, the physician inserts six thin flexible plastic rods under the skin in the upper arm. Each capsule has a powdered crystal of the hormone levonorgestrol, a progestin (synthetic version of prodesterone). By releasing a small amount, the drug keeps the ovaries from releasing eggs and thickens cervical mucus. Norplant is an effective method of protecting against unwanted pregnancy for five years. Litigation, however, took its toll. The first year sales were $141 million. When lawsuits entered the picture, sales dropped to $3.7 million the next year.

A landmark U.S. Supreme Court decision in 1973 called *Roe v. Wade* declared a Texas law making abortion a crime unconstitutional. The constitution was interpreted to protect a woman's right to choose to have an abortion during the first three months of pregnancy. State regulations of abortion were acceptable in the last six months. Forces on both sides mobilized. Some groups sought to liberalize this decision. Others determined to overthrow *Roe v. Wade*. Clinics performing abortion came under fire with picketing and even violence. By the turn of the twenty-first century, the law making abortion legal was as divisive as ever in U.S. society, with many groups still advocating for the repeal of the law.

The story of RU-486, a French abortion pill, is full of subterfuge and intrigue. The drug answers the search of women even from the time of ancient Egypt for a simple and effective way to end pregnancy. In 1982 Group Roussel Uclaf, a subsidiary of Hoechst, Germany, stunned the world with the announcement of the creation of an abortion pill. As the uterus prepares for pregnancy, it receives progesterone by special receptors, but RU-486 binds to the receiving cells and blocks the hormones. Since the uterine lining does not get the hormone, it sloughs off, taking the embryo with it. In 1987 trials indicated a success rate of 95% when used with a second drug. Like a miscarriage, it produced heavy bleeding, which could last up to two weeks; yet, some scientists stated that it was much safer than mechanical abortion. In 1988, in response to serious pressure from anti-abortion advocates, Roussel announced it would no longer market RU-486. Only 48 hours later the French government ordered the company to reverse the position. In 1992, ten years after the discovery, Roussel announced it would not seek to market the drug in the United States because of American anti-abortion forces but would sell only in France, Britain, and Sweden. The Chinese created their own version in 1988 in response to the one child policy—a law mandating that married couples could have no more than one child.

When Bill Clinton became U.S. president in 1993, the FDA sought to have Hoechst market RU-486 in the United States, but the company decided against it but instead donated the rights to the pill to the Population Council, a non-profit research group. The group now had the patent but no one to market it. In 1992 another nonprofit group, ARM (Abortion Rights Mobilization), growing tired of Hoechst and the Population Council, found a little-known law that allowed researchers to copy and distribute the drug as long as it was not for profit. Lawrence Lader, a 78-year-old New York writer and head of the group, built a warehouse in Westchester. The group hired a scientist to produce a copy and distributed it through an underground organization. ARM found a manufacturer (although the name is not revealed) and distributes it to clinics in several cities.

Another abortion drug, methotrexate, stops the cells of the embryo from dividing. Like RU-486, it is given with a second drug, misoprostrol. The jury is still out on what will happen to drugs like RU-486. With legal liability and threats of anti-abortion activists to boycott all products made from a company producing the drugs, as of 1997 only five companies were engaged in contraceptive research. Societal forces related to birth control are complex. The interplay of birth control advocates, scientists, the churches and other conservative groups, feminists, drug companies, the courts, trial lawyers, the government, consumers, and consumer advocates made the issues of contraception and abortion as controversial as ever as the twenty-first century dawned.

EVELYN B. KELLY

Further Reading

Asbell, Bernard. *The Pill: A Biography of the Drug That Changed the World.* New York: Random House, 1995.

Baulieu, Etienne-Emile with Mort Rosenblum. *The Abortion Pill.* New York: Simon and Schuster, 1990.

Knowles, Jon and Marcus Ringel. *All about Birth Control: A Personal Guide.* New York: Three Rivers Press, 1998.

Rubin, Eva R. *The Abortion Controversy: A Documentary History.* Westport, CT: Greenwood Press, 1994.

The Discovery of Genetic Markers for Disease

Overview

Genetic markers are sequences of DNA located near defective or disease-causing genes that can be used to indicate the presence or absence of these genes. Genetic markers are always at the same place on a chromosome.

Several scientists hit upon the idea of markers at the same time. The association of a gene with a particular chromosome forms the basis of the field of cytogenetics. Cytogentics is a subdiscipline within genetics that links chromosomal variations to specific traits. Beginning with single-gene diseases, such as Duchenne muscular dystrophy, Huntington's disease, and cystic fibrosis, the search for genetic markers has mushroomed to an all-out hunt.

Along with these discoveries have come social and ethical debates over the use of genetic markers. To these debates has been added the idea of changing the problematic gene using gene therapy, creating one of the most controversial issues of the twentieth century.

Background

The idea of mapping the human genome actually began in 1950 with the goal of associating a particular chromosome with a specific physical trait. Geneticists created linkage maps, looking especially at large families with an aberration in a chromosome.

The development of technology has underlain the search for genetic markers. Restriction enzymes were first used to map DNA in 1971. In 1975 two simple techniques sequenced DNA in gels and in transformed bacteria. E.M. Southern developed a technique for gel electrophoresis, in which, using an electric field, light bases travel farther along a gel plate than heavier bases. More complicated techniques were later developed for protein eletrophoresis. For example, high throughput analysis combines many techniques involving microchips and spectrographic analysis. In addition, each October the magazine *Science* produces a full-color page of known gene locations on the genome. The map is thus slowly becoming complete.

Impact

In 1978 two scientists, Y.M. Kan and Andrea Dozy, from the University of California Los Angeles (UCLA) noted that a harmless piece of DNA was always inherited along with sickle cell anemia a blood condition. This variation marked the possibility of a faulty gene, aiding researchers to develop the idea of "reverse genetics." Instead of starting with a product such as a protein and then searching for the gene, members of a family who have the gene are compared with relatives who do not. This method was still time consuming because of the massive amount of DNA to be searched through.

Beginning in 1979, Raymond White began work on restriction fragment length polymorphisms or RFLPs. Certain enzymes, called restriction enzymes, cut DNA out of a chromosome at specific sites. For example, one may cut at a site with the sequence G-A-A-T-T-C and not cut if the sequence has mutated to G-A-C-T-T-C. The fragments can thus be cut into different lengths and separated through gel electrophoresis. Variations in their lengths are called RFLPs. These RFLPs are scattered across the 23 human chromosomes, which have over 100,000 genes. White began by eliminating parts of the genome that were identical. This procedure narrowed the search down to five sequences of DNA between 15,000 to 20,000 bases. Using DNA from 56 donors and radioactive labels, he was able to find eight RFLPs.

Throughout the 1980s, White worked on the theory that RFLPs could be used as genetic markers. In 1981 a team of British researchers studying Duchenne muscular dystrophy (DMD), a condition that affects the muscles of young boys (usually killing them before age 20), found the first RFLP on the X chromosome. In 1986 Louis Kunkel found the gene responsible for Duchenne muscular dystrophy. White also worked with markers for neurofibromatosis 1, a gene related to colon cancer, and the BRCA1 genes related to breast cancer.

In 1982 Canadians Lap-Chee Tsui and Manuel Buchwald started looking at a statistical analysis of factors from blood samples of over 50 families with two or more children with cystic fibrosis (CF). CF is a fatal disease in which the mucus lining thickens, causing major lung congestion and other problems. For two years they searched without finding a marker. A biotechnology firm in Massachusetts offered to provide them with labeled fragments of single-strand DNA involving 200 additional markers. Within a few weeks, a link was found between one of the markers and the DNA from the sample. The firm traced the marker to chromosome 7. White then found a close link between CF and "met," a cancer-causing gene. The English researcher Robert Williamson discovered another marker on chromosome 7. Seven teams had been looking for the CF gene; they decided to meet in Toronto and pool their knowledge. In 1986 Jean-Mark Laloud and White determined that two markers were on either side of the CF gene. In April 1987 Willamson's team announced that they had found a candidate gene but additional tests revealed an error.

Tsui continued to bombard chromosome 7 with markers. Hei obtained membranes from sweat glands, which are associated with CF, and began to compare CF cells with normal cells. In the September 1989 issue of *Science,* they reported that their new gene had 27 sequences of DNA; only three base pairs were missing—a small number for such a lethal condition.

In 1872 a physician named George Huntington (1850-1916) had described a disease that starts in middle age and then destroys the person both physically and mentally. The disease was later called Huntington's disease (HD). Over one hundred years later, Nancy Wexler, a Columbia University psychologist whose mother had HD, saw a photo of natives in a village on Lake Maricaibo in Venezuela, whose gait resembled that of individuals with HD. Traveling to the remote area, she heard the story of a Portuguese sailor in the 1800s who always walked as if he were drunk. He had married a local woman and had many children. Seven generations later, 250 out of 5,000 of his descendants now had HD. This large family was a boon to geneticists, who could now map a huge family tree. (Wexler gave them blue jeans in exchange for 2,000 blood and skin samples.)

Back in the United States, James Gusella and a team at Massachusetts General Hospital had been sampling tissue from an Iowa family with HD to look for an RFLP. When he received the blood from the Venezuelans, Gusella expected to look for years for the RFLP. A team in Indiana led by P. Michael Conneally, however, joined the information from the Iowan family and the Venezuelan families. In May 1983 the computer found an RFLP that matched; this RFLP was the genetic marker for HD.

Although people walk around carrying genes for lethal diseases, few realize it. In 1989, after the gene for CF was found, some clinics developed a test for the disease, but many problems were encountered. Actually, general screening of the population identifies only 70-75% of CF carriers because many mutations cause the disease. The National Institutes of Health (NIH) recommended not to test until there was a 95% accuracy level. What to do with genetic markers has presented many practical and ethical problems. Having a test that will tell if a person carries a defective gene involves a question regarding benefit. What purpose will it serve and what if the test is wrong? For example, there is a prenatal test for Down's syndrome, a condition involving mental retardation and physical disabilities that is caused by an extra chromosome 21. The test, however, does not reveal how severe the condition is and what—if any—physical impairments will develop.

Another problem with genetic testing involves detecting a gene for a disease that has no cure. Would knowing about the gene help, especially given the possibility for error or misinterpretation? In the summer of 1999, a panel at Stanford University convened to consider a genetic test for Alzheimer's disease. Because there is no cure for the disease, a disease which has major social and personal ramifications, the committee refused to support the initiative.

If screening does become available for a disease, the question then becomes who should be tested. Should ethnic groups displaying a higher proportion of the condition be targeted? Another potential problem is misuse of the information.

For example, one fear is that insurance companies may deny health coverage to a person who carries a certain gene. In 1970 the United States pushed an aggressive program to test for sickle cell anemia, a blood condition found primarily in individuals of African descent. The results of this program were a disaster. Misunderstanding was rampant as people who were carriers of the gene thought that they were going to die. Some people were denied jobs or health insurance. Couples who carried the gene were told not to have children. A screening program without counseling and follow-up information thus courts disaster.

In 1985 the NIH approved guidelines for gene therapy experiments in humans. In 1990 the first such treatment was used on a four-year-old girl with adenosine deaminase deficiency. While the therapy appeared to work, its use set off a storm of ethical debate.

Gene therapy uses a vector to send the corrective DNA to cells. Viruses may be used as vectors. Because of the complexity of the procedure and the variety of individual reactions, the progress in gene therapy has been slow. Additionally, in the fall of 1999, the death of a healthy-looking eighteen-year-old with a rare genetic disorder involving ammonia metabolism caused an ethical fury to erupt. Nevertheless, gene therapy remains an area of promise.

The search for genetic markers continues as does the work to complete the human genome. Ethical questions surrounding genetic testing and gene therapy will no doubt likewise continue.

EVELYN B. KELLY

Further Reading

Bains, William. *Biotechnology: From A to Z.* New York: Oxford University Press, 1998.

Lewis, Ricki. *Human Genetics: Concepts and Applications.* Dubuque: Wm. C. Brown, 1997.

Rabinow, Paul. *Making PCR: A Story of Biotechnology.* Chicago: University of Chicago Press, 1997.

Wexler, Alice. *Mapping Fate.* New York: Random House, 1995.

The Development of High-Tech Medical Diagnostic Tools

Overview

Near the end of the twentieth century, high-tech diagnostic imaging techniques became powerful medical tools that allowed physicians to explore bodily structures and functions with a minimum of invasion to the patient. Advances in diagnostic technology allowed physicians the ability to evaluate processes and events as they occurred in vivo (in the living body). During the 1970s, advances in computer technologies allowed the development of accurate, accessible, and relatively inexpensive (when compared to surgical explorations) non-invasive technologies. Although relying on different physical principles (i.e., electromagnetism vs. sound waves), all of the high-tech methods relied on computers to construct visual images from a set of indirect measurements. The development of high-tech diagnostic tools was the direct result of the clinical application of developments in physics and mathematics. These technological advances allowed the creation of a number of tools that made diagnosis more accurate, less invasive, and more economical.

Background

The use of non-invasive imaging traces its roots to the tremendous advances in the understanding of electromagnetism during the nineteenth century. By 1900, physicist Wilhelm Konrad Roentgen's (1845-1923) discovery of high energy electromagnetic radiation in the form of x rays were used in medical diagnosis. Developments in radiology progressed throughout the first half of the twentieth century, finding extensive use in the treatment of soldiers during WWII.

Technological innovations in computing opened the door for the subsequent development and widespread use of sonic and magnetic resonance imaging (MRI). Better equipment and procedures also made diagnostic procedures more accurate, less invasive, and safer for patients. During the course of the twentieth century x-ray images that once took minutes of exposure could be formed in milliseconds with doses less than 1/50 of those used at the dawn of the century. Because of increased resolution that allowed physicians to see more clearly and with greater detail, physicians were increasingly able to make diagnosis of serious pathology (e.g., tumors) earlier. Earlier diagnosis often translates to a more favorable outcome for the patient.

During the early 1950s, the use of fluorescent screens allowed physicians their first real-time look into the body. Used in connection with improved contrast mediums (dyes that allow physicians to see vessels and internal organs), real-time diagnosis became a way for physicians to see the dynamics of disease.

Although nuclear medicine traces its clinical origins to the 1930s, the invention of the scintillation camera by American engineer Hal Anger in the 1950s brought nuclear medical imaging to the forefront of diagnostics. Nuclear studies involve the introduction of low-level radioactive chemicals that are consequently transported throughout the body to target organs and tissues. The scintillation camera measured and created images from the gamma rays emitted from these radioactive chemicals, enabling physicians to detect tumors or other disease processes. The basic techniques of the scintillation, or Anger camera, are the most widely used tools in nuclear medicine today, and gave rise to modern imaging systems such as the P.E.T. (positive emission tomography) scanner.

The development of noninvasive diagnostic techniques allowed physicians the opportunity to probe the body with safety and precision. Medical imaging encompasses a number of subspecialties, including x ray, computed tomography, magnetic resonance imaging, ultra-sound, scanning, and endoscopy. In conjunction with the injection of various dyes and markers, these techniques allow physicians to view the dynamic workings of the body (e.g., the flow of blood through arteries and veins).

The development of powerful, high-tech diagnostic tools in the latter half of the twentieth century was initially the result of fundamental advances in the study of the reactions that take place in excited atomic nuclei. Applications of what were termed nuclear spectroscopic principles became directly linked to the development of non-invasive diagnostic tools used by physicians.

In particular, Nuclear Magnetic Resonance (NMR) was one such form of nuclear spectroscopy that eventually found widespread use in the clinical laboratory and medical imaging. NMR is based on the observation that a proton in a magnetic field has two quantized spin states. Accordingly, NMR distinguishes between different types of organic molecules and, although there are complications due to interactions between atoms, in simple terms, NMR allowed physicians to see pictures representing the larger structures of molecules and compounds (i.e., bones, tissues, and organs) obtained as a result of measuring differences between the expected and actual numbers of photons absorbed by a target tissue.

Groups of nuclei brought into resonance, that is, nuclei absorbing and emitting photons of similar electromagnetic radiation (e.g., radio waves), make subtle yet distinguishable changes when the resonance is forced to change by altering the energy of impacting photons. The speed and extent of the resonance changes permit a non-destructive determination of anatomical structures. This form of NMR is used by physicians as the physical and chemical basis of a powerful diagnostic technique termed Magnetic Resonance Imaging (MRI).

Magnetic Resonance (MR) technologies relied on advances in physics during the 1950s. The development of MR imaging, attributed to English scientist Paul Lauterbur, was used in clinical trials during 1980. By the mid-1980s MR imaging was an accepted and widely used diagnostic technique.

MRI scanners rely on the principles of atomic nuclear-spin resonance. Using strong magnetic fields and radio waves, MRI collect and correlate deflections caused by atoms into images of amazing detail. The resolution of MRI scanner is so high that they can be used to observe the individual plaques in multiple sclerosis.

Impact

Late in the twentieth century, diagnostic tools moved into the operating room. So called "image-guided surgical methods" have allowed surgeons to more accurately determine the locations of tumors, lesions, and a host of vascular abnormalities. A corollary benefit of these techniques allows surgeons to track the positions of surgical instruments. Rapidly advancing computer technology and imaging, when used in conjunction with optical, electromagnetic, or ultrasound sensors, allow physicians to make real-time diagnosis a part of surgical procedures.

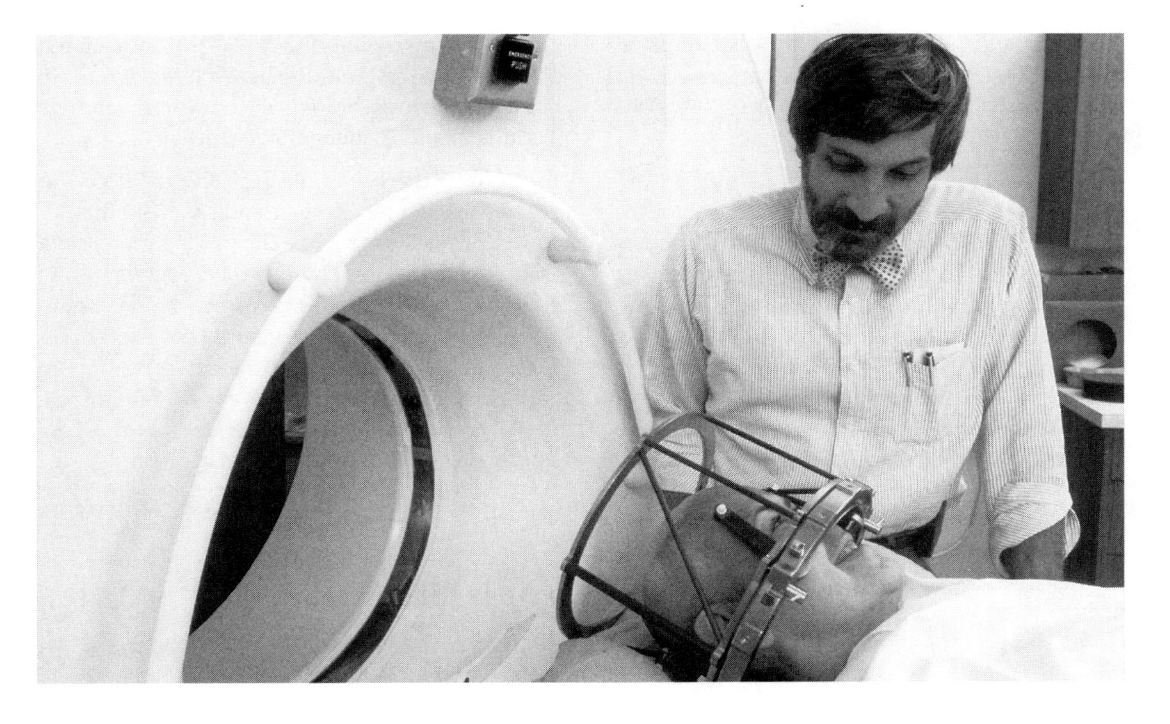

A cancer patient and her doctor before a CAT scan. *(RigerRessemy/CORBIS. Reproduced with permission.)*

Principals of SONAR technology (originally developed for military use) found clinical diagnostic application with the development of ultrasound during the 1960s. A sonic production device termed a transducer was placed against the skin of a patient to produce high frequency sound waves that were able to penetrate the skin and reflect off internal target structures. Modern ultrasound techniques using monitors allow physicians real-time diagnostic capabilities. By the 1980s, ultrasound examinations became commonplace in the examination of fetal development.

The advent of other imaging techniques allowed for less potentially damaging forms of diagnosis. Ultrasonic Doppler techniques are used to identify pathology related to blood flow (e.g., arteriosclerosis). Specialized types of scanning using Doppler techniques are also used to identify heart valve defects.

Microscopes using ultrasound can be used to study cell structures without subjecting them to lethal staining procedures that can also impede diagnosis. Ultrasonic microscopes differentiate structures based on underlying differences in pathology. Ultrasonics are also the least expensive of the latest high-tech innovations in diagnostic imaging.

Mammography, or x-ray visualization of the breast, became a common diagnostic procedure in the 1960s, when physicians demonstrated its usefulness in the diagnosis of breast cancer. Initially, mammograms were used to aid in diagnosis only, and the x-ray dose to the tissue was relatively high. In 1973, the Breast Cancer Detection Demonstration Project, a five-year study of over 250,000 women, helped to establish mammography as an effective screening tool for breast cancer detection. Today, modern mammography machines allow greater precision of breast tissue imaging with less radiation exposure to the patient. Mammography is a standard recommended screening procedure, and is credited with reducing breast cancer mortality in women over 50.

During the early 1970s, enhanced digital capabilities spurred the development of Computed Tomography (derived from the Greek *Tomos* meaning slice) imaging, also called CT, Computed Axial Tomography, or CAT scans, invented by English physician Godfrey Hounsfield. CT scans use advanced computer-based mathematical algorithms to combine different reading or views of a patient into a coherent picture usable for diagnosis. Hounsfield's innovative use of high-energy electromagnetic beams, a sensitive detector mounted on a rotating frame, and digital computing to create detailed images earned him the Nobel Prize. As with x rays, CT scan technology progressed to allow the use of less energetic beams and vastly decreased exposure times. CT scans increased the scope and safety of imaging procedures, allowing physicians to view the arrangement and functioning of the body's internal structures on a small scale.

American chemist Peter Alfred Wolf's (1923-1998) work with positron emission tomography (PET) led to the clinical diagnostic use of the PET scan, allowing physicians to measure cell activity in organs. PET scans use rings of detectors that surround the patient to track the movements and concentrations of radioactive tracers. The detectors measure gamma radiation produced when positrons emitted by tracers are annihilated during collisions with electrons. PET scans have attracted the interest of psychiatrists for their potential to study the underlying metabolic changes associated with mental diseases such as schizophrenia and depression. During the 1990s PET scans found clinical use in the diagnosis and characterizations of certain cancers and heart disease, as well as clinical studies of the brain.

MRI and PET scans, both examples of functional imaging, are the subject of increased research and clinical application. They are used to measure reactions of the brain when challenged with sensory input (e.g., hearing, sight, smell), activities associated with processing information (e.g., learning functions), physiological reactions to addiction, metabolic processes associated with osteoporosis and atherosclerosis, and to shed light on pathological conditions such as Parkinson and Alzheimer's disease.

During the 1990s, the explosive development of information technologies and the Internet allowed physicians to make diagnosis from remote locations and to tele-conference over real-time data. Multiple imaging is becoming an increasingly important tool to physicians. These and other high-tech innovations may one day contribute to the speed and accuracy of diagnosis, minimizing the need for invasive surgeries, and allowing exquisite precision when surgery is necessary.

BRENDA WILMOTH LERNER

Further Reading

Books

Kevles, Bettyann Holtzmann. *Naked to the Bone: Medical Imaging in the Twentieth Century.* Rutgers University Press, 1997.

Institute of Medicine Board on Biobehavorial Sciences and Mental Disorders. *Mathematics and Physics of Emerging Biomedical Imaging.* National Academy Press, 1996.

Mettler, F.A. and M.J. Guiberteau. *Essentials of Nuclear Medicine Imaging.* 3rd ed. WB Saunders Company, 1991.

Sorenson, J.A. and M.E. Phelps. *Physics in Nuclear Medicine.* 2nd ed. Grune & Stratton Inc, 1987.

Advances in Understanding Cancer

Overview

By the second half of the twentieth century, heart disease, cancer, and stroke had replaced infectious, epidemic diseases as the leading causes of death in the United States. After centuries of confusion about the basic causes of cancer, scientists are gaining insights into the factors that cause the loss of control of cell multiplication that is the fundamental characteristic of cancer. A cure for cancer remains elusive, but many promising experimental approaches to diagnosis and treatment, based on developments in immunology and molecular biology, were undergoing clinical trials by the end of the 1990s.

Background

Cancer is basically defined as a process that leads to the loss of control of normal cell division and multiplication. Cancer cells can form tumors that invade neighboring tissues and metastasize to establish tumors at distant sites in the body. The term "cancer" actually refers to several different forms of the disorder, which are defined in terms of the tissue of origin. A cancer that originates in connective tissues, bone, or muscle is called a "sarcoma." The most common form of cancer originates in epithelial tissues and organs, such as the breast, lungs, or stomach, and is known as a "carcinoma." The World Health Organization has classified cancers into about 100 different kinds, depending on their sites of origin, but cancer specialists believe that there are actually many more forms of cancer.

Cancer has been attributed to causes as varied as "unbalanced humors," diets rich in fat or sugar, metabolic wastes and toxins, aging, hormonal imbalance, parasitic infestations, viruses, genetic predisposition, mechanical irritation, electricity, tobacco, x rays, "melancholy," stress, and unnatural urban living. Other suggested—and much debated—causes include industrial pollutants, pesticides, occupational hazards, dietary factors, carcinogens that occur naturally in foods, genetic susceptibility, stress, radon gas, electromagnetic fields, low level ionizing radiation, and even cell phones. Viruses are associated with several human cancers: hepatitis B virus with liver carcinoma, human papilloma virus with cervical cancer, HTLV-I with adult T-cell leukemia, and Epstein-Barr virus with both Burkitt's lymphoma and nasopharyngeal carcinoma. A naturally occurring agent called interferon that was known to inhibit viral replication seemed to offer hope for preventing tumor growth. In 1979 the American Cancer Society allocated 2 million dollars for clinical trials of interferon but the results were ambiguous.

The correlation between exposure to certain chemicals and the development of cancer goes back to Percival Pott's eighteenth-century observations of scrotal cancers in chimney sweeps. Minimizing exposure to known carcinogens is an important public health goal. Analyzing the possibility that particular substances in the environment act as carcinogens, however, is complicated by a long latent period that precedes the appearance of the cancer. Another complication is the additive, or synergistic, effect of co-carcinogens, such as cigarette smoke and asbestos in the induction of lung cancer. In 1964 Surgeon General Luther Terry published his landmark report, *Smoking and Health: Report of the Advisory Committee to the Surgeon General of the Public Health Service,* which concluded that cigarettes caused cancer and were a major public health threat. Currently, many public health experts believe that tobacco products are responsible for about 30% of the cancer deaths in the United States.

Cancer research was very limited until the twentieth century, by which time infectious, epidemic diseases became less of a threat to the wealthy, industrialized world. Epidemiologists estimated that in the 1990s that one-third of the population of the industrialized world would develop some form of cancer. In contrast, for the 1970s the estimated rate was one in four. Because of major changes in age distribution in the population and the decrease in deadly epidemic diseases, however, the extent to which the incidence of cancer is increasing is actually a complicated question. Some apparent changes in the incidence of cancer are actually artifacts of improvements in diagnostic techniques. In the United States, the most common cancers in men are those originating in the prostate, lung, and colon. In women, the most common cancers are those of the breast, colon, and lung. Early diagnosis is considered important and oncologists call for the widespread use of diagnostic screening.

Conventional cancer treatment generally consists of chemotherapy, radiation, or surgery. Attempting to remove cancers through surgery is quite ancient but, as Hippocrates observed, such operations seldom cured the patient. In the case of breast cancer, for example, doctors had little evidence to show that survival rates improved as surgeons became more aggressive. Some surgeons, therefore, contended that less radical surgery combined with radiotherapy might lessen trauma and improve survival rates. George Crile, Jr., author of *What Women Should Know About the Breast Cancer Controversy* (1973), led the battle for less mutilating breast cancer surgery. In the 1990s some physicians, however, recommended bilateral breast removal for women with high risk factors, such as family patterns or specific marker genes, to prevent the disease.

Chemotherapy in the modern sense goes back to observations made of mustard gas during World War II. Autopsies of soldiers who were killed by mustard gas (dichloroethylsulfide) revealed an extreme inhibition in the ability of white blood cells to reproduce. Because cancer cells divide at a faster rate than normal cells, this finding suggested using mustard gas in the treatment of leukemia, Hodgkin's disease, and cancers of the lymphatic system. After the war, research on chemotherapy was expanded to include other "antimetabolites" in hopes of finding a penicillin-like "miracle drug" for cancer. Before the 1970s, the standard treatments for cancer—surgery, radiation, or chemotherapy—were generally administered separately. Combinations of chemotherapy and radiation were eventually found to be more effective in treating certain cancers, such as Hodgkin's disease and childhood leukemia.

Impact

An apparent increase in cancer, especially statistical increases among "civilized" populations, was hotly debated in the 1920s. In 1928 Senator Matthew M. Neely spoke to the U.S. Congress about "Cancer—Humanity's Greatest Scourge." *Mortality from Cancer Throughout the World,* a book by Frederick L. Hoffman, chief statistician for the Prudential Insurance Company, was the major source for Neely's account of cancer as a monstrous plague that threatened the very existence of the human race. The real number of cancer deaths at this time was unknown because autopsies were rarely done and the disease was often misdiagnosed or deliberated concealed. Inspired by Neely's call for a cancer crusade, Congress called for the establishment of an independent agency to lead a war on cancer. Officials, however, at the National Institutes of Health (NIH), which had been created in 1930, lobbied against the idea. In 1937 the National Cancer Institute (NCI) was created as a division of the NIH. In the 1970s Congress again attempted to launch a cancer crusade and established an independent National Cancer Authority. In his State of the Union address, President Richard Nixon promised to dedicate new funds to the conquest of cancer. As a compromise, the NCI was kept within the NIH but, unlike other components of the NIH, its director was directly appointed by the White House.

The House-Senate compromise bill, the Cancer Act of 1971, was signed by Nixon on December 23. The bill provided for 1.59 billion dollars in cancer research funds for the next 3 years, a National Cancer Advisory Board of government officials, plus a large number of panelists appointed by the president. In addition, a panel of laymen, doctors, and scientists would keep the president informed of progress and the president would control the budget for cancer research.

Although the impact the "war on cancer" might have had on saving lives is controversial, there is general agreement that major advances in understanding the origins of cancer have occurred. Cancer research has also been revolutionized by studies of oncogenic viruses, oncogenes, anti-oncogenes (tumor suppressor genes), and the sequencing of the human genome. (Oncogenes and tumor suppressor genes are genes that affect cell proliferation and the transformation of normal cells into malignant cells). John Michael Bishop and Harold Elliot Varmus, who shared the Nobel Prize in 1989, discovered cellular oncogenes while studying retroviruses. This discovery suggested that all human cells may contain genes that have the potential to cause cancer. Therefore, despite the complexity of cancer, there may be a fundamental common pathway leading to malignancy.

Despite intensive efforts—largely made possible by 25 billion dollars awarded for research by the National Cancer Institute—and optimistic pronouncements from the American Cancer Society, cures for cancer remain elusive. Five-year survival rates for the most common cancers have changed little since President Nixon launched the "war on cancer." Advances in science in the 1980s, however, have made it possible for biotechnology and pharmaceutical companies to develop such experimental products as cancer vaccines and diagnostic assays. Strategies for development involve integrating research in gene discovery, gene therapy, and small-molecule drug discovery. New treatments based on immunotherapies appear to be valuable. Genes and proteins linked to tumors as well as drug-response patterns are being identified by "molecular fingerprinting" technology. Drug manufacturers have called the new gene-based approach to the treatment and diagnosis of cancer "pharmacogenomics." These advances have suggested ways of tailoring existing therapies to individual patients as a means of improving efficacy and diminishing toxic reactions. By 1999, clinical trials for the use of gene-modified tumor cell vaccines in the treatment of prostate, melanoma, breast, and lung cancer were underway. Research on cancer pathways has led to promising studies of the p53 tumor suppressor gene pathway and studies of gene-delivery vectors carrying anti-cancer genes. In addition, progress is being made in finding genetic markers and mutations that appear to make people more vulnerable to certain cancers.

In March 1996 the Food and Drug Administration unveiled a plan to speed new oncology drugs to market. The angiogenesis inhibitors thalidomide, angiostatin, endostatin, and 2-methoxyestradiol could be candidates for such "fast-track" approval. New blood vessel growth, a process known as angiogenesis, is critical to normal pregnancy and wound healing as well as the growth of pathologic tissues such as cancers. Tumors, like other cells, depend on networks of new blood vessels to supply them with nutrients and oxygen. Therefore, agents that interfere with their blood supply may inhibit the growth of cancer cells. Thalidomide, a drug that was prescribed as a sedative for pregnant women in the 1950s and caused severe birth defects because it prevented the development of new blood vessels in the fetus, is being tested as a treatment for macular degeneration and for cancer involving the prostate, breast, skin (Kaposi's sarcoma), and brain (glioblastoma multiforma).

The possibility that some cancers could be prevented by attention to carcinogens in our food, water, and general environment is also of interest but, unfortunately, preventive approaches are less dramatic than innovative medical treatments. Some epidemiologists estimate that if major risk factors were identified and protective strategies implemented, 70% of human cancers could be prevented. Such conclusions are drawn from comparisons of cancer rates in different parts of the world and correlations between different lifestyles. Dealing with carcinogenic hazards in the environment, however, would require serious modifications of individual behavior and restrictions on the production of carcinogens.

LOIS N. MAGNER

Further Reading

Breslow, Lester. *A History of Cancer Control in the United States, 1946-1971.* Washington, DC: GPO, 1977.

Doll, Richard. *Prevention of Cancer: Pointers from Epidemiology.* London: Nuffield Provincial Hospitals Trust, 1967.

Maulitz, Russell C., ed. *Unnatural Causes: The Three Leading Killer Diseases in America.* New Brunswick, NJ: Rutgers University Press, 1989.

Patterson, James T. *The Dread Disease: Cancer and Modern American Culture.* Cambridge, MA: Harvard University Press, 1987.

Proctor, Robert N. *Cancer Wars: How Politics Shapes What We Know & Don't Know About Cancer.* New York: Basic Books, 1995.

Rather, L. J. *The Genesis of Cancer: A Study in the History of Ideas.* Baltimore, MD: Johns Hopkins University Press, 1978.

Rettig, Richard A. *Cancer Crusade: The Story of the National Cancer Act of 1971.* Princeton, NJ: Princeton University Press, 1977.

Strickland, Stephen. *Politics, Science, and Dread Disease: A Short History of United States Medical Research Policy.* Cambridge, MA: Harvard University Press, 1971.

Weinberg, Robert A. *Racing to the Beginning of the Road: The Search for the Origin of Cancer.* New York: Harmony Books, 1996.

The Global Eradication of Smallpox

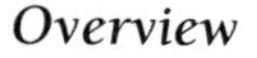

Overview

Smallpox is an acute, highly contagious, and often lethal disease caused by a virus that may be airborne but that can also be spread by direct contact or by clothing and bedding contaminated by pus and scabs. In 1979 after two years without a reported case, the World Health Organization (WHO) announced the global eradication of smallpox. The World Health Assembly continued to monitor the status of smallpox and in 1980 confirmed that the world was indeed free of the disease. The international campaign to eradicate smallpox, launched by WHO in 1967, is one of the greatest achievements of twentieth-century public health medicine and a model of international cooperation. The first steps toward the eradication of smallpox were taken in the eighteenth century, when the possibility of controlling the threat of smallpox through inoculation and vaccination was first seriously considered. By the 1970s, wealthy nations were generally abandoning routine vaccination as the threat of epidemic smallpox diminished. Because the smallpox virus could still become an agent of biological warfare or a terrorist threat, to remember and analyze the devastation formerly caused by this ancient epidemic disease is essential.

Background

The origin of smallpox is unknown, but it may have evolved from one of the pox viruses of wild or domesticated animals in Africa or Asia. Eventually, migration, warfare, conquest, and commerce carried the virus to all parts of the world. Once an individual has been infected, the virus multiplies rapidly and spreads to the internal organs. Accurate diagnosis is virtually impossible during the early stages of the disease while the patient suffers from fever, aches, sneezing, nausea, fatigue, and other flu-like symptoms. By the time the characteristic skin lesions and high fever appear, the victim may have transmitted the infection to many others. Blood poisoning, pneumonia, blindness, and deafness were not uncommon complications but, even in mild cases, smallpox pustules generally caused scarring. The form of the disease known as black, or hemorrhagic, smallpox is almost always fatal.

Various folk practices, such as exposing children to a person with a mild case of the disease, taking material from smallpox pustules and inserting it into a cut in the skin, or inhaling a powder made from dried smallpox scabs, evolved as attempts to mitigate the threat of the disease. Credit for bringing smallpox inoculation to the attention of the medical world is generally

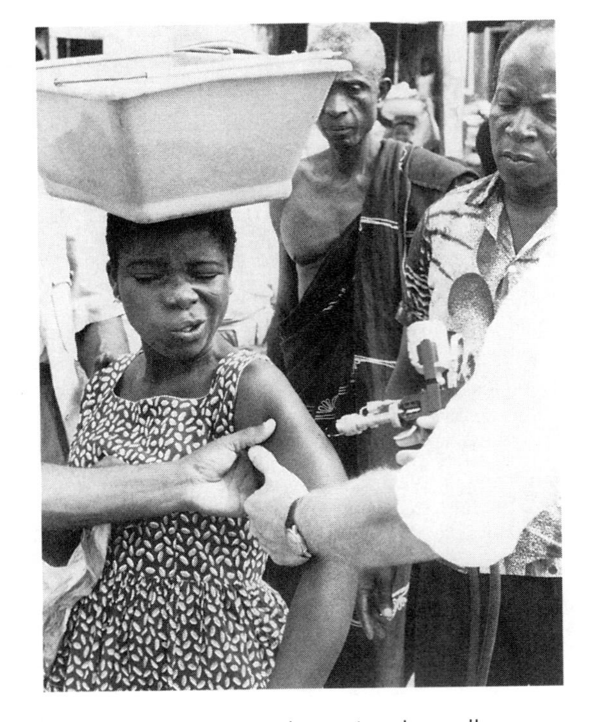

A Nigerian girl winces as she receives the smallpox vaccine. Africa was devastated by smallpox before the disease was eradicated in the late 1970s. *(Bettmann/CORBIS. Reproduced with permission.)*

attributed to Lady Mary Wortley Montagu (1689-1762) in England and Cotton Mather (1663-1728) in colonial New England. Lady Mary observed the Turkish method of inoculation while her husband was ambassador to the Turkish court at Constantinople. Fluid was collected from pox lesions on the 12th day of the illness. The fluid was then kept warm until a patient was inoculated with it through a scratch on the skin. This produced a milder, but usually survivable, form of the disease. The Reverend Cotton Mather learned about the practice from a young African slave. He also read an account of the Turkish method in the *Philosophical Transactions* (1714) of the Royal Society. When a smallpox epidemic struck Boston in 1721, Dr. Zabdiel Boylston and Mather experimented with inoculation. Although the procedure was very controversial, Boylston's statistical evidence demonstrated that the mortality rate for inoculated smallpox was significantly lower than that for the naturally acquired disease. Inoculation had important ramifications for medical practitioners, public health officials, and parents who had to assess the risks and benefits of inoculation.

By the second half of the eighteenth century, inoculation was a generally accepted medical practice. Even a person with inoculated smallpox, however, was a danger to others and needed isolation and medical attention. In 1798 Edward Jenner (1749-1823) published his studies of vaccination, a safer method of gaining immunity to smallpox. Jenner explained that English milkmaids sometimes acquired the cowpox lesions and experienced minor illness. This experience provided immunity against both natural and inoculated smallpox. Jenner proved that immunity could also be transmitted directly from person to person. To distinguish between the old practice of inoculation with smallpox material and his new method, Jenner coined the term "vaccination" (from the Latin *vacca,* for "cow").

Although vaccination rapidly replaced inoculation, debates about the safety and efficacy of preventive vaccines have raged ever since the first experiments with smallpox. Nineteenth-century critics objected to the enactment of compulsory vaccination laws, while advocates predicted that if all states adopted compulsory vaccination, smallpox would soon disappear. At the beginning of the twentieth century, epidemiologists warned that the rate of vaccination in the United States was worse than that of all other civilized nations. After World War II, the enforcement of vaccination laws improved dramatically and smallpox was no longer considered endemic in either Britain or the United States. Nevertheless, imported cases continued to touch off minor epidemics as well as major panics. Because the disease was so rarely seen in England, the rest of Europe, or the United States, smallpox patients often infected hospital personnel, patients, and visitors before the proper diagnosis was made. Once an outbreak was identified, some cities launched heroic vaccination campaigns. For example, after a smallpox outbreak in 1947, five million New Yorkers were vaccinated within two weeks.

Impact

By the 1970s, the odds of suffering ill effects from vaccination for most residents of the wealthy industrialized nations became greater than the chance of encountering smallpox itself. In 1971 the U.S. Public Health Service recommended ending routine vaccination because six to eight children died each year from vaccination-related complications. Given the extensive and rapid movement of people made possible by modern transportation, however, the danger of outbreaks triggered by imported smallpox could not be ignored as long as smallpox existed anywhere in the world. For the United States, Great Britain, and the Soviet Union, then, the worldwide eradication of smallpox offered a humane and economical solution to the vaccination dilemma.

The Smallpox Eradication Program was first adopted by the World Health Organization in 1958, but the intensive campaign for global eradication was not launched until 1967, at which time smallpox was still endemic in 33 countries. (In 1966 smallpox had killed 10 to 15 million people.) Epidemiologists agreed that smallpox was a good candidate for eradication because the disease has no animal reservoir, humans are the only hosts, and there is no carrier state or subclinical infection. Nevertheless, public health specialists were generally pessimistic about the possibility of eradicating smallpox from the world's poorest nations. Surprisingly, within a short period eradication programs in West and Central Africa were successful. During this phase of the global campaign, public health workers learned to modify their strategy in ways appropriate to special challenges. Originally, epidemiologists thought that vaccinating almost the entire population of endemic areas would be necessary. As a result of shortages of personnel and equipment in eastern Nigeria, however, public health workers there discovered that a strategy called "surveillance-containment" effectively broke the chain of transmission. By concentrating limited resources on the most infected areas and replacing expensive jet injectors with a simple bifurcated needle, the new strategy proved effective, even when only 50% of the population had been vaccinated. Finally, in October 1977, Ali Maow Maalin of Somalia became the last person outside a laboratory to contract smallpox.

Although humanitarian motives were not absent from the decision to declare war against smallpox, there is no doubt that economic factors loomed large in the choice of this target. For developing nations, malaria and other tropical diseases caused more serious problems than smallpox. Most victims of smallpox die or recover within a matter of weeks and in areas where the disease was endemic it was usually just one of many "childhood illnesses." In contrast, malaria is a debilitating, recurrent illness that reduces productivity, resistance to other infections, and the live birth rate. Global eradication of smallpox cost billions of dollars but wealthy smallpox-free nations could save even more by eliminating the threat of imported smallpox.

When the World Health Organization announced the victory over smallpox, Dr. Donald A. Henderson, who had led the campaign, urged all nations to use the lessons they had learned in the battle against smallpox to formulate and implement global immunization programs against diphtheria, whooping cough, tetanus, measles, poliomyelitis, and tuberculosis. Henderson later served as a White House science advisor and dean of Johns Hopkins School of Public Health. In the 1990s Henderson felt that it was necessary to warn politicians, scientists, and public health authorities about the potential use of the smallpox virus and other dangerous microbes for germ warfare and acts of biological terrorism. The idea that smallpox could be used in such a destructive manner is quite ancient; in colonial America, settlers and British agents were accused of deliberate attempts to cause smallpox outbreaks among Indian tribes.

With the global eradication of naturally occurring smallpox, the only reservoirs of the virus were those remaining in research laboratories. The danger of maintaining such laboratory stocks was emphasized in 1978, when a photographer working at the Birmingham University Medical School died of smallpox. The director of the laboratory committed suicide after admitting that safety precautions had been ignored. Subsequently, according to WHO, the only remaining stocks of smallpox virus were in high-containment laboratories at the Centers for Disease Control and Prevention in Atlanta and the Institute for Viral Preparations in Moscow.

At its annual meeting in 1995, the World Federation of Public Health Associations called for the destruction of all remaining smallpox virus and urged the World Health Assembly to require the destruction of all remaining stocks of smallpox virus and related materials. The following year, the World Health Assembly recommended that the last smallpox stocks be destroyed by 1999.

The smallpox virus has been called the ideal agent for germ warfare because it is simple to grow, stable, easily aerosolized, highly contagious, and causes a terrifying, often fatal illness. Moreover, the threat would be greatest to the wealthy nations that had abandoned vaccination in the 1970s. A scenario commonly envisioned by those who have studied the potential for bioterrorism involves the contamination of an airplane with smallpox virus. Many of the passengers would become infected and would carry the disease with them to their various destinations. Few American doctors would be able to diagnose smallpox, even when patients were in the advanced stages of the disease. Smallpox outbreaks could thereby overwhelm medical facilities.

Speaking at the International Conference on Emerging Infectious Diseases in March 1998,

Henderson warned his audience that the United States lacked the infrastructure, planning, and funding to deal with the threat of biological terrorism. Henderson pointed out that the kinds of microbes that might be used by bioterrorists—smallpox, anthrax, and plague—would be virtually unknown to the American medical community. Experts in bioterrorism argue that preparation would act as a deterrent to the use of biological weapons and have called for the development and implementation of strategic plans for coping with bioterrorism and biological warfare.

Moved by such warnings, President Bill Clinton approved new directives to improve the country's ability to prevent as well as respond to chemical and biological attacks, and the Department of Health and Human Welfare announced that it was working with other federal and military agencies to address the threat of bioterrorism. The Institute of Medicine and the National Research Council established a committee to investigate research and development strategies that would be useful in minimizing the damage caused by potential bioterrorist attacks. New drugs and vaccines to combat anthrax and smallpox were among the projects noted in the committee's report, *Chemical and Biological Terrorism: Research and Development to Improve Civilian Medical Response,* which was released in 1998.

LOIS N. MAGNER

Further Reading

Books

Alibek, Ken, and Stephen Handelman. *Biohazard.* New York: Random House, 1999.

Barnaby, Wendy. *The Plague Makers: The Secret World of Biological Warfare.* UK: Vision Paperbacks, 1997.

Baxby, Derrick. *Jenner's Smallpox Vaccine: The Riddle of Vaccinia Virus and Its Origins.* London: Heinemann Educational Books, 1981.

Behbehani, Abbas M. *The Smallpox Story in Words and Pictures.* Kansas City, KS: University of Kansas Medical Center Bookstore, 1988.

Brock, Thomas D. *Microorganisms from Smallpox to Lyme Disease.* New York: W.H. Freedman and Co., 1990.

Fenner, F., D. A. Henderson, I. Arita, Z. Jezek, and I.D. Ladnyi. *Smallpox and its Eradication.* Geneva: WHO, 1988.

Hopkins, Donald R. *Princes and Peasants: Smallpox in History.* Chicago: University Chicago Press, 1983.

Razzell, Peter. *Edward Jenner's Cowpox Vaccine: The History of a Medical Myth.* Sussex, England: Caliban Books, 1977.

Razzell, Peter. *The Conquest of Smallpox.* Firle: Caliban, 1977.

Shurkin, Joel N. *The Invisible Fire: The Story of Mankind's Triumph Over the Ancient Scourge of Smallpox.* New York: Putnam, 1979.

Winslow, Ola E. *A Destroying Angel: The Conquest of Smallpox in Colonial Boston.* Boston: Houghton Mifflin, 1974.

World Health Organization. *The Global Eradication of Smallpox: Final Report of the Global Commission for the Certification of Smallpox Eradication, Geneva, December, 1979.* Geneva: WHO, 1980.

Periodicals

Perkus, Marion E. , A. Piccini, B.R. Lipinkas, and E. Paoletti. "Recombinant Vaccinia Virus: Immunization Against Multiple Pathogens." *Science* 229 (1985): 981-984.

Wade, Nicholas. "Biological Warfare Fears May Impede Last Goal of Smallpox Eradicators." *Science 201* (1978): 329-330.

Internet

http://seercom.com/bluto/smallpox/index.html Smallpox Homepage.

http://www.pbs.org/wgbh/pages/frontline/shows/plague PBS Frontline "Plague Wars/Bioterrorism Report."

http://www.outbreak.org/cgiunreg/dynaserve.exe/cb/bionews.html Bioterrorism.

www.abc.net.au/m/talks/bbingstories/s10790.htm Iraq's Germ Warfare.

www.darpa.mil U.S. Defense Advanced Research Projects Agency.

www.hopkins-biodefense.org Johns Hopkins Center for Civilian Biodefense Studies.

www.nas.edu U.S. National Academy of Sciences.

http://www.apha.org/text/wfphatxt/about_wfpha.html Resolution of the World Federation of Public Health Associations Concerning the Destruction of Smallpox Virus.

The Advent of Cardiopulmonary Resuscitation (CPR)

Overview

Cardiopulmonary Resuscitation (CPR) is an emergency first aid procedure designed to re-establish or simulate heart and lung action. During cardiac arrest, CPR provides a percentage of oxygenated blood to the heart and brain, helping to keep these organs alive until advanced life support is provided. CPR techniques focus on the

"ABCs" of resuscitation—Airway, Breathing, and Circulation. Modern CPR was introduced in the medical community in the late 1950s, and is delivered using rescue breathing and chest compressions. The advent of CPR contributed to fundamental changes in the delivery of medical care, created ethical end-of-life considerations, and influenced outlook on health and fitness.

Background

During the 1950s, American physicians James Elam and Peter Safar were the first contemporary researchers of the airway "A" and breathing "B" components of CPR. Elam doubted the methods then-in-use for artificial respiration, which included the use of arm lifts and slow pressure applied to the chest. Elam proposed that expired air given directly to the victim through mouth-to-nose breathing would provide more oxygen to the body. Elam then demonstrated that his own expired air could maintain normal arterial oxygen levels in surgical patients when blown through their endotracheal tubes. Elam published his findings in 1954, and traveled to lobby governmental agencies and colleagues, championing the simplicity of his technique.

Safar, while sharing a long car ride home from an anesthesia conference with Elam, also became interested in resuscitation research. In a series of experiments on human volunteers begun in 1956, Safar focused on mouth-to mouth resuscitation. He found that mouth-to-mouth resuscitation techniques provided superior lung ventilation than the existing manual maneuvers. Safar demonstrated the ease of learning and performing mouth-to-mouth resuscitation techniques. Additionally, Safar addressed the issue of airway obstruction which often accompanies unconsciousness, showing that simply tilting the head backward and the jaw upward usually opens the airway. In 1958, Safar's research was published, and the medical community adopted Safar's mouth-to-mouth resuscitation methods. The American Medical Association called mouth-to-mouth resuscitation "an easily learned life-saving procedure" and encouraged that "information about expired air breathing be disseminated as widely as possible."

The discovery of artificial circulation was accidental, and actually a rediscovery of techniques similar to external cardiac massage described in late nineteenth-century medical literature. At that time, external cardiac massage was not accepted, and remained forgotten for over half a century. Instead, surgeons practiced open cardiac massage, in which the physician simulated circulation by squeezing the heart with his hand, as the preferred treatment for cardiac arrest until the 1950s. The circulation "C" component of CPR was rediscovered by two engineering scientists at Johns Hopkins University in 1958. William Kouwenhoven and G. Guy Knickerbocker studied ventricular fibrillation (an inefficient, "quivering" heart rhythm which leads to cardiac arrest) in anesthetized dogs. Knickerbocker noticed a slight momentary rise in blood pressure when defibrillation paddles (an instrument which delivers an electric shock to return the fibrillating heart to an efficient rhythm) were rested on the dog. Through a series of experiments, Kouwenhoven and Knickerbocker, along with surgeon James Jude, learned that when applying pressure straight downward with the heel of the hand to the sternal area of the chest, artificial circulation was most efficient. Through trial and error they also determined the most effective rate of chest compressions (60-80 per minute) and the optimal depth of compressions (1.5-2 inches). After successfully applying their techniques to human patients, the three scientists published their findings in 1960, asserting that cardiac resuscitation could be performed anywhere and "all that is needed is two hands."

Impact

Modern CPR was created in 1960 when the three techniques, mouth-to-mouth resuscitation, the head tilt, and chest compressions, were united. Safar joined Kouwenhoven and Jude to study the merger of the techniques, and together they conducted a world-wide tour to promote CPR to the medical community. American physician Archer S. Gordon, who defined the early manual artificial respiration methods, embraced CPR and contributed the "A, B, C" mnemonic for its airway, breathing, and circulation components. Gordon also helped produce a training film in 1962 which educated millions of medical personnel and students in CPR methods. The American Heart Association formally endorsed CPR in 1963, and founded a committee dedicated to its study. By 1966, the National Academy of Sciences reported recommendations to standardize the performance and training of CPR, allowing CPR to become the standard first-line treatment for hospitalized victims of cardiac arrest. The same year, deaths from cardiac arrest in hospitals began to decline.

Heart disease was the leading cause of death in America by 1960, due in part to Americans' changing lifestyles. Nutritional habits changed as processed, convenience, and fast foods contributed a higher fat and salt content to the diet. Simultaneously, increased urbanization led to a more sedentary lifestyle. As a result, more Americans suffered myocardial infarctions (heart attacks), and most of these occurred outside the hospital. Coupled with trauma suffered from automobile accidents on increasingly crowded roadways, the need for CPR outside the hospital was quickly realized.

In 1966, a national standard for training ambulance personnel was created, which included CPR, and led to the rise of emergency medical technicians (EMTs). EMTs could provide CPR at the scene and en route to the hospital. Often, however, victims of cardiac arrest, despite receiving CPR, arrived at the hospital too late to be saved by advanced life support techniques, including defibrillation. CPR delivery outside of the hospital made defibrillation and advanced life support necessary outside of the hospital as well. The first mobile cardiac care unit in the United States was established in New York in 1968, based on a similar program in Belfast, Northern Ireland. Staffed with doctors, nurses, and technicians, the unit provided medications, CPR, and defibrillation to victims with cardiac emergencies at the scene and during transport to the hospital. Although successful, the program was not practical. As it required a physician to be in attendance, often the unit experienced delays arriving at the scene after waiting for the physician to extricate himself from hospital duties and jump aboard the ambulance. American physician and engineer Eugene Nagel devised a method to deliver advanced emergency care in the field while allowing the physician to monitor the patient from the hospital in 1968. Nagel developed the first portable telemetry machine (used to determine the heart rhythm and relay it to a physician at the hospital) and trained rescue firemen in Miami in its use. In 1969, Nagel's "paramedics" performed CPR on a 60-year-old man, then used the portable defibrillator to shock the fibrillating man's heart back to a normal rhythm. Today, the paramedic system of providing advanced emergency medical care is in use throughout the developed world.

CPR was introduced into the community in 1972 in Seattle. Under Seattle's "tiered response" system, all emergency personnel, including volunteer firefighters, were trained in CPR. The first available personnel unit was assigned to reach the emergency site quickly and provide CPR, followed by the paramedics who initiated advanced life support and transported the patient to the hospital. Seattle physician Leonard Cobb, after analyzing Seattle's emergency response data, found the sooner CPR was started, the more favorable were the chances for survival. Cobb embarked upon an ambitious program to train 100,000 Seattle area citizens in a modified version of CPR, reasoning that bystanders can become the first responders to a medical emergency. Although the medical community was skeptical at first, by 1973 the American Heart Association defined the standards for basic CPR, and by 1974 , the American Red Cross conducted popular CPR classes for the public throughout the Unites States. Community-based CPR continues today as one of America's most far-reaching public health initiatives of the twentieth century.

Empowered with the new knowledge of CPR, Americans in the mid-1970s began a period of increased health consciousness. Aerobics, exercising in a manner to build cardio-vascular endurance and fitness, was embraced by many Americans who included aerobic exercise into their regular routine for the first time. Running and, particularly, jogging became popular aerobic pastimes. Health and exercise clubs increased in number, and athletic wear became a fashion trend. Americans became familiar with their individual "cholesterol counts" (fatty acids in the blood which are a risk factor for heart disease), and sought ways to improve them through a plethora of self-help publications on the market. The natural foods trend of the 1960s reemerged, with an emphasis on heart-healthy foods. These trends continue to evolve today. The rate of American deaths from heart disease reached a plateau, then experienced a decline during the 1980s and 1990s. Further innovations in life-sustaining treatment of heart disease as well as lifestyle improvements are credited with the reduction.

The advent of CPR gave rise to ethical questions, both in the hospital and the community. Prior to 1950, cardiac arrest almost always resulted in death. With CPR and advanced life support, medical personnel have greater options and responsibility to deter or determine when death occurs. The precise incidence of CPR effectiveness is not known, and at times those who are saved by CPR require intensive long-term life support. For these reasons, many with advanced disease or age choose not to undergo CPR and advanced life support. Legislation was passed in most states urging medical personnel to honor a patient's "living will" expressing end-of-life choices. Legislation was also refined to protect and encourage bystanders to render first aid, including CPR if needed, to victims in emergency situations. Most hospitals have ethics committees to define standards of life-sustaining and end-of-life care, and assist medical personnel and patients dealing with this issue.

The American Heart Association continues to set the standards for CPR training and performance. Since its inception, revisions to CPR include considerations for administering CPR to a child, an infant, or an injured person. The Heimlich maneuver for aiding a choking victim is included along with the breathing section of CPR training. Following the CPR example, health officials are encouraging many sectors of the public to receive training in advanced life support using the newly developed automatic defibrillator. New methods are under study to increase oxygen to the brain during CPR. The most significant aspect of CPR, however, remains its ability to be easily learned, and performed in almost any location, enabling trained citizens as well as professionals to give aid during critical emergencies.

BRENDA WILMOTH LERNER

Further Reading

Books

Eisenberg, Mickey S. *Life in the Balance: Emergency Medicine and the Quest to Reverse Sudden Death.* New York: Oxford University Press, 1997.

Paradis, Norman A., Henry R. Halperin, and Richard M. Nowak, et. al. *Cardiac Arrest—The Science and Practice of Resuscitation Medicine.* Baltimore: Williams & Wilkins, 1996.

Periodical Articles

Kouwenhoven, W.B., J.R. Jude, and G.G. Knickerbocker. "Closed Chest Cardiac Massage." *Journal of the American Medical Association* 173 (1960): 1064-67.

Safar, P. "Ventilatory Efficacy of Mouth-to-Mouth Artificial Respiration." *Journal of the American Medical Association* (May 1958): 335-41.

The Advent of Total Hip Replacement

Overview

In 1961 John Charnley (1911-1982) developed successful surgery for replacing diseased or injured hip joints with artificial joints. Many surgeons consider total hip replacement (THR) the greatest surgical advance of the second half of the twentieth century, because it has dramatically improved the quality of millions of lives and is relatively free of complications.

Background

The skeletal system gives the body shape, strength, and structure. Generally, it is durable enough to do this, except for a few troublesome areas, particularly the movable joints, which are the weakest parts of the skeleton. Joints, especially the hips and knees but also the shoulders, elbows, wrists, knuckles, and ankles, deteriorate with age, are easily injured, and are commonly subject to debilitating chronic diseases such as arthritis. Joint diseases and injuries have caused agony and disability since the beginning of time, but until the 1950s there was little that doctors could do to help. The best remedy for a diseased or injured joint is to replace it altogether, but that became technologically possible only through Charnley's work.

The human pelvis on each side contains a cup or socket called the "acetabulum." The ball-and-socket hip joint is formed by the head of the femur rotating in the acetabulum, lubricated by synovial fluid, supported by ligaments and muscles, and cushioned by cartilage. The head of the femur is connected to the shaft by the neck. Between the neck and shaft of the femur are two protuberances, the greater (outer) and lesser (inner) trochanters.

Any artificial body part is a "prosthesis." This could be anything from a wooden leg to a plastic heart valve to a set of dentures. Surgery to replace a natural joint with a prosthesis is "arthroplasty," from the Greek words *arthron* ("joint") and *plassein* ("to form, to shape, or to create"). The surgeon literally creates a new joint. Arthroplasty can also mean rebuilding a joint without using a prosthesis.

The first successful arthroplasty was performed at Pennsylvania Hospital, Philadelphia, in 1826 by John Rhea Barton (1794-1871), who reported the case as "On the Treatment of Anchylosis by the Formation of Artificial Joints," in *The North American Medical and Surgical Journal.* Ankylosis is the pathological stiffening of a joint. A 21-year-old sailor had fallen on his hip. Within a year of the injury the hip joint had become immovable and he had lost the use of his leg. The head of the femur had knitted into the acetabulum so that the pelvis and the femur were as one bone, with the leg turned inward.

Barton operated 20 months after the injury. He made a large cross-shaped incision centered over the greater trochanter, retracted the soft tissue, sawed through the femur transversely between the two trochanters, then returned the soft tissues to their place and closed the wound. Four months later, after intense physical therapy, the sailor was walking short distances without crutches.

Barton's arthroplasty did not involve the insertion of any prosthesis. He turned a disabled ball-and-socket joint into a functional hinge joint using only the bone, muscle, and ligaments of the disabled joint—and did it all in seven minutes. In the era before anesthesia surgeons had to work fast.

Barton's success was miraculous and isolated. The permanence of a fabricated joint with no natural connectives or lubrication could not be guaranteed, although Barton's sailor apparently overcame the improbable odds. Successful hip arthroplasty without prosthesis was reported by Royal Whitman (1857-1946) in 1924, but surgeons gradually came to understand that prostheses would be needed to achieve their goal of artificial joints with more inherent strength, more dependable structure, less friction, and greater endurance.

Modern hip prosthetic arthroplasty began in the 1920s. The femoral head was left alone, but an artificial cup was inserted into the acetabulum, mainly as a barrier against ankylosis or reankylosis. The first material used for the cup was glass (1923), but that was abandoned because it abraded or broke under the body's weight. A form of celluloid, viscaloid, was tried (1925), but the body tended to reject it. Pyrex then became the standard (1933), but problems remained with abrasion and breakage. Bakelite was used briefly (1937).

In 1937 at Harvard, Marius Nygaard Smith-Petersen (1886-1953) began experimenting with metals for acetabular cups. He built upon the work of Charles Scott Venable (1877-1961), Walter G. Stuck, and Asa Beach. He reported his design of a successful cup made of "Vitallium," a relatively inert, generally biocompatible alloy of cobalt, chromium, and molybdenum, patented by Charles H. Prange for Austenal Laboratories in 1934 and first used in dentistry.

The next major advance occurred when Jean Judet and Robert Louis Judet of Paris, France, developed a steel-reinforced acrylic femoral head. They reported in 1950 that this device reduced pain for 70% of patients and increased mobility for 90%, but two thirds of patients still needed a cane after the operation. The main problem with the Judet prosthesis was friction. The new joint sometimes even squeaked.

In 1938 Roy J. Plunkett invented polytetrafluorethylene (PTFE), a low-friction, nearly inert substance commonly known as "Teflon." In the 1950s Charnley experimented with Teflon acetabular cups and metal femoral heads to produce slippery, well-lubricated, and therefore longer lasting and less painful joints. His 1961 report of a successful operation marked the long-awaited breakthrough in low-friction prosthetic arthroplasty. For the first time both the acetabulum and the head of the femur could be replaced with artificial materials. This was the first genuine THR.

Impact

Total Hip Replacement has had a profound impact on the ability of the medical community to treat hip joint problems. As of 1994, 120,000 THR operations were being performed each year in the United States. Millions have been performed worldwide. Almost all patients receive distinct relief from their hip conditions. Sometimes re-operations are necessary to replace defective, deteriorated, or improperly installed implants, but the long-term prognosis for THR remains among the most optimistic for all surgical procedures. Rates of complication such as infection or induced thrombosis (blood clots) dropped significantly after the 1980s because of more effective prophylactic and anticoagulant drugs, earlier and more thorough physical therapy, and shorter hospital stays.

In 1950 40-year-old patients with degenerative hip disease might end up in wheelchairs by age 60. In 2000 similar patients, after THR, could be walking, playing golf, or cycling in their seventies and eighties.

One of the main problems with THR is the gradual disintegration of the materials of the artificial ball-and-socket joint. Early in the 1960s Charnley realized that Teflon was not adequate for the socket because it would erode and discharge irritating debris into the joint and surrounding tissues. His experiments with other synthetic low-friction materials led him in 1963 to high density polyethylene, which is still preferred by most THR surgeons.

Meanwhile, research in prostheses for other joints was advancing. In 1959 Earl W. Brannon and Gerold Klein reported the first successful prosthesis for a finger joint. A.B. Swanson introduced flexible silicone finger joint prostheses in 1966. Frank H. Gunston made major breakthroughs in knee replacement surgery in the 1960s and 1970s. Each type of joint presents its unique set of problems for researchers. The hip and knee are significantly different from other joints because of the tremendous load of weight and stress they must bear.

One controversy among THR surgeons is how to attach the prosthesis to the bone. The acetabulum is reamed and the cup prosthesis is inserted. The head and neck of the femur are removed either above or through the trochanters and the stem of the ball prosthesis is inserted into the shaft of the femur. The question is whether these components should be cemented to the residual bone or adhered in some cementless way. Bone will usually grow into a porous or roughened surface and create a strong bond without cement. On the other hand, bone cement may provide a stronger and more durable bond but may be more likely to cause infection or injury to surrounding tissues. Polymethylmethacrylate (PMMA) bone cement, introduced by Charnley in the early 1960s, is still the standard, but it sometimes loosens and generates debris. Whether to cement or not to cement often depends upon the age of the patient. Older patients, with thinner and weaker bones, often do better with cemented prostheses.

Debris control, friction control, and the bone/prosthesis bond remain areas of concern. Research into these problems generally centers around the materials used. Titanium, for example, was tried in the 1980s but proved too soft. David G. Murray, who chaired the prestigious U.S. National Institutes of Health Consensus Development Conference on Total Hip Replacement in 1994, prefers for the ball component an alloy of cobalt, chromium, and molybdenum with a porous or roughened femoral stem. Surgeons also use polished hardened ceramics such as alumina or zirconia for the head. The femoral stem is sometimes coated with another ceramic, hydroxyapatite, because it is chemically similar to bone and therefore adheres well without cement. With many minor improvements, ultrahigh molecular weight polyethylene for the socket has been used since Charnley's time, but methods of attaching it to the acetabulum vary. Acetabular components backed with porous ceramic-coated metal have so far shown the best results.

ERIC V.D. LUFT

Further Reading

Books

Klapper, Robert, and Lynda Huey. *Heal Your Hips: How to Prevent Hip Surgery and What to Do if You Need It.* New York: Wiley, 1999.

National Institutes of Health. *Total Hip Replacement.* Bethesda, MD: NIH, 1994.

Trahair, Richard. *All about Hip Replacement: A Patient's Guide.* Melbourne: Oxford University Press, 1998.

Waugh, William. *John Charnley: The Man and the Hip.* London: Springer, 1990.

Other

American Academy of Orthopaedic Surgeons. *Understanding Total Hip Replacement.* Videocassette. Chicago: AAOS, 1989.

Aging Issues since 1950

Overview

The study of aging, or gerontology, is a unique discipline that has emerged during the last half of the twentieth century. Advances in gerontology have been the result of a growing understanding of biomedical functions, behavior, and societal problems. Gerontology cuts across the major disciplines of biology, medicine, psychology, sociology, and even law to form a distinct discipline.

At the end of the twentieth century, only about 70,000 out of the 273 million people in the United States lived to be 100 years of age. Demographers, who study populations and trends, tell us that by 2050 the number of centenarians could swell to 4.2 million. The twenty-first century could thus be known as the age of longevity. A child born today in the United States may live well into the twenty-second century. On May 31, 1974, Public Law 93-296 authorized the National Institute of Aging (NIA) to address the multi-faceted subjects of longevity and the process of aging.

Background

Fear of aging, or gerontophobia, is nothing new. The Greek poet Homer pointed out that even the gods detest old age. In her book *Coming of Age,* Simone deBeauvoir described a vast literature by writers—ranging from Aristotle to William Butler Yeats—who decried growing old. That people reach a certain point and then slowly deteriorate until death is a myth that persists; youth culture is alive and well today.

A child born in 1776 might have expected to live to age 35; the median age at the time was 16. By the beginning of the 1900s, life expectancy had increased to 40 with a median age of 21. The term "average life expectancy" refers to the age at which, out of a group of people born at the same time, half are alive and the other half have died. For example, in 1999 a person who was born in 1922 had an average life expectancy of 77, meaning that half the people born in 1922 were alive and half had died.

In the minds of many people the year of decline is the age of 65. How this age was chosen is a story that begins in 1885. Bismarck, the chancellor of Germany, had some younger generals whom he wanted to promote but could not because of a group of older officers. To move these older ones out, Bismarck set an arbitrary age of 65 and offered the generals a pension to retire. In 1933, when President Franklin D. Roosevelt started looking at social security as part of the New Deal, he also chose the age of 65, a reasonable age considering that in the 1930s life expectancy was about age 60. What no one had planned for, however, was the intervention of science, which made life expectancy soar as well as the number of Americans living well past the age of 65.

Impact

In 1940 several far-sighted scientists such as Thomas W. Parran, the U.S. Surgeon General, recognized that to improve a nation's health research must focus on aging. Nathan W. Shock came to the NIA/Gerontology Research Center (GRC) in 1941 and set up a program, beginning in 1958, that recruited 650 males, ages 26 to 90, to participate in a study on aging. That program became known as the Baltimore Longitudinal Study of Aging. (A longitudinal study uses the same subjects over a period of years.) Volunteers reported to the center every two years for extensive tests involving physical, mental, behavioral, and social factors. About twenty years later, women were added to the study. Shock, often called "the father of aging research," was the primary influence in making gerontology an independent discipline.

When the Research on Aging Act was signed into law in 1974, the NIA/GRC went into action, appointing Robert N. Butler as permanent director in 1976. Many foundations have also been involved in the study of aging as well as groups dedicated to understanding certain diseases related to aging such as Alzheimer's disease.

The maximum human life span refers to the longest period of time a human could live if untouched by disease. At present, scientists believe that age to be about 120 years. The oldest documented person was Jeanne Calment (of France), who died in 1997 at the age of 120. While the average life expectancy has increased over the years, a debate exists as to whether life span can be extended. Roy Wolford, a professor of pathology at the UCLA School of Medicine, believes that the outer limit of the human life span has not budged. In contrast, John Wilmoth, a demographer at the University of California, thinks that there may be no fixed limit if disease is conquered. For example, Thomas E. Johnson and his colleagues at the University of Colorado have extended the life span of a roundworm, *C. elegans;* they have found eight genes that quadruple the worm's life span, which would equate to a human life span of 340 years.

A perhaps more reasonable goal for science is the "rectangularization of the curve." This term refers to the use of quality medicine to increase the average life expectancy so that average life expectancy will increase to that of life span. In other words, people would remain healthy to the end of their life span, at which point they would die suddenly. Conquering cancer, heart disease, and genetic diseases could achieve this goal. As John H. Bland stated in his book *Live Long, Die Fast:* "My goal is to die young as late as possible." Instead of tampering with the biological clock, the goal of aging research is to increase life satisfaction and productivity. Research is grouped into the following areas: cellular and genetic, body systems, mental and cognitive, and psychosocial.

Most gerontologists believe the answer lies with cells and genetics. Aging at the cellular level involves cell membranes and their receptors, growth factors, the skin and cartilage, and cellular communication. Leonard Hayflick, a professor of anatomy at the University of California, San Francisco is an authority on the cellular biology of aging and originator of the "Hayflick limit." Hayflick devised this theory in the 1960s after demonstrating that cells divide only a limited number of times and then die. Hayflick states that aging in biological terms defies definition, involving not just the passage of time but what happens over a period of time. Each person has a separate biological clock, and because of that fact biological age does not necessarily fit chronological age. A group of seventy-year-olds, therefore, are much more diverse in age than a group of sixteen-year-olds—biologically.

Some biologists believe that aging is genetic chaos. As one ages, they argue, many genes suddenly stop working. Hayflick thinks we are asking the wrong question when we ask why we age. The correct question, he maintains, should be why do we live as long as we do. Therefore, we should be searching for genes that protect vital life processes, not searching for genes that cause aging.

Hayflick's limit has been verified in the test tube. A cell will divide about fifty times and then die. At the end of the DNA chain is a minute unit called a telomere. Each time DNA replicates, a little bit of the telomere is broken off, and the cell ultimately loses its ability to divide. The telomere gene is turned on in cancer cells, allowing them to multiply wildly. A corporation in Menlo Park, California has isolated the telomere gene and has used it to see if other cells could also live longer.

Mapping the human genome has led to a mad rush to develop drugs to conquer disease at the genetic level. Gene therapy, while now in its infancy, is targeting such diseases as Alzheimer's, an insidious condition in which neurons develop tangles, called neurofibrillary tangles. Early onset Alzheimer's disease has been traced to genes on chromosomes 21, 14, and 1. Late onset Alzheimer's is related to chromosome 19. In 1999 a pharmaceutical company announced success with a vaccine against Alzheimer's in mice that are bred to have Alzheimer's-like symptoms. Researchers are also investigating diseases that resemble premature aging, such as progeria, Werner's syndrome, and Down's syndrome.

Another aspect of the biology of aging is the study of different body systems. For example, if the heart is free of disease, the heart of an older person is as efficient as a younger person's. According to the Centers for Disease Control, death resulting from heart disease has dropped by 60% since 1950. Also, stroke deaths dropped form 88.8 per 100,000 to 26.5 per 100,000 individuals in 1996. Attention to better nutrition, exercise, and quitting smoking account for these improvements. According to longitudinal studies, however, one system that is directly affected by aging is the urinary system. Likewise, the immune system seems to be affected by aging. The ability of the body to produce antibodies diminishes with age and the immune system becomes less able to function. Many ailments of older adults are immune system problems. Arthritis, which has over 100 forms, is an auto-immune disease in which the body turns on itself.

An item related to lifestyle involves the production of free radicals, highly reactive chemicals that combine with other body chemicals, causing them to change. As free agents, they roam about the bloodstream, pairing with other electrons in proteins, lipids, or DNA. Diets high in anti-oxidants help to search out these damaging free radicals. The free-radical theory was developed by Denham Harmon of the University of Nebraska.

One major myth about aging is that older people lose their memory and intellectual abilities. Long-term studies have contradicted these myths. While people do vary according to social and physical conditions, longitudinal studies show that one's level of competence improves as one grows older. Crystallized intelligence (intelligence that depends on learning and experience) increases, although fluid intelligence declines. (Fluid intelligence relates to quickness and rote memory.) The primary effect of aging on intelligence seems to be a decline in the amount of material that a person can learn quickly during a finite period of time. Personality traits also relate to learning; as one researcher has said, having the mental toughness not to succumb to the stereotypes of old age is the most powerful tool against aging. The decline of cognitive skills in older adults is more related to disease (Alzheimer's, tumors, stroke), poor health habits, or depression. Experiments in rats show that in the absence of disease or stress the aging brain does not decline in cognitive intelligence.

The social aspects of aging are also a major area of research. Social Security and Medicare are bounced around in political discussions. Moreover, several groups, such as the American Association of Retired Persons (AARP), the Older Women's League (OWL), and the Gray Panthers, are activists in fighting "ageism," discrimination against people because of their age. Other age-related issues include elder abuse, long-term care of the elderly, and custodial care.

EVELYN B. KELLY

Further Reading

Bland, John. *Live Long, Die Fast.* Minneapolis: Fairview, 1997.

Chopra, Deepak. *Ageless Body, Timeless Mind.* New York: Crown, 1993.

Hayflick, Leonard. *How and Why We Age.* New York: Ballantine, 1994.

Roizen, Michael, and Elizabeth Stephenson. *Real Age: Are You As Young As You Can Be?* New York: HarperTrade, 1999.

Rosenfeld, Isadore. *Live Now, Age Later.* New York: Warner, 1999.

The Evolution of the U.S. Healthcare System

Overview

Between the years 1750 and 2000, healthcare in the United States evolved from a simple system of home remedies and itinerant doctors with little training to a complex, scientific, technological, and bureaucratic system often called the "medical industrial complex." The complex is built on medical science and technology and the authority of medical professionals. The evolution of this complex includes the acceptance of the "germ theory" as the cause of disease, professionalization of doctors, technological advancements in treating disease, the rise of great institutions of medical training and healing, and the advent of medical insurance. Governmental institutions, controls, health care programs, drug regulations, and medical insurance also evolved during this period. Most recently, the healthcare system has seen the growth of corporations whose business is making a profit from healthcare.

Background

Prior to 1800, medicine in the United States was a "family affair." Women were expected to take care of illnesses within the family and only on those occasions of very serious, life threatening illnesses were doctors summoned. Called "domestic medicine," early American medical practice was a combination of home remedies and a few scientifically practiced procedures carried out by doctors who, without the kind of credentials they must now have, traveled extensively as they practiced medicine.

The practice of midwifery—attending women in childbirth and delivering babies—was a common profession for women, since most births took place at home. Until the mid-eighteenth century Western medicine was based on the ancient Greek principle of "four humors"—blood, phlegm, black bile, and yellow bile. Balance among the humors was the key to health; disease was thought to be caused by too much or too little of the fluids. The healing power of hot, cold, dry, and wet preparations, and a variety of plants and herbs, were also highly regarded. When needed, people called on "bone-setters" and surgeons, most of whom had no formal training.

Physicians with medical degrees and scientific training began showing up on the American landscape in the late colonial period. The University of Pennsylvania opened the first medical college in 1765 and the Massachusetts Medical Society (publishers of today's *New England Journal of Medicine*), incorporated in 1781, sought to license physicians. Medical schools were often opened by physicians who wanted to improve American medicine and raise the medical profession to the high status it enjoyed in Europe and in England. With scientific training, doctors became more authoritative and practiced medicine as small entrepreneurs, charging a fee for their services.

In the early 1800s, both in Europe and in the United States, physicians with formal medical training began to stress the idea that germs and social conditions might cause and spread disease, especially in cities. Many municipalities created "dispensaries" that dispensed medicines to the poor and offered free physician services. Epidemics of cholera, diphtheria, tuberculosis, and yellow fever, and concerns about sanitation and hygiene, led many city governments to create departments of health. New advances in studying bacteria were put to practical use as "germ theory" became the accepted cause for illness. It was in the face of epidemics and poor sanitation, government-sponsored public health, and healthcare that private healthcare began to systematically diverge.

Impact

As America became increasingly urbanized in the mid 1800s, hospitals, first built by city governments to treat the poor, began treating the not-so-poor. Doctors, with increased authority and power, stopped traveling to their sickest patients and began treating them all under one roof. Unlike hospitals in Europe where patients were treated in large wards, American patients who could pay were treated in smaller, often private rooms.

In the years following the Civil War (1865), hospitals became either public or private. More medical schools and institutions devoted to medical research emerged. A trend toward physicians needing more training led to the Johns Hopkins University's medical school's requirement in 1893 that all medical students arrive with a four-year degree and spend another four years becoming a physician.

Earliest efforts of doctors to create a unified professional organization started in the mid 1800s and, in 1846, the American Medical Association (AMA) was established. With little early impact on American medicine, by the next century the AMA had great influence over the politics and practice of medicine. An early AMA victory was the regulation of drugs.

Just after the Civil War, nursing became professionalized with the establishment of three training schools for nurses. While nursing began as a gender-based and female stereotyped "nurturing" occupation, over the next 100 years nursing would become more professionalized. By the late twentieth century, more nurses were receiving advanced degrees and playing a greater role in the administration of health care. Rarely trained as doctors even in the early twentieth century, by the 1980s women comprised up to half of medical school student admissions.

As the nineteenth century ended, advancements in biology, chemistry and related medical sciences meant that the great diseases—tuberculosis, yellow fever, diphtheria, cholera, and others—were practically eliminated with the development of diagnostic tests and vaccines. Extensive public health projects, aimed at fighting the causes of disease or to prevent their spreading, raised the levels of public health. Healthcare extended into the schools through school nurses.

By the early part of the twentieth century, doctors had more authority and were better paid than ever before. Associations, such as the AMA and the American Hospital Association (AHA), founded in 1899, became stronger. Employers and labor unions began to offer a range of benefits to workers, including paid medical care. National health insurance, such as provided by many European nations, became associated with socialism and the concept became unpopular in the United States, opening doors for private health insurance to cover the rising costs of medical care.

While private health insurance emerged prior to World War I, it was not until well after the War and toward the end of the 1920s that the first large medical insurance company, Blue Cross, was established.

The 1930s saw rising healthcare costs and an increasing number of health insurance plans. At this time, doctors were paid by a system called "fee-for-service." New insurance plans, such as Blue Cross and Blue Shield, allowed its members to pay both the costs of hospitalization and for treatment by physicians. The AHA in the 1930s took an active role in supporting group hospitalization plans. During World War II, a medical plan started by Henry J. Kaiser for his employees featured a pre-paid program that paved the way for Health Maintenance Organizations (HMOs) 40 years later.

The post-World War II era saw great expansions in the workforce, advancements in medical science and medical care, and increasing healthcare costs. The Baby Boom generation, the name given to the large numbers of children born just after World War II, received ever-higher levels of medical and preventive care during the 1950s. Advances in medicine in diagnostic techniques, such as x rays, life saving drugs, such as penicillin, and inoculations against diseases, such as polio, had created an ever-deepening scientific culture that included laboratory technicians, therapists, widening roles for nurses, and increasing specialization among physicians.

These post-World War II technological advances professionalized the roles of non-physician therapists and technicians, including respiratory therapists, physical therapists, x-ray technicians, and laboratory technicians. Improved technology and increasingly sophisticated treatments and therapies also pushed up cost of health care during the same period.

U.S. government research and health institutions and programs, such as the National Institutes of Health and the Centers for Disease Control, were established. The 1960s saw the initiation of social programs to aid in the medical care of the aged (Medicare) and poor (Medicaid). Prior to the founding of these institutions, the U.S. government had founded other health programs and institutions, such as the Indian Health Service, the U.S. Public Health Service, the Food and Drug Administration, and established an executive cabinet-level agency, the Department of Health and Human Services.

Between the end of World War II and the late 1980s, most doctors were still independent and compensated through fee-for-service. Through the powerful AMA and other organizations, doctors had fought off political attempts at creating a nationalized, universal coverage medical systems, such as those in Canada, the United Kingdom, and in Europe.

Doctors did not apparently notice, however, the growth of Health Maintenance Organizations (HMOs). By the mid 1980s, HMOs began to dominate both the organization of health care and reimbursement to physicians. In the 1990s, HMOs and their varieties would revolutionize the organization of health care in the United States and provoke controversy among recipients of healthcare as well as doctors, who came to find themselves in less control of their practices. Fee-for-service began to fade as doctors increasingly found themselves working for corporations that made profits from pre-paid healthcare by reducing the costs of healthcare, carefully restricting services, and focusing on preventive healthcare.

Fee-for-service was slowly being replaced by "capitation," a system that paid doctors a set fee from which they had to care for all of their patients, the sick and the well. Called "managed care," this system also produced changes in the consumers' role in healthcare as greater emphasis was placed on preventive medicine, consumer choice, and being accountable for one's own health and healthcare. Communications advancements such as the Internet and the World Wide Web in the 1990s added to the health information available to consumers. Also at this time, consumer interest grew in "alternative medicine," such as acupuncture, herbal preparations, and vitamin therapies. These interests could be seen as a reaction against the medical industrial complex.

Computer and communications advancements also allowed for such practices such as "telemedicine," a system utilizing the Internet by which patients could be diagnosed and often treated by physicians at a distance.

Twenty-first-century technology promises to continue changing the nature, complexity, and costs of healthcare. As knowledge increases about the genetic bases of disease, the healthcare system will make greater use of gene therapies, developing ways to prevent genetically caused diseases. Just as the impact of new technologies, such as x rays, antibiotics, vaccines, and surgical advances changed early and mid-twentieth-century medicine socially and scientifically, scientific and medical innovations, as well as social movements and economic realities, will continue to shape twenty-first-century medicine and health care.

RANDOLPH FILLMORE

Further Reading

Books

Biddle, Wayne. *A Field Guide to Germs.* New York: Henry Holt and Company, 1991.

Ehrenreich, John, ed. *The Cultural Crisis of Modern Medicine.* New York: Monthly Review Press, 1978.

Inlander, Charles B. and Michael A. Donio. *Medicare Made Easy: The People's Medical Society.* New York: MJF Books, 1999.

Muff, Janet, ed. *Women's Issues in Nursing: Socialization, Sexism and Stereotyping.* St. Louis: The C.V. Cosby Company, 1982.

Starr, Paul. *The Social Transformation of American Medicine.* New York: Basic Books, 1982.

Periodical Articles

Mark, David, M.D., M.P.H and Richard M. Glass, M.D. "Impact of New Technologies in Medicine: A Global Theme." *The Journal of the American Medical Association* (17 Nov., 1999).

McGinnis, J. Michael, MD and Philip R. Lee, MD. "Healthy People 2000 at Mid Decade." *Journal of the American Medical Association* 273, no. 14 (1996).

Trends in Alternative Medicine

Overview

"Alternative medicine" may be defined as medicine that is different from "conventional" treatment. These treatments are not based on "modern" concepts of disease. The Office of Alternative Medicine (OAM) at the National Institutes of Health (NIH) defines alternative, or "complementary," medicine as "those treatments not taught widely in medical schools, not generally used in hospitals, and not usually reimbursed by insurance." With developments in the last years of the 1900s, this definition will probably change.

In the latter part of the twentieth century, the number of various practitioners of alternative medicine had increased in Western Europe and the United States. It is estimated that half of the U.S. population may go to an alternative practitioner. This figure does not reflect the number of individuals who venture into self-help practices.

One factor in the development of the alternative medicine movement was the emerging idea of self-help, which perhaps came about as a reaction to the image of practitioners of Western medicine as doctors who dehumanize patients and who are concerned only with disease and not about the prevention of pain or illness. In the West in the 1960s the movement dubbed "counterculture" questioned all kinds of traditions, including medicine. In fact, in 1976 Ivan Illich, a guru of the counterculture, argued that Western medicine had reached a watershed and had become a major threat to health.

In the early 1970s President Richard Nixon began "Ping-Pong diplomacy," which helped draw attention to Eastern philosophies. Publicity regarding the failures of Western medicine, with its drugs and technology, added to the public's cynicism. (The thalidomide scare of the early 1960s exemplified this situation.) At the time, a number of countries in Europe joined in a project called the European Council of Science and Technology to examine the potential of alternative medicine. The United States also opened the Office of Alternative Medicine for funding purposes.

Many therapies in alternative medicine are holistic, emphasizing the physical, mental, emotional, and spiritual aspects of the whole person. Some of the treatments are preventive, stressing good health practices.

Background

To understand alternative medicine, one must look at traditional Chinese medicine and the Indian system of Ayurvedic medicine. In these philosophies, the body is treated as a whole. It is a unity, but one also affected by outside forces. In Indian medicine, the body interacts not just with the physical environment, but also the social structure, changing mental needs, and the accumulated karma of the soul that inhabits it. All these factors have vital energy, which strengthens or weakens the body and affects consciousness.

Indian medicine is called *Ayurvedic,* a system that teaches life-enhancing practices. Yoga, a series of meditative exercises, comes from the Siddha tradition, which believes in balancing the body's innate vital energies.

Chinese medicine combines physical movement and thought, called *Dao Yin,* as both preventive and healing arts. Central to Chinese medicine is *ch'i,* which Taoist philosophies describe as the vital force in the body; control of ch'i promotes power and longevity. Philosophers of the Sung dynasty (960-1279 A.D.) describe ch'i as composed of a balance between *yin* and *yang.* Yin, the female principle, is dark and passive and represented by the earth. Yang, the male part, is active and light and represented by the heavens. According to their beliefs, the forces of yin and yang act in an individual's body just as in nature. Disease is caused by a disharmony between the two and bringing balance back will restore health. The five elements of wood, metal, earth, water, and fire are balanced to ward off disease.

Both Chinese and Indian medicine draw upon animals, plants, and minerals to provide treatment for disease. Eastern medicine is contrasted to Western science, whose history reflects years of fighting to destroy "vitalism" and has reduced life to molecules and atoms. The Western reliance on technology and its concentration on disease also contrasts with the principles of holistic medicine, or the treatment of the whole person.

Impact

While many forms of alternative medicine have roots in the Eastern tradition, the West has changed or adapted many ideas. Growth in the

West, is after all, is consumer driven. Popular forms of alternative, or complementary, medicine include acupuncture, chiropractic therapy, osteopathy, homeopathy, herbalism, hydrotherapy, and meditation.

Acupuncture, the insertion of one or more small metal needles into the body for therapeutic purposes, is of Chinese origin. Devised before 2500 B.C., the practice grew out of the cosmic theory of yin and yang. A yin-yang imbalance causes a disruption of the life force, or ch'i, which flows through twelve meridians, or pathways, in the body. Practitioners of acupuncture insert arrowhead-shaped needles over the twelve bases or other specialized points. When inserted, the practitioner may twist or twirl the needle or apply an electrical current.

Used in China as an anesthesia during surgery, Western visitors have witnessed how normally painful operations have been carried on with fully conscious patients. Western explanations include such speculations as that the needle insertions activate certain natural painkillers (endorphins or enkephalins, for example) or that the stimulation of acupuncture closes pain impulses to the brain. Other Western observers have rationalized that there is a placebo effect.

Applications of acupuncture, however, have supplemented Western medical practice. For example, preliminary studies from the University of Maryland indicate that acupuncture reduced pain and improved joint function in osteoarthritis patients, although it did not cure the underlying cause of inflammation.

Two popular therapies, chiropractic medicine and osteopathy, do not have a clear connection with either the East or West but have a holistic and preventive component. Chiropractic, a form of manipulation of spinal vertebrae and other joints to stimulate the nervous system, was developed in 1895 by Iowa merchant David D. Palmer (1845-1913). Doctors of chiropractic are trained in accredited chiropractic colleges, which teach an interrelation between the musculoskeletal structure and nervous system. Chiropractic techniques include heat therapy, traction, and nutritional counseling.

Osteopathy, which also manipulates the spine and other joints, was founded by Andrew Taylor Still (1828-1917). One difference between chiropractic and osteopathy is that osteopathic doctors can prescribe medication. Major insurance companies in the United States are beginning to cover payments for these types of therapies.

Homeopathy is based on the idea that "like cures like,"—the more dilute a homeopathic substance is the greater the medicinal effect. This system was introduced in 1796 by the German physician Samuel Hahnemann (1755-1843), who noted that large doses of quinine, a treatment for malaria, produced effects similar to the symptoms of malaria. He theorized that if large doses aggravate the illness, then minute doses will treat it. When one thinks of the treatments of the day, such as bleeding, purging, and using strong doses of drugs (polypharmacy), patients welcomed homeopathy. In the twentieth century, homeopathy has been criticized because it focuses on disease and not health. Nevertheless, homeopathy does have adherents, including the International Homeopathic Medical League in Bloemendaal, Netherlands.

Herbalism, the use of plants to prevent or treat illness, draws heavily on Eastern tradition. Many herbs, such as ginkgo biloba and ginseng, are popular and greatly advertised. The Chinese medical tradition has an extensive folklore regarding herbals. Herbal remedies, however, have not met the test of scientific studies. The alleged benefits of herbs come from anecdotal testimonies given by people who have taken the herbs and declared that they have been helped. The FDA therefore has not approved herbal remedies, noting a lack of standardization and proven effectiveness. Some of the herbal remedies are beginning to be investigated using scientific methods. For example, ginkgo biloba, used for memory improvement, was found somewhat effective in relieving symptoms of Alzheimer's disease but caused major side effects, such as gastrointestinal disturbances, in some people. Also, an herbal mixture called Maharishi Amrit Kalish, from the Indian tradition, is a potent antioxidant and decreases atherosclerosis.

Hydrotherapy, a therapy involving health-giving baths and the drinking of spa water, came into vogue in the nineteenth century through the efforts of Vinzenz Priessnitz (1799-1851), a farmer who believed the water on his land could make one well. This belief in the healing power of water has also been connected to religious shrines. Many Western practitioners agree that disease and injuries may directly improve from the relaxing effect of being immersed in water but do not given credence to water's healing powers.

The main tenet of mind/body medicine is that thoughts and emotions are central to health. The mind and body are considered to be united and practices include prayer and meditation.

Boston's Deaconess Hospital, for example, has a large Mind/Body Institute for research. Spiritual healing, or the laying on of hands, supposedly transmits energy to the client and marshals the subject's energy for physical healing. This technique is an old one that links Eastern practices to traditional Western beliefs.

Other alternative, or complementary, therapies include: aromatherapy, the use of essential oils from plants for massage or inhalation; hypnosis, the use of an artificially induced state of semi-consciousness for therapy; massage, the manipulating of the body to improve health; naturopathy, an approach to prevention and treatment based on enhancing lifestyle, diet, and exercise; radisthesis, the use of dowsing with a pendulum to diagnose and treat illness; reflexology, the therapeutic massage of the feet; shiatsu, a Japanese technique bringing together finger pressure and massage; and biofeedback, self-monitoring using an electronic apparatus designed to give information so one may gain control one's over body or mind. Some Western practitioners use a form of biofeedback to assist patients. For example, after prostate surgery, biofeedback is used to teach patients how to regain bladder control.

Books based on Eastern therapies are best-sellers. Writers such as Deepak Chopra (1946-), who blends Ayurvedic medicine with Western science, has sold millions of copies of his books.

The debate over alternative medicine is alive and well. Hard-liners of the scientific method demand proof rather than philosophy—a major difference between the East and West. Driven by medical consumers, however, their position is softening. In 1992 Congress mandated establishment of the OAM and awarded grants to six U.S. universities for various studies. In 1996 Congress allotted 7.4 million dollars to the OAM and 12 million in 1997. Moreover, in 1997, 30% of American medical schools were offering courses in complementary medicine. In 1998 Congress gave the NIH 10 million dollars a year for five years for mind/body research.

EVELYN B. KELLY

Further Reading

Benson, Herbert. *Timeless Healing.* New York: Scribner, 1996.

Chopra, Deepak. *The Complete Mind-Body Guide.* New York: Crown, 1991.

Goleman, Daniel, and Joel Gurin. *Mind Body Medicine: How To Use Your Body for Better Health.* Yonkers, NY: Consumers Union of the United States, 1993.

Porter, Roy. *Medicine: A History of Healing.* New York: Barnes and Noble, 1997.

Trends in Epidemiology since 1950

Overview

Epidemiology is the branch of medicine that deals with the investigation of the causes, distribution, and control of disease in the general population, rather than at the level of individual cases. Over the past 50 years there have been significant changes in disease patterns throughout the world. Epidemiology uses statistics not only to explain present disease patterns but also to help predict how they may change in the future. The explosion in the world's population that has taken place in the last half century and the vast environmental and lifestyle changes experienced worldwide have created significant shifts in the causes of mortality as well as in morbidity, the rate of incidence of diseases. The development and more widespread use of vaccines to prevent infectious diseases and of antibiotics (antibacterial drugs) also have had an important impact on morbidity and mortality. This essay looks at these changes and their consequences for the worldwide battle against disease.

Background

One of the most important factors in epidemiology in the last 50 years has been a greater emphasis on using statistics to track disease trends and to attempt to discover their causes. For example, in the 1970s a large-scale study on cancer rates throughout the United States revealed that cancers were related to a number of environmental conditions and to lifestyle choices. The relationship of smoking to lung cancer was made more obvious by this study, as was the link between high-fat diets and colon cancer. When American cancer rates were compared with those in other parts of the world, it became even more obvious that differences in lifestyle were important. While breast cancer is relatively rare among Japanese women, it is the most common cancer among women in the United States, even Japanese-American women, which indicates that lifestyle—in this case the American high-fat diet—plays a more decisive role than does genetics in the development of many cases of this cancer.

In drug therapy the second half of the twentieth century saw the development and widespread use of many antibiotics. The first effective antibiotic, penicillin, became widely available in the late 1940s, so that by the 1950s its effect on the incidence of infectious diseases such as bacterial pneumonia, strep infections, syphilis, and gonorrhea became obvious. These drugs reduced both morbidity and mortality and made a significant contribution to rising population rates in developing nations, where infection was a more significant cause of death than in developed countries. In the latter, where there was a longer life expectancy, most people lived long enough to develop degenerative diseases such as atherosclerosis, the build-up of fatty deposits in the walls of the arteries that can lead to heart attack and stroke. Most types of cancer, though they may occur in younger people, also become more common with age.

The World Health Organization (WHO) was created in 1948 by the United Nations. In its mission to improve the health of people everywhere, especially in developing nations where healthcare was usually scarce, WHO set out to attempt to find cost-effective ways to reduce morbidity and mortality while at the same time developing better statistics on disease incidence worldwide. One of WHO's most successful projects was the elimination of smallpox , an often fatal infection that left the skin of survivors seriously scarred from the pustules that covered the body. WHO mounted a large-scale international vaccination program coupled with aggressive treatment and isolation of those with the infection. The last smallpox case was treated and cured in 1977, making smallpox the first known infectious disease to be completely eradicated.

The elimination of smallpox led to the hope that other infectious diseases could also be eradicated, but at the end of the twentieth century smallpox remains the only such success. In economically advanced countries polio was brought under control in the 1950s and 1960s with the vaccines developed by Jonas Salk (1914-1995) and Albert Sabin (1906-1993), but the battle against polio is still being waged in developing countries. Malaria remains one of biggest health problems worldwide despite a period in the 1950s during which its incidence declined rapidly. Malaria is caused by a one-celled parasite transmitted by mosquitoes, and in the 1950s the pesticide DDT was used extensively to eliminate mosquito populations. This led to a significant drop in the incidence of malaria, but within a few years mosquitoes became resistant to DDT and their populations rose again, and the scourge of malaria returned in developing nations in the tropics. Though anti-malarial drugs were developed, the parasite eventually became resistance to them, rendering them ineffective. In the 1990s, substantial progress was made on a vaccine, but malaria remains a particularly difficult health problem. In many tropical areas it is endemic, meaning that it is likely the people there will contract the disease, usually early in life. Since the parasite remains in the body, it periodically continues to damage the red blood cells, sapping its victims' energy and lowering resistance to other infections.

Impact

The history of malaria points up the relationship between the environment and disease, a relationship that can often be extremely complex. For example, it might seem that economic improvement projects that bring increased employment and other benefits to underdeveloped areas would also lead to improved health. But in fact the incidence of schistosomiasis, which is caused by a snail-borne parasite, has increased in areas of such projects because snails breed in drainage and irrigation canals. There is also evidence that the Ebola virus—a dreaded virus that causes rapid illness and death in most of its victims—becomes more likely to spread from apes to humans as human populations replace forested areas with farmland, bringing the apes into more direct contact with humans.

Human immunodeficiency virus (HIV) may also have spread from an isolated indigenous population in Africa that had developed a resistance to it. Individuals in this group may have had more contact with people from other areas, as African nations in the 1950s and 1960s became more urbanized and as their transportation systems improved. HIV is an example of how, despite the conquering of some infectious diseases with vaccines and antibiotics, humans still remain vulnerable to infections. For example, acquired immune deficiency syndrome (AIDS) was first identified by U.S. physicians working with male homosexual patients in the early 1980s. Soon it was also discovered among U.S. intravenous drug users and those who had received tainted transfusions and blood products. It was found to be a global rather than a national problem as cases were identified all over the world, but particularly in Africa, making it apparent that this continent was the likely source of the infection. By 1983 HIV was identified as the cause of AIDS, and by the mid-1990s relatively effective drugs were developed that slow the virus's destruction of the immune system. But these drugs are expensive and their effective use requires the kind of healthcare systems that only exist in developed nations. In developing nations such as those in Africa, where 69% of the people with HIV or AIDS live and where 84% of the deaths from the disease in 1999 occurred, such drugs and healthcare are completely lacking. It is estimated that by 2010, life expectancy in Africa will drop from 59 years to 45 because of AIDS. The incidence of HIV has also dramatically increased in the 1990s in Asia and Latin America.

While emerging infectious diseases such as HIV pose a critical health problem in the world today, many more common infections that have long been on the scene continue to have significant impact. Particularly in developing nations, diarrheal infections such as dysentery still claim many lives, particularly among infants and young children. These infections are often water-borne and are common in areas where a clean water supply and sewer systems are nonexistent. Respiratory infections also kill many children in developing countries, in part because antibiotics are not as readily available as they are in the developed world. While these problems have long plagued these areas, new threats are emerging with increased urbanization and industrialization. These often bring increased air and water pollution as well as exposure to many toxic industrial wastes.

In the developed world the patterns of disease are quite different, but nonetheless disturbing. While over the past 50 years the death rate due to heart disease has decreased by more than a third in the United States, the death rates for most cancers have remained relatively steady and in some cases have actually increased. While the decrease in the incidence of and death from heart disease is very clear, it is still difficult to pinpoint its causes. Most experts attribute it to a variety of factors involving both lifestyle changes and improved healthcare. For instance, fewer Americans smoke today than they did 50 years ago. Also, more people are weight conscious and are aware of the importance of a low-fat diet and exercise to a healthy heart. Fifty years ago, there was little that could be done to prevent or treat heart disease, but now there are cholesterol-lowering drugs and a wide variety of blood-pressure reducing medications, both of which combat the risk factors for heart disease. There are also a number of medical techniques that were developed, including bypass surgery, to counteract the effect of clogged arteries that lead to heart damage.

With cancer there have also been definite improvements in treatment and prevention over the past 50 years. Childhood leukemias that were almost always fatal 50 years ago now have a survival rate of more than 50%, a change largely due to the development of effective chemotherapy. The low-fat diet that helps to prevent heart disease will also lower one's risk of colon and breast cancer. Yet statistics indicate that one out of nine American women will develop breast cancer. Lung cancer is also becoming more common among women, since less women than men are giving up cigarettes and more young women than men are taking up smoking. In other parts of the world smoking-related illnesses are also on the rise, since the anti-smoking campaigns that have been effective in the United States are quite uncommon elsewhere.

This brief survey of the world health trends over the past 50 years indicates how important epidemiology is to our understanding of disease patterns and how they are changing. In the next 50 years disease patterns are likely to continue to change as new treatments as well as new diseases emerge. The median age of the population in developed countries will continue to increase, which means that degenerative diseases will become an even bigger problem, while in most developing nations the population will continue to increase significantly despite the development and increased use of contraceptives. This means that in these nations the median age will continue to remain low and that infant mortality, problems of malnutrition, and poor sanitation are likely to remain crucial factors in causing disease.

ROBERT HENDRICK

Further Reading

Desowitz, Robert S. *New Guinea Tapeworms and Jewish Grandmothers: Tales of Parasites and People.* New York: Norton, 1981.

Dubos, René. *Mirage of Health: Utopias, Progress, and Biological Change.* New York: Harper & Row, 1959.

Epstein, Helen. "Something Happened." *New York Review of Books* (December 2, 1999): 14-18.

Garrett, Laurie. *The Coming Plague: Newly Emerging Diseases in a World Out of Balance*. New York: Farrar, Straus and Giroux, 1994.

Hooper, Edward. *The River: A Journey to the Source of HIV and AIDS*. Boston: Little, Brown, 1999.

Kalipeni, Ezekiel, and Philip Thiuri, eds. *Issues and Perspectives on Health Care in Contemporary Sub-Saharan Africa*. Lewiston, NY: Edwin Mellen Press, 1997.

Lambo, Thomas, and Stacey Day, eds. *Issues in Contemporary International Health*. New York: Plenum, 1990.

Mann, Jonathan, and Daniel Tarantola, eds. *AIDS in the World II: Global Dimensions, Social Roots, and Responses*. New York: Oxford University Press, 1996.

Siddiqi, Javed. *World Health and World Politics: The World Health Organization and the UN System*. Columbus: University of South Carolina Press, 1995.

Public Health Efforts since 1950

Overview

Public health is a branch of preventative medicine concerned with the physical, mental, social, and environmental health of the community as a whole. From everyday lifestyle needs such as food and workplace safety to emerging infectious diseases, public health professionals apply scientific principles to analyze a community environment and institute measures which promote community well-being. In the latter half of the twentieth century, public health achievements such as the eradication of smallpox and institution of vaccination and health education programs have benefited millions throughout the world.

Background

Most developed countries support national public health institutions which research current health trends, and rely on information provided to them by smaller local public health centers throughout the country. For instance, the Center for Disease Control and Prevention (CDC) in the Unites States, among its other functions, collects and analyzes data provided by state and local health departments. Special methods of information-gathering regarding disease prevalence, along with methods to recognize patterns and institute control measures for significant findings, form the basis of epidemiology. Epidemiologists use statistics to find relationships between the incidence of disease and correlating factors, such as diet, lifestyle patterns, and the environment.

International public health goals are set by agencies such as the World Health Organization (WHO), established in 1948 as a special adjunct of the United Nations. During the 1950s and 1960s, the WHO successfully conducted mass campaigns to minimize endemic diseases (diseases normally occurring in a population at a relatively unchanged rate), as well as epidemic diseases (diseases with a rapid increase in the rate of infection) in the developing world. Additionally, the WHO helped teach the methods of epidemiology and public health principles to developing countries so that they could establish their own public health programs. Regardless of the level of the organization—local, national, or international—public health agencies assist the government in passing legislation to protect the health of the population at large.

Impact

Twentieth-century gains in public health dramatically impacted life expectancy. Many public health officials credit the chlorination of water with the almost 50% jump in life expectancy experienced by the end of the century. Waterborne diseases such as cholera and typhoid are almost non-existent in developed countries, but remain a threat in some countries where nearly half of the population still drinks untreated water. Poverty, overcrowded living conditions, and lack of infrastructure to treat and deliver water contribute to waterborne diseases which kill approximately 25,000 people in developing nations a day. Waterborne-diarrheal disease outbreaks in the early 1990s were particularly deadly to children, killing more than three million infants and children in a five-year period. The World Health Organization identified the availability of safe drinking water through chlorine decontamination as its number one priority of the 1990s, and, along with other international assistance groups, is providing developing countries with technical assistance and education to help improve the quality of drinking water.

Infectious diseases have spurred some of twentieth century's most spectacular public health accomplishments, as well as occupied public health scientists with some of its most frustrating mysteries. Almost all bacterial infections were considered curable with the advent of powerful and specialized antibiotics, until the 1970s, when bacteria resistant to standard antibiotic treatments began to emerge. The liberal over-usage of antibiotics, and failure to complete courses of antibiotics when prescribed, presumably encouraged bacterial evolution which created resistant strains.

Poliomyelitis, an infectious viral disease of childhood that often resulted in severe muscle paralysis, created a decade-long atmosphere of panic in the United States—with mothers isolating their children from both playmates and public locations fearing exposure to the disease. By 1954, American physician Jonas Salk (1914-1955) tested his polio vaccine made from the killed virus, and Americans eagerly waited in lines at makeshift public health centers by the thousands to inoculate their children. Polio is now on the brink of extinction, with vaccination routine in the developed world, and efforts to vaccinate those in remote areas continuing.

Other infectious diseases were the target of successful public health interventions. The last known case of smallpox occurred in Africa in 1977. Smallpox vaccinations are no longer routinely necessary as the disease is considered eradicated. As a result of global vaccination efforts, 80% of the world's children are vaccinated against major childhood diseases such as diphtheria, pertussis, measles, polio, and tetanus. Subsequently, child mortality dramatically decreased since 1980.

The emergence of the Human Immunodeficiency Virus (HIV) which progresses to AIDS (Acquired Immunodeficiency Syndrome) in the early 1980s sparked a public health revolution. As the epidemic took hold across the world, public health officials responded to intense pressure to quickly disseminate the latest knowledge and preventative strategies. Previous social taboos vanished as public health officials spoke openly of the connection between AIDS and sexuality in churches, schools, and other public forums. As cities debated needle exchange programs and distribution of condoms, public health officials continued its education strategies to minimize public misconception of the bloodborne illness. Today, as the epidemic continues, public health agencies world wide participate in the search for a vaccine to prevent AIDS, to document the epidemic's evolution, and to educate the public in preventing its spread.

In the 1990s, bloodborne types of hepatitis, hemorrhagic fevers, and the re-emergence of tuberculosis (including resistant strains) all join AIDS in challenging public health professionals to educate and protect the public, as well as healthcare workers caring for victims of these evolving diseases.

Preventative services for chronic conditions brought about by age, genetics, or lifestyle are also within the scope of public health. In the last half of the twentieth century, public health officials identified risk factors for chronic conditions, then planned interventions to reduce mortality. The reduction of deaths from heart disease, due in part to a diet- and exercise-conscious public in the 1970s and 1980s, is hailed as one of the great public health accomplishments of the century. Deaths from stroke also decreased in the United States, as a televised public health campaign labeling hypertension (high blood pressure) as the "silent killer" encouraged citizens to seek screening and treatment. Infant mortality decreased as pregnant women were encouraged to seek prenatal treatment.

Public health dramatically affected women's issues during the 1960s and 1970s. Sexually transmitted disease screening was included in family planning care, and the advent of the oral contraceptive enabled women to exercise more control of their bodies. The sexual revolution of the 1960s evolved to the women's movement of the 1970s, and women entered the workforce in record numbers. During these two decades, mortality from cervical cancer plummeted, as women were encouraged to seek regular pap smears for early cervical cancer detection.

The environment in which people live and work is a public health concern. Pressure from an increased population and increased industrial activity in the twentieth century contributed to a higher volume of pollutants threatening the air and water. Public health officials monitor air quality, post alerts on high pollution days, and support legislation enforcing clean air emission standards from industry and automobiles. Additionally, public health organizations monitor bodies of water, as well as municipal water supplies for biological and chemical contaminates. Public health officials also plan for environmental disasters which may affect the water supply or air quality, such as radioactive discharges, severe flooding, or biological terrorism.

The effects of tobacco use in the United States were a major public health concern as the twentieth century ended. Although per capita consumption of cigarettes was down in 1998 to its lowest level since the 1940s, smoking among young people has been on the rise since 1991. Many states successfully litigated against tobacco companies, claiming teens were especially targeted by tobacco advertisements. Some tobacco companies were ordered to pay for anti-smoking advertising aimed at teens, and to fund public health campaigns stating the dangers of smoking. Stricter policies to protect the public from second-hand smoke have been implemented. Many states passed laws restricting smoking in governmental and private worksites. Since the Surgeon General's 1964 report which first spoke of the dangers of cigarette smoking, strong clinical evidence has been compiled by the medical community supporting the link between smoking and lung cancer. Although public health agencies encouraged "smoke-free" days each year and engaged in anti-smoking advertising campaigns aimed at teens, smoking remained a frustrating addiction resulting in 30% of all cancer deaths in the United States. World-wide, smoking grew at exponential rates, especially in developing countries.

Smoking restrictions prompted some to claim that governmental public health policies interfered with individual citizen rights, though few argued that laws prohibiting the sale of tobacco to children were oppressive. This public health double-edged sword can also be illustrated by the decision of some mothers not to vaccinate their children against childhood diseases, claiming that the risk of exposure was low since most other children were vaccinated. In most cases the public good was eventually satisfied by enacting laws—for example, requiring children to be vaccinated before conferring eligibility to enter school. Instances of exotic or unusual burial requests sometimes made news, and were almost always refuted by strict burial and tissue disposal laws enacted to protect the public from the shedding of potentially harmful or infected cells from the deceased. One exception to this was the Ebola outbreak of the 1990s in Africa. New cases of the deadly virus continued to emerge until control measures banned customary mourning practices of handling the dead victim's body. Some legislative insistence to public health conformity is less controversial, such as requirements to buckle the seat belt while driving a car.

Public health official still face many challenges. At the close of the twentieth century, overcrowding and growth pressures expanded as the Earth's population reached six billion. Gains in mental health care eroded as the result of budgetary cuts. Environmental concerns such as global warming and the depletion of the ozone layer may affect the world's ability to grow crops. Hunger persisted in large parts of the world. With surveillance and applied principles of epidemiology, public health professionals hope to tackle these problems in the twentieth-first century, and assist the peoples of the world to achieve better states of health and well-being.

BRENDA WILMOTH LERNER

THE CAMPAIGN TO ERADICATE POLIO

In 1988 the World Health Organization launched a global drive to eradicate polio by the year 2000. Joined by Rotary International, the De Beers diamond company, UNICEF, and the governments of Australia, Canada, Denmark, Germany, Japan, the United Kingdom and the United States of America, the campaign has been a spectacular success. Most countries are now free of the disease, and the goal appears to be within reach. The WHO has targeted the Indian subcontinent and sub-Saharan Africa in their final push to eliminate the disease.

When WHO announced their eradication drive in 1988, there were 35,000 cases of polio worldwide. Using a strategy similar to the one that had eliminated smallpox a decade earlier, the number of cases dropped to 6,000 within 10 years. An outbreak in Angola pushed the numbers higher in 1999, but in the first quarter of 2000, there were only 39 confirmed cases. The WHO plans to target measles for their next eradication campaign.

For more information see the WHO web site: http://www.who.int/vaccines-polio/eradconcept.htm

Further Reading

Books

Garrett, Laurie. *The Coming Plague—Newly Emerging Diseases in a World Out of Balance.* Penguin, 1995.

Rose, Geoffrey. *The Strategy of Preventative Medicine.* Oxford University Press, 1992.

Virginia Apgar
1909-1974
American Obstetric Anesthesiologist

Virginia Apgar was an obstetric anesthesiologist best known for developing a scoring system to evaluate newborn babies that is now used around the world. Apgar published more than 60 papers in professional journals on topics related to anesthesiology, birth defects, and resuscitation. During her career she trained about 250 anesthesiologists and administered anesthesia in over 20,000 operations and over 17,000 births.

Virginia Apgar was born in Westfield, New Jersey, on June 7, 1909. She graduated from Mount Holyoke College in Massachusetts in 1929. During her undergraduate days she played in the college orchestra and on several athletic teams and reported for the college newspaper. Her energetic pursuit of diverse interests continued beyond college training. Apgar completed medical school at Columbia University's College of Physicians and Surgeons in New York City in 1933 and spent 1934 and 1935 completing a surgical residency at the city's Presbyterian Hospital.

Although Apgar did very well in the residency, the Chairman of Surgery, Dr. Alan Whipple, encouraged her to pursue anesthesia as a career. At that time the U.S. was deep in the Great Depression, and Whipple knew that even male surgeons were having difficulty succeeding in New York, where numerous surgeons wanted to practice. Whipple also knew that anesthesia, which was an underdeveloped specialty at the time, must advance for surgery to advance as well and that anesthesia needed doctors like Virginia Apgar.

Apgar took her mentor's advice and in early 1936 began anesthesia training at Presbyterian Hospital. In 1937 she spent six months in Madison, Wisconsin, to study anesthesia with Dr. Ralph Waters, who had created the first academic anesthesia training program in the United States more than a decade earlier. When she returned to New York City, Apgar spent six more months training under Dr. E.A. Rovenstine—another student of Waters's—at Bellvue Hospital. Thus Apgar had the finest anesthesia education available to her at the time.

In 1938 Apgar was named Director of the Anesthesia Division in the Department of Surgery at Columbia. She faced numerous obstacles in attempting to run her department. The clinical case load was enormous, and Apgar had trouble recruiting other physician-anesthesiologists in an era when anesthesia was still considered a nurse's job. Apgar faced conflicts with surgeons over who was in charge during operations and was poorly compensated for her work; at first, she was not allowed to bill patients for her work. Yet Apgar prevailed, and remained in this department until 1959. She managed the clinical load and attempted to add a research component to her division. In 1949 Dr. Emmanuel Papper, an anesthesiologist at Bellvue with a strong research interest, became chief of the division, and he and Apgar were both named full professors in the medical school. Apgar was the first woman to reach that position at Columbia.

Virginia Apgar. *(AP/Wide World Photos. Reproduced with permission.)*

Since she was no longer involved in the administrative duties of the division, Apgar could develop her interest in obstetric anesthesia. At mid-century evaluation of newborn babies was a haphazard process. In 1949 during breakfast in the hospital cafeteria, a medical student remarked on the need to evaluate newborns. Apgar quickly wrote down five items needed in such an evaluation and began to test her ideas. Her scoring method was presented at a medical meeting in 1952 and published the following year. The evaluation of newborns at one- and five-minute intervals quickly became standard delivery room practice, and today the Apgar Score is known around the world.

For the next decade Apgar continued to research the application of her scoring system to the condition of newborns. She also studied the effects on the baby of anesthetics given to the mother during delivery. By 1959 Apgar was ready for a change. She left Columbia and moved to Johns Hopkins University in Baltimore, Maryland, and earned a masters degree in public health. When the March of Dimes organization asked her to lead a new division of birth defects, she accepted and spent the last 15 years of her life in that position.

Apgar never married or had children of her own. She did maintain throughout her life the interest in music demonstrated in her undergraduate college years. Whenever she traveled, she carried along her viola and played with local musicians as often as possible. In October, 1994, a 20-cent stamp honoring Apgar was issued by the U.S. Postal Service. At the ceremony releasing the stamp, four musicians played a viola, cello, violin, and mezzo violin that Apgar herself had made. In 1996 these instruments were donated to the Columbia University College of Physicians and Surgeons, where this remarkable woman did so much to improve the health of newborn infants.

A. J. WRIGHT

Christiaan Neethling Barnard
1922-
South African Surgeon

Christiaan Barnard performed the first human heart transplant in 1967 in South Africa. Barnard's surgery focused attention on the rapidly developing field of organ transplantation and sparked social and philosophical debates concerning the ethics of transplantation. Following Barnard's pioneering efforts, more than 30,000 human heart transplants have been completed worldwide.

Barnard was born in Beaufort West, which is in the Cape of Good Hope province of South Africa. Barnard studied medicine at The University of Cape Town, where he received an M.D. in 1953. Shortly afterwards, Barnard left South Africa to continue his studies at the University of Minnesota, where he earned a Ph.D. in 1958. While in Minnesota, Barnard was trained by C. Wallton Lillehei (1918-1999), who is considered the "father" of open-heart surgery. Barnard changed his specialty from general to cardiothoracic surgery and assisted Lillehei's team with research, which led to the development of the first heart-lung machine. The development of the heart-lung machine, a device that pumps oxygenated blood throughout the patient's body, allowed Barnard and his colleagues to immobilize the heart during surgery long enough to complete complex repairs.

Christiaan Barnard. *(The Library of Congress. Reproduced with permission.)*

In 1958 Barnard returned to South Africa and taught surgery at the University of Cape Town. Subsequently, Barnard became director of surgical research at the Groote Schuur Hospital, where he introduced open-heart surgery, designed artificial heart valves, and engineered surgical protocols for several heart procedures. After years of research and experimentation with canine heart transplantation, Barnard, who, by 1967, was senior cardiothoracic surgeon at Groote Schuur, felt confident that the time was right to attempt the first human heart transplant.

Barnard's confidence was based on more than a decade of advances in organ transplantation. In 1954 surgeons in Boston had broken the human transplantation barrier when they transplanted a kidney from one identical twin to another. By 1963, success with kidney transplants had increased dramatically with the development of drugs that suppressed the body's immune response. These innovations, along with the development of the heart-lung machine, set the stage for Barnard's attempt to transplant a human heart.

On December 3, 1967, Barnard removed the diseased heart of 55-year-old Louis Washkansky in a five-hour operation. Barnard and his team of physicians and nurses then transplanted a healthy heart into Washkansky. The donor heart was obtained from 25-year-old Denise Darvall, who had died earlier at the hospital from injuries sustained in an automobile accident. Washkansky survived 18 days after the transplant, dying from pneumonia as a result of an immune system that had been suppressed to prevent rejection of the donor heart.

Barnard performed his second heart transplant on Philip Blaiberg on January 2, 1968. Blaiberg achieved fame as a symbol of hope for victims of heart disease. He survived 563 days after the operation. Based on this success, Barnard continued to develop and refine surgical techniques for the burgeoning field of heart transplantation. It was not until the early 1980s, however, with the advent of cyclosporin and other "next-generation" antirejection drugs, that the heart transplantation procedure became widely accepted. Research into cardiothoracic surgery continues in the laboratories of the Christiaan Barnard Building at the University of Cape Town.

In 1983, as rheumatoid arthritis interfered with the dexterity of his hands, Barnard retired from surgery. He has authored a cardiology text and several novels. In addition, he is the author of *Heart Attack: You Don't Have to Die* and an autobiography , *One Life*. Barnard lives near his boyhood home on a 32,000-acre ranch on South Africa's Karroo plateau.

BRENDA WILMOTH LERNER

Michael S. Brown

1941-

American Geneticist

One of America's leading experts on cholesterol metabolism in the human body, geneticist Michael S. Brown shared the 1985 Nobel Prize in physiology or medicine with Joseph Goldstein (1940-). Brown and Goldstein, who began working together in the 1970s, discovered the LDL receptor, a protein in the membranes of a cell that plays a central role in the body's ability to regulate cholesterol levels.

Brown was born on April 13, 1941, in New York City. He performed his undergraduate studies at the University of Pennsylvania, graduating in 1962, and went on to enroll in the university's medical school. There his research work earned him the Frederick Packard Prize in Internal Medicine. After earning his M.D. in 1966, he served as an intern and resident at Massachusetts General in Boston, where he first met Goldstein.

In 1968 Brown became clinical associate at the National Institutes of Health (NIH) in Bethesda, Maryland, where he worked in the biochemistry lab alongside Earl Stadtman, head of the laboratory for the National Heart, Lung, and Blood Institute. Brown focused on research in gastroenterology, particularly on the function of enzymes in digestive chemistry.

In 1971 Brown was appointed assistant professor at Texas Southwestern Medical School in Dallas. Goldstein joined the school's faculty in 1972, and soon afterward the two began working together. Their study involved familial hypercholesterolemia, an inherited disorder that causes dangerously elevated levels of cholesterol in the blood. To discover the genetic causes of this condition, they studied skin samples from a variety of patients.

Of particular interest to Brown and Goldstein were those rare subjects who had not one, but two defective genes. From childhood, they discovered, these people had exhibited extremely high levels of low-density lipoprotein (LDL), which carries cholesterol to the cells. In large quantities LDL can clog arteries and encourage heart disease. These patients, Brown and Goldstein observed, were missing a crucial type of protein known as a receptor, which binds with and regulates the levels of LDL. Without these receptors the body is unable to break down LDL, which then accumulates in the blood.

Having located the LDL receptor, Brown and Goldstein soon identified the gene responsible for its production. They then sequenced and cloned the gene, enabling them to isolate the genetic mutations that cause familial hypercholesterolemia and other inherited disorders involving cholesterol metabolism. By giving their patients specific combinations of drugs that inhibited their livers' ability to synthesize cholesterol, Brown and Goldstein increased their need for cholesterol from outside sources. The patients' bodies produced more LDL receptors, which in turn lowered their cholesterol levels.

Brown and Goldstein later performed a series of experiments that tracked the body's processing of cholesterol from first production to final dissolution. They noted that a liver transplant could correct genetic deficiencies in the production of LDL receptors, and demonstrated the means whereby a combination of low-fat diet and regular exercise could decrease cholesterol levels. In addition, they genetically engineered a mouse with an abnormally high number of LDL receptors, which allowed it to eat a high-fat diet without a significant rise in LDL.

Brown and Goldstein, in the words of the Nobel Prize Committee, "revolutionized our knowledge about the regulation of cholesterol metabolism and the treatment of diseases caused by abnormally elevated cholesterol levels in the blood." Brown has received a number of other awards in addition to the Nobel Prize, including the National Medal of Science in 1988. He has served as Paul J. Thomas Professor of Genetics and director of the Center for Genetic Diseases at the University of Texas Southwestern Medical School since 1977. Brown married Alice Lapin in 1964, and they have two daughters.

JUDSON KNIGHT

Benjamin S. Carson
1951–
African-American Neurosurgeon

In 1987 Dr. Ben Carson marked a milestone in neurosurgery when he successfully separated a pair of Siamese twins conjoined at the head, an operation that is arguably one of medicine's most significant accomplishments. Carson also made surgical history when he performed a successful hemispherectomy (removal of half the brain) on a young girl who suffered interminable seizures. In yet another groundbreaking procedure, he operated on an unborn twin suffering from hydrocephaly—cerebrospinal fluid on the brain—that caused an abnormal expansion of the baby's head. Carson was able to relieve the swelling and remove the surplus fluid—all while the infant remained in its mother's uterus.

Carson was born on September 18, 1951, in Detroit, the son of Robert Solomon and Sonya Copeland Carson. His father left when Carson was eight, forcing his mother, a domestic worker, to raise Carson and his brother Curtis on her own. In his 1990 autobiography *Gifted Hands*, cowritten with Cecil Murphey, Carson described himself as a hot-tempered child and a poor student. However, his mother was able to improve his grades dramatically by prohibiting him from watching television, and by requiring that he read two books a week. Carson ended up graduating at the top of his high school class.

With his outstanding academic record, Carson was in demand among the nation's highest-ranking colleges and universities, and he enrolled at Yale University, from which he graduated with a B.A. in 1973. From Yale he went to the University of Michigan at Ann Arbor, where he enrolled in the School of Medicine. In 1975 he married Lacena Rustin, whom he had met at Yale; they eventually had three children. Carson earned his medical degree in 1977, and the young couple moved to Maryland, where he became a resident at Johns Hopkins University. By 1982 he was the chief resident in neurosurgery in Johns Hopkins.

Carson spent 1983 at Sir Charles Gairdner Hospital in Perth, Australia, returning to Johns Hopkins in 1984, where he became director of pediatric neurosurgery in 1985. In the same year he performed the first of several significant neurosurgical operations, a hemispherectomy, or removal of half the brain, on an 18-month-old girl who had been having more than 100 seizures a day. The girl was healed, and Carson went on to perform numerous other successful hemispherectomies.

On September 5, 1987, in an operation that required 22 hours and a team of 70 people, Carson separated Patrick and Benjamin Binder, seven-month-old twins joined at the head. Although the twins sustained some brain damage, both survived the separation, making Carson's the first successful operation of its kind. Part of the operation's success was due to Carson's application of a technique he had seen used in cardiac surgery: a drastic cooling of the patients' bodies to stop blood flow. This ensured their survival during the delicate period when the surgeons were separating their blood vessels.

Carson performs some 500 operations a year, three times as many as most neurosurgeons. Because his career has represented a triumph over circumstances, he has become a well known inspirational writer and speaker, with several books in addition to *Gifted Hands*. He has received a number of awards, including the American Black Achievement Award from *Ebony* magazine in 1988; the Paul Harris Fellow from Rotary International (1988); the Candle Award from Morehouse College in 1989; and numerous honorary doctorates. In 1994 Carson and his wife (who goes by the name of Candy) established the Carson Scholars Fund, Inc., which offers scholarships to students, including those in elementary school, whose scholarship money is held in trust until they are ready for college.

JUDSON KNIGHT

Sir John Charnley
1911-1982
British Surgeon and Inventor

In the late 1950s and early 1960s John Charnley developed successful techniques and materials for total hip replacement (THR) surgery. This operation is now common, and is responsible for giving long-term pain relief and restoring mobility, functionality, and high quality of life to millions of patients.

Charnley was born into a middle class Methodist family in Bury, a northwestern suburb of Manchester, England. His father, Arthur, was what is called a "chemist" in Britain but a "pharmacist" in America. His mother, Lily, was a nurse. He had a younger sister, Mary, who received her degree in English literature from Girton College, Cambridge University, and had a distinguished career as a teacher and school administrator.

Charnley went through the usual boys' course of study at Bury Grammar School from 1919 to 1929. He was not a diligent student, but did well in science. In the fall of 1929 he entered the Victoria University of Manchester School of Medicine. In 1935 he received both the Bachelor of Medicine (M.B.) and Bachelor of Surgery (Ch.B.) degrees. He became a Fellow of the Royal College of Surgeons in 1936.

When World War II began in 1939, Charnley already had plenty of surgical experience in several prominent British hospitals. He volunteered for military service immediately, and was commissioned a lieutenant in the Royal Army Medical Corps in May 1940. That same month he participated in the evacuation of trapped British soldiers from Dunkirk, France. From 1941 to 1944 he was an orthopedic surgeon to the British forces in North Africa. While stationed in Cairo he began inventing and improving orthopedic devices and instruments. He arrived back in England just before D-Day, and returned to civilian life in February 1946.

His first two books, *Closed Treatment of Common Fractures* (1950) and *Compression Arthrodesis* (1953), established his reputation as an innovative and thoughtful biomechanical engineer as well as a surgeon. In 1958 the Manchester Royal Infirmary allowed Charnley to create his own hip surgery facility at Wrightington Hospital. This was the great turning point in his career, because it gave him the resources and staff to perform the kinds of experiments and operations he envisioned.

At Wrightington Charnley was incredibly productive. He used polytetrafluorethylene (PTFE), better known as Teflon, and stainless steel to realize his ideas for "low friction arthroplasty," that is, manufacturing and safely implanting strong, durable, biochemically inert artificial joints. In 1961 he published his basic results in *Lancet,* an influential British journal of medicine. By the mid-1960s THR had become a routine surgical procedure.

Charnley's major book, *Low Friction Arthroplasty of the Hip: Theory and Practice* (1979), described his further refinements of THR. His methods of attaching the artificial hip joint or "prosthesis" to the inside of the femur reduced the need for second operations. His clean-air surgical systems and sterile operating techniques decreased surgical wound infection.

Charnley was almost 46 when he married Jill Heaver, 20 years his junior, in 1957. Their son Tristram was born in 1959 and their daughter Henrietta in 1960. Charnley tried hard to be a good father, but could not cope with what William Waugh termed, in his standard biography *John Charnley: The Man and the Hip,* the "two-generation gap." Charnley especially could not tolerate the youth culture of the 1970s, and grew distant from his children, but reconciled with them when they became adults. They, along with his wife and sister, were nearby when he died in Manchester of complications from two heart attacks.

Charnley received several prestigious honors for his work, including Companion of the British Empire (1970), Fellow of the Royal Society (1975), and Knight Bachelor (1977).

ERIC V.D. LUFT

Denton Arthur Cooley
1920-
American Cardiovascular Surgeon

Denton A. Cooley is a pioneer in the evolution of modern cardiovascular (heart-circulatory system) surgery. In 1969 Cooley was the first surgeon to implant an artificial heart in a human awaiting heart transplantation. He also performed the first successful human heart transplant in the United States. Cooley is the founder of the world-renowned Texas Heart Institute, where more open-heart surgeries and heart diagnostic procedures have been performed than in any other facility in the world. Cooley's high-profile surgeries have sparked both praise and criticism in the medical community as well as broadened the boundaries of surgical practice.

Born and raised in Houston, Texas, Cooley was the son of a successful dentist and real estate investor. He excelled academically in school and overcame his shyness by playing basketball. Cooley majored in zoology at the University of Texas, where he also played on the varsity basketball team. While taking pre-med courses, Cooley became fascinated by surgery. Cooley earned his M.D. degree from Johns Hopkins University in 1944. During his residency there, Cooley assisted Alfred Blalock in the first "blue baby" operation, a milestone surgical procedure that corrected a congenital (inborn) heart defect in an infant. Blalock's influence and pioneering work on the frontiers of open-heart surgery inspired Cooley to specialize in the field. Cooley interrupted his residency in 1946 to fulfill an obligation with the U.S. Army, serving for two years in the Army Medical Corps. He was made Chief of Surgical Services at the Army Hospital in Linz, Austria. Upon his discharge, Cooley returned to Hopkins, where he finished his residency and became an instructor of surgery. Cooley earned a reputation for simplifying complex surgical techniques while exercising speed and dexterity in the operating room.

In 1950 Cooley spent a year in London at the Brompton Hospital for Chest Diseases, where he studied and worked with Lord Russel Brock, an eminent British surgeon, and participated in England's first intracardiac operation. Cooley returned home to Texas the following year to become an associate professor of surgery at Baylor University College of Medicine. At Baylor and its sister institution, Houston Methodist Hospital, Cooley began a collaboration with another cardiovascular surgeon, Michael E. Debakey, one which resulted in major innovations in heart surgery throughout the 1950s. Together the two developed and perfected a heart-lung bypass machine that used at Methodist to allow for the immobilization of the heart during surgical repairs. They also collaborated on surgical techniques to remove aneurysms (weakened areas of the arterial wall) from the aorta and to repair damaged heart valves.

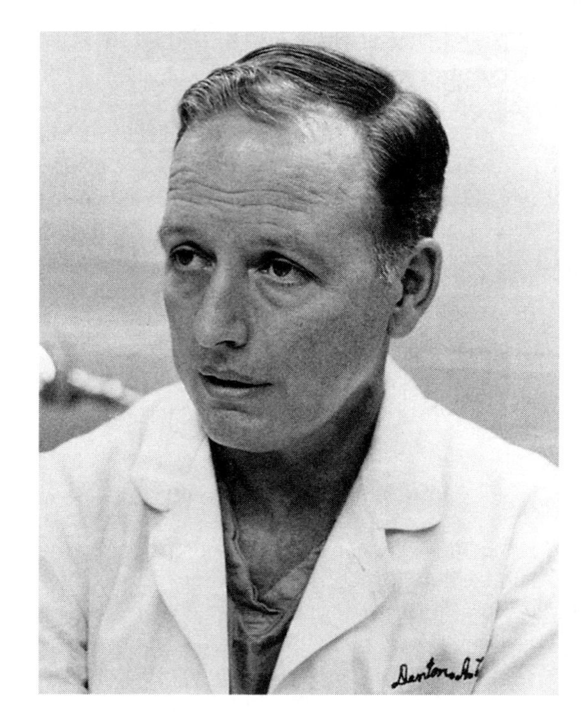

Denton A. Cooley. *(Archive Photos. Reproduced with permission.)*

In 1962 Cooley founded the Texas Heart Institute at St. Luke's Episcopal Hospital in Houston. Here, Cooley and his colleagues encouraged and refined the development of artificial heart valves, dramatically reducing the mortality rate for valve transplant patients. Cooley's team was the first to remove successfully a pulmonary embolism (blood clot in the lungs) and pioneered delicate procedures to correct congenital heart defects in infants and children. In 1967 the International Surgical Society awarded Cooley its highest accolade, the Renée Lebiche Prize.

Cooley performed the first successful human heart transplant in the United States on May 3, 1968. His patient, a 47-year-old man, received the heart of a 15-year-old girl who had committed suicide. (The donor's brain function had ceased but her heart was still beating.) Although the recipient lived for 204 days and Cooley received praise from the medical community, an ethical debate ensued as Americans wrestled with the issue of determining when the moment of death occurs. In 1969 Cooley implanted the first artificial heart in a human. This mechanical heart served as a temporary bridge to a human donor heart, which became available approximately three days later. Although the patient died the day after the human heart was transplanted, this pioneering effort sparked intensive worldwide research into artificial hearts. The procedures and laws governing organ donation, organ harvesting, and allocation of organs for transplant continue as a subject for debate and revision today.

During the 1970s, Cooley turned his attention to the study of coronary artery disease, which was by then the leading cause of death in the United States. Cooley severed his relationship with Baylor to assume his present post as Surgeon-in-Chief at the Texas Heart Institute, where Cooley pioneered many techniques for the coronary bypass operation. Cooley and his associates at the Heart Institute have performed approximately 100,000 open-heart operations—more than any other institution in the world. In 1998 President Bill Clinton presented Cooley with the Medal of Technology, the nation's highest honor for technical innovation.

BRENDA WILMOTH LERNER

William DeVries. *(Bettmann/Corbis. Reproduced with permission.)*

William Castle DeVries

1943-

American Surgeon

On December 2, 1982, Dr. William C. DeVries successfully implanted the Jarvik-7 artificial heart, a plastic and titanium pump powered by compressed air, in the first total heart replacement intended for permanent use. The recipient survived only a few months with the device, illustrating the limited potential of the Jarvik-7 and other artificial replacement hearts as they currently exist. Efforts to develop an improved artificial heart continue.

DeVries was born in Brooklyn, New York, on December 19, 1943, the son of Hendrik and Cathryn Castle DeVries. He earned his B.S. degree (1966) and his M.D. (1970) from the University of Utah Utah. He served his internship at Duke University Medical Center in North Carolina, and went on to an eight-year residency in general and thoracic surgery. From 1979 to 1984 he worked as professor of surgery at the University of Utah. During this time, he continued to perform surgery—including the historic implantation of the Jarvik-7 heart.

The Jarvik-7 artificial heart, designed by American physician Robert K. Jarvik, used a pump made of plastic and titanium to deliver compressed air through two tubes inserted in the patient's abdomen. On December 2, 1982, DeVries performed an operation on Barney Clark, who was gravely ill and running out of options. Clark and his surgeons hoped that the Jarvik-7 would successfully replicate the functions of a natural heart, and for nearly four months it appeared that it had. After 112 days, however, Clark died. Later, other surgeons implanted Jarvik-7 hearts in four patients, each of whom died—though one, William Schroeder, lived for 620 days, or nearly two years. Artificial hearts have since been used primarily to assist weakened and damaged hearts, allowing them to recover, or as a temporary devices that allow patients to survive until a suitable organ is found for transplantation. Research and development on artificial hearts continues, however, because there are many more candidates for heart transplants than there are organs that can be used.

7In 1989, DeVries left his teaching position in Utah to form DeVries & Associates of Elizabethtown, Kentucky. He sits on the boards of numerous Louisville, Kentucky, hospitals, and continues to teach at the University of Louisville. A recipient of the Wintrobe award in 1970, DeVries belongs to a number of professional associations, including the American College of Chest Physicians. He has seven children.

JUDSON KNIGHT

Ian Donald
1910-1987
English Obstetrician

Obstetrician Ian Donald helped develop the first successful diagnostic ultrasound machine in the 1950s. His innovation—applying sonar techniques to diagnosis—was initially greeted with skepticism, but eventually adopted and adapted for widespread use. Brown also invented a respirator for newborn infants, and in his later years worked to develop a birth-control device that would predict ovulation.

Donald was born in Liskeard, Cornwall, England, on December 27, 1910, the eldest of four children. His father, John Donald, was a doctor, and his mother, Helen Wilson Donald, a concert pianist. Donald attended Fettes College in Edinburgh, then went to South Africa, where he studied at the Diocesan College in Rondebosch, and later at Capetown University. There he obtained a B.A. in French, Greek, English, and music. Upon his return to England, he enrolled at St. Thomas's Hospital Medical School, earning his M.B. (bachelor of medicine) and B.S. in 1937. In the same year he married Alix Richards, with whom he eventually had four daughters.

During World War II, Donald served in the Royal Air Force, and after four years of service in 1946, was named Member of the British Empire. In 1947 he received his M.D. from St. Thomas's. Five years later, he became an instructor at Hammersmith Hospital, where he developed a device for resuscitating newborn infants with respiratory problems. He was appointed to the regius chair (a chair endowed by the British Crown) of midwifery at the University of Glasgow in 1954. The following year he published the textbook *Practical Obstetric Problems.*

Donald's search for an instrument that would help diagnose problems in unborn children was aided by his wartime experience with radar and sonar. With help from T. G. Brown of Kelvin Hughes, an electronics corporation, he created the world's first ultrasound machine. In 1958 Donald and John MacVicar published their findings in the *Lancet*. At first the idea of using sonar for diagnosis was greeted skeptically by the medical community. Over the course of his career, however, Donald saw his device—which underwent considerable development over time—gain wide acceptance.

Donald began to suffer from ill health in 1961, but he continued to work for several years. His department moved from Glasgow Royal Maternity Hospital in 1964 to the new Queen Mother's Hospital, which he helped design. A devout Christian, he was an outspoken opponent of the Abortion Act of 1967, and also opposed experiments on embryos. In the 1970s and 1980s he received a number of honors, including the Eardley Holland gold medal (1970), the Blair Bell gold medal (1970), Commander of the Order of the British Empire, (1973), the Order of the Yugoslav Flag (1982), and many others.

During his later years, Donald worked on a birth-control method that would alert women when they were approaching ovulation. Plagued by heart problems, he underwent three major operations, and died at his home in Paglesham, Essex, on June 19, 1987.

JUDSON KNIGHT

Robert Geoffrey Edwards
1925-
British Physiologist

Robert Geoffrey Edwards and Patrick Christopher Steptoe (1913-1988) pioneered in vitro fertilization (IVF), making the birth of the first "test-tube baby" possible in 1978. By quickly transferring the oocyte (the egg prior to maturation) to an optimal cultural medium, Edwards was able to replicate the conditions necessary for an egg and sperm to survive outside the womb.

Edwards was born on September 27, 1925, the son of Samuel and Margaret Edwards. He attended the University of Wales from 1948 to 1951, and the University of Edinburgh from 1951 to 1957. He then worked for a year as research fellow at the California Institute of Technology (Cal Tech) before joining the staff at the National Institute of Medical Research in Mill Hill, England, in 1958.

Edwards took a position at the University of Glasgow in 1962, but moved to Cambridge University the following year. In 1965 he served as visiting scientist at Johns Hopkins University, and in 1966 at the University of North Carolina. Later, he returned to Cambridge, where he became Ford Foundation reader (instructor) in physiology in 1969, a position he held until 1985. While at Cambridge in 1968, Edwards met P. C. Steptoe, with whom he began an important collaboration.

By analyzing the conditions necessary for an egg and sperm to survive outside the womb, Ed-

Robert Edwards (right), shown here with collaborator Patrick Steptoe. *(Bettmann/CORBIS Reproduced with permission.)*

wards was able to develop an appropriate medium—which he called "a magic culture fluid"—in which to achieve fertilization. In 1971 he and Steptoe performed their first attempt to implant a fertilized egg in a patient. They were not successful, however, until the birth of Louise Brown, dubbed the first "test-tube" baby, in July 1978. The IVF method developed by Edwards and Steptoe soon gained wide acceptance, and proved successful in dealing with a number of types of infertility.

After serving as visiting scientist at the Free University of Brussels in 1984, Edwards became professor of human reproduction at Cambridge in 1985. He remained in that position until his retirement in 1989, after which he became a professor emeritus. Together with Steptoe, he established the Bourne Hallam Clinic, and served as its scientific director from 1988 to 1991.

Edwards is Extraordinary Fellow of Churchill College, and served as chair of the European Society of Human Reproduction and Embryology from 1984 to 1986. He is a member and honorary member of several other professional societies, and has received awards from around the world. His publications include *A Matter of Life* (1980), written with Steptoe, and several other works.

JUDSON KNIGHT

René Geronimo Favaloro

1923-

Argentinean Surgeon

In 1967 René Favaloro performed the first successful, fully documented coronary bypass, substituting a vein from the patient's leg for a damaged artery in the heart. The procedure had been attempted unsuccessfully before, and successfully in an emergency situation, but Favaloro's was the first planned bypass, complete with supporting literature, which he published in 1968. Favaloro also established one of Latin America's leading medical training facilities, the Institute of Cardiology and Cardiovascular Surgery in Buenos Aires, Argentina.

Favaloro was born in 1923 in La Plata, Argentina. His father, Juan Favaloro, was a carpenter, and his mother, Ida Raffaeli Favaloro, a dressmaker. Favaloro attended the National College and Medical School at the University of La Plata, from which he received his M.D. degree in 1949. He underwent his internship and residency at the Instituto General San Martin in La Plata, where he also held his first staff position.

In 1961 Favaloro traveled to Cleveland, Ohio, to observe the latest techniques in myocardial revascularization (increasing a restricted blood supply to the heart) at the Cleveland Clinic. He asked for a job, and joined the hospital's

Medicine

1950-present

thoracic and cardiovascular team in 1962. Five years earlier, F. Mason Sones, Jr., of the Cleveland Clinic had developed angiography, in which dye is inserted via catheter into the arteries, exposing on X-rays the exact location of blockages. This tool had greatly improved conditions for bypass surgery by facilitating the selection of candidates for such operations.

In 1962 David Sabiston of Duke University in North Carolina performed the first human bypass, but it had been unsuccessful. Two years later, in 1964, H. Edward Garrett performed the first successful bypass operation, but he did so in an emergency situation, and did not publish documentation for another decade. Favaloro's 1967 bypass was a groundbreaking operation, and so successful that in the following three years Favaloro and others on his team performed more than 1,000 such operations. A quarter were multiple bypasses, and the death rate was an impressively low 4.2 percent.

Favaloro left the Cleveland Clinic in 1971 and returned to Buenos Aires, where he established the Favaloro Foundation to teach bypass surgery. He also began work on a dream he finally realized in 1992, with the completion of a ten-story, $55-million facility for the Institute of Cardiology and Cardiovascular Surgery, one of the finest medical teaching programs in Latin America. Favaloro's programs have trained hundreds of heart surgeons, and he and his team have performed thousands of bypass operations.

Favaloro, who served in the Army of the Republic of Argentina as a lieutenant, is considered a hero in his homeland, where he is often suggested as a presidential candidate. Favaloro himself, however, has expressed little interest in a political career. He is married to Maria A. Delgado, and is a member of several professional societies.

JUDSON KNIGHT

Daniel Carelton Gajdusek

1923-

American Virologist and Pediatrician

In 1976 Daniel Carleton Gajdusek shared the Nobel Prize in Medicine with Baruch S. Blumberg (1925-) for their discoveries concerning "new mechanisms for the origin and dissemination of infectious diseases." Gajdusek had suggested the existence of "slow viruses," novel viruses that seemed to remain dormant for long periods of time before attacking the body. The concept of slow viruses emerged from Gajdusek's studies of kuru, a degenerative brain disease found among the Fore people of Papua New Guinea. The slow viruses were eventually implicated as the causative agents of other diseases, including Creutzfeldt-Jakob disease and mad-cow disease.

Daniel Carleton Gajdusek. *(Bettmann/CORBIS. Reproduced with permission.)*

Gajdusek was born in Yonkers, New York. His father was born in Slovakia and his mother was a first-generation Hungarian American. His aunt, Irene Dobrozcky, an entomologist, who arranged for him to obtain a summer job at the Boyce Thompson Institute for Plant Research, encouraged his interest in science. Gajdusek attended the University of Rochester and was awarded his bachelor's degree summa cum laude in 1943. Three year later he was awarded the M.D. from the Harvard Medical School, where he had specialized in pediatrics. To complete his training, Gajdusek spent the years from 1946 to 1951 doing an internship at Columbia Presbyterian in New York, and residencies at Children's Hospital in Cincinnati and Children's Medical Center in Boston. He also participated in a postwar medical mission to Germany, studied physical chemistry at Caltech with Linus Pauling (1901-1994), and carried out research in the virology laboratory of John Enders (1897-1985) at Harvard.

In 1952 Gajdusek was drafted and assigned to the Walter Reed Army Institute for Research.

Joseph Smadel, who supervised his virology research at the Institute, suggested that Gajdusek might be more suited to field epidemiology work. After he left the service in 1954, Gajdusek worked with Marcel Baltazard at the Pasteur Institute in Teheran studying the epidemic diseases of the Middle East. Two years later Gajdusek moved to Melbourne, Australia to work with Sir MacFarlane Burnet (1899-1985). Gajdusek met Vincent Zigas, district medical officer of the Public Health Department in Port Moresby, Australia, during a field excursion to New Guinea. Zigas introduced him to a strange neurological disorder, known as kuru, which was found only among the Fore people, who lived in isolation in the eastern highlands of New Guinea. Most of the victims of the disease were women and children. Gajdusek spent almost a year living with and studying the Fore, and collecting tissue samples from victims of kuru.

When Gajdusek returned to the United States in 1958, he was able to carry out laboratory studies of kuru at the National Institute of Neurological and Communicative Disorders and Stroke, where he eventually established the Laboratories of Slow, Latent, and Temperate Virus Infections and of Central Nervous System Studies. The brain tissue obtained from victims of kuru revealed many lesions and atrophy, but not the expected signs of infection. William Hadlow of the NIH Rocky Mountain laboratory pointed out the similarity between kuru and a viral disease of sheep known as scrapie. To test the possibility that kuru was a viral disease, Gajdusek and his associate, Clarence J. Gibbs, inoculated chimpanzees with brain tissue from kuru victims. Two years later the experimental animals showed symptoms of the disease. Brain issue from these animals produced the same disease in previously healthy chimpanzees. Studies of the Fore people suggested that the disease was transmitted by ritual cannibalism, in which women and children ate the brains of those who had died of kuru. After the ritual was abandoned, the disease eventually disappeared. Gajdusek and Gibbs were able to transmit other brain diseases, such as Creutzfeldt-Jakob disease, in a similar manner.

Gajdusek adopted many children from New Guinea and brought them to the United States. He used much of his Nobel Prize money for the education of his adopted children. He retired in 1997 from the National Institutes of Health in Bethesda, where he had been chief of the Laboratory for Central Nervous System Studies.

LOIS N. MAGNER

Robert Charles Gallo

1937-

American AIDS Researcher

Robert Charles Gallo is known for his discovery of the AIDS virus, for developing a test for the virus that not only diagnoses cases but screens the blood supply, for development of AZT, the only effective AIDS medication, and for his relentless pursuit of an AIDS vaccine.

Gallo's father was a northern Italian immigrant who worked very hard at his metallurgy business in Waterbury, Connecticut, and had abandoned the religious beliefs of the Roman Catholic Church. His mother Louise, a very religious and happy person, was part of a big southern Italian family. When Gallo's sister Judith became ill with leukemia, the family was literally torn apart. Gallo blamed God for stealing his childhood and began at 14 to abandon his family for the streets.

Through the entire ordeal of his sister's illness, he had talked with Judith's pathologist, Dr. Marcus Cox, and admired him greatly. Cox enticed him away from the streets to the adventures of medicine. Gallo attended the Thomas Jefferson University School of Medicine in Philadelphia. He disliked being around sick people but was very interested in microbiology, genes, and viruses. After a residency at the University of Chicago, he went to the National Institute of Health (NIH) in Bethesda in 1965. Gallo won recognition at the institute for his tremendous ego and ability to generated large amounts of money for the government.

As head of the National Cancer Institute Tumor Laboratory, he discovered the first human leukemia virus, the condition that had killed his sister. He also found a number of oncogenes—cancer-causing genetic material—and established that interleukin-2 helps white blood cells fight tumors.

During the early 1980s, a strange new virus threat was emerging. The unknown affliction was characterized by an immune deficiency, along with a group of other conditions. It was called "acquired immunodeficiency syndrome" or AIDS. On April 3, 1984, U.S. Health and Human Services Secretary Margaret Heckler stood along side of Gallo as she announced that the probable cause of AIDS had been found—a virus that was called human immunodeficiency virus or HIV—and that a blood test had been developed. Attributing these discoveries to

Robert C. Gallo (right). *(Archive Photos. Reproduced with permission.)*

Gallo, she also predicted that a vaccine would be possible in a few years.

A firestorm erupted immediately with charges that Gallo had stolen the discovery of the virus from French scientist Luc Montagnier (1932-), of the Pasteur Institute in Paris. Indeed, the French researcher had taken the virus from the lymph nodes of a patient with AIDS, but did not show it caused AIDS. Gallo had actually shown the virus to be the cause. Another controversy erupted when Gallo was charged with using a strain of the French virus to develop the test for AIDS. The story was recreated in the film *And the Band Played On,* with actor Alan Alda cast as Gallo. The film portrayed Gallo as a self-seeking, pompous scientist, and the image stuck. While it was established in 1991 that the sample used by Gallo's lab to isolate the virus was indeed contaminated by a sample from Montagnier's lab, Gallo was cleared of any wrongdoing in 1993.

Gallo's research led to the formulation of AZT for the treatment of AIDS. Money began pouring in for AIDS research, and with the enormous sales of AZT, Gallo became a respectable researcher in great demand in industry. Maryland Governor Parris Glendening, whose brother who died of AIDS, lobbied intently. When Maryland and Baltimore offered to build a multimillion-dollar institute around Gallo, he accepted. The research center, called the Institute of Human Virology, is part of the University of Maryland, but Gallo's private company was authorized to commercialize any marketable results.

Many promising projects are being developed toward the treatment of AIDS. Scientists have discovered naturally occurring molecules called chemokines that suppress HIV in vitro. Gallo has also undertaken work on a gene therapy vector. However, a vaccine for AIDS is a major goal. Although controversial, Gallo is committed to AIDS research. He holds or shares 79 patents and his discoveries have generated more that one million dollars in private sector research.

EVELYN B. KELLY

John Heysham Gibbon Jr.

1903-1973

American Surgeon

Inventor of the heart-lung machine, John H. Gibbon, Jr., first demonstrated that life can be maintained by an external pump acting as an artificial heart during an operation on a cat in 1935. Eighteen years later, in 1953, Gibbon performed the first successful open-heart operation using a heart-lung machine.

Born in Philadelphia on September 29, 1903, Gibbon was the second of four children born to John Heysham Gibbon (a surgeon) and Marjorie Young Gibbon. He earned his B.A. from Princeton University in 1923, and his M.D. from Jefferson Medical College in Philadelphia in 1927. He completed his internship at Pennsylvania Hospital in 1929, and in 1930 went to Harvard Medical School as a research fellow in surgery.

In 1931 Gibbon married Mary Hopkinson, a surgical researcher, and took a position as fellow in medicine at the University of Pennsylvania School of Medicine. In the same year he became assistant surgeon at Pennsylvania Hospital. In 1934 he went to work as a research fellow in surgery at Harvard Medical School, where he developed the first generation of his heart-lung machine using a rotating blood-film oxygenator. Then, on May 10, 1935, he made his first successful use of the heart-lung machine during an operation on a cat in which the device took over cardiac and respiratory functions.

Gibbon began work as a surgeon at Pennsylvania Hospital in 1937, and in the following year developed a second-generation version of his machine using DeBakey roller pumps. In 1940, before America entered World War II, he joined the U.S. Army Reserves as a major in the medical corps, and in 1942 was sent on active duty to New Caledonia in the south Pacific. He was promoted to the rank of lieutenant colonel in 1944, and in 1945 acted as chief of surgical service at Mayo General Hospital in Galensburg, Illinois.

From 1945 to 1946, Gibbon worked as assistant professor of surgery at the University of Pennsylvania School of Medicine, and served as a member of council for the American Association of Thoracic Surgery. He became professor of surgery and director of surgical research at Jefferson Medical College, as well as attending surgeon at Jefferson Medical College Hospital, in 1946. Gibbon held both positions for 10 years, and served as chairman of the editorial board for *Annals of Surgery* from 1947 to 1957.

In 1949 IBM developed the model I heart-lung machine with a revolving film oxygenator, and in 1951 produced the model II oxygenator. On May 6, 1953, Gibbon performed the first successful bypass surgery, the repair of an atrial septal defect on an 18-year-old female, using the heart-lung machine. In 1954 IBM produced its model III, which cost $79,650—or about $320,000 in 1990s dollars.

In 1956 Gibbon became the Samuel D. Gross professor of surgery and head of the surgery department at Jefferson Medical College and Hospital, positions he would hold until 1967. During this same period, he worked as attending surgeon in chief. Gibbon also worked as consulting surgeon at Pennsylvania Hospital and as consultant in general surgery for the Veterans Administration Hospital in Philadelphia from 1950 to 1967. In 1959 he was appointed Taub visiting professor of surgery at Baylor Medical College, and in 1960 visiting professor of surgery at Harvard Medical School.

Gibbon served as a member, fellow, and officer of numerous professional organizations, ranging from the board of governors for the American College of Surgeons (1950-64) to the Pennsylvania Medical Society, for which he served as delegate at large from 1961-63. He held posts and memberships on some two dozen other organizations. His awards included the Distinguished Service Award from the International College of Surgery (1959), an honorary fellowship from the Royal College of Surgeons in England (1959), the Gairdner Foundation International Award from the University of Toronto (1960), an honorary Sc.D. from Princeton University (1961) and the University of Pennsylvania (1965), the Research Achievement Award from the American Heart Association (1965); the Albert Lasker Clinical

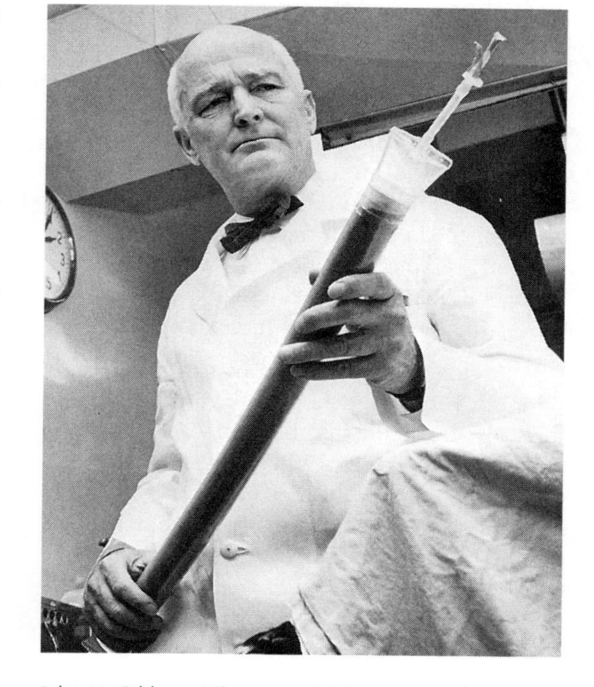

John H. Gibbon. *(Cbettmann/CORBIS. Reproduced with permission.)*

Research Award (1968), the Dixon Prize in Medicine from the University of Pennsylvania (1972), and many others.

Gibbon himself suffered from heart trouble in his later years. He had his first myocardial infarction in July 1972, and died of a massive myocardial infarction while playing tennis on February 5, 1973.

JUDSON KNIGHT

Joseph Leonard Goldstein
1940-
American Geneticist

Nobel laureate Joseph Goldstein was the first person to describe the process by which cholesterol is metabolized and accumulates in the human body. The understanding of the relationship between cholesterol, blood lipids, and heart disease has been one of the most important health discoveries of the twentieth century, enabling the development of new drugs and diet regimens to lower blood cholesterol levels.

Goldstein was born April 18, 1940, in Sumter, South Carolina. His family owned a clothing store in Kingstree, South Carolina. Goldstein graduated from Washington and Lee University in Lexington, Virginia, in 1962. He received the doctor of medicine from Southwestern Medical School of the University of Texas Science Center in 1966. Goldstein was offered a faculty position by Donald Seldin, who was impressed by Goldstein, if he would study genetics and return to Dallas to establish a division of medical genetics in the department of internal medicine. Goldstein declined at that time and went on to do an internship at Massachusetts General Hospital in Boston, where he met Michael Brown (1941-), another intern who became a life-long friend and future collaborator.

Moving to the National Institute of Health (NIH) in 1962, Goldstein worked in the laboratory of Marshall Nirenberg (1927-), a biochemist whose great challenge was to decipher the genetic code. Goldstein also served as an associate at the National Heart Institute, where he described the action of several proteins related to protein synthesis. While at NIH he launched a study into cholesterol metabolism in the human body, a topic that would become a lifelong pursuit. Noting that cholesterol accumulation in the blood was related to heart attacks, he began a study of the risk factors, leading him to the idea of a genetic component.

From 1970-72 Goldstein worked at the University of Washington as a Special NIH Fellow in the laboratory of Dr. Arno G. Motulsky, who helped establish the relationship between medicine and human genetics. There Goldstein studied tissue culture techniques and population genetics strategies. He and his colleagues initiated and completed a study to determine the frequency of hereditary lipid disorders in heart attack survivors. They found that 20 percent of heart attack survivors have one of three single-gene types for hyperlipidemia, or the presence of excess fat and cholesterol in the blood. Familial hypercholesterolemia affects one out of 25 heart attack victims, and it is present in one out of every 500 people.

Taking up Dr. Seldin's earlier offer, in 1972 Goldstein returned to Southwestern Medical as an assistant professor in Internal Medicine and led the school's first Division of Medical Genetics. He also convinced them to bring Michael Brown, by then an expert in metabolic disease, to Dallas. The two combined their expertise in the study of genetic regulation of cholesterol metabolism.

Concentrating on the problem of accumulation of cholesterol, they discovered that low-density lipoproteins (LDLs) are the major carrier of cholesterol. Special receptors on the cell's surface take the LDLs out of the bloodstream. They wondered what would occur if these receptors were not present. They found that under such circumstances LDLs build up in the blood vessels, causing them to eventually clog. Absence of the LDL receptors are found in family hypercholesterolemia. The body cannot remove LDL from the bloodstream, allowing it to build up. This discovery led to the use of drugs to combat the buildup and new understanding of the importance of low cholesterol diets.

Goldstein was elected to the National Academy of Science in 1980 and has won numerous awards and lectureships. In 1985 he and Brown won the Nobel Prize for Physiology or Medicine. Goldstein has since continued his research at the Southwestern Texas Medical School in Dallas, where he has worked to unravel the mechanisms by which the sterol regulatory element binding proteins (SREBPs) pathway regulates cholesterol metabolism at the molecular, cellular, and whole body levels.

EVELYN B. KELLY

Roger Charles Louis Guillemin
1924-
French Physiologist and Endocrinologist

Roger Guillemin's studies of the hormonal control of the pituitary gland, and his work in isolating several hormones, have led to a greater understanding of the disease diabetes, as well as female sexual development. He was awarded the Nobel Prize for Medicine, along with Andrew Schally (1926-) and Rosalyn Yalow (1921-), in 1977.

Guillemin was born on January 11, 1924, in the small French town of Dijon, the capital of the Burgundy region. His education began at the local public schools and progressed to Dijon's medical school in 1943. Six years later, he received his M.D. from the Faculté de Médecine of Lyons, with his studies based on clinical training, including a three-year "rotating internship." While Guillemin completed his studies, a dark pall settled over his hometown. France was in the throes of World War II, and the German army had taken control of Dijon.

Guillemin soon gained an interest in endocrinology, the study of the glands and hormones of the body, thanks to two of his favorite teachers, P. Etienne-Martin and J. Charpy. The two professors had conducted some of the earliest research in the field. Although Guillemin yearned to pursue a career as a laboratory researcher, there were no such facilities in Dijon, save those reserved for the field of gross anatomy.

While visiting Paris, he attended a lecture on endocrinology by Hans Selye, and was fascinated by the man and his research. A few months later, he joined Selye's newly created Institute of Experimental Medicine and Surgery at the University of Montreal, where he completed his doctoral dissertation in 1949. Rather than submit to the formalities of the French research profession, he returned to Montreal in 1953 to complete his Ph.D. in physiology at Selye's Institute and to begin his career in a less rigid setting. While there, he developed an interest in the process by which physiological controls effect the secretions of the pituitary gland in response to stress.

In 1953 Guillemin was hired as an assistant professor at the Baylor University College of Medicine in Houston, Texas. He continued his research there, becoming more and more involved in the search for clues to the hormonal control of the pituitary gland, particularly by the hormones secreted by the hypothalamus. In 1970 he moved to the Salk Institute at La Jolla, California, where he established his Laboratories for Neuroendocrinology. He later moved to the Whittier Institute for Diabetes and Endocrinology in La Jolla, and also served as Adjunct Professor of Medicine at the University of California at San Diego.

Guillemin has been honored by several scientific establishments. He was elected to the National Academy of Sciences in 1974, and became a member of the American Academy of Arts and Sciences in 1976. He has received several awards, most notably the 1977 Nobel Prize in Medicine.

In addition to his work in the sciences, Guillemin is known as an art collector, and as a talented artist himself. His computer-generated paintings have been exhibited in galleries throughout the United States, Mexico, and Italy.

Guillemin's work in the field of endocrinology led to the isolation of many hormones, including somatocrinin, or growth-hormone releasing factor, which is crucial to our understanding of diabetes. He also discovered endorphins and identified many of the hormones that regulate female sexual development.

STEPHANIE WATSON

Charles Brenton Huggins
1901-1997
American Surgeon

Charles B. Huggins won the Nobel Prize for physiology or medicine in 1966 for his discovery, made three decades earlier, of the relation between hormones and prostate and breast cancer. His research yielded a number of valuable ideas and techniques, most notably hormone therapy, the first nontoxic and nonradioactive chemical treatment for cancer.

Charles B. Huggins. *(The Library of Congress. Reproduced with permission.)*

Born on September 22, 1901, in Halifax, Nova Scotia, Huggins was the elder of two sons born to pharmacist Charles Edward Huggins and wife Bessie Spencer Huggins. In 1920 Huggins earned his B.A. degree from Acadia University in Wolfville, Nova Scotia, graduating in a class of just 25 students. Later that year he entered Harvard Medical School, from which he graduated in 1924 with both M.A. and an M.D. degrees. Huggins spent his internship at the University of Michigan Hospital, and in 1926 became instructor in surgery at the university's medical school.

A year later, in 1927, Huggins became a surgery instructor for the newly opened University of Chicago Medical School, and married Margaret Wellman; the couple had two children. Huggins became an assistant professor in 1929, then an associate professor in 1933—the same year he became an American citizen. In 1936 he attained the rank of full professor.

Though urology had originally been his focus, Huggins became interested in cancer research as early as 1930, when he met the German cancer researcher and Nobel laureate Otto

Warburg (1883-1970). Using cells from the male urinary tract and bladder, Huggins began to experiment with changing normal connective tissue elements into bone. His research soon focused on the role played by chemicals and hormones in the prostate gland, the male accessory reproductive gland located at the base of the urethra. Using dogs as experimental subjects, since they are the only animal other than man known to develop cancer of the prostate, Huggins developed, in 1939, a procedure for isolating the dogs' prostate glands. This allowed him to analyze and measure glandular secretions, and Huggins went on to study human subjects with prostate cancer.

He soon noted high levels of the male sexual hormone testosterone in the secretions of cancerous prostate glands. This offered two possibilities of treatment for prostate cancer, though the first—orchiectomy, or castration—was a radical one. And though the levels of androgens, or male sex hormones, dropped markedly after orchiectomy, they often rose again, in some cases higher than before.

The other possibility for reducing testosterone levels appeared to be injecting estrogen, a female hormone; this method, however, appeared effective on only a small proportion of patients. Huggins then discovered why: the adrenal glands produced their own androgens, apparently to compensate for the lower levels induced by hormone therapy, and these androgens helped spread the cancer. Following this discovery, Huggins performed the world's first bilateral adrenalectomy in 1944, removing the two adrenal glands located above the patient's kidneys. Nine years later, he reported that a combination of adrenalectomy and cortisone therapy—another radical treatment—proved effective on about 50 percent of patients suffering with prostate or breast cancer.

Huggins had meanwhile served briefly as professor of urological surgery and director of the department of urology at Johns Hopkins University in 1946. He soon returned to the University of Chicago, and in 1951 became director of its Ben May Laboratory for Cancer Research, where he served for the next 18 years. In a 1958 speech before the University of Glasgow, he referred to breast cancer as "one of the noblest of the problems of medicine;" it was around that time he and students D. M. Bergenstal and Thomas Dao developed a treatment for it that involved removal of both ovaries and both adrenal glands.

During the 1960s, when scientists debated whether birth-control pills encouraged breast cancer, Huggins analyzed his own work over the preceding decades and found that it did not. Some researchers, in fact, later suggested that the pill actually worked to discourage some forms of cancer. In 1969, three years after winning his Nobel Prize, Huggins left the Ben May Laboratory, and in 1972 returned to Acadia University as chancellor. He retired from that post and moved to Chicago, where he devoted his time to his family and his favorite music, Bach and Mozart. He died in 1997.

JUDSON KNIGHT

Robert Koffler Jarvik

1946-

American Physician and Bioengineer

Robert Jarvik is known for his design of an artificial heart. This mechanical device, known as the Jarvik-7, was used with some success during the mid-1980s to permanently replace a diseased heart or to be used temporarily with individuals waiting for a human donor. Although the Jarvik-7 is no longer in use, this mechanical heart prolonged the lives of nearly 50 people and contributed significant information to our understanding of how artificial organs work and how they can be improved upon.

Jarvik, the son of a physician, was a child who enjoyed mentally disassembling things and putting them back together. At age 8 he was building elaborate model boats and airplanes. When he was 16 years old, after watching several surgical procedures with his surgeon father, he invented an automatic surgical stapler that simplified clamping and tying blood vessels. This was to be the first of his five patented inventions.

Jarvik initially enrolled in architecture school at Syracuse University and did not become interested in pre-med courses until his father developed cardiovascular disease. He was rejected from 15 medical schools in the United States before eventually being accepted at the University of Bologna in Spain. He studied there but did not earn a degree. Later, in 1971, he earned a master's degree in occupational biomechanics at New York University. After graduating he joined Dr. Willem Kolff, a pioneer in the field of artificial organ research at the University of Utah. With the encouragement of Kolff and his own personal interest in designing and developing the artificial heart, he once again entered medical school and in 1976 earned his M.D. degree.

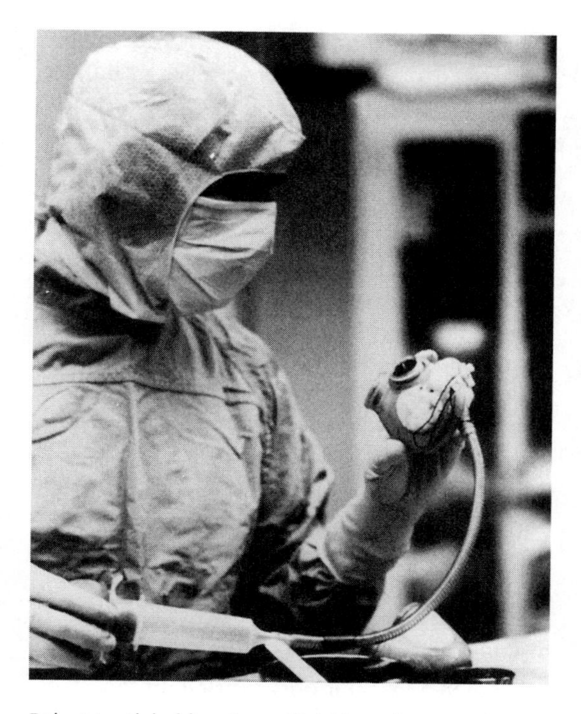

Robert Jarvik holding the artificial heart he created. *(AP/Wide World Photos. Reproduced with permission.)*

Kolff and Jarvik joined their expertise in 1977 and together launched a bioengineering company, Symbion, to research and develop artificial hearts. Kolff, a Dutch-born physician, had been involved in artificial heart research since the mid-1950s. Kolff's artificial heart design successfully kept a human patient alive for more than 60 hours in 1969. Building on Kolff's research and design, the two men worked together for nearly 10 years to re-design an effective artificial heart.

The artificial heart that the two developed was identical to the structure and action of a natural heart. The Jarvik-7 artificial heart was constructed of plastic, aluminum, and polyester and required an external source of power supplied through a system of compressed air hoses that entered the heart through the chest. This compressed air energized and regulated pumping action of the device. The artificial heart was designed to function like a natural heart and had two pumps (ventricles), each with a disc-shaped mechanism that pushed the blood from the inlet valve to the outlet valve.

The completion of the device was generally met with some skepticism within the medical community. At the time, human heart transplants were experimental, and sufficient numbers of human hearts were unavailable. Therefore, the artificial heart held great promise for thousands of people with life-threatening heart disease even though the procedure was experimental.

In 1982 a team led by cardiac surgeon William DeVries (1943-) of the University of Utah implanted the first Jarvik-7 device into Barney Clark. Clark lived for 112 days until he died from complications of a stroke. In 1984 William Schroeder received a Jarvik-7 and lived for a record 620 days before he died of similar complications. The artificial heart was implanted three additional times, and unfortunately all of the recipients died. It appeared that the man-made metal and plastic materials triggered a physiological response as clots formed on pump surfaces and broke free and lodged in the brain.

After these deaths, the Jarvik-7 was used only for individuals awaiting a heart transplant. In 1986 the federal government requested that the artificial heart not be used on human beings until additional research was conducted.

LESLIE HUTCHINSON

Clarence Walton Lillehei
1918-1999
American Surgeon

Clarence Walton Lillehei was one of the leading innovators of modern cardiac surgery. Through his pioneering advances in open-heart surgery and the use of the pacemaker, countless lives have been saved. Open-heart operations that once seemed like science fiction are now commonplace in hospitals around the world.

Lillehei was born in Minneapolis on October 23, 1918. He attended a local high school, and went on to graduate with honors from the University of Minnesota in 1939. In 1942 he graduated from the University of Minnesota Medical School.

During World War II, Lillehei served from 1942-46 in the United States Army, enlisting as a first lieutenant and eventually rising to the rank of lieutenant colonel. He was awarded the Bronze Star for "meritorious services in support of combat operations" while serving in Anzio, Italy. After returning from the war, he received his Ph.D. in surgery from the University of Minnesota Graduate School and became a clinical instructor in the University of Minnesota Medical School's Department of Surgery in 1951. He taught medical students as a professor in that department in 1956, and was later appointed its chairman, a position he held for seven years.

Lillehei was known by his peers and admirers as the "father of open-heart surgery," a title he earned by developing the techniques that first made surgery on the human heart a possibility. Until the 1950s, there was no surgical replacement for the natural process by which oxygen was pushed into the bloodstream and circulated throughout the body. Opening the heart and interrupting this process without creating an alternate oxygen pathway was impossible. The only heart ailments that could be remedied, therefore, were those that could be treated without direct invasive surgery.

Two of Lillehei's predecessors in the field, John F. Lewis and John H. Gibbon, Jr. (1903-1973), had in the early 1950s successfully used artificial pump oxygenators to circulate oxygen while they repaired atrial heart defects. After their landmark operations, however, progress in heart surgery was stalled after the oxygenators failed on several occasions. Also, many surgeons at the time held the misconception that open-heart surgery was impossible without the availability of an artificial heart to support post-operative recovery.

To overcome these obstacles, Lillehei pioneered the use of "cross circulation," in which the patient being operated upon was hooked up to the bloodstream of a healthy donor by tubes. Lillehei and his team used this method for the first time at the University of Minnesota on March 26, 1954. The operation was a success, representing the first step in the development of open-heart surgery.

"Cross circulation" ultimately proved too risky to the donor and was soon abandoned, but a new method Lillehei developed proved even more effective. In 1955 he and colleague Richard A. DeWall used a heart-lung machine, called a helix reservoir bubble oxygenator, to 'bubble' oxygen throughout the bloodstream during an operation.

Lillehei made several other breakthroughs in the 1950s, which made the treatment of once-fatal heart conditions possible. One such condition is known as "heart block," in which the body is unable to produce the small electrical signals that regulate heartbeat. Heart block was one of the leading causes of death among patients undergoing repair of ventricular defects.

In 1957 Lillehei and his colleagues were able to successfully link wires into the human heart, electronically replicating the missing signals and causing the heart to beat in a steady rhythm. The signals were generated by a battery-powered pacemaker—which was about the size of a pack of cigarettes—small enough to be easily hidden under the patient's clothing. Lillehei introduced his pacemaker at a conference on cardiovascular surgery in January 1958. The pacemaker ultimately created a billion-dollar industry, allowing those with heart defects to live a healthy, normal life.

Lillehei also made advancements in the management of heart valve disease. For the first time, diseased valves could be replaced with artificial valves, like the widely used St. Jude Mechanical Heart Valve. In 1967 he introduced a new type of artificial heart device consisting of layers of plastic sheets and silicone rubber membranes that helped strengthen hearts weakened by disease.

While heading up the surgical team at New York Hospital in 1969, Lillehei performed the first heart and double-lung transplant at that hospital, placing the organs of a 50-year-old woman into a 43-year-old man suffering from terminal lung and heart failure.

Lillehei was not only himself a pioneer, he trained hundreds of doctors in the precise art of heart surgery, and wrote numerous papers and articles on the subject. His protegees included Norman Shumway, who went on to develop the surgical technique used in the heart transplant operation, and Christiaan N. Barnard (1922-), the South African surgeon who performed the first successful heart transplant in 1967.

Lillehei's achievements have been recognized with numerous international honors. He was nominated for the Nobel Prize in Medicine several times, received the 1955 Lasker Award for "Outstanding Contributions to Cardiac Surgery," the Harvey Prize in Science and Technology, and the American Medical Association's Hektoen Gold Medal. Lillehei was President of the American College of Cardiology, and held honorary memberships in 33 foreign societies, as well as honorary doctorates at five international universities.

Lillehei succumbed to cancer in July 1999. He was 80 years old. Though he is no longer living, his great contributions to cardiac surgery continue to improve and prolong the lives of many others.

STEPHANIE WATSON

Luc Montagnier
1932-
French Virologist

Since 1972 Luc Montagnier has headed the Viral Oncology Unit of the Pasteur Institute.

Under his leadership, in 1983 this unit discovered the retrovirus later named the human immunodeficiency virus (HIV-1) and identified it as the causative agent of acquired immune deficiency syndrome (AIDS). Two years later Montagnier's team isolated the second-known human AIDS virus, HIV-2. Montagnier has established an international network of researchers to study the AIDS pathogenesis (how the disease originates and develops). He is convinced that such knowledge will lead to the development of a vaccine against this fatal disease, which now effects millions of individuals worldwide.

Montagnier studied both medicine and biology at the University of Poitiers and then at the Sorbonne in Paris. In 1957 he decided on a career in virology and spent four years doing research in the United Kingdom. His early work focused on the relationship of viruses and oncology, the formation of cancer tumors. He was the first to show that during the reproduction of a virus inside a cell, its RNA takes the form of a double helix very similar to the DNA double helix. In early 1964 Montagnier developed a test *in vitro* (outside the body) showing the carcinogenic power of a virus.

Such successes led to his appointment in 1972 as the head of the newly created Viral Oncology Unit of the Pasteur Institute in Paris, where he concentrated his research on antiviral defenses. Viruses are molecular parasites in the cell and use the cell's mechanisms for transmitting their genetic messages. Throughout the 1970s and 1980s Montagnier continued to work on RNA tumor viruses. He did extensive research on the interferons (small proteins) that are produced in viral-infected cells and that signal neighboring cells to produce enzymes that curb viral multiplication.

Montagnier also began research on human retroviruses. A retrovirus is a RNA virus that makes a DNA copy of itself in the host cell. This copy is incorporated into the host's DNA. The cell then makes more copies of the virus that spread to other cells. Some retroviruses cause very slow, degenerative diseases; they are known as lentiviruses or retrolentiviruses. This research put Montagnier in a key position when AIDS became an epidemic in the mid-1980s, since AIDS is a disease caused by a retrolentivirus that devastates the body's immune system.

In 1983 Montagnier's team isolated the retrovirus that causes AIDS. He also discovered that the virus at first does not openly attack the infected individual but apparently remains dormant for a lengthy period. After that dormancy the virus suddenly becomes active and destroys the body's immune system. The patient dies from a wide variety of diseases, such as pneumonia and tuberculosis, against which his/her body no longer has defenses. In 1985 Montagnier's Pasteur team discovered a second AIDS virus.

Luc Montagnier (right), with Jonas Salk. *(Bettmann/CORBIS. Reproduced with permission.)*

Montagnier became involved in a bitter controversy with Dr. Robert Gallo (1937-) of America's National Institutes of Health (NIH) over who first identified the AIDS virus and which country should receive royalties from AIDS-related tests and vaccines that might be developed. It was established in 1991 that the sample containing the virus Gallo claimed his laboratory had isolated, and which was used to develop a blood test for the disease, had been accidentally contaminated by a virus from a sample sent to him by the French. Montagnier is therefore recognized as the discoverer of the virus. After a long legal battle the Pasteur Institute and the NIH agreed in 1987 to joint ownership. In the previous year the initially identified virus was named HIV-1, and the second one isolated by Montagnier HIV-2. A person whose blood samples contain the virus but who has not yet developed the symptoms of AIDS is termed HIV-positive.

During the 1990s researchers learned that developing an effective vaccine against HIV will not be easy. HIV mutates rapidly when attacked, so that drugs quickly lose their power as the virus mutates to a less vulnerable form. This mutation makes finding an HIV vaccine almost impossible. Montagnier thinks that, since bacterial factors increase the virulence of some animal lentiviruses, perhaps such a "cofactor" exists in the case of HIV that can be identified and destroyed. He has been concentrating his research efforts on mycoplasmas, tiny bacteria often associated with animal lentiviruses.

Each day, 16,000 people become infected with HIV and 7,000 people die because of AIDS. The United Nations estimates that AIDS is responsible for 5,500 deaths a day in Africa alone. Thirty-three million people are HIV-positive and at least sixteen million have died of AIDS. As the epidemic spreads to underdeveloped areas of Asia and Africa, Montagnier has become a world leader in encouraging AIDS prevention and treatment efforts. In 1993 he co-founded and became the president of the World Foundation for AIDS Research and Prevention.

ROBERT HENDRICK

Joseph Edward Murray

1919-

American Surgeon

Joseph Murray began his career as a plastic surgeon, but soon became interested in transplantation—surgically moving tissue from one part of the body to another or from one person to another. Murray eventually performed the first successful transplant of an entire human organ.

Murray was born in 1919 in Milford, Massachusetts. He did not take premedical courses in college, but instead earned a liberal arts degree from the College of the Holy Cross. His studies focused on philosophy, English, Latin, and Greek. After graduating, however, he attended Harvard Medical School and completed a medical degree in 1943.

Soon afterward, Murray entered the army and was assigned to Valley Forge Army Hospital in Pennsylvania. This hospital specialized in plastic surgery, and Murray worked there to treat burn victims. One treatment for burn victims is the use of skin grafts—the removal of healthy skin from one part of a patient's body to cover a burned or otherwise injured area. Sometimes, however, the patients at Valley Forge were so severely burned that they had no undamaged skin available for grafting. As a temporary measure, skin was transplanted to the patients from other people. These "foreign" grafts could not be left in place permanently, however, because the patient's body eventually rejected them. Murray became interested in why this rejection occurred. (It is now known that the immune system can recognize a unique set of marker proteins on the foreign cells.) At Valley Forge, Murray was taught that the more closely related two people are, the less likelihood of rejection. (Closely related people are more likely to have similar marker proteins.)

After three years, Murray left the army and returned to Boston to finish his surgical training. He became a plastic surgeon, specializing in the treatment of facial cancer and birth defects. In 1951, he joined a group of doctors who were attempting to develop procedures for transplanting kidneys. Kidneys remove waste from the bloodstream and help to maintain water balance. Although humans have a pair of these organs, they can survive with only one. The doctors reasoned that a healthy person could theoretically donate one of his or her kidneys to someone whose kidneys were damaged or diseased.

Before attempting the surgery on humans, Murray experimented with transplanting kidneys in dogs. Then, in 1954, the case of Richard Herrick was brought to his attention. Herrick was near death due to kidney disease. He had an identical twin brother, and a kidney transplant was proposed. Murray believed that Herrick's immune system would not recognize his twin's kidney as foreign and so would not reject it. First, Murray performed a skin graft from one twin to the other as a test. When the graft proved successful, Murray led a team of doctors in the kidney transplant operation. The operation worked, and Herrick lived for an additional eight years.

Murray then began to examine ways in which kidneys could be transplanted between people who were not identical twins. Other scientists had developed ways of suppressing a patient's immune system so that it would not attack foreign tissue. One of these methods was with the use of x rays. In 1959, Murray transplanted a kidney between fraternal, or nonidentical, twins. The kidney recipient was irradiated with x rays and went on to live for an additional 26 years. Other such operations, however, were not as successful. The x rays often so reduced the ability of the patients' immune systems to function that their bodies were susceptible to lethal infections.

As transplanting technology improved, Murray began testing a drug that suppressed the immune system called azathioprine. He became the first doctor to transplant kidneys from both unrelated and recently deceased donors, and he treated the recipients with this drug. About half of the first group of patients who received azathioprine survived. (It was thought that only 20% would do so.) Today, azathioprine is still given to the recipients of kidney transplants, and thousands of these operations are performed each year. Seventy percent of patients survive for at least a decade.

In addition to his work in the operating room, Murray helped to establish a worldwide registry of patients who had received kidney transplants so that their progress could be tracked and monitored. In 1990, he won the Nobel Prize in medicine for his work in kidney transplantation. His successes helped to initiate the advances in other transplant surgery such as liver, heart, and pancreas transplant operations.

STACEY R. MURRAY

John Rock
1890-1984
American Gynecologist and Obstetrician

American gynecologist, obstetrician, and researcher John Rock played a leading role in two landmark developments in the field of human reproduction. He is best known for his work in producing and testing the oral contraceptive, a device which revolutionized the modern birth control movement. Rock's earlier research, motivated by the desire to help childless women, culminated in the first successful fertilization of a human ovum in a test tube—in vitro fertilization.

John Rock was born on March 24, 1890, in Marlborough, Massachusetts, where his father was a businessman. After being fired twice as an accountant, Rock abandoned the business world to pursue an education, obtaining his M.D. degree in 1918 at Harvard Medical School. In 1925 he married Anna Thorndike, and eventually became the father of five and the grandfather of 19. Appointed Assistant Professor of Obstetrics at Harvard in 1922, Rock began practicing in Boston, and in 1924 opened the Fertility and Endocrine Clinic, one of the nation's first facilities to focus on the problems of infertility.

In the summer of 1944, Rock and Harvard scientist Miriam F. Menkin announced a breakthrough in the history of assisted reproductive technology—the first successful fertilization of a human ovum, or egg, outside the human body. The two researchers had been collecting ripened ova from female surgical patients, maintaining them in test tubes filled with the donor's blood serum, and then exposing them to sperm. After six years of failure, they succeeded in bringing fertilized ova to the three-cell stage of development, publishing their results in the prestigious journal *Science*. Although Rock himself doubted that it would ever happen, his experiments did in fact lead to the birth of the world's first "test tube baby," Louise Brown, born 34 years later in London in 1978.

FROM KIDNEYS TO ARMS

In 1984 an Australian man severed his hand in an accident. Although doctors attempted to reattach the hand, the operation eventually proved unsuccessful, and the man's arm was amputated just below the elbow. Then, in 1998, fourteen years after the accident, a team of doctors led by Jean-Michel Dubernard (1941-) of France and Earl Owen of Australia transplanted a hand and forearm to the patient from a recently deceased donor. This complicated operation was the first of its kind to be successful. The first part of the procedure involved attaching the bone of the transplant to the patient's upper forearm. Then, the blood vessels, nerves, muscles, and skin of the donated arm were sewn to those of the patient, a process that took more than 13 hours. To keep the transplant from being rejected, the patient was given drugs that suppressed his immune system. A year after the operation, the patient had some feeling in all of his new fingers and was even able to write and use a fork with his new hand. In 2000 Dubernard and Owen were part of the team that performed the first double-arm transplant, in which one patient received two arms from a single donor.

Still searching for ways to overcome sterility, Rock experimented in the early 1950s with injections of progesterone (the female hormone that suppresses ovulation) in childless women, surmising that fertility might increase when the injections were stopped. He contacted scientists Gregory Pincus (1903-1967) and Min Chueh Chang (1908-1991) for assistance with developing a synthetic hormone that could be taken orally. Pincus and Chang were more interested in the contraceptive properties of progesterone, and research proceeded simultaneously for its use as a treatment for sterility and as a contraceptive.

Upon his retirement, Rock started the Rock Reproductive Clinic and devoted himself full time to investigations with progesterone. Clinical trials were successful, and the first report of the use of hormones to suppress ovulation in women was published by Pincus and Rock in 1956. In 1960, the U.S. Food and Drug Administration approved the manufacture and sale of the synthetic hormone Enovid.

By 1964 four million women were on "the pill," and by 1998 the estimated number was 100 million women throughout the world. Since that time, oral contraceptives have been the focus of medical controversy and media attention, as the relative risks and benefits have been investigated and debated. According to a 1998 report from the U.S. Centers for Disease Control and Prevention, the impact of reliable family planning afforded by oral contraceptives has been immense: as couples chose to have fewer children, child mortality declined, people moved from farms to cities, the age of marriage increased, and the health of children and women improved, along with the social and economic status of women.

Rock was an outspoken activist for the use of contraceptives for family planning and population control despite his lifelong membership in the Roman Catholic Church, which forbade the practice. In 1963 his book, *The Time Has Come: A Catholic Doctor's Proposal to End the Battle Over Birth Control*, helped to legitimize this form of birth control among people of all faiths.

John Rock died on December 4, 1984, in Peterborough, New Hampshire, at 94 years of age.

DIANE K. HAWKINS

ROCK VS. THE ROMAN CATHOLIC CHURCH

Pope Pius XI's 1931 encyclical *Casti Connubii* articulated the Church's traditional stand on birth control, condemning all methods except periodic abstinence, also called the rhythm method, as unnatural and constituting grave sin. The Second Vatican Council (1962-65), however, allowed that sex within marriage served the dual purposes of procreation and the expression of love. Rock, uniquely positioned as a devout Roman Catholic and leading gynecologist, urged the Church to reexamine its stand in the light of emerging science. He argued that the pill, created from natural hormones, extended the time when a woman was infertile, thereby acting in concert with the rhythm method.

Conservative Catholic theologians harshly criticized Rock, but the media painted him as David facing Goliath. In response to public pressure, Pope Paul VI appointed the Commission for the Study of Population and Family Life to study the issue. In 1966 the Commission submitted its report, with medical experts recommending by a vote of 60 to four, and cardinals by a vote of nine to six, to liberalize the teaching on birth control. Pope Paul rejected that recommendation, reasserting in his 1968 encyclical *Humanae Vitae* that every conjugal act must be open to the transmission of life. Six hundred Catholic scholars signed a protest, great numbers of priests left the Church, and the laity were left in a painful conflict between obedience and conscience.

Albert Bruce Sabin
1906-1993
Polish-American Microbiologist

Albert Sabin is best known for his pioneering research on poliomyelitis ("infantile paralysis") and his development of an orally administered live attenuated vaccine for the prevention of the disease. Poliomyelitis is an acute viral infection that can invade the nervous system and cause paralysis. Where the disease is common, most infections probably go unnoticed, or result in mild symptoms.

Sabin was born in Bialystok, Poland. At the end of World War I, he immigrated to the United States to escape religious persecution. His family settled in New Jersey in 1921, where he attended Patterson High School. He enrolled in the dental school at New York University, but switched to microbiology. After working in Dr. William H. Park's laboratory, Sabin was admitted to the medical school of New York University. He received his M.D. in 1931, then pursued further training as an intern at Bellevue Hospital and a year at the Lister Institute in London. He was a member of the scientific staff at the Rockefeller Institute (now University) in New York from 1935 to 1939. Most of his medical career was spent at the Cincinnati College of Medicine and the Children's Hospital Research Foundation in Ohio. During Word War II, from 1943 to 1946, he worked with the U.S. Army as an epidemic disease investigator. He was appointed president of the Weizmann Institute in Israel in 1972. Although he left that position two years later, he remained active in various national and international medical agencies.

Dr. Park, his mentor at New York University, stimulated Sabin's interest in virology and viral diseases. One of Sabin's first successful efforts to isolate a dangerous virus occurred in 1932, when an associate at NYU died after being bitten by a laboratory monkey. The new viral agent, which was called the B virus, had caused the acute ascending myelitis. Poliomyelitis, one of the most feared epidemic diseases of the time, became Sabin's major research interest. Contrary to prevailing views about the means of transmission of the poliovirus, Sabin proved that the virus was spread by the fecal-oral route rather than the nasal route. The virus multiplied in the human intestinal tract. By 1936 Sabin and his associates had been able to isolate and propagate the poliomyelitis virus in laboratory cultures of human embryonic nervous tissues.

Although the natural history of poliomyelitis was still generally obscure, it was known that protective antibodies could be found in the body of survivors many years after infection. Sabin believed that long lasting immunity could best be established with a live attenuated vaccine. A killed vaccine might be easier to develop, but its effects might not be as long lasting. Another advantage of the live vaccine was that it could be administered by mouth. The attenuated virus might also be spread from those who had been vaccinated to others. During a period when unvaccinated people might easily encounter the wild type virus, the benefits of spreading the attenuated virus might outweigh the dangers. When the disease was almost eradicated, the transmission of even an attenuated form of the poliovirus to non-immune people might, however, be dangerous. Eventually, American epidemiologists found that the use of the live vaccine was associated with a small, but real risk of paralytic polio.

Before Sabin had perfected his oral vaccine, Jonas Salk (1914-1995) developed a successful killed vaccine that had to be administered by injection. A nationwide trial of the Salk vaccine in 1954 was successful. As a result of the Salk vaccine program, the incidence of paralytic polio in the United States decreased dramatically by 1961. The National Foundation for Infantile

Albert Sabin. *(Archive Photos. Reproduced with permission.)*

Paralysis was committed to the Salk vaccine and unwilling to sponsor other vaccines. The World Health Organization, however, supported tests of oral polio vaccines. Sabin was an active participant in the vaccine trials carried out in Mexico, Czechoslovakia, and the Soviet Union. The highly successful vaccination tests in the Soviet Union reached about 145 million people by 1960. In 1985 the World Health Organization began an effort to eradicate polio worldwide. Sabin was active in the battle to lift the burden of infectious diseases throughout the world by bringing immunization to Third World children.

LOIS N. MAGNER

Jonas Edward Salk

1914-1995

American Virologist and Physician

Jonas Salk is best known for his pioneering research on poliomyelitis ("infantile paralysis") and his development of an injectable killed virus vaccine for prevention of the disease. Wide scale testing in 1954 quickly led to the national distribution of the Salk vaccine. Immunization campaigns resulted in a major reduction in the incidence of poliomyelitis in the United States. Work by Salk's rival, Albert Sabin (1906-1993), led to the development of a live attenuated vaccine that could be administered orally.

Jonas Salk. *(AP/Wide World Photos. Reproduced with permission.)*

Salk was born in New York City. He was the oldest son of Russian-Jewish immigrants who encouraged their children to pursue higher educational and professional goals. Salk attended Townsend Harris High School and the City College of New York. His original intention was to study law, but he was attracted to the medical sciences and decided to become a doctor. While a student at the medical school of New York University, Salk began research on the recently discovered influenza virus. This work served as the basis for his later research on the poliovirus and convinced him that he would prefer research to medical practice. He received his medical degree in 1939 and began a two-year internship at Mount Sinai Hospital, New York. He applied to Dr. Thomas Francis, Jr., a respected virologist and epidemiologist, for a research position in his laboratory at the University of Ann Arbor, Michigan. During World War II, Salk worked with Francis on influenza, a problem that was considered significant to the war effort. Salk moved to Ann Arbor in 1942 and spent the next six years working on various was of inactivating the influenza virus and producing a safe and effective vaccine. In 1947 Salk accepted an offer from the University of Pittsburgh Medical School, where he became the only full-time member of the medical school faculty. The Sarah Mellon Scaife Foundation provided funds for renovation of Salk's laboratory. To secure funds for research, Salk applied to the National Foundation for Infantile Paralysis and began research on the poliovirus. The next eight years were devoted to the development of a vaccine against polio.

During the 1950s, poliomyelitis was one of the most feared epidemic childhood diseases, because infection could lead to paralysis or even death. Poliomyelitis is an acute viral infection that can invade the nervous system, but where the disease is common, most infections probably go unnoticed, or result in a mild febrile illness, with sore throat, headache, vomiting, and stiffness of the neck and back. Before the introduction of the polio vaccine, the annual incidence in the United States during certain epidemic years reached over 10,000 paralytic cases. Such epidemics formed the basis of the image of polio as the great crippler of children, and exerted a profound influence on the direction of medical research.

Although the natural history of poliomyelitis was still generally obscure, Albert Sabin had proved that the virus was spread by the fecal-oral route rather than the nasal route. Before Sabin had perfected his oral vaccine, Salk developed a successful killed vaccine that was administered by injection. A nationwide trial of the Salk vaccine in 1954 was successful. The results of the trial were announced in April 1955 and led to widespread acclaim for Salk, who was further lauded for his refusal to patent the vaccine. As a result of the Salk vaccine program, and the efforts of the National Foundation for Infantile Paralysis, the incidence of paralytic polio in the United States decreased dramatically by 1961. In 1963 Salk established the Jonas Salk Institute for Biological Studies in La Jolla, California, to encourage innovative scientific and medical research. During the last years of his life, Salk remained active in research dealing with AIDS.

LOIS N. MAGNER

Thomas E. Starzl

1926-

American Surgeon and Researcher

Thomas E. Starzl is the father of liver transplantation and a pioneer in the field of organ transplants. In a career spanning nearly five decades, his research and surgical techniques have set the standard for organ transplants worldwide. Starzl stands as one of the greatest medical minds of the twentieth century.

Born in the small town of LeMars, Iowa ("corn and hog capital of the world") in 1926,

Starzl grew up in a strict German Catholic family. His father, Roman (R.F.), was editor and publisher of the daily paper. His mother, Anna Laura Fitzgerald, was a surgical nurse and his inspiration in pursuing a career in medicine. Reportedly, the driven young man knew he wanted to be a surgeon by age 11.

Starzl was an outstanding student in high school and worked summers for his father at the newspaper. He was on the debating, football, and basketball teams and even played the trumpet in the band. Next, Starzl majored in biology at Westminster College in Fulton, Missouri, then went on to Northwestern University Medical School in Chicago. By 1952 he had earned both a Ph.D. in neurophysiology and an M.D. with distinction.

Starzl held surgical residencies and fellowships at Johns Hopkins, the University of Miami, and the Veterans Administration Research Hospital in Chicago. In 1958 he returned to Northwestern, where he performed surgery and conducted research. At the time Starzl searched for a subject, as he recalls, "difficult and complex enough to invest a lifetime in." It turned out to be the nascent field of liver transplant. "Transplanting was hardly even thought of as a possibility then," Starzl remembers. "I was working blind."

Transplant surgery and research attracted Starzl because it was a new field and fulfilled his commitment to saving lives. In 1962 Starzl, with wife, Barbara, and three children, joined the University of Colorado School of Medicine. While at Colorado, Starzl made his mark in organ transplanting. Liver transplant surgeries, Starzl soon discovered, were grueling tests of endurance and stamina, but the more difficult aspect was that the body's own immune system fought against the new organ. Starzl spent long hours of research trying to overcome this challenge.

Although few colleagues agreed with him, Starzl's early research successes bolstered his efforts and convinced him that liver transplant would work in humans. On March 1, 1963, Starzl performed the world's first human liver transplant, but the patient bled to death during the operation. Starzl's work continued, to the chagrin of the legal establishment. In 1967, the same year as the first heart, pancreas, and lung transplants, Starzl met success when a young girl lived for 13 months after having her liver replaced. By the time he left Colorado, Starzl and his surgical team had transplanted over 1,000 livers.

In 1980 Starzl accepted a position at the University of Pittsburgh Medical Center (UPMC). Already a medical superstar, Starzl and his second wife, Joy, settled into the Steel City and built the hospital into the world's foremost transplant institute, re-named the Thomas E. Starzl Transplantation Institute in 1996. Starzl did not stop at research and surgery, either. He founded the United Network of Organ Sharing, which prioritizes the recipient list nationwide. Hopping planes and helicopters through adverse weather conditions and operating for up to 24 hours straight, Starzl and his team also established the logistical system of retrieving donated organs, then transporting them for surgery.

After two heart surgeries of his own in 1990 and reaching age 65 a year later, Starzl stopped operating. It was a relief for him, as he described in his 1992 memoir, *The Puzzle People,* "I was not emotionally equipped to be a surgeon or to deal with its brutality." As a researcher, Starzl is the most cited scientist in clinical medicine and has averaged one paper every 7.3 days, making him one of the most prolific scientist in the world.

Starzl's amazing career has laid the foundation for nearly all the work done today in liver transplantation. His research into anti-rejection drugs and his pioneering spirit set the stage for the high success rate liver transplant patients can now expect. His most important legacy may be that he and his team from UPMC have taught other surgeons around the globe, thus making liver transplants accessible worldwide. Recently, Starzl placed 213th in the book, *1,000 Years, 1,000 People: The Men and Women Who Charted the Course of History for the last Millennium.*

BOB BATCHELOR

Patrick Christopher Steptoe
1913-1988

British Obstetrician, Gynecologist, and Reproductive Biologist

Patrick Steptoe pioneered the technique of human in vitro fertilization for the treatment of infertility. The technique involves removing eggs from the ovary of a woman, fertilizing them with sperm in the laboratory, and returning the developing embryo to the mother's uterus, where pregnancy proceeds. The process, first done successfully in 1978, sparked widespread ethical debates that continue today.

Steptoe was born in Whitney, Oxfordshire, in England. He studied medicine at King's College, London, and St. George's Hospital, where he qualified as a medical doctor in 1939. Steptoe

Patrick Steptoe. *(AP/Wide World Photos. Reproduced with permission.)*

entered the Royal Navy in 1939, serving as a surgeon during World War II. He was a prisoner of war in Italy from 1941 to 1943. After the demobilization in 1946, Steptoe began his practice of obstetrics and gynecology. In 1951 he moved to the northern city of Oldham, where he became senior obstetrician and gynecologist at the state Oldham Hospitals.

In the early 1950s Steptoe noted the excessive number of laparotomies (abdominal surgeries) performed on women who were being treated for infertility. He searched for a simpler method of visualizing internal female reproductive organs. Steptoe accomplished this task by pioneering the technique of laparoscopy. In this procedure, a thin, flexible tube tipped with a fiber optic light is inserted into the abdomen, allowing the visualization of blocked Fallopian tubes and other causes of infertility. Steptoe spent nearly a decade researching and perfecting laparoscopic techniques, including laparoscopic sterilization.

In 1967 Steptoe teamed up with Robert Edwards, a Cambridge University physiologist, to continue infertility research. Steptoe devised a way to retrieve eggs from the ovaries of women who had been made infertile by blocked Fallopian tubes. Edwards's research concentrated on fertilizing the retrieved eggs outside of the body with sperm in laboratory conditions that would allow the fertilized egg to grow into a zygote. In 1968 Edwards achieved the first such successful fertilization and, by 1970, normal zygote growth to the sixteen-cell stage was achieved.

In 1972 Steptoe first returned a fertilized and dividing egg to the uterus of a woman from whom the egg had been removed. Implantation, however, was not successful. Steptoe and Edwards endured the scientific skepticism of their peers as well as a difficult relationship with the mass media. Finally, in 1977 successful implantation and pregnancy was achieved through in vitro fertilization. The mother, Leslie Brown, was unable to conceive naturally because of blocked Fallopian tubes. On July 25, 1978, at Oldham Hospital, Brown gave birth to a healthy baby girl, named Louise. The media reported the birth of the world's first "test-tube baby" with a zeal not seen since man first landed on the moon.

The event inspired debate as well as praise. Many infertile women volunteered for further research, hailing in vitro fertilization as the miraculous answer to their quest for motherhood. Others, including the Roman Catholic Church, saw in vitro fertilization as a morally questionable use of scientific techniques. This debate continues today as improved technologies allow for the storage of fertilized zygotes for extended periods of time.

Steptoe and Edwards were initially reluctant to publish their research. With the birth of baby Louise Brown, however, skepticism faded and fellow scientists eagerly received Steptoe and Edwards's work, which they presented in 1979. Advances in what has become known as ART—assisted reproductive technologies— have enabled couples with various fertility problems to have children through in vitro fertilization. Today, over 300 clinics in the United States alone are dedicated to the practice of ART, with in vitro fertilization serving as the primary means of achieving pregnancy for infertile couples. Worldwide, an estimated 500,000 babies have been born using in vitro fertilization technologies.

Steptoe, along with Edwards, founded the Bourn Hall Clinic in 1980, which serves as a research and treatment center for infertility. Together, they authored their scientific memoirs, *A Matter of Life,* which was published in 1981. Steptoe received knighthood in 1987 and was made a Fellow of the Royal Society. He died of cancer in 1988 and was survived by his wife, son, and daughter.

BRENDA WILMOTH LERNER

Thomas H. Weller. *(The Library of Congress. Reproduced with permission.)*

Thomas Huckle Weller

1915-

American Physician, Virologist, and Bacteriologist

Thomas Huckle Weller is an American virologist and bacteriologist who, along with Drs. John Franklin Enders (1897-1985) and Fredrick C. Robbins (1916-), was awarded the Nobel Prize in physiology and medicine in 1954 for the discovery that the polio-causing virus could be cultivated in a test tube. This was the breakthrough required in order to mass-produce a polio vaccine.

Weller was born in Ann Arbor, Michigan, on June 15, 1915. His father, Carl V. Weller, was a pathologist and chairman of the department of pathology at the University of Michigan Medical School. Weller attended the University of Michigan, obtaining both his bachelor's and master's degrees there. He followed his father's footsteps into medicine by attending Harvard Medical School. In his senior year, he had the opportunity to work with Dr. John Franklin Enders on test tube (in vitro) virus cultivation. This work was important because during this time period the study of viruses was conducted primarily on monkeys, which was expensive and time consuming. In some cases mice, or in rare cases human patients, were used to study viruses and so it was imperative that some other method of study be developed that was cheaper and less time consuming. This was the focus of the experiments Enders was conducting, and Weller had the opportunity to assist with this groundbreaking work. Weller graduated in 1940 with his medical degree.

In the years that followed, Weller developed his career first as an intern of medicine, then a teaching fellow of bacteriology at Harvard, followed by a research fellow of tropical medicine and pediatrics, and an intern of bacteriology and pathology at Children's Hospital in Boston. During World War II, Weller served in the Army at the Antilles Department Medical Laboratory, where he was in charge of studies in parasitology, bacteriology, and virology.

Following the war, Weller reunited with Enders at Harvard and became the assistant director of the infectious disease laboratory at Children's Medical Center. At this time, polio cases were occurring in epidemic numbers across the United States and the world. The polio virus infects the nerve cells in the spinal cord that are responsible for supplying nerve impulses to the muscles. By doing so, the signals to the muscles are interrupted and the result is paralysis. This often results in the death of the nerve cell, which leads to permanent paralysis. Many people were crippled by this disease. During this period of polio epidemics, Dr. Weller, along with Drs. Enders and Robbins, continued to experiment with in vitro virus cultivation. They discovered that the polio virus could be successfully cultivated and recognized in non-nerve tissue. This made it possible to provide a virtually endless supply of virus for vaccine production and finally control the polio epidemics. It also made it unnecessary to experiment with monkeys, mice, and human patients. For this the three doctors were awarded the Nobel Prize in physiology and medicine in 1954.

In the same year he was awarded the Nobel Prize, Weller was made Richard Pearson Strong professor and the chairman of the tropical public health department at the Harvard School of Public Health. Weller went on to experiment with other diseases and later, with Franklin Neva, demonstrated the common causes of shingles, a skin condition marked with blisters and itching on the trunk of the body, and chicken pox.

MICHAEL T. YANCEY

Paul Maurice Zoll
1911-1999
American Cardiologist

Paul M. Zoll introduced the first cardiac pacemaker in 1952. He was also a pioneer in the use of heart monitors, which he introduced in 1955, and of external countershock defibrillators, which he first used in 1956.

Zoll was born in Boston on July 15, 1911, to Hyman and Molly Homsky Zoll. He earned his B.A. from Harvard, graduating summa cum laude in 1932, and went on to earn his M.D. at Harvard Medical School in 1936. Zoll performed his internship at Beth Israel Hospital in Boston, and later at Bellevue Hospital in New York City, from 1936 to 1939.

Zoll then began his medical practice in Boston, specializing in cardiology. He also began working as a medical researcher at Beth Israel Hospital, where he became assistant in medicine in 1947. During World War II, Zoll served in the U.S. Army, joining in 1941 as a first lieutenant and ending his term of service in 1945 as a major. During that time, he received the Legion of Merit medal.

At Beth Israel, Zoll conducted his groundbreaking work with heart machines. After showing for the first time that electrical stimulation can revive a heart that has stopped beating, he went on to conduct research that led to the creation of the first pacemaker in 1952. He was also instrumental in the development of electronic heart monitors in 1955, and of external countershock defibrillators, which are capable of restarting a heart through electrical stimulation, in 1956.

Zoll also worked as a medical researcher at Harvard Medical School from 1939, and as a research fellow in medicine from 1941. He was a consultant in cardiology to Boston hospitals, associate editor of *Circulation* (1956-65), and contributor of articles to other medical journals. He received a number of honors, most notably the Albert Lasker award for clinical medical research in 1973. Zoll also belonged to a number of professional organizations, among them the American Heart Association, which gave him its award of merit in 1974 and 1992.

Married twice, to Janet F. Jones in 1939, and to Ann Blumgart Gurewich in 1981, Zoll had two children, Ross and Mary Janet. He retired in 1994, and died of respiratory arrest in Chestnut Hill, Massachusetts, on January 5, 1999.

JUDSON KNIGHT

Biographical Mentions

Murray Llewellyn Barr
1908-1995

Canadian physician who sparked interest in genetic disorders when he discovered sex chromatin masses within the nuclei of normal female body cells. These masses are called Barr bodies. The presence or absence of Barr bodies may indicate genetic defects involving X-chromosomes, such as Down syndrome or Turner's syndrome. Barr received many honors, including the Joseph P. Kennedy Foundation International Award, and an appointment to the Order of Canada. He was also nominated for a Nobel Prize.

Baruj Benacerraf
1920-

Venezuelan-born American immunologist who received the 1980 Nobel Prize for Physiology or Medicine, along with Jean Dausset and George Snell, for research involving immune response genes and major histocompatibility complex (MHC). While working to understand the mechanisms and genetics of immunology, Benacerraf helped identify the way in which the immune system recognizes its own cells and attacks foreign substances. This discovery explained the genetic basis for blood and organ compatibility, leading to advances in tissue and organ transplantation.

Frank Milan Berger
1913-

Czech-born American physician recognized for his research in neuropharmacology. Berger's studies focused of how drugs affect psychomotor behavior and emotional states. He was especially interested in muscle relaxants, tranquilizers, and anti-epileptic mediations.

Douglas Bevis
1919-1994

British physician who in 1950 first described how amniocentesis can be used to test fetuses for Rh factor incompatibility. The prenatal test later was used to screen a battery of genetic disorders.

R. Michael Blaese
1939-

American geneticist noted for his innovative work in gene therapy and applied genomics. In the early 1980s Blaese and his colleagues hit upon the idea that defective genes could be changed, and subsequently devised a strategy to deliver engineered viruses to correct such defective genes. In 1990 the team had their first success when they treated a young girl with adenosine deaminase (ADA). This breakthrough led them to apply the same principles of gene therapy to other metabolic disorders and cancer.

Baruch Blumberg
1925-

American physician and researcher noted for his research on the spread of virus infections, especially hepatitis B. In 1963, while examining samples of different ethic groups and their response to disease, Blumberg found an antigen in the serum of an Australian aborigine that he later related to a virus that causes hepatitis B, a serious liver disease. The discovery led to the screening for the virus and, in 1982, a safe and effect vaccine. Blumberg shared the 1976 Nobel Prize for Physiology or Medicine with Carleton Gajdusek for their work on viruses.

Murray B. Bornstein
1917-1995

American physician noted for his work on multiple sclerosis (MS). Realizing a strong immune system was essential for treatment of MS, Bornstein sought a drug that would treat the disease without weakening the immune system. While at Albert Einstein College of Medicine in New York City, he found the drug Copolymer 1 to be effective, offering an entirely new approach to the treatment of MS.

Denis Parsos Burkitt
1911-1993

British surgeon and researcher noted for his research on tropical medicine and high-fiber diets. After World War II, Burkitt went to Uganda as a surgeon in the British colonial service, where he studied a lethal cancer of the lymphatic system in children. Recognizing that the cancer occurred in areas where malaria and yellow fever were endemic, Burkitt suspected an insect was the vector. He linked the disease to a virus that emerges when the immune system is suppressed. Later, he developed chemotherapy for this lymphoma, named Burkitt's lymphoma. Burkitt also advocated a high fiber diet to protect against colon cancer and other ailments, popularized by his book *Don't Forget Fiber in Your Diet* (1979).

John Edmond Buster
1941-

American physician noted for his pioneering research in reproductive endocrinology and infertility. In 1983 Buster and Maria Bustillo were the first to successfully conduct a human embryo transfer. Buster has been involved in many aspects of obstetrics and gynecologic research, ranging from menopause to advanced methods of treating infertile women.

Maria Bustillo
1951-

Cuban-born American noted for her work in reproductive endocrinology. Bustillo pioneered methods of ultra-sound-guided egg retrieval and uterine lavage (washing) for oocyte donation. She has advanced reproductive medicine by introducing innovative concepts and has performed over 4,000 assisted reproductive technology (ART) procedures. In 1983 Bustillo and John Buster were the first to successfully conduct a human embryo transfer.

Alexa I. Canady
1950-

First African-American woman to be recognized in neurosurgery. Born on November 7 in Lansing, Michigan, she was recognized throughout her school years for her academic ability. Receiving the doctor of medicine in 1975 from the University of Michigan, she completed her residency in neurosurgery at the University of Minnesota in 1981. She has written extensively on pediatric neurosurgery, her specialty. She became director of the Children's Hospital in Detroit in 1987 and a clinical professor at Wayne State University School of Medicine.

Min Chueh Chang
1908-1991

Chinese-American biologist who is known for his work in the field of animal reproduction. He has written numerous articles on artificial insemination and has conducted research in which mammal's eggs were successfully transplanted. He has also written on early embryonic development and oral contraceptives. He received his Ph.D. from Cambridge University in 1941; he moved from China to the United States in 1945.

William H. Clewell
1943-

American obstetrician and gynecologist who pioneered the treatment of fetal diseases and malformations. Born in Newport Township, Pennsylvania, he received his medical degree from Stanford School of Medicine in 1970 and completed a residency in Obstetrics and Gynecology there in 1974. Majoring in fetal medicine, he went to the University of Colorado where he pioneered the treatment of fetal hydrocephalus. In 1987 Clewell joined Phoenix Perinatal Associates, serving also as clinical professor at the University of Arizona College of Medicine and Chief of Service at Good Samaritan Hospital.

Irving S. Cooper

American surgeon who pioneered the practice of cryosurgery (destroying tissues by freezing them) in the removal of brain tumors. Although variations of cryosurgery had been attempted for more than 100 years, Cooper's method, using liquid nitrogen, was the first to work well with internal tissues. Cryosurgery is now used to remove skin lesions (such as warts), skin tumors, gynecologic tumors, ophthalmic lens removals, and other procedures. It is considered to be safe and reliable in both the operating room and in a family doctor's office.

Allan MacLeod Cormack
1924-1998

American physicist who shared the 1979 Nobel Prize for Medicine with Sir Godfrey Newbold Hounsfield for the development of the Computerized Axial Tomography (CAT) scanning system, which revolutionized noninvasive medical imaging and diagnosis. Cormack calculated the mathematical equations required for generating the three-dimensional CAT scan images of body organs. He was a professor of physics at Tufts University in Massachusetts, and served as the department's chairman from 1968 to 1976.

Eric Courchesne

American physician whose research into autism has provided valuable insights into the origins of the condition. In 1988 Courchesne and his colleagues showed that changes to the brain that lead to autism begin either *in utero* or shortly after birth. This finding tended to confirm a theory that autism is caused in part by a lack of certain cells that help to regulate brain activity and communication between parts of the brain.

Fernand Daffos

Physician who in 1983 became the first to extract fetal blood from the umbilical cord and to use it for diagnosing disease in the fetus. Daffos's technique, along with that of amniocentesis, uses samples of fetal cells to help determine the current and future fetal health. In addition, by withdrawing a sample of blood, Daffos's technique allows additional testing, for example, for enzyme activity, organ function, and so forth, making it in some respects a more sensitive test than amniocentesis.

Jean Baptiste Gabriel Joachim Dausset
1916-

French immunologist who discovered the genetic cause of blood and tissue incompatibility and the mechanism by which the body recognizes its own cells. Dausset hypothesized that patients who had immunological reactions to blood transfusions did so because of specific variations they carried in their genes. He eventually isolated that gene, which he termed HLA (human leucocyte antigen). Dausset was awarded 1980 Nobel Prize for Medicine, which he shared with George Snell and Baruj Benacerraf.

Kay Davies

Internationally renowned geneticist who developed the first prenatal and carrier test for the muscle wasting disease Duchenne muscular dystrophy, paving the way for identification of the mutated gene. Her research has also led to the mapping and characterization of other disease genes, including spinal muscular atrophy, the leading form of infant mortality. In 1995 Davies was elected chair of the department of biochemistry at Oxford University.

Morris E. Davis

American physician who in 1965 was the first to show that estrogen therapy could prevent osteoporosis and arteriosclerosis in post-menopausal women. This discovery led the way to recognition of these problems as well as suggesting ways to help reduce their effects in older women. It also led to increasing use of preventative measures, such as measuring bone density and taking calcium supplements.

Michael Ellis DeBakey
1908-

American cardiologist who pioneered the fields of heart transplantation and telemedicine. An internationally renowned surgeon, teacher, and medical statesman, DeBakey is recognized for a number of medical accomplishments. He was the first to successfully perform a coronary artery bypass and to implant a partial artificial heart in a human patient. He also helped NASA develop a miniaturized artificial heart. DeBakey revolutionized medical training and treatment by developing a satellite-based system that electronically linked remote sites of the world to the Texas Medical Center in Houston.

Forest D. Dodrill
1902-1997

American surgeon who performed the first successful open-heart surgery using a heart-lung machine he helped to invent. Dodrill's work helped pave the way for virtually all heart surgery, including coronary bypass operations, heart valve replacement, and heart transplants. The heart-lung machine both pumps blood through the body and oxygenates it, allowing both heart and lungs to be stopped. His machine, developed in conjunction with General Motors, is on display at the Smithsonian Institute in Washington, D.C.

Gertrude Belle Elion
1918-1999

American chemist who with George H. Hitchings and Sir James Black won the Nobel Prize for medicine in 1988. Elion began her career as an instructor of high school science students and nursing students. During World War II many professions experienced a shortage of workers. This shortage enabled Elion to enter into the male-dominated world of academia. Here Elion's research led to the discovery of many drug treatments for diseases such as leukemia, malaria, herpes, gout, and AIDS. Elion's drug developments also improved the field of organ transplanting.

John Franklin Enders
1897-1985

American microbiologist who shared the 1954 Nobel Prize in physiology and medicine with F. C. Robbins and T. H. Weller for their discovery that the virus which causes polio can be grown in a test tube (*in vitro*) on various tissues. This discovery allowed for the mass production of a successful polio vaccine. He also cultivated the measles virus in 1954 and developed a vaccine for measles in 1962.

Moses Judah Folkman
1933-

American surgeon and professor of surgery at Harvard Medical School who suggested that a tumor, which is an abnormal mass of tissue in the body, might release a substance that promotes the growth of blood vessels, a substance he named tumor angiogenesis factor. Folkman received his B.S. from Ohio State University in 1953, and he graduated with an M.D. from Harvard Medical School in 1957.

Michael S. Gazzinga

American neurosurgeon who was the first to show that both hemispheres of the brain function more or less independently of each other. Gazzinga noticed that in patients in whom the corpus callosum (the bundle of nerves connecting the two brain hemispheres) was severed, each hemisphere continued to function independently of the other. This finding, in turn, led to more involved studies of brain function and the mapping of various functions to different parts of the brain.

Jacob Gershon-Cohen

American physician and radiologist whose work in the detection of breast cancer culminated with the development of mammography in 1964. Mammography, a technique that uses periodic x-ray examinations to detect breast cancer, has been a significant development in the early detection of this disease, leading to more effective treatment at a stage when the disease is often more treatable. Although some controversy exists about its use, mammography is generally credited with helping increase the survival rate for women diagnosed early.

Evarts A. Graham

Physician who with Ernest Wydner was the first to show that tars from tobacco smoke could cause cancer in mice. This was the first of many studies that became progressively more convincing at showing a direct link between cigarette smoking and many diseases, including cancer and heart disease. These studies, originally vigorously attacked by the tobacco industry, were eventually acknowledged to be accurate by the industry.

Andreas R. Gruentzig

Swiss surgeon who developed angioplasty in 1977. This procedure, performed with great regularity today, inserts a small catheter containing a balloon into blocked coronary arteries. The balloon, when inflated, compresses the plaque against the arterial wall, opening the way for improved blood flow. Recent innovations have included placing small stents to help prop the arteries open and using radiation to help keep them from re-closing after the procedure (a relatively common occurrence).

James F. Gusella

American geneticist who was the first to find a genetic marker for Huntington's disease. This was one of the first diseases for which a positive genetic marker was located, making it possible to identify people likely to develop the disease in the future. This information could also be used, via amniocentesis, to identify fetuses with the potential to develop Huntington's disease later in life.

Henry J. Heimlich

American surgeon best known for inventing the Heimlich maneuver to expel food that might otherwise cause a person to choke to death. Heimlich developed this procedure in the early 1970s after noting that the accepted method of hitting a person on the back was usually unsuccessful. It has been widely taught, and both the procedure and the name are well-known throughout most of the world.

Eric P. Hoffman

Researcher who helped discover a major feature of Duchenne muscular dystrophy. Along with Louis Kunkel and other colleagues, Hoffman in 1988 found that sufferers of the disease lack the protein dystrophin. The finding opened the way to earlier diagnosis of the disease.

Frank A. Horsfall, Jr.

American physician who was the first to announce that all cancer results from genetic change (or mutation) in cellular DNA. Earlier researchers had shown that genetic change *could* cause cancer, but had not stated that all cancer originated in this fashion. Horsfall's discovery opened the door to more intensive study of cancers and its genetic origin, leading in turn to a closer examination of the role that mutagens (such as certain chemicals, viruses, and radiation) can play in carcinogenesis.

Godfrey Newbold Hounsfield
1919-

British physicist, medical engineer, and inventor who in 1971 at Atkinson Morley's Hospital in Wimbledon built the first clinically successful computerized axial tomography (CAT), or computed tomography (CT), scanner. In 1951 he received his undergraduate degree from Faraday House Electrical Engineering College in London and became a scientist for Electrical and Musical Instruments (EMI) Ltd. He shared the Nobel Prize in medicine or physiology with Allan Cormack in 1979 and was knighted in 1981.

Charles Anthony Hufnagel

American surgeon who developed the first artificial heart valve. The valve, a flimsy-looking cage containing a small ball, worked nearly as well as a natural heart valve and set the stage for rapid strides in the field of biomedical engineering, the science of designing artificial parts for the body.

Alick Isaacs
1921-1967

Scottish virologist whose work on the effects of interferon on viral infections suggested entirely new ways to fight disease. Interferons are substances produced by the body that interfere with a virus's ability to function normally. For a time, they appeared to be a "magic bullet" that would help to cure many viral diseases. Now, although apparently less powerful than originally thought, interferons are still thought to hold a great deal of promise in the fight against disease.

P. A. Jacobs

American physician who with J. A. Strong was the first to show that Klinefelter's syndrome was caused by a genetic defect on one of the sex chromosomes. This, along with Turner's syndrome, was among the first diseases to be localized to a specific chromosome. In addition, by helping to localize the chromosome on which the faulty genes are carried, Jacobs and Strong helped set the stage for a better understanding of genetic disease in general.

David S. Janowsky

American physician who first showed that bipolar disorder (also called "manic-depression") is caused by an imbalance between two types of neurotransmitters in the brain. Made in 1972, this discovery was one of the first to show that some mental illness has an organic basis—that it is due to a physical problem in the brain and is analogous to diabetes and other similar diseases. This, in turn, has helped remove some of the stigma associated with mental illness.

Alec J. Jeffreys
1950-

British biologist who discovered the process of genetic fingerprinting and who helped discover the genetic marker for colon cancer. Genetic fingerprinting has become increasingly useful in law enforcement and in establishing paternity because it identifies certain stretches of DNA that, although unique to each individual, also show patterns that run in families. This, in turn, has allowed genetic testing to be used to show guilt in many cases as well as proving the innocence of many who had been falsely accused. See long biography on p. 157.

Bruno Kirsh

Physician who in 1956 discovered strange, dense bodies in guinea pig heart cells. Although Kirsh did not at first understand the purpose or significance of these bodies, later research showed them to be involved in the release of hormones that help to regulate the circulatory system. This, in turn, led to a better understanding of the circulatory system in general, including the chemical processes that help to keep it operating properly.

Louis M. Kunkel

Physician who in the mid- and late-1980s made a number of discoveries pertaining to the genetics and mechanisms by which Duchenne muscular dystrophy works. In successive years, Kunkel and his collaborators began to understand the genetic basis for this often-fatal disease. From that start, they were able to determine that the absence of the protein dystrophin helps to cause the disease and to develop an early screening test to identify persons at risk.

Rita Levi-Montalcini
1909-

Italian-born American biologist who with Stanley Cohen won the Nobel Prize for medicine in 1986. At the age of 20 Levi-Montalcini convinced her father that she should pursue a professional career. She studied medicine at the University of Turin and began her research on nerve cells in 1936. Working with Viktor Hamburger in the United States, Levi-Montalcini studied a substance that caused nerve cells to grow. She and Stanley Cohen successfully isolated this substance, called nerve-growth factor.

Jack Lippes

American physician who introduced the intrauterine device (IUD) for birth control. The IUD is a small, inert, plastic device that is inserted into the uterus to prevent pregnancy. Although the manner in which the IUD helped prevent pregnancy was not fully understood for some time, its effectiveness could not be argued. More certain than many forms of birth control and less expensive than others, the IUD remained very popular for many years.

Ignacio Navarro Madrazo

Physician who showed that implanting adrenal cells into a part of the brain could help to cure or alleviate the symptoms of Parkinson's disease. This procedure had been attempted earlier, but to no avail. Navarro Madrazo was able to show that, by altering the location of implantation slightly, this treatment became much more effective. While other treatments have been attempted recently, many of them use fetal cells, and the ethics of these treatments are currently hotly debated.

Barry J. Marshall
1951-

Australian-born medical scientist and gastroenterologist who co-discovered Helicobacter pylori, a spiral shaped bacterium responsible for stomach inflammation, including peptic ulcers and gastritis. Barry's discovery helped dispel myths that the stomach was actually sterile and resistant to bacterial infection. He was so convinced that the medical community was wrong that he actually ingested the bacteria to prove his case. Because of his research, the bacteria is now widely recognized as the most common chronic infection in the world.

Harry Meyer Jr.

American physician who with Paul Parman developed the first rubella vaccine in 1966. Rubella, also known as German measles, was a common disease that affected many children and adults prior to introduction of the vaccine. In pregnant women rubella is particularly dangerous as it causes a wide variety of birth defects in developing fetuses.

Gregory Goodwin Pincus
1903-1967

American endocrinologist who, with Min Chueh Chang and John Rock, developed the first effective birth-control pill. Pincus co-founded the Worcester Foundation for Experimental Biology in 1944, where he studied steroid hormones and reproduction. At the encouragement of Margaret Sanger, founder of the birth-control movement in the United States, Pincus turned his attention to the use of synthesized hormones to prevent pregnancy. After experimenting with as many as

200 potential substances, Pincus and his collaborators derived a steroid from the wild Mexican yam that could inhibit ovulation without serious side effects. The "pill" was first made available to the public in 1960.

Bruce A. Reitz

American surgeon who in 1981 performed the first successful heart-lung transplant in the world. This operation was important because many patients with failing hearts have lung problems, too, requiring additional surgery. The ability to replace both of these organs at the same time helps to reduce stress on the patient while simultaneously reducing stress on the transplanted organs. However, the surgery can be very delicate and, until Reitz pioneered it, had not succeeded.

Frederick Chapman Robbins
1916-

American physiologist and pediatrician who cultivated viruses to combat polio. Earning a Harvard medical degree, Robbins worked as a pediatrician at Boston's Children's Hospital. With Thomas H. Weller and John F. Enders, Robbins grew the poliomyelitis virus in various tissue cultures. The trio won the 1954 Nobel Prize in medicine and physiology because their work enabled the development of a polio vaccine. Robbins also served as dean of the Case Western Reserve School of Medicine and as president of the Institute of Medicine.

Francis Peyton Rous
1879-1970

American pathologist and oncologist who discovered cancer-causing viruses in chicken. In 1909, while working as a scientist at the Rockefeller Institute for Medical Research, Rous discovered that chicken tumors were derived from viruses that could be transmitted to other fowl of the same stock by injecting an agent extracted from the tumor cells. His work was not well received in the scientific community, and he eventually abandoned his research and instead turned to the study of liver and gallbladder pathology. However, his theory of viral cancer causation was later validated, and Rous was awarded the 1966 Nobel Prize for Physiology or Medicine, which he shared with Charles Huggins.

Mary Jo Schmerr

American research chemist who has worked at the National Animal Disease Center since 1975, when she received her Ph.D. in biochemistry from Iowa State University. Schmerr's research interests include the application of microtechniques to the study of animal diseases and the development of new methods to define viral interactions with host cells. She is a member of several professional societies, including the American Association for the Advancement of Science, the American Chemical Society, and the International Association for Comparative Research on Leukemia and Related Diseases.

Norman E. Shumway

American surgeon who in 1968 performed the first heart transplant on an adult patient in the United States. He also conducted the first heart-lung transplant and pioneered a procedure for correcting birth defects through bypass surgery. Shumway was instrumental in urging the government and medical community to adopt a definition of death based on the cessation of brain activity rather than on the absence of a heartbeat.

George D. Snell
1903-1996

American geneticist whose pioneering research helped decipher the complex genetics of the immune system. His work in defining the genes that determine whether a body accepts or rejects organs paved the way for modern transplants. Snell was co-recipient of the 1980 Nobel Prize for Medicine, along with Baruj Benacerraf and Jean Dausset, for discoveries in genetics and immunological reactions.

Ellen Solomon

Research scientist in the area of the genetic aspects of human cancer. Solomon held the position of Research Scientist at the Imperial Cancer Research Fund in the United Kingdom until 1996. She then became the Prince Philip Professor of Human Genetics at Guy's Hospital in London, where she worked in the Division of Medical/Molecular Genetics. Solomon is currently Vice-Chairman of the British Society for Human Genetics.

J. A. Strong

American physician who with P. A. Jacobs helped show that Klinefelter's syndrome was linked to one of the sex chromosomes. This was one of the first diseases shown to be either "x-linked" or "y-linked." Linkage to a sex chromosome is important because, in males, there is only one of each sex chromosome. This means that a genetically recessive disease or trait will not appear in women but will appear in men because men lack a second x chromosome to mask the disease.

Bert L. Vallee

American physician who in 1985 was the first to find the tumor angiogenesis factor predicted in 1961 by Judah Folkman. This factor is important to growing tumors because it stimulates the formation of new blood vessels that are needed to feed a growing tumor. One approach to combating cancer may be to inactivate this in growing tumors, causing them to starve for lack of food- and oxygen-carrying blood. Valle renamed this compound angiogenin.

Patrick Walsh

Surgeon who in 1982 announced a new surgical technique for removing cancerous prostate glands. This technique greatly reduced the risk of post-operative impotence in men receiving it, a major advantage from the standpoint of men diagnosed with this disease. Since prostate cancer will eventually affect virtually all men who live long enough, this technique had a significant impact. Today, this technique is often replaced by the implantation of radioactive "seeds" to destroy the tumor, but Walsh's technique is still popular.

J. Robin Warren

American physician who with Barry Marshall showed that a bacterium may be responsible for many gastric and intestinal ulcers. This bacterium, *Helicobacter pylori,* is the first to be linked to ulcers, and its discovery suggests that conventional treatment methods may be inappropriate. This, in turn, has led to investigations in antibiotic methods to control ulcers. However, as of this writing the traditional treatments of lowered stress and bland diet are still common.

Nancy Wexler
1945-

American geneticist who researches hereditary diseases. Wexler earned a clinical neuropsychology Ph.D. from the University of Michigan and is a professor at Columbia University's College of Physicians and Surgeons. Since 1969 she has served as president of the Hereditary Disease Foundation. After Wexler's mother died from Huntington's disease, Wexler located the gene that causes that condition and developed a chromosomal test to identify potential sufferers. Wexler has won numerous awards, recognizing her efforts to cure genetic diseases.

Robert Wallace Wilkins
1906-

American physician who began using the drug reserpine after observing its use as snakeroot in India. Reserpine was first used as a medicine for high blood pressure, one of the first successful drugs to be so used. It was later found to be a tranquilizer, the first to be found, and it was subsequently used in psychiatric treatment, as well. Although eventually displaced by more effective drugs in both arenas, many patients were served well by reserpine.

Robert Williamson

American researcher who with Kay Davies found the first genetic marker indicating the likelihood of developing Duchenne muscular dystrophy. As one of the first genetic markers found, this opened the possibility of diagnosing the disease early as well as the hope of finding a cure for the ailment. It also opened a host of ethical questions affecting the medical and insurance industries as well as those who, having a genetic marker, may not have developed the disease yet.

Ernest L. Wydner

Physician who with Evarts Graham first showed the carcinogenic properties of the tar from tobacco smoke in mice. Their mouse studies were followed with studies in other animals, all showing that cigarette smoking could cause cancer and other diseases. This led to a statement by the Surgeon General of the United States linking tobacco smoking with these diseases and to several decades of near-constant fighting between tobacco companies and medical researchers.

Rosalyn Sussman Yalow
1921-

American physicist who won the 1977 Nobel Prize for the development of radioimmunoassays for peptide hormones. As founder of the radioisotope laboratory at the Bronx Veterans Administration Hospital, she began a very fruitful collaboration with Dr. Solomon Berson. The revolutionary radioimmunoassay developed by Yalow and Berson made it possible to measure minute amounts of almost any substance of biologic interest, such as hormones, drugs, enzymes, and antibodies, in blood and body tissues.

Michael Zasloff

American physician who helped discover a possible treatment for some types of brain cancer using compounds from the liver of a type of shark. The compound, called squalamine, is an angiogenesis inhibitor, meaning it interferes with a tumor's ability to cause the growth of new blood vessels necessary to feed it as it expands. This, in turn, may help to limit the growth and spread of tumors, holding the promise of a treatment to limit the severity of some types of cancer.

Bibliography of Primary Sources

Books

Gallo, Robert C. *Virus Hunting: AIDS, Cancer, and the Human Retrovirus: A Story of Scientific Discovery* (1991). Gallo's personal defense of his work on the discovery of the AIDS virus, HIV. Although Gallo is credited as co-discoverer of the virus, French researcher Luc Montagnier was the first to isolate the virus, and Gallo's reputation was damaged in the ensuing controversy about where Gallo's viral samples had originated.

Steptoe, Patrick and Robert Edwards. *A Matter of Life* (1981). Memoirs of the acclaimed researchers who pioneered human in vitro fertilization.

Periodical Articles

Charnley, John. "Arthroplasty of the Hip: A New Operation" (1961). The publication of this article in the medical journal *Lancet* signaled the long-awaited breakthrough in low-friction prosthetic arthroplasty. For the first time both the acetabulum and the head of the femur could be replaced with artificial materials. This was the first genuine total hip replacement.

Rock, John. "In Vitro Fertilization and Cleavage of Human Ovarian Eggs" (1944). This was the first published account of in vitro fertilization of a human ovum.

Rock, John. "Effects of Certain 19-Nor Steroids on the Normal Human Menstrual Cycle" (1956). This article was the first clinical report of the use of hormones to suppress ovulation in humans.

Physical Science

Chronology

1952 Walter Baade, a German-American astronomer, discovers that the Andromeda galaxy is more than 2 million light-years away, not the 800,000 that Edwin Hubble had estimated; this results in a dramatic increase in the estimated size and age of the universe.

1963 Dutch-American astronomer Maarten Schmidt discovers quasi-stellar radio sources, which are later named quasars.

1964 Murray Gell-Mann, an American physicist, first postulates the existence of unusual particles—which he dubs "quarks"—that carry fractional electrical charges.

1967 Jocelyn Bell Burnell, a graduate student working with English astronomer Antony Hewish, discovers a rapidly fluctuating but unusually regular radio signal between the stars Vega and Altair; the phenomenon is later identified as a pulsar.

1969 Dutch-American astronomer Peter Van de Kamp postulates the existence of solar systems other than Earth's, an idea that implies that some form of life may exist elsewhere in the universe.

1974 English physicist Stephen Hawking spurs efforts to delineate the properties of black holes when he proposes that, in accordance with the predictions of quantum theory, these emit subatomic particles until they exhaust their energy and finally explode.

1974 Howard Georgi and Sheldon Glashow develop the first grand unified theory to account for strong, weak, and electromagnetic forces as parts of a single force that broke apart when the universe cooled down after the Big Bang.

1980 The father-and-son team of Luis and Walter Alvarez speculate that a giant asteroid collided with Earth, causing a prolonged dust blackout and mass extinctions—including the disappearance of the dinosaurs.

1986 Laverne Kulm and colleagues publish the first detailed report regarding deep-sea vents, and demonstrate that fluids and gases emerge not only from continents but also from the ocean floor.

1992 American astronomer George Smoot discovers what he calls "the handwriting of God," evidence of the Big Bang: small fluctuations in the microwave background radiation detected by the Cosmic Background Explorer (CBE) satellite.

1995 Astronomers Michel Mayor and Didier Queloz at Geneva Observatory discover two new planets: one, roughly the mass of Jupiter, orbits the star 51 Pegasi, 42 light-years from Earth; the other, 20 times Jupiter's mass, revolves around the star GL229, 30 light-years from Earth.

1991 Research concerning fullerenes, a recently discovered from of elemental carbon which may enable the advance of high-temperature superconductors, increases dramatically.

Overview: Physical Sciences 1950-present

The Legacy of 1900-1949

In the first half of the twentieth century, modern physics emerged out of a watershed of theory and discovery concerning the realm of the atomic nucleus and subatomic energies and speeds. From the profound insights of Albert Einstein's (1879-1955) relativity to the compartmental order of Max Planck's (1858-1947) quantum theory and its antecedent quantum mechanics, humanity was upon the threshold of an atomic age. Likewise, chemists also delved into the study of atomic structure following groundbreaking investigations of chemical reactions and equilibrium laws, new elemental gases, radioactive elements, their isotopes and chemistry, and atomic weights. Shared areas of study included surface chemistry, molecular structure, chemical bonding, and the synthesis of new radioactive elements.

In contrast to study of the physically infinitesimal, study of the vast universe advanced through the use of large aperture telescopes, beginning at Mount Wilson in California. These celestial studies confirmed early relativistic physics and provided new evidence, put forth by Edwin Hubble (1889-1953), suggesting that the universe is expanding. Or was it static or contracting? Complementing these questions, early astrophysics, such as Ejnar Hertzsprung (1873-1967), analyzed the physical and chemical makeup of the stars and their life cycles. New cosmological theories also emerged, such as genesis via the Big Bang, in late 1940s. Between the two world wars, researchers in the earth sciences were defining atmospheric motion mathematically as a subset of fluid and hydrodynamics, represented by the work of Vilhelm Bjerknes (1862-1951), mapping the ocean floors, and peeling away the structure of the earth to its very core.

The Further Rise of Applied Physical Science

It is perhaps not surprising that after World War II, with its negative application of science and technology for destructive purposes, that subsequent scientific endeavors focused on Mother Earth, threatened rather than embraced by the realities of the atomic age. The postwar peace quickly evolved into a Cold War world of polarized superpowers, of bigger government accommodating bigger science and military interests. However, in 1957-58 an extraordinary, six-year cooperative international effort to study the total physical earth system came to fruition with the International Geophysical Year (IGY). The design of the ICBM missile served just as well for IGY rockets lifting the first satellites into near space orbit. A mammoth IGY database supported many scientific advances, including the development of plate tectonic theory, new understanding of charged particles in the upper atmosphere (called Van Allen Belts), and a new perspective of the unity of the solar system.

By 1950 several high-speed computers were already half a decade old and pushing the envelope of evolving mass data needs. Princeton's MANIAC pointed to the potential of modeling physical phenomena through vast scientific data banks. It demonstrated its power in 1952 with the first numerical weather forecasts. By about 1961, using computer models of atmospheric flow, MIT theoretical meteorologist Edward Lorenz (1917-) formulated chaotic system theory to show that extended forecasting of atmospheric dynamics had its limits. He thus introduced the applications of determinate chaos theory to the sciences. Computer modeling surged from the 1970s through the 1990s, with a progression of applications to study most physical and chemical phenomena of nature. Sophisticated submersible technology and robotics made possible visits to the deepest of ocean floors in 1960, and enabled the collection of oceanographic and geological data encompassing complex ocean currents, volcanic activity, and deep sea vents, as well as the biology around such vents. The planetary sciences of the solar system graduated from complex sensory satellites to space probes equipped with computer-controlled cameras and sensors, making flyby and orbiting data missions to nearly all of our planetary neighbors. As part of this effort, chemists designed special chemical analysis instruments and automatic experimental devices for probe landings.

During the second half of the twentieth century, astronomers looked to the stars with clearer eyes and a perspective further into cosmic origins with the Hale 200-inch (5-meter) reflector on Mount Palomar (1948), the largest in the world for decades to come. The simplistic idea of glimpsing a horizon of the universe with the Hale, as anticipated by some, led to more complex cosmic spatial quandaries as the 1960s approached, including new curving and multidimensional models of the universe. Other radiative wavelengths beyond the visual became cosmos-searching tools. The radio telescope's steady refinements, including the use of several together, or aperture synthesis, proved that the Milky Way is a spiral galaxy and detected the first pulsars (neutron stars) in 1968. In addition to cosmic ray and infrared detectors, the use of other wavelength detectors followed in the 1950s, including gamma rays (*Explorer 11*, 1961), x rays (rocket detection, 1962), cosmic background microwave remnants of the Big Bang (1965, 1966), cosmic ray detection of quasars (energetic distant galactic nuclei, 1963) and cosmic sources of infrared radiation (IRAS, 1983).

During the 1990s, new large telescopes and observatories were built that sharpened delineation of the origins of our galaxy and the universe. By 1986 the idea of an orbiting telescope (first theorized in 1947) placed above the visual barriers of the earth's atmosphere became reality with the Hubble Space Telescope (HST). Optical corrections to the HST improved its clarity and resulted in pictures that revealed a changing cosmos. There were indirect and then direct (1998) observations of not only proto-planetary areas of stellar dust but actual extra-solar planetary objects. Such glimpses, clues to our own primordial solar origins, add to persistent questions of cosmic order. Since the concept of black holes emerged in 1968, posited by John Wheeler (1911-), the heart of its theory has been the nature of gravity and relativity. Black holes are physical objects, the most massive, yet spatially limited objects, in the universe, and their gravitational forces massive enough to hold even light prisoner. A span of analysis in the full electromagnetic spectrum has revealed stellar binary and galactic nuclear black hole varieties—our own Milky Way belonging to the latter variety—with energetic characters classed as dormant and active. Other theoretical by-products of black hole research have come from astrophysicists such as Stephen Hawking (1942-), with his application of concepts including charged holes, Hawking radiation, and D-branes to Big Bang theory. The Big Bang theory itself is continually updated or compromised by new databases, so that a slightly expanding universe is challenged by various steady and quasi-steady state universes of which the outcome still lies in the future.

Mainstream Science

In addition to such theoretical and applied advances in earth and space science since 1950, mainstream physics and chemistry have provided the constant underpinning to these areas. Delving into the nuclear heart of the atom for peaceful advance of physical knowledge was foremost in the collective physics and chemistry minds at the end of World War II. The science of particle physics was just beginning, already equipped with the cloud chamber and early particle accelerators foretelling the energies needed to proceed. Framed by a quantum field theory of quantum electrodynamics (QED), the discovery of the pion (1948) and "strange" particles (1950) anticipated a whole revolutionary world of subatomic particles. Cyclotron, betatron, bevatron, and many others were all "atom smashers" to the general public. However, from the 1950s through the 1970s, in ever greater numbers, these provided the ammunition for the bubble chambers used to obtain countless photos of subnuclear collisions in the 1960s, identifying new particles and the various mesons and neutrinos. In 1962 Murray Gell-Man (1929-) of Caltech, who coined "strangeness" as a characteristic of unstable elementary particles (1959), theorized the "eightfold way" subatomic particle interrelationship to predict particle collisions based on the eight quantum numbers describing elementary particles. This also led to his defining early "quark" model particles.

Predictions of an expanding roll call of missing particles ensued as accelerator energies continued to grow. The Greek alphabet was exhausted as a source of names, and the addition of leptons and further defined characteristics of quarks (up, down, strange, charm, bottom, and tau) resulted in reorganizations of a Standard Model of particles and the weak and strong forces defining them. This deeper inspection of the limits of the elemental world, including the discovery of the so-called top quark in 1994, gave pause to further maturing of Einstein's ideas about unifying the forces of nature from the elemental level to cosmological space-time theory, a so called "theory of everything" (TOE) or more formerly, grand-unified theory (GUT).

Yet the anomalies between the macro and subatomic levels of the physical world reflect the discontinuities between quantum theory and Einstein's general relativity. And for many scientists, so-called "string theory" holds promise as the concept that explains everything from quarks to black holes. Strings have emerged in various theories as unknown, elemental vibrating strands making up the matter of the universe and reacting by merging rather than, as the accepted atomic particles, colliding in space and time. Perhaps our familiar concepts of space and time will eventually become obsolete. As a new century dawns, our understanding of the physical world remains in a constant state of change.

WILLIAM MCPEAK

Plate Tectonic Theory and the Unification of the Earth Sciences

Overview

It took nearly a century for scientists to accept the idea that continents were not forever fixed in their places, but had, in fact, slowly drifted to their current locations. In the 1960s plate tectonics, a further refinement of this concept bolstered by irrefutable geologic proof, burst into widespread acceptance in less than a decade.

Plate tectonic theory holds that continents ride atop thin plates of crust that are constantly moving across the face of the Earth. These plates break apart at midocean ridges, such as the mid-Atlantic Ridge; when they come together, one plate dives beneath the other to be recycled into the mantle—a process called subduction. These subduction zones, appearing as deep-sea trenches, are the sites of most of the world's earthquakes. As the plates descend into the Earth, they heat up and start to melt. The rising magma reaches the surface, forming volcanoes at the surface, usually within about 100 kilometers of the subduction zone.

Plate tectonic theory has become the single unifying factor in the earth sciences. In the words of John Tuzo Wilson (1908-1993), one of its founders: "The acceptance of continental drift has transformed the Earth sciences from a group of rather unimaginative studies based upon pedestrian interpretations of natural phenomena into a unified science that holds the promise of great intellectual and practical advances." This theory, for the first time, gave a single mechanism to explain the locations of mineral and ore deposits, the origins of volcanoes, the reason for the "Ring of Fire," the origin of many earthquakes, the origin of seafloor magnetic anomalies, the formation of mountains, and much more. Plate tectonics may not have the emotional and theological impact of evolutionary theory, but it helps explain phenomena that evolution alone could not. It is the workhorse of theories because it serves many sciences.

The impact of plate tectonics is not limited to the earth sciences, however. Paleontological evidence first helped to prove plate tectonics and, later, continental drift helped explain otherwise impossible fossil evidence, such as the presence of strictly land-based animals in Antarctica and South America. This theory was a true scientific revolution because of its wide-ranging applicability and the sweeping changes it forced in both scientists' and nonscientists' view of the Earth. Put more simply, before plate tectonics we believed that we lived on a static, dead world enlivened with only an occasional spasm. We discovered instead that we live on a vibrant, active world, constantly changing and in continual motion.

Background

Ever since the world was mapped, scientists noticed that the coastlines of Europe, Africa, and the Americas seemed as if they could fit together like puzzle pieces. Eduard Suess (1831-1914) proposed in the late nineteenth century that large ancient continents had broken into smaller ones. He believed, however, that instead of drifting apart, large parts of these giant continents had sunk beneath the ocean. This, along with other land bridges that had disappeared, explained similarities between continents that were no longer connected. Not until the first part of the twentieth century, however, did anyone suggest that the continents had separated and moved to their present locations. Alfred Wegener (1880-1930), who held a Ph.D. in astronomy and worked as a meteorologist, presented his findings at a lecture in 1912; he published his studies in 1915 in *Die Entstehung der Kontinente und Ozeane* (The Origin of Continents and Oceans). Scientists of the day scoffed at Wegener and criticized his idea.

Wegener was not ridiculed simply because his idea was thought to be absurd, though that did play a part. His ideas were dismissed because he could not suggest a motivating force for continental motion and because the very concept of something as huge as a continent moving was simply incredible to his contemporaries. Geology in 1912 was far different than it is today. Its practitioners were not far removed from the great debates over evolution, whether or not fossils represented animals lost in the biblical flood, and similar concepts. Here are some other "facts" believed by early-twentieth century geologists:

> Mountains and valleys were caused by the wrinkling of the Earth's crust as it cooled.
>
> Similar animals in the Americas and Africa crossed from one continent to the other via land bridges that later sank into the ocean.
>
> Continents are fixed and unmoving on the Earth's surface.
>
> The Earth occasionally shifted on its axis, causing the poles to "wander."

Wegener had synthesized information from a number of fields, including stratigraphy, zoology, paleontology, and others to support his theory. Despite this, his theories were rejected. One reason is that he was an outsider dabbling in areas in which he was not trained. Unfortunately, specialists often focused only on their own particular problems and failed to see, as Wegener did, more than one possible answer. Another factor was that while no single scientist knew enough about all fields to criticize Wegener's theory as a whole, specialists knew more than he did about their particular fields, allowing them to carp about small inaccuracies. This combination of ridicule and scientific rejection caused continental drift theory to be discarded for nearly a half century.

After World War II, oceanographers began to conduct magnetic surveys of the ocean floor. They saw that when molten rocks cool they "freeze" into mineral crystals that reflect the pattern of the Earth's magnetic field. By studying magnetic fossils, geophysicists learned that the polarity of the Earth's magnetic field had undergone many past reversals. Oceanographers were surprised to discover that these magnetic field traces changed as they moved across the ocean floor. They encountered stripes running the length of the Atlantic Ocean in which the magnetic field polarity was identical. Then they found another stripe with opposite polarity, and yet another stripe with polarity reversed again. This pattern was repeated on the European and African side of the mid-Atlantic Ridge. This suggested that the oceanic crust was formed continuously and, as the molten rock forming the crust solidified, it froze into place whatever magnetic field existed at that time. The stripes simply revealed the history of oceanic crust formation over time.

In the meantime, other observations continued to trouble and puzzle geologists. First, there was the still-nagging presence of nearly identical fossils on opposite sides of oceans, something that had led Wegener to propose his continental drift theory. Specifically, the plant *Glossopteris,* a fern, was found in South America, Africa, India, and Australia. A number of nonmarine animals were also found in similar locations. In addition, geologists were becoming increasingly uncomfortable with the concept of a land bridge a few thousand miles long that just happened to vanish without a trace. Plus, there were well-documented matching rock formations in England and New England, Antarctica and Australia, South America and Africa, and other places; too many places to be simple coincidence. Some earlier geologists had postulated a mid-Atlantic continent, now vanished, that could explain some of these similarities in rock type, but this, again, stretched credulity to the breaking point.

Next came the realization that the Hawaiian Islands were likely formed by a single "hot spot" in the Earth's mantle, and that the Emperor Seamounts, a chain of submarine mountains, were probably a continuation of the Hawaiian Island chain that had eroded beneath the waves. Scientists also discovered that no oceanic crust on Earth was more than a few hundred million years old, yet 3 billion-year-old rocks could be found on the surface. Geologists were also puzzled by unmistakable evidence of ancient glaciers in Australia and Africa; they also found evidence of swamps in the Antarctic. The more they looked, the more they realized that continental drift must be causing these, and perhaps other, phenomena. The question then became *how?*

The motivating force for continental drift—the explanation that Wegener lacked—turned out to be plumes of hot rock in the Earth's mantel. Although a solid material, hot rock can deform and flow, just as hot glass can. Over geologic time, the rock of the mantel forms rising plumes that can move the continental crust. Under this scenario, hot mantle pushes its way up at the midocean ridges, where convection cells rise to the bottom of the crust. Here, it melts the oceanic crust, perhaps contributing some mantel rock to the mix. The crust buckles up and is forced to the sides, pushing the entire plate with it, away from the ocean ridge—at the breakneck rate of an inch (2.5 cm) or so per year. That translates to about 25 kilometers (16 miles) every million years—enough to open the Atlantic Ocean to its present width in about 100 million years. Geologically speaking, this is about 2% of the age of the Earth, so the Atlantic could have opened to its present width and closed again 25 times over the history of the Earth.

If oceans open up, they must close, too. This means that at various times in the Earth's past, the continents must have been assembled into supercontinents—huge single landmasses. This explains how the land plant, *Glossopteris* could appear simultaneously in so many locations and how worms could move from South America to South Africa. Scientists now envision a "supercontinent cycle" in which, over hundreds of millions of years, the continents assembled themselves into supercontinents that stayed together for awhile and then broke up again. The last supercontinent, Pangaea, broke up into Gondwanaland and Laurasia about 200 million years ago. These two continents then broke up to form the seven present continents beginning about 120 million years ago.

By 1971 all of these pieces had come together into a coherent theory that, while not fully accepted, had won acceptance by a large number of prominent geologists, paleontologists, and others. From that point on, rather than fighting for acceptance, adherents to plate tectonic theory refined and expanded their theory, finding it ever more valuable in explaining the workings of the Earth. It would be fairly accurate to say that by the 1980s the theory of plate tectonics was almost universally accepted.

Impact

The impact of plate tectonic theory on science has been enormous and may not yet be fully realized. The impact on the general public has been somewhat less obvious, but significant nonetheless. Plate tectonics has allowed us to:

1) Better understand volcanoes and earthquakes, giving us insight into their causes, mechanisms, and risks.

2) Improve theories for locating mineral deposits, allowing more efficient prospecting and recovery of mineral resources.

3) Reconstruct earlier climates and land positions at those times, giving more detailed information about the Earth's climate history.

4) Extend our knowledge of the Earth to other planets and moons in the solar system as we try to better understand them.

5) Develop a unified theory of the Earth, instead of a number of piecemeal and ad hoc theories explaining individual features.

Volcanoes and Earthquakes

Most earthquakes and virtually all volcanoes originate at the margins of tectonic plates. As slabs of oceanic crust descend into the lower crust and mantel, they catch, build up stress, and then slip. As plates move past one another, they catch and release. This is what causes earthquakes. As one moves further from a subduction zone, the earthquakes move steadily deeper, following the descending plate into the Earth. Further from the surface, the earthquakes cause less damage at the surface and the area becomes seismically safer.

Most volcanoes are formed when molten rock rises from descending plates to the surface. Again, this means that living further from subduction zones generally reduces the hazards associated with volcanic eruptions. Notable exceptions to this general rule are "hot spot" volcanoes such as the Hawaiian Islands or the volcanic caldera beneath Yellowstone. In these cases, a plume of mantel material is apparently hitting the bottom of the crust as it moves past, melting crustal rocks and causing volcanoes to form. However, knowing that this is the cause leads us to predict that, while Yellowstone will likely erupt again someday, Cincinnati will probably not do so because it is not over any known hot spot (the mantel hot spots have been pretty well mapped by measuring the temperature and heat flow rate of the crust over the Earth).

Understanding how volcanoes erupt, earthquakes occur, and why they are where they are can help in many ways. Knowing where to find them lets us construct maps showing the seismic and volcano risk of various parts of the world. These maps are used to determine insurance rates, to help find locations for hazardous and radioactive waste disposal sites, to calculate the risk of damage to power plants, and so forth. This knowledge can also help people to decide where to live. On the other hand, many geologists purposely move toward such zones to be closer to the events they study.

Mineral Deposits

Continental rifting creates stress in rocks and is accompanied by the movement of magma and hot liquids (geothermal fluids). Volcanoes are often associated with mineral deposits that vary depending on the chemistry of the subducting slab and the overlying "country" rock. Much work has been done to correlate seismically and tectonically active regions and mineral deposits, leading to a greater potential for economically viable mineral extraction.

In addition, many rock formations were simply ripped asunder during continental rifting. A rich mineral deposit on one side of an ocean or sea implies a corollary deposit on the other side, if one can determine continental motion in the intervening millions (or tens or hundreds of millions) of years. A careful examination of the rock record near good mineral deposits may indicate areas of similar geology several thousand miles away that would be worth exploring for economic mineral deposits.

Despite the arguments against environmental degradation, maintaining and (for less-developed nations) building a modern industrial society requires ready access to large amounts of relatively inexpensive energy and raw materials. Much human progress has depended on the use these commodities. The lack of raw materials can even push a country into war, as Japan did in World War II to assure continued access to raw materials and energy sources. Understanding the location of valuable mineral and energy deposits can help the world to continue progress towards goals of greater wealth and a higher standard of living for all.

Climatological Studies

Earth sciences can also help study past climates. For example, the composition of rocks formed at the Earth's surface indicate the temperature at which they formed and the rough chemical composition of the atmosphere at that time. In addition, certain plants and animals, which may be found as fossils, are known to have been associated with specific types of climate. However, simply knowing what the weather was like at some time in the past is not sufficient to reconstruct a past climate. For example, a scientist 30 million years from now, finding the fossil of a palm tree beneath arctic ice may conclude that, at this time, the polar caps were melted and the poles were warm. However, that same scientist, looking at continental motions over this time period, would realize that the palm tree grew near the equator and had been subsequently carried to its frigid new home.

There is great interest in studying ancient climates because of the debate currently raging over whether and by how much humans are changing the Earth's climate. To know if we're having an effect, we must know what's "normal." To do that, paleoclimate records must be adjusted for ancient position on the globe. This, in turn, will help us better understand our effect on the world. As it turns out, the past few million years have been uncommonly cool. Ironically, the situation known as "global warming" more closely resembles the typical climate on Earth. However, that must be balanced by the fact that, in the more recent past, we have been in an ice age that does not seem to be over yet. Therefore, although our current climate is "unseasonably" cold, any warming trends can't yet be said to reflect what the Earth would be doing without our help, or if it a departure from the expected. Research in this area continues.

Other Worlds

In our exploration of the other planets and moons in our solar system we have seen evidence of tectonic activity in a number of places. Io, a moon of Jupiter, is the most volcanically active body known. Ganymede, another Jupiter moon, shows evidence of "ice tectonics" that suggests an underlying ocean of liquid water. Mars, probably devoid of tectonic activity at present, has the largest known volcanoes and rift valley, possible signs of early tectonic activity. Venus appears to have been subject to plate tectonics at one point, if not still today. Other worlds are not yet sufficiently well explored, but we are likely to find more tectonic activity in the future as we learn more about our planetary neighbors.

This is important for a number of reasons. First, it confirms that plate tectonics is a viable and important theory. In addition, we can observe other planets and virtually all of their features because, with the possible exception of Ganymede, they are devoid of the oceans that hide so many of Earth's tectonics. By studying other worlds, we can better understand our own. And, finally, we are learning more about some of the mechanisms that can cause plate tectonics. Plate movement requires energy. On Earth, this energy comes from the radioactive decay of uranium, thorium, and an isotope of potassium in rocks. On Io, the energy likely comes from tidal flexing of the moon as it moves through Jupiter's intense gravitational field. Mars probably lost its internal heat long ago, which is why it's now tectonically dead. As we learn more about the "styles" of tectonic activity elsewhere, again, we learn more about our Earth.

Interdisciplinary Unification of the Sciences

As mentioned above, plate tectonics was a unifying theory. It helped bring together many disparate parts of the earth sciences, uniting them with evolutionary theory, paleontology, and some aspects of biology. In fact, this may be its single-biggest contribution to science because, more than anything, it shows how a few well chosen tools can solve seemingly unrelated problems in a number of disciplines. The more we look, the greater the number of interconnections we see, and we start to realize that all problems are interdisciplinary to some degree, so that the answer to any problem lies with no single discipline or person. It both broadens and unifies science as few other theories have ever done.

P. ANDREW KARAM

Further Reading

Continents Adrift and Continents Aground: Readings from Scientific American. San Francisco: W. H. Freeman and Company, 1976.

Hallam, Anthony. *Great Geological Controversies.* Oxford: Oxford University Press, 1983.

Hellman, Hal. *Great Feuds in Science.* New York: John Wiley & Sons, 1998.

McPhee, John. *Assembling California.* New York: Farrar, Straus, & Giroux, 1993.

McPhee, John. *Basin and Range.* New York: Farrar, Straus, & Giroux, 1980.

McPhee, John. *In Suspect Terrain.* New York: Farrar, Straus, & Giroux, 1983.

McPhee, John. *Rising from the Plains.* New York: Farrar, Straus, & Giroux, 1986.

Quasars: Beacons in the Cosmic Night

Overview

The term quasar is used to describe quasi-stellar radio sources that are the most distant, energetic objects ever observed. Quasars are enigmatic. Despite their great distance from Earth, some are actually brighter than hundreds of galaxies combined, yet are physically smaller in size than our own solar system. Astronomers calculate that the first quasar identified, 3C273 (3rd Cambridge catalog, 273rd radio source) located in the constellation Virgo, is moving at the incredible speed of one-tenth the speed of light and, although dim to optical astronomers, is actually five trillion times as bright as the Sun. Many astronomers theorize that very distant quasars represent the earliest stages of galactic evolution. The observations and interpretation of quasars remain controversial and challenge many theories regarding the origin and age of the Universe. In particular, studies of the evolution and distribution of quasars boosted acceptance of Big Bang-based models of cosmology (i.e., theories concerning the creation of the Universe) over other scientific and philosophical arguments that relied on steady-state models of the Universe.

Background

In 1931 American engineer Karl G. Jansky's (1905-1950) discovery of radio waves emanating from the central region of the Milky Way Galaxy laid the foundations for the development of modern radio astronomy. Six years later, another American engineer, Grote Reber, constructed the first radio telescope in his own backyard and over the course of the next decade, radio telescopes increasingly were used to explore the Cosmos. The information they provided astronomers served to shape one of the greatest mysteries of astronomy, the discovery of emission of strong radio waves from dim stellar sources. By the mid-1950s an increasing number of astronomers using radio telescopes sought explanations for these energetic radio emissions.

Puzzled astronomers suspected a variety of star-like objects might be the source of radio waves. In 1960 astronomers Allan Sandage

The Hubble Space Telescope views of the distant quasar 120+101 indicate that its image has been split by gravitational lensing, a phenomenon by which the pull of a massive object, such as a galaxy, can bend the light of another object when the light passes near or through the massive object. *(NASA. Reproduced with permission.)*

(1926-) and Thomas Matthews identified a source of strong radio and ultraviolet emissions that appeared to originate from an optically faint star. In 1962 British radio astronomer Cyril Hazard, in an attempt to fix the location of radio emissions, used the Moon as a occultive shield to measure the duration of eclipsed radio waves. With the help of other astronomers, Hazard discovered strong radio source traceable to a seemingly ordinary single star-like object. Optical telescopes looking at that same region of space pinpointed a faint star-like object (subsequently designated quasar 3C273) with the same unusual spectral emissions.

In 1963 Maarten Schmidt (1929-), an astronomer working at the Mount Palomar Observatory, explained the abnormal spectrum from 3C273 by suggesting that its seemingly bizarre wide emission lines were really the highly red-shifted spectral lines normally associated with hydrogen. Astronomers Jesse Greenstein and Thomas Matthews, conducting an independent study of what was later known to be quasar 3C48, also noted that their object's strange spectrum was explainable by an extremely large red shift.

Red shift describes the shift of spectral emission lines toward longer wavelengths. Determination of an object's red shift allows the calculation of a object's recession velocity. Because the rate of recession increases with distance, the velocity is a function (known as the Hubble relation) of the distance to the object.

Schmidt's keen analysis made a profound impact because it meant that 3C273 had to be three billion light-years away from Earth. Because 3C273 was so far away, it must also be thousands of times more luminous than a normal galaxy to appear as optically bright as it did. Ascribing the 3C273 spectrum to Doppler-like red shift also implied an immense velocity of recession (a spreading of space in all directions).

Impact

In the wake of Schmidt's unraveling of the mystery of 3C273, subsequent studies of the evolution and distribution of quasars indicate that they were more abundant when the Universe was younger. Although astronomers have yet to fully unravel the mystifying radiation of such enormous energy from objects no larger than our solar system, the observable and measurable differences between the younger Universe and the Universe as it exists now heralded the decline of steady state models of the Universe.

In addition, prior to more direct observations, the discovery of quasars provided tacit proof of the existence of black holes. Russian scientist Yakov Zeldovich proposed that only the theoretical properties associated with black holes could explain how quasars outshine galaxies composed of hundreds of billions of stars. Black holes are created when giant stars collapse to tremendous density and thereby create a gravitational field so intense that not even light can reach the required escape velocity. Such a black hole, if located near the center of a galaxy, would begin to consume the galaxy's gas and stars and, just outside the black hole in an area called the accretion disk, intense radiation would be emitted as matter accelerated toward oblivion in the black hole.

Subsequent studies of Quasar 3C273 showed that it blasts jets of visible and x-ray energy tens of thousands of light-years into space, a phenomena that could only be explained by the presence of one rotating, supermassive object, with galactic matter orbiting in an accretion disk. One explanation for the emission of radio waves postulates that as electrons in the accretion disk are accelerated to speeds near the speed of light, they also move in the presence of a magnetic field along helical paths and thereby emit radio waves by a process termed synchrotron radiation. Waves similar to the waves emanating from quasars are observed on Earth when physicists shoot high energy electrons through synchrotron particle accelerators.

After decades of observation there is a growing consensus that quasars represent a class of galaxies with extremely energetic nuclei. Large radio emissions seem most likely associated with large black holes with a large amounts of matter available to enter the accretion disk. Less vigorous radio sources (e.g., Seyfert galaxies or QSOs—"quasi-stellar objects"—that offer a similar optical appearance to quasars but that are radio quiet or silent) may be accounted for by smaller black holes or by black holes in smaller galaxies with less matter available for their consumption.

The nature and location of quasars, as well as the existence of objects that might be associated with them, garnered and consumed considerable research attention. The limitations of ground-based telescopes, however, frustrated astronomers searching for clues to unravel the mysteries. In fact, the need and ability to study quasars was often cited as one of the principal reasons to build the Hubble Space Telescope launched by the United States in 1990. In particular, astronomers wished to determine whether quasars were associated with galaxies.

Photos from the Hubble Space Telescope subsequently determined that quasars did reside in galaxies. In addition, studies of brighter quasars that act as a powerful electromagnetic back lighting allow astronomers to examine intervening absorbing material. Using such quasars, astronomers are able to study the primitive gas clouds found in the early Universe.

The impact of quasars upon cosmology (the study of the nature and origin of the cosmos) cannot be understated. The discovery of quasars provided an enormous boost to cosmological models based on the Big Bang theory. Because of the finite speed of light, the discovery of quasars allowed astronomers to look back into the history of the Universe. Although challenged by American astronomer Halton Christian Alp, most astronomers now accept that the large red shifts of quasars indicate their great distance and that, correspondingly, their light emissions present us a picture of the early Universe as it existed within a few billion years after the Big Bang.

Other compatible interpretations of quasars assert that they were formed in large numbers in the young Universe and that these quasars evolved into normal galaxies. Accordingly, they are visible only in distant, ancient light.

Prior to the discovery of quasars, a rival cosmological model (i.e., a model describing the creation of the cosmos) was termed the steady state model. The steady state model relied on a Universe that was the same in all directions (when averaged over a large span of space) and at all times. To account for Hubble's discovery of an expanding Universe the steady state model postulated a continuous creation of matter in the space between the stars and galaxies so the density of the Universe was maintained in a steady state. Although many astrophysicists rejected the steady state model because it would violate the law of mass-energy conservation, the model had many eloquent and capable defenders, including British physicist Fred Hoyle. In addition, the steady model was seen to be compatible with many philosophical, social, and religious concepts centered on the concept of an unchanging Universe that has existed in much the same state as it had been created.

Problematic for skeptics of the steady state model was the fact that the proposed rate of "mass creation" required to support the steady state model was far too small to be detected by experimental observation. To maintain the Universe in a steady density state, slightly less than one hydrogen atom per cubic centimeter would have to be created every millennium. In essence, there was no way to absolutely disprove the steady state model of the Universe, and thus the steady state theory and the Big Bang theory competed with each other for scientific and philosophical favor.

As astronomers studied quasars they discovered an early Universe that contained many more quasars than exist now. This change in the Universe over time (e.g., specifically the rate of formation and existence of quasars) contradicted the steady state model. Along with the discovery of a permeating cosmic background radiation, the discovery of quasars tilted the cosmological argument in favor of Big Bang-based models.

Building on the work of Schmidt and others, recent discoveries concerning quasars aid astronomers, cosmologists, and philosophers refine models concerning the history, structure, and future of the Cosmos. As observation and detection techniques improved, so did the discovery of new quasars that stretch the frontiers of the Universe. In 1979, a x-ray quasar was found with a red shift of 3.2, meaning that the velocity of recession was 97% the speed of light. Another enigma surrounding quasars came with the apparent discovery of quasars approximately at a distance of 17 billion light years. Although seemingly fantastic, these measured distances would make these quasars older than the current upper estimates for the age of the Universe and thereby challenge existing cosmological theory.

K. LEE LERNER

Further Reading

Books

Kaufmann, William. *Black Holes and Warped Spacetime.* W. H. Freeman and Company, 1979.

Shipman, Harry. *Black Holes, Quasars, and the Universe.* Houghton Mifflin, 1976.

Periodical Articles

Schmidt, M. and R. Green. "Counts, Evolution, and Background Contribution of X-Ray Quasars and Other Extragalactic X-Ray Sources." *Ap. J.* 305, 68 (1986).

Talcott, Richard. "A Quasar Lights up the Universe." *Astronomy* (September 1991): 42.

The Discovery of Pulsars

Overview

The discovery of pulsars in 1967 can be said to have been almost accidental. Pulsars were discovered by Jocelyn Bell Burnell (1934-), then a graduate student at the University of Cambridge who was using her advisor's radio telescope to search for quasars. Her discovery had considerable impact, both for astronomers in general and women scientists in particular. Since their discovery, pulsars have been recognized by astronomers as crucial to understanding the nature of stars, especially exotic stars like black holes. For women scientists, Bell Burnell's discovery was to be an inspiration. Rarely had a female scientist gained so much fame for a scientific discovery. And although she did not share the Nobel Prize in Physics given to her advisor for the discovery of pulsars, she has since come to be recognized for something perhaps more significant: for helping to pave the way for women in all fields of science.

Background

The discovery of pulsars has as its background the whole of radio astronomy, and the discovery of

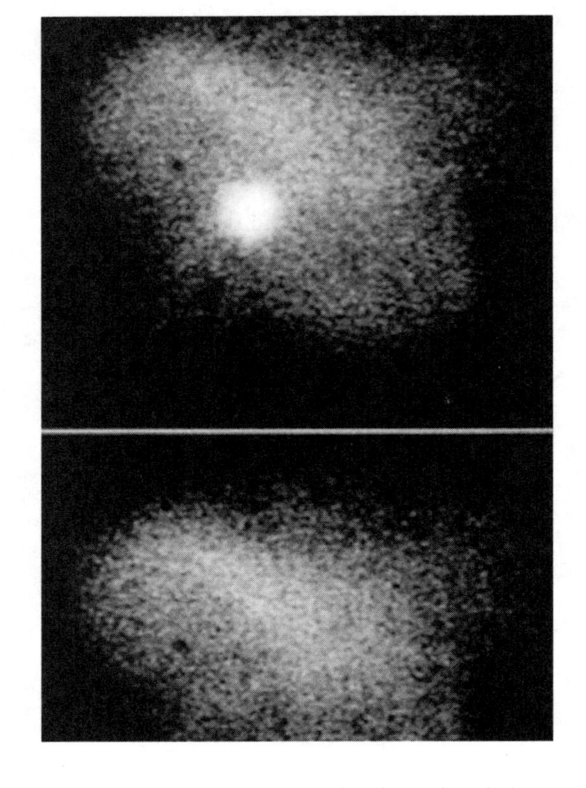

A very fast pulsar was discovered in the Crab Nebula in 1968.*(NASA. Reproduced with permission.)*

quasars in particular. This is because the use of radio telescopes to search for quasars led to the discovery of pulsars. The history of radio astronomy and the development of radio telescopes is important to the discovery of both quasars and pulsars.

Radio telescopes receive radio waves, not light. Therefore they are not like the optical telescopes we normally associate with astronomy. Radio telescopes do not have lenses and they are not shaped like tubes. Instead, radio telescopes typically consist of radar dishes, or of very large arrays of wires suspended above the ground. These "telescopes" receive radio waves from space. Unlike optical telescopes they can operate night and day, and also during cloudy weather. They can amplify the signals they receive so that they are made stronger, and these signals can then be transformed into both audio and video signals that are interpreted by astronomers. One problem that radio telescopes have is that they often pick up human-made radio signals from Earth. This can cause considerable confusion. Such confusion was part of the story of the discovery of pulsars, and will be discussed below. First, however, we must consider the discovery of quasars.

Quasars were discovered in 1960 with a type of radio telescope called an interferometer. The radio astronomer Thomas Matthews was using this telescope to get an accurate position for an object referred to as "3C 48." Earlier this object had been observed as a blue-colored star. Matthews demonstrated that this star was a source of large quantities of radio waves. In the next few years, other such radio wave-emitting objects were discovered. One of these objects, called "3C 273," was closely studied in 1962. It was demonstrated to be both very distant and very bright. So bright, in fact, that astronomers estimated this single object to be as bright as 100 galaxies, the equivalent of one trillion stars. Further studies of these objects revealed that they all shared the traits of being extremely bright, large (each being roughly the size of our solar system), and radiating vast quantities of energy in the form of radio waves. They were called quasi-stellar radio objects, or *quasars*.

The best way to detect quasars was to use a technique called "interplanetary scintillation." Radio waves coming to the Earth from objects in space, like quasars, will be slightly disrupted by the solar wind (ionized gas) that "blows" off of our Sun. While radio signals from space are affected by the solar wind, radio signals from Earth are not. The technique of "interplanetary scintillation" detects radio signals from space by looking for the disruption of these signals by the solar wind; this disruption is detected as a twinkling or "scintillation." In order to detect such scintillations, unique radio telescopes had to be built.

In July 1967 radio astronomers at the university of Cambridge in England finished building such a radio telescope. The director of this project was Antony Hewish (1924-). He was aided by Jocelyn Bell Burnell, who was then a graduate student, and other volunteers. This radio telescope took two years to build and consisted of 120 miles (193 km) of cable suspended on 128 pairs of poles. The entire telescope covered roughly 4.5 acres of ground. As part of her Ph.D. work, Bell Burnell analyzed the charts of data produced by the telescope's computer. Her job was simply to review the numerous data charts, find scintillations like those produced by quasars, and then plot their positions on maps of the heavens. She could not have known that this seemingly mundane task would lead to a most remarkable discovery.

Impact

Jocelyn Bell Burnell's work with the radio telescope was routine for about two months, until August of 1967. On 6 August the telescope picked up a radio source whose signals came in pulses. At first Bell Burnell thought the pulses were just "scruff," as they did not appear to be the quasars for which she was looking. After a while she realized that these pulses of " scruff" came with extreme regularity. Initially, neither Bell Burnell nor her advisor Hewish thought they had discovered anything new. They believed it was a human-made radio signal, perhaps reflected back to their telescope off of the Moon or a satellite, or even a nearby building. But by November they realized this was not the case, that their mysterious signal did in fact come from a location outside of our solar system. Astoundingly, its radio-wave pulses came with such rapid regularity—once every 1-1/3 seconds—that Bell Burnell and Hewish thought the source might not be natural. As a joke, they said the signal must come from "Little Green Men," and so they called the pulsing radio source LGM1.

In the next month, December of 1967, Bell Burnell was analyzing data from a different part of the sky and found another regularly pulsing radio source with a slightly shorter period of 1-1/5 seconds. And then, over the Christmas holiday, she discovered two more such pulsing sources. So by January 1968, Bell Burnell and Hewish knew they had discovered a new class of objects in space. They announced their discovery in February of 1968 in a paper in the journal *Nature*. The announcement was sensational, and soon afterwards the objects were given the name pulsars.

But what sort of objects were these pulsars? A few months before Bell Burnell's discovery, the astronomer Franco Pacini, then at Cornell University in New York, published a paper arguing that a rapidly rotating neutron star, if one were found to exist, would have a very strong magnetic field and would therefore be a powerful source of radiation. In June of 1968, soon after the discovery of pulsars was announced, Thomas Gold (1920-), also at Cornell University, published a paper in *Nature* in which he identified the pulsars discovered by Bell Burnell with the theoretical rotating neutron stars indicated by Pacini. Thus it was shown that the pulsars were rapidly rotating neutron stars. They emitted high-intensity radio waves from their magnetic poles. Because of their rapid rotation, pulsars' radio waves are detected as "pulses" much like the way light is seen "pulsing" from a light-house.

Perhaps one of the most interesting results of the discovery of pulsars was the controversy over who actually discovered them. In 1974, Antony Hewish and Sir Martin Ryle (1918-1984) received the Nobel Prize in physics for their work in radio astronomy. Hewish was recognized for his role in the discovery of pulsars. Jocelyn Bell Burnell did not share the prize. She was not considered the discoverer of pulsars; at the time she had merely been a graduate student and the Nobel Prize committee felt that the award should go to a scientist with a long and established record of research. Her exclusion from the Nobel Prize led many distinguished astronomers, including Thomas Gold, to complain that Bell Burnell was in fact the discoverer of pulsars and therefore should have shared the prize.

In all of this Bell Burnell did not complain. She said "Nobel Prizes are based on long-standing research, not on a flash-in-the-pan observation of a research student." She did win many other prizes, medals, and honors for her discovery of pulsars and became an inspiration for women scientists. Living in England, she considers herself "a role model, a spokeswoman, a representative, and a promoter of women in science in the United Kingdom." And she has undoubtedly inspired women scientists throughout the world.

The discovery of pulsars has impacted science and society in two significant ways. First, it was an incredible discovery for astronomers. It not only confirmed the existence of the theoretical neutron star, but it also enabled scientists to make advances in astrophysics, particularly in their theories of stellar collapse and the formation of black holes. Furthermore, pulsars are the most regular "clocks" in the universe. They have enabled scientists to make important tests of Albert Einstein's theory of general relativity.

Second, the discovery of pulsars shed light on the important role of women in science. Perhaps more surprising than the fact that a new type of star was discovered was that a woman had discovered it. In 1967 there were relatively few established women in science. Jocelyn Bell Burnell was then and continues to be an important example for women scientists. In 1991 she was made a Professor of Physics at the Open University in England. Soon after her appointment, the number of women physics professors in the United Kingdom doubled.

STEVE RUSKIN

Further Reading

Books

Lyne, A. G. and F. Graham-Smith. *Pulsar Astronomy.* Cambridge: Cambridge University Press, 1990.

North, John. *The Norton History of Astronomy and Cosmology.* New York: W. W. Norton, 1995, pp. 563-66.

Periodical Articles

Bell Burnell, Jocelyn. "Little Green Men, White Dwarfs, or What?" *Sky & Telescope* (March 1978): 218-21.

Reed, George. "The Discovery of Pulsars: Was Credit Given Where it Was Due?" *Astronomy* (December 1983): 24-28.

Woolgar, S.W. "Writing an Intellectual History of Scientific Achievement: The Use of Discovery Accounts." *Social Studies of Science* 6 (1976): 395- 422.

Advances in Radio Astronomy Revolutionize Man's View of the Universe and its Origin

Overview

In the 1960s a Cambridge University radio astronomer named Martin Ryle (1918-1984) discovered ways to combine the information from several radio telescopes to simulate a much larger instrument. These advances in radio astronomy helped pave the way for instruments such as the Very Large Array in New Mexico and techniques such as very long baseline interferometry. These helped to turn radio astronomy into an incredibly valuable scientific tool that has given profound insights into some of the most exciting phenomena in the universe, including supernovae, quasars, the Big Bang, and star formation.

Background

In the early 1930s engineers from Bell Laboratories set up large antennae to study ship-to-shore communications. When these antennae consistently picked up static and hissing that could not be correlated with any known sources, an engineer named Karl Jansky (1905-1950) was asked to assist in the investigation. He discovered that the stray noise came from the sky, starting the field of radio astronomy. The first radio telescope consisted of eight large metal hoops set in a wooden frame that was mounted on a set of wheels from a Model T Ford automobile.

The following 20 years or so saw minor improvements in radio astronomy, but astronomers remained limited by many factors. Radio telescopes, because of the longer wavelength of radio waves compared to that of visible light, gave a "fuzzier" view of the universe than visible-light telescopes. They could not be made large enough to have a similar resolution because of engineering constraints, so our view of the "radio cosmos" remained blurry for many years. In addition, some radio bands could not be investigated because of radio "noise" from terrestrial sources. To observe, for example, at a frequency used by a local radio station would be like trying to observe a faint star with a spotlight shining in your eyes. Finally, not all radio wavelengths are visible on the Earth's surface because longer wavelengths are reflected back into space by the ionosphere and shorter ones are absorbed by the air. In spite of these limitations, however, radio astronomy continued to be used and became steadily more valuable to astronomers.

The biggest engineering breakthroughs in radio astronomy came with the development of interferometry techniques. These techniques, similar to those used in the Michelson-Morley experiment, allowed astronomers to combine the images obtained by radio telescopes that were widely separated to simulate a much larger telescope with correspondingly greater resolution. (In other words, they can see finer detail than single dishes.) The technique of interferometry has been extended in two fundamentally different ways. First, very long baseline interferometry (VLBI) uses data from one or more radio telescopes that may be separated by the diameter of the Earth to simulate a telescope with a diameter of 8,000 miles (12,875 km). Second, the Very Large Array (VLA) is a collection of 27 radio dishes, each 25 meters (82 feet) in diameter, that are arranged in a Y shape along railroad tracks. These dishes can be extended to a total diameter of 27 kilometers (17 miles) to achieve a resolution up to 1,000 times better than is possible with visible-light telescopes, or they can be clustered together to collect a greater number of radio waves.

Both of these techniques rely on Ryle's technique, called aperture synthesis, a method for gathering and combining the data collected simultaneously from multiple sources and synthesizing it into data from a simulated single, large telescope. (Space here does not permit a description of the technical details of this technique,

The Very Large Array radio telescope near Socorro, New Mexico, consists of 27 dish antennas, each 82 feet across. The dishes are linked together by aperture synthesis. *(JLM Visuals. Reproduced with permission.)*

but it is described in many popular books on astronomy, including those noted below.)

With the "new" radio telescopes created using aperture synthesis, detailed views of structures in this galaxy and others were suddenly possible. This development, in turn, helped astronomers gain a better appreciation of the mechanisms that power quasars (in fact, quasars were first discovered via radio astronomy) as well as shed new light on pulsars (neutron stars), supernovae, and other astronomical phenomena.

Meanwhile, other radio astronomers were noting other phenomena with their instruments. In the 1960s British astronomer Tony Hewish (1924-) and his student, Jocelyn Bell Burnell (1943-), discovered regular radio signals coming from the sky. These signals were determined to be of natural, not artificial, origin and were later found to be from neutron stars, stars that collapsed at the end of their lives but were not massive enough to form black holes. In 1965 American astronomers Arno Penzias (1933-) and Robert Wilson (1936-) discovered a persistent radio signal coming, seemingly, from all directions. They eventually determined it was an "echo" of the Big Bang in which the Universe was born. These discoveries were later rewarded with the Nobel Prize because they fundamentally changed the way we look at the universe.

Impact

These advances in the technology of radio astronomy have resulted in a significant impact on the sciences of physics and astronomy. This scientific impact, in turn, has led to a number of discoveries that have affected the general public in several ways. For example, the scientific theory that the universe originated in a "Big Bang" is deeply troubling to many whose religious beliefs are contradicted by this theory. (In fact, the Big Bang theory was only one of several competing hypotheses describing the origin of the universe; this issue was only settled scientifically with the discovery of the cosmic background radiation.) Additionally, since we do not see in radio wavelengths, all the data collected by radio astronomers must necessarily be rendered in such a way that it can be viewed in visible light. Such false-color images have tended to excite the public interest because of their novelty and beauty. This public interest, in turn, has helped to persuade funding agencies (and the politicians who approve their funding) that astronomy is worth continued public support. (Each impact will be discussed in further detail in the following paragraphs.)

The scientific impact of radio astronomy was immediate and obvious. As one astronomer put it, "Nothing interesting happens in the visible spectrum." The first scientific impact was the realization that many objects emit radio-frequency radiation. This led to the discovery of galactic magnetic fields (which affect the paths taken by cosmic rays), quasars (galaxies whose centers seem to house super-massive black holes), pulsars (the first evidence of stars collapsing at the end of their lives), and "radio jets" emanating from the centers of some galaxies.

All of these phenomena required the development of new physical theories, or the application of physics in extreme conditions, to explain. In addition, the discovery of pulsars allowed certain tests to be performed that have confirmed the accuracy of the general relativity theory while examinations of the cosmic microwave background have given deep insights into the origin of the chemical elements, the formation of structure in the universe, and further confirmation of general relativity predictions. Some of these findings, such as the exploration and mapping of radio jets, supernova remnants, and radio-bright galaxies have led to a much deeper appreciation and understanding of the violent events that unfold in the universe and that might have, at some time in the past, affected life on Earth. Notably, this precision mapping of such structures has depended entirely on the development of aperture synthesis techniques by Ryle.

Confirmation of the Big Bang theory in the formation of the universe is arguably the most significant impact advances in radio astronomy have had from the standpoint of public interest. Before launching into a discussion of this topic, however, a popular misunderstanding regarding the word "theory" with respect to scientific inquiry must be dispelled. As opposed to popular usage, scientists use the word "theory" to refer to something that is accepted as fact by the majority of the scientific community. In effect, scientists use the word "theory" in a manner similar to what most people mean when they say something is a "fact." From this standpoint, the great majority of physicists and astronomers do not doubt that the Big Bang occurred, although they may disagree about some of the details of the event. Penzias and Wilson's discovery of the cosmic microwave background radiation was the single discovery that turned the Big Bang from hypothesis (proposed but not proven) into theory (proposed and proven).

This discovery, in turn, has proved threatening to many religious beliefs because science appears to be invading territory that was previously in the domain of religion, the formation of the universe. Many have felt that science is trying to address questions already answered by God, while others have feared that the Big Bang theory might be a portent of future scientific advances that would convince many that God either did not exist or was not necessary. In spite of widespread acceptance of the Big Bang theory, however, there is still general disagreement over what sparked the Big Bang. In other words, the Big Bang theory still leaves room for religion. (See the following book by Stephen Jay Gould for more about the relationship between science and religion.)

Finally, scientific developments from radio astronomy, as well as astronomy in other wavelengths (x-ray, gamma-ray, infrared, ultraviolet, and so on) have excited public interest in the field of astronomy. Part of this interest no doubt results from the sheer novelty of the phenomena described and part is because of the fact that the pictures are beautiful to look at. False-color images of nearby galaxies, radio jets, x-ray emissions from black hole accretion disks, and colliding galaxies seem to appeal to more than just scientists and the scientifically inclined. NASA's World Wide Web site logs many "hits" and a great number of them are for images, including radio wavelengths and others that are not visible. In fact, entire books have been published that are devoted solely to astronomy in radio, x-ray, and other "invisible" wavelengths. Moreover, this public interest seems to be reflected to some extent in support for further research. In spite of NASA's budgetary difficulties and cuts in science funding in general through the 1980s and much of the 1990s, observatories—radio and otherwise—continued to receive funding to continue their operations and to construct new instruments. This fact can only be explained by continued public interest in astronomy, an interest not only in beautiful astronomical images but also an interest in how stars and galaxies, as well as the universe, live and die.

P. ANDREW KARAM

Further Reading

Bartusiak, Marcia. *Thursday's Universe.* New York: Times Books, 1986.

Davies, Paul, ed. *The New Physics.* New York: Cambridge University Press, 1989.

Gould, Stephen Jay. *Rocks of Ages.* New York: Ballantine Publishing Group, 1999.

Henbest, Nigel, and Michael Marten. *The New Astronomy.* New York: Cambridge University Press, 1983.

The Debate Between "Big Science" and "Small Science"

Overview

As the twentieth century began, scientists became increasingly aware of fundamental new forces and scientific phenomena that were observed fleetingly, if at all, but that were required by emerging theories. This sparked interest in scientific apparatus to investigate the nature of the atom, the structure of the universe, and more. This quest for more detailed knowledge continues, but the questions are growing larger, requiring increasingly expensive equipment to address. However, the vast majority of scientists work on relatively small projects, and most of scientific research takes place in relatively small laboratories with relatively small budgets. These small budgets are often threatened when huge scientific projects threaten to cut funding for other, less visible projects. This, in turn, has led to the tension that often exists between "big science" and "small science."

Background

One of the first large science projects was the construction of the 100 inch (2.5 meter) Mount Wilson Observatory's telescope in California. This telescope was designed to probe the universe and, for nearly 40 years, remained the largest and most powerful telescope in the world. Mount Wilson was superseded by the 200 inch (five meter) Hale telescope in 1948 but, by then, the Manhattan Project had forever changed the way that large scientific projects would be tackled.

During this time, many other fields of scientific endeavor grew to demand increasingly large expenditures because, as our knowledge of the sciences grew, the questions to be answered became ever more intractable. For example, early research into atomic structure was performed on a cyclotron that could fit on a small kitchen table. This provided ample energy for the atomic structure research done in the 1930s. Direct descendents of this device include the 27 km-circumference (17 miles) CERN accelerator and the 53 km-circumference (33 miles) Superconducting Supercollider (SSC), the latter of which failed to receive funding and was cancelled. By another measure of comparison, the first cyclotron could accelerate particles to an energy of nearly 100,000 electron volts, while several accelerators today reach energies of over one trillion volts on a routine basis. Other areas of science have seen similar increases in required funding as the price of continuing advancement grows. The Hubble Space Telescope cost over $1 billion, several interplanetary probes (*Galileo, Cassini,* Mars Observer, for example) have cost over $1 billion, and the Human Genome Project will cost several billion dollars when completed.

While huge scientific projects are becoming necessary, small science is flourishing as a result of breakthroughs in engineering, computer power, and scientific understanding. For example, sequencing a gene used to take years of laborious work by a laboratory full of people. This same job can now be performed automatically by any of a number of labs for a very small cost. Similarly, it is now possible to construct a "poor man's supercomputer" for a few tens of thousands of dollars, simply by networking a dozen or so desktop computers to give the same performance as a 1990-vintage Cray supercomputer costing several million dollars.

Thus, we are led to an apparently growing dichotomy between a relatively small number of high-energy physicists, astronomers, and geneticists whose work on some of the fundamental concepts of the universe and life require ever-increasing sums of money, while the great majority of scientists, many of them doing less glamorous but no less necessary follow-on work, are starved for funding. In science, as in most other human endeavors, it is the follow-on research that helps determine the validity and eventual utility of major breakthroughs. On the one hand, we run the risk of being unable to capitalize on scientific discoveries if small science is neglected in favor of large projects, but, on the other hand, without the large projects, there will be no new scientific territory to explore and exploit.

Of course, the issue of national pride enters into the mix. Many nations, the United States included, are willing to spend vast amounts of money to build the newest, most powerful, or most advanced piece of scientific equipment in the world. However, they fail to see the rationale behind funding the thousands of scientists laboring to fill in the blanks, confirm the results, and to do the other work necessary to make sense of the findings of large projects. This has led many scientists to feel that there is an impending crisis in the sciences.

Impact

Large science projects have captured the public's attention for decades, and continue to do so. Photos from the Hubble Space Telescope and probes to the outer Solar System appear on television regularly and the intricacies of subatomic physics are broadcast on network news shows when a new particle is discovered or created in the laboratory. The creation of new chemical elements receives similar attention, and the sequencing of additional chromosomes or genomes is often front-page news. There is no doubt that large science projects such as these are a major factor behind the public's interest in science, and that interest is one large factor that keeps science funding alive in the United States and many other nations.

If huge science projects are exciting, promote public interest in science, and add to national pride, there is also no doubt that they are necessary. For example, in the realm of atomic physics, scientists have simply looked at just about every process that can be examined at relatively low energies. All of the predicted particles that exist at current accelerator energies have been found and the only way to move on is to construct larger accelerators that cost billions of dollars. By this reasoning, spending a few billion dollars now may take funds away from smaller science for the next decade as a new accelerator is built. However, the new machine will then make possible the next decade's worth of work for the smaller scientists, with the funds generated by public interest supporting their work.

A similar argument was made to promote the construction and launch of the Hubble Space Telescope. Ground-based observatories, it was argued, had fundamental constraints placed on them due to atmospheric turbulence. The only way to see the universe in greater detail was to raise a telescope above the atmosphere, giving an unhindered view of all. However, shortly after Hubble was launched, advances in what is known as adaptive optics (ironically made possible by yet another massive project, the Strategic Defense Initiative, or "Star Wars") made it possible for much larger ground-based telescopes to achieve the same resolution as Hubble with greater light-gathering capability and at a fraction of the cost. There are still many things Hubble can do that current ground-based telescopes cannot, but the fundamental premise behind building and launching Hubble was overtaken by events that were unforeseen when Hubble was first designed and funded.

It is also important to note that, in spite of the positive publicity engendered by successful large science projects, public opinion cannot always be counted on and a reversal of public opinion is often only a headline away. For example, cost overruns caused the cancellation of the Superconducting Supercollider when partially completed. Negative publicity surrounding the flaws in Hubble's main mirror tarnished NASA's image, even though Hubble remained perfectly capable of performing over 70 percent of its original science objectives. Public opinion of NASA was further diminished when the Mars Observer was lost and the high-gain antenna on *Galileo* failed to open, hampering the probe's ability to relay data to the Earth. Large projects tend to be hyped by their promoters in order to attract funding approval; when results fail to live up to initial promises, or when the public tires of it, the backlash can be devastating. This diminution of public confidence can carry over to science in general, leading to a general reluctance to spend money for any scientific research that is not of immediate and obvious utility.

Small science, by comparison, is neither as widely appealing nor as likely to arouse public disdain or distrust. Most people view smaller science projects as doing little more than filling in the gaps between landmark, important discoveries made by "big science." However, in spite of the romance of working on a huge project that is on the cutting edge of science and technology, it must be remembered that the laser, the polymerase chain reaction, discovery of the cosmic microwave background radiation field, plate tectonics, discovery of the asteroid impact that ended the Cretaceous Period, unraveling the structure of DNA, and other major advances in scientific understanding came as the result of "small" scientific projects.

The great majority of scientists are never going to announce a breakthrough in understanding that rivals plate tectonics or DNA. Nonetheless, their contributions are important and there is no way to tell in advance which avenues of inquiry may someday prove valuable. For this reason, advocates of small science fear the increasing tendency towards large, expensive projects that often threaten to pauperize national and international science funding. For example, a typical scientist with a modest laboratory and one or two assistants can operate for a year on about $200,000-$300,000. This may seem a large amount, but when one considers that it includes two salaries, rent for laboratory space (which is often leased to researchers by universities), supplies, travel to conferences, other necessary expenses, and university overhead (which can run to 60 percent of the total budget), it is apparent that even a budget of $300,000 annually is not much to run a lab on.

By comparison, if we assume that Hubble cost $1.5 billion, it seems that the costs of building and launching Hubble could completely support 5,000 "small" scientists with annual budgets of $300,000 each. Therefore, unless large science projects such as Hubble are funded from a new and completely separate source of financial backing, every large project drains the collective "pot" of science funding and can have a dramatic impact on scientific research as a whole. This is the concern of small science advocates—that the returns from gargantuan projects, while important to furthering our understanding, are not sufficiently dramatic to offset the loss of funding and scientists that result. And so, the debate continues.

P. ANDREW KARAM

Further Reading

Chaisson, Eric J. *The Hubble Wars: Astrophysics Meets Astropolitics in the Two-Billion-Dollar Struggle Over the Hubble Space Telescope.* Cambridge: Harvard University Press, 1998.

Lederman, Leon. *The God Particle: If the Universe Is the Answer, What Is the Question?* Boston: Houghton Mifflin, 1993.

Taubes, Gary. *Nobel Dreams.*New York: Random House, 1986.

Scientists Get Closer to Determining the Age of the Universe

Overview

In the first half of the twentieth century, astronomers demonstrated that the universe was expanding, thus providing support for the idea that it had arisen from a tremendous explosion billions of years ago. This explosion came to be known as the Big Bang. Depending on the methods used and the assumptions that are made, different values are obtained for the age of the universe. Today, scientists are coming closer to agreement on an age of 12-16 billion years.

Background

The nineteenth century invention of spectroscopy allowed astronomers to break up the light coming from distant stars and galaxies into its constituent wavelengths. Different types of stars and galaxies have characteristic patterns, or spectra. According to the Doppler effect, the spectra are shifted towards longer wavelengths, or red-shifted, when the object is receding. The greater the red shift, the faster the object is moving away from us. This corresponds to the lowered pitch you hear as a train whistle or ambulance siren fades into the distance.

During the 1920s scientists began using spectroscopy to compile data on the velocities of galaxies. Meanwhile, American astronomer Edwin Powell Hubble (1889-1953) was studying their distances. The powerful new telescope on Mt. Wilson in California, with its 100-inch (2.5 m) mirror, allowed him to observe individual stars in many galaxies. In particular, Hubble was interested in Cepheids.

Cepheids are a type of variable star; that is, their brightness changes over time in a regular way. The special feature that makes Cepheids so valuable for determining distances is that the period of their variability is related to their brightness. The longer the period, the brighter the star. Knowing how bright the star actually was, and comparing this to how bright it appeared from Earth, enabled Hubble to calculate its distance. He employed this method for the galaxies in which he could observe Cepheids. For other galaxies he used the brightest star he could find and assumed its actual brightness was comparable to the brightest stars in our own galaxy. If a galaxy was too faint for Hubble to measure any individual stars, he measured the brightness of the galaxy as a whole to estimate its distance.

In 1929, Hubble announced that the farther away a galaxy was from Earth, the faster it was receding. This result, since called *Hubble's law,* is written $v = H_0 d$, where v is the velocity, d is the distance, and H_0 is the Hubble constant, pronounced "H nought."

At first it might seem odd that everything appears to be moving away from us. The Solar System does not occupy any special place in the cosmos; we circle an average star in a sort of celestial suburb, out on one of the spiral arms of the Milky Way. What is actually happening is that on a large scale *everything* is moving away from everything else, like the raisins in a loaf of rising bread. The entire universe is expanding, and the galaxies are receding into the distance at rates of hundreds of millions of miles per hour.

Such an expansion could have been caused by an enormous explosion at the beginning of time. Russian-born American physicist George Gamow (1904-1968) coined the term *Big Bang* to describe this explosion, and the name stuck. The Big Bang implied an actual beginning to the universe. Many scientists of the time resisted the idea. They preferred the steady state theory, which held that the universe had no beginning or end, and any expansion was simply a temporary phase, perhaps one stage of a periodic oscillation.

However, in 1948 Gamow realized that a Big Bang would have generated extremely high energy radiation that would have gradually cooled down over the eons. It should still exist as cosmic background radiation, coming equally from every direction. In 1964 Arno Penzias (1933-) and Robert Wilson (1936-) detected this cosmic background radiation, an achievement for which they received the 1978 Nobel Prize in physics. At this point almost all scientists agreed that the Big Bang had actually happened. But there was still the problem of determining when it had occurred.

Impact

Looking at the equation for Hubble's law, it is easy to see that if the rate of the expansion of the universe is constant, its age is simply the inverse of H_0, often called the *Hubble time*. Based on Hubble's data, the calculated age was about 2 billion years. This presented an immediate problem, because by the 1930s geologists using radioactive dating methods had found rocks on Earth that were almost twice as old.

Of course, it was by no means a foregone conclusion that the expansion speed was actually constant. If the expansion is slowing down, then the universe would be younger than the Hubble time; if it is speeding up, the universe would be older. Furthermore, H_0 was determined by estimating distances. These estimates could very well be off, and the true value for H_0 has been vigorously debated for decades.

In 1952 German-American astronomer Walter Baade (1893-1960) discovered that there were two varieties of Cepheid variables. Hubble had used the equations for the wrong type in some of his measurements. In distant galaxies he had occasionally mistaken glowing hydrogen gas clouds for especially bright stars. Correcting these errors pushed the date of origin back far enough to satisfy the geologists, to something on the order of 10 billion years. Data taken in the 1960s and 1970s supported various ages ranging from about 10 to 20 billion years.

A problem with the smallest estimates for the age of the universe is that some of the oldest stars in our own galaxy, seen in globular clusters, have been tentatively dated at 14 to 18 billion years old. Obviously, the universe itself can not be younger than its oldest stars. However, measurements in the late 1990s seemed to indicate that the clusters were further away than had been previously thought, implying that their stars were more luminous. This would put them at a different stage in their life cycle, with ages closer to 11 or 12 billion years.

For some time there have been two main camps in the age debate, with Allan Sandage (1926-) of the Carnegie Institution supporting an older universe, and other astronomers proposing a younger one. At one point Sandage's expansion rate was half the value considered accurate by some of his colleagues, and there were many estimates in between.

The Hubble Space Telescope, launched by NASA in 1990, has provided new measurements that have narrowed the gap. The ability to see distant clusters of galaxies is important for accuracy. For relatively nearby objects, local motion, such as two galaxies moving toward one another due to gravitational attraction, may be more obvious than the overall expansion of the Hubble flow. Most scientists now believe the universe is between 12 and 16 billion years old; Sandage continues to maintain that ages up to about 18 billion years old are possible.

Accurate values for the universe's age, rate of expansion, and how that rate changes will provide clues to its future. The ultimate fate of the universe depends on whether its expansion can counteract the gravitational forces tending to pull it in upon itself. At present the evidence seems to favor a universe that expands forever, rather than collapsing in what has been called the "Big Crunch."

SHERRI CHASIN CALVO

Further Reading

Linder, Eric V. *First Principles of Cosmology.* Harlow, England: Addison-Wesley, 1997.

Overbye, Dennis. *Lonely Hearts of the Cosmos: The Story of the Scientific Quest for the Secret of the Universe.* New York: HarperCollins, 1991.

Smoot, George and Keay Davidson. *Wrinkles in Time.* New York: William Morrow and Company, 1993.

Advances Related to Quantum Electrodynamics (QED)

Overview

Quantum electrodynamics (QED) is a scientific theory that is also known as the quantum theory of light. QED describes the quantum properties (properties that are conserved and that occur in discrete amounts called quanta) and mechanics associated with the interaction of light (i.e., electromagnetic radiation) with matter. The practical value of QED rests upon its ability, as set of equations, to allow calculations related to the absorption and emission of light by atoms and to allow scientists to make very accurate predictions regarding the result of the interactions between photons and charged atomic particles such as electrons. QED is a fundamentally important scientific theory because it accounts for all observed physical phenomena except those associated with aspects of relativity theory and radioactive decay.

Background

QED is a complex and highly mathematical theory that paints a picture of light that is counterintuitive to everyday human experience. According to QED theory, light exists in a duality consisting of both particle and wave-like properties. More specifically, QED asserts that electromagnetism results from the quantum behavior of the photon, the fundamental "particle" responsible for the transmission electromagnetic radiation. According to QED theory, a seeming particle vacuum actually consists of electron-positron fields. An electron-positron pair (positrons are the positively charged antiparticle to electrons) comes into existence when photons interact with these fields. In turn, QED also accounts for the subsequent interactions of these electrons, positrons, and photons.

Photons, unlike other "solid" particles, are according to QED theory, "virtual particles" constantly exchanged between charged particles such as electrons. Indeed, according to QED theory the forces of electricity and magnetism (i.e., the fundamental electromagnetic force) stem from the common exchange of virtual photons between particles and only under special circumstances do photons become observable as light.

According to QED theory, "virtual photons" are more like the wave-like disturbances on the surface of water after it is touched. The virtual photons are passed back and forth between the charged particles much like basketball players might pass a ball between them as they run down the court. As virtual particles, photons cannot be observed because they would violate the laws regarding the conservation of energy and momentum. Only in their veiled or hidden state do photons act as mediators of force between particles. The "force" caused by the exchange of virtual photons causes charged particles to change their velocity (speed and/or direction of travel) as they absorb or emit virtual photons.

Only under limited conditions do the photons escape the charge particles and thereby become observable as electromagnetic radiation. Observable photons are created by perturbances (i.e., wave-like disruptions) of electrons and other charged particles. According to QED theory the process also works in reverse as photons can create a particle and its antiparticle.

In QED dynamics the simplest interactions involve only two charged particles. The application of QED is, however, not limited to these simple interactions, and interactions involving an infinite number of photons are described by increasingly complex processes termed second-order (or higher) processes. Although QED can account for an infinite number of processes (i.e., an infinite number of interactions), the theory also dictates that more interactions also become increasingly rare as they become increasingly complex.

The genesis of QED was the need for physicists to reconcile theories initially advanced by British physicist James Clerk Maxwell (1831-1879) regarding electromagnetism in the later half of the nineteenth century (i.e., that electricity and magnetism are two aspects of a single force) with quantum theory developed during the early decades of the twentieth century. Prior to WWII, British physicist Paul Dirac (1902-1984), German physicist Werner Heisenberg (1901-1976), and Austrian-born American physicist Wolfgang Pauli (1900-1958) all made significant contributions to the mathematical foundations related to QED. Even for these experienced physicists, however, working with QED posed formidable obstacles because of the presence of "infinities" (infinite values) in the mathematical calculations (e.g., for emission rates or determinations of mass). It was often difficult to make predictions match observed phenomena, and early attempts at using QED theory often gave physicists wrong or incomprehensible answers.

The calculations used to define QED were made more accessible and reliable by a process termed renormalization, independently developed by American physicist Richard Feynman (1918-1988), American physicist Julian Schwinger (1918-1994), and Japanese physicist Shin'ichirô Tomonaga (1906-1979). In essence the work of these three renowned scientists concentrated on making the needed corrections to Dirac's infinity problems, and his advancement of QED theory helped reconcile quantum mechanics with Albert Einstein's special theory of relativity. Their "renormalization" allowed positive infinities to cancel out negative infinities and thus allowed measured values of mass and charge to be used in QED calculations.

The use of renormalization initially allowed QED to accurately predict the observed interactions of electrons and photons. During the later half of the twentieth century, based principally on the work of Feynman, Schwinger, Tomonga, and another influential physicist, Freeman Dyson, QED became an important model used to explain the structure, properties, and reactions of quarks, gluons, and other subatomic particles. Although Feynman, Schwinger, and Tomonga each worked separately on the refinement of different aspects of QED theory, in 1965 these physicists jointly shared the Nobel Prize for their work.

Because QED is compatible with special relativity theory and special relativity equations are part of QED equations, QED is termed a relativistic theory. QED is also termed a gauge-invariant theory, meaning that it makes accurate predictions regardless of where it is applied in space or time. Like gravity, QED mathematically describes a force that becomes weaker as the distance between charged particles increases—reducing in strength as the inverse square of the distance between particles. Although the photons themselves are electrically neutral, the predictions of interactions made possible by QED would not be possible between uncharged or electrically neutral particles. Accordingly, in QED theory there are two values for electric charge on particles: positive and negative.

Impact

QED theory was revolutionary in physics. In contrast to theories that strove to explain natural phenomena in terms of direct causes and effects, the development of QED stemmed from a growing awareness of the limitations on scientist's ability to make predictions regarding the subatomic realm. In fact, QED was unique precisely because QED did not always make specific predictions. QED relied instead on developing an understanding of the properties and behavior of subatomic particles characterized by probabilities rather than by traditional cause-and-effect certainties. Instead of allowing scientists to make specific predictions regarding the outcome of certain interactions—predictions often mystifyingly incompatible with human experience (e.g., that an electron could be in two places at once), QED allowed the calculation of probabilities regarding outcomes (e.g., the probability that an electron would take one path as opposed to another).

In particular, Feynman's work, teaching, and contributions to QED theory reached near legendary status within the physics community. In 1986, Feynman published a book titled *QED: The Strange Theory of Light and Matter.* In his book, Feynman attempted to explain QED theory in much the same manner as Einstein's writing on relativity theory a half century earlier. In fact, although Feynman's profound contributions to QED theory were well beyond the understanding of the general public, no other physicist since Einstein and Robert Oppenheimer (1904-1967) had so captured the attention of the lay public. In addition, Feynman also became somewhat of a celebrity for chronicles relating to his life and studies.

Feynman's work redefined QED theory, quantum mechanics, and electrodynamics, and Feynman's writings remain the definitive explanation of QED theory. With regard to QED theory, Feynman is perhaps best remembered for his invention of simple diagrams, now widely known among physicists as "Feynman diagrams," to portray the complex interactions of atomic particles. The diagrams allow visual representation of the ways in which particles can interact by the exchange of virtual photons. In addition to providing a tangible picture of processes outside the human capacity for observation, Feynman's diagrams precisely portray the interactions of variables used in the complex QED mathematical calculations.

Schwinger and Tomonga also refined the mathematical methodology of QED theory so that predications became increasingly consistent with predictions of phenomena made by the special theory of relativity. Tomonga also solved a perplexing inconsistency that vexed Dirac's work (e.g., that an electron could be calculated to have a seemingly infinite amount of energy). Tomonga's mathematical improvements—along with refinements made by Schwinger and Feynman—resolved this incompatibility and allowed for the calculation of finite energies for electrons. In a master stroke, Tomonga renormalized and made more accurate the prediction of particle properties (e.g., magnetic properties) and the process of radiation.

QED went on to become, arguably, the best tested theory in science history. Most atomic interactions are electromagnetic in nature and, no matter how accurate the equipment yet devised, the predictions made by renormalized QED theory hold true. Some tests of QED, for example, predictions of the mass of some subatomic particles, offer results accurate to six significant figures or more. Even with the improvements made by the renormalization of QED, however, the calculations often remain difficult. Although some predictions can be made using one Feynman diagram and a few pages of calculations, others may take hundreds of Feynman diagrams and the access to supercomputing facilities to complete the necessary calculations.

The development of QED theory allowed scientists to predict how subatomic particles are created or destroyed. Just as Feynman, Schwinger, and Tomonga's renormalization of QED allowed for calculation of finite properties relating to mass, energy, and charge-related properties of electrons, physicists hope that such improvements offer a model to improve other gauge theories (i.e., theories which explain how forces, such as the electroweak force, arise from underlying symmetries). The concept of forces such as electromagnetism arising from the exchange of virtual particles has intriguing ramifications for the advancement of theories regarding the working mechanisms underlying the strong, weak, and gravitational forces.

Many scientists assert that if a unified theory can be found, it will rest on the foundations established during the development of QED theory. Without speculation, however, is the fact that the development of QED theory was, and remains today, an essential element in the verification and development of quantum field theory.

K. LEE LERNER

Further Reading

Books

Feynman, Richard P. *The Character of Physical Law.* MIT Press, 1965.

Feynman, Richard P. *QED: The Strange Theory of Light and Matter.* New Jersey: Princeton University Press, 1985.

Schweber, Silvan S. *QED and the Men Who Made It: Dyson, Feynman, Schwinger, and Tomonga.* New Jersey: Princeton University Press, 1994.

Finding Order among the Particles

Overview

In 1961 the physicist Murray Gell-Mann (1929-) developed a method of organizing the dozens of subatomic particles that had been discovered to date. This classification system led him to realize that many of these particles were made up of even smaller particles. Gell-Mann named these smaller particles *quarks*. Eventually, physicists used Gell-Mann's work to formulate the standard model, a way of describing the known types of matter in the universe.

Background

Atoms are composed of particles called protons, neutrons, and electrons. These *subatomic particles,* because they are smaller than atoms. During the 1950s and 1960s, physicists discovered many new types of subatomic particles, which were given names such as *positrons, pions,* and *muons.* These particles do not generally exist as parts of atoms, but are observed in cosmic rays—high-energy radiation that reaches Earth from space—or created in machines called particle accelerators. Accelerators are used to create high-speed collisions among particles. Such collisions often result in the production of new types of particles.

Physicists began to look for a way to organize the many kinds of subatomic particles that would show how the different types were related. In other words, they were searching for a physics version of the periodic table of the chemical elements. They first divided subatomic particles into two broad groups based on whether the particles are affected by the strong nuclear force. The strong nuclear force is what holds the nucleus of an atom together; it acts over very short distances (10^{-15} meters) and allows similarly charged protons to exist close to each other. It is one of the four basic forces; the other three are gravity, electromagnetism, and the weak nuclear force. Particles affected by the strong nuclear force are called *hadrons.* Examples of hadrons are protons and neutrons. Particles that are not affected by the strong nuclear force are called *leptons.* There are six types of leptons, one of which is the electron. Leptons are said to be fundamental particles because they do not seem to be made up of simpler structures.

Unlike leptons, there are hundreds of types of hadrons. These particles can be subdivided into two groups called *mesons* and *baryons.* These groups are based on the particles' spin. Subatomic particles behave as though they are spinning balls carrying an electric charge. *Spin* is a measure of both the speed at which the particle is spinning and the mass of the particle. Some particles have a greater spin than others. For instance, protons are assigned a spin of ½, *kaons*—particles discovered in cosmic rays—have a spin of zero. The spin of mesons is equal to zero or whole number integers. The spin of baryons is equal to half-integers, such as ½. Kaons therefore belong to the group of mesons, and protons belong to the group of baryons.

In 1961 Murray Gell-Mann, a physicist at the California Institute of Technology, introduced a method of further organizing hadrons, which he called the Eightfold Way. He grouped similar particles into families. Each family contains 1, 8, 10, or 27 particles. (Most families contain eight particles, as indicated by the name Eightfold Way.) Particles with the same spin belong to the same group. For each family of hadrons, Gell-Mann made a graph in which he plotted the particles' electric charge on one axis and their strangeness on the other. The *strangeness* of a particle is related to the speed with which it decays (if it decays at all). The data points of these graphs formed geometrical patterns. For instance, families of eight mesons formed hexagonal patterns (six data points arranged in a hexagon with two data points at the center). Families of ten baryons formed triangular patterns (a row of four points, a row of three points, a row of two points, and a row of one point). The relationships revealed by these graphs could be used to predict the mass, magnetic properties, and interactions of the particles. The Israeli physicist Yuval Ne'eman (1925-) suggested a similar organization system at about the same time.

Impact

Gell-Mann was able to use the Eightfold Way to predict the existence of particles that had not yet been discovered. For instance, he predicted the discovery of a baryon that he called the *omega-minus particle,* which belongs to a 10-particle family. Gell-Mann had made a graph of this family, and the data points formed a triangular shape. At the time, however, one of the data points was missing. He anticipated that the particle corresponding to this data point would soon be found. In 1964 the omega-minus particle was discovered with the aid of particle accelerator at the Brookhaven National Laboratory in Upton, New York. The properties of this particle corresponded to those that Gell-Mann had predicted, offering proof of the Eightfold Way.

However, the Eightfold Way did not explain *why* there are so many hadrons. In fact, it suggested to physicists that hadrons were composed of even smaller particles. A similar situation occurred after the creation of the periodic table. In the table, the chemical elements are organized into families with similar properties. Scientists eventually discovered that the patterns exhibited by the chemical families are based on the subatomic particles—the protons, neutrons, and electrons—that made up each element. Similarly, in 1964, Gell-Mann and the physicist George Zweig (1937-) suggested that hadrons are not fundamental particles—that they are in fact made up of even smaller particles that are responsible for the patterns seen in the families of the Eightfold Way. Gell-Mann called these particles quarks.

Gell-Mann proposed the existence of three types of quarks. Each type is called a flavor, and he named the three flavors up, down, and strange. Unlike all previously observed particles, quarks are thought to have fractional electric charges. Up quarks have a charge of $+^2/_3$ and down and strange quarks each have a charge of $-^1/_3$. Each flavor of quark also has an antiquark with an opposite charge. (An antiquark is a type of antimatter. Antimatter is identical to matter except that its particles have charges that are opposite to those of matter. Every particle of matter has a corresponding particle of antimatter. For instance, there are antiprotons and antineutrons.) Gell-Mann proposed that all baryons are composed of three quarks and that all mesons are composed of a quark and an antiquark.

A proton, which is a baryon, consists of two up quarks and one down quark. The charges on these three quarks add up to +1 $[+^2/_3[+^2/_3 + (-^1/_3) = +1]$, which is the charge on a proton. A neutron, which is also a baryon, consists of one up quark and two down quarks. The charges on these quarks add up to zero $[+^2/_3 + (-^1/_3) + (-^1/_3)= 0]$, which is the charge on a neutron. Gell-Mann's rules could account for every hadron that had been discovered up to that time.

However, some scientists argued against Gell-Mann's ideas because no one had been able to observe quarks individually. It seemed that it was impossible to knock quarks loose from the hadrons of which they were a part. Therefore, a team of physicists led by Jerome I. Friedman (1930-), Henry W. Kendall (1926-1999), and Richard E. Taylor (1929-) set out to observe quarks indirectly. They did so by firing high-speed electrons at protons with the aid of a particle accelerator. Most of the electrons passed straight through the protons, indicating that protons are mostly empty space. However, a few of the electrons were reflected from the protons at sharp angles. The pattern formed by the scattering of these electrons indicated that the proton contained three point-like structures. It is believed that these structures are the quarks proposed by Gell-Mann.

In 1965 the physicists Oscar Greenberg (1932-) and Yoichiro Nambu (1921-) described a property of quarks called color. In this case, *color* does not refer to an actual color. Instead, is similar to an electrical charge. Each flavor of quark comes in the colors red, green, and blue. Every baryon must contain one red quark, one blue quark, and one green quark. Every meson must contain one quark and one antiquark of the same color. These color rules support the idea that individual quarks cannot exist in isolation and cannot be observed directly.

More evidence of the existence of quarks occurred when Samuel Chao Chung Ting (1936-) and Burton Richter (1931-) discovered the psi meson in 1974. To explain the properties of this particle, the physicists realized that a fourth flavor of quark was needed, which they called the charm quark. Additional particles containing

THE LITERATURE OF PARTICLE PHYSICS

It may seem that particle physicists have a fondness for rather bizarre (and even silly) language. Some of their terms are taken directly from everyday speech and given new meanings. Examples include *color, flavor,* and *strangeness.* Other terms have their roots in literature. For instance, Gell-Mann took the word *quark* from a line in James Joyce's 1939 novel *Finnegan's Wake.* The line, which reads, "Three quarks for Muster Mark!," is part of a dream description involving a character who has passed out drunk on the floor. Joyce invented much of the language in this experimental novel, including the word *quark.*

The term *Eightfold Way* derives from Buddhist teaching on the way to nirvana (a state of perfect peace or bliss): "Now this, O monks, is noble truth that leads to the cessation of pain; this is the noble Eightfold Way: namely, right views, right intention, right speech, right action, right living, right effort, right mindfulness, right concentration."

A third example of the blending of physics and literature involves top and bottom quarks, which were initially called truth and beauty quarks. These names refer to the final lines of the poem "Ode on a Grecian Urn," written by John Keats in 1819: "Beauty is truth, truth beauty,—that is all / Ye know on earth, and all ye need to know."

charm quarks were soon found. Fifth and sixth flavors were eventually discovered as well—the bottom quark in 1997 and the top quark in 1994.

Scientists now believe that all matter is composed of three fundamental types of particles: leptons, quarks, and *force carriers*. Force carriers can be thought of as the "glue" that binds the other particles together. For instance, the force carriers that bind quarks together are called *gluons*. This description of matter is called the standard model.

The quarks and leptons of the standard model can be organized into three generations, or groups. Each generation consists of a pair of quarks and a pair of leptons. The greater the number of the generation, the heavier the particles. The first generation contains up and down quarks, electrons, and electron *neutrinos*. (Electron neutrinos have no charge and are believed to have very little or no mass. They play a role in the radioactive decay of some elements.) The particles of the first generation make up the ordinary matter found on Earth. The second generation contains strange and charmed quarks, and leptons called *muons* and *muon neutrinos*. The third generation contains bottom and top quarks, and leptons called *tau particles* and *tau neutrinos*. The particles of the second and third generations are seen in cosmic rays or are produced in particle accelerators.

Physicists can see patterns in the three generations of particles, but presently there is no explanation as to *why* these patterns exist. Just as subatomic particles accounted for the patterns seen in the periodic table and quarks accounted for the patterns seen in the Eightfold Way, physicists hope to someday discover what causes the patterns seen in the generations of fundamental particles.

STACEY R. MURRAY

Further Reading

Chester, Michael. *Particles: An Introduction to Particle Physics*. New York: Macmillan, 1978.

Gell-Mann, Murray. *The Quark and the Jaguar: Adventures in the Simple and the Complex*. New York: W. H. Freeman and Company, 1994.

Stwertka, Albert. *The World of Atoms and Quarks*. New York: Twenty-First Century Books, 1995.

Stephen Hawking Makes Pioneering Discoveries in Gravitational Field Theory

Overview

Black holes, formed by stars that have collapsed under their own weight, are perhaps the strangest physical objects in the universe. Their gravitational fields are so strong that neither light nor matter can escape. At their center are thought to be *singularities*, where the laws of physics cease to function normally. Within singularities, scientists believe they can find the keys to the universe's birth and death.

Background

The concept of black holes was first described in 1783 by the Reverend John Mitchell, a British amateur astronomer. Using the simple concept of escape velocity, the speed at which an object must move to escape a planet or star's gravitational field, he theorized that light might also be held to these rules.

Albert Einstein's (1879-1955) theory of general relativity laid the keystone for the discovery of black holes. His system for describing gravity was quickly explored by many other scientists. One of these was Karl Schwarzschild (1873-1916). While serving the German army in the trenches of the Russian front during World War I, Schwarzschild theorized that a star could collapse under its own weight, crushing itself down until its gravitational field became so strong that neither light nor matter could escape. Physicist John Wheeler named these objects "black holes" in 1969. The point to which an object must be crushed to hold back light is now called the "Schwarzschild radius." (The Earth, for instance, would have to be crushed to a diameter of about 2 centimeters to hold back light.)

Stars normally burn hydrogen and helium, the two lightest elements, producing energy in a process called nuclear fusion. As the energy they emit radiates outwards, it prevents the star from collapsing under its own gravity. After a star has exhausted its hydrogen and helium, it begins to burn heavier elements such as carbon and oxygen. Soon this fuel is exhausted as well. Then, when the radiated energy can no longer hold up against gravity, the star quickly collapses inward.

Black holes form at this point—when a massive star reaches the end of its life. Two theories have evolved to explain their formation. In the first, a large star (more than 25 times the mass of the Sun) collapses under its own weight and crushes itself down to a black hole. In the second, a star explodes violently in a type II supernova. The star, having exhausted its nuclear fuel, then quickly collapses. The energy from this compression triggers one last massive burst of fusion reactions as the star crushes itself. The resulting energy blows apart the star, letting off almost as much energy as it radiated in its entire lifetime in just a few seconds. What is left at the center collapses inward to form a neutron star. As nearby matter falls back on to the neutron star and increases its mass, the star will form a black hole if it becomes heavy enough to collapse further.

Black holes may also have been formed shortly after the Big Bang. These can range from microscopic to massive sizes. Astrophysicists now believe that the centers of many galaxies contain black holes with masses equivalent to that of millions of suns. They are fed continually by rich fields of gas and dust as well as entire stars that fall into the center.

Since black holes cannot be detected by any telescope or sensor, they can only be observed indirectly. Accordingly, astronomers look for the effects black holes have on surrounding gas, dust, and stars. A black hole's tremendous gravity can pull in nearby gas and dust. Dust gives off increasing amounts of energy as it falls toward the hole, energy that can be detected as x rays. The first such black holes were found in the 1990s. Since then, many more have been detected. Black holes at the center of galaxies are identified by observing rapidly swirling stars near the galactic center as well as bipolar jets of x rays moving away from the plane of the galaxy.

Impact

Until the 1960s black holes remained interesting quirks of physics. Only after Stephen Hawking discovered that black holes radiate heat did their importance in the evolution of the universe come to be appreciated. Until then it seemed as if they were eventually going to sop up all the matter in the universe, leaving nothing but black holes.

Physicists then faced a paradox: how could black holes seemingly violate the second law of thermodynamics (all objects lose energy as heat)? Black holes grew larger and larger whether or not they were absorbing matter, without losing energy. A series of discoveries by Hawking and others showed that this assumption was false.

In 1970 Jakob Bekenstein, a graduate student of John Wheeler, was the first to suggest that black holes had a temperature. This meant that they must also emit particles, something most physicists believed a black hole was unable to do. Bekenstein's ideas were roundly dismissed by the physics community, especially Hawking. Shortly thereafter, Russian physicist Yakov Zeldovich published a similar paper showing that a rotating sphere should emit particles, at least until it stopped rotating.

In 1972 Hawking, with Brandon Carter and Jim Bardeen, wrote a contradictory paper explaining the mechanics of black holes. They were trying to show how a black hole's surface area could increase without violating certain physical laws. They realized that their equations matched perfectly with the nineteenth-century equations governing thermodynamics, but at the time they noted that the similarities were only superficial, not identical. Later they would realize their mistake.

The turning point came in 1973 when Hawking decided to examine Zeldovich's equations on his own. To his "surprise and annoyance" he found that black holes *do* emit particles, and would do so even after they stopped rotating. Apparently, Bekenstein had been right—black holes do indeed have a temperature.

Hawking's discovery showed that black holes are, in fact, dynamic objects. Instead of acting like mindless vacuums sucking up matter, black holes actually have a temperature, given by this tiny amount of radiation. The temperature is minute, only a few millionths of a degree above absolute zero. But it is enough—this small amount of radiation means that black holes will slowly but surely lose mass until they evaporate into a puff of x and (maybe) gamma rays.

These emitted particles are created by a property of quantum mechanics in which subatomic particles flash in and out of existence in fractions of a second. Normally they disappear just as quickly as they are formed. But Hawking found that if these particles form just inside the event horizon of a black hole, they can tunnel across the event horizon and travel away from the black hole. The energy of these particles gives the black hole its temperature. In his honor, this radiation is now called "Hawking radiation." (Tunneling is another quantum property by which particles can cross barriers that they normally do not have enough energy for.)

This discovery was one of the most important physical theories of the century. Not only was it a brilliant solution to a difficult problem, it was the first physical theory to merge thermodynamics, quantum mechanics, and general relativity into a single system. It was the first step towards a "theory of everything" that would simultaneously describe the microscopic world of atoms and the macroscopic world of planets, galaxies, and the universe. Since then, many physicists have taken on the challenge of developing theories that describe both of these worlds. The biggest problem they face is how to reconcile "smoothness" and "roughness." Einstein's general relativity theory treats the four dimensions of space and time and their curvature by gravity as a simple, smooth force. Quantum mechanics, in contrast, sees the world in discrete, bumpy jumps called quanta.

In addition to searching for a theory of everything, Hawking and others also tackled the "information puzzle" hypothesis, first brought up by Hawking in 1976. The problem involves what happens to the information given into a black hole. All particles that fall into a black hole have "information" as defined by certain quantum mechanical properties. Physicists are still debating and theorizing about what happens to this information: does a black hole completely absorb it and keep it hidden from the universe forever, or is this information somehow reexpressed by the Hawking radiation the black hole emits?

The solution to this problem may rely upon the fate of the black hole. At the center of the black hole resides a singularity, where the curvature of space and time become infinite and Einstein's general relativity is of no use in describing space and time. The properties of this singularity, which are being studied by Hawking and others, may resolve the information puzzle-the singularity may hold all the information inside itself. Or, conversely, the information may be gone.

Hawking had a long-standing bet regarding this singularity with physicist Kip S. Thorne and John Preskill of the California Institute of Technology. Hawking bet that we will never be able to observe the singularity, that it will vanish with the rest of the black hole as it evaporates. This view is the "cosmic censorship" hypothesis, first stated by mathematician Roger Penrose (1931-). Thorne took the other side, believing that we will be able to observe the naked singularity. The bet was partially settled in 1997 when Matthew Choptuik showed with supercomputer simulations that under certain circumstances the singularity could be revealed.

Studying singularities is important because the universe may have begun in a similar state. The Big Bang theory states that the universe began as a tightly compressed object that exploded outward to create everything we can see today. The properties of this initial ball of matter may have been just like those of a singularity. Understanding the properties of singularities better will help us define the original state of the universe.

PHILIP DOWNEY

Further Reading

Hawking, Stephen. *A Brief History of Time.* New York: Bantam, 1988.

Ferris, Timothy. *The Whole Shebang.* New York: Simon and Schuster, 1997.

Toward the Unification of Forces

Overview

Scientists develop theories as attempts to understand the physical world. The more observed phenomena a theory can account for, the more promising it is. One of the main goals of modern physics is to construct a single theory that would explain the four known forces of nature: electromagnetism, gravity, and the strong and weak nuclear forces. The philosophical implications of such a fundamental understanding of the workings of the universe have led many physicists to speak metaphorically of "reading the mind of God." But the ultimate "theory of everything" has yet to be found.

Background

The main objective of science is to find the underlying explanations for the phenomena we observe in nature. The wonderful variety of these phenomena might lead one to conclude that the underlying causes are just as numerous. Who would imagine, for example, that the long neck of the giraffe has anything to do with the sweet smell of flowers? Yet the theory of evolution explains them both quite nicely.

In the physical sciences, the behavior of objects is understood in terms of *forces*. A force is a vector quantity, meaning it has both *magnitude* (amount) and *direction*. A force is an impetus that tends to cause an object to move. The study of astronomy proceeded for thousands of years without anyone realizing that the motions of the planets are governed by the same gravitational force that keeps our feet on the ground. Finally, Isaac Newton (1642-1727) made this great leap in a realization symbolized by the famous legend of the falling apple.

Similarly, until the nineteenth century, electricity, magnetism, and optics were studied as three completely separate disciplines. There was no obvious reason for early experimenters to suppose that electrical charge, magnetized iron, and lenses were related in any way. However, James Clerk Maxwell's (1831-1879) theory of electromagnetic waves traveling at the speed of light swallowed up the entire field of optics and explained electricity and magnetism as different manifestations of the same force.

Thus the two known forces of nature at the end of the nineteenth century were the electromagnetic force and the gravitational force. Both of these forces presented the same philosophical problem: that of action at a distance. If you use a hammer to exert a force directly upon a nail, you would not be surprised to find the nail moving into the wood. On the other hand, the hammer couldn't cause the nail to move just by being somewhere in its vicinity. Yet the gravitational force holds the Earth fast in its orbit around the Sun, despite the 93 million miles (150 million km) of space in between. Likewise, there will be an attraction between a positive electrical charge and a negative one, whether they are in a metal wire or in empty space.

To get around this difficulty, nineteenth-century physicists began to use a new terminology to describe forces. Instead of regarding them solely in terms of the interaction between two objects, such as the Sun and the Earth, they began thinking in terms of force fields. Any object with mass has a gravitational field, whether or not another object is present. The force field interacts with other objects; for example, the gravitational field of the Sun keeps the Earth in its thrall. A charged particle such as an electron has an electrical field around it that acts on other charges.

Impact

The early twentieth century saw a revolution in physics. Albert Einstein (1879-1955), in his general theory of relativity, described gravitational fields in terms of the curvature of space and time. Not satisfied with merely superseding Newtonian mechanics, Einstein strove to develop a unified field theory that would include electromagnetism as well. He spent the last 25 years of his life working on this project, and he worried in his final years that he had failed either to construct such a theory or prove that one could not exist.

Einstein's blind spot was quantum mechanics. By the 1930s two more forces of nature had been discovered, both operating at the scale of the atomic nucleus. The *strong force* holds the nuclear particles together. The *weak force* participates in their decay. Quantum mechanics describes the behavior of atomic particles in terms of probabilities and maintains that it is impossible to determine both position and velocity precisely. Such a formulation was profoundly offensive to Einstein's idea of an ordered universe. His famous response was that "God does not play dice with the universe." By refusing to consider quantum mechanics, the great man condemned his unified field theory efforts to failure.

Physicists had better luck applying Einstein's special theory of relativity to electromagnetism. The mathematics of special relativity becomes important for objects moving very quickly, at a velocity comparable to the speed of light, which is 186,000 miles per second, or *c*. While nothing can go faster than *c*, massless particles can achieve this speed, and subatomic particles can come close. In quantum mechanics, particles' positions are expressed as probabilistic wave functions, leading to the idea of wave-particle duality. Light, or sub-atomic particles like electrons, become quantities that sometimes act like particles and sometimes act like waves.

In relativistic quantum field theories (RQFTs), every field has an associated particle, or quantum. The quantum of the electromagnetic field is the photon, a massless packet of energy. According to the theory of quantum electrodynamics (QED), the most successful relativistic quantum field theory, the photon is the carrier of the electromagnetic force.

Grand unified theories (GUTs) attempt to reconcile three of the four known forces: electromagnetism, the strong force, and the weak force. One important step in this direction has been the *standard model* of twentieth-century particle physics. While it continues to be extremely useful, it does have its deficiencies, including a number of parameters that must be juggled to keep the model in line with experimental values. However, during the last decades of the twentieth century, there were a number of important experimental confirmations of the standard model's predictions, and no experiments that contradicted them.

In the standard model, electromagnetism and the weak force are unified and called the *electroweak force.* Sheldon Glashow (1932-) first proposed this unification in 1961. The quanta of the electroweak field, and thus the carriers of the electromagnetic and weak forces, are called *intermediate vector bosons* (*IVBs*). A problem with Glashow's theory was that the weak force required massive carriers to account for its short range, while the photon, carrier of the long-range electromagnetic force, is massless.

This conundrum was resolved independently by Steven Weinberg (1933-) and Abdus Salam (1926-). At low energies, these forces appear separate because of a field called the *Higgs field.* The Higgs field requires carriers with mass for the weak force and massless photons to carry the electromagnetic force. However, above a high threshold energy called the *electroweak scale,* about 100 billion electron volts, the Higgs field vanishes and all the IVBs are massless. Glashow, Weinberg, and Salaam shared the 1979 Nobel Prize in physics for this work.

The standard model also concerns itself with the strong force. The strong force holds together the *hadrons* (protons and neutrons) in the atomic nucleus. Each hadron is made of three fundamental particles called *quarks*. Quarks come in three categories called "colors," and so the theory of the strong force is called *quantum chromodynamics* (*QCD*). The carrier of the strong force between quarks is called the *gluon.*

GUTs generally hold that all three forces will be unified at some grand unification scale, perhaps at energies on the order of a quadrillion electron volts. This picture, in which increasing energies reveal first one force and then the next to be part of a unified whole, motivate particle physicists to propose ever more powerful accelerators. Unfortunately it is often difficult to induce non-physicists to share their enthusiasm, at least on a consistent basis. The much-anticipated Superconducting Super Collider was cancelled in 1993 after 14 miles of tunnels had already been dug in Texas, and $2 billion spent on the project. On a more theoretical level, particle physicists and cosmologists have found common interests in discussing the forces that existed in the extremely high energies of the early universe.

Supersymmetry theories unify matter particles, called *fermions,* with the force carrying bosons. It offers a framework for including not only the strong, weak, and electromagnetic forces of the grand unified theories, but also gravity, with its hypothetical force carrier, the graviton. Thus, supersymmetry may be a part of any four-force "theory of everything." Unfortunately, the current supersymmetry theories, constructed based on relativistic quantum field theories, have yielded mathematical infinities in inconvenient places. This may be a fault of the supersymmetry theories themselves, but is just as likely to be a problem with conventional RQFTs.

In string theory, first discussed in the late 1960s, the hadrons were understood as different oscillation modes of a "string" 10^{20} times smaller than the diameter of the atomic nucleus. The strings hold the quarks together but are too small to be observed, and their effects become apparent only at extremely high energies. At lower energies, string theory reduces to the standard model. However, troublesome features of early string theories included as many as 26 spatial dimensions, and faster-than-light particles called *tachyons.*

In the 1980s string theory was combined with supersymmetry to yield superstring theory. The problematic tachyons disappeared, and the inclusion of supersymmetry pointed the way to bringing in gravity as well. Superstring theory doesn't solve all the problems, though; it introduces particles and interactions that have not been observed and does not seem to result in a solution as unique as one might expect a "theory of everything" to be.

The conception of gravity as a property of the very fabric of space-time continues to cause problems in developing a universal theory. Some physicists have suggested that gravity is not a separate force at all, but rather a byproduct of the other three. It is also possible that Einstein's formulation may break down at the Planck scale, corresponding to energies of approximately 10^{19} GeV.

As physicists consider the basic questions of the universe, many have the sense of treading on hallowed ground. The past few decades have seen an unprecedented number of books on physics invoking the name of God in their titles. Perhaps it is not surprising, then, that the universe does not give up its secrets easily. However, it is in the nature of humanity to keep looking for the ultimate answers. "I want to know how God created this world," wrote Einstein. "I want to know His thoughts. The rest are details."

SHERRI CHASIN CALVO

Further Reading

Davies, Paul. *Superforce: The Search for a Grand Unified Theory of Nature.* New York: Simon and Schuster, 1984.

Weinberg, Steve. *Dreams of a Final Theory: The Search for the Fundamental Laws of Nature.* New York: Random House, 1992.

Zee, Anthony. *Fearful Symmetry: the Search for Beauty in Modern Physics.* New York: Macmillan, 1986.

Edward Lorenz's Groundbreaking Study of Weather Patterns Leads in Part to the Development of Chaotic Dynamics

Overview

In 1961 Edward Norton Lorenz (1917-) demonstrated that as nonlinear deterministic systems evolve, they exhibit sensitive dependence on their initial conditions. This means that small changes in those conditions have large, unpredictable consequences. Lorenz was the first to emphasize the importance of identifying and studying such systems. His original research initiated the development of chaos theory, which has applications in fields ranging from astronomy and engineering to economics and medicine.

Background

Physical systems whose later states evolve from earlier ones according to fixed laws are deterministic systems. Isaac Newton's (1642-1727) *Principia Mathematica* embodied the belief that the natural world was a deterministic system—one whose behavior was governed by his equations of motion and law of gravitation. When Newtonian mechanics were successfully applied to natural phenomena in the eighteenth and nineteenth centuries, their predictive accuracy was directly related to the accuracy of initial conditions. Minor inaccuracies in the initial states translated into small prediction errors. Consequently, extremely weak effects could be neglected without significantly affecting predictability.

Certain phenomena, though, proved particularly recalcitrant to Newtonian methods, the most notable being weather and turbulence, both of which are complex and apparently random processes. Meteorologists did develop a set of equations they believed governed changes in weather patterns, but employing them successfully required a vast number of computations. Only with the introduction of self-programming digital computers in the 1940s could the solutions to these governing equations be approximated.

Even with the power of computers, numerical weather forecasts were only accurate for a few days. These limitations were thought to be caused by insufficient knowledge of initial conditions and the physical approximations dictated by the computational constraints of available computers. But in 1961 Edward Lorenz demonstrated Newtonian predictability to be a fantasy.

Lorenz formulated a simple atmospheric model that allowed him to simulate recognizable weather patterns. To track variations in these patterns, he printed out the values of prognostic variables after each simulated day. During the winter of 1961 he decided to investigate a particular computer run in greater detail. He used a line of numbers from a previous printout as the initial conditions for the new run. The new results differed significantly from the original. Lorenz quickly traced the problem to his truncation of the initial condition values. The numbers stored in the computer were accurate to six decimal places, but Lorenz had them rounded off to three decimal places before printing. Assuming that such small differences could have no significant effect, Lorenz used the printout values. What he found was that extremely small differences can generate widely varying outcomes, making long-term prediction impossible. Lorenz referred to this as sensitive dependence on initial conditions— known today as chaos.

Sensitive dependence is a direct result of the nonlinearity of chaotic systems. In linear systems, changes to initial states result in proportional changes to later states. Nonlinear systems like the atmosphere are not so well behaved. Variables in these systems are coupled such that changes result in complex cumulative feedback effects. For example, increases in temperature raise humidity. This causes more solar radiation to be absorbed, which raises the temperature further, which then increases humidity and so on.

Lorenz studied the behavior of nonlinear equations in greater detail. In particular, he investigated the *attractor* for such a system. An attractor is a set of states a system can occupy once transient effects have dissipated. The simplest attractor is a fixed point that represents the stable states of a pendulum in an unwound clock. Regardless of how it is set in motion, the pendulum will always come to rest in the same position. The attractor for a pendulum in a continuously wound clock will be a closed elliptical curve. The points of the curve represent the states of the system. In both cases future states can be accurately predicted.

The attractor Lorenz discovered proved to be the first *strange attractor.* As with any attractor there is a set of states to which the nonlinear system is attracted. However, once the system is in the strange attractor, nearby states diverge from each other exponentially fast, making long-range prediction impossible. Strange attractors, however, are stable geometric objects with definite structure. This gives order to the apparent random behavior of nonlinear systems. For instance, the underlying order of the weather's strange attractor—the climate—explains why similar weather patterns occur over and over.

Impact

Many physical phenomena besides weather are nonlinear deterministic systems and exhibit chaotic behavior. While this is generally viewed as undesirable, chaotic properties can be exploited to analyze unresolved problems and generate practical applications.

As already mentioned, the most immediate consequence of sensitive dependence is the impossibility of ever making perfect or even approximate long-range predictions. However, by studying the structure of strange attractors it may be possible to determine the probability distribution of allowed events. This would then allow long-term probabilistic weather and other predictions to be made with mathematical precision.

Economics is another area where forecasting is of primary importance. Some economists now believe many economic data series—such as gross national product, unemployment, stock indexes, and industrial production—are chaotic. Chaos theory has given insight into the forces driving economic fluctuations, allowing better model specification and improved predictive capabilities.

Transition to chaos is another area of rapidly growing research. This is particularly so in fluid dynamics, where turbulence is of great concern, because smooth-flowing fluids often break-up into wild swirls and eddies. Fluid dynamicists try to minimize this turbulence, which controls drag around ships and planes and affects turbine engine and propeller efficiency. Chaotic attractors are now believed to underlie turbulence. Investigating these attractors has increased our understanding of how and when turbulence emerges. This in turn may lead to better designs.

Chaotic transition is also important in physiology and medicine. At present scientists are debating whether healthy biological systems are regular and predictable or chaotic. Interestingly, periodic and chaotic phenomena are present in both healthy and diseased conditions. Research is trying to detect transitions from one regime to the other in the hopes of developing predictive and diagnostic tools. Heart rhythms have been studied under normal and abnormal conditions in an attempt to identify impending arrhythmia and cardiac arrest. Brain wave activity has also been scrutinized in an attempt to understand epilepsy, manic depression, and other illnesses.

Chaotic dynamics might also be exploited in situations where great flexibility is beneficial. Small changes in control variables could be used to change system dynamics rapidly. This would allow quick responses to changing conditions. Practical applications are yet to be realized, but this idea has been used to explain the evolution of life on Earth.

Control through small changes also suggests connections between the theory of chaotic systems and information theory. For instance, by manipulating various control variables it might be possible to encode information into chaotic oscillator signals. Chaotic dynamics may also be exhibited by the brain's electrical activity. Tantalizing research in this area suggests that the cortex supports a global attractor that provides for rapid dissemination of information.

Chaos theory is also extensively used in astronomy, because the dynamical structures and evolution of the solar system exhibit chaos. The Kirkwood gaps in the asteroid belts are generally thought to be chaotic regions where asteroid orbital elements are altered drastically. The Cassini divide in Saturn's rings is similarly thought to be a chaotic region, and the spin-orbit of Saturn's satellite Hyperion displays characteristically chaotic behavior. The orbits of the planets also show some evidence of chaotic motion over extremely long periods of time. In addition, chaos theory has been fruitfully applied to the analysis of galactic dynamics, stellar oscillations, and sun spots.

Electrical engineers confront chaos regularly. Many electrical systems involved in the generation and transmission of power are susceptible to chaotic responses. Voltage collapse, the principal threat to the stability and reliability of power grid systems, is believed to be chaotically induced. Flow-induced vibrations in condenser tubes and transmission wires are also chaotic. Chaos is believed to occur at the physical data level in computer networks as well. Efforts remain focused on developing procedures for ensuring optimal operation of electrical systems and electronic devices fundamentally plagued by chaotic dynamics.

This review is only a glimpse of the wide range of fields in which chaos theory is used. The explosive growth this discipline experienced during the last two decades of the twentieth century shows no signs of abating, and many important applications are as yet expected.

STEPHEN D. NORTON

Further Reading

Books

Cramer, F. *Chaos and Order: The Complex Structure of Living Systems*. New York: VCH, 1993.

Cvitanovic, Predrag, ed. *Universality in Chaos*. Second ed. Bristol: Adam Hilger, 1989.

Davies, Brian. *Exploring Chaos: Theory and Experiment*. Reading, MA: Perseus Books, 1999.

Favre, Alexandre, Henri Guitton, Jean Guitton, André Lichnerowicz. *Chaos and Determinism*. Baltimore, MD: Johns Hopkins University Press, 1995.

Gleick, James. *Chaos: Making a New Science*. New York: Viking Penguin, 1987.

Kellert, Stephen H. *In the Wake of Chaos*. Chicago, IL: University of Chicago Press, 1993.

Kim, Jong Hyun and John Stringer. *Applied Chaos*. New York: John Wiley & Sons, Inc., 1992.

Lorenz, Edward N. *The Essence of Chaos*. Seattle, WA: University of Washington Press, 1993.

Ott, Edward. *Chaos in Dynamical System*. New York: Cambridge University Press, 1993.

Periodical Articles

Lorenz, Edward N. "Deterministic Nonperiodic Flow." *Journal of Atmospheric Science* 20 (1963): 130-141.

Asteriods, Dinosaurs, and Geology: Catastrophic Events and the Theory of Mass Extinction

Overview

Fossils were first determined to be the remains of extinct animals in the nineteenth century. This realization carried with it the concept of extinction, something that did not play a major part in the religion-dominated science of the day. As the science of geology developed, so did the understanding that, every so often, a very large percentage of species on the Earth vanished for no apparent reason. There have been at least 10 mass extinctions recorded in the fossil record in the past 600 million years. While many reasons for this have been proposed, there is only one such event for which the cause is known, that which precipitated the extinction of the dinosaurs. This catastrophic event, caused by the impact of a large body such as a comet or small asteroid on the Earth's surface, posed a completely novel ap-

Luis and Walter Alvarez view a sample of an iridium-layer deposit. *(Roger Ressmeyer/CORBIS. Reproduced with permission.)*

proach to understanding our geologic past. It also helped spur awareness of events ranging from nuclear disarmament to the effects of potential asteroid impacts on the modern world.

Background

Dinosaurs were first brought to the public's attention in the nineteenth century when it was realized that their fossils represented the remains of huge reptilian creatures that no longer existed. For over 100 years they have captured the public's attention, both for what they might have been like as well as how they might have died out. Over this time, speculation abounded with suggestions that they fell to evolutionary degeneracy, overspecialization, egg predation by mammals, climate change, disease, or other factors.

In the late 1970s, Nobel laureate Luis Alvarez (1911-1988), his son Walter, Frank Asaro, and Helen Michel were conducting investigations of rock layers in Italy, looking for iridium that might have come from cosmic dust settling to the surface. They felt that measuring the accumulation of this dust could help to correlate sections of rock over the planet, showing which sediments were deposited at the same time. They decided to look for iridium because it is more common in meteorites than on the surface of the Earth (most terrestrial iridium now resides in the Earth's core) and because iridium is more easily detected than are other likely metals.

What they found was a layer of clay that marked the boundary between the Cretaceous period (the latest geologic period in which dinosaurs were known to have lived) and the subsequent Tertiary period. This "K-T" boundary (so called because of the geologic shorthand for these two periods) contained a layer of very distinctive clay with iridium levels that were significantly higher than those in neighboring sediments. After eliminating other possible reasons for this geologic anomaly, they concluded that a large object had struck the Earth, causing the dinosaurs' extinction. This finding was published by Alvarez in the premier scientific journal *Science* in June 1980. Science fiction writers had for years speculated about such impacts, and Eugene Shoemaker had convincingly demonstrated that the Earth still bore impact craters. This was, however, the first time that a major geologic event was directly tied to an impact.

Initially viewed with a high degree of skepticism, further discoveries, combined with a lack of counterevidence, led scientists to realize that such a large impact must have happened. The final piece of evidence was the discovery of a large impact crater just off the Yucatan Peninsula. The date and physical characteristics of this crater were such a close match to that predicted by Alvarez that virtually no doubt remains about this event, although the role of impacts in other mass extinctions is still hotly debated.

Impact

Alvarez's paper presented results that seemed more like science fiction than scientific theory. His findings affected science and society in several ways:

1) Scientists had to consider and, later, accept the role that catastrophe can play in geologic and biologic events. At first, this seemed to herald a retreat from uniformitarianism, the doctrine that the forces of geological change that shaped the prehistoric world are the same processes at work today. In addition, scientists such as Stephen Jay Gould (1941-) began to think increasingly about the role of contingency, or independent phenomena, in evolutionary theory.

2) Further research showed that nuclear war could mimic some of the effects of large impacts, particularly a "nuclear winter" induced by clouds of sun-obscuring dust raised after a massive explosion. This vivid picture of the potential outcome of nuclear weapons use may have changed the way nuclear weapons were viewed toward by end of the Cold War.

3) The public, always fascinated with dinosaurs and their extinction, started thinking about the chance that such events could happen during their own lifetime. This led to several programs to search for near-earth asteroids and ways to divert asteroids or comets that might hit the Earth.

Charles Lyell (1797-1875) led one of the first great revolutions in geology by advocating the theory of uniformitarianism, the view that geologic change occurs at a steady, slow rate over millions of years, an idea first proposed by James Hutton (1726-1797). Though Georges Cuvier (1769-1832) proposed that catastrophic events and extinctions had occurred in the earth's past, the notion that they might have played a major role in the history of life was largely discounted. Recent geological evidence of a massive asteroid impact, however, has prompted modern scientists to revisit their views.

Largely as a result of Alvarez's paper, scientists increasingly accept the view that catastrophic events do indeed play a major role on Earth. Whether that event is a massive volcanic eruption, a nearby supernova, or a meteor impact, random catastrophes can occur. Uniformitarianism has been recast somewhat in the wake of the K-T impact; scientists now feel that catastrophes happen, but the causes for them and the mechanisms by which they wreak havoc can be understood; furthermore, these causes and mechanisms remain the same over time. For example, meteors fall to Earth—they are not thrown by angry deities. Further, the damage that a meteor

METEORITES ON ICE: HUNTING FOR METEORITES IN ANTARCTICA

Shooting stars have captured man's attention for millennia. Most meteorites vaporize high in the atmosphere because they are generally specks of dust or grains of sand, and the high temperatures caused by friction with the air heats them to the point of disintegration. Some, however, reach the ground, giving our only direct evidence of what lies beyond the Moon. Until recently, meteors were found more or less by luck. The bottoms of impact craters were mined, some found their way to museums, and some were bought by researchers. Meteors helped to establish the age of the Solar System, Earth's age, and the chemical composition of asteroids and comets. However, this wealth of information was doled out only sparingly because confirmed meteorites were rare and hard to find. In the late 1980s researchers realized that the ice caps might hold many meteorites, and expeditions were sent to Antarctica to look for them. Meteorites were found almost immediately. In particular, some areas were discovered where, due to the ice flow and lack of native rocks, virtually every rock seen seemed to be a meteorite.

Most meteorites are thought to originate from comets and asteroids. However, several Antarctic meteorites seem to share key chemical signatures with Mars, leading to the supposition that they were blasted from Mars by a giant impact at some time in the past. One in particular was examined and thought to contain signs of ancient Martian microbes, the first signs of life elsewhere in the universe. Unfortunately, this was shown to be an unlikely interpretation of the data, and the search for extraterrestrial life continues.

will do can be calculated, and is the same as an identical meteor would have done 100 million years ago. In other words, the natural and physical laws that govern our world and universe are unchanging, even if they sometimes manifest themselves through random catastrophe.

Another victim of large impacts is the concept of steady, gradual survival of the fittest. Replacing it is a more dynamic theory of evolution that recognizes long periods of evolutionary quiet, punctuated with brief intervals of rapid change. Among the events that can usher in this change is a mass extinction event such as the K-T impact, which opened up a tremendous number of evolutionary niches. Many organisms moved in to fill them, evolving at accelerated rates. Once the niches were filled, a quieter, steady-state condition resumed until the next major event. Survival of the fittest in this scheme is tempered by the concept of contingency. Some organisms died out, not because they were less fit, but because they were unlucky enough to live underneath a future asteroid impact site. Other organisms lived because they happened to be on the other side of the world when the asteroid fell, regardless of their degree of evolutionary fitness.

In addition to the scientific impact of this discovery, there were social and political effects as well. People began to realize that such events could happen again and that, even if they didn't, nuclear weapons could produce the same results. This realization, coupled with the wide coverage of Comet Shoemaker-Levy 9's massive 1994 impact on Jupiter, made people aware that "it could happen here." One result of this was a spate of movies and novels about comets hitting the Earth. News stories about near-Earth asteroids gained more attention, and there was some discussion (still continuing in a haphazard fashion) about deploying systems to detect and either divert or destroy potentially deadly comets and asteroids. Largely as a result of Alvarez's paper, people increasingly view the solar system as a giant shooting gallery in which the Earth is bound to be hit again at some point.

Many assume that, because the K-T extinction was caused by an asteroidal impact, then all mass extinctions must have been caused in the same way. Other explanations for mass extinctions include continental drift, massive volcanic eruptions, nearby supernovae, and nearby gamma ray bursts. Interestingly, virtually every paper suggesting a mechanism by which a mass extinction may have occurred tends to suggest that *all* extinctions occurred for the same reason. This is known as conceptual simplification. The K-T extinction has taught us only that this single extinction was caused by an asteroid impact; others *might* have been caused in the same way. More importantly, it has shown us that large, catastrophic events, of which impacts are one, can profoundly affect life on Earth. Other, similarly catastrophic events may have caused other mass extinctions. However, there are at least as many potential causes for mass extinctions as there are extinctions recorded in the fossil record. It is not realistic to think that a single type of event caused them all.

P. ANDREW KARAM

Further Reading

Gould, Stephen Jay. *Wonderful Life: The Burgess Shale and the Nature of History.* New York: W. W. Norton, 1989.

Hallam, Anthony. *Great Geological Controversies.* Oxford: Oxford University Press, 1983.

Hellman, Hal. *Great Feuds in Science.* New York: John Wiley & Sons, 1998.

Wignall, P. B., and Anthony Hallam. *Mass Extinctions and their Aftermath.* Oxford: Oxford University Press, 1997.

Solar System Exploration: 1970-2000

Overview

Between 1970 and 2000, solar system exploration included major missions to most of the planets. The United States sent out many unmanned spacecraft that studied the planets, their moons, and even asteroids. The Soviet Union's planetary missions included the *Venera* spacecraft series, some of which landed on Venus and sent back pictures from the surface. The Soviets, the Japanese, and the European Space Agency all sent probes to study Halley's Comet in 1985-86.

The space probe *Mariner 10,* launched November 3, 1973, was the first to travel to Mercury. *(NASA. Reproduced with permission.)*

Background

By 1970 humans had landed spacecraft on the moon but not on any planet. By 1999, however, spacecraft had landed on and mapped Mars and Venus, our nearest planetary neighbors. *Mariner 10* visited Mercury, and the *Voyager* missions visited all four of the gas giant planets. The *Galileo* probe spent nearly four years studying Jupiter and its four largest moons. Even some of the smallest bodies in the solar system, asteroids and comets, were studied by unmanned missions.

Impact

In 1974 *Mariner 10* visited Mercury, the closest planet to the sun. Mercury's surface is heavily cratered and barren, and scientists originally thought it was similar to the moon. *Mariner 10* data, however, revealed that Mercury began as a mostly molten planet that was deformed by the tidal influences of the Sun. As it cooled and shrank, its surface cracked and shifted as its shape became rounder. *Mariner 10* revealed the cracks and ridges that resulted, and also indicated that Mercury must have a huge iron core not far below the surface.

Mariner 10 was also the first spacecraft to get close-up pictures of the thick atmosphere that surrounds Venus and hides its surface. *Pioneer Venus 1* began orbiting the planet in 1978, using radar to map the surface through the clouds. It also discovered significantly more sulfur dioxide in the planet's atmosphere than scientists had measured from Earth. The sulfur dioxide levels slowly dropped, and astronomers concluded that a volcano had erupted on Venus just before *Pioneer 1* arrived, pumping sulfur dioxide into the atmosphere. *Pioneer Venus 2* sent four small probes through the atmosphere to measure it. One survived its landing and sent back data from the surface for an hour.

The Soviet *Venera* series of spacecraft visited Venus in the 1970s and sent back photographs and other data. Several landed on the surface, where they briefly studied surface rocks and took photographs before being crushed by the pressure and heat of the atmosphere.

The *Magellan* missions in 1990 and 1994 used radar to map Venus's surface in great detail. *Magellan* was able to see features as small as 400 feet across. It revealed an utterly dry surface containing large volcanic mountaintops and few craters. The tectonic forces that shape Earth are nonexistent on Venus, which has a single thick plate rather than many thinner plates that can move across its surface.

Earth's moon was the object of two missions in the 1990s. In 1994 *Clementine* mapped most of the moon's surface. It was followed by lunar prospector, which orbited the moon for 1.5 years, returning data about the moon's resources, magnetic field, and gravity. The prospector was able to confirm the existence of water ice in permanently shaded regions near the moon's south pole.

Probes to Mars in the 1960s examined mostly the southern hemisphere and seemed to discover a cratered, dead world. *Mariner 9,* which landed in a global dust storm in 1971, studied the northern hemisphere after the dust began to settle, and discovered four huge volcanoes, winding valleys that resembled dried riverbeds, and an enormous canyon complex that was named Valles Marineris after the spacecraft.

The Viking missions in the 1970s included two orbiters and two landers. The latter established a base from which data was returned about the surface of Mars and the weather in its thin atmosphere. *Viking* included an experiment that scooped up a small sample of the Martian soil and looked for indications of living things; it found none. However, if water had ever existed on the surface of Mars, as the *Mariner* results seemed to indicate, it was possible that primitive life had existed there as well. In the 1990s, NASA examined the planet's geological history, in particular the possible past existence of surface water, with several missions.

In July 1997, *Pathfinder* bounced to a landing and released the *Sojourner* rover. *Pathfinder* and *Sojourner* sent back data for nearly three months, well beyond their expected lifetimes. They returned a wealth of scientific data about the rocks on the martian surface, including a particular area that looks like it was once a floodplain. In March 1999, *Global Surveyor* began a Mars mapping mission. It has sent back stunning pictures revealing previously unseen details of the martian surface.

The Voyager missions to the outer planets began in the 1970s. *Voyagers 1* and *2* neared Jupiter in 1979. The *Voyagers* studied weather systems in Jupiter's atmosphere, including rotation of the Great Red Spot, and observed lightning and auroras in the Jupiter's clouds. The *Voyagers* also observed the four largest moons of Jupiter, discovering the first nonterrestrial volcano as it erupted on the moon Io, and discovered three new moons. The discoveries made by the *Voyager* spacecraft were expanded by the *Galileo* mission in the 1990s.

The *Galileo* mission, beginning in December 1995, studied Jupiter and its moons for four years. After a two-year primary mission, its two-year extended mission included many flybys of its four largest moons. As a daring finale to the extended mission, *Galileo* made two very close passes of the volcanic moon Io, coming within 200 miles (322 km) of the surface. Io harbors the hottest volcanoes in the solar system, and the latest images show both active volcanoes and past lava formations. *Galileo* greatly extended our knowledge of Jupiter's moons, including evidence suggesting the presence of subsurface oceans on Europa and Callisto, and a thin atmosphere and magnetic field at Ganymede (the largest satellite in the solar system and so far the only one known to have its own magnetic field).

Voyagers 1 and *2* continued on to Saturn after studying Jupiter, arriving in 1979 and 1981, respectively. They examined the weather systems on Saturn (which were similar to those on Jupiter) and mapped the rings in exquisite and initially baffling detail. Scientists were able to unravel the evidence and figure out some of the intricate mechanisms by which the gravity of Saturn and some of its moons produced features like the braided effect observed in some of the rings.

Voyager 1 was scheduled to take many photographs of Saturn's largest moon, Titan. However, *Voyager* showed that the moon has a thick cloud layer that prohibits visual observation. The atmosphere itself turned out to be interesting, probably resembling that of the Earth in its earliest years. This makes Titan a possible candidate for the future development of simple life forms like those that first arose on Earth.

Voyager 2 continued on to Uranus, arriving in 1986. Uranus itself presented a bland appearance, with a thick atmosphere covering a warm ocean surrounding a hot rocky core. *Voyager 2* was able to measure the temperature in different areas of the planet's atmosphere, but the most exciting find of its Uranus mission was the surface of the moon Miranda. Miranda is tiny, and scientists expected it to be too small for the kind of geological forces that create interesting surface features. However, Miranda shows some of the most dramatic topography in the outer solar system, with strange grooves, white markings, and a very distinctive chevron pattern.

In 1989 *Voyager 2* arrived at Neptune, which, like Jupiter and Saturn, shows weather patterns in its atmosphere. Neptune's atmosphere is a deep blue, a sign that it contains large amounts of methane. Neptune has darker blue spots (some with white clouds) that contain giant storm systems. In examining these weather systems, *Voyager* found Neptune to be the solar system's windiest planet. It also revealed that Neptune's moon, Triton, is one of the very few moons (along with Titan) that have an atmosphere, although a very thin one. *Voyager* 2 also made the surprising discovery that this icy moon appears to have geysers of nitrogen or methane gas that erupt nearly five miles up from the surface before being carried sideways by the wind.

It wasn't until 1978 that astronomers discovered that Pluto had a satellite, the moon Charon, and even then the only evidence they had was a bulge on one side of images of Pluto. This tiny, distant, and icy world is still largely a mystery. No spacecraft have yet visited Pluto, though the Hubble Space Telescope has provided pictures that show Pluto and Charon as two distinct bodies, as well as patterns of light and dark areas on the surface of Pluto.

In 1985-86 the return of Halley's Comet to the inner solar system gave scientists an opportunity to study a comet closely. The Soviet *Vega* mission took photographs of the nucleus at the center of the comet, and the *Giotto* mission flew only 400 miles (644 km) away, returning images of dark, carbon-rich rock. *Giotto* was redirected to Comet Grigg-Skjellerup, passing within 125 miles (201 km) of that comet's center in 1992. Early in the twenty-first century the *Stardust* mission is scheduled to fly close to the Comet Wild 2 and return of a small sample of the comet's gases and dust to Earth.

In 1991, on its way to Jupiter, *Galileo* observed the asteroid Gaspra, providing the first close-up view of one of these irregularly shaped chunks of rock. *Galileo* also flew by the asteroid Ida and discovered that it had a tiny companion moon, the first time such a thing had been observed.

The twentieth century saw dramatic advances in our knowledge of the solar system. As the century drew to a close, new solar system missions were beginning. The *Cassini* probe is on its way to Saturn, *Stardust* is on its way to a comet, and NASA is planning a series of Mars missions for the early years of the twenty-first century.

MARY HROVAT

Further Reading

Books

Greeley, Ronald, and Raymond Batson. *The NASA Atlas of the Solar System.* Cambridge, England: Cambridge University Press, 1997.

Henbest, Nigel. *The Planets.* New York: Viking, 1992.

Periodicals

Chaikin, A. "Magellan Pierces the Venusian Veil." *Discover* 13, No. 1 (January 1992): 22-26.

Dowling, T. "Big, Blue: The Twin Worlds of Uranus and Neptune." *Astronomy* 18, No. 10 (October 1990): 42-53.

Newcott, William R. "In the Court of King Jupiter." *National Geographic* 196, No. 3 (September 1999): 126-139.

McLaughlin, W. I. "Voyager's Decade of Wonder." *Sky & Telescope* 79, No. 1 (July 1989): 16-20.

Newcott, William R. "Return to Mars." *National Geographic* 194, No. 2 (August 1998): 2-29.

Robinson, Cordula. "Magellan Reveals Venus." *Astronomy* 23, No. 2 (February 1995): 32-41.

Strom, R. G. "Mercury: The Forgotten Planet." *Sky & Telescope* 80, No. 3 (September 1990): 256-260.

Planets Beyond Our Solar System

Overview

Until the 1990s, no one had evidence of planets around Sun-like stars anywhere in the universe, except for Earth and the other eight planets circling the Sun. Then in October 1995, Michel Mayor and Didier Queloz of the Geneva Observatory changed history. They located a planet orbiting 51 Pegasi, a star located just 40 light years away from Earth. Shortly thereafter, additional astronomers chimed in, announcing evidence for other extrasolar planets. The announcements triggered a great deal of excitement in the scientific community and among the public as well. The announcements finally brought confirmation that the nine planets within our solar system are not the only ones in the universe, and begged the question, "Is there life out there?"

Background

Astronomers had long suspected that Sun-like stars had orbiting planetary systems, but they never had data to back up their assumption. They wondered whether planets were common to other stars or rare, whether other planetary systems were similar to our own, and ultimately whether Earth was the only planet in the universe that could support life as we know it.

In 1995 Michel Mayor and Didier Queloz of the Geneva Observatory announced that they had collected data verifying the existence of a planet around a star other than the Sun. The star, 51 Pegasi, is visible from Earth with the naked eye in the constellation Pegasus. The planet is at least half as large as Jupiter, and orbiting 51 Pegasi so closely that a full revolution takes only four days. With a proximity of 7 million kilometers (4,350,000 miles), the extrasolar planet travels much nearer to 51 Pegasi—closer than Mercury's trek some 58 million kilometers (36 million miles) from the Sun. Scientists have estimated the planet's surface temperature at 1,300° Celsius (2,372° Fahrenheit), much too high to support life.

Although some astronomers were skeptical, most doubts were erased when Geoffrey W. Marcy and R. Paul Butler of San Francisco State University and the University of California at Berkeley, verified the planet and shortly thereafter reported evidence of four others. Two of the planets have masses about 80 percent that of Jupiter. One orbits at 25 million kilometers (16 million miles) from its parent star and the second at about 14.5 million kilometers (9 million miles). Another of the planets is at least twice as large as Jupiter and orbits at a much less extreme 300 million kilometers (186 million miles) from its star. The fourth is the largest: a likely gaseous planet at least 6.5 times larger than Jupiter and with a great elliptical orbit around its star, 70 Virginis, in the constellation Virgo.

The evidence for each of these four planets came from measurements of the motion of the "parent" stars. According to these astronomers, the gravitational pull of the orbiting planets causes the stars to exhibit a wobbling phenomenon, which is visible from Earth as a so-called Doppler shift in the star's spectrum. They assert that this method of detection can determine the presence of a planet, its size, and its orbit.

In 1996 George Gatewood of the University of Pittsburgh, announced another extrasolar planet using a different technique but the same basic idea. Instead of scanning spectral data, he sought out the Doppler shift by watching for tiny aberrations in the path of the star Lalande 21185. Gatewood's first planet is about the same size as Jupiter, with an orbit of about 300,000 million kilometers from its parent star. The Earth, in contrast, maintains a distance of about 150 million kilometers (93 million miles) from the Sun.

During the following year astronomers continued to report extrasolar planets, including a Jupiter-sized planet in the Northern Crown constellation. The scientists suggested that other planets might share the parent star, called Rho Coronae Borealis, giving hope that a planetary system would soon be discovered. In 1999 Marcy, along with Debra Fischer of San Francisco State University, identified a multi-planet system orbiting the Sun-sized Upsilon Andromedae in the constellation Andromeda. Harvard's Robert Noyes confirmed the discovery. The three-planet system brought the total aggregate of extrasolar planets to about two dozen, a number astronomers believe will increase dramatically in the coming years.

Along with the rapid succession of planet discoveries, a scientific team led by Jane Luu of the Harvard Smithsonian Centre for Astrophysics in Massachusetts reported in 1997 that even our own solar system might contain other planets. The researchers found a 480-kilometer-wide (300 miles) object orbiting the Sun from beyond the orbit of Neptune. This miniature "planet" is larger than the comets orbiting the Sun in the Kuiper Belt, but much smaller than the system's nine known planets. Even the smallest planet, Mercury, is more than 10 times as wide as the newfound object.

Impact

The discoveries of extrasolar planets confirmed that our solar system is not unique, and apparently not even rare. These discoveries, coupled with data collected in recent years about the universe and the galaxies within it, are helping astronomers to understand how the cosmos evolved, and specifically how the Earth came to be. The discoveries have also sparked the imaginations of scientists and the lay public who wonder if life exists beyond Earth.

When Mayor and Queloz announced evidence of the first planet orbiting a Sun-like star, and other astronomers began reporting additional planets, the excitement in the scientific community was palpable. In an article in the May 1996 issue of *Scientific American,* Alan P. Boss of the Carnegie Institution of Washington, declared, "It is the most exciting thing I've seen in my scientific career."

The earliest discoveries described large planets mainly orbiting very close to their parent stars—a pattern much different than our solar system. Large planets with tight orbits didn't mesh well with current planet-formation hypotheses, and gave astronomers new food for thought. Some scientists proposed that the large planets were formed further from the star, but were actually migrating inward until finding stable orbits closer to the parent star. Since the discoveries, astronomical theorists have been busy reviewing data and analyzing potential planet-forming strategies to figure out what options are available in the creation of extrasolar systems.

A great deal of research also began to determine whether any of the planets could harbor life. The large planets with close orbits were poor candidates. By traveling so closely to their parent stars, their surface temperatures would be much too high to support life. The planet around 51 Pegasi, for example, actually orbits within the star's outer corona where temperatures are almost eight times that required to boil water. Planets with extreme elliptical orbits were also poor candidates, because their temperatures would vary too greatly as they circled their parent stars.

Some of the other planets, including those orbiting 70 Virginis in the constellation Virgo and 47 Ursae Majoris in Ursa Major (which contains the Big Dipper), are cool enough to allow liquid water to persist. Similar to Jupiter, these two planets are mostly gaseous, leaving little probability for the existence of life like that on Earth. The moons, however, are another story, according to Marcy, who discovered the planets. In the January 29, 1996, issue of *Time,* he remarked, "If they are comparable in size to the moons of Jupiter and Saturn, they could easily have rain and oceans." In the view of most scientists, water extends at least the possibility for life.

In late 1999 a research team using the Keck I Telescope in Hawaii reported finding five giant gaseous planets with similarly temperate climates. The researchers likewise suggested that their moons might support life-giving conditions.

The very presence of extrasolar planets also brings up the possibility that extrasolar systems exist. The Doppler-shift technique favors the discovery of larger planets, because their gravitational pull is strong enough to create the parent star's "wobble" that tips off astronomers to an orbiting planet. Many astronomers believe that some of the newly found planets could be part of larger systems, and that they may be part of systems with a half dozen or dozen planets.

In 1998 David E. Trilling and Robert H. Brown of the University of Arizona, detected what they believe is a Kuiper Belt around a star. They studied the star 55 Rho1 Cancri, which has an orbiting planet, and located the dusty ring of comets at about the same distance from the star as our solar system's Kuiper Belt is from the Sun. They contended that the belt suggests the system contains more than one planet.

The following year, another scientific team found evidence of the first extrasolar system surrounding a Sun-like star. Again using the Doppler-shift technique, the researchers found evidence of three planets orbiting Upsilon Andromedae in the constellation Andromeda. The scientists scrutinized the star's erratic "wobble" and found it was actually the combined effects of three large planets tugging at the star.

These discoveries have prompted the National Aeronautics and Space Administration (NASA) to begin its own investigations. NASA is planning to launch a five-year mission to search for planets as part of its medium-class Explorer (MIDEX) program. Scheduled for launch in 2004, the Full-Sky Astrometric Mapping Explorer is a space telescope that will check some 40 million stars for large, orbiting planets.

As the planetary discoveries continue, they are providing scientists with unexpected and enlightening insights into the formation of planets and planetary systems, and injecting a new air of excitement into the question: "Are we alone?"

LESLIE A. MERTZ

Further Reading

Boss, Alan. *Looking for Earths: The Race to Find New Solar Systems.* New York: John Wiley, 1998.

Clark, Stuart. *Extrasolar Planets: The Search for New Worlds.* New York: John Wiley, 1998.

Croswell, Ken. *Planet Quest: The Epic Discovery of Alien Solar Systems.* New York: Oxford University Press, 1999.

Lemonick, Michael. *Other Worlds: The Search for Life in the Universe.* New York: Simon & Schuster, 1998.

Mariotti, J. M., and D. Alloin, eds. *Planets Outside the Solar System: Theory and Observations.* Boston: Kluwer Academic Publishers, 1999.

Physical Science

1950-present

Deep-Sea Hydrothermal Vents: New World under the Ocean

Overview

In 1977, scientists began to search for deep-sea hydrothermal vents, 2,500 to 2,600 meters (8,000 to 8,500 feet) below the sea, using submersible vehicles such as *Alvin* and *Angus*. It was on the Galapagos Rift, an area southwest of Ecuador, that the search for hydrothermal vents was first undertaken in an effort to examine the metal deposits surrounding the vents. While exploring these vents, researchers made an amazing discovery. Pictures taken in 1975 revealed the presence of large white clams, but in 1977 an expedition led by Dr. John B. Corliss of the Oregon State University exposed an entire community of animals thriving in the environment around the vents. Other expeditions ensued, as many were intrigued by this unique ecosystem.

Background

Geologists had hypothesized about the existence of deep-sea vents, but it was impossible for them to examine deeper regions of the ocean since the technology did not exist to venture that deep into the ocean. When equipment, such as *Alvin* from the Woods Hole Oceanographic Institution in Oregon and France's Cyana were developed, deep-sea explorers were able to reach these areas of the ocean. *Alvin* was capable of descending to depths of 4,000 meters (13,000 feet) and was used in 1977 by Corliss and his team to study the vents on the Galapagos Rift.

These vents were first proposed to exist in 1965 by geologists like Harry Hess and Robert Dietz, as a part of the expanding theory of plate tectonics. In this theory there are 13 semi-rigid plates making up the Earth's crust, which float on the fluid asthenosphere or mantle. The constantly moving plates push together and pull apart, shaping the Earth's surface for billions of years. Individual vents are not thought to last more than 60 years, but the phenomenon of hydrothermal venting is believed to predate life on Earth and may have existed for 3.5 to 4 billion years.

Hydrothermal vents are often found on the mid-ocean ridge, the most volcanically active continuous zone on the Earth's surface. Over 60,000 kilometers (37,282 miles) long, the mid-ocean rift is the largest feature of the Earth's surface and extends from the Arctic Ocean to the Atlantic and crosses the Pacific Ocean to the west coast of North America. A rift is a giant spreading center where tectonic plates are pulling away from one another. As the plates spread apart, hot magma rises up to create new crust. The new crust hardens as it comes into contact with the cold ocean water, fissures develop, and hot water trapped within the crust is released. Water seeping through the ocean's crust becomes heated when travelling through the warm interior of the Earth. The warm water rises and makes its way back into the ocean through the cracks found along zones of tectonic activity. Created along crests in the mid-ocean rift, vent fields form linear zones that can be a few kilometers long and 100 meters wide. Water exiting the vents is rich in minerals and can be as hot as 400 degrees C (752 degrees F), but cools rapidly as the hot water mixes with the cooler seawater. Minerals and sulfides precipitate out when the warm water exits the vents and creates the appearance of billowing white or black smoke. These precipitates accumulate, and create chimneys around the vents.

The vents found on spreading centers in the Galapagos are characterized by hot spewing water and white or black plumes, but vents in areas of convergence, where plates come together, do not have water velocities or temperatures high enough to create these spectacular visuals. Along areas of convergence, such as the Oregon subduction zone located off the coasts of Oregon and Washington, are vents similar to those at spreading centers. More subtle and difficult to detect, these vents are characterized by radon anomalies and elevated concentrations of methane gas.

Impact

The findings of the 1977 expedition prompted the National Science Foundation and the Office of Naval Research to complete a more thorough examination of the hydrothermal vents in 1979. A team of scientists led by geologist Dr. Robert Ballard once again used *Alvin* to descend to the depths of the Galapagos Rift. What they discovered was truly remarkable. Large white clams, *Calyptogena magnifica*, that were a foot (0.3 meters) long and large tubeworms, *Riftia Pachyptila*, that were as much as two meters (6.5 feet) long were seen. Living in basally closed tubes, the worms with their bright red plumes were clustered around the vent openings. Also observed were a mytilid mussel and smaller alvinellid worms, including the pompeii worm, *Alvinella pompejana*. The area was home to scavenging crabs, such as *Bythograea thermydron*, and a bythitid fish that was seen with its head down in the vents, presumably feeding. Species of shrimp, limpets, and whelks were in the thickets created by the masses of the worm *Riftia* and in the vents themselves. These animals are endemic to the vents and do not exist outside the vent fields; indeed the areas surrounding the vent fields are sparse and barren in comparison. Octopus that feed on the clams and mussels may venture into the vent fields, but are not endemic to the vent environment and are the exception. Elevated temperatures at the vents as well as the chemistry of the water make them unsuitable to other marine life, where vent animals have uniquely adapted themselves to this toxic environment.

Sunlight is unable to penetrate so deep into the ocean, and the creatures living around the deep-sea vents survive in an environment devoid of solar inputs. Previously, it had been thought that solar energy and photosynthesis were the basis for all life on Earth; the discovery of the vents challenged that view. Instead, the animals living in the vent fields depend on the energy from the Earth, and a process called chemosynthesis to survive. Living in the vents, and in dense mats surrounding the vents, bacteria oxidize hydrogen sulfide or methane and use the energy to create organic carbon from carbon dioxide. Serving as primary producers in these ecosystems, bacteria are the source of food for primary consumers like the clams, mussels, and worms.

The large white clam and *Riftia* use symbiotic relationships with bacteria to derive their nutrition. Bacteria have been found in the gills of the clams and in a part of *Riftia* called the trophosome. Concentrated in the trophosome, bacteria live on sulfides and oxygen transported to this organ by specialized hemoglobin in the worm's blood. In turn, the worm can digest these bacteria. Some alvinellid worms also use symbiotic relationships and are covered by hair-like strands of bacteria, which produce eurythermal enzymes that can withstand dramatic temperature changes and serve to protect the worm from the extreme temperatures experienced at the vents.

The discovery of the vents and the vent ecosystems is significant for many reasons. Animals and bacteria found at the vent sites provide researchers and industry with new concepts and could help develop technology that may help improve our quality of life. Enzymes produced by the bacteria might be used to dislodge oil inside wells, convert cornstarch to sugar, and speed up biological and chemical reactions, possibly useful in other industrial processes. The enzymes can withstand high temperatures, toxic chemicals, and total darkness, and genetic engineers hope to culture bacteria that could decompose toxic waste. Potentially important in the development of new drugs and medicine, vent creatures continue to be investigated.

Rich in metals such as copper, zinc, iron, and gold, the vent chimneys are of interest to the mining industry as well, but the exploitation of this resource is the subject of debate. Some question the ability to safely mine the vents without damaging the unique ecosystems. Despite this, deep-sea vents continue to be of interest to industry, as sulfide ore deposits at the vents might be used when terrestrial resources have been depleted. Sulfide ore is economically important as this class of minerals includes ores of metals such as lead, copper, and silver.

The exploration of extreme environments deep in the ocean also gives researchers an opportunity to develop methods for work in places like Mars, and other dangerous environments. Remote technology and photography are used in deep-sea exploration, along with robotic arms and other specialized parts. Deep-sea vehicles are equipped with advanced sampling devices just as those on Mars might be and give researchers an opportunity to become more familiar with this kind of technology.

Methanococcos jannaschii, a unicellular, microscopic and chemosynthetic organism, found at a vent on the Eastern Pacific Rise in 1982, was confirmed to be part of a newly recognized third domain of life call *Archaea* by Dr. Carl Woese and others at the University of Illinois. Previously just two domains of life were recognized: 1) Bacteria and 2) Eukaryota. *Archaea* do not have a nucleus, but one half to two thirds of the genes are unlike anything else on Earth and distinguishes them from the other domains of life. *Archaea* are believed to most closely resemble the organisms contended to be the first forms of life on this planet.

Important for their potential commercial and scientific value, hydrothermal vents also play an important role in the regulation of the temperature and chemical balance of the oceans. Geochemists, such as John Edmund, have proposed that all the world's oceans circulate through the crust once every 10 million years. When seawater circulates through the crust, magnesium and sulfates are removed, while calcium, potassium, and gases such as hydrogen sulfide and methane are added. Previously it was held that the chemical balance of the oceans was determined by run-off from the continents, but now hydrothermal infusion, or circulation, and continental run-off are considered equally important in this function.

Vent ecosystems continue to be of interest as they provide insights about the origin of life on Earth. While some believe the bacteria thought to be the origin of life arrived on meteors from outer space, called the Panspermia theory, others believe the hydrothermal vents at the bottom of the ocean may be the origin of life on Earth. Useful in the study of life on Earth and life elsewhere in the universe, vents ecosystems are unique. The uncovering of an ecosystem that is based on chemosynthesis and geothermal energy is tremendously important. It challenges the traditional thinking about the physical parameters that will support life and opens new doors for speculation.

KYLA MASLANIEC

Further Reading

Books

Rona, Peter A., et al. *Hydrothermal Processes at Seafloor Spreading Centers.* New York: Plenum Press, 1983.

Periodical Articles

Ballard, Robert D. and Frederick J. Grassle. "Return to Oases of the Deep." *National Geographic* Vol. 156 (December 1981): 689-703.

Kulm, L.D., et al. "Oregon Subduction Zone." *Science* Vol. 231 (February 1986): 561-66.

Other

Fearon, Bertha and Jonathan Chu. "What Are Deep-Sea Hydrothermal Vents?" http://www.geneseo.edu/~jc99/whatarethey.html

National Geographic Society. *Dive to the Edge of Creation.* Video. 1980.

A World Within: The Search for Subatomic Particles

Overview

Although the concept of the atom dates back to ancient Greece and the scientist/philosopher Democritus (c. 460-370 B.C.) who defined atoms as matter "unable to be cut," and since English scientist John Dalton's (1766-1844) articulation of atomic theory in 1803, the concept that matter consists of atoms and that atoms, in turn, are composed of smaller, subatomic particles has become an evolving but fundamental postulate of physical science. In the later half of the twentieth century as physicists explored the composition and properties of subatomic world, the impact of discoveries regarding subatomic particles bounded into the vastness of cosmological theory.

Background

The elegant simplicity of the indivisible atom was first shaken in the 1890s with English physicist J. J. Thomson's (1856-1940) discovery of the electron. Because electrons are negatively charged and atoms are electrically neutral, the existence of the electron implied the existence of at least one other subatomic particle to balance the electron's negative charge. That particle, the proton, was discovered by English physicist Ernest Rutherford (1871-1937) in 1919. The composition of the atom was far from settled, however, as evidence also existed for the presence of a third subatomic particle. Discrepancies between atomic number (number of protons in an atom) and atomic weight indicated that atoms must contain a subatomic particle that was electrically neutral yet approximately equal in mass to the proton. Subsequently, the neutron was discovered by James Chadwick (1891-1974) in 1932.

Scientists, and the lay public, became comfortable with a new atomic model that included protons, neutrons, and electrons. The model was simple and attractive because it explained most of the known physical and chemical phenomena. Yet, already looming were predictions that the subatomic world might harbor other strange and exotic particles.

Early in the 1930s French mathematician and physicist Paul Dirac's (1902-1984) prediction of a positively charged electron was confirmed when positrons were discovered in a cosmic ray shower by physicist Carl David Anderson (1905-1991). Anderson's discovery not only garnered him the 1936 Nobel Prize in Physics, it touched off a seemingly endless hunt for subatomic species. Although nearly 30 years elapsed before the existence of the antiproton was confirmed, the number of subatomic particles predicted and discovered seemed to grow at an increasing tempo.

To confound matters further, Albert Einstein's (1879-1955) use of the concept of photons in 1905 postulated the existence of particles without mass, and in 1931 Wolfgang Pauli (1900-1958) predicted the existence of the massless neutrino.

During a search for the mediator of the strong force, in 1938 the mu meson, or muon, was discovered. Although the new particle turned out not to be the carrier of the strong force, the findings established a pattern of "accidental" discoveries made while scientists explored the subatomic world. In 1947 English physicist Cecil Powell (1903-1969) discovered another type of meson in cosmic ray showers, the pi meson, or pion, that was found to interact with protons and neutrons.

Impact

The finding of the pi meson cemented a relationship between theoretical physicists and particle "experimental" physicists that had existed since Dirac and Anderson. Japanese physicist Hideki Yukawa (1907-1981) was awarded the 1949 Nobel Prize in Physics for his nuclear force theories that predicted the existence of mesons, and English physicist Cecil Powell won the 1950 Nobel Prize in Physics for his work developing the photographic methods used in the study of nuclear processes.

A pattern was established whereby theoretical physicists' predictions of the existence of subatomic particles would lay out the need for technological breakthroughs that would, in turn, allow experimentalists the tools to confirm the existence of the predicted particles.

One technological breakthrough driven by the quest for subatomic particles was the invention and construction of machines that could accelerate particles to very great energies. So powerful was this new technology that experiments conducted with these particle accelerators soon swelled the family of subatomic particles to more than 100 new species.

Victims of their own success, physicists were overwhelmed with the increasing numbers of newly discovered subatomic particles—and with it a taxonomy ill equipped to handle the bewildering variety of properties these subatomic particles exhibited. Physicists and modern scientific community traditionally uneasy with disorganization and uncertainty in nomenclature proposed many classification schemes to organize the seemingly chaotic and voluminous data regarding subatomic particles.

Some proposals were relatively simple, and merely grouped and organized particles according to common properties. Some schemes, for example, classified particles according to their spin. Those with half-integral spins (e.g., 1/2, 3/2, 5/2, or 7/2) were termed fermions, while those with integral spins (0,1,2) were classified as bosons. Particles acted upon by the strong force were grouped together as hadrons.

Some particles were grouped with others that existed only in neutral states, while others (e.g., pion and delta particles) existed in one or more charged (or neutral) states. Other schemes relied on the observed life expectancy of subatomic particles, some having lives as short as 10^{-23} second. Particles with unusually long lifetimes were said to have a property labeled "strangeness." Additional schemes relied on the decay properties of particles.

Subatomic particles were also grouped according to mass—the lightest particles being leptons, those of medium mass being mesons, the heavier particles named baryons, and the extremely heavy particles termed hyperons.

In contrast to the often unexplainable taxon of particles with mass, the search for and classification of subatomic particles without mass, however, provided exciting new vistas to confirm some of the leading theoretical interpretations of both quantum and relativity theory. In fact, the very concept of subatomic particles without mass could only be understood in terms of the interconvertability of mass and energy.

At times confusion seemed to reign supreme in the fast-paced world of particle physics. If not bewildered, many scientists and the general public seemed bemused at the whimsical non-descriptive names sometimes ascribed to particles. Hadrons, (e.g., protons and neutrons) for example, were determined to be composed of various combinations of three more elementary particles named "quarks." The name quark was derived from Irish writer James

Joyce's novel *Finnegan's Wake* and a character's cry, "Three quarks for Muster Mark." The brilliant American physicist Murray Gell-Mann (1929-) who named the quark defended the name as appropriate because quarks always occur together in nature in sets of three.

Further havoc to the conventions of nomenclature came with the delineation of three types of quarks eventually labeled the up, down, and strange quarks. The terms "up, down, and strange" were not at all descriptive. In fact there is nothing up about the "up" quark, nor is there anything strange in the conventional sense about the "strange" quark.

In 1961, a taxonomic scheme dubbed the "eight-fold way" was proposed for organizing subatomic particles. Particles were grouped according to properties such as spin and strangeness. In much the same manner that the early periodic table was validated by the discovery of elements that filled in holes in the table, the validity of the "eight-fold way" was enhanced with the 1964 discovery of the W particle.

During the last decades of the twentieth century, physicists continuously refined their methods of classifying subatomic particles. The resulting system of classification based on particle properties is termed the Standard Model. According to the Standard Model, six types of each particle are arranged in pairs according to mass and energy level. In addition to quarks and leptons, there are other fundamental particles. These mediating particles include the undiscovered graviton (proposed carrier of the gravitational force), photons (carriers of electromagnetic or light energy), gluons (mediators of the strong force), and the W+, W-, and the Z bosons (carriers of the weak force further associated with radioactivity and nuclear decay). Not to be without its own eccentricities in nomenclature, the weak force, for instance, is postulated to allow quarks to change to another type of quark, or a lepton to another type of lepton. When these transformations occur, quarks and leptons are said to have changed "flavor." The discovery in 1994 of the top quark added validity to the Standard Model.

The seemingly stable matter of everyday experience (energy level one) contains only electrons, neutrinos, and the up and down quarks that interact to form protons and neutrons. Particles associated with energy level two (e.g., the strange and charm quarks, leptons, muons, and the muon neutrinos) are found at the energy level of cosmic rays. Particles associated with energy level three (e.g., the bottom and top quarks) can be demonstrated only in accelerators that can obtain the power characteristic of fundamental cosmological processes (i.e., the energy levels that existed near the Big Bang).

Because the Standard Model accounts for the strong, weak, and electromagnetic fields and forces (including the electroweak theory force), it is a powerful theoretical model. The discovery of subatomic particles as means to understanding the forces ruling the quantum world, however, has become inexorably entwined with relativity-dominated cosmological theory regarding the origin and structure of the Universe.

By the end of the twentieth century more than 200 subatomic particles had been identified. In general the particles exist as pairs of particles and antiparticles with identical mass but with opposite properties of charge, magnetic moment, or spin. More powerful accelerators will be needed to determine if there is a more elemental composition to quarks and leptons. The construction of the expensive accelerator is not without controversy, however. After a decade of controversy and the expenditure of over 200 million dollars, construction on the United States Superconducting Supercollider was stopped after funding was cut off by a budget-conscious United States Congress. The termination of the SSC project left CERN (*Organisation Europeene pour la Recherche Nucleaire*) in Geneva, Switzerland, as the focal center of the particle physics world.

Because there are important questions left unanswered by the Standard Model and because of defects in the model (e.g., at the extremely high temperatures thought to have existed during the Big Bang, the model breaks down), most physicists consider the Standard Model to be but a stepping stone toward a grand unification theory that can unite incompatible quantum and relativistic theories. Articulation of such a theory will have the profound scientific and philosophical impact comparable to, and perhaps surpassing, Newton's theories of gravitation, Einstein's theories of relativity, and Planck's contribution to quantum theory.

Such a unified theory may allow for a further understanding of the nature and existence of a critically important hypothesized subatomic particle that continues to elude physicists—the proposed mediator of the gravitational force, the graviton.

K. LEE LERNER

Further Reading

Books

Ezhela, V.V., et. al. *Particle Physics: One Hundred Years of Discoveries: An Annotated Chronological Bibliography.* American Institute of Physics, 1996.

Feynman, R., et. al. *Elementary Particles and the Laws of Physics: The 1986 Dirac Memorial Lectures.* Cambridge University Press, 1999.

Periodical Articles

Davies, P. "Particle Physics for Everybody." *Sky & Telescope* (December 1987).

Freedman, D. "Particle Hunters." *Discover* (Dec. 1992).

Huth, J. "The Search for the Top Quark." *American Scientist* (Sept./Oct. 1992).

Hubble Space Telescope and Its Influence on Astronomy

Overview

From the age of the universe to the existence of black holes and the discovery of new galaxies, the Hubble Space Telescope (HST) is providing a multitude of data for astronomers eager to understand the cosmos. While the telescope is providing images and data about deep-space objects, stellar systems, and other curiosities, it actually travels no farther than 380 miles (612 km) from Earth. A cooperative program of the European Space Agency and the National Aeronautics and Space Administration (NASA), the Hubble Space Telescope is an orbiting observatory, revolving around the Earth once every 93 minutes. It is able to collect data unavailable to Earth-based observatories, because the HST's location is unhampered by distortion from our planet's atmosphere. Since its 1990 deployment, the telescope has collected information and images that are helping astronomers generate a better picture of the universe and how it works.

Background

The $1.5 billion Hubble Space Telescope was deployed into orbit by the space shuttle *Discovery* on April 25, 1990. The launch culminated two decades of work by researchers at NASA and other laboratories. Because the HST received its transportation into orbit via shuttle, it had to be lightweight and compact. The 13-ton [43 feet] HST is about the size of a railroad car (13.1 meters [43 feet] long and 4.3 meters [14 feet] in diameter at its widest point), considerably smaller than major Earth-based observatories, such as the 12-story Mount Palomar observatory in California. Its mirror is also much lighter than the traditional variety. By using a honeycomb design, engineers were able to eliminate two-thirds of the weight of a conventional mirror.

In addition, scientists had to design an intricate pointing system that would allow the solar-powered telescope to lock onto a target and maintain that lock even as the telescope circles the Earth at 8 kilometers per second (17,895 mph). The telescope accomplishes this feat through a complex of sensors and computer-controlled wheels. The Fine Guidance Sensors verify the telescope's target position 40 times a second, while the wheels correct for any deviation.

Protection of the HST from extreme heat change was another consideration. The HST circles the Earth every 93 minutes, going from the light side of Earth to the dark side—about a 120° difference in temperature—during every revolution. The solution was an exterior thermal blanket, which maintains a constant temperature for the telescope.

Despite all of the technology and abilities available aboard the HST, scientists, engineers, and computer programmers on Earth still are essential to the telescope's operation. Men and women at the Space Telescope Science Institute and Goddard Space Flight Center in Maryland carefully plan the telescope's operation down to the last detail, and radio instructions to the orbiting observatory. The telescope can then make observations in the visible, near-ultraviolet, and near-infrared spectra.

The HST follows the instructions, which may be to take a picture of or collect data on a distant star or galaxy. The pictures and data are translated into signals, radioed to Earth, then retranslated into pictures and data for use by astronomers and other scientists.

Throughout the telescope's tour of duty, astronauts have periodically visited the HST to upgrade its capabilities, perform tune-ups, and

The Hubble Space Telescope. *(UPI/Bettmann. Reproduced with permission.)*

make repairs. The most well-publicized repair occurred after scientists and engineers discovered a flaw in the telescope's primary mirror, which blurred the HST's pictures. The solar panels also experienced problems when the telescope moved from the light to the dark side of Earth. The panels flexed enough to affect the accuracy of the pointing system. In 1993 astronauts aboard the space shuttle *Endeavor* replaced the solar panels and installed two devices with corrective optics.

The Hubble Space Telescope is named for American astronomer Edwin Powell Hubble (1889-1953), who discovered in 1929 that the universe is expanding. This idea formed the basis for the widely held Big Bang Theory, which states that a huge explosion created the universe, and the universe is even now continuing to expand out from the point of the explosion. Hubble also proved that galaxies exist beyond the Milky Way.

Impact

The data and images collected by the Hubble Space Telescope have yielded key evidence for a variety of scientific hypotheses, provided information that is helping to unlock some of the mysteries of the cosmos, and changed the way scientists and the public look at our universe.

For example, the HST has repeatedly provided views of previously unseen galaxies. In 1999 the HST yielded images of a spiral galaxy much like the Milky Way in size but 60 million light-years away. (One light year is the distance that light travels in a year, or approximately 6 trillion miles. [over 9 trillion km]) The spiral arms of the galaxy, dubbed NGC 4414 and located in the constellation Coma Berenices, are signs of ongoing star formation.

The telescope's data and images have also provided information about how quickly different types of galaxies formed following the Big Bang and about how rapidly the universe is expanding from that initial cosmic event. Wendy L. Freedman, head of the Hubble Space Telescope Key Project on the Extragalactic Distance Scale, and Allan R. Sandage of Carnegie Observatories, analyzed HST-collected data for Cepheid variable stars, including three dozen located in the spiral galaxy NGC 4603 in Centaurus, and put the expansion speed, known as the Hubble constant, at 58-70 kilometers per second per megaparsec.

While the HST was viewing yet another galaxy, the giant radio galaxy M87, astronomers observed subatomic particles appearing to move at six times the speed of light within a jet emanating from the galaxy's core. The movement was actually an optical illusion, called superluminal motion, produced because of the orientation of the jet to Earth. In reality, the particles likely move at a velocity slightly below the speed of light. As reported in the May 1999 issue of *Astronomy,* "According to John Biretta of the Space Telescope Science Institute, the discovery of superluminal speeds in M87 practically confirms the belief that radio galaxies, quasars and blazars are essentially the same, and differ only in orientation with respect to the observer."

Astronomers provided a glimpse back in time by producing an image called the Hubble Deep Field. The image is a melange of 342 separate exposures photographed by HST's Wide Field and Planetary Camera 2 during a 10-day period in 1995. Taken inside a small sliver of the universe near the handle of the Big Dipper (part of the constellation Ursa Major), the Hubble Deep Field image contained some 1,500 galaxies, most of which had never before been seen. In a Space Telescope Science Institute press release, institute director Robert Williams remarked, "We are clearly seeing some of the galaxies as they were more than 10 billion years ago, in the process of formation. As the images have come up on our screens, we have not been able to keep from wondering if we might somehow be seeing our own origins in all of this."

The data and images collected about galaxies are also providing distances between cosmic neighbors. With this information, astronomers are essentially working backward to try to determine how far the galaxies have traveled out from the origin of the Big Bang, and, thus, the age of the universe. In addition, the HST has detected gases and elements that give further credence to the Big Bang Theory.

In some of its most dramatic images, the HST offered views of the birth of stars. Images taken in 1999 depicted young stars surrounded by disks of dust and gas that astronomers believe may be the stuff of future planets. The images not only confirmed indirect evidence that the disks existed, but provided a level of detail previously unknown. Estimated to be from eight to 16 times the size of Neptune's orbit, the disks were located around newly forming stars 450 light years from Earth and in the constellation Taurus. Scientists believe the images might also provide clues to the formation of our solar system.

The Hubble took other revealing pictures of a star-forming nebula within the bright winter constellation of Orion. The images again showed embryonic stars surrounded by disks of potentially planet-forming dust. The Hubble took some of its most-used images of star birth in columns of dust and gas in the 2-million-year-old Eagle Nebula, which is located about 7,000 light years from Earth in the constellation Serpens. The images show so-called evaporating gaseous globules (EGGs) at the tips of the columns. Astronomers believe that the EGGs surround newly forming stars, created when the gas in the columns collapses on itself.

Closer to home, the HST has also provided the first-ever direct view of the surface of Pluto, unparalleled pictures of Jupiter as it was struck by pieces of the comet Shoemaker-Levy 9, images of Mars minerals containing signs that the Red Planet once held water, and additional evidence for a debris- and comet-filled ring that surrounds the solar system. The Hubble determined that the ring holds at least 100,000 comets.

Beyond the scientific impact, the Hubble Space Telescope is promoting education through the Space Telescope Science Institute. The institute merges images and data from the HST into online educational opportunities and other resources for students from elementary through high school levels.

LESLIE A. MERTZ

Further Reading

Barbree, Jay, and Martin Caidin. *A Journey Through Time: Exploring the Universe With the Hubble Space Telescope.* New York: Penguin Studio, 1995.

Fischer, Daniel, and Hilmar Duerbeck. *Hubble Revisited: New Images From the Discovery Machine.* New York: Copernicus, 1998.

Mitton, Jacqueline, and Stephen P. Maran. *Gems of Hubble.* New York: Cambridge University Press, 1996.

Petersen, Carolyn C., and John C. Brandt. *Hubble Vision: Further Adventures With the Hubble Space Telescope.* New York: Cambridge University Press, 1998.

Buckyballs: Carbon Goes 3-D

Overview

In 1985 Harold Kroto (1939-), Robert Curl (1933-), and Richard Smalley (1943-) discovered a novel form of pure carbon, called fullerenes, and opened up a new field of chemistry and materials science. Buckminsterfullerene, named after American engineer Richard Buckminster Fuller (1895-1983), consists of 60 carbon atoms joined together into the shape of a soccer ball. It and related fullerenes are hollow, highly stable, and have unusual physical and chemical properties, including optical activity and superconductivity. Nanotubes, which are "buckyballs" rolled into cylindrical forms, are being tested for a variety of applications and are considered to be the most promising material for nanotechnology, the building of new materials at the molecular scale.

Background

Fullerenes represent a new form of the element carbon in which the atoms are arranged to create closed shells. Much of the potential of these new compounds is rooted in the vast experience scientists and engineers have in carbon chemistry, but the first fullerenes, clusters of 60 carbon atoms (C_{60}) and 70 carbon atoms (C_{70}), have their origins in radioastronomy and cluster chemistry.

In 1985 Kroto, Curl, and Smalley came together at Rice University to perform an experiment. Kroto was an expert in microwave spectroscopy at the University of Sussex. He studied the formation of carbon in stars and had already established the existence of long carbon chains in the atmospheres of red giant stars and in interstellar space. He was interested in understanding the structure of these chains.

Curl, also a microwave spectroscopist, had told Kroto that his colleague, Smalley had a device that could provide data he needed. Smalley was using a laser-supersonic cluster beam apparatus to study the design and distribution of chemical clusters. Clusters are aggregations of molecules at a scale between the macroscopic and microscopic worlds. In the gas phase, atoms tend to condense to form clusters of various sizes. Smalley was trying to understand why cluster components of certain sizes were much more common that those others. His apparatus, which could vaporize materials into plasmas of atoms, seemed ideal to simulate the conditions in the atmospheres of red giant stars.

The scientists put a sample of graphite into Smalley's apparatus. An intense pulse of laser light quickly vaporized the carbon, and its atoms were drawn into a stream of helium, where they combined to form clusters. These were then captured in a vacuum chamber maintained at only a few degrees above absolute zero.

The scientists analyzed the carbon clusters using mass spectrometry. As they fine-tuned the experiment, two strong peaks, representing compounds C_{60} and C_{70}, showed up. It was soon clear that these compounds were highly stable because their spectrographic peaks persisted even when a half dozen different gases were added to the mix. A new, highly symmetrical, cage-like form of carbon fit the data. Before the scientists finished the second week of experiments, they had mailed off their report to *Nature*. Because the C_{60} cluster resembled Buckminster Fuller's geodetic dome, it was name buckminsterfullerene. It and related compounds are now popularly known as "buckyballs."

Within a month, the three scientists' results were in print. The scientific community reacted both with excitement and skepticism. While applications suggested themselves from the beginning, the new compounds needed to undergo extensive chemical testing to confirm the surprising structure. The analytic community jumped on buckyballs almost immediately, building evidence that the cage-like structures postulated were correct, but final proof required larger (gram) quantities. This came in 1990,

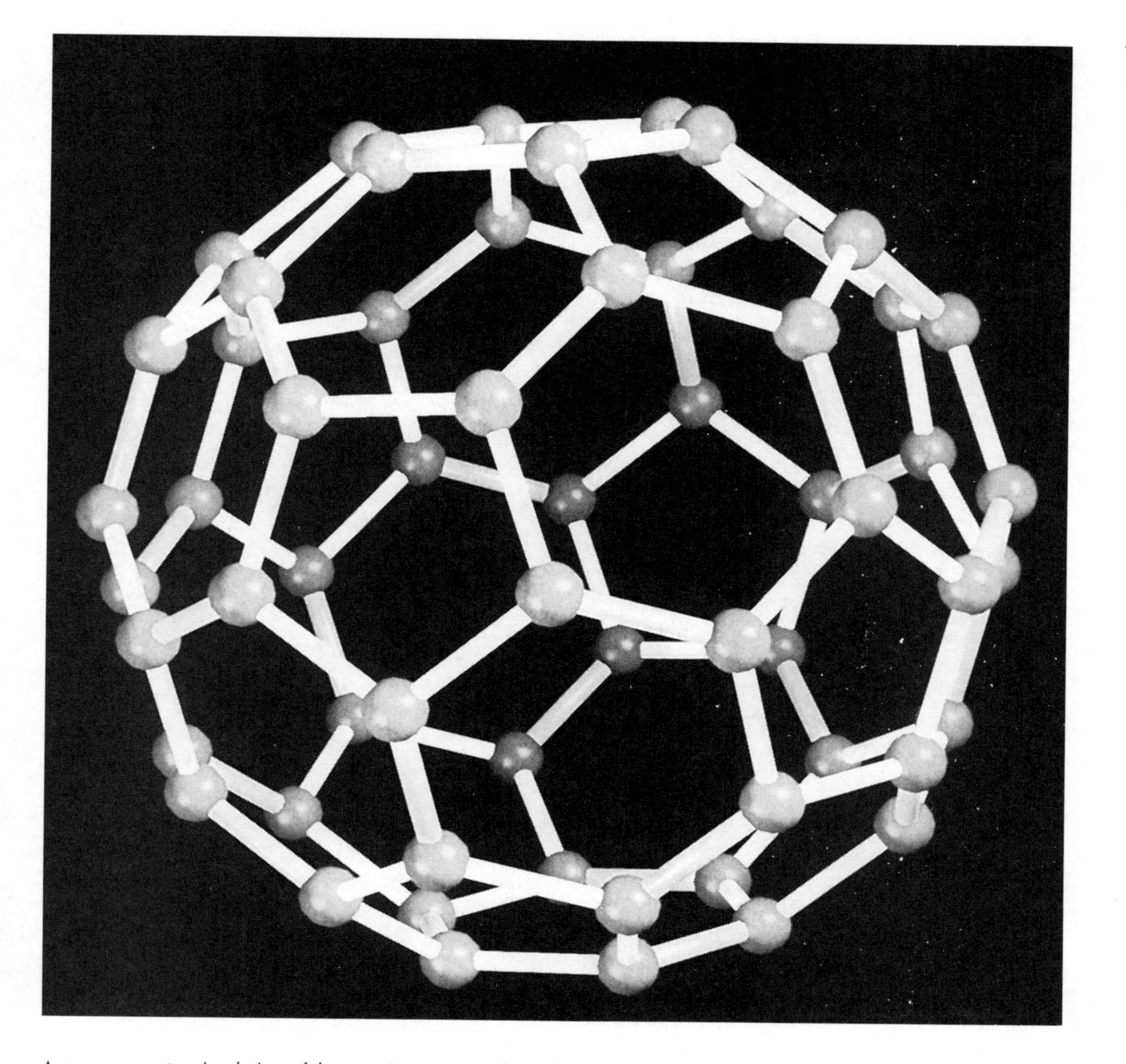

A supercomputer simulation of the atomic structure of a molecule of Buckminsterfullerene. Carbon molecules appear as small spheres; double bonds between them are darker than single bonds. *(J. Bernholc et al., North Carolina State Univ./Photo Researchers. Reproduced with permission.)*

when Donald Huffman and Wolfgang Kraetschmer used carbon arc vaporization between two graphite electrodes to produce macroscopic quantities.

The result was a black powder that was highly stable to both pressure and temperature. (Breaking a fullerene requires temperatures over 1,000°C [1,832°F]). As a cubic crystal, the material is extremely hard and electrically insulating. Buckyballs have strong optical effects, linking together into chains upon irradiation with intense ultraviolet laser light. With large quantities to test, C_{60} passed every test to prove its soccer-ball shape (20 hexagons and 12 pentagons in a spherical configuration). In fact, Kroto himself verified the structure with nuclear magnetic resonance.

Impact

In 1996 Kroto, Curl, and Smalley were awarded the Nobel Prize for Chemistry. Since then, fullerene chemistry has become one of the most dynamic and fastest growing areas of investigation in chemistry and materials science. Thanks to the work of Huffman and Kraetschmer, fullerenes can be worked on with the full toolbox of organic chemistry, and thousands of scientific papers on fullerenes and their derivatives have been published. Specialty chemical companies even sell a variety of buckyballs for a few hundred dollars a gram.

For most commercial applications of fullerenes, lower prices and simple methods of mass production are required. However, their properties—and the variety of additional properties that are achievable with known organic chemical reactions—promise a number of important uses that are actively being investigated.

The shapes of fullerenes immediately suggests ball bearings, so these compounds may find a use as lubricants. Their cage-like structure offers the possibility of using them to enclose drugs or radioactive tracers for medical imaging. The electrical and optical properties of fullerenes also make them good candidates for transistors, memory bits, photoconductors, tunnel diodes, and sensors.

Organic chemistry—the chemistry of manipulating carbon compounds—is a well-established field and has supported the development of chemical methods to manipulate the structure and properties of fullerenes. The stability and the aromaticity (alternating single and double bonds) of fullerenes make them available to chemical reactions that do not destroy their hollow, cage-like structure. Chemists can make specific modifications, such as adding chemically polar groups to make the compounds water soluble, and can even begin to work in three dimensions to create catalysts and other useful materials.

Because of the relatively large size of the individual fullerene molecules, big, empty spaces exist between them when they are in crystalline form, much like the space between gum balls in a gum ball machine. It's possible to put small molecules and atoms into these spaces (intercalating them). With certain metal compounds this has an interesting result: the materials become superconductors. While the practical value of this effect may be limited since the presence of air destroys these materials, fullerene-based superconductors do provide a new and flexible model for understanding the phenomenon of high temperature superconductivity.

Astrochemistry has also benefited from fullerene research, with better explanations of the galactic carbon cycle. Nature has been producing buckyballs for eons, and spectroscopy confirms their presence in space.

Buckyballs have an intellectual appeal because the shape, a truncated icosahedron, is one of the Archimedian polyhedra (which, like Platonic solids, were once thought to have magical properties). This, along with their aromatic bond structure, makes them an object of theoretical interest. The simplicity and symmetry of fullerenes make them nearly ideal for testing chemical theories in the lab. And since fullerenes can be fabricated as both thin films and crystals, they are available for a wide variety of experimental and analytical techniques.

Perhaps the most intriguing work that has come from the discovery of buckyballs is the creation of thin tubes with closed ends, nanotubes, that arrange carbon atoms in the same cage-like way as fullerenes. These molecular drinking straws can be made into an abundant variety of sizes, lengths, and diameters. They can be twisted into helices, with their conductivity determined by their shapes. They can be empty or filled with other materials, and they offer the possibility of insulating metallic and semiconducting wires. They could provide the molecular electronics for nanomachines and are being looked into as a basis for creating artificial muscles. Nanotubes also promise applications in energy, including fuel cell technology, lithium ion batteries, and hydrogen storage.

Already, nanotubes are used to strengthen polymers and as nearly indestructible tips and cantilevers for atomic force microscopes. A string of metal atoms down the center could provide a nanowire, which could be extremely strong. A cable made up of nanowires would have 100 times the weight-bearing strength of steel. The strength and stability of nanotube fibers could make them a potential replacement for graphite fibers in body armor, helmets, airframes and rocket nozzles.

The comparisons of buckyballs to soccer balls and the structures of architect Buckminster Fuller have created a public awareness that few other chemical compounds enjoy. Less directly, buckyballs have provided justification for speculations in science fiction, particularly in those stories that exploit the possibilities of nanotechnology. Star Trek has referenced buckyballs as the focus of a character's chemistry lab and as a key component of communicators. The combination of the material properties of fullerenes is inspiring engineers. For instance, Arthur C. Clarke's novel *The Fountains of Paradise* proposes a skyhook, a cable connecting an orbiting satellite to the ground that allows an elevator into space. Such a structure would greatly reduce the costs of traveling and living in space. On paper, nanotubes seem to provide the material properties for constructing such a skyhook, but they will first need to get beyond the current length limit of one millimeter.

While thousands of patents have already been filed for the production and application of fullerenes and nanotubes, there are no products that have any serious commercial or social impact. There are no obviously successful uses on the horizon, but these materials are still early in their development and show great potential for providing future benefits.

PETER J. ANDREWS

Further Reading

Aldersley-Williams, Hugh. *The Most Beautiful Molecule: The Discovery of the Buckyball.* New York: John Wiley & Sons, 1995.

Amato, Ivan. *Stuff: The Materials the World Is Made Of.* Austin, TX: Bard Books, 1998.

En Route to a Grand Unified Theory: The Unification of Electromagnetism and the Weak Nuclear Force at the Turn of the 1970s.

Overview

Sheldon L. Glashow (1932-), Steven Weinberg (1933-), and Abdus Salam (1926-1996) jointly received the 1979 Nobel Prize in physics "for their contributions to the theory of the unified weak and electromagnetic interaction between elementary particles, including inter alia the prediction of the weak neutral current." The Glashow-Weinberg-Salam theory of electroweak interactions is, in fact, the first experimentally proven scheme at unifying in a single set of fundamental laws two of nature's four basic forces. This major step in theoretical physics greatly enhanced our comprehension of the universe and lies at the foundation of subsequent attempts at unifying all natural forces.

Background

Aristotle thought that all matter in the universe was made of four basic elements—earth, air, fire, and water—upon which only two forces acted: gravity—the tendency for earth and water to sink—and levity—the tendency for air and fire to rise. While levity lost its "scientific" credentials over the centuries, gravity became recognized as one of nature's universal forces. In *Philosophiae Naturalis Principia Mathematica* (1687), Isaac Newton (1642-1727) postulated a law based on the principle of action at a distance in order to explain how (though not why) every particle of matter in the universe is attracted toward one another. The more massive and closer to each other two material bodies are, the more they will feel the strength of that ubiquitous force. Today gravity is best understood by Albert Einstein's (1879-1955) general theory of relativity.

During the nineteenth century electricity and magnetism were synthesized into a complete and unique set of equations that now stand for electromagnetism, the second of the natural forces. This achievement was described in a *Treatise on Electricity and Magnetism* (1873), written by one of the most influential physicists of the last century, James Clerk Maxwell (1831-1879). Electromagnetism is the force that binds, for instance, negatively charged electrons to a positively charged nucleus to create what is called an atom. As with gravity, the distance over which the electromagnetic force acts is infinite though, contrary to the former, it can be either attractive or repulsive. Nowadays the theory that explains the nature of this force is called quantum electrodynamics (QED), one of the most successful theoretical concepts ever designed by physicists.

The weak nuclear force, third on our list of basic forces, is responsible for radioactivity, which is the property of unstable atoms to release energetic particles after the spontaneous disintegration of their nuclei. An example of the weak nuclear force is beta decay, which briefly converts a neutron into a proton by emitting an electron and an antineutrino. In 1933 the Italian physicist Enrico Fermi (1901-1954) developed a theory of beta decay that tried to explain what was at the time a not-so-well understood phenomenon. Only in the late 1960s, however, did a field theory proposed by Glashow, Weinberg, and Salam come to be viewed as the solution to the problem. (We will discuss this first field theory, also known as the electroweak force, later on.)

The fourth and final natural force is the strong nuclear force, which holds quarks together in the proton and neutron and glues the latter two into atomic nuclei. It is also responsible for nuclear fusion (which powers the Sun) and fission (used in atomic bombs). The strong nuclear force was the last of the basic forces to receive some kind of formal theoretical description in what is now called quantum chromodynamics (QCD), which was put forward during the mid-1970s. (The "chromo" prefix is explained by the fact that quarks composing atomic nuclei come in three different virtual "colors": red, green, and blue by analogy with the three primary colors of light.)

These four natural forces constitute our physical world; lying at the foundation of matter and energy and therefore of life itself. Their unification could lead to nothing less than the complete understanding of the origin of the universe. This is one of the reasons why the Glashow-Weinberg-Salam theory is so fundamental to our continual quest for knowledge.

Impact

In the early 1970s the Glashow-Weinberg-Salam theory of electroweak interactions was brought to the forefront of the physics community by experiments that were conducted at the Fermi National Accelerator Laboratory (Fermilab) in Batavia, Illinois, and at the European Organization for Nuclear Research (CERN) near Geneva, Switzerland. These experiments had one essential aim—finding whether weak neutral currents existed or not. The answer was given in 1973; it was, for some at least, an astonishing yes.

The neutral current is predicted by the Glashow-Weinberg-Salam theory, which was advanced independently by the three authors late in the 1960s. Their theory was first designed to give a complete account of the weak nuclear force in a conceptual framework akin to QED. In accordance with the latter theory, the attractive or repulsive force experienced between two charged particles is transmitted by a so-called force particle, or vector boson—a photon in the particular instance of QED. Hence, when an electron feels the repulsion of another electron, the phenomenon manifests itself through the exchange of a photon. Similarly, to make use of the aforementioned beta decay, the archetype of weak interactions, the Glashow-Weinberg-Salam theory assumes that the neutron radiates a force particle called W-, causing the former particle to turn into a proton; that W- particle then rapidly decays into an electron and an antineutrino.

Glashow, Weinberg, and Salam did more than explain all known weak nuclear interactions by suggesting the existence of two force particles, W- and W+. Their theory predicted yet another vector boson, Z_0, which describes, for instance, how a neutrino (a neutral particle) scatters an electron by emitting such a Z_0. The acknowledgement of this neutral current in 1973 launched a new era of modern physics. According to this original gauge theory, the electroweak force, which unifies electromagnetism and the weak nuclear force—since then recognized as two different aspects of the same interaction—can today be described as being mediated by a set of four force particles: the photon, W-, W+, and Z_0. All three of the new vector bosons were discovered experimentally at CERN in 1983.

Following the explanatory success of the theory of electroweak interaction, quantum chromodynamics (QCD) was developed during the mid-1970s in order to find a concise conceptual scheme for the strong nuclear force. (The strong nuclear force is mediated by force particles called gluons.) These two theories nowadays form what is labeled the "Standard Model." As it stands, the Standard Model is completely successful in its description of subatomic interactions—so successful, in fact, that experimental evidence that would contradict it has yet to be detected. But can this Standard Model actually be the final physical reality scientists are able to propose? Since the Standard Model is, in fact, composed of two distinct theories, why not try, some have asked, to unify them both into a unique theory?

Many attempts at a Grand Unified Theory (GUT) that would unite the electroweak force to the strong nuclear force have failed. In fact, a number of GUTs do exist but they are, to say the least, highly speculative. But what about gravity? Since the beginning of our discussion, gravity has been completely left out of the theoretical picture. The main reason for this situation is that no one, not even Einstein himself, was ever able to combine gravity (the theory of the very big) and quantum mechanics (the theory of the very small) into a single theory in which the principle of action at a distance would ultimately be replaced by a force particle called the graviton. Nevertheless, mathematical physicists continue to hope to find not only a GUT but a TOE a "Theory of Everything" as it is sometimes comically referred to, that would encompass the four natural forces. Many have put their faith of a TOE in superstring theories but, unfortunately for now, to no avail.

How could a Grand Unified Theory be corroborated experimentally? The energy required to verify any of the GUTs would call for a particle accelerator as big as the solar system! These machines have been at the forefront of particle physics since the 1930s. The latest prototype to be proposed would have been a gigantic 53-mile- (85 km) diameter machine built in Texas at a cost of at least five billion dollars. Called the Superconducting Supercollider (SSC), it would have been used to probe the structure of matter in regions a hundred thousand times smaller than the diameter of a proton. The U.S. government canceled the project a few years ago.

The ultimate theoretical unification of all natural forces will face in the coming years extraordinary conceptual and technical challenges but also—and perhaps essentially—tremendous pressure from society. How can we agree, people could argue, to build a machine costing billions of dollars when there is so much poverty and social injustice in the world? This question will have to be answered someday. One thing, however, is certain; science has always relied on observation and measurement to further its understanding of the macrocosm and the microcosm. The dialogue between theory and experiment is as vital as the two sides of a coin; one does not go without the other. During the twentieth century, the theories and the apparatus for practicing science have grown ever more complex and the fact is that if we do not build those gargantuan machines needed to probe nature, nothing excitingly new will come out from science. Weinberg said to a journalist a few years ago that today is a "terrible time for particle physics." Does this mean, therefore, that rather than reaching the edges of science during the next millennium, we will be facing the end of science instead? The Glashow-Weinberg-Salam theory of electroweak interaction opened new vistas in theoretical and experimental physics in the late 1960s and early 1970s. In the coming decades, however, if humankind wants to take the next step in its current journey to the last frontiers of science, it will have to answer ethical, scientific, and economic questions of the utmost importance.

JEAN-FRANÇOIS GAUVIN

Further Reading

Books

Galison, Peter. *How Experiments End.* Chicago: The University of Chicago Press, 1987.

Weinberg, Steven. *The First Three Minutes.* New York: Bantam, 1984.

Weinberg, Steven. *Dreams of a Final Theory.* New York: Pantheon Books, 1992.

Periodicals

Weinberg, Steven. "Unified Theories of Elementary-Particle Interaction." *Scientific American 231* 1 (July 1974): 50-9.

Weinberg, Steven. "A Unified Physics by 2050?" *Scientific American 281* 6 (December 1999): 68-75.

Internet

particleadventure.org

The International Geophysical Year (IGY), 1957-58

Overview

The International Geophysical Year (IGY) was prompted by a lack of concise data about Earth and its land, oceans, and atmosphere. Post-World War II technology had become available to launch a worldwide research effort to solve questions concerning the physical processes, patterns, and cycles of the forces of nature—the geophysics of the earth. During the IGY, global cooperative research expanded knowledge of Earth itself, the lower and upper atmosphere, the oceans, the polar regions, and near space, where the first satellites orbited successfully.

Background

Through the nineteenth century marine and polar exploration had highlighted the earth as a natural laboratory for research and brought new emphasis to the development of modern Earth sciences—geology, meteorology, and oceanography. The mystery of the polar regions stimulated the first cooperative, multinational scientific effort, the International Polar Year (1882-1883), but it was hardly international and consisted of failed expeditions. The world's scientific community tried again during the Depression but many countries could not honor their commitments. The end of World War II, however, brought a new technological age.

In 1950 a group of geophysicists met at the home of American physicist James Van Allen (1914-). Present were the world-renowned English geophysicist Sydney Chapman (1888-1979) and two other Americans, physicist and engineer Lloyd Berkner (1905-1967) and geophysicist S. Fred Singer (1924-). They discussed acquiring comprehensive upper-atmospheric data from the application of such post-war technology as radar, radiation sensors, sonar, rockets, high-performance aircraft, and the computer. Berkner suggested another polar year was needed; instead, they agreed the whole Earth should be studied. This meeting was the seed for the IGY. The idea was worked on for the next two years by a four-nation committee (the United States, Britain, France, and the Soviet Union) headed by Chapman and guided by the International Council of Scientific Unions.

Chapman came up with an appropriate new name, the International Geophysical Year. All IGY projects were to focus on "specific planetary problems of the earth" with emphasis on the Arctic and Antarctica, the equatorial regions, pole-to-pole profiles of atmospheric circulation, electrical layers of the ionosphere, the ocean depths, geodesy, seismic research, and geomagnetism. The ambitious showcase projects were to orbit the first artificial satellites and to explore Antarctica thoroughly. The massive organizational logistics would take some years. The period 1957 to 1958 (a year of maximum solar activity and several eclipses) was far enough ahead to make the preparations feasible. The ambitious research agenda actually required that the IGY be lengthened to 18 months (July 1, 1957, to December 31, 1958).

The near-space research was much anticipated. It was the beginning of the exploration of the "last frontier." Both the United States and the Soviet Union had taken advantage of German rocketry technology; few, however, knew that the Soviets were advanced in the design of large rocket engines. With the capture of the German V-2 rocket arsenal and under the leadership of Wernher von Braun (1912-1977), from 1946 to 1951 U.S. scientists such as Van Allen launched V-2s with payloads, bolstering American military rocket designs. By the late 1940s Singer suggested that it was possible to launch "artificial moons"—satellites—into orbit around Earth.

The satellite addition to the IGY agenda called for both American and Soviet satellites to be launched simultaneously to orbit the earth, recording and transmitting data on the geomagnetic field, the layers of the ionosphere, solar radiation, and cosmic rays. (The U.S. Navy and Air Force arrogantly competed with the Army for the launching of America's first satellite, believing the Soviets were far behind the West.) Satellite data would augment atmospheric data obtained using instrument packages on high-altitude balloons as well as small sounding rockets and earthbound sensors. Van Allen had designed a "rockoon," a balloon-born rocket, for the IGY to study cosmic rays, which were known to be the most energetic radiation in the universe and were an IGY emphasis.

Comprehensive worldwide ground observations of many geophysical phenomena were also an IGY goal. A study of the aurora (located about 70 miles [113 km] up) involving dense ground observations of the upper latitudes was organized. The airglow (low-energy ionization of oxygen) was to be studied using ground and rocket-mounted photometers. The obtainment of the first exact data on stratospheric ozone (tri-atomic oxygen) was slated. For the lower atmosphere, the troposphere, there would be extensive airborne and ground analyses of air masses (particularly the origins of Antarctic weather systems), the structure of the high-velocity jet stream, and wind patterns in the Pacific Ocean and southern hemisphere. And lower still, on Earth and its oceans, an extensive data base would be created: geologists and seismologists would gather geothermal, gravitational, and geodetic data to delineate Earth's structure (with special emphasis on earthquake data); and oceanographers would obtain data on deep-ocean basins, protein sources, ocean chemistry, currents and countercurrents, the effects of climate change, and sea-level variation.

The greatest earthbound challenge to the IGY was the study of the polar regions. Since the Arctic was a more familiar region, a greater emphasis was placed on Antarctica. While the Arctic is basically an ice cap over the ocean, Antarctica is a landmass covered with snow and ice. Was Antarctica an extension of South America, two islands, or a continent? The IGY was the time to find out. For the Arctic expedition, the United States and Soviet Union each committed to camping on a large Arctic ice floe, literally floating through the Arctic Ocean to collect all manner of data. Long-term planning for Antarctica called for at least 20 research stations, both on the coast and in the interior, with 179 people scheduled for research work. The actual numbers would involve 48 stations and 912 people, requiring huge airlift operations to build the stations and stockpile supplies as well as the establishment of routes in the interior. The military led the U.S. effort, called Operation Deep Freeze, beginning in 1954.

Impact

By the start of IGY research, 67 countries were involved with over 60,000 scientists and technicians. With data compiled at three data collection centers, the momentous project became a reality. A more sobering reality, however, was to become clear.

The first artificial satellites were launched—but with ironic results. On October 4, 1957, the West woke up to the Soviet's *Sputnik I* in orbit; the Soviets had not waited for a simultaneous launch. Though the Soviets tried to downplay the situation, the United States was angry and suspicious of a Communist agenda. The Soviet Union then launched its even larger *Sputnik II* on November 3 with a dog aboard. The U.S. political establishment, galled by its complacency over the technological superiority of the United States, drove American IGY satellite officials to launch more satellites than the Soviets. The U.S. Navy had won launching rights for its small *Vanguard* satellite atop a new rocket design, but technical problems ended in two humiliating failed launches. Von Braun stepped in with the simpler *Explorer I* satellite, which was essentially a small rocket fitted with instruments. Riding atop the big Army *Redstone* (to become the Jupiter C) on January 31, 1958, it became the first U.S. satellite in orbit. *Vanguard* was finally sent into orbit on March 17, 1958, followed about a week later by *Explorer III* (along with *Explorer IV* and *V* in the summer). On May 15 Russia put its last IGY satellite, *Sputnik III,* into orbit.

The irony of these events was that, in fact, U.S. technology was superior to the Soviet's, even though they had won the race into space. The Soviets actually had to use large rockets to get large satellites into orbit because their electronic instrumentation was antiquated, whereas the United States was far ahead in electronic miniaturization of instruments, affording the use of smaller satellites. *Explorer I* telemetered the data Van Allen used to identify the belts of charged particles (wrongly called radiation) trapped by Earth's magnetic field and named for him. Tiny but sophisticated, *Vanguard* amassed far more data than all the Russian satellites put together. Yet, the period was the Cold War and the Space Race was on. (The United States announced it would put a man on the moon by the end of the 1960s—a race it did win.) The dawn of the artificial satellite became controversial; were there national rights to orbital air space and what of non-satellite nations' rights? The satellite as both "spy in the sky" and weapon joined nuclear war as another threat.

Though it was never on the IGY agenda, a secret project took place. This project was "Argus," a U.S. military experiment in which three rocket-propelled nuclear electron shells were detonated in near space (above fallout levels) by the Navy in the South Atlantic in the summer of 1958. The resulting auroral displays and communication blackouts were recorded by unsuspecting IGY stations as astonishing anomalies—but many scientists had their suspicions. At a time when nuclear testing was under heated negotiations, Argus was a potential embarrassment, which explains why the U.S. government did not admit to the project until 1959. Nevertheless, nuclear testing was carried out during the IGY—137 detonations—with the purpose of studying fallout using different altitudes and wind conditions. The results showed that fallout fell more quickly than had naively been supposed, helping the later ban on atmospheric nuclear testing.

The call for a dense network of seismic stations revitalized defunct stations, and instruments that had been reserved for monitoring nuclear testing were put to a more humane purpose. At least 100 additional stations were activated, including in China, which, though withdrawing from the IGY, invited the IGY seismic coordinator, the Canadian geophysicist J. Tuzo Wilson (1908-1993), to survey their installations. Mother Nature cooperated admirably with IGY research. A huge solar flare in November 1958 provided the greatest number of magnetic storms ever recorded. Along with many volcanic eruptions during the IGY, in December 1957 one of the largest earthquakes on record occurred in central Asia, the Gobi-Altai Quake, which flattened a mountain range and drained a lake. Coupled with a great store of data on ocean-floor mountains and faults, the geological and seismic data helped formulate the new plate tectonic theory of Earth.

Some 20 nations operating on 86 vessels combed the oceans, mapping deep-water circulation. Sounding for the deepest ocean floors was tedious—the Pacific's was the deepest with trenches tens of thousands of feet deep. The

Marianas Trench was the deepest at over 36,000 feet (11,000 m). The idea of "ancient water" (water so deep it does not circulate) was found to be false. America's most active oceanographers—Columbus Iselin (1904-1971), Maurice Ewing (1906-1974), and Roger Revelle (1909-1991)—led important IGY projects in the study of (respectively): salinity, temperature, and currents; ocean basins and sediments; and the warming climate's relation to oceanic carbon dioxide balance. English oceanographer John Swallow (1923-) used his neutral buoyancy float in the Atlantic Gulf Stream, finding a countercurrent beneath it.

SATELLITES AND ANXIETY DURING THE IGY

Though reactions were mixed following Russia's unannounced launch of *Sputnik I* in October 1957, curiosity about the first "artificial moon" or satellite was tremendous. Cold War fears before the International Geophysical Year (IGY) had already put the anticipated launch of satellites into a negative perspective. Some armchair political analysts predicted world domination would be played out in the near space of Earth by those with the most sophisticated satellites.

But people worldwide stepped outside an hour or so after sunset—or before sunrise—to catch a glimpse of *Sputnik I,* the polished metal 185-pound (84 kg) sphere that streaked overhead every 95 minutes, lit like a star from the sun's reflection. American gazers watched in awe, but wondered if it carried spy cameras or maybe a bomb on board. The U.S. satellite effort was behind, and the anxiety of failure was heightened by the lack of technical cooperation from the Russians in trying to track *Sputnik.*

The U.S. visual tracking program for *Vanguard* was called Moonwatch (the Russians also had a program, but not as organized), under the supervision of the Smithsonian Astrophysical Observatory with volunteers at home and around the globe. Official observers were given a special short, rich-field telescope and tasked with scanning a set area overhead (along the meridian). Unofficial watchers gathered in large groups, using their own binoculars and telescopes. Moonwatch tracked *Sputnik I* and then the larger *Sputnik II,* when it was launched in early November 1957. They turned to tracking the successful U.S. *Explorer I* and *Vanguard I* in early 1958. These satellites were much smaller, so more challenging to observe. The original IGY satellites fell back to earth over time-spans measured in months—except tiny *Vanguard I,* which still orbits Earth.

Knowledge of the North and South Poles was advanced by the IGY. Innovations in resupplying research stations were made when two U.S. submarines surfaced through the ice at the North Pole to deliver supplies for the ice floe research stations. The measurement of ice areas in the Arctic, the continental mountains, and the Antarctic indicated that the world's ice was in a melting stage and thus that the climate was warming. The research conducted in Antarctica had been spectacular, including all manner of geophysical observations, such as climatic information reaching back thousands of years taken from drilled ice cores. The United States and Soviet Union volunteered for the most isolated observation stations, the South Pole and the Pole of Inaccessibility (the exact center of Antarctica), respectively. Traverses, overland scientific trips, involved the United States, Soviet Union, Britain and Commonwealth nations such as Australia and New Zealand, and France. Some of these traverses were not completed until 1959. The longest, most publicized traverse was the British Trans-Antarctic Commonwealth Traverse led by geologist Sir Vivian Fuchs (1908-), which crossed the whole of the landmass in 99 days, proving via detailed seismic and geodetic surveying that Antarctica was one continent, a great basin with central highlands. Antarctica was the real test of IGY cooperation. Historically, there had been many claims and disputes over the area. Worldwide cooperation, however, resulted in a seven-article scientific treaty, making Antarctica a natural laboratory to be shared by all nations—which it still is.

Another significant result of the IGY was that data on the influence of the Sun's radiation on the planets of the solar system fostered a new perspective on Earth's integration with the solar system and the cosmos beyond. The IGY data storehouse, moreover, was a springboard to the delineation of further specialized geophysical disciplines, such as geochemistry, tectonophysics, space physics, and planetology. As a humble epitaph to the IGY, Chapman said in a 1958 lecture, "The time will come when the International Geophysical Year will be viewed as an important but primitive contribution to the exploration of the cosmos." Instead, the IGY should be remembered as a great deal more.

WILLIAM J. MCPEAK

Further Reading

Antarctica in the International Geophysical Year. Geophysical Monographs No. 1, American Geophysical Union, 1956.

Bates, D. R. *The Earth and Its Atmosphere.* New York: Basic Books, 1957.

Berkner, Lloyd. *Rockets and Satellites.* New York: McGraw-Hill, 1958.

Berkner, Lloyd, and Hugh Odishaw, eds. *Science in Space.* New York: McGraw-Hill, 1961.

Chapman, Sydney. *IGY: Year of Discovery.* Ann Arbor: University of Michigan Press, 1959.

Odishaw, Hugh, ed. *Earth in Space.* New York: McGraw-Hill, 1962.

Siple, Paul. *90 Degrees South.*New York: Putnam & Sons, 1959.

Sullivan, Walter. *Assault on the Unknown.* New York: McGraw-Hill, 1961.

Wilson, J. Tuzo. *IGY, The Year of the New Moons.* New York: Alfred A. Knopf, 1961.

Biographical Sketches

Luis Walter Alvarez

1911-1988

American Physicist and Engineer

Luis W. Alvarez was perhaps the best-known Hispanic scientist of the twentieth century, excelling in experimental physics and engineering. In the 1940s he worked on the Manhattan Project designing the detonating device for the first atomic bomb. In 1968 he won the Nobel Prize in physics for work that included the discovery of resonance particles—subatomic particles that have very short lifetimes and that occur primarily in high energy nuclear collisions, and the design of an experimental bubble chamber to measure these particles. In the late 1970s he theorized that the dinosaurs became extinct as a result of a meteor crashing into Earth 65 million years ago.

Alvarez was born June 13, 1911, in San Francisco. His father was a physician who enjoyed research, and his mother taught school. As a child Luis often traveled to work with his father, and he liked to play with the equipment in the lab. He showed no interest in the medical aspects of his father's medical practice, but by the time he was 10 years old he was capable of using all of the small tools in the lab and wiring electrical circuits. In 1925 the family relocated to the Mayo Clinic in Rochester, Minnesota, in order for the elder Alvarez to take a position there. During high school summers, Luis was employed in the instrument shop at the clinic.

Initially, he enrolled at the University of Chicago in chemistry, but he discovered that he had a natural aptitude for physics and optics. He took 12 physics courses in 5 quarters and spent all of his spare time in the physics library, beginning a long and productive acquaintanceship with research and libraries. He earned both his bachelor's (1932) and doctoral degrees (1936) at the University of Chicago. He joined the faculty of the University of California—Berkeley in 1936 and taught there until his retirement in 1978.

Alvarez's first scientific discovery was the phenomenon of orbital electron capture. He discovered that in certain elements, the nucleus could capture an electron that is in an inner orbit. In 1939 Alvarez was able to measure the magnetic moment of the neutron.

During the World War II years, Alvarez, worked at the Massachusetts Institute of Technology (MIT) and was responsible for the design of three important radar systems: the microwave early warning system, the Eagle high altitude bombing system, and a landing radar system for civilian and military planes. After he left MIT he worked on the first nuclear reactor with Enrico Fermi (1901-1954). He was recruited by the United States government with the outbreak of World War II to work on the top-secret Manhattan Project. While in Los Alamos he developed the detonating device for the atomic bomb. He was on board the aircraft that dropped the atomic bomb over Hiroshima, Japan. Alvarez was shocked and sickened by what he saw, but he continued to support the United States in weapons research.

After the war ended Alvarez continued his work in high energy theoretical physics. He was involved in the design and construction of the first proton linear accelerator completed in 1947. This led to the development of the liquid bubble chamber, in which subatomic particles and their reactions are detected and photographed. The new particles produced by these accelerators and measured by the bubble chamber gave scientists exciting new information on the behavior and lifespan of subatomic particles, particles so small that they can be identified only by the tracks they leave. Alvarez in the course of this work was able to detect, record, and analyze 70 subatomic particles. He received the Nobel Prize in 1968 for this body of work.

In the late 1970s Alvarez, in collaboration with his son Walter, a respected geologist, announced a theory of the extinction of dinosaurs. They suggested that a giant asteroid, perhaps six miles (10 km) in diameter, crashed into Earth and created inhospitable conditions for the dinosaurs. The theory was controversial and many scientists rejected it. However, fifteen years later the theory is widely accepted.

LESLIE HUTCHINSON

Jocelyn Bell Burnell, the discoverer of the first pulsar. *(Bettman. Reproduced with permission.)*

Jocelyn Bell Burnell

1943-

English Astronomer

Jocelyn Bell Burnell discovered pulsars when she was a graduate student. Despite this accomplishment, Burnell's professional life as an astrophysicist has been erratic. A researcher, teacher, and administrator, Burnell's most significant position has been as an astronomical observer who provided new information regarding the life cycle of stars. Her work on pulsars allowed other astronomers to comprehend previously unknown aspects of celestial physics. Burnell's research also benefitted male colleagues who won the Nobel Prize for pulsar-related achievements.

Born on July 15, 1943, in Belfast, Northern Ireland, Burnell, the daughter of G. Philip and M. Allison Bell, became interested in science when her father took her to the Armagh Observatory. She graduated from the University of Glasgow in 1965, completing a doctorate in radio astronomy at Cambridge University three years later. Here, Burnell built a radio telescope for her graduate advisor, Anthony Hewish (1924-), to look for quasars. She operated the instrument and scrutinized data, noting possible quasars and identifying occurrences of human-generated interference. Burnell was using this instrument in 1967 when she recognized that another, distinctive type of signal had been recorded according to sidereal time, which is time based on Earth's daily cycle relative to the stars instead of to the Sun. Burnell began to focus on determining the source of these mysterious pulses that were rapidly being emitted in predictable patterns.

Burnell believed that the signals came from a star because they were sent consistently from the same place that she observed with the telescope at different times. In order to study the source's structure, Burnell made a high-speed recording and determined that the pulses occurred approximately every one and a third seconds, which was faster and more constant than other stars' signals. For several months she monitored this phenomenon, finding three additional radio signal sources that seemed to be pulsating—thus inspiring the term "pulsar," which was coined by a journalist.

Burnell and Hewish published articles about pulsars, declaring that these objects were located far outside the solar system but inside the galaxy. Using this information, other astronomers also found pulsars, increasing knowledge of their properties and of how they produce radio signals. Pulsars were defined as dense, burned-out neutron stars that rotate quickly and emit radio waves from magnetic and electric fields.

Burnell and Hewish received the Franklin Institute of Philadelphia's Albert A. Michelson Prize in 1973, but Burnell was excluded from

Nobel Prize honors for physics the next year. Hewish and Sir Martin Ryle (1918-1984) were awarded the Nobel for their contributions to radio astrophysics, particularly in Hewish's case his involvement in the discovery of pulsars. Burnell, however, was ignored as the person who originally found pulsars. Prominent astronomers argued that Burnell should have been included with the Nobel winners in recognition of her pioneering work.

Burnell also became frustrated when many male colleagues did not understand how her career was affected by her family. She moved frequently as her husband, Martin Burnell, accepted new employment, and her diabetic son required constant supervision. Working in part-time and temporary positions, Burnell focused on gamma-ray and x-ray astronomy using available resources. She held research and teaching positions at the University of Southampton in England, served as editor of *The Observatory,* and was affiliated with the Mullard Space Science Laboratory at London's University College.

During the 1980s, Burnell, as a senior research fellow at The Royal Observatory in Edinburgh, Scotland, directed the James Clerk Maxwell Telescope section, which oversaw British contributions to a multi-national telescope project located in Hawaii. She also managed EDISON, an infrared observatory. In 1991 Burnell received the physics department chair at Open University. She has published significant papers in leading scholarly journals and edited a book about infrared space observations. Burnell has received numerous honorary degrees and awards, including the Herschel Medal, presented to her by the Royal Astronomical Society of London in 1989.

ELIZABETH D. SCHAFER

William Maurice Ewing

1906-1974

American Geologist, Geophysicist and Physicist

William Maurice Ewing, known as "Doc" to his associates, was a pioneer in the development of techniques to study ocean basins, which at that time were still essentially a mystery. He also made fundamental contributions to the study of earthquake origins.

With a working scholarship, Ewing studied at the Rice Institute in Houston (1926-1931), earning a Ph.D. in physics. He had already become interested in ocean crust and sediment study during summer jobs prospecting for oil in the Gulf of Mexico with other future figures in marine geophysics—Albert D. Crary and H. N. Rutherford. In 1924 they were on a 22-foot (6.7 m) whaleboat throwing packages of explosive blasting gelatin over the side—a way of obtaining seismic reflection profiles of the continental shelf that was improved by Ewing. He taught physics in Pennsylvania, first at the University of Pittsburgh, then Lehigh, but was turning more toward marine geological research, conducting

PROMOTING GENDER EQUITY IN SCIENCE EDUCATION

During the last decades of the twentieth century there was an increasing call to promote gender equity in college and secondary school science and mathematics classrooms. Compared to their male counterparts, many studies continued to show that disproportionately fewer female students took elective science courses at the secondary school or college level. Consequently, disproportionately fewer women pursued university degrees or careers in mathematics, science and engineering.

Despite continual gains started in the early 1970s, near the end of the twentieth century women still comprised a relatively small percentage (15%) of the scientists and engineers in the United States and only about 10% of university-level science faculty. In 1991 the American Association of University Women began a series of studies and reports investigating the causes of the gender gap in science and science education. The AAUW studies set out, in part, to debunk disputed data that suggested a biological basis to the gender gap in scientific achievement. In 1996 the National Science Foundation also undertook studies of student proficiency in science. The NSF report of its findings ultimately suggested significant differences between male and female secondary school students with regard to achievement in, and attitudes toward, math and science regardless of race, socioeconomic status, and other demographic factors. The AAUW claimed that its studies substantiated perceptions that classroom level gender bias damaged female students' self-esteem, school achievement, and career aspirations. Among the factors consistently identified in the AAUW reports as perpetuating the gender bias was differential treatment by teachers that steered girls away from math and science courses.

William Maurice Ewing. *(AP/Wide World Photos. Reproduced with permission.)*

part-time work at the Woods Hole Oceanographic Institute in Cape Cod, Massachusetts. Because of his research in the travel of seismic waves through strata he was asked in 1934 by the U.S. Coast and Geodetic Survey and the American Geophysical Union to study the continental shelf of North America. At Woods Hole he took to the sea, developing methodical seismic survey procedures using seismic reflection and refraction techniques for the open ocean. From data gathered from the Atlantic basins, Mid-Atlantic Ridge, Norwegian, and Mediterranean bottom profiles, he showed the ocean floor crust to be much thinner than the continental crust. He also improved methods for studying gravitational anomalies over the ocean.

After World War II, during which he conducted important marine acoustical research for the U.S. Navy, Ewing accepted a position in the geology department at Columbia University in New York. At that time, money and land were donated to the school to start the prestigious Lamont Geological Laboratory (today the Lamont-Doherty Earth Observatory) in Palisades, New York. Ewing became its first director, remaining at the post for 25 years. One of his early wishes was that Lamont be an oceanographic institution as well. Ewing introduced the use of deep underwater photography and developed more seagoing seismic methods, involving earthquake waves to obtain oceanic crust topography. He introduced the use of sonar (propagated sound waves) for topographic soundings and the use of a submersible coring machine to take longer ocean sediment samples (the Ewing corer).

By 1952 Ewing was conducting extensive ocean sediment sampling. He directed research vessels to tow magnetometers to obtain measurements of changing magnetic fields. Sediment cores validated the phenomenon of trapped magnetic fields. During the International Geophysical Year (1957-1958) Ewing continued this research with his colleagues; they discovered a central rift valley associated with the Mid-Atlantic Ridge and confirmed their hypothesis that great mid-ocean ridges encircle Earth. Others took these leads to form the foundation for plate tectonic theory, which Ewing did not accept until relatively late. He did propose with other researchers, however, that earthquakes were associated with central oceanic rifts around the globe. He further hypothesized that sea-floor spreading was probably global and episodic.

Ewing devoted himself to data acquisition rather than theory—the first essential, the latter overabundant—yet his efforts were paramount to both. Many of his published works were collaborations with colleagues, such as *Propagation of Sound in the Ocean* (1948) and *The Floors of the Oceans* (1959). In 1972 he returned to his native Texas, joining the University of Texas at Galveston. He received many scientific awards. Ewing was honored on his death by the American Geophysical Union, when it instituted the Maurice Ewing Medal in 1974, honoring individuals for work in marine geophysics and technology.

WILLIAM J. MCPEAK

Richard Phillips Feynman
1918-1988
American Physicist

Richard Feynman's brilliance is legendary. His method was once described as: "You write down the problem. You think very hard. Then you write down the answer." Feynman contributed significantly to our understanding of quantum mechanics, which describes the behavior of minuscule particles. He gave science a way to visualize the motion of these particles and a deeper understanding of why and how they behave the way they do. For this work he shared a Nobel Prize in 1965. His own quirks and eccentricities, however, were almost as important as those of the particles he studied; Feynman's in-

Richard Feynman. *(The Library of Congress. Reproduced with permission.)*

subordinate spirit gave the physics world a lesson in the value of questioning authority.

Feynman was introduced to science at an early age, growing up in Far Rockaway, New York. In his books, he fondly remembers his father's explanations of the world—not always perfectly accurate, sometimes fantastic, but full of the spirit of curiosity and investigation. His family's support and encouragement of his scientific pursuits allowed him to attend the Massachusetts Institute of Technology (MIT), from which he graduated in 1939. From there he went on to Princeton University, where he earned his Ph.D.

In 1942, just after receiving his degree from Princeton, Feynman and his bride moved to Los Alamos, New Mexico, where he was one of the youngest physicists working on the Manhattan Project. His wife died of tuberculosis while they were there; when the project was disbanded after the war, Feynman relocated to Cornell University. There he began the work that would later win him—along with Julian Schwinger (1918-1994) and Shin'ichiro Tomonaga (1906-1979)—a Nobel Prize.

Feynman's work at Cornell was crucial to the understanding of how particles interact. He calculated the probability of each path a particle could take between two points, then figured out the summation of all these possible paths—a path integral. This was mathematically identical to wave function but related to the particles themselves, rather than their function within an electromagnetic field. Once a particle's path integral is determined, a Feynman diagram can be produced. This is essentially a space-time graph with the x-axis representing particles in space and the y-axis representing time. If the diagram is covered with a sheet of paper that is slowly pulled upwards, exposing one instant of the y-axis at a time, the particle's movement can be viewed. The Feynman diagram was the first visual model of quantum motion; Schwinger called it "bringing computation to the masses."

After leaving Cornell, Feynman carried on his theoretical physics work at Caltech. He also bolstered his quirky reputation by teaching introductory physics courses that drove away some unsuspecting freshmen (but were heavily attended by graduate students and professors), practiced his drumming and bongo-playing, frequented topless bars, dated heavily, married and divorced twice, and acted in a student production of *South Pacific*.

Feynman's diagnosis of abdominal cancer in 1979 slowed down his irrepressible nature, but by then he was already a celebrity. In 1986 during Congressional hearings on the *Challenger* disaster, Feynman gave a simple but dramatic demonstration of the tragedy's cause. He put a clamp around some of the space shuttle's gasket material then dropped it into the ice water he had been drinking, showing that the material lost its resiliency under freezing conditions. This was the peak of Feynman's fame; he died two years later. His books, however, are still widely read and enjoyed and his contributions to the understanding of quantum particles are indispensable to anyone in the physics field.

JESSICA BRYN HENIG

Murray Gell-Mann
1929-

American Physicist

Following World War II, a new level of governmental support for high energy physics research in the United States and Europe led to the discovery of numerous new elementary particles, most being substantially heavier than the electron and many heavier than the proton. Murray Gell-Mann played a key role in bringing order to this "zoo" of newly discovered particles, first through the introduction of a new quantum

Murray Gell-Mann. *(AP/Wide World Photos. Reproduced with permission.)*

number called "strangeness" and then through a unification named (partly in jest) the "eightfold way," which led to the "quark" model of the more massive subatomic particles. In his later career, Gell-Mann played an essential role in interesting physicists in a new interdisciplinary field devoted to the study of complex phenomena and in setting up the Santa Fe Institute as a center for research in this area.

Murray Gell-Mann was the son of Arthur Gell-Mann, the owner of an unsuccessful language school and a man learned in several scientific disciplines. A polymath and child prodigy, Gell-Mann entered Yale University at the age of 15 and received his Ph.D. in physics from the Massachusetts Institute of Technology at the age of 21. As Gell-Mann began his scientific career, the generation of physicists who had built the atomic bomb under the Manhattan Project were returning to academic research or working in new national laboratories. A major thrust of the research was understanding the forces at work within the atomic nucleus. Particle accelerators were built to create highly energetic beams of protons or electrons that could create substantial numbers of unstable particles when they collided with a target. These accelerators complemented a very limited number of observations made on cosmic rays, which are very energetic particles that collide with Earth's atmosphere.

Gell-Mann's first major contribution to physics involved the identification of a new quantum number, strangeness, needed to explain the unusual stability of newly discovered particles somewhat heavier than a proton. Gell-Mann correctly reasoned that strangeness was a property conserved under the strong nuclear force but not concerned with the weak nuclear force, which operates on a much slower time scale. Gell-Mann's main contribution to physics, however, is the classification of the heavier baryons (particles more massive than the proton) and the lower-mass meson according to an abstract set of symmetries described by Lie Groups, mathematical structures introduced almost a century earlier by the Norwegian mathematician, Marius Sophus Lie (1842-1899). Application of the Lie Group designated SU(3) suggested that elementary particles would occur in groups of eight or ten and made possible the prediction of new particles, particularly the omega-minus, which was discovered in 1964. The SU(3) symmetry also allowed for a set of three fundamental particles, particles Gell-Mann named "quarks" after a nonsensical line in the novel *Ulysses* by James Joyce. Although initially unpopular, the quark model has become an accepted part of quantum chromodynamics, the theory of the strong nuclear force.

Gell-Mann received the Nobel Prize for physics in 1969. Unlike most recent Nobel awards, this one was not shared with other scientists, thus recognizing Gell-Mann's overall career rather than any one particular achievement. He has maintained a broad range of intellectual interests, including biology and languages, and as many physicists became interested in the nature of complex and nonlinear phenomena in the 1980s he assumed a leadership role. In 1983 Gell-Mann was a featured speaker at a meeting intended to set the agenda for a new research institution, the Santa Fe Institute, which was designed to bring together researchers from a broad range of the social and natural sciences to study the common features of "complex systems." He continues to be a strong advocate ofinterdisciplinary research.

DONALD R. FRANCESCHETTI

Stephen William Hawking
1942-
English Theoretical Physicist

Stephen Hawking is arguably the most famous scientist of the second half of the twentieth century. His work on black holes, his best-selling book *A Brief History of Time*, and his long struggle with Lou Gehrig's disease have made him a celebrity.

Hawking was trained as a physicist at Oxford University. He wanted to study mathematics, but his father wanted him to study medicine. Since University College didn't offer mathematics, he ended up in physics. After completing his undergraduate degree he went to Cambridge University for his Ph.D., where he was supervised by Dennis Sciama. He remained at Cambridge, where he is now in the department of Applied Mathematics and Theoretical Physics, and is the Lucasian Professor of Mathematics, a position also held by Sir Isaac Newton (1642-1727).

Hawking's greatest scientific achievements came in the 1960s and 1970s with his advancements of Albert Einstein's (1879-1955) general relativity as it relates to black holes and the Big Bang theory. He worked out many of the characteristics of black holes, such as shape and temperature. His greatest discovery is that black holes radiate tiny amounts of heat, meaning they lose mass and energy. He showed that black holes will eventually shrink down to sub-molecular sizes and evaporate into nothingness. In his honor this radiation is now called Hawking radiation. This discovery solved many problems related to the evolution of black holes.

But this evaporation also created new problems. What exists at the center of the black hole seems to be a "singularity" where space and time cease to have any normal meaning. The English mathematician Sir Roger Penrose (1931-) and Hawking also showed that rewinding the history of the universe must lead one to conclude that the universe began as a singularity. This singularity could have started the Big Bang, creating our entire universe. This and other findings showed that Einstein's theory of general relativity and quantum mechanics would have to be linked together into a "theory of everything." This search has occupied Hawking and many other scientists since then and has led in part to the development of string theory.

Hawking is a victim of amyotrophic lateral sclerosis, popularly known as Lou Gehrig's disease. The disease robs victims of control of their muscles. Hawking was diagnosed with the illness in 1963. He was given only a few years to live but has since defied that prediction. He began using a wheelchair as he lost many motor skills. Then, during a 1985 bout with pneumonia he had to undergo a tracheotomy that left

Stephen Hawking. *(AP/Wide World Photos. Reproduced with permission.)*

him almost without speech. He now speaks in a faint, slurred mumble that only those who are with him regularly can understand. Soon after that operation he began to use one of his trademarks—a speech synthesis program that speaks in a robotic, monotone voice. With the two fingers he can still control he slowly and laboriously pecks out text for the machine to read. He claims he can communicate more effectively with this system than with his voice. Today he can use the Internet, make cellular phone calls, and control the lights and doors in his house from his wheelchair.

His 1988 book *A Brief History of Time* became a surprise best-seller, staying on the New York Times Bestseller list for 100 weeks and selling over one million hardcover copies. It was also adapted into a television series called *Stephen Hawking's Universe.* He followed that up with *Black Holes and Baby Universes and Other Essays* in 1993. He has also written, edited, and contributed to many academic texts and popular books.

Hawking has three children—Robert, Lucy, and Timothy—by his first wife, Jane. Hawking and his wife endured a bitter divorce in 1991 and relations with her and his children remain strained. In 1995 he married his second wife, Elaine Mason.

PHILIP DOWNEY

Harry Hammond Hess
1906-1969
American Geologist

Born in New York in 1906, Harry Hammond Hess was considered a "very promising student" even in his high school years. He had no difficulty gaining acceptance into the electrical engineering major group at Yale University in New Haven, Connecticut. However, after two years, he elected to change his major to geology and received his B.S. in that field in 1927.

Prior to his graduate studies at Princeton University, Hess spent two years in south central Africa (formerly Rhodesia) working as an exploration geologist. He returned to America and earned his doctorate at Princeton. This led to a year of teaching at Rutgers University and another year as a research associate at the Geophysical Laboratory in Washington, D.C. He then rejoined the faculty at Princeton, where he remained for the balance of his career.

His academic career was interrupted by World War II, during which he enlisted in the U.S. Navy and attained the rank of captain. Even while he was engaged in the conflict in the Marianas, Leyte, Linguayan, and Iwo Jima, Captain Hess (with the cooperation of several crew members aboard the U.S.S. *Cape Johnson*) conducted echo-sounding surveys while en route to various battles. These soundings gave Hess the data he needed to form ocean floor profiles across the North Pacific Ocean, resulting in his finding a series of flat-topped volcanoes beneath the surface. His loyalty to Princeton influenced his calling the volcanoes "guyots" after the university Geology Building.

Later, in 1957, Hess supported the Mohole Project (initiated by Walter Munk), to drill through the oceanic crust, deep into the earth's mantle. The pair arranged support and settled on the initial venture off Guadalupe Island, Mexico. By the time the second phase of the project was cancelled by Congress in 1966, many of the problems of ocean drilling had been solved, making the task much easier for subsequent scientific ocean drilling teams.

In 1959 Hess published an informal manuscript that was widely circulated and introduced a groundbreaking process that would later be known as seafloor spreading. By 1962 these ideas were formalized in a paper that became one of the most important documents in the science of plate tectonics—*History of Ocean Basins.*

Its importance was based on the question geologists had pondered for many years: If the oceans were at least 4 billion years old, why was there so little sediment found on the floors? Hess hypothesized that the sediment had been accumulating for about 300 million years. He reasoned that this would be the approximate time needed for the floor of the ocean to move from the ridge crest to the trenches.

Hess's ideas were particularly interesting to followers of an earlier geologist, Alfred Lothar Wegener (1880-1930), who proposed that pieces of the original continents may have broken up and drifted apart to form the areas we know today as continental formations. Several new theories were developed but, finally, improved seismic data confirmed that portions of the ocean's crust were sinking into trenches, lending credence to prove Hess's hypothesis, which originally was based on intuitive geologic reasoning.

Hess continued to head the Geology Department at Princeton until his death in 1969. Unlike his predecessor, Wegener, he lived to see his work accepted, confirmed, and expanded. In recognition of his contributions to global knowledge and better understanding of our planet, in 1962 President John F. Kennedy appointed Hess to serve as Chairman of the Space Science Board of the National Academy of Sciences. From this position he played a significant part in the design and development of the United States space program.

BROOK ELLEN HALL

Robert Hofstadter
1915-1990
American Physicist

Corecipient of the 1961 Nobel Prize in physics, Robert Hofstadter discovered that protons and neutrons are not indivisible particles, as was previously believed, but are complex components of the atom. He also developed the sodium iodide-thallium scintillator, a device for measuring radiation still used in particle accelerators.

Hofstadter was born in New York City on February 5, 1915, the third of four children born to Louis (a salesman) and Henrietta Koenigsberg Hofstadter. He went to public schools, then enrolled at the City College of New York (now the City University of New York). Hofstadter graduated magna cum laude with a B.S. in physics in 1935, and earned the Kenyon Prize for exceptional achievement in mathematics and physics.

Robert Hofstadter. *(The Library of Congress. Reproduced with permission.)*

He later earned M.A. and Ph.D. degrees from Princeton University, completing his work in 1938 but remaining at Princeton for postdoctoral studies under a Proctor fellowship. During this time, Hofstadter concerned himself with photoconductivity in crystals. In 1939 he took a position as instructor in physics at the University of Pennsylvania, then became Harrison research fellow before returning to the City College of New York, this time as an instructor in physics.

After the beginning of World War II, Hofstadter went to work as a research assistant at the National Bureau of Standards (NBS) in Washington, D.C. There he assisted American physicist James Van Allen in creating a proximity fuse, used to detonate a bomb just before it reached its target. He remained at NBS for a year, then went to Norden Laboratories, home of the Norden bombsight, in New York City. In 1942 he married Nancy Givan, and they had three children, including son Douglas, author of several science-related books, including *Gödel, Escher, and Bach.*)

After the war was over, Hofstadter went back to Princeton as an assistant professor of physics, and began work on problems in radiation detection. In 1948 he developed the sodium iodide-thallium scintillator, which emitted light whenever radiation passed through it. The intensity of the light varied with the degree of radiation, and the device proved so effective that it remains in use in particle accelerators.

In 1950 Hofstadter became associate professor of physics at Stanford University, where he would remain for the rest of his career. There he began work on the questions that would lead to his Nobel Prize a decade later. The first issue he approached was simply how to look at atoms. Because they are smaller than light rays, atoms cannot be seen even under the most high-powered microscopes. X rays, gamma rays, and high-speed electrons, however, all have wavelengths smaller than that of light. Hofstadter soon discovered that the high-energy electrons in Stanford's linear accelerator (linac) made it possible to view atomic particles.

Thus equipped, he made a number of significant discoveries. Although the nuclei of atoms varied in size, their density was nearly uniform. As he began to look deeper into the structure of the atom, Hofstadter found that protons and neutrons were not solid and indivisible like little ball bearings, but detailed in structure—more like cells. Each proton or neutron, he discovered, contained a dense, positively charged core, surrounded by two shells of mesonic material—one of which was negatively charged in the neutron.

Promoted to a full professorship in 1954, Hofstadter became Max H. Stein professor of physics at Stanford in 1971. He retired in 1985, after which he was made emeritus professor of physics. In addition to the Nobel Prize, which he shared with Rudolf Mössbauer (1929-), Hofstadter earned the Roentgen Medal (1985), the National Science Medal (1986), and numerous other awards. He died of heart disease on November 17, 1990.

JUDSON KNIGHT

Edward Norton Lorenz
1917-
American Mathematician and Meteorologist

Edward Norton Lorenz is known for his pioneering work on chaos theory as a way of explaining atmospheric science. He established the theoretical basis of weather and climate predictability as well as the tools used for computer-aided atmospheric physics and meteorology. His "butterfly theory" influenced a wide-range of basic sciences and has contributed to significant changes in the ways that scientists view nature, the universe, and the future direction of mathematics and science.

Lorenz was born May 23, 1917, in West Hartford, Connecticut. He graduated from Dartmouth College with an undergraduate degree in mathematics (1938) and later earned a master's from Harvard (1940) and a Ph.D. from the Massachusetts Institute of Technology (1948). He taught mathematics at Harvard during the 1941-42 academic year and served as a weather forecaster with the U.S. Army Corp from 1942-46.

In 1946 Lorenz began his meteorology career at the Massachusetts Institute of Technology (MIT) in the department of meteorology. He was influenced by MIT meteorologists Carl-Gustav Rossby (1898-1957) and Vilhelm Bjerknes (1862-1951), both highly regarded weather scientists. Their work emphasized a physical-dynamical interpretation of atmospheric movement and provided the first systematic model of theoretical and practical meteorology. This system of weather forecasting was used extensively by the U.S. Weather Bureau and the military during World War II.

While a visiting scientist at Harvard (1950), his research focused on the idea of atmospheric energy balance. Later, this research on the theoretical basis of dynamic and statistical systems became part of a joint research effort between UCLA and MIT in statistical weather prediction ("MIT Statistical Forecasting Project"). At the conclusion of the forecasting project in 1959, Lorenz was promoted to associate professor, and his research began to focus on weather prediction using computer modeling.

Lorenz saw the chance to combine mathematics, the recently developed computer, and meteorology and set out to design a mathematical model of the weather—a set of differential equations that represented changes in temperature, pressure, wind velocity, etc. He developed a model containing 12 equations and in 1961 began a project to simulate weather patterns using a computer. Lacking much memory, the computer was unable to create complex patterns, but it was able to show the interaction between major meteorological events such as tornadoes, hurricanes, easterlies, and westerlies. Lorenz began to see patterns emerge and was able to predict with some degree of accuracy what weather pattern would occur.

While carrying out an experiment, Lorenz made an accidental discovery that eventually became known as "deterministic chaos" or the "Butterfly Effect." His experiment began using an identical set of numbers, but when the data was inputted for the second time, the new run was not identical to the first as expected. Although initially the same, the data slowly changed after many computations and became an entirely different set of numbers representing a totally different weather system. Lorenz realized this occurred because of minute errors that had been unintentionally programmed into the computer when the numbers were rounded off. Therefore, Lorenz had unexpectedly discovered that no matter how much information he gathered, his weather prediction could be wrong.

In the early 1970s he wrote a paper examining the hypothesis that "the flap of a butterfly's wings in Brazil can set off a tornado in Texas." This paper, presented at the convention of the Global Atmospheric Research Program at MIT in 1972, suggested that some very minor, small undetectable influence could lead to something detectable after a sufficient period of time. With this *Butterfly Effect*, Lorenz concluded that weather is generally unpredictable beyond a 14-day period and even then subject to change.

In 1983 Lorenz and colleague Henry Stommel received the Crafoord Prize from the Royal Academy of Sciences in Sweden. In 1991 Lorenz received the Kyoto Prize and in 1992 he was the first recipient of the American Geophysical Union's Revelle Medal for achievements in understanding atmospheric process and determining climate.

LESLIE HUTCHINSON

Drummond Hoyle Matthews

1931-1997

English Geologist and Geophysicist

Together with his colleague and former student Frederick J. Vine (1939-1988), Drummond H. Matthews helped establish the theory of plate tectonics in the 1960s. Prior to that time, geologists and geophysicists had remained skeptical about claims that Earth's crust was divided into shifting plates. The findings of Matthews and Vine revolutionized thinking on the subject, and plate tectonics theory later proved useful in areas such as oil exploration and the assessment of volcanic risk.

Born in 1931, Matthews was educated at King's College, Cambridge, and in 1955 went to work as a geologist on the Falkland Islands Dependencies Survey. In 1957 he returned to Cambridge, where his students included a promising doctoral candidate named Frederick J. Vine.

While taking part in a 1962 mapping expedition aboard the H.M.S. *Owen* in the northwestern Indian Ocean, Matthews noted a pattern of magnetic bands or stripes as much as 20 miles or 30 kilometers wide. Such magnetic stripes had been found on the floors of other oceans, but Matthews's was the first such readings in that particular region.

Upon his return to Cambridge, Matthews—along with Vine— examined these findings and discovered curious reversals in polarity, or the direction of the magnetic field, among the stripes. In particular, the two men noted that the polarity of the stripes varied symmetrically: for instance, if the third stripe to the west of a ridge was a wide one with a north magnetic pole near the north geographic pole, the same was true of the third stripe to the east of a ridge. Matthews and Vine presented their findings in the September 1963 issue of *Nature*.

Half a century earlier, Alfred Wegener (1880-1930) had put forward the theory of continental drift, but his ideas had come under attack in the years preceding the publication of Matthews's and Vine's data. Now the two offered compelling evidence to support not only continental drift, but plate tectonics theory, which began to gain wide acceptance following the presentation of their findings at the December 1966 meeting of the Geological Society of America in San Francisco. This led to a significant change in geological thinking, and ultimately plate tectonics theory found practical application in a wide variety of areas.

Matthews was appointed assistant director of research in the department of geophysics at Cambridge in 1966, and became reader (instructor) in geology in 1971. During the 1970s he published a number of influential papers on the subject of marine geophysics. He was elected Fellow of the Royal Society in 1974.

Matthews later left his position at Cambridge but continued his research, focusing on the continental crust. He became director of the British Institutions Reflections Profiling Syndicate, or BIRPS, and retired in 1990. Matthews died in July 1997.

JUDSON KNIGHT

Jan Hendrik Oort

1900-1992

Dutch Astronomer

Jan Oort is popularly known for his description of the Oort Cloud that now bears his name. This cloud is located far outside of our solar system and is the source of the comets that orbit our sun. However, among professional astronomers, Oort is best known for his many discoveries concerning the structure of the Milky Way and other galaxies.

Oort was born in the Netherlands in 1900. During his lifetime he saw our view of the cosmos transformed from a single galaxy to an entire universe filled with galaxies. Many of his discoveries drastically changed our perceptions of our home galaxy, the Milky Way.

Oort's first major discovery came in 1927. His analysis of the motions of stars in our galaxy provided support for Bertil Lindblad's 1926 theory that our galaxy was rotating. Oort showed that the motion of different streams of stars could be explained easily if the galaxy was rotating. He also showed that it would take our Sun 200 million years to orbit the center. This was a revolutionary discovery, since it had only been recently discovered that the Milky Way did not comprise the entire universe. Oort's discovery was also one of the first demonstrations that gravity didn't act only in our solar system or in nearby binary solar systems, but that the force was felt throughout the entire galaxy. Oort also calculated that most of the galaxy's mass would be concentrated toward its center.

Oort began his studies of comets in the late 1940s when he began supervising a Ph.D. student whose advisor had died. At that time many researchers were grappling with the origin of comets. Measurements of a typical comet's path show that it orbits our Sun and is not merely passing through our solar system a single time. But they seemed to originate from far outside the solar system. Secondly, comets seem to enter our solar system from all directions. Thirdly, comets lose a large percentage of their mass each time they pass near the Sun, which heats and boils them away. Therefore comets have a relatively short lifetime. Given that we still see them today, how is the supply of comets being replenished?

Oort proposed that comets are debris left over from the initial formation of our solar system. Icy clouds of dust and water that were over 30 times the distance of the Sun to the Earth (1 astronomical unit or AU) were not absorbed into the planets. Oort suggested that these clouds of debris merged to form comets and drifted away until they were 30,000-100,000 AU away from the Sun. Chance collisions and gravitational interactions with passing stars would occasionally nudge a comet into motion towards our Sun. This cloud of debris,

Jan Oort proposed that comets exist in an enormous cluster, called the Oort cloud, at the edge of the solar system. *(Dr. Seth Shostak/Science Photo Library/Photo Researchers. Reproduced with permission.)*

which surrounds our solar system like a giant ring, is now named the Oort Cloud in his honor.

Oort was also an active leader of the International Astronomical Union. He served as General Secretary and President. As General Secretary during and after World War II Oort was often in the position of mediator between scientists whose countries had formerly been enemies. He was also one of the guiding forces behind the European Southern Observatory. He first attempted organizing this collaboration between many of the nations of Europe in 1954. By 1962 these nations had agreed to begin construction of an optical telescope in the Southern Hemisphere.

Oort was one of the first astronomers to recognize the effect of the interstellar medium on the structure of galaxies. He was also instrumental in advancing radio astronomy in the Netherlands. He used radio measurements to confirm our galaxy's spiral structure, map its center, and track the motion of vast clouds of gas inside and outside our galaxy. After his retirement from active research in 1970 he continued to publish many important review papers on the structure of galaxies, galactic nuclei, and the organization of superclusters of galaxies throughout the universe.

PHILIP DOWNEY

Sir Martin Ryle
1918-1984
English Radio Astronomer

Martin Ryle did not establish radio astronomy, but he made it practical by overcoming difficulties astronomers encountered when trying to study stars by analyzing their radio emissions. Due to the long wavelengths of radio waves, it was thought that a radio telescope would have to be impossibly large and too expensive to build. It was Ryle's achievement to develop a giant "phantom" telescope, actually a series of measurements made from smaller telescopes, linked by computer. He shared the 1974 Nobel Prize in physics with Antony Hewish (1924-).

Born on September 27, 1918, in Brighton, England, Ryle was the son of physician John A. Ryle and Miriam Scully Ryle. His was an exceedingly distinguished family: not only was his father the director of the Institute of Social Medicine at Oxford University, as well as Oxford's first professor of social medicine, his uncle Gilbert Ryle was a well-known philosopher. Ryle attended Bradfield College and Christ Church at Oxford, earning first-class honors in the latter's school of natural sciences.

Just after Ryle graduated from Christ Church in 1939, World War II broke out, and he went to work for the British government in the Telecommunications Research Establishment. (The latter was later renamed the Royal Radar Establishment.) While there, he met his fellow future Nobel laureate Antony Hewish, and worked on developing countermeasures against German radar.

In 1947 Ryle married Ella Rowena Palmer, a nurse and psychotherapist. By that time the war had ended, and Ryle had returned to Cambridge, where he had worked briefly before it began. It was there that he began his first important work in radio astronomy. The latter had been developed during the 1930s by Karl Guthe Jansky (1905-1950), an American engineer who noted that certain stars emitted very short radio waves. These, Jansky postulated, might carry useful information in the same way that light waves do, but the idea was a controversial one.

The first great contribution to the field of radio astronomy made by Ryle was the development of a map showing the radio-emitting sources in the sky. His first map, made in 1950, identified 50 of these, but a second map made five years later showed nearly 2,000.

These investigations led Ryle face-to-face with the problem that would win him the Nobel more than two decades later. Because radio waves have a much longer wavelength than light waves, a radio telescope must be much larger than a light telescope—so big, in fact, that it would be too costly to build. Ryle attacked this problem by envisioning a phantom giant telescope; then he designed a number of telescope parts that could be moved to different places along the imaginary apparatus. He took readings from various spots, and tied these together using a master computer. The latter then generated a picture of the type that might have been obtained by a single large radio telescope.

In addition to the Nobel, Ryle earned the Hughes Medal from the Royal Society in 1954, the Gold Medal of the Royal Astronomical Society in 1964, the Henry Draper Medal of the United States National Academy of Sciences in 1965, and the Royal Medal of the Royal Society in 1973. He was knighted in 1966. Ryle devoted the last decade of his life to developing wind power as a renewable source of energy. He retired in 1982, and died of lung cancer in Cambridge on October 16, 1984.

JUDSON KNIGHT

Carl Edward Sagan

1934-1996

American Astronomer and Biologist

For most teenagers and adults living in the 1970s, the idea of space exploration was synonymous with the name Carl Sagan. His enormously popular television series *Cosmos* and his many appearances on the "Tonight" show made him a recognizable figure wherever he went. Corey S. Powell, staff writer for the *Scientific American,* estimated that *Cosmos* reached an audience of over 500 million people.

Sagan's parents, Samuel and Rachel Sagan, were living in the Bensonhurst area of Brooklyn, New York, when their son, Carl, was born in 1934. Although the elder Sagan hoped that his son would follow him in the coat business, young Carl spent his early years immersed in reading science fiction and related topics: astronomy, chemistry, physics, and the likelihood of space travel.

The family moved to Rahway, New Jersey, just before Sagan entered high school, but he elected to attend college at the University of Chicago and received both his undergraduate and graduate degrees there. Part of his Ph.D. dissertation was a "greenhouse model" for Venus that indicated the possibility of sustaining human life there as well as on the Moon. This theme would run throughout his sparkling career; he

Carl Sagan. *(The Library of Congress. Reproduced with permission.)*

argued constantly for funding and academic research for any signs of extraterrestrial life.

His first book, *The Cosmic Connection,* enjoyed wild success. He wrote about exceedingly complex subject matter and made it both understandable and intriguing to people in all walks of life. Later, in 1985, he published two books: *Nuclear Winter* and *Contact.* The latter was made into a successful motion picture in 1998.

Although he was well-known in scientific circles for his valuable contributions to the U.S. space program, outsiders were not often aware of the specific projects in which he was deeply involved. While NASA was working on *Pioneer 10* and *11* spacecraft intended to reach Jupiter and Saturn, the project engineers called on Sagan to devise a series of messages that might be picked up by other possible civilizations as the vehicles drifted into other galaxies over millions of years. The messages Sagan devised took the form of a plaque that showed a galaxy map and the location of Earth on the map as well as drawings of a man and woman.

When NASA went on to develop the two *Voyager* interstellar ships, Sagan led a group that updated and improved the original messages carried in the *Pioneer* ships. Along with written signals, the new package included photographs of our planet and some of the people who live here, samples of our music, and multi-language greetings for those who might encounter *Voyager* over the eons to come.

Carl Sagan's intellectual gifts ranged far beyond astronomy and astrophysics. He was also a competent biologist (his first wife, Lynn Margulis (1938-), is eminent in this field), a Pulitzer prize-winning, best-selling author, and a compelling personality in the world of television, where the competition was not only heavy but constant. He was also a charismatic professor at Cornell University, where he was Associate Director of the Center for Radio Physics and Space Research. (He had worked earlier at the Smithsonian Astrophysical Observatory.) Until the day he died, he was David Duncan Professor of Astronomy and Space Sciences and director of the Laboratory for Planetary Studies, both at Cornell. During his tenure at the university, he developed what was then an entirely new field: exobiology—the study of possible alien biochemistry and life forms. Sagan died in 1996.

BROOK ELLEN HALL

Abdus Salam

1926-

Pakistani Physicist

For his work on the relationship between the electromagnetic and weak forces—two of the four known basic forces governing nature—Abdus Salam shared the 1979 Nobel Prize in physics with Sheldon Glashow and Steven Weinberg. During the 1960s, all three men, working independently of one another, developed similar theories. Salam's, published in 1968, showed that the electromagnetic and weak forces might be two parts of a single, more fundamental force, the electroweak force. Salam has been active in the movement to improve scientific education in the Third World and to this end helped establish the International Center for Theoretical Physics in Trieste, Italy, in 1964.

Born on January 29, 1926, in the rural Pakistani village of Jhang, Salam was born to Muhammad Hussain and Hajira Salam. His father worked for the local department of education. Salam entered the Government College at Punjab University in Lahore when he was 16 years old, and by the age of 20 had earned his master's degree in mathematics. He then received a scholarship to St. John's College at Cambridge University, where he earned a bachelor's degree in mathematics and physics in 1949.

Abdus Salam. *(The Bettmann Archive/Newsphotos. Reproduced with permission.)*

After several more years at Cambridge, where he also received his Ph.D. in theoretical physics (in 1952), Salam felt it was time to give his talents to his homeland. Therefore, he returned to Pakistan and accepted a joint appointment as professor of mathematics at the Government College of Lahore and chairman of the department of mathematics at Punjab University. Unfortunately, he had no opportunity to conduct research and found, as he later recalled, that he was "the only practicing theoretical physicist in the entire nation." Moreover, university administrators expected him to "look after" the school's soccer team. Therefore, Salam returned to Cambridge, where he taught mathematics for two years. In 1957 he accepted an appointment at the Imperial College of Science and Technology in London as professor of theoretical physics, a position he has held ever since.

By the mid-1950s, scientists had recognized that there were four fundamental physical forces: gravitational, electromagnetic, strong, and weak. Many scientists posited that these forces were manifestations of a single basic force, but the unity of these forces could only be demonstrated under high-energy conditions—in cosmic radiation, for instance, or in the most powerful particle accelerators. Salam developed a mathematical theory to unify the electromagnetic and weak forces; meanwhile, Steven Weinberg and Sheldon Glashow, working independently of Salam and one another, arrived at similar results.

The synchronicity of these findings seemed to confirm the theory, and further confirmation appeared in 1973 with the discovery of weak "neutral currents," which had been predicted by both Salam and Weinberg. The electroweak theories had also posited the existence of certain force-carrying particles, a fact verified by a series of experiments in 1983.

In addition to the Nobel Prize, Salam has received numerous other awards and has sat on a wide array of scientific councils both for the government of Pakistan and for the United Nations. He has also devoted much of his time to creating opportunities for other scientists from the Third World. Remembering his own frustrating experience in Pakistan in the early 1950s, he helped establish the International Center for Theoretical Physics as a place where scientists from developing nations can receive support, encouragement, and instruction.

JUDSON KNIGHT

Maarten Schmidt
1929-
Dutch-American Astronomer

In 1963 Dutch-American astronomer Maarten Schmidt discovered what are now called quasars, or quasi-stellar radio sources (QSOs). Having found what appeared to be invisible stars that emitted radio waves, he observed that the objects' light spectrum was like nothing ever seen. These source of ultra-intense radio emissions remain somewhat mysterious; they are no bigger than stars, yet they produce more energy than an entire galaxy. In later years, this fact has led to speculation that black holes are the source of quasars; as for Schmidt, he has continued to investigate the spectral qualities of quasars.

Born on December 28, 1929, Schmidt was raised in Groningen, Holland. He attended the University of Leiden, where he studied under Jan H. Oort and earned his Ph.D. in 1956. In 1959 he moved to the California Institute of Technology (Cal Tech), where he was initially concerned with questions involving the galaxy's mass distribution and dynamics. In the early 1960s, however, astronomer Rudolph Minkowski retired from Cal Tech, and Schmidt assumed leadership of the project Minkowski had directed, the analysis of spectra of radio-emitting objects.

With the advent of radio astronomy in the 1930s and 1940s, scientists were able to observe stars not only in terms of their light waves but also their radio emissions—a development that greatly expanded the reach of the known universe. At this time radio astronomers first became aware of faint blue star-like objects that emitted extraordinary amounts of radio waves and ultraviolet radiation. While observing these objects at California's Mount Palomar Observatory in 1963, Schmidt had a great insight.

The light spectrum associated with these objects contained a few strangely wide lines, which initially looked like nothing ever seen, but Schmidt soon realized that these lines were normal spectral lines for hydrogen that had been shifted toward the red end of the visible spectrum, and this shift had made the configuration difficult to recognize. The red coloring indicated that the object was moving away from Earth—at a rate of about 30,000 miles per second (48,280 km/s).

These odd red-shifting objects came to be known as quasi- stellar radio sources, or quasars. Since 1963, thousands more have been discovered, and the majority of quasars are now believed to emit no radio waves at all—only an intense ultraviolet light. These "radio-quiet" quasars now constitute about 99% of all known quasars.

Since the mid-1960s, Schmidt has turned his attention to a number of subjects involving quasars. Analyzing their evolution and distribution, he found that there were far more quasars in an earlier epoch. This finding helped lead to the decline of the steady-state theory, which had once competed with the Big Bang theory as a model for how the universe came into being.

Schmidt, who retired as a professor in 1996, has served in a variety of roles at Cal Tech: executive officer for astronomy from 1972 to 1975; chairman of the Division of Physics, Mathematics, and Astronomy from 1976 to 1978; and director of the Hale Observatories from 1978 to 1980. Winner of the 1992 Bruce Medal and other honors, he continues to investigate quasars with a high red shift, his aim being to find the "red shift cutoff," above which no quasars exist.

JUDSON KNIGHT

Julian Seymour Schwinger
1918-1994
American Physicist

Julian Schwinger was one of a small number of physicists to be involved in the creation of quantum electrodynamics, the physical theory of electrons, positrons, and photons. The theory is unique in providing a complete description of a well-defined set of physical phenomena that is in complete agreement with experiment. As such, quantum electrodynamics has strongly influenced the search for comprehensive theories of the less-well understood nuclear interactions. It also represents a new level of mathematical sophistication to be required of theoretical physicists in the future.

Schwinger was a child prodigy, the son of a businessman and his wife, who taught himself physics by reading library books and the *Encyclopedia Britannica,* and, skipping several grades, graduated from high school at the age of 14. While attending the City College of New York, he published a scientific paper in the *Physical Review* that attracted the attention of I. I. Rabi, who then arranged a scholarship for Schwinger at Columbia University. He received his Ph.D. degree from Columbia in 1939. Like many American physicists, Schwinger was drawn into military research, working on both the atomic bomb and radar. After the war, he joined the faculty at Harvard University, becoming a full professor at the age of 29.

The quantum theory developed by Erwin Schroödinger and Werner Heisenberg in the 1920s provided a description of the wavelike properties of electrons moving at low velocities. A form of this theory consistent with Einstein's theory of relativity was published by Paul Dirac soon after. While the new quantum mechanics made it possible to explain the absorption and emission of light by atoms and many aspects of the chemical bonds formed by atoms, the theory was unsatisfactory in that it did not provide a treatment of the particle-like properties of light in a manner consistent with its treatment of electrons.

The need for a more comprehensive theory was highlighted at one of the first postwar conferences on theoretical physics, a meeting held at Shelter Island, New York, in 1947. Schwinger and another former child prodigy, Richard Feynmann, attended. Much of the discussion at the Shelter Island conference focused on two new experimental results: a precise measurement of the difference between two energy levels of the hydrogen atom, the simplest atom; and new precise measurements of the magnetic moments of the electron. Both of these results disagreed slightly but indisputably with what was expected from the existing quantum theory. A consensus thus developed that a proper treatment of the

Julian Schwinger. *(The Library of Congress. Reproduced with permission.)*

electron interaction with the electromagnetic field would explain the new experimental results.

Because the electron has an electrical charge, it interacts with the electromagnetic field. Attempts to compute the energy of this interaction were beset with serious mathematical difficulties. Chief among them was the fact that traditional approaches gave results that involved adding together an infinite number of small quantities, and, depending on the order in which the terms were added, partial sums of the terms could be infinitely large. An effective way to deal with these mathematical problems, using measured quantities to replace some of the mathematics, was found by three individuals, the Americans Schwinger and Feynmann and the Japanese Sin-Itero Tomonoga, who would share the Nobel Prize for physics in 1965 for their contributions. Schwinger's contribution, which was characterized by logical clarity and mathematical elegance, was published in a series of papers in the *Physical Review* beginning in 1948. Feynman's approach was more intuitive, based on a graphical representation that has come to be known as the Feynman diagram. Tomonaga had actually anticipated some of the Americans' results but was prohibited from publication by wartime conditions in Japan.

Schwinger remained active in the study of particle physics until the time of his death, influencing research on the weak force and attempts to unify the fundamental forces. He is remembered as a brilliant lecturer who often spoke without notes and as a reserved individual with strong opinions but an open mind.

DONALD R. FRANCESCHETTI

Edward Teller
1908-
Hungarian Physicist

Teller is one of the greatest physicists of the twentieth century. Until the end of the Cold War, he was one of the leading members of the military industrial complex.

As a young boy, Edward Teller was introduced to extreme nationalism, prejudice, and authoritarian government. When he was six, his native Hungary became deeply involved in World War I. Severe wartime shortages and unprecedented carnage gave Teller an early lesson in the importance of peace and security.

After the war, clashes between Communist and anti-Communist forces, frequent executions, and anti-Semitism hardened Teller's hatred of all types of authoritarianism. These early childhood experiences would influence his stand against both fascism and Communism. Edward Teller came to believe that democracy was the best form of government. He used his scientific expertise to fight all threats to its existence.

At an early age Teller showed formidable intellectual capacity, especially in mathematics. His excellent grades opened the doors to the great European universities. After graduating from secondary school, Teller chose to study in Germany, eventually receiving a Ph.D. from the University of Leipzig. When Hitler came to power in 1933, Teller decided to emigrate to America, where he eventually accepted a position at George Washington University. Here Teller became part of the American scientific establishment.

On August 2, 1939, Albert Einstein (1879-1955) reported to President Franklin Delano Roosevelt that the development of nuclear weapons was possible. Because this research had been pioneered in Germany and many scientists were working within the Nazi organization, the probability was high that the Germans had a bomb program in operation. Teller and other prominent American scientists were given permission to develop a bomb on October 9, 1941, an undertaking called the Manhattan Project.

The first atomic weapons were used against Japan to end World War II in August 1945.

Soon another conflict was at hand. The Cold War dominated international relations for half a century. Teller saw this as another battle against totalitarianism. His fears intensified when, on August 20, 1949, the Soviets exploded their first atomic weapon. He was also afraid that the Russians would develop an even more powerful hydrogen bomb. After intense secret debate spurred by Soviet advances in nuclear weapons, President Truman announced in January 1950 that an American hydrogen bomb project was underway.

Teller, known as the father of the American hydrogen bomb, accelerated research efforts. On May 8, 1951, a small test explosion proved the feasibility of Teller's concept, a theory called radiation implosion. On November 1, 1952, the United States detonated a device that was 500 times more powerful than the Nagasaki bomb. The final test of the hydrogen bomb came on March 1, 1954. The explosion produced a yield of 15 megatons, 1,000 times as large as the Hiroshima bomb. Although the Soviets were also working to develop thermonuclear weapons, their program lagged consistently behind that of the United States; they did not test a megaton weapon until 1955.

With both superpowers able to launch thermonuclear weapons, a horrific stalemate known as mutual assured destruction (MAD) ensued. Each side knew that a launch against its adversary would end its own existence. Teller, however, viewed MAD as an unacceptable means of maintaining peace. He believed that to hold the world's population hostage to nuclear destruction was unjustifiable. His prominent role in the development of the Strategic Defense Initiative (SDI) in the 1980s, was an attempt to end this nightmarish scenario. This project, known popularly as "Star Wars," was the most controversial of Teller's career. It was based on the theory that a combination of lasers and ballistic missiles could shoot down incoming missiles. The project has languished since the end of the Cold War.

RICHARD D. FITZGERALD

Sin-Itiro Tomonaga

1906-1979

Japanese Physicist

A pioneer in the field of quantum electrodynamics, Sin-Itiro Tomonaga developed a theory about subatomic particles that resolved earlier difficulties encountered by physicists seeking to bring together principles of quantum mechanics and special relativity. He shared the 1965 Nobel Prize for Physics with Richard Feynman (1918-1988) and Julian Schwinger (1918-1994).

Tomonaga was born in Tokyo on March 31, 1906, to Sanjuro and Hide Tomonaga. His father became a professor of philosophy at Kyoto Imperial University when Tomonaga was a boy, and the family moved to Kyoto. Later Tomonaga enrolled in Kyoto's renowned Third High School, and there he studied alongside Hideki Yukawa (1907-1981), who would later become Japan's first Nobel Prize winner (also in physics) in 1949. Both men later majored in physics at Kyoto Imperial University, earning their bachelor's degrees in 1929, and both remained as research assistants to physicist Kajuro Tamaki.

In 1932 Tomonaga took a job as research assistant to Yoshio Nishina at the Institute of Physical and Chemical Research in Tokyo. Five years later, he went to the University of Leipzig in Germany, where he studied under the great Werner Heisenberg (1901-1976). At Leipzig Tomonaga wrote his dissertation, on the atomic nucleus, and in 1939 earned his Ph.D. from Kyoto Imperial.

Tomonaga married Ryoko Sekiguchi, daughter of the director of the Tokyo Metropolitan Observatory, in 1940. The couple later had two sons, Atsushi and Makoto, and a daughter, Shigeko. In 1941 Tomonaga became professor of physics at Bunrika University (now Tokyo University of Education), and during this period conducted some of his most important research.

Quantum electrodynamics (QED) had emerged in the 1920s, as physicists sought answers from both quantum mechanics and relativity theory as a means of explaining the behavior of particles and their interaction with energy. English physicist Paul Dirac (1902-1984) had developed a theory of QED that initially seemed effective in explaining these questions, but in time Dirac's theory ran into difficulties. Among these was the fact that, according to Dirac's predictions, particles under certain circumstances would have infinite mass and infinite electrical charge—which was clearly impossible.

Tomonaga, however, was convinced that he could resolve the "divergence difficulties" Dirac had encountered. Using a mathematical technique called renormalization, he was able to demonstrate that although a particle could theoretically assume infinite mass and charge, this would never occur in the real world. He published a paper on his findings in 1943, but because of World War II, this information did not reach the scientific community at large until 1947. Around the same time, Schwinger and Feynman, working at the California Institute of Technology (Cal Tech), published similar findings. In 1965 the three physicists shared the Nobel Prize in recognition of their independently derived solutions to the "divergence difficulties" problem.

In 1949 Tomonaga accepted an invitation to go to the Institute of Advanced Studies (IAS) in Princeton, New Jersey, as a visiting scholar. Two years later he returned to Tokyo as director of the Institute for Scientific Research, and in 1955 he helped establish the Institute for Nuclear Studies at the University of Tokyo. A year later, he became president of the university.

In addition to the Nobel Prize, Tomonaga received the Japan Academy Prize in 1948, the Order of Culture of Japan in 1952, and the Lomonosov Medal of the Soviet Academy of Sciences in 1964. He retired in 1962, but shortly afterward became president of the Science Council of Japan and director of the Institute for Optical Research. Tomonaga held these posts until 1969. He died in Tokyo on July 8, 1979.

JUDSON KNIGHT

Frederick John Vine
1939-
English Geophysicist

In the 1960s Frederick J. Vine and his colleague Drummond Matthews (1931-) emerged as foremost proponents of the theory of plate tectonics—the theory that the earth's crust is divided into shifting plates that include the continents embedded on their surface. Vine's research, presented in 1966, helped sway the opinions of geologists toward the idea that the ocean floor was created at the mid-ocean ridges and slowly spread apart as large individual plates.

Vine was born in Chiswick, England, a suburb of west London, to Frederick Royston Vine, an accountant, and Ivy Bryant Vine, a personal secretary. He studied natural sciences at St. John's College, Cambridge, and went on to obtain a Ph.D. in marine geophysics under the tutelage of Drummond Matthews.

In 1962, while undergoing his graduate studies at St. John's, Vine became intrigued, as many before him had, by the apparent fit between the great gulf of West Africa and the bulge of Brazil. This seemed to confirm the theory of continental drift put forth half a century earlier by Alfred Wegener (1880-1930). However, this theory had come under attack in recent years.

Also in 1962, Matthews took part in a mapping expedition aboard the H.M.S. *Owen* in the northwestern Indian Ocean, and noted a pattern of magnetic bands or stripes as much as 20 miles or 30 kilometers wide. Upon Matthews's return to England, he and Vine examined this data—such magnetic stripes had been found on the floors of other oceans, but not the region studied by Matthews—in light of the then-new hypothesis of sea-floor spreading put forth by Harry Hammond Hess (1906-1969).

Vine and Matthews began to examine the bands for variations in polarity, or the direction of the magnetic field, taking note of research conducted by Allan Cox and other American geologists who had shown that Earth's magnetic field reverses its polarity every hundred thousand years or so. They subsequently discovered that the polarity of the stripes varied symmetrically: for instance, if the third stripe to the west of a ridge was a wide one with a north magnetic pole near the north geographic pole, the same was true of the third stripe to the east of a ridge.

The theories of Hess, as well as those of Cox and others, had not gained wide acceptance. But Vine and Matthews, who presented their findings in the September 1963 issue of *Nature,* offered compelling evidence of magnetic reversals in the ocean floor.

Vine further confirmed his findings in 1965, when American geologist Brent Dalrymple told him about a previously unrecognized geomagnetic reversal that had occurred about a million years ago near Jamarillo Creek in Mexico. Shortly afterward, while visiting the Lamont Geological Observatory in New York, Vine discovered data confirming a geomagnetic reversal on the South Pacific floor at the same time as the Jamarillo event. Not only were the geomagnetic reversals real, but they were clearly linked.

At the December 1966 meeting of the Geological Society of America in San Francisco, Vine presented a paper entitled "Proof of Ocean-Floor Spreading." Thereafter sea-floor spreading, continental drift, and plate tectonics theory began to gain wide acceptance.

In 1967 Vine took a teaching and research position at Princeton University in New Jersey, but in 1970 he returned to England to work in the environmental sciences department of the University of East Anglia in Norwich. He currently serves as dean of environmental sciences for the university. He has also worked with Eldridge Moores, studying a rock formation in the Troodos Mountains of southern Cyprus that is thought to be an upthrust slice of ocean floor.

Vine married Susan McCall in 1964, and they have a son and daughter.

JUDSON KNIGHT

John Tuzo Wilson

1908-1993

Canadian Geologist

In 1963 John Tuzo Wilson revolutionized geology by suggesting that volcanic islands, such of those of Hawaii, had been formed by the movements of plates over a "hotspot" in the earth's mantle. This helped revive the theory of continental drift, which had suffered due to apparent contradictions—contradictions that Wilson's theory resolved. Two years later, in 1965, he offered another groundbreaking idea when he published a paper describing a third type of plate and plate movement, in addition to the two types already identified.

Born on October 24, 1908, in Ottawa, Ontario, Wilson was the son of a Scottish engineer who had immigrated to Canada. In 1930, when he earned his B.A. from Trinity College at the University of Toronto, he became the first student at any Canadian university to earn a degree in geophysical studies. Wilson went on to St. John's College, Cambridge, where he earned a second B.A. in 1932, then to Princeton to earn a Ph.D. in 1936.

From 1936 to 1939, Wilson worked with the Geological Survey of Canada. With the outbreak of World War II, he joined the Royal Canadian Engineers, a unit of Canada's army in which he attained the rank of colonel. Following the war's end, in 1946 Wilson became professor of geophysics at the University of Toronto. There he would remain until 1974, during which time he would perform the most significant work of his career.

By this time, the idea of continental drift, first put forward by Alfred Wegener (1880-1930) in the early 1900s, seemed to have come and gone. Geologists had returned to the belief that the continents were fixed in place, particularly in light of apparent contradictions, such as the fact that some volcanoes could be found many thousands of miles from the nearest plate boundary. Then, in 1963, Wilson published his revolutionary findings.

Perhaps the islands of Hawaii and those of other volcanic groups, Wilson suggested, had been formed not on the boundaries between two plates, but from the movement of a plate over an immovable "hotspot" in the mantle, deep beneath the plate. This resolved the contradictions in continental drift theory, and hundreds of studies since then have proven Wilson correct. However, at the time Wilson's ideas went so radically against the grain of received opinion that the major international scientific journals all rejected his manuscript. When in 1963 he finally did find a publication that would accept the paper, it was a relatively small one, the *Canadian Journal of Physics*.

Two years later, Wilson again presented a paper that contradicted prevailing opinion. Geologists knew of two types of plates and plate movement that connected trenches and ridges beneath the ocean: plates that moved apart, or divergent plates; and plates that moved toward one another, or convergent plates. Wilson now postulated a third type of plate boundary. Ridges and trenches, he observed, often end abruptly and "transform" into major faults that drop off sharply. These "transforms" or transform-fault boundaries, of which the San Andreas Fault in California is an example, offset the earth's crust horizontally, but do not create or destroy crust. As a result of Wilson's observations and those of others, continental drift theory was revived, helping to open the way for plate tectonics theory in the 1960s and 1970s.

Wilson's publications include *One Chinese Moon* (1959), *IGY: Year of the New Moons* (1961), *A Revolution in Earth Science* (1967), and *Continents Adrift and Continents Aground* (1977). He served as president of the Royal Society of Canada (1972-73) and the American Geophysical Union (1980-82), of which he was elected a fellow in 1962. His awards include the Walter H. Bucher Medal (1968) and the Mau Medal (1980). A range of mountains in Antarctica is also named for him.

Wilson remained at the University of Toronto until 1974, when he retired and became director general of the Ontario Science Centre. He served as chancellor of York University from 1983 to 1986, and died on April 15, 1993, in Toronto.

JUDSON KNIGHT

Robert Burns Woodward

1917-1979

American Chemist

Robert Woodward was a Nobel Prize-winning chemist who developed numerous techniques for producing complex chemical compounds in the laboratory. Many of the methods he developed are used today to produce compounds that were once only obtainable from living organisms.

Woodward was born in Boston, Massachusetts, in 1917. His childhood interest in science prompted his mother to give him a chemistry set. By the time he was in high school, he was conducting experiments in his home that were similar to those performed in college chemistry classes. Woodward began attending college at the Massachusetts Institute of Technology when he was only sixteen. He rarely went to class, however, usually only showing up for final exams. Instead, he spent his time in the school's chemistry laboratories and the library. He earned his degree in three years and a doctorate in chemistry in one year. After graduating, Woodward spent a summer teaching at the University of Illinois. He spent the remainder of his career teaching chemistry at Harvard University.

Woodward specialized in organic chemistry. Organic chemistry is the study of compounds containing the element carbon. Most compounds that make up plants, animals, and other organisms contain this element, so organic chemistry is largely the chemistry of living things. When Woodward began teaching at Harvard, the field of organic synthesis was still in its infancy. Synthesis involves the use of chemical reactions to produce complex compounds from simpler starting materials. Chemists use synthesis to produce useful compounds in the laboratory.

Woodward focused his attention on stereochemistry and the synthesis of specific stereoisomers. Stereochemistry is the study of the three-dimensional arrangement of the atoms of compounds. Two compounds that are identical except for the way their atoms are arranged in space are said to be stereoisomers. Each stereoisomer of a compound has unique chemical properties. Many of the compounds that make up living things are stereospecific; that is, the organism produces only one particular stereoisomer of a compound. Woodward developed methods not only for synthesizing particular organic compounds but also for isolating specific stereoisomers of the compounds.

In 1944 Woodward and fellow chemist William Doering (1917-) were the first to synthesize the compound quinine. Quinine is used to treat the mosquito-carried disease malaria. Before quinine was made in the laboratory, it had to be obtained from the bark of South American cinchona trees. Woodward went on to synthesize many other stereospecific organic compounds, including cortisone (a human hormone), cholesterol, reserpine (a tranquilizing drug that was once obtained from the roots of certain tropical plants), chlorophyll, and cephalosporin C (an antibiotic). In 1971, after working for ten years with a team of more than one hundred researchers, he completed the synthesis of vitamin B_{12}. Woodward won the Nobel Prize for chemistry in 1965 for his many syntheses of organic compounds.

Synthesizing such complex organic chemicals required many steps and intermediate compounds. (The synthesis of vitamin B12 required more than one hundred chemical reactions.) Woodward often was able to envision the entire sequence of reactions needed to produce complicated organic compounds before he ever began work in the lab. Many of the syntheses Woodward performed were among the most complicated up to that time. Several of his syntheses formed the basis for drug-manufacturing procedures, such as that used for reserpine.

In 1965 Woodward and the chemist Roald Hoffmann (1937-) proposed a set of rules that could be used to predict the stereochemistry of the products of a chemical reaction. Today these rules are known as the Woodward-Hoffmann orbital symmetry rules. They helped explain the sometimes unexpected stereochemical results Woodward had observed during his syntheses. This work won Hoffmann a Nobel Prize in 1981. Woodward most likely would have shared the prize if not for his death in 1979.

STACEY R. MURRAY

Chien-Shiung Wu

1912-1997

Chinese-American Physicist

Chien-Shiung Wu verified a theory for which her colleagues received a Nobel Prize in physics. An experimental physicist, Wu proved that the physical law of conservation of parity was invalid. Physicists had accepted this concept for thirty years before Wu's revelation, and the idea had been the foundation for theories regarding the universe's physical structure. Wu's remarkable investigation in nuclear physics demonstrated that science is dynamic and that physical laws can be fallible, reminding scientists to doubt assumed facts, to scrutinize their methodology and data to achieve accurate results, and to be receptive to new interpretations of the physical world.

Born on May 29, 1912, near Shanghai, China, Wu was born to Wu Zhongyi and Fuhua H. Fan Wu. Wu planned to become a scientist and her parents sent her to a boarding school, where she became fascinated by physics. She enrolled at the National Central University in Nanjing and prepared a thesis on x rays to earn a physics degree. While researching x-ray crystallography at Shanghai's National Academy of Sciences, Wu met a female physicist who had earned an American doctorate. Wu then decided to continue her education in the United States. In 1936 she began graduate studies at the University of California at Berkeley. Her professors included outstanding nuclear physicists, such as Robert Oppenheimer (1904-1967), who were devising new atomic theories.

Wu concentrated her research on atomic particles, writing her doctoral dissertation on two of her experiments. She had developed a method to separate two types of rays emitted during beta decay, a type of radioactivity in which the nucleus expels high-speed electrons and becomes another element. Wu observed that the electromagnetic energy discharged as a particle decelerated moving through matter, validating her hypothesis. She also investigated radioactive noble gases produced during uranium fission.

After Wu graduated in 1940, she remained at Berkeley to conduct research with Ernest O. Lawrence in the radiation laboratory. Wu became an authority on fission and was invited to present lectures nationwide. Wu married physicist Luke Cha-Liou Yuan and they moved to New Jersey, where she was the first female instructor at Princeton University. She then joined the Manhattan Project at Columbia University to design radiation detectors and to improve Geiger counters in addition to studying uranium enrichment and conducting neutron research. She stayed at Columbia after World War II. Wu returned to her research on beta decay and developed new techniques to study the shapes of particles' spectra and how they interacted during beta decay in an effort to achieve more precise experimental results.

In 1956 Wu agreed to test a theory for her colleagues Tsung Dao Lee and Chen Ning Yang. They wanted to prove that the principle of the conservation of parity, which stated that in nuclear structures an object and its mirror image behave symmetrically, was not always true. At the time, this principle was considered a fundamental law of physics. Wu examined beta rays, which are the electrons in the nuclei of radioactive substances, by chilling the nuclei to slow their thermal motion in order to observe the directions in which electrons were discharged during beta decay. She concluded that nuclear particles did not consistently behave symmetrically and, instead, often moved in the direction opposite to that of the spinning nucleus. Wu diligently verified her results and asked fellow researchers to duplicate her methodology.

On January 16, 1957, Wu announced her findings. Although Wu was co-author of articles regarding her experiments, Lee and Yang won the 1957 Nobel Prize in physics for conceptualizing the theory that she had verified. Wu, however, received other professional accolades, including the National Medal of Science, and was promoted to full professor. Research into beta decay continued to interest her and she published significant articles and books about her work, including nuclear physics applications to medicine.

ELIZABETH D. SCHAFER

Biographical Mentions

Walter Alvarez
American geologist who, with his father, Louis, discovered evidence that a large impact was the most likely cause of the mass extinction of dinosaurs. Alvarez analyzed samples from the Italian mountains that were deposited at the boundary between the Cretaceous and Triassic periods (known to geologists as the "KT boundary") and noticed high levels of iridium, commonly found in meteorites. This led to his discovery that a

large meteor or a small asteroid impact caused the dinosaurs' extinction.

William Arkell
1904-1958

English geologist and paleontologist who made major contributions to understating stratigraphy (sequence of rock layers) and invertebrate paleontology. Arkell's *Jurassic Geology of the World* helped to place stratigraphy on a more systematic basis, making that part of geology much more rigorous a science. In addition, his work on ammonites (an extinct cephalopod) was definitive and his work to correlate English geologic structure with European equivalents was both important and impressive.

Neil Bartlett
1932-

American chemist who in 1962 produced the first compound containing a noble gas. Previously, scientists had believed that the noble gases were incapable of undergoing reactions. Bartlett's discovery involved combining the noble gas xenon with the highly reactive gas fluorine. The compound that resulted was xenon hexafluoroplatinate(V); its chemical formula is $Xe(PtF_6)_x$, where *x* varies from 1 to 2. Scientists have since discovered that fluorine will also react with the noble gases argon and krypton.

Derek Harold Richard Barton
1918-1998

English chemist who won the Nobel Prize for chemistry in 1969 for his work on conformational analysis. Conformational analysis is the study of how a compound's three-dimensional shape affects its chemical properties. Barton studied the conformation, or shapes, of steroids, which are a class of compounds that includes cholesterol and many hormones. In 1958 Barton developed an efficient method for producing in the laboratory the steroid aldosterone, a hormone that helps to regulate the body's salt and water balance.

Llody Viel Berkner
1905-1967

American physicist and engineer who focused on fundamental geophysical research. His early naval duty involved Antarctic exploration and development of radar, navigation systems, and aircraft electronics. Vice president of the committee of the International Geophysical Year (1957-1958), he was the first to measure the extent and density of the ionosphere, describing all aspects of radio wave propagation. Later research centered on the origin and development of Earth's atmosphere, culminating with a theory, co-authored by L.C. Marshall, to describe the evolution of the atmospheres of the inner planets of the solar system.

J. C. Bhattacharyya

Indian astronomer who, during ground-based observations in 1984, discovered additional rings in the Saturn system. While observing changes in light intensity of a star while it passed behind Saturn's rings, Bhattacharyya and collaborators noticed light dimming unexpectedly. This indicated the presence of two rings that had not been discovered in the previous three centuries of study by ground-based telescopes, space-based observatories, and interplanetary space probes.

Adriaan Blaauw

Dutch astronomer who, in 1952, discovered that the zeta-Persei star cluster is relatively young (about 1.3 million years old). This was an important discovery because it showed that star formation is still taking place in the Milky Way galaxy and, presumably, in other similar galaxies. This, in turn, means that galaxies continue to evolve with time, rather than remaining static from the time of their formation until the last stars die billions of years later.

Felix Bloch
1905-1938

Swiss-American physicist awarded the 1952 Physics Nobel Prize with Edward Purcell for precision nuclear magnetic resonance (NMR) techniques and discoveries made therewith, including magnetic moment measurements of the proton and neutron. NMR methods are widely used in physics and analytic chemistry to study nuclear structure and behavior and as a diagnostic tool in medicine. Bloch was also a founder of the quantum theory of electrons in metals and introduced the concept of spin-waves in ferromagnetism.

Nicolaas Bloembergen
1920-

Dutch-American physicist who won the Nobel Prize for work leading to the first continuous maser. The maser (a microwave equivalent of the laser) had been invented in 1954 by Charles Townes, but operated only in pulses. Two years later, Bloembergen developed a method that generated continuous maser emissions, an important advance towards the invention of the laser in 1960. Since then, several hundred natural masers have been discovered in space.

Bloembergen was awarded the Nobel Prize in Physics in 1981.

Aage Niels Bohr
1922

Danish physicist who has made significant contributions to the understanding of atomic nuclei. Bohr, the son of Nobel Prize-winning physicist Niels Bohr, worked with his father on problems in atomic structure for much of his early life. His research into the mechanical properties of the atomic nucleus, particularly with regards to its deformation under some circumstances, led to his winning the Nobel Prize in Physics in 1975.

Dirk Brouwer
1902-1966

Dutch-American astronomer who made significant contributions to the understanding of celestial mechanics. Brouwer was also a pioneer in the use of (then) high-speed digital computers for solving problems in celestial mechanics, many of which found practical application with the launching of artificial satellites in the late 1950s and early 1960s. For his contributions to astronomy, Brouwer was awarded the Gold Medal of the Royal Astronomical Society in 1955. Other work led to mathematical techniques for precisely determining the positions of planets, setting the stage for their later exploration.

Herbert Charles Brown
1912-

English-American chemist who won the Nobel Prize in chemistry in 1979 for his work with the compounds of boron. He discovered a way to produce alcohols from compounds containing carbon-carbon double bonds. (Alcohols are a class of carbon-containing compounds important in science and industry.) Brown used diborane (B_2H_6) to catalyze, or speed up, the process, which is now called hydroboration-oxidation. Brown also discovered that sodium borohydride ($NaBH_4$) could be used to catalyze other types of reactions that produce alcohols.

Melvin Calvin
1911-1997

American biochemist who first determined many of the chemical reactions that take place during photosynthesis. Using radioactive carbon-14, paper chromatography, and patience Calvin spent many years tracing the chemical processes that take place during photosynthesis. This helped to unravel one of the most important biochemical processes on Earth because most life depends either directly or indirectly on photosynthesis for food. Calvin was awarded the 1961 Nobel Prize in Chemistry for his work.

Sydney Chapman
1888-1979

English geophysicist who invented theories and made fundamental discoveries in general geophysics, including the idea that solar radiation is the source of terrestrial magnetic phenomena: auroras, magnetic storms, and radio wave-reflecting atmospheric layers. He theorized that the atmosphere has tidal-like characteristics from gravitational and temperature changes; thus, radio wave layers would fluctuate, causing communication disturbances. He discovered lunar atmospheric tides and conceived an accurate model of earth's atmosphere. Chapman, president of the International Geophysical Year committee, coined the name for that historic year of international geophysical research (1957-1958).

James W. Christy
1938-

American astronomer who co-discovered Pluto's moon, Charon, in 1978. Pluto, the furthest planet from the Sun, was discovered in 1930 by Clyde Tombaugh. However, because of its distance from the Sun, very little was known about it except for its orbital parameters. The discovery of Charon by Christy and Robert Harrington made it possible to determine Pluto's size and mass through application of basic physics governing planet-moon systems.

John Warcup Cornforth
1917-

Australian chemist whose work in many aspects of biochemistry led to a fuller appreciation of the manner in which many hormones and other compounds affect the body. Deaf for most of his life, Cornforth studied chemistry in wartime England, first working on the chemistry of penicillin. This work was followed by research into cholesterol, particularly its chemical structure and its use in the body (cholesterol is an important part of cell membranes, for example). Cornforth won the 1975 Nobel Prize for Chemistry for his research.

Donald James Cram
1919-

American chemist whose work on molecular recognition in organic chemistry was awarded the Nobel Prize in 1987. Cram's father died during his youth, leaving him to be raised by his mother and sisters. At first encouraged to seek a career in industrial chemistry, Cram was determined to become an academic. He eventually became an outstanding researcher, entering chemistry at a time when new tools were revolutionizing the field.

James W. Cronin
1931-

American physicist who showed through his research with kaons that the CP (charge and parity) symmetry law was sometimes violated. This important discovery upset one of the principal tenets of physics and led to other significant discoveries in particle physics. Cronin has been honored by several scientific organizations and shared the 1980 Nobel Prize for Physics with Val Fitch.

Robert F. Curl
1933-

American chemist who discovered fullerenes, a new form of carbon that had been predicted but not previously found in nature or synthesized. Since its synthesis in the laboratory, fullerenes have been discovered in nature. While still primarily a laboratory tool, many uses have been suggested for fullerenes and, as their properties and manipulation become better known and more routine, fullerenes seem certain to be of great utility for a number of purposes.

Robert Henry Dicke
1916-1997

American physicist who made contributions to a number of theoretical and experimental fields, including relativity and cosmology theory, atomic physics, and radar and microwave technology. He is best known for his prediction that the universe should be filled with background radiation of microwave frequency, the remains of the energy of the Big Bang that was, theoretically, the beginning of the universe. His quantum theory of coherent radiation emission led to the development of the laser.

Leo Esaki
1925-

Japanese physicist whose research into quantum mechanical tunneling won the 1973 Nobel Prize for Physics. Tunneling, the process by which electrons (and other particles) can "escape" from an atom, has been used most recently in the scanning tunneling microscope, which can provide atomic-scale "images" of surfaces. In addition to his work on tunneling, Esaki helped to develop the first quantum electron device, the Esaki tunnel diode.

Ernst Otto Fischer
1918-

German inorganic chemist who discovered a new kind of chemical bond between metals and organic molecules and founder of organometallic chemistry. Using x-ray crystallographic methods, he showed that ferrocene is made up of an iron atom sandwiched between two five-membered carbon rings (cyclopentadienes). He subsequently investigated the molecular structure and properties of other organometallic compounds, especially those formed by various metals with cyclic organic molecules. He shared the Nobel Prize for chemistry with Geoffrey Wilkinson in 1973.

Val Logsdon Fitch
1923-

American physicist who shared the 1980 Nobel Prize for Physics with collaborator James Cronin for their study of kaons. Fitch and Cronin's research demonstrated that the kaon and antikaon (the antimatter particle equivalent to the kaon) decayed with slightly different half-lives. This highly significant discovery led directly to the concept of Charge-Parity (CP) symmetry violation, the first time any symmetry law had been observed in nature.

Paul J. Flory
1910-1985

American chemist who conducted basic research into the properties of polymers and other macromolecules. Flory first worked on developing synthetic rubber during World War II, moving to Cornell after the war and, later, to Carnegie University and finally to Stanford. During this time, he studied the properties of polymers and contributed toward better understanding of these important molecules that form the basis of plastics and similar compounds. Flory's work was recognized with the 1974 Nobel Prize for Chemistry.

Kenneth Lynn Franklin
1923-

American astronomer who was the first to detect radio emissions from Jupiter. This was a surprising discovery because, previously, no planets were known to emit radio waves. Through this discovery, much was learned about Jupiter's magnetic field and its interactions with charged particles from the Sun and elsewhere. In addition, this made it clear that planets could emit radio waves, a property previously thought limited to stars and galaxies.

Jerome Friedman
1930-

American physicist whose work in particle physics led to the discovery of quarks, for which he shared the 1990 Nobel Prize for Physics. Until Friedman's experiments in the 1960s, it was thought that the proton and neutron had no internal structure. Friedman's work, carried out at high energies, showed the existence of subnuclear particles that scattered incident electrons in unexpected directions. This discovery was analogous to Ernest Rutherford's discovery of the atomic nucleus at the beginning of the twentieth century.

Kenichi Fukui
1918-1998

Japanese chemist whose research into the nature of chemical reactions was awarded the 1981 Nobel Prize for Chemistry. Fukui's initial research was in the development of synthetic fuels, later changing to more theoretical research in the area of charge transfer as a controlling agent in chemical reactions. Fukui pursued this line of inquiry from about 1950 until his death in 1998, including important work on the theory of chemical reactions that led to a better understanding of this most fundamental aspect of chemistry.

Donald Arthur Glaser
1926-

American physicist who invented the bubble chamber, used in detecting products of atomic collisions. Glaser realized that he could heat pressurized liquid helium and it would remain a superheated liquid if the pressure were reduced slightly. Charged particles traversing the superheated liquid would create localized boiling along their paths, which were bent by a magnetic field, giving information about the particles' mass, energy, and charge. Glaser was awarded the 1960 Nobel Prize for Physics for this invention.

Sheldon Lee Glashow
1932-

American theoretical physicist who generalized and extended the electroweak theory of Steven Weinberg and Abdus Salam and shared the Nobel Prize for physics with them in 1979. Their work demonstrated that the electromagnetic force and the weak force in the nucleus are aspects of a single force, called the electroweak force. Glashow also proposed a new property of quarks (entities that make up elementary particles) which he called charm.

Jeffery Goldstone
1933-

English physicist who predicted in 1961 that symmetry breaking would produce particles with a rest mass of zero. Unfortunately, all particles meeting these criteria were thought to be known. This resulted in an apparent paradox that vexed particle physicists for several years until it was solved by Philip Anderson.

Thomas Gold
1920-

Austrian-born, British astronomer who proposed the steady-state theory of the universe in which the density of matter is kept constant by the continuous creation of matter throughout the expanding universe. An alternative explanation, the Big Bang Theory, is now more accepted, however. Gold also explained pulsars as revolving neutron stars and developed theories for the origin of the universe and for living organisms. He was an early proponent of space exploration.

Bruce Charles Heezen
1924-1977

American oceanographer and geologist who in 1956 discovered the Mid-Oceanic Ridge. Part of this global system of sub-sea volcanoes had been discovered by oceanographic expeditions in the nineteenth century, but it was not known to be global in extent until Heezen's landmark work. His discovery helped set the stage for subsequent research that gave rise to the modern theory of plate tectonics, in which the Mid-Oceanic Ridge plays a crucial role.

George Howard Herbig
1920-

American astronomer known for his studies of recently formed stars and the star-formation process. Working independently, he and Guillermo Haro discovered Herbig-Haro objects, a glowing gas cloud surrounding new stars thought to form when gas jets from the star collide with interstellar material. Herbig has also made spectroscopic observations of the gas and dust that lies between stars in the Milky Way galaxy, from which stars form.

Gerhard Herzberg
1904-

German-born, Canadian physicist who made significant contributions to a number of areas of science, including molecular spectroscopy, astrophysics, physical chemistry, and quantum mechanics. He is best known for his studies of the structure and geometry of free radicals for which he was awarded the 1971 Nobel Prize for chemistry. He determined the chemical composition of comets through spectroscopic studies, investigated chemical reactions in gases, and studied free radicals as intermediate agents in chemical reactions.

Anthony Hewish
1924-

English astronomer who was awarded the 1974 Nobel Prize for Physics for co-discovering pulsars. Pulsars were among the first of the "astronomical oddities" that, once discovered, greatly revised scientific understanding of the cosmos. In 1967 Hewish's graduate student, Jocelyn Bell Burnell, noted very regular radio signals from a part of the sky. First thought that they may be signals from another civilization, Hewish eventually understood these electromagnetic energy pulses to be very fast-spinning remnants of supernova explosions.

Roald Hoffmann
1937-

Polish-American chemist who won the Nobel Prize in 1981 for his work on the mathematical prediction of electron behavior in chemical reactions. Interested in studying the electron structures and reaction patterns of molecules, he considers himself an "applied theoretical chemist" who uses experimental results on specific molecules as the starting point for generalized models of molecular activity. A professor at Cornell University since 1965, Hoffmann is also an accomplished essayist and poet.

Columbus O'Donnell Iselin
1904-1971

American oceanographer who specialized in analysis of the Gulf Stream current and made early and fundamental contributions to research on ocean salinity and temperature distributions. Iselin became the first director of the Woods Hole Oceanographic Institute in Massachusetts, turning it into a premier institution. His research spanned the properties of underwater acoustics and involved the investigation of world current variations. During the International Geophysical Year (1957-1958) he worked with Britain's John C. Swallow on the delineation of the Gulf Stream. Iselin also taught physical oceanography at the Massachusetts Institute of Technology and Harvard.

Shirley Ann Jackson
1946-

American physicist who was the first African-American woman to earn a physics doctorate. Jackson graduated from the Massachusetts Institute of Technology. She conducted research at the Fermi National Accelerator Laboratory. In 1976 Jackson accepted a Rutgers University professorship and also served as a consultant for AT&T Bell Laboratories concerning semiconductor theory. She was named chairman of the United States Nuclear Regulatory Commission in 1995 to supervise atomic energy resources. Four years later, Jackson was selected president of Rensselaer Polytechnic Institute.

David Jewitt

American astronomer who, with Jane Luu, discovered several small, icy bodies in orbit between planets Neptune and Pluto. Prior to this discovery, the presence of many "Kuiper Belt" objects was predicted. It has even been suggested that Pluto may simply be the largest object in the Kuiper Belt and not a bona fide planet after all. The Kuiper Belt, a ring of comets and other small celestial objects beyond the orbit of Neptune, is thought to be home to many short-period comets, such as Halley's Comet.

Klaus von Klitzing

German physicist who won the 1985 Nobel Prize in Physics for his discovery of the quantized Hall effect. This is one of the few examples of quantum behavior that is directly observable and measurable in the laboratory and refers to the property in which changes in the electrical resistance of a metal plate, cooled to near absolute zero and suspended in a magnetic field, changes in discrete steps rather than continuously.

Charles T. Kowal
1940

American astronomer who in 1977 discovered the astronomical body Chiron inside the orbit of Uranus. Although early newspaper accounts wrongly called Chiron the solar system's tenth planet, Chiron is actually a very large (200-300 km /120-190 miles) diameter comet that travels between Saturn and Uranus. Chiron was among the first of such bodies to be detected, though current theories predict that thousands should be present. Chiron is also one of the smallest bodies to be detected without the benefit of either planetary probes or space-based telescopes.

Harold W. Kroto
1939-

German-born English physicist who shared the 1996 Nobel Prize for Chemistry for, the synthesis and discovery of fullerenes, a class of carbon molecules. Kroto's early interest in chemistry led to the important discovery of carbon chains in space and, later, to the synthesis of C_{60}, the chemical formula for buckminsterfullerene. Kroto is also a dedicated educator and graphic designer.

Tsung-Dao Lee
1926-

Chinese-American physicist awarded the 1957 Nobel Prize in Physics for his theoretical discovery that the conservation of parity was violated in some kaon decays. Lee's interests included astrophysics, field theory, and other exceedingly complex problems in physics before turning to the aspects of particle physics for which he is best known. Hailed by Robert Oppenheimer as one of the most brilliant theoretical physicists of that time, Lee was 31 when he won the Nobel Prize for his work.

Jean-Marie Lehn
1939-

French chemist recognized for his contributions in organic chemistry and the chemistry of large molecules. Lehn was born and studied in France, although he spent some time as a visiting professor at Harvard University. His research involved the recognition, transport, and catalytic properties of very large molecules, with the aim of designing molecular "devices" that might be made into molecular-scale electronic devices. Lehn won the 1987 Nobel Prize in Chemistry for his research in these areas.

William Nunn Lipscomb Jr.
1919-

American physical chemist who determined the structures of compounds containing boron and hydrogen, called boranes, using x-ray crystallographic and electron diffraction techniques and developed the bonding theory to explain their structures. He and his research associates also determined the molecular structure of the protein enzyme carboxypeptidase A and a number of other complex molecules, including the anti-cancer agent vincristine. He won the Nobel Prize for chemistry in 1976 for his work with the boranes.

Jane X. Luu

Vietnamese-American astronomer who, in 1992 with astronomer David C. Jewitt, was the first to discover something orbiting our Sun beyond the orbits of Pluto and Neptune. The object, called 1992 QB_1, confirmed the existence of the Kuiper Belt, a collection of comet-like objects that orbit our Sun at a distance of 30-50 astronomical units (1 AU is the distance from the Earth to the Sun). Since then many more objects in the Kuiper Belt have been discovered by Luu and others.

Theodore Harold Maiman
1927-

American physicist whose design improvements of Charles Townes's solid-state maser greatly increased its utility. He constructed and patented the first working laser in 1960 while at Hughes Research Laboratories. Maiman left Hughes and formed Korad Corporation in 1961 to develop and manufacture commercial lasers. Korad was acquired by Union Carbide in 1968, after which Maiman established Maiman Associates. He was also co-founder of Laser Video Corporation (1972), which developed large-screen laser video displays.

Ben R. Matthias
1918-1980

German physicist who made important contributions to research concerning superconductivity. Superconductivity refers to the phenomenon by which certain materials lose all resistance to the passage of an electrical current at low temperatures. Superconducting magnets are widely used in MRI (Magnetic Resonance Imaging) machines, as well as in high-energy particle accelerators. In 1967 Matthias and his research group developed a compound that became superconducting at a temperature of over 20°K, two degrees higher than the previous record-holder.

Rudolf Ludwig Mossbauer
1929-

German physicist who was awarded the Nobel Prize for physics in 1961 for the discovery of the phenomenon which bears his name. Radioactive nuclei generally emit gamma waves of varying wavelength due to the recoil of the nuclei. When such nuclei are held in a crystal lattice at low temperature, the emitted gamma rays have precise wavelengths since the recoil is absorbed by the lattice. This phenomenon is known as the Mossbauer Effect.

Benjamin Roy Mottelson
1926-

American-born Danish physicist best known for his work on the structure of the atomic nucleus. Mottelson joined the Navy during World War II and was sent for officer's training at Purdue University, where he earned his undergraduate degree. Working with Aage Bohr and others in Europe after receiving his Ph.D., Mottelson viewed the nucleus as a liquid drop, which helped explain many of the properties exhibited during atomic interactions. This work was recognized with the 1975 Nobel Prize for Physics, which he shared with Bohr and James Rainwater.

Robert Sander Mulliken
1896-1986

American chemist recognized for his research concerning the nature of chemical bonds and the electronic structure of molecules. Mulliken's work on the fundamental principles of chemistry overlapped into quantum mechanics and atomic physics, and produced results important to all three fields. He also conducted landmark studies of atomic and molecular spectra. Mulliken received many awards for his work, including the 1966 Nobel Prize for Chemistry.

T. Nakano

Japanese physicist who developed the concept of "strangeness" independently of Nobel laureate Murray Gell-Mann. During the 1950s and 1960s, as new subatomic particles were rapidly discovered, physicists faced the daunting challenge of explaining these new particles and their properties in terms of a self-consistent model. The concepts of "charm" and "strangeness" were developed to explain properties of certain quarks. Other types of quarks are known as "up," "down," "top" (or "truth"), and "bottom" (or "beauty").

Giulio Natta
1903-1979

Italian chemist recognized for his work with polymer chemistry. Natta's early work involved the examination of polymer crystal structures by means of x-ray and electron diffraction, contributing toward better understanding of their molecular properties and making possible the synthesis of improved polymers. He was also the first to create the polymer polypropylene, now widely used for many products and applications. For these and other accomplishments, Natta was awarded the 1963 Nobel Prize for Chemistry.

Kasuhiko Nishijima

Japanese physicist who, with T. Nakano, developed the concept of "strangeness" to refer to properties of some subatomic particles. Strangeness is one of the so-called "quantum numbers" that can be used to describe how these particles behave. Other examples of quantum numbers are the baryon number, hyperon number, and electrical charge. Together, these quantum numbers give a great deal of predictive and interpretive power for understanding and classifying these particles.

Lars Onsage
1903-1976

Norwegian-born, American chemist who won the Nobel Prize in 1968 for his theory of irreversible chemical reactions. He made significant contributions to statistical mechanics, the thermodynamics of solutions, superfluidity, superconductivity, and liquid crystals. He modified the solution theory to take into account Brownian Motion and the interaction of ions with solute molecules and each other. He proposed the fourth or zeroth law of thermodynamics: if two systems are in equilibrium with a third, they are in equilibrium with each other.

Eugene Norman Parker
1927-

American astronomer who first identified the solar wind. Made of hydrogen, helium, and other particles emitted by the Sun, the solar wind causes the tails of comets to point away from the Sun and constitutes a source of charged particles found in the Van Allen radiation belts surrounding Earth. Since Parker's discovery, stellar winds have been found to play a very important role in stellar evolution and the chemical evolution of galaxies.

Charles J. Pederson
1904-1989

American chemist who received the 1987 Nobel Prize for Chemistry, along with Jean Lehn and Donald Cram, for work involving the development of crown ethers and phenolic ligands. Born in Korea to his Norwegian father and Japanese mother, Pederson moved to the United States after graduating from high school. He spent his entire professional career as a research chemist for the Du Pont Company, performing research in a number of areas of chemistry, including polymer and petroleum chemistry.

Arno Allan Penzias
1933-

German-American physicist who, with R. W. Wilson, first detected cosmic background radiation. This diffuse glow coming from all over the sky is most likely a remnant of the Big Bang, and is one of the strongest pieces of evidence supporting the Big Bang theory. This work earned Penzias and Wilson the 1978 Nobel Prize in physics. Penzias worked for Bell Laboratories for many years. He has also written a book about the impact of computers on society.

Waverly J. Person

American geologist who in 1994 reported a huge earthquake that struck deep in the rock beneath Bolivia. This earthquake, measuring 8.2 on the Richter scale, caused virtually no damage, though it was felt as far away as Toronto (over 6,000 miles away). The muted impact of the earthquake may be explained by its depth and the presence of a "seismic channel" that managed to convey seismic waves a great distance, allowing a slow release of earthquake energy along this entire path.

Vladimir Prelog
1906-1998

Bosnian-born Swiss chemist best known for his work on the stereochemistry of organic molecules and reactions, for which he shared the 1975 Nobel Prize for Chemistry. Prelog studied and worked in Bosnia and Croatia until forced to leave by the German invasion during World War II. Escaping to Switzerland with his wife, Prelog continued his research, earning Swiss citizenship in 1959. Prelog eventually became director of the Organic Chemistry Laboratory of the Swiss Federal Institute of Technology.

Ilya Prigogine
1917-

Russian-born Belgian chemist who received the 1977 Nobel Prize for Chemistry for work concerning non-equilibrium thermodynamics and the theory of dissipative structures. Prigogine's family escaped revolutionary Russia shortly after his birth, resettling in Belgium, where he spent the rest of his life. Prigogine was one of the first chemists to explore the thermodynamics of chemical systems in transition between two equilibrium states, arriving at important and fundamental discoveries that were not fully appreciated until much later.

Edward Mills Purcell
1912-

American physicist whose work led to the discovery of nuclear magnetic resonance imaging (now known as magnetic resonance imaging, or MRI). Purcell was also one of the first scientists to detect the ubiquitous 21-cm radiation from neutral hydrogen in interstellar space, now used to map the distribution of hydrogen gas in galaxies and intergalactic space. Purcell shared the 1952 Nobel Prize in Physics for his work on the nuclear magnetic moment studies of liquid and solid helium.

Fredrick Reines
1918-1998

American physicist who discovered the antineutrino and was awarded the Nobel Prize in Physics in 1995. Like many young physicists in the 1940s, Reines worked on the Manhattan Project and continued working for the National Laboratories for several years after the end of World War II. Reines also had a great interest in music, at one time singing in the chorus for a concert of the Cleveland Symphony Orchestra.

Roger Randall Dougan Revelle
1909-1991

American physical oceanographer who provided fundamental theories on sea-floor spreading, the interaction between atmospheric processes and the land and ocean, the carbon dioxide balance and its effects on global climatic change (the "greenhouse effect"), and population studies. Revelle's research at the Scripps Institute of Oceanography (where he served as director from 1950 to 1964) ranged from oceanographic exploration to radiation effects on marine life. He received the U.S. National Medal of Science and the American Geophysical Union created the Revelle Medal in his honor for achievements in ocean science.

Burton Richter
1931-

American physicist who advanced particle physics research methodology. Richter earned a doctorate from the Massachusetts Institute of Technology. He served as director of the Stanford Linear Accelerator Center and directed development of the Stanford Positron Electron Asymmetric Ring. In 1976 Richter and Samuel C. C. Ting were co-recipients of the Nobel Prize for physics in recognition of their detection of a new type of heavy elementary particle, called J/psi, validating the hypothesis that matter has a quark substructure.

Bruno Benedetto Rossi
1905-

Italian astronomer known for his research into cosmic rays and for discovery of the first confirmed x-ray source outside the solar system. In 1962, Rossi and his colleagues discovered an x-ray source in the constellation of Scorpio. Previously, x-rays had been detected from the Sun and from a number of terrestrial sources, but many thought they could not reach the Earth from outside the solar system. Rossi's discovery opened the door to x-ray astronomy, which has provided a great many discoveries of utmost importance about the universe.

Carlo Rubbia
1934-

Italian physicist known for his contributions in particle physics, especially his discovery of the Z_0 particle, for which he was awarded the 1984 Noble Prize in Physics. Rubbia was initially turned down for admission at a college for physics in Italy, and was later granted admission only because another student resigned. He has spent his entire professional life exploring the realm of high energy physics, providing crucial insights into the nature of matter.

Vera Cooper Rubin
1928-

American astronomer who has advanced galactic dynamics knowledge. Rubin earned a Georgetown University astronomy doctorate and joined the staff of the Department of Terrestrial Magnetism at the Carnegie Institute of Washington. Focusing on dark matter, Rubin proved that such invisible material exists by showing that it influences observable celestial objects. She studied galaxies, particularly how they rotate internally and move through the universe. Rubin received the National Medal of Science in 1993. She published *Bright Galaxies, Dark Matters* in 1997.

Allan Rex Sandage
1926-

American astronomer who was a pioneer in radio astronomy and who, in 1961, discovered quasars (quasi-stellar radio sources). Located billions of light-years away, quasars are sources of radio waves, the first radio source discovered beyond the solar system. Quasars are thought to be early stages in the formation of galaxies. He also helped refine the method of measuring astronomical distances, studied the evolution of stars, and related the age of stars to their color.

James M. Schlatter
1942-

American chemist and researcher whose discovery of the sweetness of aspartame (in 1965) led to the development of G.D. Searle & Co.'s most famous product, NutraSweet. Company folklore tells the tale of how Schlatter licked his finger to turn a book's page and tasted a sweet flavor, which the company decided to research. The resulting sugar substitute—180 to 200 times sweeter than sucrose—became one of the world's most widely used sweeteners under the name NutraSweet.

Irwin Shapiro
1929-

American astronomer who discovered the first apparent "superluminal" sources, using very long baseline interferometry. Superluminal jets appear to be moving faster than the speed of light because they are pointed almost directly at or away from the Earth, causing the illusion of having a velocity that is impossible to achieve. This discovery was the first indication that events can take place in the universe that would create such long-lived and fast-moving beams. Superluminal jets have been noted in a number of quasars and other active galaxies as well as in a few very active star systems.

Seigfried Fred Singer
1924-

Austrian-born American physicist and geophysicist who was one of the guiding leaders of the International Geophysical Year (1957-58). Singer was an early proponent of using rockets for geophysical research and the launching of artificial satellites in the late 1940s. (He published *Geophysical Research with Artificial Earth Satellites* in 1956.) Other research by Singer dealt with the origins of cosmic rays, geomagnetic storms, and auroras. His IGY research would make a fundamental contribution to the study of the history of the solar system as well as astrogeological theory.

Richard Errett Smalley
1943-

American chemist who won the 1996 Nobel Prize in Chemistry for his discovery of fullerenes, which, along with diamond and graphite, are the third form of pure carbon. Fullerenes are circular balls of carbon where the molecules are in a soccer ball-like arrangement. Smalley has gone on to develop nanoscale tubes of carbon molecules with similar arrangements, which may see use in the creation of molecular machines.

George F. Smoot
1945-

American physicist who was the chief scientist of the Cosmic Background Explorer (COBE) satellite. COBE measured variations in the microwave background radiation of the universe, the afterglow of the "Big Bang." The results, released in 1992, showed that the background radiation had a temperature of 2.7°K—just above absolute zero—and contained small variations (around one part in one hundred thousand) that provided the strongest support ever of the Big Bang theory.

John Crossley Swallow
1923-

English physical oceanographer and geophysicist who discovered the countercurrent to the Gulf Stream during the International Geophysical Year (1957-58). After conducting extensive seismic refraction experiments and data collection, Swallow joined the British National Institute of Oceanography, where he developed his neutral buoyancy float in 1956. Sinking to a predetermined depth, the float could drift and be acoustically tracked. Swallow's float revealed the Gulf Stream countercurrent, strong mesoscale eddies at all depths in mid-ocean, the Somali current, and equatorial circulation in the Indian Ocean.

Henry Taube
1915-

Canadian-American chemist who studied the mechanisms of electron transfer in metal complexes. Taube's work on the effects of charge transfer on chemical reactivity and the specific mechanisms underlying electron transfer reactions has been of great importance in understanding these fundamental issues in chemistry. In his long career, Taube's contributions have been amply recognized through a great many awards, including the 1983 Nobel Prize in Chemistry.

Samuel Chao Chung Ting
1936-

Chinese-American physicist who has performed groundbreaking research in particle physics. Ting began experimental work looking at the production of particle-antiparticle pairs resulting from photon interactions. This led to the discovery of the J particle, for which he was awarded the 1976 Nobel Prize in Physics. Other work has explored aspects of quantum electrodynamics, production and decay of photon-like particles, and other aspects of high-energy particle physics. Meanwhile, his discovery of the J particle has led to the discovery of an entire family of new, related particles.

Alexander Robertus Todd
1907-1997

Scottish chemist best known for his work on chemicals of biological importance, such as vitamins, nucleotides, and nucleotide co-enzymes. He was awarded the 1957 Nobel Prize in Chemistry. Todd also investigated the properties of vitamins B_1, B_{12}, and E as well as the chemistry of some cannabis species. He was raised to the peerage in 1962 as Baron Todd of Trumpington.

James Van Allen
1914-

American physicist and space scientist responsible for discovering the radiation belts now named for him. The first successful American artificial satellite, *Explorer*, was sent into orbit in 1958 carrying a radiation detector. This noted extremely high radiation levels in distinct belts surrounding the Earth. Van Allen discovered the radiation belts and determined they originated from charged particles from the Sun, trapped by Earth's magnetic field.

Steven Weinberg
1933-

American theoretical physicist who, along with Sheldon Glashow and Abdus Salam, won the Nobel Prize in Physics in 1979 for his work on unifying the electromagnetic and weak forces into the electroweak force. They described how photons involved in the electromagnetic force were related to bosons, a component of the weak force, the key force behind radioactive decay. He is also the author of the 1977 book *The First Three Minutes*, one of the first popular cosmology books for the public.

John Archibald Wheeler

American physicist who coined the term "black hole" in 1969. Wheeler earlier showed how a star collapses to form a black hole. His later work with Sir Roger Penrose showed that all black holes form perfect spheres. With Niels Bohr, he published the first paper explaining nuclear fission in terms of quantum theory. He is also known for his views on the "anthropic principle," the theory that our universe requires participators and observers.

Geoffrey Wilkinson
1921-1996

Scottish chemist who shared the 1973 Nobel Prize in Chemistry for work on organo-metallic compounds. Wilkinson originally studied nuclear chemistry under the tutelage of Glenn Seaborg, changing to the study of metallic and organo-metallic compounds in the 1950s. Concentrating primarily on the transition metals, his work led to a better understanding of metal catalysts that has benefited the field of catalytic reactions, such as the hydrocarbon industry.

Robert Woodrow Wilson
1936-

American physicist, astronomer, and sculptor who was awarded the Nobel Prize for physics in 1978 for the serendipitous discovery, with Arno Penzias in 1964-65, of the cosmic background microwave radiation that had been predicted by Robert Dicke as the energy remaining from the Big Bang that occurred at the formation of the universe. He directed the building of the proton accelerator Femilab, highly regarded for its aesthetic qualities as well as its scientific usefulness.

George Wittig
1897-1987

German chemist who was awarded the Nobel Prize for chemistry in 1979 for studies of a family type of organic molecules containing phosphorus, called ylides, that he discovered. Ylides are used in an organic chemical process known as the Wittig reaction that is used to make alkenes from aldehydes and has been employed in the synthesis of organic molecules such as vitamins and steroids. He also studied structurally strained molecules and the reactions of carbanions.

Chen Ning Yang
1922-

Chinese-American physicist who, with Tsung Dao Lee, showed that parity need not always be conserved in some nuclear reactions. For this, he was awarded the 1957 Nobel Prize in physics. Yang's primary interests in physics, aside from symmetry principles, include studies in statistical mechanics and thermodynamics. Yang is currently a professor at the Institute for Advanced Studies, where he has been a professor since 1955.

Frits Zernike
1888-1966

Dutch physicist whose development of the phase-contrast method, which had a profound impact on high-performance optical microscopy, earned him the 1953 Nobel Prize in Physics. Ironically, in spite of a promising university career, Zernicke's invention of the phase-contrast microscope was virtually ignored until the Nazi invasion of the Netherlands. Seizing on the principle, Germany manufactured the first such devices. After the war, they moved into production world-wide, proving a boon to science and medicine.

Karl Ziegler
1898-1973

German chemist who conducted important research into the chemical properties and technology of polymers. Ziegler also produced significant results in the areas of organo-metallic compounds, some of which are crucial for the synthesis of high-density polyethylene that is widely used in many areas of industry. Ziegler's work earned him a number of prestigious awards and medals during his career, including the 1953 Nobel Prize in Chemistry.

Bibliography of Primary Sources

Books

Feynman, Richard. *QED: The Strange Theory of Light and Matter* (1985). This collection of lectures includes Feynman's discussion of his pioneering work on quantum electrodynamic theory (QED), work that won him a Nobel Prize in physics.

Hawking, Stephen. *A Brief History of Time* (1988). This book became a surprise best-seller and made Hawking the most-famous scientist of his day. In the work, Hawking discussed the origin of the universe and elaborated on black holes, about which he made pioneering discoveries during the 1970s and 1980s.

Hess, Harry H. *History of Ocean Basins* (1962). A central document in the science of plate tectonics, in which Hess offered explanations for the accumulation of ocean sediment, the movement of ocean floor, and creation of midocean ridges.

Sagan, Carl. *The Dragons of Eden* (1977). In this Pulitzer Prize-winning book Sagan discussed contemporary neurophysiology and the genetic origin of human intelligence. His speculation, largely concerned with the history of human behavior, focused on the interplay among three hypothetical stages of human brain development: the first and lowest is the reptilian R-complex, a vestige of our pre-mammalian progenitors that is responsible for aggression and ritual; the second is the limbic system, similar to that of birds and lower-order mammals, from which emotions and religion derive; the third and highest is the more developed neocortex or primate brain.

Sagan, Carl. *Cosmos* (1980). A companion volume to Sagan's acclaimed 13-part television series of the same title, which aired on PBS, this book featured Sagan's ambitious investigations into the design and evolution of the universe. Both the book and the TV series featured a semi-omniscient Sagan as cosmic guide and instructor aided by stunning intergalactic illustrations and photographs.

Teller, Edward. *Better a Shield than a Sword: Perspectives on Defense and Technology* (1987). In this work Teller discussed his advocacy of the Strategic Defense Initiative (SDI), also known as "Star Wars," an effort promoted by U.S. President Ronald Reagan to develop a nuclear defense strategy using satellites that would shoot down incoming missiles.

Periodical Articles

Vine, Frederick. "*Proof of Ocean-Floor Spreading*" (1966). Presented at the December 1966 meeting of the Geological Society of American (San Francisco), this paper put forth data linking the theories of sea-floor spreading, continental drift, and plate tectonics, and promoted its wide acceptance as valid.

Weinberg, Steven. "*Unified Theories of Elementary-Particle Interaction*" (1974). In this article Weinberg discussed his groundbreaking efforts to unify two of the four basic fources: the electromagnetic force and the weak nuclear force. Weinberg performed his research with Abdus Salam and Sheldon Glashow, and the trio won a Nobel Prize in physics for their efforts.

Technology

Chronology

1952 The United States explodes the first thermonuclear weapon, a hydrogen bomb, at the Eniwetok Atoll in the South Pacific.

1955 Indian physicist Narinder Kapany introduces fiber optics, demonstrating the relatively minor loss in intensity over great distances for glass fiber surrounded with cladding.

1957 The Soviet Union launches *Sputnik 1,* the first man-made Earth satellite, thus inaugurating the space age—and the space race between the U.S. and the U.S.S.R.

1958 American electrical engineer Jack St. Clair Kilby builds the first integrated circuit, a concept introduced six years earlier by G. W. A. Dummer.

1960 Seven years after Charles Hard Townes built the first maser (microwave amplification by stimulated emission of radiation), American physicist Theodore Harold Maiman builds the first laser (light amplification by stimulated emission of radiation.)

1969 The U.S. Department of Defense establishes the first packet-switched network, ARPANET—out of which will develop the Internet more than two decades later—to link computers in research facilities.

1971 Ted Hoff designs the first microprocessor, the Intel 4004, a chip smaller than a dime which has all the functions of a computer's central processing unit.

1972 Hewlett-Packard introduces the first hand-held calculator, the HP-35 "electric slide rule."

1972 Philips introduces the first home video-cassette recorder.

1973 The Universal Price Code (UPC), developed by Xerox to automate supermarket operations by signaling a price to the cashier and deducting the item from inventory, makes its first appearance.

1974 The United Nations sets the first international fax standard, allowing facsimile messages to be transmitted at the rate of about one page in six minutes.

1975 The user-assembled Altair 8800 microcomputer makes its appearance, thus inaugurating the personal computer revolution; two years later, Commodore introduces the Personal Electronic Transactor (PET), the first personal computer designed for the mass market, and Apple debuts its highly popular Apple II.

1979 The first commercial network of cellular telephones is created in Tokyo, Japan.

1981 Microsoft Corporation introduces MS-DOS (Microsoft Disk Operating System), which soon emerges as the dominant operating system and spurs a boom in software development.

1991 Linus Torvalds announces the first official version of the Linux operating system. In the subsequent decade, Torvalds makes use of numerous programmers, hackers, and other members of the general public to improve on the OS until it becomes a complete UNIX clone. By 2000 Linux emerges as a viable operating system for the Internet-based computer industry.

Overview: Technology 1950-present

Overview

The foundations of modern technology grew out of developments in electricity, communications, industrial research, and science in the first half of the twentieth century. During that period, technology became much more science-based, invention and innovation became much more deliberate, and the exploitation of electricity provided a new world of electronics for communications, entertainment, and information.

More than any other Western figure, American inventor Thomas Edison (1847-1931) epitomized the changes occurring in technology during the 1900s. His pioneering work with electrical technology in the late nineteenth century, his establishment of an industrial research and development laboratory, and his success in anticipating new markets for communications devices—such as the phonograph, the telephone, and motion picture systems—set an agenda for technological development that shaped the era. The use of electricity to encode, detect, and amplify signals grew out of the previous contributions of Edison and others. These efforts led to radio, television, long distance telephony, and the myriad of consumer products available at the end of the twentieth century for the reproduction of sight and sound.

Edison's Menlo Park research laboratory was the prototype of the commonplace industrial laboratories integral to many modern major corporations. Invention and innovation became less the domain of the lone inventor and more the product of team efforts, particularly in places such as Bell Laboratories, birthplace of the transistor, which miniaturized and transformed electronics. Research-based technological development relied not only on group projects but also on science as a source of innovation. The accelerated research efforts brought about by World War II resulted in electronic computers, jet aircraft, radar, rockets, and nuclear power, all of which demonstrated the efficacy of science in technological change and the increasingly important role of government support in creating new, costly, and complex technology.

These many forces shaping technology at the end of World War II accelerated the rate of technological change and the increased dependence on technology in the Western world. Powered chiefly by electricity and petroleum, this mid-century technology touched more and more people, and their dependence upon it grew. The world of electrical lighting and appliances, electrical power, automobiles, aircraft, and consumer culture created a century of pervasive technology.

The Electronic Age

The development of the transistor in 1947 at Bell Laboratories by William B. Shockley (1910-1989), John Bardeen (1908-1991), and Walter H. Brattain (1902-1987) transformed communications technology in the post-World War II era. This electronic device allowed for smaller, improved and more reliable products such as handheld radios, television sets, various sound reproduction systems, and computers. The availability of the semi-conductor led to the integrated circuit and the microprocessor, which permitted a new method of control and operation for a wide variety of products, most notably the computer, which began as a specialized, expensive machine but was transformed into a universal piece of technology, ranging from super computers to ubiquitous personal computers.

This electronic age, with many of its components developed by research laboratories, gave consumers a new world of choices from cellular telephones to notebook computers, making communication and access to information easier and almost instantaneous. The ongoing conversion from analog to digital systems, first widely used in computers, produced compact disks, digital videodisks, and developing digital systems for both radio and television. Lastly, the rapid expansion and use of the Internet and the World Wide Web changed both information gathering and commerce; the result was a knowledge revolution in full force during the closing years of the twentieth century.

The Nuclear World

Building upon the achievements of the Manhattan Project to produce nuclear weapons during World War II, scientists and technologists refined nuclear technology with the introduction of the hydrogen bomb in 1954 and the development of nuclear reactors in the 1950s. This new technology held the promise of peaceful uses with the introduction of nuclear reactors in the Soviet Union, Britain, and the United States. At the same time, it created the threat of nuclear destruction with refinements in nuclear weapons that played a pivotal role in Cold War diplomacy for much of the last half of the twentieth century. Nuclear technology epitomized a new world of technology based on science, while being funded and nurtured by governments. The initial hopes for the peaceful uses of this technology were overshadowed by the threat of global annihilation inherent in nuclear weaponry.

Aerospace Technology

Just as nuclear technology built upon research and development efforts during World War II, so did aerospace technology. With the marriage of scientific theory and technological experience, the jet engine emerged after the war as a faster, more efficient power source for both commercial and military aircraft. By the 1950s and 1960s jet aircraft dominated commercial air traffic, making business and leisure travel more rapid and more accessible. Faster and more aerodynamically stable aircraft made supersonic flight a reality by the 1970s and reduced global distances even more.

Also, the German efforts with V-1 and V-2 rockets under the direction of Hermann Oberth (1894-1989) and Wernher von Braun (1912-1977) demonstrated that rocketry was a viable means of controlled flight. Both the United States and the Soviet Union created extensive rocket research programs after World War II; the successful Soviet launch of Sputnik in 1957 and the concerned American response led to continuing space exploration, manned space flights, a lunar landing in 1969, space shuttles and the *Mir* space station.

Along with efforts at space exploration, both nations used this new rocket technology as part of the Cold War arms race, especially with the development of intercontinental ballistic missile systems. An ancillary development in the world of rocketry was the use of satellites, first for meteorological and military purposes and then for worldwide communications. The stuff of science fiction in the 1930s became routine late in the century with regular space travel and the transmission of images from around the globe and the cosmos.

Conclusion

Technological developments during the last half of the twentieth century moved mankind further along into a new era of technological change. The industrialism so characteristic of the pre-World War II era, while still fundamental to a technological culture, was being supplanted by an information or knowledge revolution made possible by the process of deliberate invention and innovation, by the science-technology interface, and by an electronic era with its myriad of new devices and processes. Extensive new worlds opened for people as they could access information from countless sources and as they could communicate instantaneously. This impatient culture expected instantaneous interaction and results; at the same time, the information age required thinking globally as the new technology eliminated regional and national borders for both commerce and access to knowledge.

Along with the prospects of the information age were the possibilities of fusion technology as a source of new, clean, unlimited energy. Such a result would fulfill the optimistic promise of a new era in technology envisioned in the 1950s by nuclear scientists and engineers. Continuing research efforts will play a key role in seeking a practical process for this new energy source.

The exploration of space, based on continuing developments in aerospace science and technology, have provided, and will continue to provide, new perspectives and knowledge that will reshape our sense of our world and the cosmos. These efforts could rival the impact of the information age in shaping the technological milieu of a new century.

By the year 2000 technological developments had removed traditional barriers of time, distance, and space that defined technology a century earlier. In the process, the emerging technology relied more on electronic rather than mechanical devices, more on knowledge than materials, and more on information than industrialism.

H. J. EISENMAN

Technology

1950-present

The Development of Integrated Circuits Makes Possible the Microelectronics Revolution

Overview

True revolutions in technology are relatively rare. They mark radical departures from one way of life to another. The use of tools, the invention of movable type, and the construction of the atomic bomb are examples of developments that changed society in a fundamental way. The microelectronics revolution that followed the development of the integrated circuit in 1959 has again remade the world as we know it. In our lifetimes, it has propelled us into the era we call the information age.

Background

The year 1958 marked the tenth birthday of the transistor, a tiny device used to boost electrical signals as they are transferred through it. Transistors were slow to find wide commercial application until 1954, when they were used to produce portable radios. Compared with the bulky vacuum tubes they were designed to replace, transistors were small, cheap, and consumed little power.

The problem with transistors was that the more complex the system that required them, the greater the number of transistors needed. Not only that, but each transistor was equipped with two or three connectors that needed to be attached to something else—for example, electrical components called diodes, resistors, and capacitors—and all the connections that held the packet, or circuit, together had to be made by hand. The workers, often women, who assembled the tiny pieces had to pick them up with tweezers and wire them together. A malfunction among a few of the hundreds of soldered joints in a transistor spelled ruin.

These drawbacks became increasingly apparent toward the late 1950s. At the time, the Korean War was at its height, and because the military had a particular interest in making things smaller, lighter, and more reliable, it led the drive to miniaturization. One proposal the Air Force found attractive was to do away with the individual components and the wire leads altogether. That approach would entail making everything from a single crystal, that is, a single block of a substance called a semiconductor that has properties between those of metal and glass. This concept of making a complete circuit from a single material was dubbed the "monolithic integrated circuit," from the Greek word for "single stone."

The idea of a solid block with no connecting wires was first proposed in 1952 by Geoffrey A. Dummer of Britain's Royal Radar Establishment. Although several years later the RRE awarded a contract to a company to build such a device, the project never progressed very far. Dummer later attributed the failure of the United Kingdom and Europe to exploit electronic applications to war-weariness. Only in 1960 did the RRE form a team dedicated to studying the idea. U.S. companies such as RCA and Westinghouse also attempted similar projcts, usually with support from the military. But there, too, researchers worried that the individual elements of a monolithic integrated circuit would be inferior to components made separately.

In 1958, working alone in the laboratory at Texas Instruments, a physicist named Jack St. Clair Kilby (1923-) wrote in his notebook that he thought resistors, capacitors, transistors, and diodes could all be assembled into a circuit on a single silicon wafer. As Kilby saw it, he could make all the components on one side of a piece of silicon, using a technique known as batch processing that included the interconnections as part of the manufacturing process. Kilby ran the idea by his boss, who told him to test the principle by first making the circuit in the ordinary way, using separate components—but to make them all out of silicon. Because silicon was not available, Kilby used germanium, a greyish-white element. The result was crude, but it worked. Kilby was able to show that integrated circuits could be constructed from a single piece of semiconductor material.

Kilby next turned to improving and refining the techniques he would need to make his circuits. Then in January 1959, a rumor that RCA was planning to patent an integrated circuit of its own spurred Texas Instruments to submit an application in Kilby's name, titled "Miniaturized Electronic Circuits." One month later, the company unveiled its so-called solid circuit at a press conference at the annual Institute of Radio Engineers show. The circuit was no bigger than a pencil point and performed as well as circuits many times larger.

An integrated circuit. *(Charles O'Rear/CORBIS. Reproduced by permission.)*

In the meantime, at rival company Fairchild Semiconductor in northern California, Nobel laureate Robert N. Noyce (1927-1990) was also developing a scheme for making multiple devices on a single piece of silicon with an eye to reducing size, weight, and cost. On July 30, 1959, Fairchild filed an application with the Patent Office in Noyce's name titled "Semiconductor Device-and-Lead Structure."

Nearby, William Shockley (1910-1989), head of Shockley Transistor Corporation, was working on his own integrated circuit. Shockley shared the 1956 Nobel Prize in physics with Walter H. Brattain (1902-1987) and John Bardeen (1908-1991) for the invention of the transistor. He called his circuit the Shockley diode (a valve for electrical current), and he hoped AT&T would purchase it in bulk for use in electronic switching systems for their telephone network. But the components were quirky and switched from "off" to "on" unpredictably. Unlike the methods used by Noyce and Kilby, Shockley diodes were made on both sides of the silicon slice, and constructing them was a painstaking process. For this reason, Shockley's method had been rejected earlier by Noyce and others who went on to prove they were right to abandon it. But Shockley persisted for another year, refusing advice from his staff that they try to produce a simpler device.

Integrated circuits are made by a combination of processes that use chemicals, gases, and light to create transistors by building up thin layers of semiconductor on a silicon wafer, and then etching away or adding material according to a pattern worked out in advance. In a complex chip, for example, the cycle of steps is carried out 20 times or more to form the three-dimensional elements of the chip's circuitry.

Bell Laboratories, where the transistor had been invented, also turned to producing integrated circuits. Their approach was to eliminate as many components and interconnections as possible. This approach turned out not to work, but in the attempt, a scientist named M. M. "John" Atalla realized a breakthrough that had been sought from the early days of transisters, called the *field-effect transistor.* In 1960, taking his discovery one step further, Atalla and a colleague created the first metal-oxide-silicon, or "MOS" transistor, on which most integrated circuits and microchips are now based.

Impact

The first patent to be awarded for integrated circuits went to Noyce on April 25, 1961, and Fairchild was the first company to introduce the new circuits into the market. But by October Texas Instruments, which had been turning out individual circuits by hand, produced an array of silicon circuits the size of a grain of rice that contained two dozen transistors along with the other necessary components. By the year 2000, a fingernail-size sliver of silicon was able to contain millions of transistors, and hundreds of chips could be made on a single wafer.

President John F. Kennedy's (1917-1963) call to put a man on the Moon by the end of the decade created a market for the integrated circuit overnight—nowhere would the advantages of miniaturization be more welcome than aboard spacecraft. Electronics companies such as Motorola and Westinghouse rushed to catch up with pioneers Fairchild and Texas Instruments. *Business Week* magazine announced an "impending revolution in the electronics industry."

That revolution has indeed occurred. Integrated circuits have enhanced our lives in countless ways. The microelectronics industry, to which integrated circuits gave birth, has created millions of jobs. Computers that once would have occupied a space the size of a house have become small, available, and cheap enough for almost anyone to own. Machines run more cleanly and efficiently, medical technology saves lives, and banks the world over exchange money through electronic networks, all thanks to integrated circuits. In poorer countries, technologies built on integrated circuits have decreased the cost of capital investment required for industrialization and development, allowing those countries to compete in the global marketplace.

We have adjusted very quickly to the microelectronics revolution. Washing machines, digital clocks and watches, the scoreboard in a ballpark, the bar code on your groceries, and the collar that lets only your cat to go in and out of its catflap are just a few of the mundane applications of integrated circuits that we take for granted every day.

Microelectronics is about information—ever-increasing amounts of information. And the ability of integrated circuits to store and process this information has redefined the meaning of power. Who controls information, who has access to it, how much it costs, and the uses to which people put it are questions that bear more and more on the conduct of our private and public lives.

Like the transistor from which it developed, the integrated circuit has had repercussions that were totally unpredictable. It is argued that there are limits beyond which miniaturization cannot go, but for now at least, devices have not stopped shrinking. One instance is microelectromechanical systems, a hybrid of machines and electronics the size of a speck of dust that combine the ability of computers to think and of machines to do things. Still largely at the stage of laboratory prototypes, their potential has barely been tapped. But microelectronic devices are not only growing smaller. So-called power electronics uses larger and larger transistors—about the size of a postage stamp—to handle greater amounts of electrical power, for example, to control electric motors.

GISELLE WEISS

Further Reading

Riordan, Michael, and Lillian Hoddeson. *Crystal Fire: The Birth of the Information Age*. New York: W. W. Norton, 1997.

Ryder, John D., and Donald G. Fink. *Engineers and Electrons*. New York: IEEE Press, 1984.

"The Solid-State Century: The Past, Present, and Future of the Transistor." *Scientific American* (special issue, 1997).

The Development of the Maser and Laser Leads to Widespread Commercial and Research Applications

Overview

In 1953 Charles Hard Townes (1915-) produced the first working maser. Masers generate and amplify beams of coherent microwave radiation through stimulation of excited energy states in resonant atomic or molecular systems. MASER is an acronym for this process of microwave amplification by stimulated emission of radiation. Maser principles were extended by Townes and Arthur Leonard Schawlow (1921-) to the optical portion of the electromagnetic spectrum in 1958 when they published the first

An experimental laser beam is reflected through mirrors and filters. *(Chris Rogers/Bilderberg, The Stock Market. Reproduced by permission.)*

detailed proposal for building a laser (light amplification by stimulated emission of radiation). Masers have important though limited applications while lasers are more widely used in research and industry.

Background

In 1951 Charles Townes realized that Albert Einstein's (1879-1955) theory of stimulated emission could be exploited to generate and amplify microwave radiation. According to quantum theory, atoms or molecules exist in certain discrete energy states. The lowest energy level is the ground state, with higher levels being excited states. Moving from one state to another requires absorption or emission of precise amounts of energy in the form of photons of light. Since the wavelength of light determines its energy, photons must have specific wavelengths and no others. When atoms in the ground state absorb photons, they move into excited states. Excited atoms can spontaneously emit this extra energy as a photon or, as Einstein noted in 1916, emission may be accomplished by stimulation from a photon with the same energy. This stimulated emission results in two photons (amplification) of the same wavelength that can then stimulate other atoms. However, since most atoms are in the ground state, more photons will be absorbed than emitted.

Substances containing more atoms in ground states than excited states are said to have "inverted populations." Though rare in nature, inverted populations do occur. Stimulation in such substances produces photon cascades. Townes's key insight was to realized that inverted populations could be created artificially by isolating an ensemble of excited atoms. When placed in a resonating cavity and stimulated by electromagnetic waves of the appropriate wavelength, this unstable ensemble becomes self-oscillating and generates a coherent beam of monochromatic radiation. In 1953, after two years of work with James P. Gordon and H. J. Zeiger, Townes successfully produced a working maser. Various design improvements followed, after which masers were quickly adapted for research, commercial, and military applications.

The precise frequencies generated by masers have made possible better atomic clocks. Maser amplifiers also have extremely low noise and high sensitivity, which enables detection of weak signals in radio and radar astronomy. The most important application arising from applications in astronomy has been Arno Penzias (1933-) and Robert Wilson's (1936-) discovery of the cosmic background radiation predicted by Big Bang theories. They were awarded the 1978 Nobel Prize in physics for their research. Masers are also used for military radar, microwave spectroscopy, and microwave satellite communications.

In 1957 Townes turned his attention to creating an optical maser or laser. As the name suggests, the laser operates with visible light instead of microwaves. Townes and Arthur Schawlow, having earlier collaborated on the classic *Microwave Spectroscopy* (1955), decided to work together on the optical maser. Their "Infrared and Optical Masers" paper, published in the December 1958 issue of *Physical Review,* provided the first detailed theoretical description of a laser and initiated the race to build the first working laser.

All lasers consist of (1) an active medium; (2) a pump; and (3) resonating cavity. The active medium is a collection of atoms or molecules that are raised to excited states. This is typically achieved by optical pumping of electromagnetic radiation with the appropriate wavelength into the resonating cavity. Continued pumping initiates stimulated emission. The resonating cavity usually consists of a pair of mirrors (one only partially reflective) know as a Fabry-Perot etalon. The photon beam is amplified as it is reflected back and forth between the mirrors. When amplification rises above a certain level, the beam passes through the partially reflective mirror.

While at Hughes Research Laboratories in 1960, Theodore H. Maiman (1927-) constructed the first working laser. For the active medium he chose a synthetic ruby rod. The resonator consisted of silver mirrors applied to the rod ends. Optical pumping was achieved by placing the laser rod inside the coil of a helical quartz-xenon flashlamp. The bright lamp pulse stimulated the rod to emit a short ruby florescence. Maiman's ruby laser work was quickly duplicated and employed for various industrial and research purposes.

Impact

While many different types of lasers have been and can be constructed by adopting different lasing materials, pumps, or resonators, relatively few have found widespread use. This is due to trade-offs among efficiency, ease of use, reliability, and cost. For example, the second working laser was also an optically pumped solid-state laser with a uranium lasing medium instead of synthetic ruby. However, because of its low efficiency and cryogenic cooling requirements, the uranium laser has yet to find practical applications.

The high-power output and room temperature operating conditions of Maiman's ruby laser made it the most widely used during the 1960s. One of its first applications was optical ranging, the most dramatic example of which was the determination of the distance between Earth and the Moon to within an inch. Optical ranging has also been used for military targeting and land surveying.

The first working gas laser was the helium-neon laser (1960). This was the first laser to emit a continuous beam, which for many applications is preferable to the pulsed operation of the ruby laser. The red helium-neon laser was produced a year later and, after overtaking the ruby laser in the late sixties, became the most commonly used laser up until the mid-1980s. Red helium-neon lasers are still sold by the hundreds of thousands. Their applications range from supermarket scanners for reading product bar codes to aligning construction and laboratory equipment.

The next important laser development was the CO_2 gas laser (1963). Early gas lasers had limited power, but the CO_2 laser operated in the 10 kilowatt range and is still one of the most efficient lasers. This device made possible full-scale laser welding and machining operations. Its low atmospheric absorption also made it an ideal candidate for battlefield weapons development.

Another important solid state laser is the Nd:YAG (neodymium ions embedded in a yttrium aluminum garnet crystal) laser first successfully demonstrated in 1964. The versatility of the YAG laser has made it the most common crystal solid state laser today. It emits over a range of different wavelengths and has a range of beam durations including continuous emission. YAG lasers are employed in many aspects of materials processing such as cutting, trimming, welding, marking, and laser annealing of electronic components.

These lasers have also made possible new alloys. Traditional alloying methods for surface coating have been plagued by various problems including inefficient use of key materials, structural weaknesses due to excessive heating, and discontinuities between alloy coatings and interior metal. The high power density and short pulse duration of lasers makes it possible to heat only the surface coating and small portions of the underlying metal, thus conserving resources and avoiding excessive heating. A greater degree of control is also exercised over the surface properties, allowing one to select those characteristics best suited to the design requirements. However, laser alloying is most fundamentally different from traditional coating practices in the elimination of the near-surface discontinuities. Laser alloys are continuous extensions of the interior metals, which makes for a more effective coating.

Tunable dye lasers were first operated in 1965. Their wavelengths can be continuously varied across a given range. This feature has made them important for many high-precision spectroscopic experiments. Schawlow's Stanford research group extensively used such lasers and developed advanced techniques to reveal spectra and give improved values for fundamental constants. For this work Schawlow won a share of the 1981 Nobel Prize in physics.

Semiconductor diode lasers were first built in 1962, but it was not until 1975 that the first continuous beam semiconductor diode laser, capable of operating at room temperatures, reached the commercial market. With tens of millions sold each year, they are now the most common lasers. They are used to read compact discs and CD-ROMs, for laser printing and optical data retrieval, and for amplifying attenuated laser signals in fiber optic communication cables.

Other lasers include multi-level devices that generate extremely high peak powers in short pulses for use in thermonuclear fusion research. The development of chemical and x-ray lasers has focused on their potential as high-energy sources for military weapons. Eximer lasers provide powerful sources of pulsed ultraviolet radiation and have various commercial applications. Finally, development of free-electron lasers was heavily funded as part of the Reagan administration's Strategic Defense Initiative because of their potential for generating extremely high power signals at infrared wavelengths. Small-scale free-electron lasers are used in basic research and medicine.

STEPHEN D. NORTON

Further Reading

Books

Ausubel, Jesse H. and H. Dale Langford, eds. *Lasers, Inventions to Application.* Washington, DC: National Academy Press, 1987.

Bertolotti, Mario. *Masers and Lasers: An Historical Approach.* Bristol: Adam Hilger Ltd., 1983.

Bromberg, Joan Lisa. *The Laser in America, 1950-1970.* Cambridge, MA: MIT Press, 1991.

Chiao, Raymond Y., ed. *Amazing Light.* New York: Springer-Verlag, 1996.

Duffner, Robert. *Airborne Laser: Bullet of Light.* New York: Plenum Trade, 1997.

Hecht, Jeff. *Laser Pioneers.* Revised ed. Boston, MA: Harcourt Brace Jovanovich, 1992.

Townes, Charles Hard. *How the Laser Happened: Adventures of a Scientist.* New York: Oxford University Press, 1999.

Periodical Articles

Bromberg, Joan Lisa. "The Birth of the Laser." *Physics Today* 41 (1988): 26-33.

Hecht, Jeff. "Winning the Laser Patent War." *Laser Focus World* (December 1994): 49-51.

Schawlow, Arthur L., and Charles H. Townes. "Infrared and Optical Masers." *Physical Review* 112 (1958): 1940-9.

Nuclear Weaponry

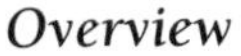

Overview

Humans have warred against one another since the being of time. Until recently, this tendency to wage war endangered hundreds, thousands, or even millions of people, but never our existence as a species. The advent of nuclear weapons, however, has changed the potential outcome of war. First envisioned simply as larger bombs, nuclear weapons ended World War II, precipitated a decades-long Cold War, and functioned as a precarious deterrent to a Third World War, while, at the same time, making all humanity fear its results. For decades, the nuclear club was limited to a handful of nations, but it expanded in the last part of the 1990s and is threatening to increase again. Paradoxically, while the end of the Cold War reduced the chance for global nuclear annihilation, it may have increased the risk of smaller-scale nuclear war between nations with recently acquired nuclear weapons.

Test explosion of a nuclear device, 1957. *(CORBIS. Reproduced by permission.)*

Background

In the final days of the World War II, while America and her allies were readying for a costly invasion of Japan, two nuclear weapons were dropped on Japanese cities, promptly ending the war. The enormous research and development effort that went into the design and construction of these weapons may well represent the greatest scientific and technical effort in human history. The Manhattan Project scientists, 15 of whom were Nobel laureates, was arguably the greatest collection of intellect ever assembled. The irony is that this tremendous effort was made to produce the first weapon powerful enough to threaten human existence and that of most other species on Earth.

Nuclear fission had an inauspicious beginning. In fact, Italian-born American physicist Enrico Fermi (1901-1954) was thought to have proved in the early 1930s that fission was impossible. He received a Nobel Prize for creating transuranic (heavier than uranium) elements by bombarding uranium with neutrons. Soon afterwards, however, Lise Meitner (1878-1968) and Otto Hahn (1879-1968) showed that Fermi's transuranics were actually fission products, caused by splitting uranium atoms. In 1935 Irene Curie (1897-1956) and her husband Frédéric Joliot-Curie (1900-1958) were awarded the Nobel Prize for showing that more neutrons are released from each nucleus during fission than are used to create fission. This sparked the idea of a self-sustaining chain reaction, if properly controlled, or the possibility of nuclear weapons if left carefully uncontrolled.

The impetus for developing American nuclear weapons was a carefully worded letter, written by a number of prominent scientists, that was delivered to President Franklin D. Roosevelt by Albert Einstein (1879-1955). This letter outlined the possibility that a nuclear weapon could be developed—and what could happen if Nazi Germany developed it first. A premier research team, called the Manhattan Project, was subsequently assembled. Charged with creating the first atomic weapon, they worked in strict secrecy. The German atomic weapons program, later revealed to be misguided and misdirected, never reached its goal, and the European war ended before the American's new weapon was completed. The war with Japan, however, lasted long enough and promised to have a sufficiently bloody end that atomic bombs were called into use, bringing that war to a rapid conclusion with devastating effect on Japan.

The immediate postwar period saw the American monopoly in nuclear weapons vanish. The Soviet Union, followed by Britain and France, developed nuclear (and then thermonuclear) weapons. Later, China, South Africa, and a number of other nations followed suit (although South Africa abandoned its nuclear weapons program in 1989). A generally successful international program to halt nuclear weapons proliferation helped limit their spread, although India, Pakistan, and (many suspect) Israel did successfully design and build their own nuclear arsenals. Other so-called rogue states have attempted to either design, purchase, or steal nuclear weapons or the technology to make them. Such attempts by Iraq, under the leadership of Saddam Hussein during the 1990s, raised the stakes of the Persian Gulf War and resulted in the imposition of contentious United Nations inspections after the war was concluded.

Impact

The development and fear of nuclear weapons was a defining aspect of the second half of the twentieth century. For many years, every scientific advance, athletic victory, family vacation, or school class existed in the shadow of imminent nuclear war. The Cold War spurred the space program, drove technological development, financed scientific research, invaded popular culture, and more. For nearly 50 years, people worldwide wondered if, or when, a nuclear war might occur, and many seriously debated whether it would be better to survive such a war or to be among the hundreds of millions killed.

As first conceived, nuclear weapons were simply a more efficient form of artillery. Little thought was given to fallout, cancer, genetic damage, or nuclear winter because these issues were beyond the experience of anyone on earth. Gradually, awareness grew that nuclear war would bring with it all of these things. This led to the atmospheric test-ban treaty, for which American chemist Linus Pauling (1901-1994) was awarded the Nobel Peace Prize. Massive demonstrations against nuclear weapons in Europe and the United States also led eventually to several treaties that first limited and later reduced the size of the superpowers' nuclear arsenals.

At the same time, research by Luis Alvarez (1911-1988) and Walter Alvarez showed that a major asteroidal impact had likely caused the mass extinction of the dinosaurs. Additional research by Carl Sagan (1934-1996) and others showed that such impacts could create a global winter by raising clouds of dust that would block out the sun for several years. When people realized that nuclear war could do the same thing, nuclear winter became something more worry about.

In many ways, the Cold War and nuclear weapons are inseparable. Only the fear of nuclear devastation kept the Cold War from turning hot, and even with that, nuclear war nearly broke out on several occasions. Yet, fears of mutually assured destruction (MAD) also worked to prevent a seemingly inevitable conflict between the two antagonistic ideologies of democracy and communism. This checkmate between the two opposed camps helped to keep war smaller in scale and regional in scope.

THE DISCOVERY OF GAMMA RAY BURSTS

In an effort to better monitor nuclear weapons testing by the Soviet Union in the early 1960s, the United States began sending satellites into orbit that could detect gamma rays given off by nuclear reactions in atomic bombs. In conjunction with seismic readings obtained from stations built to detect atomic explosions, the satellites would let the United States know if the Soviet Union was testing weapons larger than allowed by treaty.

Shortly after being activated, these satellites, code named Vela, began detecting gamma ray bursts (GRBs). However, these bursts were not associated with any seismic events that resembled nuclear detonations, and the characteristics of the burst were far different than those expected from nuclear weapons tests. Scientists finally decided that these bursts of gamma rays had to come from space, since there was no reasonable terrestrial source. Since the Vela program was top secret this information was not released to the scientific community for many years.

Eventually, these gamma ray bursts were discovered by other scientists, who launched their own studies to investigate them. Although a number of theories were floated, in the late 1990s the riddle began to be solved. Astronomers now believe that GRBs are caused by an extreme class of supernovae at faraway points in the galaxy. Because they are detectable at great distances across most of the observable universe, they may also be the most energetic events known to occur.

The amount of scientific research and technological advance wrought by the Cold War is truly incredible. Both the American and Soviet space programs were the result of national pride and fear of the other's missiles. When the Soviet Union launched *Sputnik,* the U.S. military saw, instead of a small satellite in orbit, a potential bomb flying over the pole toward New York. The Cold War and the space program brought the world miniature electronics, home computers, wireless telephones, Velcro, fuel cells, communications and weather satellites, and more. In addition, scientists in any number of specialties benefited, even if their research was only peripherally related to war. Astronomers, for example, benefited from advances in optics spurred by the Strategic Defense Initiative (SDI) (popularly known as "Star Wars"); the Human Genome Project, initially conceived as a way to better understand the genetic effects of radiation, was initiated by the Department of Energy.

The SDI program warrants special mention. Begun under the Reagan administration in the early 1980s, this ambitious program aimed to design a space-based system that could detect and destroy a full-blown Soviet nuclear attack. It is widely thought that Soviet fear of American technological prowess and the rapid development of Star Wars induced the Soviet government to spend more than it could afford on its military efforts. This may well have hastened the collapse of the Soviet Union and the overthrow of Communist governments throughout Eastern Europe at the Cold War's close. Though SDI's initial goals may have been ambitious, subsequent efforts have turned to intercepting small-scale launches by rogue nations rather than stopping a full-scale attack.

The cultural aspects of nuclear weapons were important, too. Movies like the original *Godzilla* and *Them* postulated monsters created by exposure to radiation from nuclear-weapons testing. American school children were taught to "duck and cover" in case of a nuclear attack. People dug bomb shelters beneath or beside their homes in case of a nuclear attack, and other movies, such as *On the Beach, Dr. Strangelove,* and *Failsafe* looked at nuclear war from a number of aspects. In some parts of the world, particularly Europe, these fears may have been even greater, as they were positioned between the two nuclear-armed superpowers.

The end of the Cold War removed many of these fears. Although the United States and Russia still possess enough nuclear weapons to destroy human civilization, both nations have agreed to reduce their stockpiles. That and other political events have since made an all-out nuclear war less likely. Replacing the fear of global nuclear catastrophe, however, is the fear of local nuclear war or of nuclear terrorism. The collapse and impoverishment of the former Soviet Union raised many fears that a combination of desperation, hunger, and poverty could lead former Soviet nuclear scientists to work for rogue governments, helping them obtain the materials and expertise needed to build their own weapons. Where Cold War movies explored total nuclear war, post-Cold War movies imagine nuclear terrorism.

We now find ourselves in a world dominated not by fears of global catastrophe, but, rather, by fears of nuclear terrorism or small-scale nuclear war between minor nuclear powers. It remains to be seen how and to what extent nuclear weapons will shape the geopolitical balance and social consciousness around the world.

P. ANDREW KARAM

Further Reading

Books

McPhee, John. *The Curve of Binding Energy.* New York: Noonday Press, 1994.

Rhodes, Richard. *Dark Sun: The Making of the Hydrogen Bomb.* New York: Simon and Schuster, 1995.

Rhodes, Richard. *The Making of the Atomic Bomb.* New York: Simon and Schuster, 1986.

Serber, Robert. *The Los Alamos Primer: The First Lectures on How to Build an Atomic Bomb.* Berkeley: University of California Press, 1992.

Other

Trinity and Beyond. 1997. Goldhill Videos. Videocassette.

Harnessing Solar Power and Earth's Renewable Energy Sources

Overview

In 1839 a young physicist experimenting with light discovered the photovoltaic effect, which would be exploited to create the world's first solar cells. The resulting photovoltaic cells were slowly perfected over the following century, and in 1954 Bell Laboratories developed the first practical solar cells, made of silicon. By the late twentieth century scientists had embraced these clean, renewable (and modular) energy sources, finding numerous applications to take advantage of the Sun's power. In the course of finding applications for solar power, scientists also searched for alternatives to fossil fuels, turning to other "green" technologies such as wind, hydro, and geothermal power.

Background

From man's earliest days on Earth, he has investigated his natural surroundings, exploring infinite uses for the byproducts of nature—water, stone, wood, metals. As this scientific exploration evolved, man's interest in his world produced a curiosity for the heavens, and he began studying the Sun and the stars in the sky, discovering, in the process, that sunlight provided energy. While the largest solar energy system on Earth occurs naturally in its green vegetation (where the sunlight causes chlorophyll to combine the air's carbon dioxide with water supplied by plant roots, producing carbohydrates—sugars, starch, and cellulose), man searched for synthetic methods to harness the sunlight. In 1839 French physicist Alexandre Edmond Becquerel (1820-1891) observed that light falling on certain materials could produce electricity, a phenomenon he termed the "photovoltaic effect."

Building on Becquerel's observations, scientist Willoughby Smith experimented with other materials, and in 1873 he discovered the photoconductivity of selenium. Soon afterwards, in 1877, William G. Adams and R. E. Day observed the photovoltaic effect in solid selenium, then developed a primitive (it converted less than one percent of the Sun's energy into electricity) photovoltaic cell from the selenium. In 1883 Charles Fritts, an American inventor, improved upon Adams and Day's design with his solar cells made from selenium wafers. As a result of the new selenium cells, and experiments in the 1920s with solar cells made of copper and copper oxide, the photovoltaic cells found their first practical application. Cells of both selenium and copper oxide were adopted in the emerging field of photography for use in light-measuring devices. (In the late twentieth century, light sensors for cameras were still made from selenium.)

In their search for a more practical and efficient solar cell, scientists continued to experiment with photovoltaic materials in the 1940s. In the early 1950s major progress was achieved when a sophisticated crystal growing method, invented in 1918 by Polish scientist Jan Czochralski, was used to produce highly pure crystalline silicon. In 1954 at Bell Laboratories in New Jersey, researchers Daryl M. Chapin, Calvin S. Fuller, and Gerald Leondus Pearson (1905 -) used the Czochralski process to create the world's first practical photovoltaic cell, made from crystalline silicon. (The patent for their "Solar Energy Converting Apparatus" was issued in 1957.) at first, the cells could perform with an efficiency of only 4%, but, within a few months, the researchers had raised efficiency to 6%. Soon, the efficiency was 11%. The researchers' next obstacle was in producing a silicon cell that was less costly. This seemed an insurmountable problem. Had it not been for mankind's growing interest in space, the technological advancement of photovoltaic cells might have been stalled.

Impact

In July 1958 the U.S. Congress passed the National Aeronautics and Space Act, which created the National Aeronautics and Space Administration (NASA) and jumpstarted U.S. space science and exploration. Before that, however, U.S. scientists had been researching the launch of a satellite to collect geophysical data in and above the earth's atmosphere as part of the International Geophysical Year (IGY) research program. A key aspect of this satellite project was a search for a lightweight, long-lasting power source. The answer was found in the silicon solar cells originally designed by the team at Bell Laboratories.

In March 1958 the U.S. launched the world's first photovoltaic-powered satellite, *Vanguard 1*, which was the second U.S. satellite de-

A photovoltaic system at the National Renewable Energy Laboratory in Golden, Colorado. *(U.S. Department of Energy. Reproduced by permission.)*

ployed into orbit above the earth. The satellite power system operated for eight years aboard *Vanguard 1*, a breakthrough for solar cell technology as its success led to large contracts from NASA and the expansion of the photovoltaic industry. By the late 1960s at least four U.S. companies were producing hundreds of thousands of solar cells each year, many earmarked for the growing satellite industry. (In the late twentieth century, solar cells power virtually all Earth-orbiting satellites.)

Achievements in solar cell development during the buildup of the space program included a major increase in cell efficiency and reductions in cost, but photovoltaic cells had still not found many terrestrial applications. Then, the mid-1970s oil embargo saw fossil-fuel costs soar dramatically. As a product of the oil crisis, the U.S. Department of Energy funded the Federal Photovoltaic Utilization Program, resulting in the installation and testing of thousands of solar power systems. Other applications followed.

In 1973 the University of Delaware built one of the world's first photovoltaic residences, Solar One. In 1982 the first photovoltaic power station was brought online in Hysperia, California. By the early 1990s electric utilities in 20 U.S. states had installed nearly 2,000 cost-effective photovoltaic systems. The Gulf War of 1990 again triggered interest in non-fossil fuel energy alternatives, and by the end of the 1990s there were several hundred thousand photovoltaic systems installed worldwide. In 1998 a team of Harvard University researchers led by physicist Eric Mazur discovered a new "spiky" black silicon whose future in photovoltaic technology was still being researched at the end of the century. As solar cell technology continued to progress, new discoveries like this brought on even more possibilities.

While the emerging technology of photovoltaics converted the Sun's energy into electricity, progress was made investigating other solar energy alternatives—mainly addressing the transfer of the Sun's heat to a fluid, which was then used to warm buildings, heat water, and even generate electricity. Solar thermal activities such as these were nothing new. In 100 A.D. historian Pliny the Younger (62-113) built a summer home in Northern Italy that featured thin sheets of mica windows in one room, which got hotter than the others, saving on short supplies of wood. By the sixth century sunrooms on houses and public buildings were common. Advances in glass technology, such as double and triple pane windows and low emissivity glass that employed a coating to allow heat in but not out, were a large contributor to further building efficiencies.

Just as buildings harnessed the Sun's energy for centuries, solar heating for water and other functions had also been practiced for many years. Solar thermal collectors, first built in 1767 by Swiss scientist Horace de Saussure (1740-1799), found many applications. In 1891 Baltimore inventor Clarence Kemp patented the first commercial solar water heater. After photovoltaic cells were perfected, the 1970s saw President Jimmy Carter (1924 -) authorizing the installation of solar water heating panels on the White House. Solar pool heating technology wasn't far behind. Solar thermal electric power systems also developed that focused the Sun's power on heating water to make steam that was subsequently used to rotate a turbine attached to an electric generator. In 1998 in Southern California, these thermoelectric systems met the energy needs of over 350,000 people and were part of the world's largest solar energy power plant.

Besides the expansion of solar energy systems, the energy dilemma spotlighted by the 1973 oil embargo and the 1990 Gulf War brought about advancements in other renewable energy technologies. Wind power, actually generated as a result of temperature differences on the earth's surface caused by the Sun, had been one of the first natural energy sources utilized by man. Windmills, said to be invented by the Chinese and used as early as 200 B.C. by the Persians, were updated in the 1970s. Government-sponsored wind turbine development programs flourished in the United States in the twentieth century. Tax credits for investors (which expired in 1985) resulted in nearly 7,000 turbines being installed in California alone between 1981 and 1984. Advancements in design reduced the costs of wind power, making it cost-competitive in many electric power applications.

Another inexpensive renewable energy source explored was that of hydropower, which harnesses falling water, usually in rivers. The most common form of hydropower was found in dams on these rivers. By the mid-1990s hydropower was generating about 10% of the United States' electricity—supplying enough energy to power whole towns, cities, and even entire regions of the country. Water also played a part in early usage of geothermal power. For at least 10,000 years mankind had used hot springs for cooking, refuge (they were neutral zones where members of warring nations could bathe together in peace), and, of course, relaxation. The exploitation of geothermal resources deepened beyond hot springs in 1892, when the world's first district heating system in Boise, Idaho, was served by pipes from their hot springs. In 1922 John D. Grant founded the United States' first geothermal power plant near The Geysers in California. More plants followed and alternative uses of geothermal energy, including agribusinesses such as crop drying (introduced by Geothermal Food Processors, Inc. in 1978), were initiated. By the 1990s engineers and scientists were developing technologies that would allow mankind to probe more than 10 miles below the earth's surface in search of geothermal energy. Along with photovoltaic cells and solar thermal technologies, hydro and geothermal power became viable alternatives to fossil fuel by the end of the twentieth century.

From the early efforts of researchers to harness the Sun's energy through photovoltaic cells, to discoveries such as a "spiky" black silicon, from windmills to sophisticated turbines, scientists endeavored to learn more about the myriad uses for the byproducts of nature. As scientists continue to explore all of these renewable energy technologies, constantly improving efficiencies and reducing costs, it may one day be possible to eliminate most of the world's fossil-fuel consumption.

ANN T. MARSDEN

Further Reading

Books

Asimov, Isaac. *How Did We Find Out About Solar Power?* New York: Walker & Co. Library, 1981.

Butti, Ken, and John Perlin. *A Golden Thread: 2500 Years of Solar Architecture and Technology.* Palo Alto: Cheshire Books, 1980.

Perlin, John. *From Space to Earth: The Story of Solar Electricity.* Ann Arbor, MI: Aatec Publications, 1999.

Internet Sites

U.S. Department of Energy. "About Photovoltaics." http://www.eren.doe.gov/pv/

U.S. Department of Energy. "Learning about Renewable Energy. http://www.eren.doe.gov/erec/factsheets/rnwenrgy.html. National Renewable Energy Library, 1985.

Advances Related to Silicon Transistors Spur the Microelectronics Revolution

Overview

In 1895, using principles of electricity worked out in the earlier part of the century, the Italian engineer Guglielmo Marconi (1874-1937) sent a radio signal more than a mile (1.6 km). The work of Marconi, and of those before him, laid the groundwork for investigations into the electrical properties of solids that opened ways of manipulating electrons to do useful work. The result of these endeavors was a humble device called the transistor, which formed the basis for the revolution in microelectronics that is now reshaping the world. It won for its inventors the 1956 Nobel Prize in physics.

Background

Radio waves are easy to send, but they are difficult to detect. They are difficult to detect because the current in an antenna oscillates back and forth, or alternates, and earphones require one-way bursts of current. Twenty years before Marconi's transmission, Ferdinand Braun (1850-1918), a German physicist, discovered that current in certain crystals flowed more easily in one direction than another. Moreover, pressing one of two metal contacts into the crystal face caused the current to flow in a single direction. Braun called this phenomenon *rectification,* and he created inventions based on it, including crystal sets, that made it possible to transmit voices and music by radio. In 1909 he shared the Nobel Prize in physics with Marconi for developments in wireless telegraphy.

In 1904 an English physicist named John Ambrose Fleming (1849-1945) created a rectifying vacuum tube. Fleming used this device—a glass bulb with the air sucked out and equipped with two charged parts called *electrodes*—to successfully improve the incoming signal in a radio-receiving system. American inventor Lee de Forest (1873-1961) took Fleming's device a step further by adding a third electrode to control the current. He showed his invention, which he called the *audion*, to AT&T, which was looking for a way to boost electrical signals along its long-distance telephone lines. Any call over a distance of more than 2,000 miles (3,219) dissolved into static. At the time de Forest's device worked better for low voltages than for the high voltages required to amplify telephone currents. But A&T was impressed enough with the potential of the gadget to invest in research to develop it. In January 1915 the company successfully tested a coast-to-coast telephone call over its 3,400-mile (5,472 km) transcontinental line.

During World War I AT&T adapted the vacuum tube to wireless communications. It became known as the "electron tube," and in the first part of the twentieth century it came to dominate the new technology of electronics, which was the name given the branch of physics that studied the behavior of electrons in a vacuum. An early limitation of the vacuum tube, however, was that it was not useful for short wavelengths, which were of interest in electrical communications. So during the 1920s and 1930s, physicists turned again to the kinds of crystals used in crystal sets, which were a class of substances called *semiconductors.*

Semiconductors are neither electrical conductors like metals nor insulators like rubber and glass. They fall in between, and they also have unique properties of their own. Unlike metals, which are more resistant to electrical conductivity as they heat up, semiconductors become less resistant as temperature rises. Moreover, they are very sensitive to light. The study of metals, insulators, and semiconductors is known as *solid-state physics*. It concerns itself with the arrangement and behavior of the atoms that compose solids.

On January 1, 1925, AT&T's research department incorporated into a separate entity called the Bell Telephone Laboratories. By the mid-1930s Walter H. Brattain (1902-1987) and William Shockley (1910-1989) had both been hired to work at Bell Labs on the physics of semiconductors. They were joined in their efforts after World War II by John Bardeen (1908-1991).

The United States' entry into the war in 1941 redirected research priorities toward the war effort. But almost as soon as it was over in 1945, understanding how and why semiconductors worked became a focus of research aimed at developing commercial uses for semiconductors in radio and communications. Vacuum tubes had made radio transmission possible and long-distance calling practical, but their disadvantages were increasingly apparent. They were big, expensive to manufacture, hot, and unreliable. In early computers the glow they gave off attracted bugs and moths, which short-circuited the computers. (Ever since, the term "debugging" has been used to describe fixes for computer problems.) And the mechanical relays in telephone systems required regular cleaning and adjusting. Solid-state solutions to these problems would have the advantage that they did not require mechanical parts.

Brattain, Shockley, and Bardeen, along with a few others, were known as the semiconductor group. Interdisciplinary but closely knit, the group worked in a spirit of give and take that was later credited in large measure with the success of their effort. In beginning their investigations of alternative amplifiers and replacements for mechanical relays, the group made two major decisions: to focus on silicon and germanium, the simplest semiconductor materials, and to focus particularly on the surface properties of the semiconductors. Using these materials, they experimented with switching and modulating electric currents.

On December 23, 1947, the three successfully demonstrated within Bell Labs the product of two years of labor essentially performed by Brattain and Bardeen: a crude-looking contraption less than an inch (2.5 cm) high, assembled from a paper clip, a chunk of germanium, a thin plastic wedge, and two slivers of gold foil. When they powered up the device, the oscilloscope to which it was connected showed that, as it was being transferred, the electrical input signal was being amplified many times over. They called the invention the *point-contact transistor.*

A transistor is essentially an electronic switch. It has two functions, to turn on and off, like a light, and to run high and low, or modulate, in the same way that a dimmer switch allows a light to grow dimmer or brighter. Unlike a light switch, which is controlled by your hand, a transistor is controlled by an electrical current. A remarkable ability of the transistor is that it can conduct current or block it, whichever is needed. When you speak into a microphone, for instance, the transistor adjusts the mix of electrical signals coming from your voice and the signals from the power supply to boost the sound of your voice coming out of the speaker.

The invention was kept secret for months afterward while patent applications were being filed. The public first heard of the transistor at a press conference held July 1, 1948. Spelling out the word "transistor," the director of research at Bell Labs explained to reporters that it was a "little bitty thing," the equivalent of a vacuum tube, only better. Moreover, it was "composed entirely of cold, solid substances." The account in the *New York Times* ran on page 46 in the "News of Radio" column, and stated simply that a device called a transistor with several applications in radio had been demonstrated for the first time Bell Labs.

Bardeen and Brattain's gadget proved the principle of the transistor very well. But it was unwieldy, which meant that manufacturing it would be difficult. Even before the world learned of the existence of the transistor, Shockley set to work developing a device that would be easier to make and to use. His solution was called a *junction transistor*, and it consisted of a small strip of either silicon or germanium with three wires attached, one at each end and one in the middle. The Bell Labs press release dated July 4, 1951, announced the tiny structure as "a radically new type of transistor which has astonishing properties never before achieved in any amplifying device." Although in 1952 point-contact transistors entered commercial production for use in telephone switching equipment, they never really established themselves in the marketplace. When manufacturers finally decided to produce transistors, the junction transistor was the one they chose.

The first transistors were made from germanium. Although silicon is the second most plentiful element in the earth's crust, it is hard and brittle and has other properties that make it more difficult to use in making transistors than germanium. But silicon-processing techniques developed in the 1950s exploited the advantages of these properties. And in 1954 Texas Instruments produced the first commercial transistors made from silicon. Silicon eventually became the semiconductor material of choice.

On November 1, 1956, Shockley, Brattain, and Bardeen were awarded the Nobel Prize in physics for the invention of the transistor. *Newsweek* magazine printed a picture of the three scientists and reported, "This year the prize went, for a change, to an eminently practical little American invention."

Impact

No one could have predicted the impact of the transistor. To be sure, its advantages were many. It was small, needed no warmup time, and required about as much power as that generated by a flea jumping once a minute.

But by the mid-1950s the transistor had still not caught on in the popular culture. It was too expensive for use outside of such applications as hearing aids and military communications. All that changed in 1954, when the first portable transistor radios appeared, and consumer electronics was born. Although U.S. leadership in consumer applications would wane in the 1960s, transistors had created a billion-dollar semiconductor industry.

The trend to miniaturization did not start with the transistor age. Already in 1948, as a result of the military's emphasis on smaller and lighter applications, vacuum tubes had shrunk to the size of a pencil head. But in 1948 a transistor whose performance was comparable to that of a vacuum tube was half the size of a paper clip. Twenty years later, with the efficiency made possible by an innovation called the integrated circuit, a chip could pass through the eye of a needle. By the year 2000 a modern processor chip the size of a fingernail was able to carry 3 to 5 million transistors and cost virtually nothing to make.

The information age and what we call the global village would be unimaginable without the transistor. We chat to our friends and families on the Internet or over mobile phones across long distances, we witness on television events taking place on the other side of the globe beamed via satellite, the Hubble telescope sends back pictures of storms on Mars—these things as well as the myriad devices that are ordinary parts of our lives, for example, digital wristwatches, pocket calculators, children's games, and video cameras—are all based on the transistor.

GISELLE WEISS

Further Reading

Books

Riordan, Michael, and Lillian Hoddeson. *Crystal Fire: The Birth of the Information Age.* New York: W. W. Norton, 1997.

Ryder, John D., and Donald G. Fink. *Engineers and Electrons.* New York: IEEE Press, 1984.

Other

"50 Years of the Transistor." Internet article. http://www.lucent.com/ideas/heritage/transistor/

"The Solid-State Century: The Past, Present, and Future of the Transistor." *Scientific American* (special issue, 1997).

Nuclear Power

Overview

Nuclear power is a valuable and controversial source of energy in many nations around the world. Originally heralded as a reliable source of clean, safe power, two accidents raised concerns about the wisdom of continued reliance on nuclear power. However, with a limited store of fossil fuels, concerns about greenhouse gas emissions, drawbacks in alternative power sources, and the limited natural resources of many nations, nuclear power is likely to remain an important source of power for many years.

Background

The first nuclear reactor of which there is a record began operation about two billion years ago in a sandstone bed in what is now Gabon, in the western part of Africa. This natural reactor operated intermittently for tens of millions of years, unheralded and unknown until discovered by French scientists in the 1970s. It is not known if the Oklo reactor (named after the part of Gabon where it lies) was unique in the world or if other natural nuclear reactors operated at other times, but there is no doubt that this body of uranium ore did indeed operate as a nuclear reactor when algae was the most advanced life on earth.

Research into nuclear fission began in the 1930s, when Lise Meitner (1878-1968) and Otto Hahn (1879-1968) first showed that uranium atoms could be made to break apart, or fission, when struck by a neutron. At about the same time, Leo Szilard (1898-1964), a Hungarian-born American physicist realized how a nuclear chain reaction could be made to be self-sustaining, producing power. These ideas and many more came together during the Manhattan Project when Enrico Fermi (1901-1954) led the effort to design and build the world's first artificial nuclear reactor at the University of Chicago.

The atomic-electric generating station at Shippingport, Pennsylvania. *(The Library of Congress. Reproduced by permission.)*

In a short period of time, more nuclear reactors were constructed to help make plutonium for atomic weapons.

After the Second World War, nuclear reactors were used primarily for research and plutonium production at first. However, nuclear fission is much more efficient than burning fossil fuels, thus the development of nuclear reactors for power generation was seemingly inevitable. The first use of nuclear power for peacetime energy production occurred on a small scale in the Soviet Union in 1954. England followed in 1956, and the United States in 1957. The *Nautilus,* the world's first nuclear submarine began operation in 1954, inaugurating Naval nuclear power.

Since then, the use of nuclear power has continued to grow worldwide, in spite of highly publicized accidents at Three Mile Island, Chernobyl, and in Japan. However, this growth has not been without controversy, as various groups argue over the risks that nuclear power generation may entail. Opposed to nuclear power are those who feel that the resultant waste cannot be safely disposed of and that every nuclear reactor is a potential disaster. On the other side of this debate are those who point out that nuclear power has cost far fewer lives than any other form of widely used power generation, that radioactive waste can be safely disposed of, and that we are less sensitive the possible ill effects of radiation than popular opinion might believe. However, this debate remains unresolved and is likely to continue for some time.

Impact

Power production is perhaps the most visible function served by nuclear reactors, in both civilian and military use. But nuclear reactors are also used to manufacture radioactive isotopes for medical treatment and scientific research, and to manufacture plutonium and tritium for nuclear weapons. They are used in scientific research in branches of engineering, geology, biology, and other fields, and some are used to make drinking water out of seawater. The use of nuclear power has generated widespread debate, as noted above, about the role of science and technology in society. In addition, for many reasons the use of nuclear energy for power production has become almost inextricably linked with nuclear weapons. Although this is no more appropriate than linking the petroleum industry with fire bombings, the linkage persists, coloring any debate about nuclear energy.

Power Generation and Desalinization

A nuclear reactor and its cooling towers is perhaps the most recognizable symbol used by environmentalists to represent technology run amok. However, in spite of these objections, it must be admitted that nuclear reactors provide, and will continue to provide an important source of power to the world. France, Japan, Belgium, the United States, and other countries all depend on nuclear power to some extent. This use of nuclear power has contributed to lower use of fossil fuels and reduced emissions of many pollutants in the atmosphere and water. In addition to the civilian use of nuclear power, several nations use nuclear reactors to provide a reliable source of power for naval vessels, which eliminates their dependence on refueling and, in the case of submarines, their ties to the ocean's surface.

Nuclear reactors can also be used to desalinate seawater for use as drinking water or for irrigation. This is done routinely in nuclear-powered naval vessels, all of which generate steam using nuclear power and then use the steam to boil seawater. The steam is pure water that is then condensed and used for drinking, cooking, or other uses. In a similar manner, large nuclear power plants can be used to produce many millions of liters of fresh water per day in places such as the Middle East, where deserts stretch to the edge of the sea.

THE COLD FUSION DEBACLE

In 1989 two relatively unknown scientists in Utah announced that they had discovered a way to generate energy using hydrogen fusion in a small and simple setup that fit on a benchtop. Called "cold fusion" because it did not require the very high temperatures of conventional fusion research, it took the world by surprise.

The promise of fusion power is that it is a nearly limitless source of energy, such as that which powers the Sun. For several decades scientists and engineers have been inching steadily closer to the goal of "ignition," the point at which a fusion reaction will sustain itself long enough to produce more energy than it consumes. Until ignition is reached, fusion will not be a viable source of energy for humanity. Current thinking is that, to reach ignition, temperatures of millions of degrees are needed. To reach these temperatures, facilities have been built costing tens and hundreds of millions of dollars. Many researchers feel that commercial fusion power may be a reality by 2050 or so, but nobody is at all certain because of the great theoretical and engineering difficulties yet to be overcome.

Cold fusion held the promise of changing this paradigm, perhaps making fusion power available within a decade. For several weeks researchers around the world tried to replicate the Utah experiment. To verify fusion is taking place, one would need to see production of both heat and neutrons. Some experiments seemed to show neutron generation but no heat, while others showed the opposite. Finally, after several months, the overwhelming scientific consensus was that cold fusion had not been discovered.

Isotope Manufacture for Scientific, Medical, or Military Use

All chemical elements have a given number of protons that give them their unique chemical properties. Most elements can contain varying numbers of neutrons, giving them different atomic weights. For example, uranium 238 has 92 protons and 146 neutrons, while uranium 235 has 92 protons and only 143 neutrons. Atoms of an element with different numbers of neutrons are called nuclides or isotopes. Some elements, such as uranium, naturally occur with several isotopes, while other, such as technetium or promethium, do not even occur in nature. It was recognized early in the twentieth century that radiation might be successfully used in research or for medical treatments. After World War II, scientists realized that nuclear reactors could be used to manufacture radioactive isotopes of elements that could be used for scientific research or to diagnose or treat disease. Today, radioactive isotopes of sulfur, hydrogen, carbon, iodine, and phosphorus are routinely used in research, and isotopes of technetium, iodine, gallium, thallium, palladium, and others are routinely used in medical care. Some isotopes, such as iodine 131, can be given to patients to treat cancer or to help image diseased organs. Others, such as radioactive gold, can be inserted into a tumor to irradiate and kill it. Many of these isotopes are generated in nuclear reactors designed and built specifically for this purpose. Recently, particle accelerators have been used to fabricate these radionuclides, too. Nuclear reactors are also used, of course, in the manufacture of isotopes of plutonium and hydrogen for nuclear weapons. It is a peculiar irony that the same process can be used to make isotopes for either war or medicine. To date, however, the number of medical reactors is greater than that of reactors used for military isotope production.

Environmental and Health Concerns

Much of the debate over nuclear energy centers on the environmental effects of continued use of nuclear power. In related arguments, many feel that the risks of using large sources of radiation are simply too great. The response from industry professionals is that radiation is a tool and, like many other tools, it must be treated with respect but not with fear. However, "radiation phobia" has stirred public interest like few other subjects. This fear is reflected in the case of magnetic resonance imaging (MRI), a diagnostic medical apparatus originally called nuclear magnetic resonance imaging (NMR). Since so many people are frightened by anything containing the word "nuclear," a name change was made.

Although radiation professionals disagree with the arguments and fears that are used to justify anti-nuclear power rhetoric, these arguments have been generally effective, regardless of their factual accuracy or validity. One of the largest battlegrounds in this controversy is in the arena of health effects. Some feel that exposure to any amount of radiation is harmful and should be avoided, while others point out that we live in a background radiation field that is higher in some places than others, raising the possibility that some radiation exposure may not cause harm. This group also notes that all living organisms have mutation repair mechanisms that are quite adept at repairing radiation damage, further suggesting that exposure to some levels of radiation may be safe. This is, in fact, a major scientific controversy that remains unresolved. Only time will tell which side of this controversy is correct.

These arguments have successfully hampered efforts to construct repositories for both low- and high-level radioactive wastes in the United States, and have caused some governments to decide to phase out the use of nuclear power altogether. In part, this success may be traced to the continuing association of nuclear power and nuclear weapons in the mind of the general public. However, in many cases, the anti-nuclear power activists have simply been better organized and more vocal than their opponents, winning their arguments almost by default. In any event, this argument is likely to continue for some time, as the long design life and high cost of nuclear power plants makes them unlikely to be shut down on short notice.

P. ANDREW KARAM

Further Reading

Knief, Ronald. *Nuclear Engineering: Theory and Technology of Commercial Nuclear Power.* New York: Hemisphere Publishing Corporation, 1992.

Murray, Raymond. *Understanding Radioactive Waste.* Edited by Judith A. Powell. Columbus: Battelle Press, 1994.

Nero, Anthony V. *A Guidebook to Nuclear Reactors.* Berkeley: University of California Press, 1979.

The Development of Computer Languages and Programmers

Overview

One of the principle strengths of the modern electronic digital computer is its ability to be programmed to perform a wide variety of useful and disparate functions. Originally designed as "super-calculators" for limited use in military and scientific computation, computers have become one of the most ubiquitous technologies of late twentieth-century society. What makes the computer so powerful is its enormous flexibility: given the appropriate software, an inexpensive and mass-produced computer chip can emulate the function of many more costly, special-purpose devices. The remarkable success of the computer industry in the United States is in large part due to the ability of programmers to develop software applications that appeal to a broad range of corporate consumers. The cornerstones of this "software revolution" are the computer programming languages used to create versatile and efficient software.

Background

The earliest electronic digital computers were designed and constructed for military or scientific purposes and were generally large, expensive, and designed for speed and reliability rather than ease of use. Programmers used numeric machine codes to communicate directly with the computer's hardware in order to achieve the high level of performance required by repetitive scientific computations. Since programming costs represented only a small percentage of the total cost of owning and operating these computers, and the amount of software development that occurred in this period was small, there was little incentive to develop expensive programming tools. A number of organizations developed assembler programs that allowed programmers to write software using simplified mnemonic codes instead of esoteric machine language, but for the most part programming in the early 1950s required extensive knowledge and a painstaking attention to detail. Individual programmers often developed reputations for their idiosyncratic styles and displays of virtuoso programming technique.

As commercial computers became less expensive and more widely used by corporations interested more in processing data than crunching numbers, the need for new programming methods and techniques became increasingly apparent. Computer manufacturers wanted to ensure that their devices were accessible to the broadest range of corporate consumers. Programmers hoped to eliminate some of the tedious clerical work associated with machine code and assembly language. Corporate managers wanted to free themselves from a dependence on apparently eccentric programmers. A whole host of new products appeared aimed at making programming less difficult and time consuming. Most of these so-called "auto-coders," however, simply exchanged one confusing and incomprehensible set of mnemonic shortcuts for another. As programming projects became larger and more complex, the costs of software development increased dramatically; by the middle of the 1950s it was estimated that programming and debugging accounted for as much as three-quarters of the cost of operating a computer.

The first widely used programming language, called FORTRAN (FORmula TRANslator), was developed by the IBM Corporation in response to the rising costs of software development. The head of the FORTRAN development team, mathematician John Backus (1924-), was outspoken in his belief that programming in the 1950s was "a black art" lacking in generally accepted standards and principles and overly dependent on the individual programmer's "private techniques and inventions." FORTRAN allowed programmers to describe their programs using relatively comprehensible algebraic expressions, rather than in cryptic assembly code. The FORTRAN compiler translated these algebraic expressions into the machine-level code required by the underlying hardware. One of the principle reasons behind the widespread adoption of FORTRAN was its ability to produce efficient machine code that would run almost as fast as that produced by the experienced programmers. Another is that it was supported by IBM, by then an industry giant. FORTRAN was an essential component of IBM's successful line of Model 704 computers, and by the end of the 1960s a version of FORTAN was available on almost every computer ever made up until that point.

FORTRAN was by no means the only high-level programming language developed in the 1950s, however. By the end of the decade, the proliferation of new computers and programming tools had created a veritable Tower of Babel of competing languages and dialects. A demand for a universal programming language developed among user groups and industry consortiums. In 1957 the Association for Computing Machinery (ACM) began work on a universal programming language called ALGOL (ALGOrithmic Language). Although ALGOL was never widely adopted outside of academia, the ACM effort highlighted the benefits of language standardization. In 1959 an influential group of government, military, and industry leaders held a meeting at the Pentagon to discuss the need for a common business-oriented programming language. The outcome of this and other meetings was the development of COBOL (COmmon Business Oriented Language). COBOL rapidly became a de facto standard within the business and defense community, largely as a result of a 1960 Department of Defense decree that the military would longer purchase or lease any computer that did not have a COBOL compiler available. Despite the many changes that have occurred in computer and programming technology since the early 1960s, COBOL remains the world's most widely used programming language.

The same developments in the computer industry that created a need for new programming languages also led to a demand for more and better programmers. By the end of the 1950s a number of universities had established programs in computer science and computer engineering. At Dartmouth University in New Hampshire, Professors John Kemeny (1926-1992) and Thomas Kurtz (1928-) began work on what was to become the BASIC (Beginner's All-purpose Symbolic Instruction Code) programming language. Their purpose was to create a language that could be easily learned by non-technical undergraduates. BASIC was part of a whole system of technologies (including time-shared computers and teletype terminals) designed to allow Dartmouth students immediate and intimate access to computers. Although not as fast or powerful as FORTRAN or COBOL, BASIC was nevertheless a useful and capable language, especially for the specific pedagogical purposes for which Kemeny and Kurtz designed it. Millions of students who otherwise would have been intimidated by programming have learned the fundamentals of computing through their exposure to BASIC.

By the end of the 1960s a "software crisis" was brewing in the computer industry. Despite the widespread adoption of standardized languages like FORTRAN and COBOL, software development costs continued to skyrocket and numerous large programming projects either exceeded their budgets or failed altogether. A number of related developments contributed to the growing sense of crisis: the increase in the size and complexity of software applications; an influx of inexperienced and poorly trained programmers; and the failure of corporate executives to recognize that programming was a highly creative activity difficult to predict and control using traditional management methods. Whatever the underlying causes, the "software crisis" provoked a new interest in developing programming languages that would speed development and encourage good programming practice. The PASCAL programming language, named after the seventeenth-century French mathematician and philosopher Blaise Pascal (1623-1662), was particularly suited to a "structured programming" approach to software engineering. First developed in 1970, PASCAL was widely adopted by the academic community as the ideal tool for teaching proper technique to aspiring computer scientists and programmers.

As the commercial computing industry expanded and new computer technologies developed, the needs of the programming community evolved accordingly. In 1965 the Digital Equipment Corporation (DEC) introduced a new class of relatively small and inexpensive "mini-computers." These computers were less powerful than traditional mainframes but were accessible to a broader range of organization and were extremely popular. Many of these mini-computers did not have the resources required by highly structured languages such as PASCAL. Innovative computer scientists developed languages specifically suited to this new technology. Ken Thompson (1943-) and Denis Ritchie, two researchers at Bell Telephone Laboratories in New Jersey, created the UNIX operating system and the C programming language on a DEC PDP-8 mini-computer. Since the UNIX operating system was written in C, the success of the one contributed to the adoption of the other. The C language and its successors (C+ and C++) are the most widely used programming languages for workstation and personal computer software development.

Impact

The development of useful and efficient programming languages has had a dramatic influence on the adoption of computer technologies by contemporary society. In the early 1950s it was not at all clear to many corporate observers what computers were good for or how they could best be utilized. Standardized programming languages such as FORTRAN and COBOL allowed these companies to share their software and experience with others, thereby encouraging the spread of information, personnel, and techniques. Each of the many programming languages that were developed in this period served a different purpose: FORTRAN enabled scientists to make more efficient use of computers; COBOL provided features peculiar to the computing needs of businesses; BASIC and PASCAL allowed universities to train the next generation of programmers and computer-savvy executives. The move away from machine code and assembly language towards more algebraic and English-like statements expanded the user base of computers and made computers more accessible and understandable by a broad public audience. Finally, the accumulation of programming expertise and techniques allowed for the further development of more sophisticated and imaginative software.

NATHAN L. ENSMENGER

Further Reading

Books

Campbell-Kelly, Martin and William Aspray. *Computer: A History of the Information Machine.* New York: Basic Books, 1996.

Ceruzzi, Paul. *A History of Modern Computing.* Cambridge, MA: MIT Press, 1998.

Kraft, Philip. *Programmers and Managers: The Routinization of Computer Programmers in the United States.* New York: Springer-Verlag, 1977.

Sammet, Jean. *Programming Languages: History and Fundamentals*. Englewood Cliffs, NJ: Prentice-Hall, 1969.

Weinberg, Gerald. *The Psychology of Computer Programming*. New York: Dorset House Publishing, 1998.

Wexelblat, Richard, ed. *History of Programming Languages*. New York: Academic Press, 1981.

Periodical Articles

Fritz, W. Barklay. "The Women of ENIAC." *Annals of the History of Computing* 18 (3): 13-23.

Xerox Introduces the First Photocopier

Overview

The first fully automatic photocopier was introduced by the Xerox Corporation in 1959. It was called the 914, and it could make 7.5 copies per minute on any kind of paper. At that time few companies realized how important the photocopier would be, how many millions of dollars Xerox would make, and how it would become an essential tool of every modern business.

Background

The history of the photocopier is largely the history of the Xerox Corporation. The driving force behind the company was Chester Carlson (1906-1968), an American physicist and patent lawyer. In the 1930s, while he was working as a patent clerk he found that there were never enough copies of a patent. He wanted a better method than copying by hand or sending it out for photographic duplication.

Carlson began to study the problem, and his researches led him to the field of photoconductivity. He found that certain metals' and alloys' electrical conductivity will change after being exposed to light. Carlson's flash of inspiration was quite simple: if you shone an image onto a photoconductive surface, there would be different degrees of electrical current through the light and dark regions. If there were a way to attract some sort of ink to the different regions of conduction, a reproduction of the image could be made.

In 1938 Carlson and an assistant, Otto Kornei, made the first photocopy by projecting a slide image onto a piece of sulfur-coated zinc. This charged up the surface. After that they covered it with a powder of lycopodium (the spores of a moss). They blew off the powder and saw that it had glued itself to the sulfur, showing the words "10-22-38 Astoria," a fuzzy but true reproduction of their slide.

Carlson took his invention to Kodak, IBM, General Electric, and other companies, all of which turned him down. In 1944 Carlson met a researcher from the Battelle Memorial Institute, a non-profit group which funded his research. Carlson and Battelle joined up with the Haloid Company, a maker of photographic paper, in 1946. Eventually the two merged and changed the company name to XeroX in 1961. (The X at the end was in imitation of Kodak, and was later changed to a lower case x.) The word came from their term for photocopying, xerography—*xero* was Greek for dry, and *graphos* was writing. Today the generic term photocopying has replaced xerography. (At one time the Japanese used the term ricohing, after the name of that country's best-selling photocopier.) Xerox produced the first manual photocopier in 1949. This machine, the Model A, was difficult and messy to use and was not very successful.

After 10 years of redesigning their photocopier Xerox produced the first fully automatic photocopier, the 914. With a massive advertising campaign the copier first went on sale in late 1959, with delivery beginning in 1960. Xerox leased their products, and replaced and repaired all their machines. Xerox was quickly filling orders as fast as it could. Two years after the 914's introduction they had sold $60 million worth of photocopiers. By the middle of the 1960s Xerox had revenues of nearly half a billion dollars.

Impact

Before the 914 there were four ways of copying documents: by hand, photography, carbon copies, which transferred impressions through multiple sheets of paper held in a typewriter, or the mimeograph, a machine that made copies with ink from a specially prepared master document. Any company that wanted to make thousands of copies of a document had to contract out to printing companies. The costs associated with this were large and were beyond the reach of most smaller companies and individuals. Xerox's 914 changed all of this, because companies leased their machine and were charged according to the number of copies made.

After the successful introduction of the 914, Xerox was looking for ways to expand its market. The first way was to make a smaller desktop copier—the 914 weighed 650 pounds (295 kg). The 813, the first desktop copier, was developed during the early 1960s and was released at the end of 1963. Like the 914, it sold extremely well. From the early 60s to the early 70s Xerox was one of the fastest-growing stocks in the world.

Modern photocopiers work on Carlson's original principles, but the innards are much different. The first stage is to give the photoconductive surface, a hollow cylinder called the drum, a charge by running a small electric current through it. This surface, usually made from selenium, is kept in the dark to preserve its charge. Next, the document to be copied is illuminated by passing a light over it. Mirrors reflect the light coming off the document onto the rotating drum. Anywhere the light hits the drum, the electrical charge is dispersed. Wherever there is text or an image, the charge is conserved. Then toner, microscopic particles of black dust with an opposite charge, is passed over the drum by a series of belts: it sticks only to the charged (dark) areas. Now that the toner is stuck to the drum, a sheet of paper with a small static charge is passed over the drum. The toner is transferred from the drum to the paper by static electricity. The paper is pressed to ensure adhesion and heated to dry the toner. The copying process is now complete and the paper is ejected from the photocopier.

With the successes of Xerox it was inevitable that other companies would enter this market. Many Japanese companies, including Canon, Ricoh, Minolta, and others soon had competing products in the market. They introduced their first models during the 1970s. But the products were of an extremely low quality: some even caught fire. They had little chance of making a dent in Xerox's domination of the market. Over time the quality improved and many of these companies were able to erode Xerox's market share by entering the low-end of the market and working their way up. These copiers were less expensive than Xerox's, and weren't high-performance machines. But with companies that made few copies, or where speed was not a priority, the Japanese competitors were successful.

One example is the Ricoh Corporation, which introduced the Savin 750 in the summer of 1970. It cost two and a half times less than Xerox's closest comparable model, and had one other striking advantage: it used liquid toner. Xerox's toner was a powder that needed to be melted, sprayed onto the paper, then cooled to harden. All these steps forced the copying process to require at least a few seconds. The Savin 750 had liquid toner, which reduced this to a one-step process. It was simpler than Xerox's method, which made it faster and cheaper.

Today photocopiers are found everywhere, ranging from offices to schools to libraries to convenience stores. Millions of copies are made daily throughout the world with the touch of a single button. This ease of use has also brought with it some problems. Many publishers worry about plagiarism and copyright infringement. When it is easy to copy 30 pages out of a textbook many publishers fear they are losing revenue when pages and chapters are extracted from books by individuals and groups to avoid paying for entire, often costly, textbooks and reference works. The law governing this in most countries is called fair use, where individuals are allowed to make limited copies for their own personal use. As well, many universities and schools have permission to make limited copies of portions of academic journals and textbooks for use by students.

Modern copiers are versatile machines with many features. Photocopiers can collate multiple pages into any order, staple papers, fold them, punch binder-holes, and copy onto both sides of a sheet of paper. Photocopiers can accommodate more varieties of paper; images can be placed on transparencies and other materials. Almost any size paper can be used, sometimes measuring over four feet square. The speed of copying has also gone up—high-end copiers can now make up to 150 copies per minute of a single page.

Color copiers, which began to arrive on the market in the 1970s, work on similar principles to black and white photocopiers. They work more slowly because they make a copy in stages, analyzing how different primary colors mix to form the final image. Color images are created by scanning the image multiple times; each time the document is scanned it is seen through different color filters. After breaking the document down to its components colors, four different colored toners (yellow, cyan, magenta, and black) are used to build up a color image by layers.

Digital technologies have changed the process of making copies. Digital photocopiers can store the image of a page in memory and then print as many copies as required by using the stored copy instead of the original. This allows a user to walk away with his original while the copying process continues. A digital copier also has other features that give people more control over the quality of the copy. Using controls on the copier, hole-punch marks, margin notes, or other defects can be removed. Images can also be moved, magnified, or centered on the finished copy. Xerox also invented the laser printer, which uses a laser to trace an image onto a photoconductive surface, instead of reflected light. Their first model, the 9700, was released in 1977.

PHILIP DOWNEY

Further Reading

Books

Dessauer, John H. *My Years With Xerox: The Billions Nobody Wanted.* Garden City, NY: Doubleday, 1971.

Jacobson, Gary and John Hillkirk. *Xerox: American Samurai.* New York: Macmillan, 1986.

Kearns, David T and David A. Nadler. *Prophets in the Dark.* New York: HarperCollins, 1992.

Other

Scientific American: Working Knowledge: Photocopiers. http://www.sciam.com/1096issue/1096working.html

Stanford University Libraries: Copyright and Fair Use. http://www-sul.stanford.edu/cpyright.html

Xerox—The Document Company. http://www.xerox.com

A Brief History of Robotics since 1950

Overview

The term "robot" comes from a Czechoslovakian word for "work" used in the 1921 play by Karel Capek called *R.U.R.* ("Rossum's Universal Robots") to describe an army of manufactured industrial slaves. Since then, we have come to think of robots as the mechanical men or "androids" of modern science fiction. In reality, technical manuscripts from as early as 300-400 B.C. reveal that human beings have been trying to build automated machines or "automata" for centuries.

The development of modern robotics was precipitated by the advent of steam power and electricity during the Industrial Revolution. A growing market for consumer products drove engineers to devise ways of producing automatic machines to speed up production, do tasks that humans could not do, and to replace humans in dangerous situations. In 1893 Canadian professor George Moore produced "Steam Man," a prototype for a humanoid robot made of steel and powered by a 0.5 horse-power steam engine. Essentially a gas boiler housed in what looked like a mechanical suit of armor, it could walk independently at a rate of 9 miles per hour (14.5 kph) and pull light loads. In 1898 inventor Nikola Tesla (1856-1943) demonstrated a model for a remotely operated submersible boat at Madison Square Garden. Tesla also wrote that he believed it possible to someday build an intelligent, autonomous humanoid robot. Tesla's ideas were not taken seriously until well into the twentieth century. In fact, the robotics industry as we know it emerged only around the mid-twentieth century. Once research and development teams began to work in earnest, however, robots were integrated into manufacturing and gradually adapted to the military, aeronautics and space, medical, and entertainment industries.

Background

By the 1950s engineers were developing machines to handle difficult or dangerous repetitive tasks for both defense and consumer manufacturing—particularly the booming automotive industry. Because robots were meant to replicate the pattern of movement that a human would make while lifting, pulling, pressing, or pushing, designs were based upon the anatomical structure and movement of a human arm. These were modified versions of the first patents for robotic arms filed over a decade earlier. For example, patents for both the "Position Controlling Apparatus," filed in 1938 by Willard V. Pollard, and a spray-painting apparatus by Harold A. Roselund, filed in 1939, were modeled on human shoulder-arm-wrist configuration and dexterity. Roselund's design patent was granted to the DeVilbiss Company, which would later become a major supplier of robotic arms in the United States. These early prototypes were not mass produced. However, once electronic controllers came into use after the Second World War, similar but more efficient designs were developed, including the first computer-controlled revolute arms from Case Western Reserve and

Robots assemble automobiles at a Ford Motor Company plant in Detroit. *(Bettmann/CORBIS. Reproduced by permission.)*

General Mills in 1950, and a complex, hydraulically powered robotic arm by the British inventor Cyril W. Kenward, who filed his patent in 1954 and published it in 1957.

"Planetbot," one of the first commercial service robots in production, was a hydraulically powered robotic arm first used by a division of General Motors in the production of radiators during the mid-1950s. Eventually, approximately eight Planetbots were sold. The company claimed its robot could easily perform 25 individual movements and could be reset to perform a different set of operations in only minutes. However, this early model proved unsuccessful because it was controlled by a cumbersome mechanical computer, and it behaved erratically when the hydraulic fluid was cool. By the 1980s the Planet Corporation had developed a more sophisticated and efficient hydraulic arm, which has been successfully used for forging operations.

"Unimate," designed by George Devol and patented by Devol and Joe Engelberger, was originally used to automate the production of television picture tubes. The movement of Unimate's 4000-pound (1,816 kg) arm was controlled by commands stored on a magnetic drum. In 1962 it was integrated into General Motors Corporation production to sequence and stack hot, die-cast metal components. After the bugs were worked out of its design, Unimate became a popular feature in assembly lines. Of the approximately 8500 machines originally sold, more than half of them were used in automotive manufacturing plants. Today, there are approximately eight models of Unimate available, boasting payloads of from 50 to 500 pounds (23 to 227 kg). They have been adapted to such applications as material handling, spot welding, die casting, and machine tool loading, with an advertised 98 percent reliability rate.

Some of the most significant early contributions to robotics were sponsored by agencies outside of consumer manufacturing. The Case Western Reserve arm developed by Norman F. Diedrich was supported by the Space Nuclear Propulsion Office. A similar design, the "Programmable Universal Manipulator for Assembly" (PUMA) developed in the 1960s by Victor Sheinman, a graduate student at Stanford University, was intended for microsurgery. PUMA was eventually licensed and improved by the Unimation Corporation in 1978.

Taking advantage of advances in other technologies, developers gradually integrated more sophisticated computer controls and precision components into their models. Higher degrees of freedom—how far a component could move away from "home" position—led to greater versatility. The addition of finger joints in some models created manual dexterity for grasping, holding, and positioning objects. In 1960 a General Electric research team headed by Ralph Mosher developed "Handyman" and "Man-Mate"—two remotely operated robotic arms. Handyman, a two-arm electro-hydraulic teleoperated robot, had two hinge-type jointed fingers that could grasp objects via a single command from the human operator wearing the Handyman apparatus. Other robotic "arms" were actually based on the human spine. The multiple joints in these "serpentine" arms allowed for flexibility required in product-inspection operations.

A different kind of robotic arm configuration was used in NASA's Viking mission to Mars during 1975-76. The Viking landers, designed by the Martin Marietta Corporation, had to be designed with the extreme environmental conditions of Mars in mind. Instead of the heavy, jointed industrial arm, the Viking arms were made of two light, ribbon-like extenders rolled onto a drum. The two halves unfurled and connected, creating a tube to scoop samples from the planet's surface. Although the arm control mechanism had some bugs in it, operators on Earth were able to guide the robot through a repair procedure, making researchers optimistic about aerospace telerobotics—human control of robots from a remote location.

The desire to have machines that could follow commands or operate by themselves through many complex operations required special programming and controls for the machines. Some researchers came to believe that the best way to control what machines did for people was to find some artificial means to simulate the way that humans thought, remembered, and responded to their environments. Thus the study of artificial intelligence (AI) grew up alongside robotics engineering. The term "Cybernetics"—the study of the relationship between human and machine intelligence—was coined by scientist Norbert Wiener (1894-1964) in the 1940s, while he and a colleague, Dr. Arturo Rosenbluth, were working on ways to improve the automatic controllers in military aircraft. Since that time researchers have been experimenting with computer simulations of human thought. Although researchers disagreed about how the human brain works, the AI projects that arose from their debate have become a significant factor in the development of robots.

Another trend in robotics was to create mobile robots that could operate independently of humans. In order to do this, the robot must be able to avoid stationary and moving objects in its path. "Shakey," a primitive version of a mobile service robot based on this idea, was built at the Stanford Research Center in 1968. It had a vision sensor (a motorized camera and range-finder) positioned above a central processing unit (computer) on a cart. Shakey was propelled by two motorized wheels and two obstacle-sensing bumpers. It could apply logic-based methods of problem solving that allowed it to recognize the shape of objects, push them, and negotiate a ramp. Like Shakey, service robots of the 1980s were essentially "brains in a box" that ran along pre-mapped pathways laid out on the floor and could recognize and maneuver around obstacles. More recent models, like the RoboKent(r) SweeperVacs and Recycling ScrubberVacs from the Servus Robots Company, learn the area to be cleaned by themselves. They operate independently, without reference targets, and have built-in obstacle detection and avoidance protocols. Robots have cleaned more than offices: they have been used for larger, environmental hazard jobs, such as the cleanup at the Chernobyl nuclear power plant in the Soviet Union, following a major radioactive explosion there in 1986.

Shakey's ability to learn from interaction of its sensors with the environment also became the basis for the small, insect-like robots developed by researchers such as Rodney Brooks of the Massachusetts Institute of Technology (MIT) beginning in the 1980s. Brooks believed that the interaction of machine sensors with their environment creates a learning situation for robots similar to that of a human infant. According to this theory, it is not necessary to build complex computers with thousands of stored facts to control the robot. Instead, simple motor and sensor elements are combined to create robots that learn from experience. Brooks built robots modeled on multi-legged insects such as "Genghis"—designed to negotiate the kind of rough terrain that would be encountered on other planets. NASA planned to utilize "Hermes," a smaller version of Genghis, to explore the surface of Mars.

Many researchers argue that a humanoid robot is best for adapting to the human environment. When Rodney Brooks and Andrea Stein co-founded the COG humanoid robot project at MIT, their goal of producing an android that could behave like a human being and interact with human beings was considered controversial. Soon, other similar projects were in development. Vanderbilt University School of Engineering has been working on a humanoid robot called "ISAC." ISAC, like COG, is learning to interact with human people in a natural way. The projected use for ISAC is to perform as an in-home care-giver. Other university-based robotics research in the United States, including projects at the University of Southern California, the University of California-Berkeley, and Georgia Tech are destined for use in medical, private and military security capacities, or in environmental hazard situations. U.S. and European robotics research is tackling issues such as mobility, robot-human interaction, vision systems, speech imitation and recognition, and cognition.

These topics are nothing new to Japanese robot designers. Japan, which has almost twice as many robots as the United States, is developing android type robots for wide-ranging uses—to work outside of space stations and spacecraft, to interact with its growing geriatric population in hospitals and at home, and to act as civil servants in urban centers.

By 1986 the Honda Motor Company had completed "P-1," meant to "coexist and cooperate with human beings." Honda expected that by the time their humanoid robot was perfected, such robots would be used in everyday life to serve humans, not just in special operations. In the fall of 1997 Honda completed "P-3," which looks like a suited astronaut. It has a backpack with a 136-volt battery, wireless receiver, and processing unit. Commands are transmitted by the wireless ethernet modem. According to Honda, P-3's vision sensing system is able to identify stairs and other objects in a room, walk up stairs, and restabilize itself when pushed off balance.

At Waseda University, Japan, researchers have been working with another android project, "Hadaly." Like MIT's COG, Hadaly still looks much like a Leggo project with cameras for eyes, but researchers in other labs are working hard at simulating a human brain to someday operate inside of androids like Hadaly. Meanwhile, Fumio Hara and a team of researchers at the Science University in Tokyo have been working on an industrial "face robot" that can identify dozens of human expressions and make facial gestures itself. Recently, the face robot has been fitted with a simulated skin, hair, and eyes. Its designers believe that worker interaction with a humanoid face that provides emotional as well as verbal response will help reduce industrial accidents.

Impact

As with technology in general, corporate and industrial developers and independent inventors alike have enthusiastically adapted robotics to the entertainment and public relations industries. Corporations like Sony are beginning to market robotic pets that look and behave like cats or dogs, for people with allergies, or those who don't have the time to take care of a real pet. Since the 1980s novelty robots have appeared at trade shows, conference openings, and in safety programs at grammar schools.

Robotic technology has been integrated into in every facet of our lives, from manufacturing to military strategies, medicine, and other public and private service industries including environmental cleanup, space and underwater exploration, and entertainment. The integration of artificial intelligence (AI) and robotics during the last quarter of the twentieth century has resulted in predictions that androids—autonomous humanoid robots—will be a part of our everyday lives before the end of the twenty-first century.

LISA NOCKS

Further Reading

Advanced Research & Robotics Homepage. http://www.ar2.com/default.htm

Cohen, John. *Human Robots in Myth and Science.* 1st American ed. South Brunswick, NJ: A.S. Barnes, 1967.

Conrad, James M. and Jonathan W. Mills. *Stiquito for Beginners: An Introduction to Robotics.* IEEE Computer Society, 1999.

Geduld, Harry M. and Ronald Gotesman. *Robots, Robots, Robots.* Boston: New York Graphic Society, 1978.

Moravec, Hans. *Robot.* New York: Cambridge UP, 1999.

"Robonaut." Johnson Space Center, NASA, Automation, Robotics & Simulation Division. http://tommy.jsc.nasa.gov/er/er4/robonaut/Robonaut.html

Rosheim, Mark E. *Robot Evolution: The Development of Anthrobotics.* New York: Wiley, 1994.

The Advent of Modern Supertankers Facilitates the Transportation of Petroleum and Results in Environmental Catastrophe

Overview

Easily the largest movable man-made objects ever constructed, supertankers were created and designed to meet society's enormous demand for petroleum. From fueling our cars to supplying heat to our homes, supertankers have made it possible for many nations to maintain high standards of living. Although supertankers have facilitated the transportation of enormous amounts of petroleum over thousands of miles, they have also caused some of the largest environmental disasters in history.

Background

The first oceangoing tanker ever constructed was the German-designed *Glückauf*. Launched in 1866 to transport petroleum from the United States to Europe, the *Glückauf* was 300 feet (91 m) long, 37 feet (11m) wide, carried 2,300 short tons (2,088 metric tons) of oil, and had a cruising speed of about nine knots (17 kph). Today, the *Glückauf* could easily fit into the hold of a supertanker, the largest of which is the *Jahre Viking*, which is over 1,500 feet (457 m) long, 227 feet (69 m) wide, and weighs over 565,000 deadweight tons. Indeed, ships this large are no longer even called "supertankers," as that term does not adequately capture their size. Instead, they are called "very large crude carriers" and even "ultra large crude carriers."

Some of the reasons that allowed tankers to grow to so large were, of course, technological advances in engineering and ship-building material. Though these technological advances made the construction of supertankers possible, the initial impetus for constructing them largely was political. The Suez Canal in Egypt, constructed in 1869 and enlarged periodically since then, is the primary trade route for ships and supertankers traveling from the oil-rich countries around the Red Sea to the Mediterranean Sea and points beyond. Although a convenient alternative to sailing around the Cape of Good Hope at the southern tip of Africa, the Suez Canal has, on a number of occasions, been closed due to armed conflicts. Such closures forced ships to travel the great distance around the Cape in order to get to the Mediterranean, Europe, and the Americas. However, because ship-builders no longer had to consider the Canal's dimensions, larger tankers could be built. By constructing larger tankers for the longer hauls, the additional cost and time it would take to sail around the Cape would be compensated by the fact that each tanker could carry more petroleum on each trip.

Perhaps the most important reason for increasing the size of tankers involved exploiting the so-called "square/cube rule." Assume, for example, that a tanker's hold is 4 feet (1.2 m) high, 4 feet wide, and 4 feet deep. Given these dimensions, the surface area of the hold is 96 square feet (9 square meters) (4 times 4 feet per side, times 6 sides) and the volume of the hold is 64 cubic feet (1.8 stere) (4 feet high, times 4 feet wide, times 4 feet deep). If the dimensions of the tanker's hold are doubled to 8 feet in height, width, and length, however, the surface area of the hold becomes 384 square feet and the volume becomes 512 cubic feet. In other words, by doubling the hold's dimensions, the surface area increases 4-fold, but the volume increases 8-fold. Thus, even though the cost of construction increases in terms of additional material required, the carrying capacity, and therefore potential revenue, is increased by a much larger proportion. Consequently, by building larger oil tankers, profit margins increase.

Impact

Although the development of the modern supertanker was a triumph for the shipbuilding industry and a boon to oil-producing nations, it has had a serious negative side-effect: modern supertankers involved in accidents have caused enormous damage to the environment. It was not very long after the first supertankers were constructed in the early-1960s that disaster struck in a way that the shipbuilding industry had not fully anticipated.

On March 18, 1967, the 118,285 deadweight-ton tanker *Torrey Canyon* went aground on the Seven Stones rocks reef, off the coast of Cornwall, England, spilling nearly 35 million gallons (132,489,420 liters) of crude oil. The resulting oil slick was so vast that some of it

A supertanker ship transports oil. *Corbis # VC002371. Reproduced by permission.)*

spread across the entire English Channel all the way to France, covering over 260 square miles (673 square km). Although the spill took place in the ocean, it is worth noting that just one gallon (3.78 liters) of oil can ruin up to an estimated one million gallons (3785,412 liters) of fresh water—the equivalent of a year's supply of water for 50 people. Thus, a spill on the order of magnitude as the *Torrey Canyon* potentially could ruin over 35 trillion gallons (132 trillion liters) of fresh water, or equivalently, the drinking supply of 1,750,000,000 people for a year—nearly a third of Earth's population!

Since the *Torrey Canyon*, there have been even larger spills. On March 16, 1978, for example, the tanker *Amoco Cadiz* went aground near Portsall, France, spilling an estimated 65,000,000 gallons (246 million liters) of crude oil. An even larger spill resulted from the July 19, 1979, collision of the *Atlantic Empress* and the *Aegean Captain* off Trinidad and Tobago. It is estimated that an incredible 88,000,000 gallons (333 million liters) of crude oil was spilt in that accident.

Unfortunately, oil spills have continued to plague our oceans, the marine life that resides therein, and the humans whose lives are tied economically to the health of the sea. Perhaps the most infamous of recent oil spills is that of the *Exxon Valdez*. In the early morning hours of March 24, 1989, while her absentee captain lay asleep, the *Exxon Valdez* rammed into Bligh Reef in the middle of the pristine waters of Prince William Sound, Alaska. Though the grounding spilled only 11 million gallons (42 million liters) of oil (just one-fifth of the amount of oil she held), her grounding nevertheless resulted in the largest oil spill in United States history. Indeed, the spill polluted more than 1,300 miles (2,092 km) of Prince William Sound shoreline and severely degraded the economies of the surrounding fishing communities. It is no surprise that the impact of the spill was still being felt more than a decade later.

Despite the fact that the Exxon Corporation spent more than two billion dollars cleaning up the Sound, only two species—of dozens affected—have fully recovered from the effects of the spill. The Exxon Valdez Oil Spill Trustee Council, which is charged with administering the continuing clean-up efforts, has determined that contrary to initial beliefs that the oil would quickly degrade, much of the oil remains in the soil around the Sound. Thus, whenever a storm churns the Sound's water, more oil is drawn out of the soil and into the water, re-polluting the Sound. Even at seemingly infinitesimal levels, even one part of oil per billion parts of water is toxic enough to affect the reproductive cycle of salmon and herring. Consequently, the Sound likely will continue to suffer the adverse effects of the spill for decades to come. Unfortunately, what the long-term effects will be, and how and when they will manifest themselves, remains largely unknown.

It was not until the *Exxon Valdez* disaster that governments, especially the United States Government, began to closely study the long-term effects of such disasters. As a direct result of the *Valdez* grounding, the United States Congress passed the Oil Pollution Act of 1990, requiring ships transporting oil to have double hulls so as to decrease substantially the likelihood of similar disasters in the future. Essentially, the double hull acts as an additional barrier between external objects—such as rocks or other ships—and the oil carried by the tanker. For example, when the *Exxon Valdez* hit Bligh Reef, rocks from the reef penetrated more than 5 feet (1.5 m) into the ship. Had the *Exxon Valdez* been constructed with a double hull to protect its holds, however, no oil would have escaped, as there would have been 11 feet (3.35 m) separating the outer hull from the oil.

Largely because constructing tankers with double hulls increases construction costs, the oil-shipping industry avoided integrating them in tanker design. Given the enormous clean-up costs incurred by society and the long-term damage suffered by the environment after supertanker spills, requiring tankers to have this added level of security in order to prevent such spills seems not just acceptable but responsible. Indeed, this was Congress' reasoning for passing the Oil Pollution Act.

In addition to their potential for creating environmental disasters from oil spills, supertankers also contribute significantly to ocean noise pollution. The enormous engines and propellers used to move the tankers generate low-frequency noise that can carry for miles across the ocean. As low-frequency hearing is believed to be the primary sense of whales, dolphins, and other marine species for finding food, avoiding predators, and locating mates, the noise from supertankers impedes the performance of these activities. Further, at close range, a powerful sound can cause tissue in the lungs, ears, or other parts of a marine animal's body to rupture and hemorrhage. Further away, the same sound can induce hearing loss in marine animals, causing them to swim off course, abandon their habitat, and become abnormally aggressive.

Ultimately, as long as the international demand for petroleum remains high, supertankers will continue to be constructed and likely will become larger. However, as Admiral J. W. Kime, Commandant of the U.S. Coast Guard, stated in testimony before Congress, "As long as there are ships at sea, there will be accidents. We cannot alter that fact. What we can strive to do, what our goal should be, is to insure that these accidents are as infrequent as possible, and that their consequences, to the ship, the personnel onboard, and to the environment, are as harmless as possible." By requiring supertankers to be constructed with double hulls and reduce their noise output, the goal of minimizing the impact of supertankers on the environment may be realized.

MARK H. ALLENBAUGH

Further Reading

Books

Cornwell, E. L., ed. *The Illustrated History of Ships*. New York: Crescent Books, 1979.

Periodical Articles

Alcock, Tammy M. "'Ecology Tankers and the Oil Pollution Act of 1990: A History of Efforts to Require Double Hulls on Oil Tankers." *Ecology Law Quarterly* 97 19 (1992).

Allenbaugh, Mark H. "What's Your Water Worth: Why We Need Federal Fine Guidelines for Corporate Environmental Crime." *American University Law Review* 925 48 (1999).

Internet Articles

Historical Overview of the Exxon Valdez Oil Spill. http://www.oilspill.state.ak.us/mwhistory.html

National Resources Defense Counsel. "Sounding the Depths: Supertankers, Sonar, and the Rise of Undersea Noise." http://www.nrdc.org. 1999.

Modern Airplane Technology: 1950-1999

Overview

From the moment Orville and Wilbur Wright (1871-1948 and 1867-1912, respectively) took their famous flight at Kitty Hawk, North Carolina, the world fell in love with the idea of the airplane. But man's fascination with flight goes back even further. As early as ancient Greece, people gazed in wonder at birds' flight, wishing they too could reach those soaring heights. Of course, for the mythological figure Icarus that

The Boeing 747 "jumbo jet." *(Federal Aviation Administration. Reproduced by permission.)*

wish turned fatal when he flew too high and too close to the sun; the wings his father had created out of feathers and wax melted, sending him crashing to his death.

Background

Many times throughout history, man has tried to copy birds' flight and failed. Leonardo da Vinci (1452-1519) sketched flying machines in the 1500s and even made some models. The first successful flights, however, were not taken until the early 1780s, and they were not in flying machines but in hot-air balloons. In the late 1850s balloons were enhanced with steam engines to create airships.

In the 1800s the aeronautical pioneer Sir George Cayley (1773-1857) solved many of the technological questions that airplane flight posed. In 1853 his glider was the first aircraft to take a man into the air. But credit for the first sustained flight belongs solely to the Wright Brothers, who by 1905 could keep their plane airborne for a distance of more than 20 miles (32 km). The next three decades saw numerous improvements to the airplane and the distances that could be flown.

World War I called for more durable and maneuverable planes, ones capable of speeds in excess of 130 miles per hour (209 kph). In 1927 aviator Charles Lindbergh (1902-1974) took his much-celebrated flight, crossing the Atlantic from New York to Paris in just under 34 hours in his specially designed plane, *The Spirit of St. Louis*. In 1932 Amelia Earhart (1898-1937) became the first woman to cross the Atlantic alone, only to disappear five years later while attempting to cross the Pacific.

In the 1920s and 1930s planes became bigger, stronger, and faster, reaching speeds of up to 300 miles per hour (483 kph). But the turning point came in the late 1930s with British engineer Frank Whittle's (1907-1996) invention of the jet engine, which used a mixture of fuel and air to create a powerful forward thrust. The first flight of a turbojet aircraft came with the Heinkel HE-178 in 1939. Suddenly, airplanes could fly at more than 500 miles per hour (805 kph) and the aerotechnology race had begun.

Impact

Airplane production during World War II numbered in the tens of thousands across the United States, Great Britain, Germany, Russia, and Japan. But because such a large number of planes was required for the war effort, the emphasis was on building conventional propeller planes, rather than jet-powered ones. Although the jet engine was faster and more powerful, its development was essentially put on hold during the 1940s. After the war, the jet needed numerous refinements before it was ready to be used commercially.

The first jet passenger service was available in May 1952 with the launch of the British DeHavilland DH 106 Comet. The new aircraft was able to cut travel time in half from that of the previous piston-engine planes. The Comet offered a smooth, quiet flight with a pressurized cabin that allowed it to be flown in all types of weather conditions. The Comet, proved short-lived, however when it fell victim to several accidents. In 1958 a revised version, the Comet 4, was introduced. It could accommodate up to 44 passengers with a four-person crew and had more commercial success than its predecessor.

The first truly successful passenger airplane was the Boeing 707-121, which took its inaugural flight in 1954. Its earliest incarnation was powered by four jet engines, each producing 13,000 pounds of thrust. The wing span was 130 feet (40 m) across, and it could cruise at speeds of 585 miles per hour (941 kph). The 707 held 189 passengers and by 1958 was able to make nonstop flights across the Atlantic. The jet age was inaugurated on October 26, 1958, with a flight from New York to Paris on a Boeing 707-121. Six weeks later came the first commercial jet flight in the U.S., from New York to Miami.

By the mid 1950s the number of passengers traveling annually was multiplying. Part of this rise in air travel was a result of the reduced cost of flying. In 1929 the cost per passenger mile was 12 cents as compared with 5.1 cents more than a decade later. The number of people traveling jumped from 2.5 million in 1937 to 45 million in 1952 and 90 million by 1957.

There were also advancements made to jet fighter aircraft in the 1950s. Jet planes used in the Korean War (1950-1953) achieved much greater speeds than those used in World War II. Over the next decade jets were able to fly at twice the speed of sound (Mach 2). Bomber and transport jet aircraft were also able to fly at supersonic speeds. The most impressive, however, was the air-launched X-15, developed by the National Advisory Committee for Aeronautics (NACA) and debuted in the early 1960s. Dropped from the B-52 bomber, this impressive machine could travel at speeds of up to Mach 6.04—over 4,000 miles per hour (6,437 kph)—and at heights of more than 67 miles (108 km) above the earth.

Aircraft instrumentation and automation also evolved significantly throughout the years following World War I. While early pilots had to rely on a magnetic compass, barometric altimeter, and an anemometer to indicate airspeed, subsequent airplanes saw numerous technological advances. From World War II until the mid-1960s, radio altimeters, weather radar, alarms for fuel, temperature and landing-gear status, as well as airspeed and altitude indicators were added. The 1950s saw the first automatic pilots, which were able to maintain speed and direction. Between 1965 and 1980, improvements included mechanized flight directors, automatic landing systems, and digital computers for monitoring the status of hydraulic and electrical systems. Several digital displays were added in the 1980s as well as moving-map displays, collision-avoidance systems, and flight-management systems. In the 1980s and 1990s global positioning navigation systems were able to guide airplanes.

In the 1960s two more successful Boeing models were released. The 727, launched in February 1963, carried 189 passengers and went on to become the second bestseller of all time. In 1967 the most successful jet of all was introduced—the two-engine Boeing 737, which would make up close to a quarter of all U.S. commercial airplanes by the 1990s.

The introduction of large, jet-powered passenger aircraft in the 1960s ushered in a new era for air travel. In early 1969 Boeing introduced its jumbo jet airliner, the first in its 747 series. The most remarkable thing about the plane was its enormous size, making it tower over every other airplane that came before it. The 747 boasted a 185-foot-long (56 m), 20-foot-wide (6.1 m) passenger cabin, which could seat eight or ten people across with plenty of headroom and comfortable seating. The first 747s went into service on Pan American's New York-to-London route in January 1970 with 324 passengers. Over the next two decades the company added eight models to its fleet of 747s and made significant improvements to the design each year.

In 1970 the McDonnell Douglas DC-10 was introduced as a competitor to the 747. This powerful plane could carry up to 380 passengers and traveled at 587 miles per hour (945 kph). By 1974 more than 90 percent of flights were taken on jet aircraft. Then in 1976 came a dramatic new entry into the field of flight—the British-French *Concorde,* which was able to travel at twice the speed of sound (1,320 mph/2,124 kph). Unfortunately, the *Concorde,* never saw widespread commercial use because of the high cost of constructing and operating the aircraft.

The early 1980s saw the release of Boeing's twin-engine 757 and 767 models, which were more fuel efficient and filled the gap between small planes and the enormous 747. In the 1990s McDonnell Douglas introduced its MD-11, which was longer and more aerodynamic than the DC-10. In 1994 Boeing countered with its 777 model, which could carry between 292 and 500 passengers and used two Pratt and Whitney engines that could generate over 70,000 pounds of thrust. The 777 was the first plane to be built entirely by computer blueprint. It also achieved the lowest passenger-per-mile cost and the greatest fuel efficiency of any passenger jet.

In the late twentieth century the emphasis in jet building was not only to improve passenger safety but to accommodate a growing number of travelers and to meet the insatiable need for greater speed and reduced travel time. New frontiers in aeronautic technology promise to take man to greater speeds and heights than ever before experienced.

STEPHANIE WATSON

Further Reading

Braybrook, Roy. *The Aircraft Encyclopedia.* New York: J. Messner, 1985.

Ethell, Jeffrey L. *Smithsonian Frontiers of Flight.* New York: Orion Books, 1992.

Green, William, Gordon Swanborough, and John Mowinski. *Modern Commercial Aircraft.* New York: Salamander Books, 1987.

Harrison, James P. *Mastering the Sky: A History of Aviation from Ancient Times to the Present.* New York: Sarpedon, 1996.

The Development of Computer Operating Systems

Overview

For most people, understanding their computers extends no further than needing to know how to be able to install programs. However, there are a number of essential elements that allow a user to make use of the computer hardware. This essay will explore the history of one of these elements—the operating system. An operating system is a program that serves as an interface between the user of a computer and the hardware. It sets up an environment in which a user can run programs conveniently and efficiently. Examples of operating systems include DOS, UNIX, and Windows. There are four main components of a computer system: the hardware, the operating system, programs, and users. The operating system manages the hardware and software resources of the computer to best meet the diverse and sometimes conflicting needs of programs and users.

Background

The earliest computers did not have operating systems. Programmers interacted directly with the hardware through switches, tape, or punched cards. Because the computer could operate much more quickly than the programmer could load or unload tape or cards, the computer spent a great deal of time idle. To overcome this expensive idle time, the first rudimentary operating systems (OS) were devised. They were simple programs that were always in the memory of the computer and that ordered user programs by type and then automatically ran them one right after the other. The next step came about with the introduction of disk systems. Because disks are random access devices, the information on them can be accessed in any order. Disks were used to hold user input and output until the central processing unit was ready to use it. As soon as the CPU finished one task it could jump on the disk to another job that was ready to run. Time-sharing was the next logical progression. In time-shared operating systems the CPU handles many jobs at the same time by switching in between them so quickly that it is unnoticeable. Thus, while one user is typing in a command, the CPU is executing another user's program.

MULTICS and UNIX

One of the earliest formal operating systems was MULTICS, designed between 1965 and 1972 at the Massachusetts Institute of Technology. MULTICS was a time-shared system running continuously on a large complex mainframe computer with a vast file system of shared programs and data. In 1969 Ken Thompson (1943-) and Dennis Ritchie of the Research Group at Bell Laboratories began to work on UNIX, an operating system for minicomputers. Ritchie had previously worked on the MULTICS project, and UNIX was strongly influenced by MULTICS. (The name UNIX is a pun on MULTICS.) For this new OS, Ritchie and Brian Kernhagan developed the systems-programming language C to replace the assembly language previously used. By 1978 UNIX had become a product sold by AT&T (the parent organization of Bell Labs.) The size, simplicity, and clean design of the UNIX system encouraged programmers at sites other than Bell Labs to experiment with UNIX development. The most influential of these was a group at the University of California at Berkeley. The advances made by this group convinced the defense department to fund further research, leading to the development of 4BSD (Berkeley Software Distributions) UNIX. 4BSD proved to be fundamental to the development of the Internet. UNIX is a simple, highly flexible system designed to let the user build a more complex system if desired. It can run on mainframes, workstations, minicomputers, supercomputers, and even personal computers. Research and development of UNIX continued throughout the 1980s and 1990s, with special focus being placed on standardizing UNIX applications.

Apple and Microsoft

While UNIX was spreading beyond Bell Labs, the development of the Intel 4004 microprocessor in 1971 allowed the concept of a personal computer to emerge. The Intel 4004 was an entire CPU on a single microchip. Intel and other companies continued to refine the microchip, and personal computer (PC) kits that users assembled themselves became popular among computer hobbyists. Unlike mainframe computers, personal computers were not intended to have more than one user at a time and therefore were not concerned at first with time-sharing or multitasking. Instead, as the PC market grew, emphasis was placed on convenience and ease of use for the user. In 1976 Steve Jobs (1955-) and Steve Wozniak (1950-) designed and built the Apple I, which consisted of little more than a circuit board. However, by 1977 they had incorporated Apple Computer and announced the Apple II, which established a benchmark for personal computers. The Apple II had a simple operating system that came on a disk and accepted basic commands from a command line. In the same year Bill Gates (1955-) and Paul Allen (1953-) founded Microsoft Corporation.

In 1980 a computer programmer named Tim Paterson developed an operating system called 86-DOS (Disk Operating System.) Like the Apple II and the other personal computer operating systems of the time, it was a command-line interface between the user and the PC hardware. Also in 1980 IBM decided to make a personal computer and chose Microsoft Corporation to provide the operating system for the new PC. Paterson joined Microsoft in April 1981, and by July Microsoft had bought all the rights to DOS. In August IBM sold its first PC, complete with MS-DOS 1.0. In less than a year reverse engineering had allowed competitors to produce clones of the IBM personal computer. Microsoft sold MS-DOS 1.25 to these clone makers. Throughout the 1980s MS-DOS continued to develop and advance, gaining more capabilities and meeting the needs of more powerful hardware and more advanced programs.

During the Super Bowl in January 1984, Apple introduced America to a completely innovative computer in an Orwellian-themed advertisement. The Macintosh was the first commercially successful computer with a graphical user interface (GUI). The GUI style of operating system allowed users to interact with the computer through click buttons, pull-down menus, and other image options on the screen rather than through a command line. In addition to the graphical interface, the Macintosh had more advanced hardware than IBM-style PCs. Apple continued to offer both the Apple II and the Macintosh throughout the 1980s. By the end of the decade, the Macintosh offered multifinder properties that allowed it to do more than one task at once. The Macintosh proved extremely popular, especially within educational facilities. Despite the Macintosh's popularity, Apple lost ground in the PC market throughout the 1990s until its introduction of the Powermac G3 in 1997. This was followed in 1998 by the highly successful iMac computers, aimed at a low-end market.

In order to compete with the Macintosh, Microsoft produced Windows 1.0 in 1985, which brought the GUI interface and Macintosh-style features to DOS-compatible computers. In developing Windows, Microsoft signed an agreement with Apple that Windows 1.0 would not use Macintosh technology. When future versions of Windows did utilize Macintosh ideas, Apple took Microsoft to court for copying the "look and feel" of the Macintosh. Microsoft argued that the agreement only applied to Windows 1.0, and the court ruled in favor of Microsoft. In the meantime, between 1985 and 1987 Microsoft and IBM collaborated on creating a new operating system. Microsoft pulled out of the collaboration and released Windows 3.0 based on technology that had been developed jointly. IBM continued working on the new operating system and released OS/2 in 1987. Although it was a technologically advanced system, it was not a great commercial success. In 1993 Microsoft produced Windows NT, an entirely new operating system written from the ground up, designed to compete with the server market that was dominated by UNIX. The GUI on this new operating system had much the same look as Windows but with different programming underneath. Meanwhile, Windows continued to develop, with the next major advance coming with the release of Windows 95 in 1995. It included a major overhaul of the GUI, some changes to the underlying DOS, and was tested by over 50,000 individuals and companies before being released.

Microsoft was quickly taking over the PC market, and some of its practices were drawing criticism. In 1997 Microsoft was ordered to make Windows 95 available without the applications software Internet Explorer. The argument was that by automatically including Internet Explorer on Windows 95, Microsoft was using its monopoly of the PC operating system market to destroy competition in other markets, such as internet software. Microsoft appealed the order. An appeal court ruled that the 1995 injunction did not apply to Windows 98, released in 1998. However, in May of that same year the U.S. Justice Department and 20 states filed an antitrust suit against Microsoft, charging it with abusing its market powers to destroy competitors. In November 1999 Judge Thomas Jackson issued his Findings of Fact, stating that Microsoft is an illegal monopoly and that it had abused its market power in anti-competitive practices.

The Emergence of Linux

While Apple and Microsoft were battling for space in the PC market, Linus Torvalds (1970-) at the University of Helsinki in Finland was developing a freely distributed version of UNIX for personal computers called Linux. Linux began as a hobby for Torvalds, inspired by Minix, a small UNIX system developed by Andy Tanenbaum. In October 1991 Torvalds announced the first official version of Linux, 0.02. The system was still very rudimentary, but Torvalds put it out on the Internet for UNIX programmers and wizards to aid in its development. Hackers, programmers, and users of every flavor contributed, and by 1994 it had become a viable operating system, capable of running almost all UNIX programs. By 1996 Linux was a complete UNIX clone, capable of running X-windows—the UNIX version of a GUI. Because Linux was developed completely from scratch, it contains no code from AT&T or any other proprietary source. Much of the software available for Linux is from the GNU project at the Free Software Foundation in Cambridge, MA. In 1999 it was still possible to obtain Linux and a sizeable number of programs completely free of charge. Because of its flexibility, its price, and the fact that it is adaptable to most PC hardware, Linux became quite popular in the late 1990s as an Internet server. The growing interest in Linux convinced commercial software manufacturers to make their packages compatible with the Linux system. As this trend continues Linux will become even more viable and popular.

Impact

The pace of change in the computer industry makes it impossible to predict future developments in operating systems. However, this article has attempted to show that the evolution of operating systems in the second half of the twentieth century depended on a number of factors, including programmer interest, market pressures, hardware advances, and government oversight. Operating systems will continue to develop to meet the growing and changing needs of users and new hardware. It can only be hoped that the competition that gave rise to the PC revolution, the innovation that brought about GUI systems, and the free exchange of ideas that gave rise to Linux will all continue into the twenty-first century.

DANIEL BONGERT AND REBECCA B. KINRAIDE

Further Reading

Carlton, Jim. *Apple: The Inside Story of Intrigue, Egomania, and Business Blunders.* New York: Times Books, 1997.

Edstrom, Jennifer and Marlin Eller. *Barbarians Led by Bill Gates: Microsoft from the Inside.* New York: Henry Holt, 1998.

Malone, Michael S. *Infinite Loop: How the World's Most Insanely Great Computer Company Went Insane.* New York: Doubleday, 1999.

Raymond, Eric S. *The Cathedral and the Bazaar: Musings on Linux and Open Source by an Accidental Revolutionary.* O'Reilly and Associates, 1999.

Wallace, James and Jim Erickson. *Hard Drive: Bill Gates and the Making of the Microsoft Empire.* New York: John Wiley & Sons, 1992.

Wallace, James. *Overdrive: Bill Gates and the Race to Control Cyberspace.* New York: John Wiley & Sons, 1998.

Young, Robert and Wendy Goldman Rohm. *Under the Radar: How Red Hat Changed the Software Business and Took Microsoft by Surprise.* Coriolis Group, 1999.

The Explosion of Applications in Fiber Optics since 1960

Overview

Used in communication, fiber optics is the technique of sending light waves through glass fibers. It is analogous to communicating via radio waves in that signals are encoded at one end for transmission and decoded at the other. Fiber optics was developed sporadically throughout the first half of the twentieth century. Since the 1960s the field has exploded as material sciences and computer technology have made fiber optics more efficient. Today around 80% of long-distance communication is carried through fiber optic networks.

Background

Long-distance communications via optical signals has been with humanity for many centuries, ranging from smoke signals to sailing ships communicating with flags. The first mechanical optical communication systems were semaphore towers, developed in France in the 1790s.

During the nineteenth century many of these systems were made obsolete by telegraphy and the telephone. Alexander Graham Bell, the inventor of the telephone, also developed the Photophone, which communicated with light instead of electrical signals. The invention didn't get very far since light doesn't travel well through the air: a single building, cloud, or hill can effectively block any signal.

In the 1840s it was shown that light could be guided along jets of water. This was used to create fantastic, elaborate fountain displays. These scientists took advantage of the phenomenon called *total internal reflection*. Because water has a higher *refractive index* (a measure of the ability to bend light waves) than that of air, light beams will remain inside the water. This same principle was taken advantage of by dentists who used bent quartz rods to illuminate the mouth.

The first true demonstration of image transmission through glass was made by the German medical student Heinrich Lamm in 1930. He was able to transmit the image of a light bulb filament, but the quality was very poor.

The first transmissions of images through bundles of tiny drawn glass fibers were made independently in 1954 by Abraham van Heel, and the team of Harold Hopkins and Narinder Kapany. This was the discovery that broke open the field of fiber optics.

Impact

It was about 20 years from the invention of working bundles of fibers to practical communication systems. The biggest problem with the first fiber optic systems was signal strength. Signals sent through glass tend to attenuate, or die off, leaving the receiver on the end with a fuzzy, indecipherable signal. This was because the glass fibers were contaminated and light sources were not powerful or easily controllable. Two factors were indispensable in overcoming this problem: the laser provided a focused, powerful signal and new techniques and materials were used to make higher quality, pure, coated glass fibers.

The laser was the biggest step forward in providing a powerful, precise light source. It was invented in 1960. It provided a focused, coherent beam of light with an exact wavelength. Now scientists could tune a light wave transmission just as precisely as a radio transmission. Too, the shorter wavelengths of light beams theoretically allowed scientists to place much more information into a signal. But the first lasers were impractical for commercial purposes: they had to be cooled to almost -200 degrees C, and even then worked for very short periods, just seconds or minutes, before they burned out. Lasers that worked at room temperature weren't developed until 1970, and durable lasers with a lifetime of 10 years weren't made until 1976.

The other problem, signal loss, was overcome in stages. Glass fibers prior to the 1950s were not very pure, had optical aberrations and "leaked" the signal due to the poor quality or absence of coating materials. The first fix was the introduction of cladding, a coating that surrounded the glass fibers. Many researchers tried applying a plastic coating, but these had little effect on the optical quality. In 1956 Lawrence Curtiss, at that time still an undergraduate student at the University of Michigan, successfully coated glass fibers with a cladding made of a different glass. This cladding had a lower refractive index than that of air, which prevented loss of the signal into the air as well as the introduction of extraneous signals from the air. It also insulated each fiber from its neighbor, virtually eliminating *cross-talk* between fibers.

The last key transformation was a viewpoint shift as well as a technological shift. Many researchers believed that the glass itself was responsible for the rapid degradation of the signal. No matter how pure they made the glass, it would still degrade the signal. Things began to change when Charles Kao, a scientist at Standard Telecommunication Laboratories believed that if the purity of the glass was increased it would conduct light almost perfectly. His continual pushing for better fibers began a competition between many research centers.

The winner of this race was Corning Glass Works. In 1970, Robert Maurer, Donald Keck, and Peter Schultz invented a way of preparing fused silica (the glass) in an almost pure form by controlling how the fibers were drawn, the temperature needed to form the glass, and what chemicals were added. Their glass had a slightly higher refractive index than its cladding because the chemical titania (titanium oxide) was slowly doped (added) into the silica. Proper control of the amount of titania kept the refractive index high and the signal inside the glass protected by the cladding. Although the resulting fibers were fragile they had a far lower signal loss than any competing fiber. Not long after the Corning scientists found another way of improving the flexibility of the fibers by doping them with germania (germanium oxide).

The first practical fiber optic communication system was put into place in September 1975 in Dorset, England. The police in that area were looking for a new system after a current surge from a lightning strike destroyed the station's electronic systems. The police chief wanted a new system that didn't depend on electrical wires. Since fiber optic systems use light instead of electricity they are less susceptible to power surges. The English system took only a few weeks to set up and worked very well.

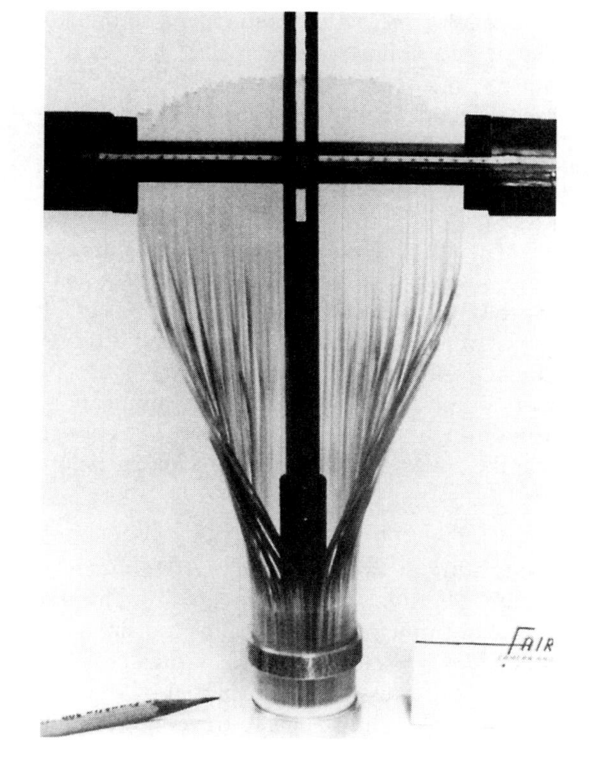

Fiber optics converter. *(The Library of Congress. Reproduced by permission.)*

AT&T tested its first fiber optic system in Chicago in 1976. They linked three office buildings in the downtown core with 2.6 kilometers (1.6 miles) of optic cable. They began tests on April 1, but were beaten to the punch by GTE, which was able to get 10 kilometers (6.2 miles) of fiber optic cable up and running for public use in Long Beach, California, on April 22. GTE's lines couldn't carry as much information as AT&T's, but it was the first working American system.

After these initial successes the first large network to be installed was the Boston-New York-Washington corridor. The previous network was composed of old microwave transmitters and carried more telephone calls and communication than almost anywhere else in the world, due to the large concentrations of industry, government, and finance companies along the American coastline. AT&T chose to line it with fiber optic cables in 1980. The line was finished in 1984, but by then changes in fiber optic transmission technology had moved so quickly that it was almost obsolete. MCI had started putting in a better system using different technology standards in 1982 with faster data transfer, more bandwidth, and fewer repeating stations. (Repeating stations amplify fading signals for transfer along the next section of cable.)

The other big goal of fiber optic communications was to stretch cables across the sea, joining the continents. The first submarine telegraph cables were laid across the Atlantic in the 1850s. None of these cables lasted very long—the first worked for 28 days—but the communications sent during those periods were valuable enough that the entrepreneurs of the time kept developing their systems.

By the 1960s there were two ways of communicating between continents with a telephone. The signal could be sent via submarine cables or through satellites. Cable transmissions tended to be static-filled, weak signals while satellite transmissions were marred by the time delay it took radio signals to travel up to orbiting satellites and back down again, plus other problems like repeating echoes, feedback, and static. Either way was extremely expensive.

The biggest problem in laying long-distance cables is that signals need to be amplified often: the TAT-6 coaxial cable, which uses electrical signals and was laid in 1976 under the Atlantic carries almost 1,000 repeaters, or about one every 9.2 kilometers (5.7 miles). These are also the parts most likely to break. Any break is extremely expensive to repair, since a ship with technicians must be dispatched to find the break, haul the cable up from the seafloor, and repair it. Therefore any way that the number of repeaters can be reduced is extremely valuable. As well, coaxial cables were approaching fundamental size limits. Anything bigger would be overly susceptible to breakage and too large to fit on cable-laying ships.

In 1978 the first fiber optic submarine cable across the Atlantic was proposed. It would be called TAT-8 and would carry fewer repeaters and more voice channels than any of the earlier coaxial cables. It was completed in 1988. Many more have been laid since then, and today 80% of worldwide long-distance communication is carried through 25 million kilometers (15,534,279 miles) of fiber optic cables.

The last problem facing fiber optics is connecting them to homes. Today fiber optic cables circle the world, delivering communications to every continent. But the final place they don't reach is one of the most common: homes. Fiber optic telephone signals reach a "neighborhood" where they stop at a hub. Here the signals are converted for transmission to groups of 100-2,000 homes via coaxial cables. This switchover slows down signals that could travel much faster both in and out of homes were the signal carried entirely through fiber optic systems. The switchover has not been made because installing fiber optic systems still costs more than coaxial cables, even though the price continues to fall. The expense isn't with the fibers themselves, but with the costs of switching—including paying skilled technicians, and installing adapters for televisions that aren't equipped to receive fiber optic signals. Still, some observers think that since the changeover must come eventually, sooner is better than later, especially with the growing use of the Internet and the increase of telecommuting workers. In Italy, the government-owned Telecom Italia has already begun this conversion process and, except for a few experimental communities scattered throughout the world, is far ahead of the rest of the world.

PHILIP DOWNEY

Further Reading

Books

Hayes, Jim Albany. *Fiber Optic Technician's Handbook.* New York: Delmar, 1996.

Hecht, Jeff. *City of Light: The Story of Fiber Optics.* New York: Oxford University Press, 1999.

Periodical Articles

Stephenson, Neal. "Mother Earth, Motherboard." *Wired* (December 1996): 97-160.

Other

2020: The Fiber-Coax Legacy http://www.wired.com/wired/archive/3.10/negroponte_pr.html

The Evolution of Satellite Communications

Overview

After the Chinese designed the first rockets, they used the "fire arrows" to repel Mongol invaders. Those early defenders would be amazed to discover the hundreds of satellites launched into orbit around Earth using technologies that developed from their invention. With the launch of the first man-made Earth satellite, *Sputnik 1*, in 1957, the world began considering the vast possibilities to be found in the "space" above the planet's surface. Satellites rapidly evolved from *Sputnik*'s simplistic beeping radio transmitters to the sophisticated communications relay stations orbiting Earth in the 1990s, revolutionizing the way mankind experienced the world.

Background

Around 300 B.C. the Chinese invented gunpowder, which was later packed into bamboo tubes to make a primitive firecracker. By 1232, when the Chinese defended their lands from Mongol invaders during the battle of Kai-Keng, an inventive native had decided to add the firecracker to an arrow, in essence creating the first bottle rocket. Ballistic weapons became commonplace in war, but rocket science didn't really emerge until the turn of the century in America. In the early 1900s American Robert H. Goddard (1882-1945) began experimenting with rocket propulsion—he received his first two U.S. patents in 1914 for a liquid-fueled rocket engine and a two or three-stage solid fuel rocket. In a technical report for the Smithsonian in 1920, Goddard outlined how a rocket might reach the moon, causing an uproar in the scientific community, who labeled him a crackpot. (His report, however, became the foundation for the early rocket program of the German army, which made further advancements in rocket science during the Second World War.) Goddard's rocketry research led to 200 patents, including important advances in liquid fuel and guidance systems. He was the first to launch a scientific payload—in 1929 he sent up a barometer and a camera. Goddard's discoveries paved the way for the modern rocket technology that would launch the first man-made satellites into space.

In October 1945 British physicist (and science fiction author) Arthur C. Clarke (1917-) published an article in *Wireless World* that described a system of manned satellites in orbit above Earth that would distribute global communications through a "relay" service. Clarke predicted that these satellites, in orbit above the equator at an altitude of 22,300 miles (36,000 km), would revolve around Earth in 24 hours, appearing motionless from the surface. While Clarke was by no means the first to theorize about a fixed orbit for satellites, his concept had a great influence on their technological development. Future satellite scientists would term Clarke's hypothesis "geostationary" orbit, and the ring 22,300 miles above Earth's surface became known as the "Clarke belt" (also called the "Clarke orbit").

In the decade following the publication of Clarke's article, satellite research was influenced by several significant scientific advancements. The Cold War between America and the Soviet Union brought about long-range, high-powered rocketry, including intercontinental ballistic missiles (ICBMs). The first solar cells were developed; these would eventually be used to power satellites. The invention of the transistor made possible the miniaturized electronic components necessary for lightweight space objects and also ushered in the age of the high-speed digital computer, which would be used to calculate and track satellite orbits. All of these scientific advances served to kick off the race into space. On October 4, 1957, the Soviet Union set an historic landmark with the launch of the first Earth-orbiting satellite, *Sputnik 1*, a 184-pound sphere about the size of a basketball. It was sent into space as part of the International Geophysical Year (IGY) research project.

Though limited (it only flew for 92 days) in success, *Sputnik 1* spurred scientists and engineers to design more sophisticated satellites. In January 1958 the United States managed, after two failed attempts, to launch its first IGY satellite, *Explorer 1*, which was instrumental in the discovery of Earth's radiation belts. In July 1958 the U.S. Congress passed the National Aeronautics and Space Act, which created the National Aeronautics and Space Administration (NASA) and served to jump-start U.S. space science and exploration. (NASA's first communications satellite, *Echo 1*, a passive communication device, was launched in August 1960.) Before the end of 1958 the world's first "active" communication

Echo 1, America's first passive communications satellite. *(NASA. Reproduced by permission.)*

satellite, *SCORE* (Signal Communication by Orbiting Relay Equipment), was launched by the U.S. Air Force. It transmitted a pre-recorded Christmas message from President Dwight D. Eisenhower (1890-1969) and lasted 12 days. Because it did not receive and retransmit a signal from Earth, *SCORE* was not a truly active communications satellite. In 1960 the U.S. Department of Defense launched *Courier*, the world's first fully active communications satellite. *Courier*, also the first solar-powered communications satellite, received and retransmitted signals, but it only functioned for 17 days.

Impact

The proven success of the early satellites stimulated a surge in private sector interest in communications satellites. By 1960 officials with AT&T had filed with the Federal Communications Commission (FCC) for authorization to launch an experimental satellite, catching the U.S. government with no policy in place to govern such requests. As a result, competitive contracts were awarded by NASA to RCA (for a medium orbit active communications satellite), AT&T (for its own medium orbit satellite), and Hughes Aircraft Company (for a high-orbit satellite). In July 1962 NASA launched the world's first private sector communications satellite, AT&T's *Telstar*, which transmitted the first live transatlantic telecast on July 10. Voice, television, facsimile, and data were transmitted between the U.S. and various sites in Europe. *Telstar* opened space to commercial users, sparking a communications revolution.

In concert with the private sector, NASA launched *Relay 1*, built by RCA, in December 1962. *Relay* was the first satellite deployed with backup systems in case of system failure, and its transmission of telephone and television signals to Europe, South America, and Japan gave the first demonstration of the possibilities of true global communications via satellite. (Because of its low orbit, however, *Relay* could only provide real-time communications for short periods, a problem that spurred research and development into higher orbiting satellites, looking towards finally realizing Arthur C. Clarke's hypothesis, the geostationary orbit.) NASA turned to their contract with Hughes and, in February 1963, launched *Syncom*, the first satellite to achieve geosynchronous orbit. Shortly afterward, *Syncom* suffered electrical equipment failure, and *Syncom II* was launched to takes its place. By 1964 *Syncom III*, the most famous satellite in the series, had been launched and used to transmit the Olympic games live from Tokyo.

The success of the NASA-Hughes satellites presaged the most significant event in modern satellite evolution—the launch of the first commercial communications satellite, *Early Bird*, in April 1965 by Comsat, the Communications Satellite Corporation. Just twenty years shy of the anniversary of Arthur C. Clarke's momentous article on satellite possibilities, global communications satellites became a reality with the launch of *Early Bird* into geosynchronous orbit over the Atlantic Ocean, providing the first continuous satellite communications link. Its siblings, *Intelsat II* and *III*, provided further links to complete the first global network, which in 1969 allowed over one half billion people to watch *Apollo 11* land on the Moon on July 20. Such live global telecasts along with international telephone transmissions became the foundation of global satellite communications.

While global communications evolved, domestic satellite systems were also learning from the experiences of the Soviet Union, which had launched the world's first domestic system, called *Molniya*, in 1965. By the early 1970s both Canada and the United States began to develop domestic satellite networks of their own. In Canada the first North American domestic communications satellite, *Anik*, was launched in November 1972. The first U.S. domestic communications satellite was Western Union's *Westar*, launched in April 1974. Early on, the North American satellites were used primarily for long-distance telephone and data communications. Then, in 1975, an American pay TV service announced it would begin using a satellite to provide its broadcasts to cable TV stations nationwide. On September 30, 1975, Home Box Office, Inc., went on the air, delivering live coverage of a world championship boxing match. Very quickly, other services joined HBO in leasing spots onboard satellites, fueling an explosion in the development of thousands of U.S. cable TV systems.

These early commercial satellite applications—international telephone, television, cable television, and data networks—were dwarfed by growth in the industry in the late twentieth century. Beginning in 1976 with the launch of *Marisat*, the first communications system to provide mobile services (to the U.S. Navy and other maritime customers), global communications grew exponentially as new technologies were introduced. Three of the most significant advances—cellular telephones, direct broadcast satellite television, and the Internet—incited rapid progress in the development of low earth orbit (LEO) satellites. Hundreds of sophisticated LEOs were deployed, providing high-speed digital communications to even the most isolated regions around the world. From Iridium's handheld global satellite telephones, first activated in November 1998, to the direct broadcast satellite technology of DirecTV, launched in 1994, to the Wireless Web service, introduced to digital telephone customers in 1999, low earth orbiting satellites brought the world closer together.

Satellite technology delivered e-mail, telemedicine, teleschooling, and telecommuting to the masses—from China to the United States to coldest Antarctica—permitting the continuous exchange of video, audio, and data information between distant locations around the earth. In the 1990s the near instant transmission of live news, sports, entertainment, and data brought the whole world to each of its inhabitants. Historic events such as the end of the Cold War (personified by the collapse of the Berlin Wall), the Gulf War, and the conflict in Kosovo were visible on world televisions as they occurred, affecting global opinion as well as the events themselves. From *Sputnik* in 1957 to the smallest low orbit device spinning out to space in 1999, satellites had changed the way the world lived and worked.

ANN T. MARSDEN

Further Reading

Books

Fthenakis, Emanuel. *Manual of Satellite Communications* New York: McGraw-Hill, 1984.

Hudson, Heather E. *Communication Satellites: Their Development and Impact.* New York: The Free Press, 1990.

Long, Mark. *1985 World Satellite Almanac: The Complete Guide to Satellite Transmission & Technology.* Boise, ID: CommTek Publishing Company, 1985.

Martin, James. *Communications Satellite Systems.* Englewood Cliffs, NJ: Prentice-Hall, Inc., 1978.

Internet Sites

Whalen, David J. "Communications Satellites: Making the Global Village Possible. http://www.hq.nasa.gov/office/pao/History/satcomhistory.html

Technology

1950-present

The Development of the Video Recorder

Overview

Video technology research began almost as soon as television was invented, as it offered distinct advantages over standard film. However, technical problems slowed development until the 1950s when video machines revolutionized television broadcasting. The home market flourished in the 1970s, although there were many setbacks, and the incompatibility of different machines scared off some consumers. In the 1980s the VHS format dominated, and consumers took to video with abandon, using it in an unforeseen variety of ways.

Background

The idea of electronically recording video images is an old one. Indeed, the first patent for the storage of television signals on magnetic tape dates back to January 4, 1927. However, fundamental technical problems limited the quality of video technology. Research blossomed after World War II, and slowly the necessary advances in amplification, noise reduction, and recording materials were made. The most challenging problems were the high recording and playback speeds needed to transfer the large amount of information in a quality image. The higher the definition of the image, the faster it needed to be retrieved from the storage medium. In the early 1950s several companies experimented with video machines that fed magnetic tape through at high speed, over 100mph in some cases, and so went through—quite literally—miles of tape for only a few minutes of recording time.

The problem of how to record and retrieve information at high speed, yet with safety, reliability, and a minimum of tape, was solved by engineers at Ampex in 1956. An engineering team that included Charles Ginsburg (b. 1920) and Ray Dolby (b.1933), later famous for his noise reduction system, came up with the idea of having the tape move slowly while rotating the recording and playback heads at high speeds in the opposite direction. This meant the relative speed between the heads and tape was very high, yet the amount of tape used was small. The Ampex team unveiled their machine at a trade convention at the Chicago Hilton in 1956. Charles Ginsburg recalled the reaction: "There must have been two or three minutes of excruciating silence, and then all hell broke loose. They were hollering and screaming and jumping out of their seats. It was a bombshell."

The reason for the enthusiasm over video was that previously the only way to record a television broadcast was to point a film camera at a television set. Quality was poor, and as the film required processing it was days before a show could be broadcast again. Video offered quality pictures available for broadcast immediately after recording. Early video machines were notoriously difficult to operate, requiring a full-time engineer to get the best out of them, but television companies quickly took to them. Breakdowns occurred often, and the phrase "normal transmission will resume shortly" became a common sight on television screens.

The impact on television production was immediate. In the United States the East Coast evening news was no longer broadcast in the afternoon on the West Coast; it was recorded live onto videotape and played back a few hours later, so that the West Coast, too, watched their evening news in the evening. Delayed broadcasts were also used for sports events, allowing the editing of games to fit within scheduling times. Video technology allowed the instant replay, which helped popularize sports broadcasting. Video changed the style and format of many television programs, which had generally been live-to-air before video. Freed from the live studio format, shows began to include location shots and quick scene changes. Editing videotape was an art form at first, as unlike film there are no visible frames on the tape.

Impact

The success of video in the broadcast industry suggested the possibilities of a home consumer market. Sales of audiotape had surprised many pundits, and videotape seemed a natural progression. In 1965 Sony introduced the CV-2000, aimed at the home user, but it did not do well as it was expensive and unreliable. However, a flood of optimism and investment into video occurred in the early 1970s. Forgotten pioneers of the home video market include the CBS EVR, the V-Cord, the TelDec, Cartrivision, and the U-Matic. Only the U-Matic, from Sony, sold in numbers, and not to home consumers, but to the education and training sectors where it remained the industry standard for many years.

In 1975 Sony introduced the Betamax, a home system developed from its successful industry models. The unit was small and cheap, relatively speaking, and quickly established itself with enthusiasts, slowly making inroads into the wider consumer market. Sony tried hard to get other manufacturers to use Betamax as a standard format, and several companies signed-up and released their own Beta machines. It was hoped that having only the one format would make everything simple for the consumer, who had been faced with a bewildering range of incompatible machines. However, such simplicity was not to be. In 1976 JVC launched its VHS format, which was similar to the Betamax with one major difference: VHS had a larger cassette size, and played a little slower than Beta tapes, which gave VHS a longer recording time. As the two formats were incompatible the now infamous Beta-VHS format war ensued, with price-cutting and fierce marketing campaigns. While many enthusiasts still make claims for the qualities of one format over the other, consumer tests at the time failed to identify a clear winner in terms of picture, sound, and other playback qualities. For the average consumer it was price and the availability of pre-recorded tapes that were the major factors in choosing a machine. Price-cutting occurred on both sides, and at first both Beta and VHS movie tapes were plentiful. However, as VHS slowly edged ahead in sales it became harder and harder to get companies to make tapes for the Beta format, and so Beta tapes became rarer, helping to reduce sales of the machines. Beta's market share grew ever smaller despite aggressive marketing. While recording time, marketing, and tape availability played important roles in the decline of Beta, many industry analysts put the result down to poor luck, and it is still a hotly debated topic.

Almost forgotten on the periphery of the VHS-Beta wars were a small host of other video formats that made little impact on the marketplace. In Europe the Phillips V-2000 offered a number of advanced recording and playback features, but failed to make a splash. In the 1980s most manufacturers abandoned their own formats and produced VHS machines. In 1988 Sony also capitulated and started producing VHS format recorders, yet Beta machines are still popular in some countries, and in the television electronic news-gathering market.

The introduction of home video recorders sparked a wave of surprise and panic in the motion picture and television industry. In the mid-seventies Disney and Universal Studios took Sony to court in the United States, arguing that taping films and programs off-air constituted copyright infringement. In 1979 Sony won the case, and the video industry breathed a sigh of relief. However, the case was appealed, and in 1981 the verdict was overturned. For a while it seemed possible that video-taping would be declared illegal. Finally, in 1984, the U.S. Supreme Court ruled that taping for private, non-commercial use was legal. In the United Kingdom the copyright laws were amended as late as 1988 to allow taping, but required tapes to be erased after 28 days. However, many video laws were only academic, as common public practice was

THE RISE OF DVDS

The pundits who predicted the success of video-discs are finally being proved right with the new generation of DVD (Digital Video Disc, sometimes called the Digital Versatile Disc) players. DVDs use similar technology to compact discs but with tighter spirals, image compression, and a two-layered system in which the laser reader can refocus itself to read either the outer or inner layer. DVD discs can hold many times the information of CDs and are so thin that they can be made double-sided. An entire motion picture can be placed on a single DVD, and the quality of sound and video is higher than tape-based systems, with a longer life span. DVDs are also used in computers, and some foresee DVD allowing the full integration of the personal entertainment system—TV, video, and computer. With re-writable DVD drives it should be possible to record television programs and transfer them to a computer, all at high definition. However, as with the introduction of other video technologies, there are problems with compatibility and copyright. Many DVD formats exist (from different companies), and they are not transferable from one to another. Even using the same brand players does not guarantee compatibility between video players and computers. Movie studios, foreseeing trouble with pirating, have included format zones across the world to limit interchange between countries, and imposed several copy protection measures on the DVD format. Unfortunately, some of the protection methods have been known to cause playback problems.

so widespread and ingrained as to make enforcement impractical.

Censorship was also challenged by the videotape revolution, with laws often lagging years behind industry developments. The ratings of motion pictures did not apply to videotapes, creating new adult markets, as well as exposing children to action and violence previously kept from television screens. Many countries imposed unenforceable censorship laws, and often it was the public who dictated the practical limits of censorship through their common rental, taping, and buying practices. Video piracy and theft troubled the industry, with an estimated 10% of all tapes sold worldwide being pirated copies. The high price and small size of video recorders made them the target of choice for thieves, creating a thriving black market.

Surprisingly many early video pundits could not see the value of recording television programs. It was expected that pre-recorded material would be the major draw for consumers, and many machines were marketed without stressing their recording capabilities. However, it was the freedom from the strict scheduling of network television that made the biggest impact on consumers. The ability to record daytime programs while at work increased the popularity of many soap operas. Going out in the evenings no longer meant missing a favorite show, and the ability to rent movies if there was "nothing on" made owning video recorders attractive to many consumers.

The videotape rental boom caught the entertainment industry napping. Manufacturers had expected to sell their pre-recorded videotapes to the home buyer. However, the high price of tapes produced low demand. The solution was to rent the tapes for overnight viewing, which keep costs down and created a new demand. While video-makers were unhappy about rental, which sometimes included the player as well as the tape, consumers took to the idea wholeheartedly. Made-for-video tapes began to appear, and unexpectedly it was "Jane Fonda's Work-Out Tape" that topped early sales figures, offering a cheap gym-style workout in the home, and spawning a host of imitators. Tape rental broadened the market for foreign and niche films, and offered failed or forgotten movies a second chance in the marketplace.

Videotape provided freedom from "official" channels, as was shown in the United Kingdom in February 1985 when a controversial documentary on MI5 (a British intelligence agency) was banned from transmission. At the same time the American soap opera *Dallas* was also taken off the air in a battle over transmission rights. Within days of these events the documentary and the next three episodes of *Dallas* were available all over the U.K. in videotape form.

Video began to be used in unexpected ways almost from its inception. The rise of video-dating agencies, where the prospective "dates" could see and hear each other before meeting face to face, bypassed the fears and stigma attached to newspaper personal ads. Video artists used the new medium for strange special effects in a new wave of art movies. The alternative press took to video to bypass the mainstream version of public events, and some alternative documentaries have enjoyed popular and, perhaps ironically, commercial success. Globally videotape allowed the fast interchange of information, and broadened the television coverage of world events.

DAVID TULLOCH

Further Reading

Books

Alvarado, Manuel, ed. *Video World-Wide: An International Study.* London, Paris: John Libbey & Company Ltd, 1988.

White, Gordon. *Video Techniques.* London: Butterworth & Co, 1982.

Wyver, John. *The Moving Image: An international History of Film, Television and Video.* London: Blackwell Publishing, 1989.

Other

The Virtual Museum of Home Video Technology. http://www.popadom.demon.co.uk/vidhist/index.htm

Videodiscovery. "DVD Frequently Asked Questions (and Answers)." http://www.videodiscovery.com/vdyweb/dvd/dvdfaq.html

The Development of Cellular Phones

Overview

The development, marketing, and resulting universal use of cellular telephones, all in less than 20 years, makes the cell phone one of the world's most popular innovations. Once merely a toy for wealthy businessmen and the rich, cell phones are now a part of everyday life. For many, the cell phone is such an omnipresent force that it is difficult to conceive of a time before its widespread utilization. In fact, a 1999 report by the United Nations predicted that cell phones will outnumber traditional land lines within the next decade. The latest statistics estimate that there are more than 400 million cell phone users worldwide, with more than 250,000 added every day, a stark contrast to the mere 11 million users in 1990.

Background

The roots of cellular phones stretch back to the crude origins of mobile radio usage in vehicles. In 1921 the Detroit, Michigan, Police Department became the first to utilize the device in the United States. Police and emergency use pushed early development, which progressed slowly. Researchers gave little thought to public applications for mobile phones. AT&T, which built the vaunted Bell System in the United States, showed little interest in mobile phones, which also hindered development.

In the late 1940s technological innovations, such as low-cost microprocessors and digital switching, made mobile telephones more practical. The first public mobile telephone system in the United States began in St. Louis in 1945 with three channels. The St. Louis experiment was made possible by the increased pool of skilled radio personnel after World War II and the use of radio communications in the armed services.

D.H. Ring, a Bell Laboratories scientist, originated the cellular concept in 1947. Ring and his colleagues realized that by using small geographic service areas (or "cells"), combined with low-powered transmitters in each area and radio spectrum frequency reuse, they could greatly increase the capacity of mobile phones. Few people, however, believed the cellular system had a commercial application, and the Federal Communications Commission (FCC) added to the problem by not allocating the necessary airwaves. AT&T asked the FCC to open the radio spectrum as an enticement for further research, but the FCC decided to limit the frequencies, thus squelching further work.

Under increasing pressure from AT&T and the general public, which had heard about advances in mobile phones, the FCC reconsidered its position in the late 1960s. The Bell Labs once again took the lead in proposing a cellular system of numerous low-powered broadcast towers, each covering a cell only a few miles in radius. By 1977 AT&T had built and operated a model cell system. The next year, after the FCC approved Illinois Bell's request, testing began in Chicago, with over 2,000 customers. In 1979 the first commercial cell phone system opened in Tokyo, Japan.

The early 1980s were crucial for the development of cellular phones. In 1981 Motorola and American Radio started a second test in the Washington, D.C., area. The next year, after dragging its feet, the FCC finally permitted commercial cellular service in the United States. The FCC's 1982 decision to break up AT&T's regulated monopoly also stymied additional research and development. Ameritech provided the first commercial service the following year in Chicago, while Motorola followed up in Baltimore and Washington, D.C.

Over the next several years consumer demand exploded. There were more than one million subscribers by 1987. The airwaves quickly became overcrowded, however, forcing the FCC to open the 800 MHz band. This decision stimulated growth in the cell phone industry and led to further research. The original analog cellular systems, Advanced Mobile Phone Service (AMPS), set the standard in North and South America, while the rest of the world used several types of analog cellular, the most prominent being Global System for Mobile Communications (GSM). Today, only a handful of U.S. carriers employ GSM. Some (AT&T) use Time-Division Multiple Access (TDMA), while others (Sprint and Bell Atlantic) utilize Code-Division Multiple Access (CDMA).

Impact

While it took decades for the cellular phone industry to develop, the nearly immediate acceptance of cell phones worldwide led to tremendous growth. In fact, cellular phones have proven to be more practical in many areas of the world than traditional wire-based phone systems. It is easier and more efficient to put up cell towers in many European, Asian, and South American countries than to string up countless miles of phone lines. Cell phones are especially important for nations that do not have strong infrastructures or whose environment is not conducive to wire lines.

In the United States the romance with cellular technology took off from the start. In 1983 the first issue of *Cellular Business* magazine was published, which helped popularize the system. A year later the Cellular Radio Communications Association (CRCA) formed as an education and legislative advocate for the cellular industry. The CRCA later changed its name to the Cellular Telecommunications Industry Association (CTIA). An earlier group, founded in 1949 and established as Telocator, later became the Personal Communication Industry Association (PCIA) and represents the broadest segments of wireless communications.

The first profile of cell phone users generated in 1987 found that they were primarily male, 35-50 years old, managers or entrepreneurs, spent a great deal of time in their cars, and had income in excess of $35,000. In fact, the average mobile phone cost $1,000 that year, with portables reaching $2,000. The biggest complaints, however, were not associated with cost. Instead, users criticized battery weight, lack of battery life, and the lack of privacy on cell phones. The FCC even denied a petition filed by the Washington Legal Foundation requesting a privacy label be placed on the devices, saying it would not serve public interest.

As the 1980s progressed, the CTIA and cellular carriers embarked on a program to raise mainstream awareness of cellular phones. In 1988 the cellular industry began testing retail sales channels, such as Sears and Kmart, and audio manufacturers like Clarion and Sanyo entered the market. It would take several years before cellular phones regularly appeared in retail stores.

Initial estimates of total cell phone users made in the early days (like AT&T's 1983 prediction of 900,000 U.S. subscribers in 2000) were dwarfed by demand, and cell phone carriers and manufacturers rushed to push the technology on consumers. The results have been phenomenal. The 1999 CTIA Wireless Industry Survey shows that more than 80 million subscribers in the United States generated over $37 billion in revenue and provided more than 141,000 jobs. As the technology has progressed, the average monthly bill has dropped from $185 a month in 1984 to a low of $39.43 in 1999.

Cellular providers are key players in the global telecommunications arena. In early 2000 Vodafone AirTouch of Britain bid more than $183 billion to acquire Mannesmann of Germany, the largest takeover in corporate history. Actually, the reaction to the deal was muted to some degree by the great number of multibillion-dollar wireless deals in the waning days of the twentieth century. The telecommunications giants are betting that someday soon wireless subscribers will be able to travel around the world without roaming problems. At stake are billions (if not trillions) of dollars in shareholder money.

Cellular technology has had other wide-ranging effects on society. It has played a key role in the information age, allowing people to telecommute and live and work where they want. Over 90% of current users say it makes them more efficient workers. Cell phones also provide a sense of safety for users. A recent study shows that two-thirds of new subscribers bought their phones for safety and security. The number of 911 calls nationwide made on cell phones has jumped from 17% in 1995 to 49% in 1998. Analysts believe this number will reach 60% as the services continue to drop in price.

The explosion of cell phone use, however, has had its critics. In 1993 a Florida man filed a lawsuit claiming his wife's brain tumor was caused by her cell phone. This case set off a slew of studies on safety issues. The CTIA formed Wireless Technology Research (WTR) and gave it more than $25 million to study the health risks. The WTR's efforts are compromised, however, by the fact that it is funded by the CTIA and several of the major carriers and manufacturers. The World Health Organization is studying radio frequency radiation (RFR), but its results are not due until 2005.

Those who use cell phones while driving are also under scrutiny, and legislatures across the United States have begun to pass legislation outlawing the practice. On September 1, 1999, Lawrence Simon became the first driver cited for using a cell phone while driving in Brooklyn, Ohio, a Cleveland suburb. The Brooklyn legislation has set off a national discussion of the issue. Bills to restrict the use of cell phones while driving are pending in eight other states. Critics of such legislation abhor the attempt to curtail their freedom to use the devices. Cell phone manufacturers and carriers, on the other hand, want to give people more options in their cars, like in-dash Internet browsers and fax machines, which all springboard off the success of cell phones.

In just over a decade, cellular communications has grown from little more than a good idea into a multibillion dollar industry. From essentially a futuristic dream, cell phones have become a common, even expected, communications tool. As cellular technology keeps pace with computer processors and wireless equipment, cell phones are becoming more than just telecommunications devices. They are being transformed into essential connections to the Internet world, strengthening the ties between people and technology.

BOB BATCHELOR

Further Reading

Books

Bedell, Paul. *Cellular/PCS Management*. New York: McGraw Hill, 1999.

Calhoun, George. *Wireless Access and the Local Telephone Network*. Boston: Artech House, 1992.

Gibson, Stephen W. *Cellular Mobile Radiotelephones*. Englewood Cliffs, NJ: Prentice-Hall, 1987.

Goodman, David J. *Wireless Personal Communications Systems*. New York: Addison-Wesley, 1997.

Other

Farley, Tom. "Digital Wireless Basics." http://www.privateline.com/PCS/PCS.htm.

Schiesel, Seth. "Europe's Megadeal: The Technology." *New York Times* (February 4, 2000): C9.

The Internet Explosion

Overview

In the early 1990s the public had no idea what the Internet was or what it could do. Just a few years later, it had exploded onto computers all over the world, revolutionizing the way we communicate, socialize, and conduct business. The Internet has created a multibillion-dollar industry and spawned a worldwide revolution. With the proliferation of e-commerce, business is no longer restricted to the traditional bricks-and-mortar operation. Now, anyone with a personal computer (PC) can create a website to sell a product. Little did the forefathers of the Internet know how far-reaching the effects of their invention would be.

Background

In 1957 America was in the throes of the Cold War with the Soviet Union. Fears of nuclear annihilation ran high, and the Soviets had just launched the first *Sputnik* into space, winning the race with the United States. In response, the U.S. Department of Defense formed the Advanced Research Projects Agency (ARPA) to boost American technology. Twelve years later, ARPA spawned ARPANET, the world's first connected computer network. ARPANET was designed to withstand a nuclear attack by routing information around the damaged areas.

At the time, there were no home PCs. Computers were massive machines that spanned entire rooms and were unable to communicate with one another. To develop a complex network of computers that could speak to one another, a whole new system of hardware, software, and connectivity had to be created. That job was undertaken by research agencies and universities such as the University of California Los Angeles, the Massachusetts Institute of Technology, Stanford, and Harvard.

By 1972 ARPANET allowed for remote login to other computers, the distribution of information by means of file transfer, and the sharing of resources between computers. The 1970s saw the introduction of electronic mail (e-mail), by which users could send messages from one computer to another, as well as Usenet newsgroups, which are essentially discussion groups focused on a single topic.

In the 1980s several unique networks were created throughout the world, including UUCP, developed by AT&T utilizing the UNIX operating system, and USENET, a decentralized newsgroup network created by the University of North Carolina. The networks, however, could not communicate with one another. This situation changed on January 1, 1983, when ARPA began using the transmission control protocol/internetwork protocol (TCP/IP), which allowed networks to com-

Lawrence Roberts, a leading figure in the development of ARPANET. *(The Library of Congress. Reproduced by permission.)*

municate, essentially creating what we now know as the Internet. Between 1984 and 1988, the number of host computers on the Internet grew from about 1,000 to over 60,000, spanning the globe from Canada to New Zealand.

In the 1980s the Internet was primarily used by universities and the government—it had not yet burst on to the public scene. This was soon to change. On June 1, 1990, ARPANET was disbanded after 21 years of service, and NSFnet took over the administration of the Internet. Created by the U.S. National Science Foundation in the mid-1980s, NSFnet linked five university supercomputers and allowed other universities to tap into its resources. NSFnet would become the central data stream for the Internet.

Impact

The Internet was a powerful new tool for sharing information. It held massive amounts of information from the world's foremost institutes of learning. Unless you knew exactly where to look, however, there was no way to find anything. Searching for information was like finding a needle in a haystack in the dark of night. Enter WAIS (Wide Area Information System), a search engine that allowed users to scan lists of the Internet's file holdings. WAIS was quickly followed by a succession of indexes, including Gopher and Veronica.

In 1989 Swiss physicist Tim Berners-Lee revolutionized the way information was shared when he proposed the creation of a seamless network that any computer would be able to access. Berners-Lee called his new system the World Wide Web (WWW).

In 1991 Congress passed the High Performance Computing Act, which created the National Research and Education Network (NREN), the successor to the NSF network. It ensured the United States primacy in the development of computer technology and high-speed networks. NSFnet had restricted Internet usage in the early years with its acceptable use policy, which forbade using the Net for profit. In 1991 those restrictions were lifted, opening the floodgates for an Internet e-commerce revolution.

In 1993 the number of Web servers sending data across the Net jumped from 50 to 250. The number of Web sites rose from 130 to 623, and the number of hosts rose to more than 1.5 million. As further proof of the Internet's success, the White House, Library of Congress, and United Nations all went online. That same year, the final, critical piece of the Internet puzzle was put in place. Berners-Lee created a software program that allowed users to browse through documents with a simple point-and-click of the computer mouse. His Mosaic browser used hyperlinks, highlighted or underlined words in a document that, when clicked, took the user immediately to that document. This development turned out to be the impetus the Internet needed to catch on like wildfire.

That same year, Internet furor spread to the University of Illinois, where a group of students (including Marc Andreessen) at NCSA (the National Center for Supercomputing Applications) decided to improve the Web's usability, giving it mass-market appeal. They did this by adding graphics to Berners-Lee's Mosaic interface and by changing the technological platform from UNIX, which only technically savvy users could understand, to the more widely used Microsoft Windows operating system. These improvements turned the tide, shifting the Internet from a system only technowizards could navigate to a platform that was clear and easy enough for anyone to use.

In 1993 Marc Andreessen, along with SGI founder Jim Clark and a group of Andreesen's former colleagues from the Mosaic development group went out on their own to build a bigger and better browser. The result was Netscape. By this time Internet traffic was expanding at almost 350,000 percent each year. More than three million hosts were now on the Internet and the number of Web servers had jumped from around 200 to 10,000.

To make navigating the enormous labyrinth of information contained on the Internet easier, two students, David Filo and Jerry Yang, created a set of bookmarks stored in a database. They made their database (which they named "Yahoo!") widely accessible on the Net and created a hugely successful search engine that would eventually be used by millions of people.

By 1994 businesses began catching on to Internet furor, and online shopping malls started to appear. The first cyberbank, First Virtual, opened for business and even the restaurant chain Pizza Hut was giving its customers the opportunity to place their orders in cyberspace.

The following year the World Wide Web became the most important service on the Internet. Online services, such as CompuServe, AOL and Prodigy, could no longer ignore the frenzy and began offering their customers complete access to the Internet for the first time.

Several new technologies also arrived in 1995, broadening the Internet's capabilities. Real Player allowed streaming audio; Sun Microsystems released a new Internet programming language called Java, which greatly improved the way applications and information could be retrieved, displayed, and used over the Internet; and Netscape released Navigator 2.0, enriching the graphical capabilities of the browser.

By 1996 the Internet had exploded with more than 10 million hosts online worldwide. Approximately 40 million people were "surfing the net," and e-commerce (the transaction of business over the Web) was booming with more than $1 billion being spent annually on Internet shopping.

Suddenly, with the proliferation of the Web, several movements began to restrict the new medium. There were clashes over copyright infringement. The Communications Decency Act tried to clean up pornography, which was easily accessible over the Net. Additionally, e-mail users were outraged that they had become the targets of spamming, or marketing aimed at consumers on the Web. Hackers were also having a field day with the new technology, breaking into secure websites and rearranging information for their own fun and profit.

But even with all of these concerns, the Internet continued to grow. Each year brought innovations and a new ways of navigating and using the Net. All over the world people were realizing the potential of the Internet. Many employees no longer had to drive to work; telecommuting allowed them direct access to the office while working from home. Students and researchers could access vast quantities of information from universities, libraries, and scientific institutions with the click of a mouse.

By 1998 the two-millionth domain name had been registered. The Internet had become a multibillion-dollar industry with consumers gaining the confidence to shop for everything from books to boats online. Projections at the end of 1999 were predicting sales to jump into the trillions of dollars in the twenty-first century.

The power of the Internet and World Wide Web cannot be ignored. No other technology has made information so accessible and has so changed the scope of business, entertainment, and society. It has absorbed print, the moving image, and sound to create a multimedia explosion that stretches around the globe.

STEPHANIE WATSON

Further Reading

Reid, Robert H. *Architects of the Web: 1,000 Days that Built the Future of Business.* New York: John Wiley & Sons, Inc., 1997.

Advances in Microprocessor Technology

Overview

In the last few decades, microprocessors have gone from being expensive and marginally useful to being the foundation of technological civilization. The importance of computers was highlighted by concerns surrounding the much anticipated "Y2K bug," the programming glitch that, if uncorrected, many feared would cause a variety

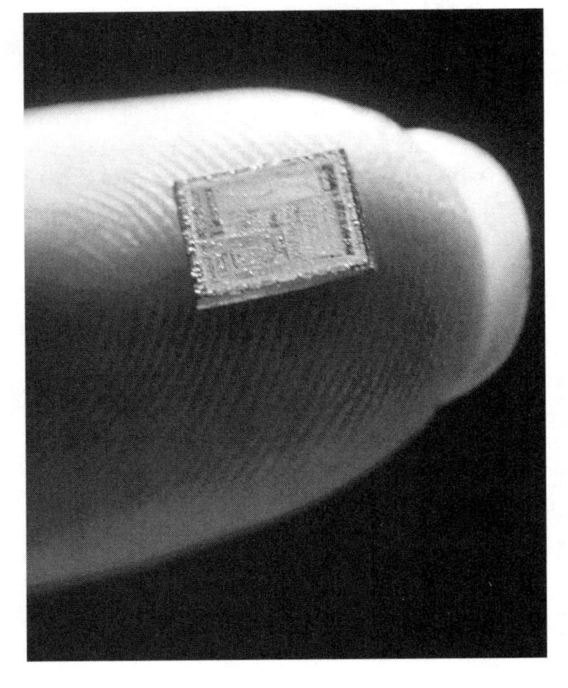

An Intel computer chip, 1982. *(Jim Sugar Photography/CORBIS. Reproduced by permission.)*

of problems, ranging from data loss and communications failures to financial crises, crashed airplanes, and jammed elevators. Indeed, the microprocessor and the information revolution it sparked have become a defining characteristic of modern society. The last few decades of the twentieth century witnessed the inauguration of a new age of inexpensive, powerful computers and ready access to an incredible array of information. While it is uncertain exactly how such technology will affect the world, the microprocessor holds great promise for those who can partake of the digital revolution.

Background

In 1952 British radar expert G. W. Dummer suggested that in the future electronic devices would no longer be constructed of a myriad of individual components stuck into sockets and wired or soldered together. He predicted, instead, that the electronics of the future would consist of all these components manufactured simultaneously in a solid block. Unfortunately, Dummer's prototype unit did not work and he failed to receive any support for further work. It was left to others to realize his prescient vision.

The first integrated circuit was invented almost simultaneously by Texas Instruments and Fairchild Semiconductor in 1959. A major advance in electronics design, the integrated circuit (often abbreviated as "IC") contained many miniature electronic components on a single small device. Incorporating transistors, diodes, resistors, and other electronic components, the IC was the first step towards development of the modern microprocessor. Early integrated circuits cost upwards of $1,000.

The next step was taken in 1971 with the unveiling of the Intel 4004, the first "computer on a chip." The Intel 4004 was actually designed for use in a calculator and was intended to operate at the same speed as a 1960s-era IBM mainframe computer. Although its specifications are laughable by today's standards, it represented a major step forward in that it was the first single-chip CPU (Central Processing Unit) manufactured and available to the public. The 4004 operated a 4-bit processor at 0.74 MHz, had 1K of data memory and 4K of program memory.

For the next several years, microprocessors gained in speed, memory, and complexity. By 1980 many single chips contained the processor, some memory, and other functions that had previously been contained elsewhere in the computer. As they became more self-sufficient, microprocessors also became more appealing for use as controllers for things ranging from automobiles to chemical processing plants to spacecraft. The next breakthrough, in the late 1970s, was the development by Motorola and Intel of the microprocessors that became the heart of the Apple/Macintosh and IBM PC computers, respectively. These processors were cheap enough to make personal computers affordable, while offering enough computing power to make them more than simply game platforms. At the same time, the increased computer power, coupled with expansions in RAM (Random Access Memory) made it possible to start writing commercial software that was worth buying, again adding to the computer's utility at home and in the office.

Since the introduction of the PC (personal computer) and Macintosh computers, personal computers have remained largely unchanged in many fundamental respects, with the personal computer world divided between the two camps—those that are PC-compatible versus those that are Macintosh-compatible. However, computing power has continued to grow at an enormous rate. In 1988, for example, a high-end computer used a 16-bit processor, contained up to 640 KB of RAM, may have had up to 20 MB on a hard drive, ran at up to 8 MHz, and cost nearly $3000. Ten years later, a 64-bit, 300 MHz processor with 64 MB of RAM, a 4 GB hard drive, video cards, sound cards, CD-ROM, and more cost less than $2500. Furthermore, the latter system, though vastly superior, was considered obsolete within two years of production. This same increase in power and cost reduction has made computing available to many people worldwide, allowing fast and reliable communication in nearly all countries and continents (including the scientific stations in Antarctica), as well as making possible the explosive growth of the Internet and the World Wide Web.

Impact

The development of the microprocessor has had, and will continue to have, an enormous impact on humanity. In fact, the microprocessor may later be regarded as one of the most important developments in the history of humanity, along with fire, the wheel, and electricity. Its impact was first felt among scientists and engineers, the wealthy, and the hobbyists, but has spread to many others as the price of computers continues to drop. The impact of microprocessors can be roughly broken into three areas:

1. Personal computers (home and business use of computers)

2. High-performance computing (scientific and military use of computers)

3. Industrial computers (microprocessor-driven controls in industrial and consumer goods)

As mentioned above, personal computers have become both powerful and affordable in the later part of the 1990s. This has made them increasingly available to groups that never before had access to computers, and has made computer literacy an increasingly important aspect of education. Use of references on CD-ROM, DVD, and the Internet has completely changed the manner in which school reports are researched and written. However, the huge amount of adult materials available on the Internet has prompted much debate and proposed legislative action, ostensibly intended to protect children from exposure to unhealthy influences, such as pornographic, violent, or bigoted sites.

At the same time, computers are ubiquitous in the workplace. Software allowing remote collaboration on documents, internet and intranet sites, e-mail, and the easy transfer of electronic files around the world has facilitated the sharing of information, while the increased power of desktop computers has made possible more capable and versatile programs. Conversely, though pundits predicted the "paperless office" with increasing computerization, the opposite has instead been true. Prior to computerization, for example, the average document would be revised only one or two times because each revision necessitated a complete re-typing. Now, however, it is not unusual for documents to have dozens of revisions, each one printed out for each person reviewing it. In fact, an argument can be made that computers in the office actually save very little time because of the collective time spent on such reports. In addition, the presence of computers on nearly every desk has resulted in a complete change in what is considered secretarial work. Since many people can type as quickly as they can write, handwriting drafts for typing by a secretary makes little sense. Instead, secretaries are increasingly asked to format documents, coordinate mass mailings, run the copier, and so forth, rather than taking dictation, learning shorthand, and other, formerly standard, skills.

This proliferation of computing power is contributing to a schism between the "haves" and the "have-nots" of the information age. While those with access to computers have access to the great benefits of widely available information resources, those without computers are being left behind. A high school student who lacks a computer is less likely to compete at a high academic level compared to peers who have and use computers regularly. When these students graduate, they are less likely to attend top-level colleges and, at any university, are less likely to perform well compared to their computer-savvy classmates. This is not to say that doing well absolutely depends on having computer skills, but those skills can have a tremendous impact on the potential for future success.

Not very long ago there were several classes of computers: personal computers, minicomputers, mainframes, and supercomputers. Sun introduced the workstation in the late 1980s, bringing high-end power to an individual desktop, primarily for use in scientific or technical fields. However, with the increasing power of the Intel and Motorola PC microprocessors, the computing world has tended to collapse towards personal computers and supercomputers, and even here the lines are blurred because many PCs can be networked together to form a virtual supercomputer.

The advent of the Cray supercomputer and follow-on machines has been a boon for many scientists. While not comprehensive, a partial list of achievements includes more accurate weather prediction, climate modeling, geophysical and hydrogeological modeling, nuclear weapons design and modeling, aircraft design, supernova simulations, investigation into the formation of the universe, particle physics data reduction, cinematic computer graphics, economic forecasting, cryptology, the availability of Global Positioning Satellites, and much more. Without going into the details of each, it is enough to say that the impact of these innovations is dramatic, from better storm predictions to the precision now possible in knowing the location of a person, vehicle, or animal that is lost. Modern science depends on computers for many reasons, including convenience, reliability, and, not least, because so many problems being addressed are simply too complex to solve manually or analytically. Not only do computers relieve scientists of the drudgery of painstaking, repetitive calculation, but they make possible many calculations and simulations that were simply impossible previously.

Microprocessors have infiltrated nearly every possible niche of modern technological society. As an example, consider the automobile. Computers continuously monitor parameters that access conditions within the engine, including temperature, oxygen levels, and exhaust gases. With this information, computers control the fuel flow, spark timing and strength, and other mechanisms to help the car operate at the highest fuel efficiency and the lowest emission of pollutants possible. In addition to the scientific and military uses noted above, microprocessors are commonly used in the entertainment industry (synthesizers, computer graphics, digital soundtracks, compact disks, and digital video disks), and many everyday electronic items, such as wristwatches, personal digital assistants, telephones, appliances, and so forth. The microprocessor is rapidly becoming an inescapable fact of life that will continue to have a profound impact on nearly everyone on in the world.

P. ANDREW KARAM

Further Reading

Banchoff, Thomas. *Beyond the Third Dimension: Geometry, Computer Graphics, and Higher Dimensions.* New York: Scientific American Library, 1990.

Computers. Alexandria, VA: Time-Life Books, 1990.

Hofstadter, Douglas. *Godel, Escher, Bach: An Eternal Golden Braid.* New York: Vintage Books, 1980.

Kaufmann, William, and Larry Smarr. *Supercomputing and the Transformation of Science.* New York: Scientific American Library, 1993.

Kidder, Tracy. *The Soul of a New Machine.* Boston: Little, Brown, and Company, 1981.

Calculators: A Pocket-Sized Revolution

Overview

The invention of the electronic pocket calculator in the 1960s ignited a world-wide microelectronics revolution. Hand-held calculating machines, portable and accurate, subsequently became valuable fixtures in science, engineering, business, and education.

Background

Calculating machines date back to ancient Babylonian devices used to perform rote mathematical operations. Although many hand-crank and machine-type devices existed (the venerable keypunch cash register is such a mechanical calculator), prior to the 1960s, calculating machines were universally heavy, cumbersome, and expensive. The more sophisticated models—those able to perform more than just fundamental mathematical operations—were available only to government, the military, and a few large businesses. Although producing accurate results, the large, fixed-site machines required specialized training and were reserved for elaborate calculations, such as those carried out during the creation of the first atomic bombs by Project Trinity at Los Alamos during the 1940s, and required teams of men to operate the expensively maintained machines. WWII also created the desperate need to perform mathematical calculations quickly, sometimes under adverse conditions. The success or failure of a bombing run or the accuracy of artillery often depended upon the nimble fingers of soldiers struggling with mechanical-type calculators.

Prior to the introduction of the electronic pocket calculator, scientists and engineers had, since the seventeenth century, relied on the slide rule (the modern version dating back to French military officer Amedee Mannheim's 1850 design based on the operation of logarithms) to make multiplication, division, extraction of roots, and raising to powers easier to perform. Although totally supplanted by the hand-held calculator, the slide rule was an important device that, it may be fairly asserted, helped take man to the Moon. The design of the rockets and spacecraft in the pioneering days of the space race was accomplished by scientists and engineers who relied on slide rules that—instead of offering precise results to many figures—relied on the operator's experience and knowledge to make correct estimations regarding the appropriate solution to a mathematical problem.

In the mid 1960s a state-of-the-art, battery-powered, transistorized portable calculating machine might weigh more than 50 pounds (23 kg) and cost in excess of $2,000.

In 1967, American scientists and engineers Jack St. Clair Kilby (1923-), Jerry Merryman, and James Van Tassel revolutionized the way mathematical calculations were thereafter performed when the Texas Instruments team invented the Pocketronic—the first pocket-sized calculator. The Pocketronic was the first successful commercial usage of the monolithic integrated circuit (generally referred to as a computer microchip). The integrated circuit is an electrical circuit consisting of resistors, capacitors, diodes, and transistors, generally made from silicon, and integrated on a single silicon chip. In fact, the development of the calculator was driven by a need to find a practical commercial use for the integrated circuit. Following the success of the pocket calculator, demand for microchip technology soared throughout the world.

Impact

Although the Pocketronic performed only the simplest of mathematical operations (enabling its users only to add, subtract, multiply, and divide), the speed and accuracy (up to 12 digit answers) of the machine proved sensational.

Kilby, whose design of the integrated circuit nosed out by few months the similar work being performed by inventors Bob Noyce (1927-1990) and Gordon Moore (1929-), is also credited with the design for the thermal printer and more than 60 patents. Kilby was named to the National Inventors Hall of Fame and, in 1970, was awarded the National Medal of Science. In 1989, the American Society of Mechanical Engineers awarded Kilby, Merryman, and James Van Tassel a medal for their engineering feat.

A modern handheld calculator, with its predecessor in the background. *(Doug Handel/The Stock Market. Reproduced by permission.)*

In 1972, the Hewlett Packard company introduced the HP-35 pocket calculator that started a steep rise in the capacity of calculators. In addition to performing basic operations, the HP-35 enabled users to utilize advanced mathematical functions. The HP-35 was the first scientific, hand-held calculator because it carried the ability to perform a wide number of logarithmic and trigonometric functions. In addition, the HP-35 was able to store intermediate solutions and to utilize scientific notations.

The drive to create the HP-35 was supplied by Hewlett Packard co-founder William Hewlett, who was interested in developing an advance calculating machine that could fit in a shirt pocket. Almost overnight, the HP-35 became the standard in computational problem solving required in scientific, industrial, and educational settings. It is interesting to note that the first calculators (including the HP-35) were often marketed as "electronic slide rules."

In addition to advanced functions the HP-35's precision was matched only by the most sophisticated computers. Because the HP-35 could handle numbers as small as 10^{-99} and as large as

10^{99}, their application to physics and engineering was unlimited. They could easily compute with far greater mathematical precision quantities far outside the range of the physical dimensions of the Cosmos.

When the HP-35 came onto the market it precipitated major changes in demand for portable calculating devices. Not only were calculators in demand by scientists and engineers but the general public clamored for the latest devices. Within a few years portable calculators replaced the use of slide rules even in high school physics and chemistry classes.

Freedom from mundane manipulations was not reserved for students, however, the HP-35 sparked interest from the business and finance communities as well. The business community soon clamored for a specialized "business computer" able to simplify often vexing financial calculations such as computing interest rates and creating amortization schedules.

On the heels of the stunning commercial success of the HP-35, the tempo of advances in calculator technology began to rise. In 1974, the first fully programmable calculator was introduced. Calculators with continuous memory functions (i.e., they maintained instructions and numbers after the units were shut off) were put on the market. In 1979 Hewlett Packard developed the first fully programmable, continuous memory, alphanumeric calculator (the HP-41 series).

The changes to education were profound and immediate. Students armed with portable calculating devices quickly and accurately performed once daunting mathematical work. Instead of being limited to highly stylized problems (i.e., problems that had whole number solutions or that contained simplified mathematical operations), teachers at all levels were, for the first time, challenged to give students more "real-world" type problems.

The introduction of the scientific calculators was not, however, without controversy. Although many teachers of mathematics viewed calculators as a tool that liberated students from pedestrian calculations, some teachers became concerned that an over-reliance on calculators was a contributing factor to sliding scores on standardized mathematics exams. Although calculators made students' work more accurate, critics of calculator usage feared a lowering of ability to perform what were once simple and rote operations. Teachers struggled with the idea that long-cherished multiplication tables might be replaced by workbooks filled with calculator exercises that ultimately became button-pushing exercises for students. During the last two decades of the twentieth century, considerable study and debate within the mathematics education community took place regarding the appropriate integration of calculator-based technology into the mathematics curriculum.

The meteoric rise in the accessibility and popularity of calculators were significantly encouraged by the low cost to the units. With mass production, prices dropped all during a time when demand, availability, and complexity of calculators rose sharply. Incredibly, Texas Instruments's Pocketronic was not even designed for mass production. In fact, Texas Instruments allowed a Japanese company (Canon, Inc. of Tokyo) to produce and market the first commercial model. Accordingly, from its quiet 1970's debut in the Japanese market, the Pocketronic's notoriety grew exponentially. The first Pocketronic, sold in the United States a few months later, carried a price tag of approximately $400. As demand rose, however, advances in technology allowed manufacturers to actually lower costs on subsequent units. By the end the end of the twentieth century, calculators able to perform far more complex math functions than the Pocketronic sold for under $10.

Seemingly near the limit in range of mathematical operations, calculator technology focused on lowering cost and improving displays through the use of Liquid Crystal Displays (LCDs). Power and durability were also targets for performance improvement (the batteries on the Pocketronic lasted for about four hours), and use of alternative power sources (e.g., rechargeable batteries and solar power) became important mechanisms to extend calculator utility. Calculators were made available to an increasingly diverse user group with the development of miniaturized Braille keyboards and talking displays.

During the 1980s improved calculator designs increased the range of operations to complex, multi-step algebraic, geometric, trigonometric, statistical, and calculus type functions. Hewlett Packard marketed alpha-numeric calculators with screens that displayed words and numbers that could be programmed to perform complex calculations within seconds. Even aboard the most sophisticated military aircraft, commercially available calculators became standard backup devices for navigators and pilots plotting vectors.

NASA placed the alpha-numeric successor to the HP-35, the HP-41 series, aboard the *Space Shuttle* spacecraft so that astronauts could, if the need arose, manually calculate the critical and exacting angles required to safely re-enter the Earth's atmosphere. By the end of the twentieth century, the hand-held pocket calculator descendants of the Pocketronic were able to rapidly graph complex equations with the power and speed once reserved to computers.

BRENDA WILMOTH LERNER

Further Reading

Books

Bronowski, J. *The Ascent of Man.* Little Brown, 1973.

Haddock, Thomas F. *A Collector's Guide to Personal Computers and Pocket Calculators.* Books Americana, 1993.

Mitchell, Robert. *Contemporary's Calculator Power: A Modern Approach to Math Skills.* NTC/Contemporary, 1996.

Packard, David, David Kirby, and Karen Lewis, eds. *The HP Way: How Bill Hewlett and I Built Our Company.* Harperbusiness, 1996.

Periodical Articles

Kim, Irene. "Functions at Your Fingerprints." *Mechanical Engineering Magazine* Vol. 112, No. 1 (January 1990).

Invention of the Bar Code Revolutionizes Retail Sales and Inventory Control

Overview

While early technological developments in bar coding were stimulated by possible retail applications, the first actual bar codes appeared in industrial settings. From a 1932 Harvard University master's thesis to imprinted codes on the side of railway cars in the 1960s to today's ubiquitous Universal Product Codes printed on manufacturers' and consumer products worldwide, bar code technology evolved from a simplistic checkout scheme to widespread inventory control and data collection systems. By the end of the twentieth century, bar code technology had found numerous worldwide applications across all industries. Hundreds of thousands of manufacturer-specific identification numbers—with no telling how many product codes based on those manufacturer codes—had been assigned not just for retail and consumer goods but also in other commercial, industrial, and government sectors in countries worldwide.

Background

In 1932 a Harvard University business student named Wallace Flint wrote his master's thesis on a proposed punched card supermarket checkout scheme. Inspired by a punched card system developed for the 1890 U.S. Census, Flint envisioned a system in which consumers would make their merchandise selections by removing corresponding punched cards from a catalog and handing the cards to a checkout attendant. The checker would place each card in a reader that would, in turn, activate an elaborate conveyor belt system (in reality, an unwieldy and expensive scheme) to deliver purchases to the customer. The consumer would receive a bill that would also give store management a record of products purchased. With Flint's thesis, a method to automatically update inventory records was born, but it went unrealized due to its conception in the midst of the Great Depression.

Progress towards the modern bar code began in 1948 when Bernard Silver (1925?-1963), a graduate student at Philadelphia's Drexel Institute of Technology, overheard the head of a local food chain petitioning a Drexel dean for research into a system that could automatically capture product data during checkout. Silver mentioned the conversation to his friend and fellow Drexel graduate student Norman Joseph Woodland. The duo began experimenting with ink patterns and ultraviolet light. Their first test device experienced difficulties due to ink instability and high printing expenses, but they were determined to succeed. Eventually, Woodland's single focus on the project led to a solution.

Frustrated with tackling the invention while at Drexel, Woodland left Pennsylvania for Florida where, after several months, he devised a linear bar code using elements from two existing technologies—Morse code and the movie sound system developed in the 1920s by Lee de Forest (1873-1961). His resulting straight line pattern was quite similar to later bar codes, but Wood-

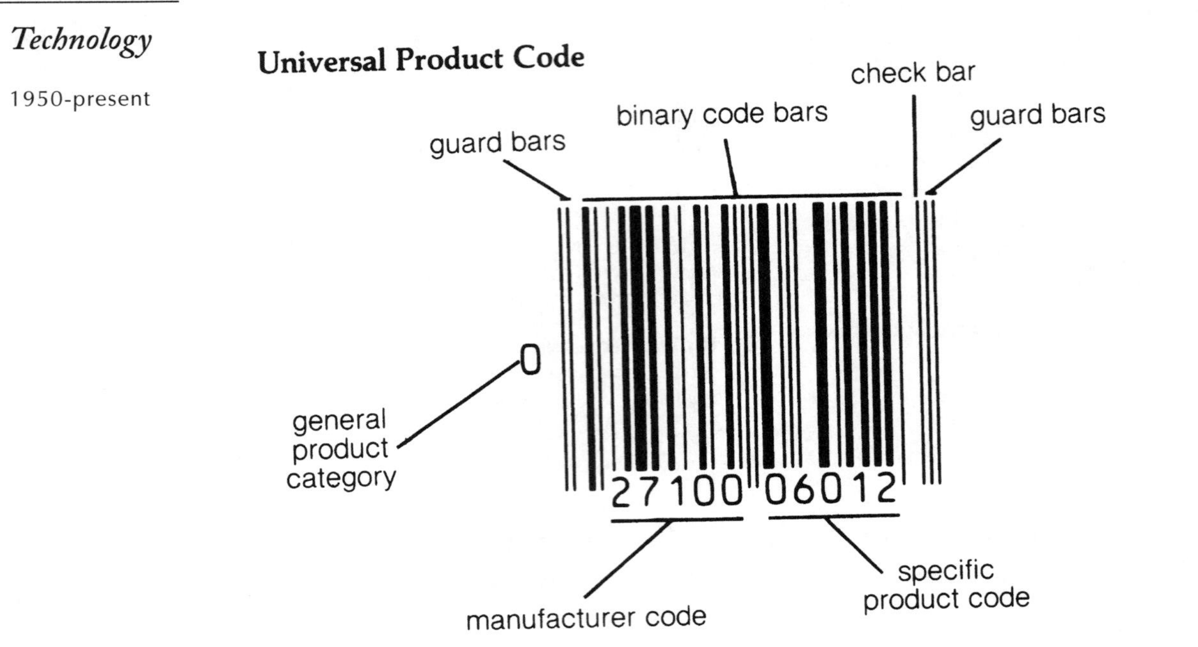

The parts of the Universal Product Code (UPC). *(Hammond, Inc. Reproduced by permission.)*

land replaced the wide and narrow vertical lines with concentric circles, known as the bull's eye code, before he and Silver filed a patent application for their "Classifying Apparatus and Method" in October 1949. (The patent was issued in October 1952.)

In 1951 Woodland accepted a job with International Business Machines Corp. (IBM), while still working with Silver to investigate the construction of a bar code reader to electronically read their printed codes. (A year later, they succeeded in building the first actual reader in the living room of Woodland's house in Binghamton, New York. It was crude, about the size of a desk, but it worked. Further advances in scanning technologies awaited the development of lasers and integrated circuit microchips in the late 1960s. These finally brought about the handheld scanners in use at the end of the twentieth century.) In 1961 Philco purchased Woodland and Silver's patent, which was later sold to RCA, which would not make any developments in bar code technologies until the early 1970s. Before then, advances in automatic data collection would result from research in the railroad industry.

In the late 1950s the Association of American Railroads was searching for a method to track railway cars. In response to this search, a Sylvania Corporation employee named David Jarrett Collins (1936-) designed a reflective, color-coded bar system read by an optical scanner, which was first tested in 1961 on gravel cars in the Boston & Maine system. Rail yards were able to use the bar code readers to automatically identify cars and supply valuable accounting data to rail industry members. In 1967 Sylvania's color bar code technology was adopted as the freight car control system throughout North America (although it was abandoned in the late 1970s in the wake of the railroad bankruptcies during the mid-1970s economic recession).

Sensing the importance of bar code technology, Collins approached management at Sylvania to develop a black and white line equivalent to the railroad's system for other industries. In a move probably regretted, Sylvania refused to fund such a program, so Collins left to co-found Computer Identics Corporation in 1968. The new company immediately began experimenting with laser beams for bar code scanning. In the spring of 1969 Computer Identics installed its first two systems—the first true bar code systems in the world—in a General Motors plant in Pontiac, Michigan, and in a distribution facility operated by General Trading Company in Carlsbad, New Jersey.

While Woodland and Silver's technological vision and Computer Identics' achievement demonstrated the feasibility of bar codes in industrial settings, it was eventually the grocery industry that, like Flint in the 1930s, explored the potential of automated data systems. By 1966 RCA owned the rights to Woodland and Silver's patent and had learned that the grocery industry's members were looking into the application of bar code technology in their stores. Research was launched, and in the spring of 1971 RCA demonstrated a bull's eye bar code system at an industry meeting. IBM executives at the meeting realized they needed to pursue the fledgling technology and transferred the bar code's inventor, Woodland, an IBM staffer, to their facilities in North Carolina, where he played a prominent role in developing the most significant version of the technology, the Universal Product Code (UPC).

While IBM was working out its technically advanced UPC system, RCA continued to test its bull's eye code. In July 1972 RCA embarked on an 18-month test in a Kroger grocery store in Cincinnati, Ohio, but experienced printing problems and scanning difficulties that hampered research and development. In April 1973 the grocery industry adopted IBM's Universal Product Code system as its standard, transforming business systems worldwide as a result. On June 26, 1974, at a Marsh Supermarket in Troy, Ohio, a package of Wrigley's chewing gum made history, as it became the first retail product sold with the help of a scanner. The UPC label on the Wrigley's package was part of IBM's fixed-length, all-numeric system designed to uniquely identify a product and its manufacturer. In addition to eliminating the necessity of putting a price label on the package, the UPC label gave both Wrigley's and Marsh Supermarkets a mechanism to track valuable consumer information while also collecting important inventory data. The resulting database could be shared electronically to allow greater efficiencies than ever experienced in the manual systems replaced.

The success of the UPC system and refinements to its basic code led to the development of other bar coding systems. For example, the Code 39 system, adopted in 1981 by the U.S. Department of Defense for labeling all products supplied to the U.S. military, was the first alphanumeric bar code. Code 39 also found applications in the automotive industry and the health industry. The variable-length Codabar system used by overnight shipping giant Federal Express was another offshoot of bar code technology, as were the postal tracking codes developed for the U.S. Postal Service. Bar codes crept into Europe with the European Article Numbering (EAN) system, also developed at IBM, which included an extra symbol to identify country of origin. Towards the end of the twentieth century, developers migrated to a stacked bar code (also known as the two-dimensional or matrix code), which found its way onto hospital patient identification bracelets and even fingerprint data encoded on identity cards.

Impact

From grocery stores to university libraries, bar code technology found a home. Aided by rapidly developing hardware (including linear image technology introduced in 1999) and software, the advanced technological systems transformed the daily functions of society. Consumers learned to scan their own groceries in self-checkout lanes (saving hours of paid labor each week per checkout stand). Students learned to scan the labels in books and resource materials they wished to take home from the library. The products and materials scanned were cataloged in innumerable databases, with the data evaluated from all angles. Everything from consumer loyalties to customer service to component inventories would be tracked and influenced. An army of bookkeepers and data-entry typists were replaced by bar codes, bar code readers, and computers. In each instance, a more effective management of resources resulted.

Despite its humble beginnings as a retail sales tracking tool, the bar code has provided immeasurable benefits across numerous industries on its way to becoming a multi-billion dollar business. From the speed and accuracy of data input to the labor savings realized through the elimination of manual systems, the bar code has far surpassed anything its early developers could have envisioned. With the increased information available to study marketplace trends, maintain inventory control, and track personnel or even ongoing projects in a workgroup, bar code technology has found its way into untold market sectors, with far-reaching impact on every aspect of our lives. It could be the single most important invention in the history of organizing mankind. The bar code is a prime example of the technological achievements of twentieth-century society.

ANN T. MARSDEN

Further Reading

Brown, Stephen A. *Revolution at the Checkout Counter.* Cambridge: Harvard University Press, 1997.

Collins, David Jarrett, and Nancy Nasuti Whipple. *Using Bar Code: Why It's Taking Over.* Duxbury, MA: Data Capture Institute, 1994.

Nelson, Benjamin. *Punched Cards to Bar Codes: A 200 Year Journey.* Peterborough, NH: Helmers Publishing Co., 1997.

The Invention of the Fax Machine

Overview

With its convenience and ease of use, the invention of facsimile communication has permanently changed business operations in the modern world. This transmission device, commonly referred to as the fax machine, is now used worldwide by businesses, governments, and households. Its ubiquitous presence in offices and homes is due to its low cost and expediency in transmitting documents. The fax machine has had a direct impact on the advancement of communications technology since its humble invention in the 1800s.

Background

The earliest form of the fax machine is attributed to Scottish inventor Alexander Bain (1818-1903). In 1843 Bain created and patented a device that simulated a two-dimensional image, making significant improvements on the telegraph. Although Bain's creation was never officially tested, in 1841 English physicist Frederick Blakewell first demonstrated the use of another type of facsimile that differed from Bain's invention in its method of transmission. In 1863 Italian abbot Giovanni Caselli (1851-1891) created the first commercial facsimile system, first used in France between the cities of Paris and Lyon. His invention continued to be used commercially in the years to come. Later, in 1902, German inventor Arthur Korn (1870-1945) proved the ability to transmit photographs through optical scanning, and in 1906 people were using his particular machines regularly in the newspaper industry. Today, nearly every business and household owns a modern type of fax machine that is capable of transmitting both colorful images and hard copy text.

The fax machine functions by scanning an image on paper and transmitting that image over telephone lines to be reproduced on paper through another fax machine. Contemporary fax machines use rolls of thermal paper with thermal printers equipped with an automatic paper cutter to separate the sheets after printing them. The thermal paper is very lightweight and inexpensive but tends to fade or turn sallow after time. Thus, it is often wise to make a photocopy of faxed documents of importance because of this problem. However, many newer fax machines are designed to print on individual sheets of standard paper much like a computer printer.

The scanning process of a document functions by taking a minimal horizontal line of the item to be faxed into the machine, where it is scanned for light and dark qualities. These lines are given a number sequence called a binary code made up of "0"s and "1"s. The quality white receives a 0, and darker qualities such as black that create the given image receive a 1. These qualities are regarded as pixels. On the average, there are 1,728 pixels per line. The facsimile that is sending the information appoints that specific binary code made up of the 0s and 1s that identifies the image made up of pixels. The scanned image is then sent, identified by the number sequence called a "bit" to the receiving facsimile, usually though telephone lines, though radio broadcast waves can also be used, and is reproduced on the other machine.

The total amount of lines per page is calculated together to create a bit sequence that can be a number that is represented in the millions. Fax machines calculate the bits per second to reproduce the image that is being transmitted from the original facsimile to the receiving facsimile according to the bits comprised of the total amount of binary codes that identify the image scanned. The images transmitted can include handwritten or typed text, graphs, charts, maps, and drawings. The images can be in black and white or color, depending on whether the receiving end is able to accommodate a colored transmission. Compatibility between each facsimile machine is necessary for the entire process to take place.

Improvements continue to be made concerning the speed of fax transmission. The process of coding the scanned information into a more compressed reading, such as using a run-length code that utilizes multiplication to cut the total amount of codes transmitted, reducing the number of bits and time spent transmitting them, is becoming standard in newer fax machines. The effect of this compressing of numbers speeds the sending and receiving of the faxed item. The speed of one's modem also impacts the rate at which an item can be sent. A good modem can transmit an item in one minute or less with elaborate images.

Facsimiles function best when a modem is the vehicle for transmission. Modems are necessary in efficient facsimile transmission because they convert digital information into analog information from a variety of telecommunication devices. Modems are either voiceband or cable operations ready. Facsimiles rely on the voiceband modem to transmit information from one machine to another in the form of number codes over telephone lines. The advancement of digital technology involves the use of digital circuitry to create an even faster printing fax machine at less than 10 seconds. This type is not common among most businesses or households at this time, however, due to limited digital transmission circuits.

Impact

Since the invention of the fax machine, improvements have been made to create better methods of transmission. The International Telegraph and Telephone Consultative Committee (CCITT), a branch no longer independent of the International Telecommunication Union (ITU) that has been an agency of the United Nations since 1947, created a standard for all voiceband modems to follow so that communication worldwide would be at its best when using a fax machine. Since both the sending and receiving fax must be compatible in order to adequately transmit information from one machine to the next, the committee assured that operations such as matching speed of transmission, encoding, and signaling would be harmonized. In 1980 the CCITT accepted the standard Group 3 type of fax machine, which calculates bits per second at a rate between 2,400 and 9,600 and produces a scanned image in around one minute using digital transmission through modems.

Because of the adoption of this standard, the growing affordability of the facsimile (due in part to mass manufacturing), and its recent compatibility with personal computers, most people and businesses own a facsimile. Prior to the CCITT decision to adopt Group 3 facsimiles, the Group 2 fax machines functioned as a much slower rate and, though they were useful in businesses for transmission of uncomplicated, plain-text documents, those machines were not suitable for home use. As a result of the adoption of the Group 3 standard fax machine, this invention has become popular not only among businesses, but in households as well because of its expedience and affordability.

The invention of the fax machine has greatly changed correspondence. The fax machine has practically made the use of the telegraph obsolete because of the facsimile's expedience, as well as its cost efficiency and its easy-to-use mechanisms. Since the adoption of Group 3 type facsimiles, these machines are easier to use because of their computer modem compatibility. It is common to have one's computer printer also convert to a fax machine and the use of the modem for online service doubles to transmit fax documents as well.

The fax machine makes the task of sending a letter to a friend halfway around the world a simple task. All one has to do to operate a fax machine is place the paper to be sent into the feeder of the fax machine, dial the number of the fax machine it is to be sent to, and wait for about minute for one's friend to receive the item on his or her fax machine. The simplicity and convenience of this machine makes sending hard copies of documents easy. They can be used 24-hours a day, seven days a week, from anywhere around the world.

The mass production of the fax machine has caused this invention to be more affordable than other methods of communication. Modern businesses have long since dispensed with their old telegraph machines and are relying on fax machines for quicker transmission of written information. The current rate of sending facsimile documents categorized in Group 3 is around one minute or less now, and that includes complex graphical text and color imaging. Although a Group 4 facsimile standard has been adopted by the CCITT, accommodating a fax machine running on pure digital networks and capable of transmitting a single page of data in under 10 seconds, the lack of availability of digital networks has caused this type of facsimile to be less popular at the present time.

Sending documents instantly is what makes the fax machine such an asset to any business. There is no more need to rely on the postal system to transport most documents. The use of couriers for paper documents is also becoming a thing of the past. Unless the documents in question must be originals and require an original signature, also referred to as a "wet signature," the need for any other communication device besides the fax machine is unnecessary. Businesses can communicate through faxing as easily as e-mail with attached documents, without the concern that one may not check one's e-mail regularly.

Facsimiles provide instant messaging in the form of a tangible document. Because of their convenience and necessity in the fast-paced modern business world, these machines will continue to be used and improved upon. The ongoing advancements in facsimile transmission, such as the development of digital circuits for practical widespread usage of Group 4 facsimiles, will have a direct impact on communications in years to come.

BROOKE COATES

Further Reading

Bodson, Dennis, Stephen Urban, and Kenneth R. McConnell. *FAX: Facsimile Technology and Systems.* 3rd ed. Boston: Artech House, 1999.

Margolis, Andrew. *The Fax Modem Sourcebook.* New York: Wiley, 1995.

The History, Development, and Importance of Personal Computers

Overview

The personal computer was introduced in 1975, a development that made the computer accessible to individuals. Up to that time computers had been very large and expensive, operated mainly by big companies. The first modern computers were created in the 1950s and have a long theoretical and technical background. The use of computers has profoundly effected our society, the way we do business, communicate, learn, and play. Its use has spread to all literate areas of the world, as have communication networks that have few limits. The personal computer has inspired new industries, new companies, and created millionaires and billionaires of their owners. It has also changed the English language and refocused the power in many businesses from the men who procure the money to those who create the product.

Background

Human beings have devised many ways to help them do calculations. Before the creation of the modern large computer and its refinement, the personal computer, a number of discoveries and inventions were necessary. The decimal system, a binary mathematical system, and Boolean algebra are required to make computers work. The eighteenth-century discovery of electricity was also essential, as was the knowledge of how to use it in the mid-nineteenth century. The first automatic calculator appeared in the seventeenth century, using wheels and gears to do calculations. In the nineteenth century, Joseph Jacquard (1752-1834) invented a loom using punch cards attached to needles to tell the loom which threads to use in what combinations and colors. With it, he wove complex patterns in cloth, still called a Jacquard design. In the same century Charles Babbage (1792-1871) designed a "Difference Engine" to calculate and print out simple math tables. He improved it with his "Analytical Engine" using punch cards to perform complex calculations, though he never had the funds to build one. Thus, by the end of the nineteenth century, many elements necessary to make a modern computer work were in place: memory cards, input devices, mathematical systems, storage capabilities, power, and input systems.

The genesis of the modern computing machine came in 1888 when Herman Hollerith (1860-1929), an American inventor, devised a calculating machine to tabulate the U.S. Census for 1890. It was mainly a card reader, but it was the first successful working computer, the grandfather of modern computers. More than 50 of them were built and sold. Hollerith's company, the Tabulating Machine Company, was the start of the computer business in the United States. When Hollerith sold out in 1911, the name was changed to the Computing- Tabulating-Recording Machine Company. In 1924, this company became International Business Machines Corporation (IBM). IBM dominated the office equipment industry for nearly 25 years with its calculators, electric typewriters, and time clocks.

Digital electronic computers appeared in 1939 and 1944, but they were only interim steps in computer development. These computers were huge and expensive, used by large companies to do bookkeeping and math quickly and accurately. They were analog computers, controlled by relays or switches, and needed huge air conditioning units to keep them cool. Because of

Steve Jobs with an Apple iMac computer. *(Reuters Newmedia Inc./CORBIS. Reproduced by permission.)*

this, and the cost of one unit, the use of computers was very limited. The first general-purpose electronic digital computer, ENIAC, was constructed in 1939. Its major components were vacuum tubes, devices that control electric currents or signals. These tubes commonly powered radios and television sets at the time. The programming of ENIAC was a long, tedious process.

A new more advanced computer was built in 1951 by Remington Rand Corporation. Called UNIVAC, it was the first commercially available computer. It was very expensive, very large, and still powered by vacuum tubes. IBM manufactured its first large mainframe computer in 1952 and offered it for sale to companies, governments, and the military. For nearly 30 years, IBM was the most successful company in information technology. The invention of the transistor in 1947 began the trend toward small computers and the personal computer. Created by three scientists at Bell Labs, for which they received Nobel prizes, the transistor is a device that does the job of a vacuum tube at a fraction of its size. It is solid with no moving parts, durable, and begins working immediately without the need to warm up like a vacuum tube. Transistors vary in size from a few centimeters in width to a thousandth of a millimeter. They are smaller, lighter, less expensive to make, cheaper to use, and more reliable than tubes. They use very little power and had replaced tubes by the early 1960s. Transistors control all of the operations in a computer as well as peripheral devices.

The first fully transistorized large computer was built by Control Data corporation in 1958, and IBM unveiled their own version in 1959. These were expensive machines, designed to work for large corporate tasks. All the essential parts of a modern personal computer had been invented by the early 1960s. A computer chip is a tiny piece of silicon, a non-metallic element, with complex electronic circuits built into it. The integrated circuit links transistors together to create a complete circuit on a single chip. Microprocessors are groups of chips that do the computing and contain the memory of a computer. With these devices, the working parts of a computer can be contained on a few computer chips. This innovation continued to shrink the size of computers. Very large computers, like the Cray and IBM machines, were called mainframes or minicomputers. By the end of the 1960s many industries and businesses had come to rely on computers and computer networks, and the personal computer was just around the corner.

Besides the hardware that makes up a computer, the most important element in making it work is the program that instructs it in what to do. The first programmer was Ada Byron (1815-1852), daughter of the British poet Lord Byron. She created theoretical steps to be used in Babbage's machines. BASIC was the first modern programming language, a simple system that almost anyone could learn. Soon it was necessary to create more complex sets of languages and instructions, and this was called software. Microsoft Corporation was started in 1976 by Bill Gates (1955-) to create and market software for personal computers. As computers increased in power, speed, and the variety of functions they performed, the size and complexity of programs also expanded. Many modern programs contain tens of millions of lines of instructions in complex codes. Some are essential to the running of the machine and are built into it. In early computers, the user had to create his own program, but it is nearly impossible to buy a computer today that can be programmed by an individual. Software is delivered to the computer by way of floppy or compact discs or is already installed in the computer. It enables a user to create written documents, display pictures, sound, play games, make charts, and gain access to the Internet.

Impact

Enormous changes have come about in the past 30 years as a result of the development of computers in general, and personal computers in particular. This creation ranks as one of the most important inventions of the twentieth century. The computer is used in government, law enforcement, banking, business, education, and commerce. It has become essential in fields of scientific, political, and social research as well as aspects of medicine and law. Everyone is affected by the manipulation and storage of data. There are negative consequences of these developments. There are those who engage in fraudulent acts, malicious mischief, and deception. These activities have spawned the need for computer security and a new category of technical crime fighters.

At first, the personal computer was defined as a machine usable and programmable by one person at a time and able to fit on a desk. It was inexpensive, accessible, simple enough for most people to use, and small enough to be transportable. The claims for the identity of the first personal computer are numerous and depend on definition. One of the first small computers was a desktop model built by Hewlett Packard in 1972. It had all the basics: a language, memory storage device, a keyboard, and a display terminal. However, because it was built for scientists and engineers, it was not available on the general market. The first personal computer available for purchase was the Altair 8800. It was introduced, described, and pictured in the January 1975 issue of *Popular Electronics* magazine. It came in kit form ready to be assembled and was aimed at hobbyists who liked to build their own radios and other electronic devices.

The first personal computer that was fully assembled and offered for sale on the general market was Apple I. It was built by 25-year-old college dropout Steven Wozniak (1950-) in his garage in Sunnyvale, California. With his friend, Steven Jobs (1955-), Wozniak showed the new machine at the first Computer Show in Atlantic City in 1976. It astonished viewers with its small, compact size and speed, but did not sell. Wozniak redesigned it. When Apple II was unveiled, encased in a plastic cover, with color graphics, BASIC, and an accounting program called VisiCalc, orders soared. No established company was willing to invest in a machine built in a garage, so Jobs and Wozniak created the Apple Computer Company in 1977. They moved out of the garage and hired people to manufacture the machine.

Soon many individuals and companies leapt into the personal computer market. Some computers were designed for the knowledgeable hobbyist, while others followed the lead of Apple. Those computers were made for those who wanted the computer to do something and didn't care how it worked. Tandy (called Radio Shack today); Texas Instruments, which had built the first electronic calculator; Commodore; and other companies began to build personal computers for sale. Some prospered, some failed.

When IBM finally got into the personal computer market in 1981, it had an immediate impact, even though it had serious limitations. Its computer had no hard disk drive and no software or graphics. But it did have the magic letters on the front—IBM. Many customers felt that if IBM, already called "Big Blue," built a computer, it had to be good. It even convinced many people that since IBM was building personal computers, then they were here to stay. IBM sold 20,000 machines in the first few months and could have sold 50,000, but they were not geared up to manufacture that many. Its design and refinements have been followed by many other manufacturers. IBM PCs, or clones, now dominate the computer market.

Just as few computer owners program their machines, few transport them. For that, a new type of personal computer has appeared: the laptop computer. It is popular among students, researchers, and business travelers, as is the new palm or hand-held computer.

Large mainframe computers changed the way businesses ran and kept records. Personal computers changed the way individuals did business, kept family records, did their taxes, entertained, and wrote letters. Even those who fear or shun computers use them or come into contact with them every day. When they use an ATM to deposit or draw out money, they are using a dedicated computer. When paying for groceries or gasoline with a credit card, a computer is involved. The internal systems of their automobiles are run by computers. Computer literacy has become a necessary skill for technical or scientific jobs and is becoming a requirement for many jobs, such as bank tellers, salesmen, librarians, and even waiters in restaurants who use computers as part of their daily work.

Today the definition of a personal computer has changed because of varied uses, new systems, and new connections to larger networks. A personal computer is now one that is used by a single operator at a business, a library, or his own home. Most personal home computers are used by individuals for accounting, playing games, or word processing. They have become an appliance that provides entertainment as well as information. They are affordable, and anyone can learn to use them. An increasing number of people do business at home on their own personal computers, or one provided by the company, and only need to travel to a place of employment a few days a week.

Personal computers are also widely used by small enterprises like restaurants, cleaning shops, motels, and repair shops. They are often linked together in networks in larger businesses like chambers of commerce, publishing companies, or schools. These computers look and behave like personal computers even when they are linked to large computers or networks. They are no longer programmable by the operators and seldom transportable.

The importance and impact of the personal computer by the beginning of the twenty-first century rests in one part on the development of the computer and in another on the creation of a new system of communications—the Internet—that depends on personal computers and could not have become so widespread without them. Together, computers and the Internet—with its attendant World Wide Web and e-mail—have made a huge impact on society, and every day radical changes are made in the way educated people all over the world communicate, shop, do business, and play.

The Internet, the World Wide Web, and e-mail are actually three distinct entities, allied and interdependent. The Internet is a network of computers that stretches around the world and consists of phone lines, servers, browsers, and clients. It began during the Cold War in a communications network linking researchers at the United States Department of Defense (DOD) and military contractors. In 1969 it was vital to be able to maintain contact in the event of a nuclear attack. When those tensions eased, the network continued as a convenient way to communicate with research groups and companies all over the world. This network was developed at the Advanced Research Projects Agency and was initially called ARPAnet.

At first ARPAnet's primary use was for electronic mail, or e-mail, beginning between 1965 and 1971. It took years of refinement and increased communication capabilities, like fiber optic cables for telephone lines, for users to be able to communicate with each other despite differing types of computers, operating languages, or speed.

ARPAnet continued to grow, still used mostly by military contractors and the DOD. In the 1970s it was opened to non-military users, mainly universities. The first host was installed at UCLA, the second at Stanford, both in California. By 1971 software was being created to enable messages to be sent to and from any computer. E-mail then became accessible to all. International connections were available by 1973. In 1983 ARPAnet was split into military and civilian sections, and the civilian network was dubbed the Internet. It is now defined as the physical structure of the network—its phone lines, servers, and clients.

The World Wide Web enhances the Internet. It is a collection of sites and information that can be accessed through those sites. Tim Berners-Lee worked at CERN in Switzerland and wrote software in 1989 to enable high-energy physicists to collaborate with physicists anywhere in the world. This was the beginning of the World Wide Web, which became an essential part of the Internet in 1991. The Web has multimedia capabilities, provides pictures, sound, movement, and text. It is made up of a series of electronic addresses or web sites. The Internet and the World Wide Web became easier and more useful when Web browsers were invented to locate, retrieve, and display this information in both text and pictures.

According to some estimates, there were approximately 40 million personal computers as the twenty-first century dawned, and most of them were connected to the Internet. No business hoping to sell products to a large audience in the new century will be able to ignore personal computers or the Internet. Any individual who wishes access to a wide range of information or to buy goods and services will need a personal computer wired to the Internet to do it.

Personal computers have changed the way we do business. Computers have created new businesses and changed others. They have altered the focus in boardrooms from the people who procure money to those who create or make decisions about new product. It has also made millionaires and billionaires of those who entered the business early. Undoubtedly, the effects of the social, economic, and cultural revolution spawned by the development of the personal computer will continue to be felt in the twenty-first century.

LYNDALL BAKER LANDAUER

Further Reading

Kidder, Tracy. *The Soul of a New Machine.* New York: Avon Books, 1981.

Oakman, Robert L. *The Computer Triangle: Hardware, Software and People.* New York: Wiley, 1995.

Shurkin, Joel. *Engines of the Mind: Evolution of Computers from Mainframe to Microprocessor.* New York: Norton Publishers, 1996.

Veit, Stan. *Stan Veit's History of the Personal Computer.* Asheville, NC: WorldComm Press, 1993.

The Invention of Compact Discs

Overview

Did you know that until the late 1800s the ability to record audio (voice or music) was just a dream? In 1877 "Mary Had a Little Lamb" was recorded and played on Thomas Edison's first experimental talking machine called a phonograph. About one hundred years (1982) after Edison's introduction of the phonograph, two companies, Sony and Philips, marketed the first digital audio 5-inch compact disc (CD). A CD-ROM (Compact Disc Read-Only Memory) disc is basically the same as a CD audio disc, except for the way the data is stored on the disc. A CD-ROM allows multimedia data (audio, text, computer graphics, and video images) to be stored on the same disc for use in a personal computer. For example, a CD-ROM can contain the complete text of the dictionary, 24 volumes of an encyclopedia, a thesaurus, a world almanac, an atlas of the world (including illustrations and animation), and a talking reference library with word pronunciations, quotes, and music. CD-R (CD-Recordable), also called CD-Write Once (CD-WO), allows the user to record using a disc and a desktop computer. The success of compact discs can be attributed to their durability, affordability, and compatibility.

Background

Durability

Compact discs are more durable than prerecorded, long-playing phonograph records (LPs) and cassette tapes. A compact disc is a laser-encoded disc that is made from the same material as bulletproof glass (polycarbonate plastic). It is almost indestructible. Unlike a record player, where a needle actually touches the record to play the music, a CD player or CD-ROM drive focuses a laser light on the reflective surface of the disc. Nothing but the laser beam touches the disc; therefore, discs do not wear out. It is estimated that a compact disc will last about 100 years.

Read-only memory (ROM) means that the computer data is permanent and un-modifiable. Files cannot be accidentally deleted from CD-ROM discs. Because the data on a CD-ROM cannot be modified, a virus cannot invade the discs.

Affordability

All types of compact discs can be mass-produced in large factories through a process called injection molding, making them cheap to produce. Since an audio compact disc was cheap to produce and much easier to carry and more durable than an LP record, its popularity increased. As a result of this increase in market demand for audio compact discs, Sony introduced its portable audio CD player in 1984. Now you could listen to your CD anywhere.

The Microsoft CD-ROM conference held in 1986 (which became known as the "Woodstock of CD-ROM") was considered the traditional starting point for CD-ROM technology. It was during this conference that Bill Gates (1955-) predicted that the CD-ROM would become the cheapest way to distribute large amounts of machine-readable financial data. Gates's prediction came true. Financial institutions in the 1990s have discovered that the cheapest way to distribute large amounts of machine-readable data is with a CD-ROM. Because CD-ROMs are easy to manufacture and cheap to produce, they have become the fastest growing consumer electronic device in history.

Compatibility

In the early 1980s Sony and Philips worked together to create the compact disc for audio players. This venture between the two companies guaranteed that any audio CD would play in any audio CD player. Together they established a standard for the physical characteristics of a compact disc. Since the color of the binder in which this standard was published was red, it became known as the *Red Book* standard. This standard is the foundation on which all other compact discs' specifications are built. Sony and Philips also established the specifications for the physical characteristics for the CD-ROM. These computer discs would use the same laser technology as the audio CD. As CD-ROM discs were introduced into the personal computer market, their specifications were created based on the CD *Red Book* standard. It was called the *Yellow Book* standard. Initially, all CD-ROMs had the same physical properties, but the filing system used for storing the information contained on the discs was not standardized.

The following generations of compact discs were each based on the previous standard. The Compact Disc Interactive (CD-I) operating system and disc layout structure was defined in the *Green Book* standard. The *Orange Book* established the specifications for allowing the user to write audio and/or data to the disc for a Compact Disc-Magnetic Optical (CD-MO) and for Compact Disc Write Once (CD-WO). This technology allowed data to be written but not erased. Initially, however, these compact disc standards did not designate how the files should be stored on the disc. In November 1985 company representatives from Apple Computer, Digital Equipment Corporation (DEC), Hitachi, LaserData, Microsoft, 3M, Philips, Reference Technology Inc., Sony Corporation, TMS Inc., VideoTools, and XEBEC met at the High Sierra Casino and Hotel in Lake Tahoe, Nevada. This committee became known as the High Sierra Group. The purpose of the meeting was to propose a standard file format structure for CD-ROM discs so that any CD-ROM could be played in any CD-ROM player. The *High Sierra* standard (as it was sometimes called) was adopted in October 1987. This proposal was submitted to the International Standards Organization (ISO) for acceptance. In 1988 the ISO, which sets international standards through committees located around the world, established universal CD-ROM standards based on the High Sierra recommendation called the *ISO 9660*. It described the directory structures and file layout for storing computer files on a compact disc. This standard stated that any CD-ROM could be used in any personal computer CD-ROM drive regardless of the type of operating system used.

In the 1990s compact discs have become the standard for storing large quantities of information for a perfect reproduction (the recording sounds the same every time you play it, no matter how many times). An audio CD holds about 74 minutes of music or text. A CD-ROM holds approximately 260,000 pages of text, or over 1,500 floppy disks. (Note: The word *disc* refers to optical storage media, such as CD audio, CD-ROM, or video disc. The word *disk* refers to magnetic storage media, such as floppy or hard disks.) Floppy disks did not have the storage capacity needed to store large amounts of computer data. A company (such as Microsoft) can produce and ship one CD-ROM that includes the software application, the online manual, and online tutorials with audio, cheaper than 36 floppy disks and printed manuals. Personal computers sold since 1995 contain a CD-ROM drive. As a result, computer software companies began producing and shipping their software on discs. The compact disc has become an optimal means of storing and delivering massive text and multimedia applications for the personal computer.

Impact

With the durability, affordability, and standardization of the compact disc industry, more industries and businesses began utilizing the compact disc. Even the United States government began using it for data distribution.

In 1983 over 800,000 audio compact discs were sold. For the first time in 1988, CD sales surpassed LP record sales. Although CD-ROM has become the primary data storage and distribution medium during the 1990s, the next generation in storage is the Digital Video Disc or Digital Versatile Disc, commonly referred to as DVD. The DVD standard was introduced in September 1997.

Initially the Digital Video Disc (DVD-Video) technology was developed to be played in a DVD player hooked up to a TV. The DVD improved movies by providing better video and audio quality. The computer industry developed the DVD-ROM, which allowed large amounts of computer data to be stored and accessed quickly. Because CD-ROM manufacturers are developing the DVD-ROM hardware, which also reads compact discs, CD-ROM drive production is expected to diminish in favor of DVD-ROM drives. The difference between the DVD-Video and DVD-ROM is similar to that between audio CD and CD-ROM. Because of the two different types of DVDs, the acronym now stands for Digital Versatile Disc.

Although the DVD physically resembles a compact disc, it can hold 7.5 times more data (audio, computer data, and video) than an audio CD. The DVD can store about 16 hours of music (audio CDs allow us to hear up to 74 minutes of music), up to 8 hours of video, or up to 17 gigabytes of data. At this time, the only DVD available is a read-only disc; however, there is a rewriteable version of DVD for computers currently being developed. The DVD is expected to replace the audio CD, videotape, laserdisc, CD-ROM, and perhaps the video game cartridge.

This new generation of optical disc storage was based on the compact disc technology. It could not have happened without the development of the compact disc. Although CDs are being replaced with DVDs, for about 15 years the compact disc was considered the best.

REBECCA J. TIMMONS

Further Reading

Parker, Dana and Starrett, Bob. *New Riders' Guide to CD-ROM*. Second ed. Indianapolis, IN: New Riders Publishing, 1994.

Sherman, Chris, ed. *The CD-ROM Handbook*. Second ed. New York: McGraw-Hill, 1994.

Technological Disasters: The Modern Challenge to the Enlightenment

Overview

The philosophers of the Enlightenment believed that humankind could maximize happiness through the application of science and technology. By the end of the twentieth century, however, several technological disasters had caused many people to rethink the role of technology in society.

Background

The seventeenth century was a turning point in the technological development of the world. Within the intellectual establishment of the West a significant change in how truth was identified took place. There was a rejection of the Aristotelian mode of syllogistic deduction in favor of the logic of induction, which is based upon individual experience. The new natural philosophers referred to the universe as the "Book of Nature," which could be understood through experiment; the universe was created by God and operated under a series of natural laws. These laws could be discovered by humankind through a process known as the scientific method. This method, based upon experience and mathematical proof, was a multi-step approach to discovering truth.

This new way of looking at the world also had a significant social impact. It created the ethical principle of utility, which stated that happiness is the highest value; all ideas are to be judged by their contribution to happiness or suffering. Empiricism, the philosophy of the scientific method, was materialistic in its orientation. Philosophers of this school believed that the only truth we could be sure of was that based upon the five senses. Thus, the new principle of utility created a view of happiness centered upon material gain. Science and technolo-

Three Mile Island, the site of a nuclear power accident in 1979. *(AP/WideWorld Photos. Reproduced by permission.)*

gy, in turn, were perceived to be tools of materialistic utility.

The first uses of science and technology in expanding utility were concentrated in the areas of food production and public health. By the eighteenth century great strides had been made in increasing the productive capacity of farmland. The scientific application of fertilizer, crop rotation, and the mechanization of farm labor had greatly increased productivity. The same held true in the area of public health. The invention of vaccines, the evolution of germ theory, and the control of drinking water and sewage greatly reduced the death rate. The success in both of these areas had an important impact on society by both increasing the population and decreasing the number of people needed to produce food. A sizable surplus of human capital was thus created for the next great technological expansion, the Industrial Revolution.

The success of industrialization was the result of another change, the energy revolution. Energy is the force or power that is needed to accomplish a particular task. Over time, energy sources have evolved significantly. The earliest form of energy used by humans was muscle power from themselves or animals. This situation progressed with successful attempts at harnessing the power of water and the wind. These two energy sources increased the amount of energy humans could utilize, but they were very unreliable. The first major breakthrough came with the invention of the steam engine.

Steam power drove the pursuit of materialistic utility until World War I, when petroleum replaced it as the energy of choice. Oil was used to drive the combustion engine. This device was small, powerful, and had both peacetime and military applications. Petroleum, a fossil fuel, has a finite supply and is unequally distributed around the world. Petroleum has thus changed the global geopolitical situation; areas such as the Middle East have become important because of their oil reserves.

In 1939 the first atomic chain reaction took place, and the world entered the nuclear age. This energy was not just viewed as a new weapon; many people believed that it could become an infinite source of power for domestic consumption.

Impact

This march toward utility, based upon a belief in humankind's ability to solve problems through the proper application of scientific and technological principles, did not operate in a vacuum. In the twentieth century ideology began to dominate the quest for utility. Two economic, political, and social philosophies dominated the world during most of the century—free market capitalism and scientific socialism.

During the Enlightenment, free market capitalism was the economic system created in the quest for happiness. Adam Smith, its founder, stated that in a free market people determine the goods and services to be produced. He believed that people would never act against their own best self-interest; thus, their economic energy would always be used in a positive way. The success of this philosophy, in turn, was to be measured in material prosperity.

The other economic model was constructed by Karl Marx as an attack against the volatile and sometimes destructive forces of the free market. Originally known as scientific socialism, it stated that the true natural law of economics was one that promoted economic equality. This equality, in turn, would be established and guided by a strong revolutionary elite.

Both economic philosophies were grounded in the production and distribution of goods. As industrialization evolved in both systems, energy became an increasingly important factor. By the last half of the twentieth century, the pursuit of material utility began to strain the financial and energy resources of the world. By the early 1960s, many technologists believed that nuclear power was the "wave of the future," while others argued that, at least initially, there would have to be a combination of nuclear and fossil fuel.

Beginning in 1979, however, a series of major technological disasters set in motion a process that would eventually begin not only a debate over the perceived infallibility of science and technology but also the ethical foundation of Enlightenment utility. March 28, 1979, was first day in this new process. On that day, in a nuclear power plant called Three Mile Island, located in Pennsylvania, a series of errors—both mechanical and human—allowed water used to cool the plant's reactor to drain. This event caused the plant's radioactive core to overheat and begin to melt down, eventually consuming a third of the core element. While proper steps were taken to avoid a major disaster, serious questions about the ultimate safety of nuclear energy took hold of the American consciousness.

Several years later, a nightmare scenario began halfway around the world from the United States in the territory of its major rival, the Soviet Union. On April 26, 1986, a major explosion occurred at the nuclear power plant at Chernobyl, which is in the Ukraine. The largest radiation disaster in history, the explosion created a cloud of cesium 137 and iodine 131 three miles high and ten times more radioactive than the one from the Hiroshima bomb. A total of 9,500 kilometers of land were contaminated in Byelorussia, Ukraine, and the Russian Federation. Twenty-eight people were killed outright, but eventually 650,000 would be affected by the radiation. This disaster was of apocalyptic proportions, and, over 13 years later, its specter still haunts that area of Central Europe. Since the disaster, there has been a dramatic increase in thyroid cancer, along with a significant increase in illness and birth defects among newborns.

These two disasters refocused much of the energy community back to the use of fossil fuels.

USING HIGH TECHNOLOGY TO SPOT ART FORGERIES

Forged artwork has been a problem for centuries, and forgers have imitated the work of virtually every important artist. Recently, however, it has become somewhat easier to detect art forgeries, thanks to modern technology. While there are a host of techniques used, there are some general categories that bear mentioning. One family of techniques is based on analyzing the materials used. Carbon dating, for example, can show if a painting is a 500 year-old artifact or a 50 year-old fake. Chemical analysis of pigments can show if the ones used even existed when a painting is said to have been painted. Other analyses of shellac, varnish, canvas, and other materials are also used to help determine authenticity. One common factor is that virtually all of these techniques require removal of a small sample of the piece of art. Another family of techniques looks at the physical properties of the artwork in question. For example, infrared analysis of a painting can show if an artist's signature was painted over the top of someone else's name or if the original signature was covered over. Non-invasive techniques can also be used to look at the chemical properties of some of the pigments, which can be compared to the properties of pigments used by authentic works by the same artist. For example, an artist who used ground lapis to make blue in one painting is unlikely to have used a chemical dye in another. This is, however, a contest likely to continue for some time as both forgers and verification technology become increasingly sophisticated.

Once again the belief that technology could create cleaner, more productive engines helped breathe new life into the use of petroleum. On March 24, 1989, however, the super-tanker Exxon *Valdez* ran aground on Bligh Reef and spilled 11 million gallons of crude oil into Alaska's Prince William Sound. This tragedy not only killed about 250,000 sea birds but also devastated the fishing industry in the entire area.

Bhopal, though not energy related, still reflects the technological drive to utility. Though the use of pesticides in the Industrial World has become suspect, its application in the developing nations has been viewed as an acceptable risk. As in Europe over two centuries earlier, a strong, stable population was regarded as a prerequisite to development. Western companies knew that large profits could be made by providing developing nations with pesticides. Human error and shoddy workmanship, however, caused 40 tons of lethal methyl isocynate and hydrogen cyanide to escape from a plant and eventually kill 16,000 people in Bhopal, India.

In reaction to these disasters, people initially focused on the ideologies of the two participants. The rigid, oppressive, controlling nature of Marxist Leninism was blamed for the Chernobyl disaster. The Soviet system proved incapable of containing the nuclear emergency. Once the accident occurred, moreover, the initial reaction of the totalitarian government was to control everything, including the flow of information. This decision allowed radiation to pollute most of the Soviet Union's neighbors.

In the United States, there was also an initial attempt to control information on the part of the government in both the Three Mile Island and Exxon disasters. A more open society with a free press eventually resulted in a release of information, but the capitalist system of the United States had also created the gluttonous appetite for material goods that had contributed to these disasters. The energy needs connected to satisfying the production levels required by this appetite have created one of the most important problems facing the American community today. Questions about how long the natural environment can withstand these types of violations are beginning to find an important place in U.S. civil debate.

The major philosophical problem in this situation revolves around the question of choice. There is a tension between the absolute right of the individual to pursue happiness (based upon the materialistic orientation of the Enlightenment's concept of utility) and the rights of society to have a safe, clean environment. The free market economy, which has over time raised a lifestyle based upon material acquisition to the level of a human right, is now being questioned. Many scientists believe that this model of mass consumption is not an environmentally sustainable one.

These questions have now entered the political arena. For example, the international political organization known as the Green Party has made the environment its sole concern; this group is in favor of instituting laws, regulations, and policies that would drastically control the amount of energy and resources the world's population expends. Technology and lifestyle will be at the center of one of the great debates of the early twenty-first century.

RICHARD D. FITZGERALD

Further Reading

Martin, Daniel. *Three Mile Island: Prologue or Epilogue.* Cambridge: Ballinger Publishing, 1980.

Medvedev, Grigori. *The Truth About Chernobyl.* New York: Basic Books, 1991.

Stevens, Mark. *Three Mile Island.* New York: Random House, 1980.

Weir, David. *The Bhopal Syndrome.* San Francisco: Sierra Club Books, 1987.

The Rise of the Appropriate Technology Movement

Overview

In the 1960s the assumption that societies should continually strive for economic growth and increasing industrialization began to be questioned. The environmental effects of unbridled consumption in the developed world had become apparent, and the impact of materialism on modern culture was of concern as well. E.F. Schumacher's book *Small is Beautiful* was among the influential works that inspired the appropriate technology movement, advocating small-scale, decentralized, environmentally sustainable enterprises.

Background

In 1955 the British economist E.F. Schumacher (1911-1977) visited Burma (now Myanmar) on a United Nations assignment. There he became acquainted with the pitfalls of introducing advanced technology to developing nations. This introduction of technology was presumed to be progress and would increase productivity. It would more than likely, however, put many laborers and artisans out of business. Schumacher viewed the dignity of work as essential to the character of individuals and the community.

Schumacher was also conscious that Earth's resources were limited. In the 1960s the deterioration of air and water quality, along with the difficulties in disposing of increasing quantities of domestic and industrial waste, were becoming all too apparent. Schumacher began to question the assumption—central to the capitalist economic system—that growth was always good. Instead, he believed, technologies that were small scale, decentralized, and not energy intensive could be used to improve a community's standard of living. At the same time, technologies should be environmentally sustainable, that is, based primarily upon renewable resources, such as growing plants and taking energy from the sun, water, and wind as opposed to such non-renewable resources as fossil fuels.

Schumacher called technologies that met his criteria "intermediate technologies" and in 1966 founded the Intermediate Technology Development Group (ITDG) in London to put his ideas into practice. Since such solutions must be tailored to the particular society in which they are applied, they are now often called "appropriate technologies."

When Schumacher published his book *Small is Beautiful: Economics as if People Mattered* in 1973, it was immediately adopted by political progressives. The book advocated production from local resources to fill local needs. Production, it argued, should be a means to an end, that of increasing wellbeing. Therefore, the goal should be to achieve that end with a minimum of resource consumption and the resultant environmental degradation. Schumacher also assumed that full employment, whether in the home or in an outside setting, was necessary for a healthy society. This philosophy runs counter to measuring wellbeing by the amount of production or consumption as is commonly done in capitalist economic analysis.

Impact

Schumacher's work was named by the *Times Literary Supplement* (London) as one of the most influential books written since World War II. It was published in more than 20 languages and gave people in the Western world a new way to think about economics. One of the most famous essays in *Small is Beautiful* is entitled "Buddhist Economics." In it Schumacher described blending spiritual values with economic progress to achieve a "Right Livelihood" that valued people over tools and production, preserved the environment, and fostered simplicity and nonviolence.

ITDG began by publishing a buyer's guide to equipment for farmers and crafts workers. Small-scale enterprises, the group maintained, should not imply isolation; in fact, the wide dissemination of information helps to avoid "reinventing the wheel." Schumacher was the movement's global ambassador, advocating production without exploitation. By encouraging local development, communities could avoid the trap of bringing in large corporations for the jobs they promise only to be left adrift when the corporations relocate again.

ITDG and other like-minded organizations continue to work on a number of projects in rural and developing areas to foster sustainable improvements. These projects often involve renewable energy but many other types of projects are possible as well. For example, ITDG—with offices in Bangladesh, Kenya, Nepal, Peru, Sri Lanka, Sudan, and Zimbabwe as well as the United Kingdom—works in the areas of energy, transportation, manufacturing, water and sanitation, construction, mining, and food production. The Development Center for Appropriate Technology (DCAT), founded in 1991, focuses on sustainable construction and restoration techniques. Of all the materials coming into the global economy each year, buildings consume about 40%. DCAT projects include building houses from straw bales and lobbying for changes in building codes that would encourage sustainable construction. The National Center for Appropriate Technology (NCAT) was established as a nonprofit corporation in 1976. NCAT is involved in a number of projects, including developing environmentally safe integrated pest management (IPM) plans for agriculture as well as energy conservation in public housing.

When Schumacher was asked what individuals could do to make his vision a reality, he suggested they begin by putting their own houses in order. "Think globally, act locally" became the slogan of a worldwide movement. Many rural homesteaders and others interested in self-reliance began experimenting with living "off the grid"—growing their own food and generating their own power.

Urban and suburban dwellers could also take part by patronizing local businesses, conserving energy, recycling their garbage, and forming economic cooperatives for buying food and other products. Today, these choices have become standard habits of the socially conscious. Community-supported agriculture is another movement that grew up in response to Schumacher's ideas. In this model, individuals purchase "shares" of a local farm's produce ahead of time, receiving a portion regularly during the harvest season. The seed money and guaranteed sales help the farm, which may be struggling to compete with large agribusinesses.

Two decades after the publication of *Small is Beautiful,* the Internet allowed for the creation of a whole new category of "cottage industries." Now, workers of the Information Age could free themselves from large institutions and sell their skills and wares on an individual basis. This development is certainly decentralized and often quite energy efficient. "Community" has been redefined to include those with whom one interacts regularly, even if those people are hundreds or thousands of miles away. Critics, however, warn of a chasm between Internet "haves" and "have nots" and remind "wired" workers not to forget the community in which they actually live.

SHERRI CHASIN CALVO

Further Reading

Betz, Matthew J., Patrick J. McGowan, and Rolf T. Wigand. *Appropriate Technology: Choice and Development.* Durham, NC: Duke University Press, 1984.

Carr, Marilyn. *The AT Reader: Theory and Practice in Appropriate Technology.* London: Intermediate Technology Publications, 1985.

Evans, Donald D., and Laurie N. Adler, editors. *Appropriate Technology for Development: a Discussion and Case Histories.* Boulder, CO: Westview Press, 1979.

Schumacher, E.F. *Small is Beautiful: Economics as if People Mattered.* New York: Harper and Row, 1973.

Thompson, William B. *Controlling Technology: Contemporary Issues.* Buffalo, NY: Prometheus Books, 1991.

Futures Imperfect: Technology and Invention in the Twenty-First Century

Overview

The most dramatic predictions for the twenty-first century can be found in the fields of biology, computers, nanotechnology, and space exploration. Biology stands ready to create the biggest changes: engineering new species, cloning, growing replacement organs, extending life, and modifying the brain could radically alter our relationship with nature and our view of what it means to be human. Computers are set to shrink until they disappear into our homes, toys, vehicles, and even our bodies. They will be combined into a worldwide network that will be always on, always sensing, and always available. Nanotechnology—building molecular devices—offers possibilities ranging from dust-sized robots to strong, light cables that can give us an elevator into space. All of our own solar system will be in reach for exploration, colonization, and even tourism.

The art of predicting is highly flawed, with a history of laughable errors. There is a tendency to be too gloomy or too optimistic. Revolutionary changes cannot be accounted for. However, predictions do help shape the future, inspiring us to create new inventions, warning us about dangers, and stretching our imaginations.

Background

In 1863 Jules Verne wrote about Paris, 1960. He imagined motor cars, computers, fax machines, and elevators, but he also had his protagonist writing up the daily accounts in a large book with a quill pen. Predictions, even "scientific" prediction, aren't easy.

Astrologers have cast horoscopes and magicians have told fortunes for millennia, but until the nineteenth century prediction was largely concerned with reconfiguring the familiar. The revolutions of the 1700s produced a sense of social change, but material change remained beyond the imaginations of most. It was only in the 1800s—the age of invention that brought us steamboats, railroads, telegraphs, telephones, cars, and even motion pictures—that the concept of changes in day-to-day life became part of the popular culture.

Verne was at the vanguard of those predicting technological change. He wrote about submarines, flights around the world, and a trip to the moon. (The rocket was launched from Florida, but it was shot from a massive cannon.) H.G. Wells wrote about the atomic bomb in 1914. Rudyard Kipling wrote about airmail in 1904.

Writers didn't have all the fun. In 1900 the magazine *Ladies Home Journal* predicted a startling set of new technologies, including refrigeration, air conditioning, tanks, and television ("Persons and things of all kinds will be brought within focus of cameras connected electronically with screens at the opposite ends of circuits, thousands of miles at a span.") But the magazine also foresaw express mail via pneumatic tubes and quiet cities where all traffic was far above or below the surface.

The errors of prediction, born of pessimism, optimism, and the inability to anticipate breakthroughs can be amusing. Historian Henry Adams foresaw the end of the world by 1950. Household robots and a cure for cancer still have not arrived. With one exception, science fiction writers, who virtually owned the future until the 1940s, missed the advent of the birth control pill and the changes it brought to society.

In 1949 science fiction's best forecaster, Robert Heinlein, predicted not only the birth control pill but also air traffic control, cell phones, answering machines, and a dozen other advances that came true by 2000. Heinlein offered his forecasts at about the same time that the discipline of futurology—first suggested by Wells in 1901—was being established.

At first, the futurologist's approach was generally to find the right theory about social change and project from accumulated data. This method was not notably successful. In the 1950s Herman Kahn pioneered the scenario technique. He looked at possible, probable, and preferable scenarios with the idea of shaping rather than predicting the future. Computer modeling followed, and one of the most famous works of futurology was *The Limits to Growth* (1972), which made extensive use of modeling to suggest future scenarios. Many of the study's models have been proven to be wrong, but the book influenced the adoption of recycling and legal controls on pollution.

In addition to developing scenarios and using computer models, futurologists ground their work in trend analysis and extrapolation. (For instance, Moore's law, which states that computing power doubles every 18 months, is the bedrock of many technical forecasts.) They discover emerging ideas by scanning publications (both technical and social) and polling experts. They combine elements of change to see if they create unexpected or larger effects.

Predicting the Future

Prediction has become a mainstay of modern life. We expect change, and we expect our experts to tell us about it. We get forecasts not just for the next five days of weather but for inflation in the next quarter and future demographics given populations. Companies depend on visions, marketing plans, and sales projections.

Still, prediction remains more of an art than a science. There is no suite of rigorous tools for prediction. Informed intuition coupled with imagination may provide the highest success rate. With the end of the twentieth century, a flurry of predictions were made for the twenty-first century. There was no shortage of robots, flying cars, and missions to Mars, but most predictions dealt with the networking of the planet and changes to the human species.

Networking the Planet

As the impact of the Web continues to grow, many predictions for 2000-2100 deal with the reinvention and reshaping of our communities expected in the first half of the century. Central to these developments are advances in computers. Current technology is expected to follow Moore's Law until at least 2017, continuing to make devices smaller, faster, and cheaper. At the same time, communications capacity is doubling every nine months.

This means that early in the 2000s, everything—televisions, shoes, vegetables—gets smart and connected. Nanotechnology—building devices on a molecular level—will enable much of this development in our networked world. Prototypes have already been made of tiny gears, motors, sensors, diodes, and transistors. Smart dust and micromachines will keep a continuing tally on where everything and everyone is and what they're up to. The huge capacity

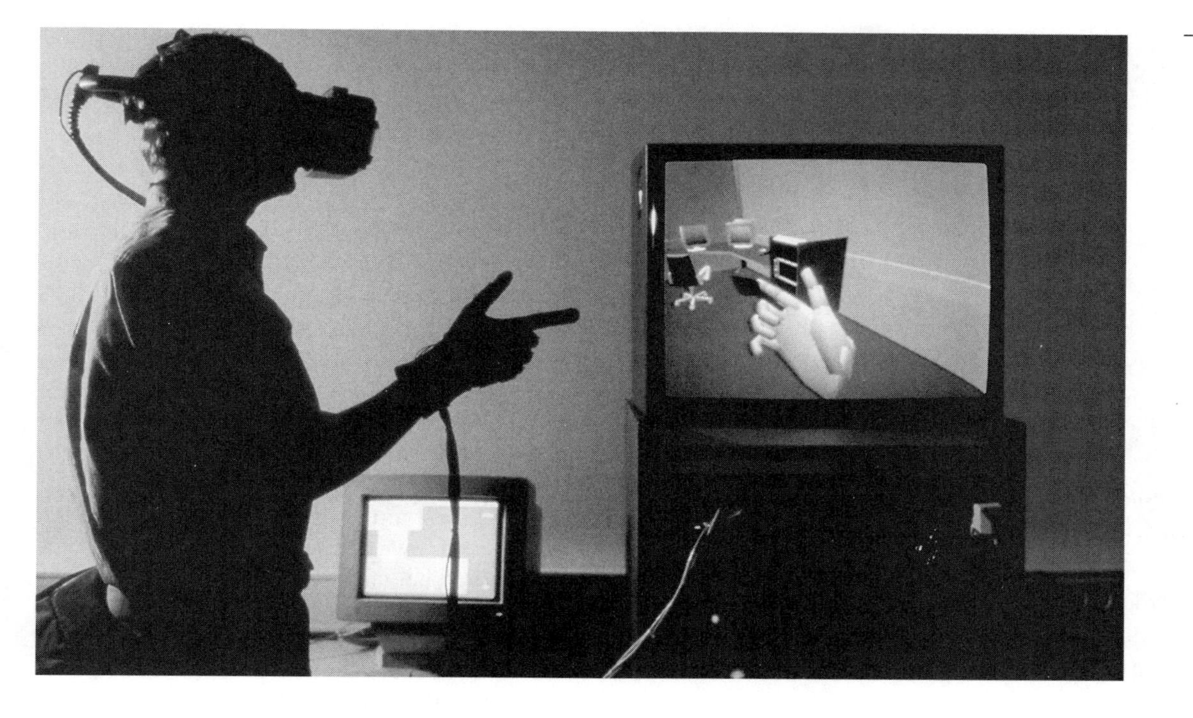

A virtual reality system. *(Thomas Ernstein/Bilderberg/The Stock Market. Reproduced by permission.)*

and power of a worldwide computer system will be both invisible (within common devices) and omnipresent (gathering and providing information everywhere).

A big piece of the computing capacity will be given over to making information more accessible and easier to turn into value. Information systems will respond to speech, expression, and gestures. They will adapt to your style and your mode of learning and communicating. They will also adapt your environment to you, adjusting lighting, heat, music, odors, and the arrangement of devices. Information you care about will be discovered, collected, and delivered based on your electronic identity and the collective judgments of your peers. It may even be possible to change hardware to suit your needs, rewiring your cell phone into a memo pad on the fly.

Virtual reality will become more compelling and invade the real world as the years go by. First, the experience will become more realistic and the gloves and goggles less cumbersome. Later, direct neural stimulation by nanomolecular machines (devices the size of viruses) may make the equipment disappear into the users. By the end of the century, and perhaps much sooner, data and interpretation will overlay our sense experience, and it will be taken for granted that every real-world experience has a virtual aspect.

Changes to the Species

While networking relies on physics to transform our experience of community, advances in biology, which may dominate the second half of the century, are likely to touch us more personally. Some developments may even challenge the concept of what it means to be human.

Cloning is on the immediate horizon. So is tissue regeneration, with simple, lab-grown organs, such as bladders, undergoing testing. And 2001 is the year of the gene, thanks to the complete sequencing of the human genome. Together, these promise a world of not just better health but control in redesigning human biology. Rejuvenation, adding talent genes, and even creating new organs to see infrared or breath under water will all be within reach.

Scientists are taking on the challenges of Alzheimer's disease, substance abuse, and mental health, and their findings are also leading to a better understanding of the brain and how it works. Drugs to prevent senility are likely to be used more broadly to improve memory. Insights into dependency and behavioral disorders could lead to medicines that enhance creativity, math skills, or verbal expression. In the 2000s it will probably be possible, within a range, to choose your personality, mood, and intelligence.

Nanotechnology may play a role here. Besides enhancing our health—by sensing and reporting on changes, hunting down pathogens and cancer cells, or repairing arteries—synthetic devices could also augment our bodies and our brains. Nanodevices could provide electronic memory or make our brains into nodes on the information network. Human brain processes are already being mapped using external tools, like Magnetic Resonance Imaging and Positron Emission Tomography. Nanodevices could make this mapping more detailed, monitoring—and possibly even altering—our thoughts.

SCIENCE AS POPULAR ENTERTAINMENT

In the last half of the twentieth century science-related entertainment moved from the mythical mad scientist's laboratory to become a mainstay backdrop for popular entertainment. Instead of literary works such as Mary Shelley's *Frankenstein* that explored man's deepest fears about himself, science-related entertainment in the post-nuclear age became a bewildering array of radioactive monsters and aliens that often portrayed our fears of each other. The development of television in the nuclear age, particularly the extensive and live coverage of the space race to the Moon in the 1960s, put science on everyone's mind and science terminology on every tongue. Science, whether in the guise of science fiction or fantasy, became a big box-office draw around the world. Films such as *ET, Close Encounters of the Third Kind, Star Wars* and *Jurassic Park* touched on subjects ranging from extraterrestrial visitations to dramas set in galaxies far, far away to genetics and the possible ramifications of DNA-based technologies. One crossover hit, the popular *Star Trek* series, became an enduring multigenerational space-based adventure phenomena.

On a more scholarly note, growing audiences argued the meanings and insights found in classical science fiction movies and books written by Arthur C. Clark. In 1980 Carl Sagan's television series *Cosmos* became the most popular public television series of its time. Millions of viewers thrilled to Sagan's explanations of profound cosmic complexities and connections. A hauntingly beautiful musical score accompanied stunning visual effects in what was then a state-of-the-art presentation of the science and the history of science. *Cosmos* also appeared in book form and quickly became the best-selling science book ever published in the English language. Spurred by Sagan's success, other mainstream scientists, used to laboring only in academia, began to produce works for lay audiences. Stephen Hawking's *Brief History of Time,* a huge critical and commercial success, provided profound insights for both scientists and non-scientists into the nature and origins of the universe.

It may be more common to have artificial brain components than electronic eyes and mechanical arms and hearts, but we will probably have plenty of cyborgs among us. An aging population may prefer more durable, versatile manufactured parts to grown organic replacements. At some point during the 2000s, the distinction between cyborg humans and more and more human robots might blur. As early as 2020 a PC could have the processing power of a brain. This will be followed by 80 more years of development, with some of those artificial brains working 24 hours a day on their own development. Will relationships with our cyber children be less valid than those with our carbon-based kin? Already, millions of people have electronic identities. These include everything from credit and purchasing profiles to avatars that may bear little resemblance to their flesh and blood antecedents, but are dealt with seriously and socially by their peers. It might not be too much of a stretch to accept artificial intelligence identities as our friends and colleagues.

Immortality may be the biggest step humans will take in the 2000s. Some scientists predict that children born in the second decade will live two or more centuries. Most of these children will be born into a world that will have ten billion inhabitants by the century's close. Population numbers are among the most predictable figures for futurologists. Disruptive changes, like the discovery of intelligent aliens, are never predictable. In fact, a radical group of futurologists, science fiction writers, and other forecasters sees the future as totally unpredictable from about 2035 on. They envision a positive feedback loop of technical development. Around that time, the combination of powerful computers, the explosion of knowledge, and new technologies, such as nanotechnology, will create a world that is beyond our imaginations and beyond the imaginations of even the best futurologists.

PETER J. ANDREWS

Further Reading

Books

Clute, John and Peter Nicholls. *The Encyclopedia of Science Fiction.* New York: St. Martin's Press, 1995.

Kahn, Herman, William Brown & Leon Martel. *The Next 200 Years: A Scenario For America and the World.* New York: William Morrow & Company, 1976.

Verne, Jules. *Paris in the Twentieth Century.* New York: Del Rey, 1997.

Sandberg, Anders. *Transhumanist Resources.* Internet site. http://www.aleph.se/Trans/

Other

"The Future Gets Fun Again." Periodical article. *Wired Magazine* (February, 2000).

Nuclear Submarines Revolutionize Naval Warfare, Intelligence Collection, and Spawn Technological Innovations

Overview

Although submarines had been used for warfare in one form or another since the American Revolution, it was not until World War I that they were ocean-capable vessels. Even then their limited battery life and diesel engines tied them to the surface. Nuclear power removed this obstacle, allowing them to operate as true undersea warships, limited only by their ability to store food and the endurance of their crews. The development and refinement of nuclear submarines led to drastic changes in the nature of naval warfare and information gathering, fostered a number of technological innovations, and fascinated the public in a manner matched only by aircraft carriers and, perhaps, battleships in their heyday.

Background

In 1620 Dutch inventor Cornelius Drebbel (1572-1633) constructed and operated a 24-man submarine that cruised five meters beneath the surface of the Thames River. It employed an unknown process that produced oxygen to keep the rowers from suffocating while they propelled the vessel. In 1755 David Bushnell (1742?-1824) constructed the first submarine designed for war, the *Turtle,* which was developed to attack British warships. Other submarines were used sporadically through the end of the nineteenth century, but until the invention of the long-duration storage battery and a diesel engine to recharge it, the submarine remained an impractical weapon. The first modern submarines were used during World War I, and the diesel-electric attack submarine reached its pinnacle of success during the Second World War. Even then, however, plans were in place for a successor that would be able to operate with complete independence from the surface.

At the start of World War II Japan, Germany, Great Britain, and the United States all knew that a nuclear chain reaction, if controlled, could produce enormous amounts of energy. Interestingly, all four countries first thought to apply this power source to submarine warfare, and thought about constructing atomic weapons later. Nuclear submarines, however, had to wait until after the war, when Hyman G. Rickover (1900-1986) persuaded the navy to let him design and construct an operable nuclear submarine. The USS *Nautilus* was completed in 1954 and immediately met all expectations. It still had the boat-shaped hull of its predecessors, lessening its underwater speed and reducing the room inside.

Later submarines rapidly became more efficient. The Skipjack class, built in the late 1950s, was built with a hydrodynamic hull for underwater efficiency. The Thresher class, built in the 1960s, had better sonar and fire-control systems, an integrated attack center, and other innovations, becoming the model for future submarines. Finally, the launch of the USS *George Washington* in 1960 put nuclear-tipped ballistic missiles on board. This created a nearly undetectable launch platform capable of rapid deployment within striking distance of any potential adversary. Subsequent designs improved on these advances as the Trident and Los Angeles, and Seawolf submarines were built and launched. Other nations followed America's lead, with Britain, the Soviet Union, France, and China building their own nuclear submarine fleets.

As submarines' capabilities expanded, so did their roles. During the Cold War, they were used to gather intelligence, train allies' navies, launch commando teams for reconnaissance or strike missions, gather oceanographic data from beneath the polar ice cap, and a host of other

The open hatches of 12 Tomahawk missile tubes on the nuclear attack submarine USS *Oklahoma City.* *(CORBIS. Reproduced by permission.)*

duties. Engineering advances improved the accuracy of submerged navigation, provided better safety systems for commercial nuclear reactors, and led to more efficient atmosphere-control systems. Rickover's insistence on safety and quality set the standard for industrial quality-control systems. Submarines' strategic importance was highlighted by the quality of their personnel: for over 40 years nuclear submarines had the best-trained sailors on their crews, enjoyed the highest priority for ordering spare parts, received the best food, and had the highest pay among any type of ship in the Navy.

Impact

Building a nuclear-powered submarine is not easy. The ship must be large enough for 120 or more sailors to work and live in for months. It has to hold a nuclear reactor and steam turbines as well as the weapons it is designed to deliver, yet still be small enough to travel at a decent speed. Its mechanical, electrical, and electronic systems must be reliable enough that the ship can safely stay at sea undetected for two or three months at a time. Finally, if it is to be truly independent of the surface, it must be able to make its own water and oxygen and remove waste gases from the atmosphere.

Virtually all of these innovations have practical applications outside the navy. Lithium hydroxide systems that remove carbon dioxide from the air are used in spacecraft; navigation systems, including global positioning satellites (GPS), are used on pleasure boats and cars; and quality-assurance programs modeled after Rickover's help keep both civilian nuclear reactors and other potentially hazardous industries safe and reliable. Improved reactor design and, as importantly, a steady stream of highly trained "ex-navy nukes" have given American nuclear power plants a safety and reliability record that, despite protests to the contrary, is virtually unmatched in any industry. Finally, onboard studies of people living in small spaces for weeks at a time have led to a better understanding of human interactions, group dynamics, and the psychology of living under stress.

One of submarines' greatest attributes is their ability to operate undetected, which makes them ideal for intelligence operations. Submarines can drop off (and retrieve) teams of specially trained personnel, who then observe potential adversaries without being observed. This knowledge can be invaluable in war, and in peacetime, can help to defuse tensions. In times of escalating tension, direct observation of an adversary's forces can show whether or not they are preparing for war. Ideally, more and better information about a potential enemy increases stability because it's less likely that innocent actions will be mistaken for malevolent intent.

The development of ballistic missile submarines has proven a mixed blessing. A certain degree of societal and political security comes from knowing that your nation maintains a fleet of submarines, almost certainly out of harm's way, fully able to launch a strike against any nation in the world. This makes it less likely that a foreign power will use weapons of mass destruction against your nation, because they presumably know that to do so would be to invite a massive and devastating counterattack. On the other hand, your adversaries have this same capability, and since sea-launched ballistic missiles have a shorter travel time than land-based missiles, a submarine strike will give less warning and may be more devastating than its land- or bomber-based counterpart. This realization can more than compensate for any feelings of security that friendly missile submarines can bring.

All things considered, the possession of ballistic missile submarines by both sides during the Cold War was ultimately a source of stability, because both understood that any nuclear strike against the other would certainly result in an equally massive counterattack that was simply not survivable. This may have helped persuade both sides that nuclear war could not be won and was not worth starting.

P. ANDREW KARAM

Further Reading

Books

Clancy, Tom. *The Hunt for Red October.* Naval Institute Press, 1985.

Clancy, Tom. *Submarine: A Guided Tour Inside a Nuclear Warship.* New York: Putnam's, 1993.

Huchthausen, Peter, Igor Kurdin, and R. Alen White. *Hostile Waters.* New York: St. Martin's Press, 1997.

Sontag, Sherry, and Christopher Drew. *Blind Man's Bluff: The Untold Story of American Submarine Espionage.* New York: Public Affairs, 1998.

Other

Ice Run. 1999. CNN Perspective. Videocassette.

Submarines, Secrets, and Spies. 1999. NOVA. Videocassette.

Submarines: Sharks of Steel. Discovery Channel. Videocassette.

Biographical Sketches

Paul Allen
1953-
American Computer Scientist

Paul Allen co-founded the Microsoft Corporation with Bill Gates (1955-) , adapting existing programming languages such as BASIC, which was originally written for mainframe computers, into software suitable for personal computers. Such innovations enabled the software industry to become established and expand as a viable business, strengthening the American economy and creating new groups of computer entrepreneurs and users. As personal computers became more affordable and accessible during the 1980s, Allen contributed to software development and distribution, enhancing the quality and usefulness of hardware technology, while major computer manufacturers incorporated Microsoft software as the primary operating system for their products.

Born in 1953 in Seattle, Washington, Allen was born to Kenneth and Faye Allen, a university library administrator and teacher, respectively. Allen grew up in Seattle's North End community and attended the private Lakeside School, where he taught his friend Bill Gates about electronics and programming languages. Allen attended Washington State University and prepared software with Gates in a campus computer laboratory for a small company. They envisioned utilizing microprocessors to perform mainframe functions in miniaturized computers. Their first business attempt, a company called Traf-O-Data, failed because of competitors who could sell products more cheaply, but in the process Allen gained awareness of how to succeed at business. He then left college to work as a programmer for Honeywell in Boston, Massachusetts. Noticing a January 1975 *Popular Mechanics* advertisement for a microcomputer kit, Allen contacted Gates, suggesting that they prepare software for this pioneering personal computer.

The pair established the Microsoft Corporation in Redmond, Washington, in 1975 with Allen serving as executive vice president, directing research to design new products. Allen believed Microsoft should promote both software and hardware but Gates disagreed. They adapted programming languages that were used for mainframes, primarily in academic and governmental institutions, for personal computer usage. BASIC (beginners' all-purpose symbolic instruction code) was their first successful program. They bought a language prototype known as Q-DOS from a local company and appropriated it for MS-DOS, an acronym for "Microsoft Disk Operating System," which became a universally adopted program used to command personal computers.

Additionally, Microsoft software had practical applications such as word processing and spreadsheet programs for business and academic uses. While Allen did not invent any of the software that Microsoft sold, he knew how to convert existing programs for personal computers. Major computer manufacturers, including IBM and Apple, selected Microsoft's software as the primary operating system for their machinery—even designing personal computers specifically to use Microsoft programs. This software was considered user friendly, that is, making personal computers more comprehensible to novice computer operators. As computer companies produced more powerful microprocessors, Allen directed Microsoft programmers to develop suitable software to meet revised technological demands.

After suffering an acute illness, Allen retired from Microsoft in 1983. He still owns, however, a significant percentage of the company's stock and serves on Microsoft's board of directors. A multibillionaire who is the third wealthiest American, Allen devotes his energy to establishing and supporting new companies interested in improving computer technology, especially Internet-related applications. He founded Starwave, a company that sold subscriptions to specially designed sports-related Internet sites. Allen also created Vulcan Northwest, the Paul Allen Group, Asymetrix Corporation, and Intervas Research. A philanthropist, Allen donates money to community groups, favoring charities that promote education, cancer research, environmental issues, and the arts. Allen focuses on improving lifestyles globally through computer technology, seeking to connect everyone to the Internet. A proponent of "distance learning," Allen has funded experimental classes through the Paul G. Allen Virtual Education Foundation at his alma mater, which presented him with a distinguished alumnus award at its 1999 commencement.

ELIZABETH D. SCHAFER

Marc Andreessen
1971-
American Computer Programmer and Inventor

A computer programming genius, at the age of 22 Marc Andreessen helped lay the foundation for the Internet and World Wide Web. His Mosaic was the first browser to come into general use. He followed it up with the commercially successful Netscape browser, which brought millions of people onto the Internet and created one of the most successful initial public offerings (IPOs) in Wall Street history.

Andreessen was born in Cedar Falls, Iowa, in July 1971, and was raised in the small town of New Lisbon, Wisconsin. His father Lowell was a seed salesman and his mother Pat worked at the mail-order clothing company Lands' End. As a child Andreessen was fascinated by the potential of personal computers. At age 8 he first showcased his genius for the technology by teaching himself the Basic programming language out of a book he had borrowed from the library. Andreessen wrote his first program (one designed to help him with his math homework) in the sixth grade using a school computer. Unfortunately, his burst of invention was short-lived; the program was erased when someone turned the power off at the end of the day. Tired of having to rely on his school for the use of a computer, he convinced his parents to buy him one of the earliest home computers, the Radio Shack TRS-80. Andreessen used his new computer to create a matchmaking program for his high school classmates.

In the early 1990s he attended the University of Illinois, studying—not surprisingly—computer science. He was not much of a student; focusing his attention instead on a burgeoning career in computers. To earn money he worked part-time as a programmer for the university's National Center for Supercomputing Applications (NCSA), a conglomeration of undergraduates, graduate students, and professors. In the early 1990s the new buzzword at NCSA was "networking," which was fueled by the Internet backbone being developed by the National Science Foundation. The NCSA group was working to expand the Internet by enabling computers to link together across the country and the world.

In 1989 Swiss physicist Tim Berners-Lee created a browser that allowed users to navigate through documents on the World Wide Web with by simply pointing and clicking with the computer mouse. The only problem was that it was text based and not very user friendly. Andreessen was convinced he could improve upon Berners-Lee's browser and convinced Eric Bina, another NCSA employee, to help him. Their new and improved browser, Mosaic, used a graphic visual interface, making it easier to use. Mosaic was introduced in March 1993 and was a huge hit. Within 18 months, it had helped the number of users on the Internet jump to 20 million.

Later that same year, Andreessen left NCSA and was contacted by the founder of Silicon Graphics, Jim Clark, who wanted to start a new company. With four million dollars of Clark's money, they formed Mosaic Communications in Mountain View, California. In the spring and early summer of 1994 Andreessen and a group of programmers began working on building a bigger and better Internet browser. Their goal was to make the new browser even easier to use in order to target a mass-market audience. In November 1994 Andreessen and Clark changed their company's name to Netscape Communications. The cross-platform, easily navigable Netscape browser was about to start an Internet revolution.

Netscape quickly became the most popular browser among consumers. In 1995 Netscape Communications's stock went public with the third-largest IPO in history, making Andreessen 50 million dollars richer virtually overnight. By 1996 the Internet had exploded worldwide with approximately 40 million people "surfing the net" on a regular basis, and millions of them relied on Netscape Navigator as their browser. While Bill Gates's Microsoft Explorer continually challenged Netscape for supremacy in the late 1990s, the company—with Andreessen at the helm—continued to reinvent itself, staying on top of the Internet market and securing its place in the history of technology.

STEPHANIE WATSON

John Backus
1924-
American Computer scientist

During the 1950s, John Backus headed the pioneering group at IBM that developed FORTRAN, the first widely used computer programming language. He was also instrumental in the development of ALGOL, which, though it did not gain wide commercial usage, had a significant influence on three languages that did: Pascal, C, and Ada. Winner of the National Medal of Science, the Turing Award, and the Charles Stark Draper Prize, Backus has continued to work on programming languages even after his retirement from IBM in 1991.

Backus was born in Philadelphia, Pennsylvania, in 1924, and grew up in Wilmington, Delaware. When he enrolled in the University of Virginia in 1942, he planned to major in chemical engineering, but these plans were stymied when, after one semester, he was expelled for skipping classes. At the time, America was embroiled in World War II, and Backus, divested of his academic commitments, was promptly drafted into the army in early 1943.

He first served in an anti-aircraft unit, but in September 1943 was sent to study engineering at the University of Pittsburgh as part of a specialized army training program. After completing his army engineering education in March 1944, Backus spent six months undergoing premedical training at a hospital in Atlantic City, New Jersey. This was followed by another six months at Flower and Fifth Avenue Medical School in New York City.

In May 1946 Backus left the army, and his interest turned from chemistry and medicine to a field that would prove a bridge into the realms of computer science and mathematics. Still living in New York City, he enrolled in the Radio Television Institute, a training school for radio and television repairman. While taking the course, he became so intrigued with mathematics that he began study at Columbia University, which granted him a B.A. in mathematics in 1949. A year later he earned his M.A. in mathematics from Columbia.

After graduating, Backus went to work for IBM, a company staking territory in the newly emerging field of computer science. Backus knew little about computers (few people did in those days) but he soon found himself on the cutting edge. In 1952 he led a group of researchers who produced the Speedcoding system for the IBM 701 computer, and a year later wrote what would prove to be a historic memo. In it he outlined for his boss, Cuthbert Hurd, the need for a general-purpose, high-level computer programming language. This was the origin of FORTRAN, an acronym for formula translator.

FORTRAN was the prototype for modern compilers—computer programs that translate high-level language statements into a form that computer hardware can read. Before FORTRAN, programmers had been forced to endure the tedium of logging in rows of zeroes and ones, the binary language of computers. FORTRAN allowed programmers much greater freedom and creativity, and enabled wide-ranging advancements—not least because, prior to its development, three-quarters of the cost of running a computer came from debugging and programming. IBM published FORTRAN I in 1954, and in the following year Backus, along with R. A. Nelson and I. Ziller, began working out the bugs in this first version of the language.

At the same time, two schools of thought about computer languages emerged: American researchers believed that programmers should develop languages as needed by users; a group of European computer scientists, led by F. Bauer of the University of Munich, Germany, held that this proliferation would create confusion. As a result, Backus, Bauer, and others met in Zurich, Switzerland, in the spring of 1958, and produced a report calling for an International Algebraic Language (ALGOL). Soon after the Zurich meeting, Backus went to work with the team developing ALGOL at IBM's Watson Research Center in Yorktown Heights, New York. The resulting language did not enjoy widespread commercial usage, but did have a great impact due to its influence on Pascal, C, and Ada.

During the decades since the development of ALGOL, Backus has worked tirelessly to refine computer languages, and in 1978 wrote a paper in which he called for their restructuring. He became an IBM Fellow in the early 1980s, which gave him an opportunity to step up the pace of his research. In 1991 Backus retired from IBM. He continued,however, to work on developing what he called a "functional" language—one that borrowed from already defined languages, eliminating the programmer's need to "reinvent the wheel" by spelling out each instruction in detail.

JUDSON KNIGHT

John Presper Eckert Jr.

1919-1995

American Computer Engineer

J. Presper Eckert was a pioneer of modern computer engineering. He served as the lead designer of the ENIAC, the world's first general purpose electronic digital computer. He founded, along with John W. Mauchly (1907-1980), the nation's first commercial computer company.

John Presper Eckert Jr. was born in 1919 in Philadelphia, Pennsylvania. He attended the William Penn Carter School and received his undergraduate degree in 1941 from Moore School of Engineering at the University of Pennsylvania. Immediately after graduation he accepted a position as an instructor at the Moore School and helped teach a government-sponsored course in defense engineering. One of his students was John W. Mauchly, a physicist from nearby Ursinus College who later became Eckert's colleague and fellow instructor.

In 1942 John Mauchly submitted a proposal to the Moore School for the construction of a general-purpose electronic digital computer. The proposal was ignored by the Moore School faculty but was picked up by the nearby Navy Ballistics Research Laboratory. One of the first engineers who Mauchly recruited for his project was J. Presper Eckert. For the next several years Mauchly and Eckert worked feverishly on the design of their electronic computer, the ENIAC (Electronic Numerical Integrator and Computer). While Mauchly provided the original vision and overall project leadership, Eckert handled most of the actual engineering. Eckert proved himself a brilliant designer, overcoming what had been considered to be an insurmountable barrier to the construction of a useful computer: the unreliability of vacuum tube circuits. Although the ENIAC required almost 18,000 of these fragile glass vacuum tubes, Eckert's inspired engineering and conservative design produced a machine that was remarkably reliable. When it was completed in 1946, the ENIAC was capable of performing 5,000 additions and 300 multiplications a second, an increase of 2-3 orders of magnitude (100-1000 percent) over existing mechanical calculators. The ENIAC played an important role in the war effort and was used for critical feasibility studies of early atomic weapons. Eckert and Mauchly were featured speakers at the famous Moore School lecture series that introduced scientists from around the globe to the principles of electronic computing.

By the end of 1946, however, Eckert and Mauchly had resigned their positions at the Moore School in a dispute over patent rights. Administrators at the Moore School attempted to force the pair to relinquish all rights to the ENIAC, but neither of them was willing to give up the opportunity to capitalize on their invention. Together they founded the Electronic Controls Company and began work on the BINAC mathematical computer for the Northrup Aircraft

Company. In 1948 they incorporated their company as the Eckert-Mauchly Computer Corporation (EMCC). Although the EMCC was awarded a major contract from the United States Census Bureau, Eckert and Mauchly found it difficult to raise adequate capital and to obtain security clearance for Mauchly (who in the pre-war period had attended a meeting of a group that, unbeknownst to him, had Communist affiliations). In 1950 they sold the company to Remington Rand (later Sperry Rand), a national manufacturer of tabulating machinery. Remington Rand delivered the first UNIVAC (UNIVersal Automatic Computer) in 1951, and for a time the name UNIVAC was synonymous with computer.

Although Mauchly left Sperry Rand in 1959, Eckert continued on with the company to become an executive. In the years between 1948 and 1966 Eckert was granted 85 patents for various electronic devices. He was elected to the National Academy of Engineering in 1967 and was awarded the National Medal of Science in 1969. He died in 1995 at the age of 76 from complications related to leukemia.

NATHAN L. ENSMENGER

Richard Buckminster Fuller

1895-1983

American Architect and Inventor and Mathematician

R. Buckminster Fuller was among the most original thinkers of the twentieth century. He is best remembered for developing the geodesic dome, the only practical type of building that has no inherent size limitations beyond which it can not support its own structure. He conceived of human beings as passengers on "Spaceship Earth" and concerned himself with finding ways to maximize the social benefits to be derived from a limited set of resources.

Fuller was born on July 12, 1895, in Milton, Massachusetts, the descendant of a long line of New England intellectuals. His great-aunt Margaret Fuller (1810-1850) was an influential philosopher of the Transcendentalist movement, a social reformer, and an ardent campaigner for women's rights.

Despite his pedigree, Fuller never completed his formal education, twice being expelled from Harvard for cutting classes. After serving in the United States Navy during World War I, he married Anne Hewlett. Anne's father, James Hewlett, an architect, had developed a modular construction method using compressed fiber blocks. The two men went into business together, and Fuller supervised the construction of several hundred houses.

Fuller was the junior partner in the business, and when it ran into financial difficulties in 1927, he found himself forced out. Impoverished and reduced to living with his family in a Chicago slum, he devoted himself to designs for the modern world. His goals were "maximum gain of advantage from minimum energy output," so that Earth's resources could be used for the benefit of all. While he became known as an inventor, he regarded his innovations as simply part of an overall strategy to do more with less, so that the needs of the entire global population could be fulfilled.

His early designs were given the name "Dymaxion," combining "dynamic" and "maximum," two key ideas of Fuller's philosophy. The factory-assembled Dymaxion house, prototyped within a year of his ouster from the construction firm, was built on a mast so that it could be easily moved from one site to another. It had its own utilities and cost no more than a new car.

Fuller then set out to build the new car. The streamlined, three-wheeled Dymaxion automobile, built in 1933, could carry 12 people at up to 120 mph, go off-road, and turn 180 degrees in its own length, while averaging 28 miles per gallon of gasoline. Tragically, a fatal accident occurred during testing. Although the fault actually lay with the other car involved, it doomed the project. Only three Dymaxion cars were built, of which one still exists at the National Auto Museum in Reno, Nevada.

Fuller developed his "energetic-synergetic geometry" based on the tetrahedron, or four-sided pyramid, because he had observed that in nature such a shape affords a maximum of strength with a minimum of structure. Applying his geometry to architecture, Fuller came up with the geodesic dome. Its lightweight polygonal facets distribute stress uniformly throughout the structure, and its strength increases logarithmically with size. Unlike other large domes, it can be prefabricated and then set into place. The geodesic dome that served as the United States pavilion during Expo 67 in Montreal was followed by "dome homes," sports facilities, theaters, and greenhouses.

Fuller joined the faculty of Southern Illinois University (Carbondale) in 1959. He died in Los Angeles on July 1, 1983. Two years later chemists

Richard Buckminster Fuller. *(The Library of Congress. Reproduced by permission.)*

Richard E. Smalley (1943-), Robert F. Curl (1933-), and Sir Harold W. Kroto (1939-) discovered the third known form of pure carbon (after diamond and graphite), consisting of 60 atoms fitted together in a hollow sphere resembling a geodesic dome. The scientists named their discovery, for which they won the 1996 Nobel Prize in chemistry, *buckminsterfullerene*. The carbon structures are colloquially known as *buckyballs*.

SHERRI CHASIN CALVO

William Henry Gates, III

1955-

American Computer Scientist and Businessman

An internationally renowned computer programming pioneer and businessman, Bill Gates has won fame and fortune as a brilliant software developer and entrepreneur. Gates cofounded Microsoft Corporation with Paul Allen (1953-) in 1974, and subsequently developed the widely popular MS-DOS operating system and Windows graphical interface. As chairman and CEO of Microsoft, Gates is one of the most influential and wealthy leaders of the computer industry. In recent years he has also won recognition as a noted philanthropist.

William Henry Gates, III, was born on October 28, 1955, in Seattle, Washington. He was raised with his two sisters in an affluent family. His father, William H. Gates, II, was a prominent attorney, and his mother, Mary, a schoolteacher. His parents nicknamed the younger William "Trey," referring to the III after his full name.

Gates struggled in public elementary school, so his parents enrolled him at private Lakeside School. There, he was first introduced to computers, and at the early age of 13 wrote a program to play tic-tac-toe.

In 1973 Gates moved across the country to attend Harvard University. While at Harvard, he developed the programming language BASIC for the world's first personal computer (PC), the MITS Altair. Two years later, at age 19, he launched the Microsoft Corporation with childhood friend Paul Allen, motivated by the belief that every business and household should have a computer. In his junior year, Gates dropped out of Harvard to devote his energies to his new company. Little did he know it would become one of the most successful companies in the world.

In the early 1980s Gates led Microsoft's evolution from a programming developer to a diversified software company producing operating systems and applications software. In 1981, under Gates' direction, Microsoft introduced MS-DOS, the operating system for the new personal computer produced by International Business Machines Corporation (IBM). MS-DOS became the standard operating system for the majority of personal computers throughout the world.

On March 13, 1986, Microsoft went public on the Stock Exchange. The initial offering for a share of Microsoft stock was $21. At age 31, Gates instantly became the richest man in the United States. His personal stock holdings exceeded $2.8 billion. In 1999 he was reported to be the richest private individual in the world, with a net worth of over $50 billion.

In the mid-1990s Gates dramatically changed the direction of his entire company and began focusing on the rapidly evolving Internet. While some of his early efforts fizzled, the company hit a milestone with its popular Internet Explorer web browser, giving the existing Netscape browser some heated competition. By June 1998 Microsoft had revenues of $14.4 billion for the fiscal year and employed more than 27,000 people in 60 countries.

Microsoft has since become the subject of federal anti-trust litigation that threatens to split Gates's corporation into several smaller companies. The suit, brought by the United States Department of Justice, accuses Microsoft of inhibiting competition from rival software companies, such as Netscape, by linking its own Internet Explorer browser with its Windows operating system, used by some 80 percent of desktop computers worldwide.

In addition to his passion for computers, Gates holds an avid interest in biotechnology. He is a shareholder in Darwin Molecular, a subsidiary of British-based Chiroscience, and sits on the board of the Icos Corporation. Gates also founded the Corbis Corporation, which is developing one of the largest resources of visual information in the world.

Gates is also a published author, having written several books about his vision for the future of information technology and society. His book, *The Road Ahead,* published in 1995, topped the *New York Times* bestseller list for seven weeks. Gates donated proceeds from the book to a non-profit fund that puts computers in classrooms.

Since Microsoft has gone public, Gates has donated more than $270 million to charities, focusing on education, population issues, and access to technology. His $6 billion donation to the Bill and Melinda Gates Foundation in August 1999 was the largest bequest ever by a living individual. The donation was earmarked to speed the development and reduce costs of vaccines for malaria, tuberculosis, and AIDS.

Gates was married on January 1, 1994, to Melinda French Gates, a former Microsoft executive. They have two children, Jennifer Katharine Gates, born in 1996, and a son, Rory John Gates, born in 1999.

KELLI A. MILLER

Gordon Gould

1920-

American Physicist

Disputes over his patents claims have made Gordon Gould one of the most controversial figures in laser history. Notwithstanding these controversies, most of the American laser industry is or has been licensed under his laser patents. The Patent Office Society named him inventor of the year for 1978.

Gordon Gould was born July 17, 1920, in New York City. As a child he idolized Thomas Edison (1847-1931) and dreamed of being an inventor. He earned a B.S. in physics from Union College (1941). Upon completing a masters degree in optical spectroscopy at Yale University (1943), he worked for the Manhattan Project—the American effort to build a nuclear bomb during World War II—separating uranium isotopes. After the war he chose to support himself with part-time jobs while he invented, designing an improved contact lens and attempting to produce synthetic diamonds.

Gould soon realized his background was inadequate for the tasks he contemplated and began taking graduate courses (1949) at Columbia University while teaching at the City College of New York. He lost his teaching position in 1954 when he refused to identify members of a Marxist study group he attended years earlier. Outraged, his thesis advisor, the Nobel laureate Polykarp Kusch (1911-1993), secured him a research assistantship allowing Gould to become a full-time student.

During this period Gould came into contact with Charles H. Townes (1915-), who had conceived and built the microwave precursor to the laser—the maser. In 1957 Townes and Arthur L. Schawlow (1921-) began seriously thinking about the possibility of an optical maser, suggesting thallium vapor as the lasing medium and optical pumping to achieve excitation. Since Gould's research under Kusch involved optically pumping thallium to observe its excited states, Townes sought him out. Gould was alarmed by their October discussions because he was also thinking about an optical masor or, as he called it, laser.

Aware of the significance of this work, Gould wrote his ideas in a notebook in early November 1957 and had it notarized, but due to bad legal advice believed he had to produce a working model before applying for a patent. As a graduate student Gould had neither the time nor the money to produce a prototype. So, in March 1958 he left Columbia to take a job with Technical Research Group (TRG).

TRG became interested in Gould's laser and in September 1958 approached the Advanced Research Projects Agency (ARPA) with a $300,000 research proposal to develop it. ARPA awarded TRG a $1 million contract. However, Gould was prevented from working directly on the project because of his past Marxist connections.

Meanwhile, Townes and Schawlow had been busy developing their optical maser. They filed a patent application in July 1958 and published the first detailed proposal for building a laser in December 1958. Gould filed his patent application in April 1959. When the Townes-Schawlow patent was granted in March 1960, Gould and TRG challenged it, claiming that although he filed later, Gould had conceived of the laser first. Gould lost the case but maintained his rights to patent coverage.

In 1977, after years of legal battles, Gould was issued a patent for optically pumped laser amplification. In 1978 Gould received a second patent covering a broad range of laser applications including machining. Gould later received patents for gas discharge lasers (1987) and for Brewster-angle windows used in laser cavities (1988). Though extensively challenged, the validity of Gould's patents have been upheld in various cases.

Gould was a Professor at the Polytechnic Institute of New York (1967-1973), where he founded a department and laser research laboratory. He co-founded (1973) the optical communications company Optelecom Inc., from which he retired in 1985.

STEPHEN D. NORTON

Grace Brewster Hopper
1906-1992

American Mathematician and Computer Programmer

Grace Hopper developed COBOL, a computer programming language, in the 1950s and her related innovations, such as compilers used for business processes, provided a foundation for more sophisticated computing systems. A computer software pioneer, Hopper encouraged her colleagues to enhance technology for data processing. She set precedents in technological professions that were considered almost exclusively male at the time, enabling other women to gain access to computing careers. Hopper's achievements helped make the production of commercial computers possible.

Born on December 9, 1906, in New York City, Hopper was encouraged by her parents, Walter Fletcher and Mary Campbell (Van Horne) Murray, to pursue her academic ambitions, especially in mathematics. Hopper resisted being restrained by conventional gender roles. Curious about how mechanical devices worked, she explored her home to investigate small machines such as clocks. Attending private girls' schools in New York and New Jersey, Hopper enrolled at Vassar College, where she majored in mathematics and physics, graduating in 1928. She married Vincent Foster Hopper in 1930. While teaching at Vassar, Hopper completed a master's degree and doctorate in mathematics from Yale University in the early 1930s.

When the Japanese bombed Pearl Harbor in 1941, Hopper decided to enlist in the U.S. Navy. Military officials, however, thought that she was too old and urged her to contribute to the war effort as a mathematics instructor. Instead, Hopper secured authorization to join the Naval Reserve in 1943, beginning a 43-year career. Hopper trained at midshipman's school and was commissioned a lieutenant for service in the Bureau of Ordnance Computation Project, located at Harvard University's Cruft Laboratories. Commander Howard Aiken directed Hopper to program the school's Mark I, an early digital computer. During World War II, the Mark I computer figured out projectile trajectories.

Using punched cards to input data, Hopper focused on preparing code for the Mark I computer. When the war concluded, Hopper resigned from active military duty but retained reserve status. She began work as a research fellow in physics and engineering sciences at Harvard's Computation Laboratory. Here, Hopper programmed the Mark II and III computers. In 1949 she accepted a mathematics position at the Eckert-Mauchley Computer Corporation. This company developed the BINAC (Binary Automatic Computer), which used code, rather than punched cards, for data entry. Hopper pro-

Grace Hopper. *(The Library of Congress. Reproduced by permission.)*

grammed BINAC with a base eight numerical system called octal that utilized the digits 0 to 7.

When the company was bought by Remington Rand Corporation (which later merged with the Sperry Corporation), Hopper was designated a systems engineer and director of Automatic Programming Development for the UNIVAC (Universal Automatic Computer) Division. She assisted John Eckert and John Mauchly, who invented the UNIVAC computer. Hopper focused on bettering compiler design in order to translate mathematical code (on tape) to machine code for performing processes. She innovated a compiler known as the Flow-Matic that understood English instructions so that the UNIVAC computers could be used for business applications. Hopper then concentrated on standardizing a universal programming language, developing COBOL (COmmon Business-Oriented Language) by 1959. This computer language recognized words—unlike its predecessors, which read numbers only.

When a moth flew into a computer that she was monitoring, Hopper developed the term "bug" to indicate a technical malfunction, and this reference was incorporated into computer terminology. Because of her age, she retired from the Naval Reserves in 1966. Navy officials, however, asked her to return to active duty to standardize computer languages that were crucial for payroll procedures. When she retired in 1986 at the rank of Rear Admiral, Hopper was the country's oldest active-duty officer. The Department of Defense presented her with its most prestigious award and she also received the National Medal of Technology. Hopper died on January 1, 1992. Five years later, a Navy destroyer was named for her.

ELIZABETH D. SCHAFER

Stephen Paul Jobs
1955-
American Computer Pioneer

Steve Jobs helped revolutionize the personal computer industry, creating the innovative Macintosh computer and developing Apple Computer, Inc., into a multibillion dollar company.

Steve Jobs was born in 1955. Orphaned as an infant, he was adopted by Paul and Clara Jobs, and grew up in the California suburbs of Mountain View and Los Altos, which would later become the heart of Silicon Valley. While in high school, Jobs was hired as a summer employee at the Hewlett-Packard electronics firm in Palo Alto. There he met Stephen Wozniak (1950-), an engineering whiz kid. Wozniak was in the process of developing his "blue box," an illegal device that allowed a user to make free long-distance calls. Jobs helped Wozniak sell his device, forging a partnership that would several years later change the face of the home-computer industry.

In 1972 Jobs entered Reed College in Portland, Oregon, but his college career was short-lived. After one semester, he dropped out to become involved with the counterculture of the 1970s. In early 1974 he signed on as a video game designer with Atari, Inc., which in the early 1980s would become famous for its Pac-Man and Space Invaders arcade games.

After only a few months at Atari, Jobs had saved up enough money to travel to India on a search for spiritual enlightenment. When he returned in the fall of 1974, Jobs again caught up with Wozniak, who was then holding regular meetings of his "Homebrew Computer Club." Unlike Wozniak, Jobs was not interested in building computers; instead, he wanted to market them. Jobs convinced Wozniak to help him create their own personal computer, one which was smaller, cheaper, and easier to use than what was currently available to consumers. In Jobs's bedroom and garage they designed and built their prototype. With $1,300 (earned from selling Jobs's Volkswagen microbus and Wozniak's scientific calculator), they started their own company. Wozniak quit his job at Hewlett-Packard to become vice president in charge of research and development. Jobs came up with the name Apple, in honor of the summer he worked in an Oregon orchard.

The first computer the pair designed and marketed, the Apple I, sold for $666 in 1976. It was the first single-board computer with built-in circuitry allowing for direct video interface, along with a central ROM, which allowed it to load programs from an external source. The Apple I earned Jobs and Wozniak $774,000. A year later, the Apple II was launched with a simple, compact design like the Apple I, plus a color monitor. Within three years, the Apple II's sales had grown by 700 percent, to $139 million.

In 1980 Apple made its move on Wall Street, its stock rising to $29 on the first day of trading, bringing the company's value to $1.2 billion. Business was good for the two Apple founders, but the fledgling computer company was not without competition. The corporate might of IBM, with its two-year-old personal computer (PC), was beginning to get the edge with consumers. Jobs realized that to compete against this industry Goliath, Apple would have to make its operating system compatible with that of IBM.

Enter the Macintosh, a powerful new computer with 128K of memory—twice that of the PC. The Mac, as it became known, had a 32-bit microprocessor, which outperformed the PC's 16-bit version. Not only was it faster, it was more versatile and easier to use than the PC, offering—as its print campaign suggested—a computer "for the rest of us." The Macintosh was introduced to the world on Super Bowl Sunday 1984 in an Orwellian-themed commercial that promised a revolutionary new wave in personal computing. The world was watching, and the Macintosh caught on like wildfire.

But just as sales of the Macintosh were taking off, president John Sculley persuaded Apple's board of directors that Jobs's emphasis on technical performance over consumer needs was hurting the company, and Jobs was sent off to a remote office that he termed "Siberia"—far from the inner workings of the company he had created. In September of 1985 Jobs resigned as chairman of Apple. In 1989 he formed a new computer company called NextStep, which he believed would compete with Apple in the personal-computer market. Eight years and $250 million later, NextStep was forced to close its hardware division. Ironically, Apple went on to acquire NextStep several years later.

By the late 1990s Jobs had moved on to a new venture as chairman and CEO of Pixar, the Academy Award–winning computer animation studio he founded in 1986. Pixar's first feature film, *Toy Story*, was released by Walt Disney Pictures in 1995 and became the third-highest-grossing animated film up to that time.

STEPHANIE WATSON

Brian David Josephson
1940-
English Physicist

Brian Josephson is as well known for his contributions to physics as he is for his beliefs that physics must also explain extrasensory perception and paranormal phenomena.

Josephson first came to the attention of the scientific community when he was still an undergraduate at Cambridge University. At the age of 20 and in his third year of undergraduate studies he developed a new, improved way of calculating how the Doppler shift, which is a change in the frequency of radiation emitted by objects moving near light speed, is affected by gravity.

Shortly thereafter in 1962 Josephson, then a graduate student at Cambridge, made several predictions concerning the behavior of superconducting circuits. When certain metals and alloys are cooled to extremely low temperatures they lose all resistance to the flow of electrons and become perfect electrical conductors, called superconductors. Josephson predicted that when two parallel superconducting wires were separated by a thin layer of a non-superconducting material (an insulator), an electric current would begin to flow between the two superconductors. This is due to an effect of quantum mechanics called tunneling. Electrons can sometimes tunnel through a material even when they don't seem to possess enough energy. Josephson predicted correctly that electrons in both superconductors could pair up. This would lower their energy state and allow them to tunnel across the barrier. As well, these electrons would be flowing simultaneously in both directions.

When these predictions were confirmed by Bell Laboratory scientists the scientific community was quick to praise him. In 1973 he was awarded the Nobel Prize for his discovery. In his honor, this physical process is now called the

Josephson effect, and these types of electrical circuits are called Josephson junctions.

Josephson junctions initially showed great promise to the computer industry. Many engineers looked for ways of applying them to computer circuits. IBM reportedly spent over $100 million before abandoning their efforts at finding broad-reaching uses for these circuits.

The single workable application to come out of these efforts are SQUIDs, or superconducting quantum interference devices. When an electrical current is traveling steadily across the insulating layer between the two superconductors, the current, called a standing wave, becomes extremely sensitive to any changes in the surrounding magnetic fields. Since electricity and magnetism are inseparably linked, small changes in electrical voltages can also be detected. SQUIDs can sense changes of picovolts (one trillionth of a volt), and are almost 1,000 times more sensitive than traditional voltmeters.

SQUID detectors have been used to study voltage changes in the brain and the heart. Detectors that analyze magnetic fields are used by geologists to monitor changes in rocks and by the U.S. Navy to detect submarines.

Since winning the Nobel Prize Josephson has turned toward more esoteric studies. He is currently a physics professor at Cambridge University's Cavendish Laboratory. There he directs the Mind-Matter Unification Project of the Theory of Condensed Matter Group. His current research interests include human consciousness, psychic phenomena, and paranormal events. He believes that some facet of quantum mechanics may be able to explain these happenings.

As well, Josephson has also been investigating the link between the mind and music. He believes that music may affect the minds of different people in the same way. Music, according to Josephson and his collaborator Tethys Carpenter, may have some sort of universal structure or pattern that gives it meaning, while cultural and individual influences play a smaller role than that attributed by other psychologists and musicologists.

PHILIP DOWNEY

Jack St. Clair Kilby

1923-

American Electrical Engineer

Jack St. Clair Kilby shares credit with the late Robert Noyce (1927-1990) as inventor of the integrated circuit, or microchip, which has been called the most influential invention of the twentieth century. Certainly it can be considered the most important invention of the century's second half, as the automobile was to the first. The world of computers, cellular phones, fax machines, satellite television, and many other fixtures of modern life could not exist without this tiny instrument, first created by Kilby and shortly afterward improved by Noyce.

Kilby was born on November 8, 1923, in Jefferson City, Missouri. He grew up in Kansas, where his father, an engineer, helped install that state's power grid. In 1947, he earned a B.S.E.E. degree from the University of Illinois, and went on to acquire an M.S.E.E. at the University of Wisconsin in 1950.

From 1947 to 1958, Kilby worked at the Centralab Division of Globe Union Inc. in Milwaukee, Wisconsin, where it was his job to design and develop thick-film integrated circuits. In 1958, he moved to Dallas, where he has lived ever since—in 1998, he was reportedly still living in the house he bought 40 years before. In Dallas, Kilby went to work for Texas Instruments, and later in 1958, he constructed the first monolithic integrated circuit.

Up to this point, computers relied on slow, ungainly vacuum tubes to do their processing. Kilby developed a single, self-contained unit about the size of a fingernail, which worked much faster than the old tubes. His chip, which was first displayed at the Institute of Radio Engineers Show in 1959, was made of germanium, and relied on external wires. It was difficult to manufacture, however, but Noyce's silicon chip, which debuted six months later, overcame this obstacle. Nonetheless, it was Kilby's chip that helped inaugurate what is sometimes called the "fourth generation" of computers.

Kilby and Noyce both went on record repeatedly stating that they should be viewed as independent co-creators; but since Kilby was working for Texas Instruments and Noyce for Fairchild Semiconductor, a company he helped establish, the issue assumed the character of an inter- company rivalry. In 1970, Kilby left Texas Instruments to work on his own, focusing on areas such as the development of a solar energy system.

Kilby has also concerned himself with promoting innovation by establishing the Kilby Awards, which annually recognizes individuals who have made outstanding contributions in science, technology, and education. In 1983, he officially went into retirement, but maintained an office at Texas Instruments headquarters, and in the late 1990s still made it a habit to stop in at least once a week.

JUDSON KNIGHT

John William Mauchly
1907-1980
American Computer Engineer

John W. Mauchly was a pioneer of modern computing. In the 1940s Mauchly headed the project that produced the ENIAC, the world's first general purpose electronic digital computer. He also co-founded the nation's first commercial computer manufacturer.

John William Mauchly was in born on August 30, 1907, in Cincinnati, Ohio. He grew up in the suburbs of Washington, D.C., where his father worked as a physicist at the prestigious Carnegie Institute. Mauchly received a solid scientific education at the McKinley Technical High School, and in 1925 was awarded the Engineering Scholarship of the State of Maryland. Two years after enrolling in an electrical engineering program at Johns Hopkins University, Mauchly transferred directly into a graduate program in physics. He was awarded his Ph.D. in 1932.

Like many depression-era academics, Mauchly had difficulty finding a position at a major research institution. He ended up accepting a job teaching physics at Ursinus, a small liberal arts college located outside of Philadelphia. Faced with a lack of resources and laboratory equipment, Mauchly turned his sights on meteorology, a field in which experimental data was inexpensive and plentiful. Mauchly soon discovered that analyzing this abundance of data required massive amounts of manual computation. He began to consider ways in which he might use electronics to automate these tedious calculations. In 1941 he enrolled in a special course in defense engineering taught at the nearby Moore School of Engineering of the University of Pennsylvania. Although the course was intended to teach young engineers to operate advanced electronic weapons and equipment, for Mauchly it served as a crash course in the logic and circuitry required for electronic computing.

Upon completion of the program, Mauchly was offered a position at the Moore School as an adjunct instructor. In 1942 he drafted a proposal outlining an electronic digital computer designed for general numeric computation. His proposal was enthusiastically received by Lieutenant Herman Goldstine of the United States Navy Ballistics Research Laboratory in Aberdeen, Maryland. Goldstine encouraged the Moore School and the Ordnance Department to fund Mauchly's proposal, and in 1943 Mauchly began work on the machine that would come to be known as the ENIAC (Electronic Numerical Integrator and Computer).

Working closely with the brilliant Moore School engineer J. Presper Eckert (1919-1985), Mauchly designed a general-purpose computer capable of performing over 5,000 additions or 300 multiplications per second. Together Eckert and Mauchly overcame what was considered to be the major barrier to the construction of a useful computer: the unreliability of vacuum tube circuits. Although the design of the ENIAC was almost certainly influenced by the earlier work of Iowa State College professor John Atanasoff (1903-1995), whom Mauchly had visited to discuss computers in 1941, the construction of a working electronic computer was, in 1946, a unique and extraordinary achievement.

In spite of the success of the ENIAC (or perhaps because of it), Mauchly and Eckert were forced to resign from the Moore School in a dispute over patent rights. In 1946 they founded the Electronic Controls Company and began work on a small mobile computer for the Northrop Aircraft Company. Two years later they received a Census Bureau contract to build a larger, general-purpose computer. The first UNIVAC (UNIVersal Automatic Computer) was completed in 1951, and for a short time the name UNIVAC was synonymous with computer. By that time, however, Mauchly and Eckert had run out of capital and sold their company to the Remington Rand Corporation. Mauchly was never really content working at Rand, and he left the company in 1959.

John Mauchly spent the remainder of his career working as a consultant and serving as the president of several small technology companies. For his role in the invention of the computer Mauchly was awarded many honors, among them the Potts Medal of the Franklin Institute and the Harry Goode Medal of the American Federation of Information Processing Societies. He died in his home in Ambler, Pennsylvania, in 1980 at the age of 72.

NATHAN L. ENSMENGER

John McCarthy

1927-

American Computer Scientist

The father of research in artificial intelligence (AI)—a term he coined—John McCarthy wrote the principal computer language for AI research, List Processing Language (LISP). He founded two of the most important AI laboratories in the world, and in latter years has been active in the movement for an "Electronic Bill of Rights" to govern electronic communications.

Born in Boston, Massachusetts, on September 4, 1927, McCarthy was the elder of two brothers. His father, John Patrick McCarthy, worked variously as a carpenter, fisherman, union organizer, and inventor; and his mother, Ida Glatt, was a Jewish Lithuanian involved in the suffrage movement, the effort to secure the vote for women. Both were members of the Communist Party of America, and McCarthy grew up in a politically charged atmosphere.

A sickly child, McCarthy turned to books for solace, and eventually his family moved to Los Angeles in hopes that his health would improve. It did, and in the meantime he proved to be a prodigious scholar, skipping three grades and entering the California Institute of Technology (Cal Tech) in 1944. Despite the fact that he took time off from school for a number of reasons, including a stint in the army as a clerk, McCarthy graduated four years later with a degree in mathematics.

He went on to Princeton, where he earned his doctorate in mathematics and took a position as an instructor in 1951. Two years later McCarthy returned to the West Coast, where he worked as an acting assistant professor of mathematics at Stanford University; but in 1955 he took a job back east again, this time at Dartmouth College. Though his time at the latter school lasted for only three years, this would prove to be a pivotal juncture in his career.

In the summer of 1956 McCarthy began working on a program that would assist a computer in playing chess. To limit the possible moves and thus speed up the game, McCarthy developed a method later termed the alpha-beta heuristic, which made it possible for a computer to quickly eliminate any moves that would benefit its opponent. This was the birth of artificial intelligence, a term McCarthy coined that year when he organized the world's first conference on modeling intelligence in computers.

John McCarthy. *(The Library of Congress. Reproduced by permission.)*

McCarthy became an associate professor of mathematics at the Massachusetts Institute of Technology (MIT) in 1958, and soon founded the first AI laboratory there. He also began creating the computer programming language that would eventually be called LISP, or List Processing Language. This remains the most commonly used language in AI research. While at MIT McCarthy began developing the means of interactive time-sharing on computers, which would make networks possible by allowing hundreds or thousands of people to share data on the same large computer. He also initiated work on the concept of giving a computer "common sense," an idea that would continue to perplex computer programmers for many decades.

McCarthy married the first of his two wives at MIT, and they had a daughter, Susan. In 1962 they moved to Stanford, where he took a professorship in computer science and inaugurated a second AI lab. At Stanford he focused on issues such as the role played by mathematical logic and common sense, which he called nonmotonic reasoning, in AI. He also examined questions such as that of a machine that could copy itself, or of an AI more intelligent than its creator.

Divorced in the 1960s, McCarthy married Vera Watson, a computer programmer and world-class mountain climber, with whom he had two more children, Sarah and Timothy. Mc-

Carthy himself was extremely active, taking part in outdoor activities such as rock- climbing, flying, and even skydiving. Tragically, his wife's own adventurous quests ended in misfortune: Watson died while taking part in a women's expedition in the Himalayas.

Though McCarthy has professed disillusionment both with the Marxism of his parents and with some of the leftist groups with which he associated in the 1960s, he has remained politically active. He was one of the first to propose that the U.S. Constitution's Bill of Rights be extended to guarantee the rights of each person to read, correct, and limit access to their own electronic files. Among the honors McCarthy has received are the Alan Mathison Turing Award (1971), the Kyoto Prize (1988), and the National Medal of Science (1990). In 1987 he took the Charles M. Pigott chair at the Stanford University School of Engineering.

JUDSON KNIGHT

Robert Norton Noyce
1927-1990
American Physicist

Although Jack St. Clair Kilby (1923-) built the world's first integrated circuit, or microchip, in 1958, Robert N. Noyce built the first practical microchip six months later. Whereas Kilby's chip had required connecting wires, Noyce used a flat transistor to replace those wires and made conducting channels printed directly on the surface of the chip. The channels were possible because Noyce had also improved on the material used: instead of germanium, as in Kilby's chip, he used silicon. Noyce later cofounded Intel Corporation.

Born on December 12, 1927, in Burlington, Iowa, Noyce was the son of a minister. He attended Grinnell College, where in 1948 he had his first encounter with a newly developed technological marvel, the transistor. Noyce later said he knew from that first moment that the transistor would change the face of electronics—but he, too, was destined to affect tremendous changes through the use of the transistor in computing.

Noyce earned his B.A. in physics and mathematics from Grinnell in 1949, and in 1953 received his Ph.D. in physical electronics from the Massachusetts Institute of Technology (MIT). Soon afterward he went to work for the Philco Corporation as a researcher, and there he remained until 1956. In the latter year William Bradford Shockley (1910-1989) recruited him to work at his new Shockley Semiconductor Laboratory. However, Shockley—another computing pioneer—had a managerial style that made him difficult to get along with, and within a year Noyce and seven others left the company to form Fairchild Semiconductor Corporation in Mountain View, California.

In early 1959 Noyce was working as research director for Fairchild, and in this capacity it was his responsibility to oversee the company's silicon chip development. Using a new chemical etching process, he was able to print transistors on the silicon wafers, which eliminated expensive wiring costs and made the chips operate much faster as well. On April 25, 1960, the U.S. Patent Office granted him a patent for a "Semiconductor Device-and-Lead Structure." These new chips, for which both Noyce and Kilby deserve credit, ended the dominance of the slow, ungainly vacuum tubes that had once powered computer processing, and inaugurated the modern electronic computing revolution.

In 1968 Noyce and Gordon Moore founded Intel, which soon developed a memory chip that made it a leader in the field. By 1974 the company was so successful that Noyce turned his attention from managing it to dealing with larger industry concerns; in the latter capacity he headed up Sematech, a consortium designed to deal with foreign competition.

In 1980 President Jimmy Carter awarded Noyce the National Medal of Science, and in 1987 he received the National Medal of Technology from President Ronald Reagan. Grinnell College named its computer center in Noyce's honor in 1984, and in 1990 he received the Charles Draper Prize from the National Academy of Engineering. Noyce died on June 3, 1990.

JUDSON KNIGHT

Admiral Hyman George Rickover
1900-1986
American Naval Officer and Engineer

At the end of the Second World War, Hyman Rickover was a relatively obscure officer in the United States Navy. He had very little experience on sea-going commands, had commanded only one auxiliary ship, and had spent most of the war in a support capacity in the United States. After the war, however, at a time when most of his naval academy classmates were retiring, Rickover almost single-handedly built the nuclear navy that, at its peak, would consist of 141 nuclear submarines and over a dozen surface ships—the largest nuclear power program in the world (except for that of the Soviet navy) and the most reliable. To do this, Rickover not only helped design a reliable and compact nuclear reactor, but also set incredibly tough standards for quality control, personnel selection and training, and ship design. Very few naval officers have had such a significant impact.

Rickover was born in the Jewish Pale of Polish Russia near the beginning of the twentieth century. His actual birth date is recorded as 1900 by navy records, though it appears as 1898 in his elementary school records. His father, a tailor, moved to the United States in 1903 to escape the persecution of Jews that was becoming increasingly common.

Rickover decided to attend the United States Naval Academy, graduating in 1922 as number 106 in a class of over 500. He spent the next 25 years primarily in engineering billets, serving aboard several ships and commanding a minesweeper in the mid-1930s. Rickover first learned of atomic energy when he was assigned to head the navy's efforts at Oak Ridge to develop a working nuclear reactor for shipboard use, with an emphasis on submarine applications. Over the next few years, Rickover became increasingly involved in reactor plant design and in naval politics. By 1947 he was directing the design of the USS *Nautilus,* the world's first nuclear-powered ship.

Over the next few years, Rickover gradually gained more control over all aspects of nuclear ship design, manning, testing, and operations. Much of this authority came from carefully cultivated contacts within Congress and industry. On at least three separate occasions, Congress directly intervened to either promote Rickover or to derail plans to force him into retirement. In the meantime, he continued to build the ships the navy needed.

Rickover knew that the navy needed nuclear reactors on submarines that were both safe and reliable. To that end, he started the world's first real quality-assurance program to ensure that "his" reactors and ships were built right and that any problems could be traced back to the responsible person or organization. Contractors were selected with care and given exacting specifications to meet. Failure to do so was simply not tolerated.

Rickover applied the same rigorous standards to nuclear-trained officers. For 30 years, every officer who served in nuclear power had to first survive a "Rickover interview." The interviews were difficult and, at times, seemingly arbitrary, but the quality of officers was almost universally high. Due to strict quality control for personnel, the naval nuclear power program quickly gained a reputation as the world's finest.

Hyman George Rickover. *(The Library of Congress. Reproduced by permission.)*

In addition to working to develop civilian nuclear power for the Atomic Energy Commission, Rickover's sailors became increasingly important in the naval and civilian world. As they went to work for regulatory bodies, nuclear power plants, and the department of energy, Rickover's principles and ideals went with them, gradually influencing most of the nuclear power industry in the United States and elsewhere.

Rickover was finally forced into retirement by the United States Navy in 1985, as he was thought by many to have outlived his usefulness to the navy and the nation. He died the following year.

P. ANDREW KARAM

Ernst Friedrich Schumacher
1911-1977
British Economist

E.F. Schumacher was an economist who argued that Earth could not afford the cultural and environmental costs accompanying large-scale capitalism. His book *Small Is Beautiful: Economics as if People Mattered* was named one of the most influential books published since World War II by the *London Times Literary Supplement.* It also made Schumacher a folk hero on the political left. Although Schumacher died in the 1970s, his opposition to excessive consumption, corporate domination, and growth for its own sake is echoed in the "simple living" movement of the late 1990s.

Schumacher, called Fritz by those who knew him, was born on August 16, 1911, in Bonn, Germany, into an academic family. After attending the universities of Berlin and Bonn, he studied at Oxford as a Rhodes Scholar, and also at Columbia University in New York. In 1937, appalled by the rise of Nazism in Germany, he settled in England. As a German national during World War II, he was sent into the country to work as a farm laborer. For three months he was confined to a detention camp, where he occupied himself with improving the sanitation and the food.

Back on the farm, he began writing on the economic requirements for peace in Europe. This brought him to the attention of William Henry Beveridge and other prominent people, and he was able to assist with British plans for full-employment policies and the post-war welfare state. He became a British citizen in 1946, and he studied the restructuring of Germany's economy as a member of the British Control Commission there.

For 20 years beginning in 1950, Schumacher served as an advisor to Britain's nationalized coal industry. In that capacity he advocated continuing British coal production while encouraging conservation, as alternatives either to depending upon Middle Eastern oil or increasing the use of nuclear energy, with its intractable problem of radioactive waste disposal. During this period he bought a house with a large backyard and became an enthusiastic proponent of organic gardening. Chemical agriculture, he wrote, worked against nature rather than in harmony with it.

Schumacher visited Burma (now Myanmar) in 1955 on an assignment for the United Nations. His time there led him to the conclusion that advanced technology was not the answer for poor countries, because it would increase productivity but not employment. Instead, he advocated intermediate technology that would allow the rural poor to improve their living conditions. This approach is often called *appropriate technology* because it must be tailored to the needs of each developing country. In Schumacher's view, employing intermediate technology while avoiding the trap of pursuing constant growth would result in a society that used both capital and energy more efficiently.

Schumacher's arguments for appropriate technology and small economic units transcended developing nations and became a critique of capitalism in general. In *Small Is Beautiful,* he argued that although capitalism brought higher living standards, the cost was environmental and cultural degradation. Large cities and large industries caused correspondingly large problems, and raised their cost beyond what Earth could bear. Small, decentralized, energy-efficient production units would better serve human needs.

Influenced by the Buddhist and Taoist thought that attracted him in Asia, the non-violent message of Mahatma Gandhi, and the Roman Catholicism in which he eventually found his spiritual home, Schumacher stressed the evils of materialism, the need for economic self-reliance, and service to others. To pursue his ideas he established the Intermediate Technology Development Group in London in 1966. His work has continued to be influential and attract new adherents at the turn of the millenium, in the face of environmental problems and concern about the influence of corporate wealth on culture and politics.

Two years after the death of his first wife, Anna Maria Peterson, in 1960, Schumacher married Verena Rosenberger. Two sons and two daughters were born of each marriage. Schumacher died on September 4, 1977, on a train near Romont, Switzerland.

SHERRI CHASIN CALVO

George Robert Stibitz
1904-1995

American Computer Scientist and Biomedical Researcher

George R. Stibitz is internationally recognized as the father of the digital computer. While working as a research mathematician for Bell Telephone Laboratories during the 1930s, he designed a binary adding machine and later developed a series of increasingly sophisticated digital computers, several of which were used during World War II. In his later years, he turned his attention to the field of biomedicine, in which he was a pioneer.

Born on April 30, 1904, in York, Pennsylvania, Stibitz was one of several children of a minister in the German Reformed Church. He spent most of his youth in Dayton, Ohio, where his father taught ancient languages at Central Theological Seminary. Stibitz attended Moraine Park School, an experimental progressive school in Dayton, and earned a full scholarship to Denison University. He graduated from Denison in 1926 and in the following year earned a master's degree in physics from Union College. Later, he went on to Cornell University, where he earned his doctorate in 1930.

After graduating, Stibitz went to work for Bell Telephone Laboratories as a mathematical consultant and in the following year married Dorothea Lamson, with whom he had two daughters, Mary and Martha. In 1937 he built his first binary adder in his kitchen using dry-cell batteries, metal strips from a tobacco can, and flashlight bulbs that he had soldered to wires from two telephone relays. A replica of this extremely early computer can be found at the Smithsonian Institution.

Together with Samuel Williams, a Bell engineer, Stibitz expanded the adder to create the Model I Complex Calculator, which was introduced in January 1940. The Model I could work faster than 100 humans using desk calculators, and given its connection to Teletypes in other Bell offices, it may also be considered a forerunner of the time-sharing system in computers. In late 1940 Stibitz presented the Model I before a joint meeting of three mathematical societies at Dartmouth. He relayed problems through a Teletype hookup from Dartmouth to a computer at Bell Labs in New York City, receiving his answers within seconds. This event is believed to be the first instance of remote computer operation.

From 1940 to 1945 Stibitz—still employed by Bell—was given on loan to the U.S. Office of Scientific Research and Development. In this capacity he created a number of more sophisticated binary computers, introducing concepts such as the excess 3 code, floating decimal arithmetic, self-checking circuits, jump program instructions, taped programs, and "table-hunting" subcomputers. With the end of the war, he became an independent consultant in applied mathematics for a number of government agencies and industrial firms. He operated from Burlington, Vermont, where in 1954 he created an inexpensive digital computer, a model for the minicomputers that followed.

Stibitz's career entered a new phase in 1964, when he joined the faculty of the Dartmouth Medical School. There he performed groundbreaking research in the new field of biomedicine, using computer applications to analyze the movement of oxygen through the lungs and to study the anatomy of brain cells. He became a professor in 1966 and a professor emeritus in 1970.

In 1965 the American Federation of Information Processing Societies honored Stibitz with its Harry Goode Award, and in 1976 he was elected to the National Academy of Engineering. He received the Emanuel R. Piore Award in 1977 and the Computer Pioneer Award from the Institute of Electrical and Electronic Engineers in 1982. In 1983 he was elected to the Inventors Hall of Fame. Stibitz published a book in 1993, *The Zeroth Generation,* its title being a reference to the fact that his computers had preceded the "first generation" of computers. He died on January 31, 1995, at his home in Hanover Center, New Hampshire.

JUDSON KNIGHT

Charles Hard Townes

1915-

American Physicist

Charles Townes conceived and built the first maser (1953), for which he won a share of the 1964 Nobel Prize in physics. Townes later worked with Arthur Schawlow (1921-) on extending maser principles to the visible portion of the spectrum, which resulted in the first detailed proposal for building a laser (1958).

Charles Hard Townes was born in Greenville, South Carolina, on July 28, 1915. Having skipped seventh grade, he graduated from high school at age 15. He graduated summa cum laude from Furman University in 1935 with degrees in science and modern languages. Townes received his physics masters degree from Duke University in 1936 before matriculating at the California Institute of Technology, where he earned his Ph.D. in 1939.

During World War II Townes worked at Bell Telephone Laboratories (1939-47) on radar-assisted bomb sights. In 1948 he joined Columbia University's physics department, where he became an expert on microwave spectroscopy—the study of interactions between microwaves and molecules. Townes worked at the Columbia Radiation Lab on producing shorter microwaves and amplifying them for use in practical applications.

In 1951 Townes realized that Albert Einstein's (1879-1955) theory of stimulated emission could be exploited to generate and amplify microwave radiation. According to quantum theory, atoms only exist in certain discrete energy states. Moving from one state to another requires the absorption or emission of fixed amounts of energy. When atoms absorb photons of light they move to higher energy levels or excited states. Excited atoms may spontaneously emit this extra energy as a photon of light or, as Einstein noted in 1916, emission may be accomplished by stimulation from another photon. This stimulated emission results in two photons of the same frequency that can then go on to stimulate other excited atoms. However, since most atoms are in lower energy states, emitted photons are generally absorbed rather than stimulating further emissions.

Townes saw that he could separate the higher-energy atoms and enclose them in a resonator cavity containing appropriate electromagnetic radiation to initiate stimulation. These emissions would be reflected back into the systems to induce further emissions, resulting in a feedback process. At sufficiently high radiation levels, the device would become self-oscillating and generate beams of coherent monochromatic radiation. In 1953, after two years of work with James P. Gordon and H. J. Zeiger, Townes successfully produced a working maser (Microwave Amplification by Stimulated Emission of Radiation). Various design improvements followed, after which masers were quickly adapted for use in radio and radar astronomy, military radar, satellite communications, and atomic clocks.

In 1957 Townes turned his attention to creating an optical maser or laser (Light Amplification by Stimulated Emission of Radiation). As the name suggests, the laser operates with visible light instead of microwaves. Townes and Arthur Schawlow, having earlier collaborated on the classic *Microwave Spectroscopy* (1955), decided to work together on the optical maser. Their "Infrared and Optical Masers" paper, published in the December 1958 *Physical Review,* provided the first detailed theoretical description of a laser. Their work initiated the race to build the first working laser, a race that was won by Theodore H. Maiman (1927-) in 1960.

Townes served as vice president and director of research at the Institute for Defense Analysis in Washington, D.C. (1959-61) before becoming provost and professor of physics at the Massachusetts Institute of Technology (MIT) between 1961 and 1966. He was awarded a share of the 1964 Nobel Prize in physics with Nicolai Basov (1922-) and Aleksandr Prokhorov (1916-), who independently produced a maser in 1955. Townes left MIT in 1966 for the University of California at Berkeley, where he remained until his retirement in 1986. Townes presently pursues research in astrophysics.

STEPHEN D. NORTON

Steven Wozniak
1950-
American Computer Designer and Electrical Engineer

Steven Wozniak designed and built the Apple I, the first complete, small, easy-to-use computer. His later Apple II added color capabilities, the simple computer language called BASIC, and came in one unit in a plastic case. Along with Steve Jobs (1955-), Wozniak co-founded the Apple Computer Company.

Steven Wozniak was born in California in 1950, the son of a Lockheed engineer. At Homestead High in Sunnyvale, where he met Jobs, his spare time was spent building computers, arcade machines, and electronic devices in his garage. He belonged to a computer club, though his friend Jobs did not.

At the University of Colorado, De Anza Community College, and the University of California at Berkeley, Wozniak majored in electrical engineering. He dropped out of college and went to work for Hewlett-Packard in 1971, while Jobs worked at Atari. Wozniak spent his spare time designing computers. He and Steve Jobs unveiled the Apple I to the public at the first Computer Show in Atlantic City in August 1976. From the beginning, the Apple computer amazed the industry but did not sell well. No one would invest in it, so Wozniak and Jobs sold their personal possessions and incorporated Apple Computer Company in 1977. The corporate logo, an apple with a stem, a bite out of it, and bright color stripes to emphasize the color capabilities of the computers, is still one of the most recognized logos in the industry.

Apple II was a distinct improvement. Introduced the next year, it came in a plastic case, had color graphics, and ran BASIC and an accounting program called VisiCalc. It was designed for those interested in what a computer

Steve Wozniak. *(Corbis Corporation. Reproduced by permission.)*

could do—not just how it worked. Orders skyrocketed and articles began to appear in magazines and newspapers about two college boys who had designed a successful computer in a garage. By the time Apple III appeared, the company had moved out of the garage, hired mid-level managers, and had several thousand employees. The machine was selling well even outside the United States. By 1981, Apple sales topped $500 million.

Success changed the company. An older, more conservative executive board wanted Apple to become more conventional. Then the personal computer market became saturated, and Apple had to lay off 40 employees. Steve Jobs, who owned 11% of Apple stock, became chairman of the board in March 1981. Steve Wozniak took a leave of absence. He wanted to design computers, not run a business.

At this time Wozniak was involved in the crash of a small airplane he was piloting to San Diego. As a result, he was afflicted with a form of amnesia in which he says he could not form new memories. He didn't remember the crash for months and could not remember what he had just done or said. Slowly his ability to remember came back, but he was changed by the experience.

In 1982 Wozniak went back to work for Apple. After disagreements with Jobs and the board, he left for good in 1985. He was worth over $100 million at the time. He still owns stock and continues to receive a small stipend from the company. He finished his degree at the University of California at Berkeley and started a company to explore expanding the uses of electronics. He became involved in a group dedicated to eliminating international dissension. In 1990 he helped establish a company to investigate the legal ramifications of computers. He also donated many Apple computers to schools. He lives with his wife and several children in Los Gatos, California. He is still involved with computers and is proud of his designs and the pioneering innovations the Apple computers represent.

LYNDALL B. LANDAUER

Biographical Mentions

Robert E. Benner

American computer scientist who contributed toward advances in the use of computers for engineering and scientific applications. Benner was involved in the development of linear and nonlinear optical systems, used in spectroscopic techniques for the characterization of materials. In addition, he has worked on optical pattern recognition, used by computerized systems to read typed or written characters or to use cameras to help interact with people or their surroundings.

Harry Bertrand

American inventor who was granted the first patent for an automobile airbag system with a crash sensor (1958), which allows the bag to inflate nearly simultaneously with an automobile's impact, cushioning the driver's impact with the steering wheel and/or dashboard. It wasn't until 1972 that any cars were produced with airbags (when the Ford Motor Co. ran a demonstration project). In 1973 General Motors began offering its first airbag. By the end of the twentieth century airbags had become standard safety equipment for automobiles.

Otis Boykin
1920-1982

American electrical engineer who invented the pacemaker, a guidance chip for missiles, and resistors for IBM computers. Boykin graduated from Fisk University and the Illinois Institute of Technology. In his career as an inventor, he created several electrical devices, including the chip used in all guided missiles and the small-component thick-film resistors used in IBM computers. Boykins's best known invention was the pacemaker, an artificial stimulator placed in the heart to keep it beating at regular intervals. Ironically, Boykin died of heart failure in 1982.

Daniel Bricklin
1951-

Computer electrical engineer who cocreated VisiCalc, the first electronic spreadsheet. Bricklin began his computer-programming career while still in high school. He received his bachelor's degree in electrical engineering and computer science from the Massachusetts Institute of Technology in 1973. While studying for his M.B.A. at Harvard University, he came up with the idea for an electronic spreadsheet, which he developed with the help of his friend, programmer Bob Frankston. Together they founded Software Arts, Inc., where Frankston served as chairman from 1979 to 1985. Their VisiCalc program helped fuel the growth of the computer industry. In 1995 Bricklin founded a new company, Trellix Corporation, to develop Internet productivity software.

Nolan Bushnell
1941-

Video-game pioneer who invented the first video arcade game and founded the Atari company. Bushnell was inspired by one of the earliest video games, Spacewar, while studying at the Massachusetts Institute of Technology. He developed his first game, Computer Space, in 1971. The game was too sophisticated for a general audience, so he went back to the drawing board and came up with Pong, a simple game based on table tennis. In 1972 Bushnell founded the Atari Corporation, which went on to lead the electronic gaming revolution. He sold Atari to Time Warner in 1977 and initiated several other ventures, including Chuck E. Cheese, the arcade-like restaurant franchise.

Wesley Clark

American computer scientist who was awarded the Eckert-Mauchly Award for his important contributions toward the development of the minicomputer, multiprocessor computers, and personal computers. Minicomputers were the first step toward bridging the gap between large mainframe computers and personal computers. They gave way to desktop workstations and, eventually, to very powerful and inexpensive personal computers. All of these types of computers have been important in bringing greater computer performance to an increasing number of people.

Sir Christopher Cockerell
1910-

English engineer who invented the hovercraft. Cockerell was born in Cambridge and studied at Cambridge University, where he initially focused his engineering skills in the fields of radio and radar before turning to hydrodynamics. In the early 1950s he discovered that air could be used to speed a boat's journey across water. His hovercraft, which "hovered" a few inches above the water as it moved, was first built in 1958. Cockerell was knighted for his work in 1969.

Edgar Frank Codd
1923-

British computer scientist who won the Alan Turing Award for his contributions to database management systems. Codd first developed the concept of a relational database, the conceptual basis of most current database systems, which was realized in programs such as Dbase, Access, FoxPro, and others. This conceptual breakthrough helped to revolutionize the computerization of records, with significant impacts in science, marketing, administration, and virtually any other field in which storing, organizing, and correlating large amounts of data is of value.

Ole-Johan Dahl
1931-

Norwegian computer scientist who, with Kirsten Nygaard, developed Simula, the first object-oriented programming language. Object-oriented programming was a completely new paradigm in computer programming and has had a profound impact on the way in which computer programs are written. Developed in the 1960s, object-oriented programming works with "objects," or predefined chunks of program that more or less stand alone and can be assembled like building blocks into new, more complex programs.

Ray Milton Dolby
1933-

American electrical engineer and physicist whose name has become synonymous with the sophisticated noise reduction system he developed in the mid-1960s. In 1963 Dolby accepted a two-year appointment as a United Nations advisor in India, before founding Dolby Laboratories in England in 1965. Among his many achievements, Dolby holds more than 50 U.S. patents and has been presented with numerous awards, including an Oscar in 1989 and the U.S. National Medal of Technology in 1997.

G. W. A. Dummer

English electrical engineer who first developed the concept of a microprocessor. In a paper written in the early 1950s, Dummer proposed a single, monolithic device that would contain a great number of circuits and electronic components. Though the inferior technology of the time prohibited Dummer from producing such a machine, it became a reality with the next two decades and spawned the computer revolution.

Harold E. Edgerton
1903-1990

Electrical engineer who developed the use of stroboscopic photography. Edgerton studied electrical engineering at the University of Nebraska, then went on to complete his graduate work at the Massachusetts Institute of Technology. In 1931 he pioneered the use of the stroboscope, a flashing strobe light that allows fast-moving objects to be photographed. His work led to the development of the modern electronic flash. Edgerton's stop-action photographs captured athletes, animals, and even bullets moving through mid-air. Edgerton was also a pioneer in underwater photography, collaborating with oceanographer Jacques-Yves Cousteau on numerous projects.

Philip Emeagwali
1955-

Nigerian computer scientist and mathematician who developed the world's fastest supercomputer. Emeagwali has made impressive contributions in a number of fields and, though forced from school at age 12, is widely regarded as one of the world's most brilliant minds. He has helped develop methods for oil recovery in nearly spent oil fields and made important contributions to the science of weather forecasting and a variety of mathematical specialties.

Douglas Carl Engelbart
1925-

Electrical engineer who invented the computer mouse and pioneered the design of modern interactive-computer environments. The grandson of early western pioneers, Engelbart grew up near Portland, Oregon, and served with the Navy during World War II as an electronics technician. He went on to work with NASA's Ames Research Laboratory and the Stanford Research Institute. Engelbart gained an interest in computers and envisioned an easily navigable interface that would allow them to be used in offices around the world. In 1963 he started his own research lab devoted to the augmentation of human intellect via technology. Throughout the 1960s his lab developed a hypermedia-groupware system called NLS (oNLine System), which debuted—along with the first computer mouse—at the 1968 Fall Joint Computer Conference.

Frederico Faggin
1941-

Italian-born American physicist and computer designer who, with Marcian Hoff and Stanley Mazor, created the first microprocessor. He developed the original silicon gate technology while at Fairchild Semiconductor. He then moved to Intel in 1970 to work on the 4004 microprocessor, improving Hoff's design architecture with Mazor. Faggin founded Zilog, Inc. in 1974, which produced the Z80 microprocessor, an early rival to Intel's 8080. He co-founded Synaptics in 1986 to develop neural-net chip technology.

Edward Albert Feigenbaum
1936-

American electrical engineer and pioneer in the field of artificial intelligence (AI) who is considered the "father" of expert systems technology. The knowledge-based applications of AI have enhanced productivity in business, science, engineering, and the military. In addition to his work as a professor of computer science at Stanford University, Feigenbaum was a co-founder of three firms in applied AI—IntelliCorp, Teknowledge, and Design Power Inc.—and served as Chief Scientist of the U.S. Air Force from 1994 to 1997.

Julian Feldman
1920?-

American information and computer science professor who with Edward Feigenbaum (1936-) published the first book on artificial intelligence (AI), a collection of papers entitled *Computers and Thought* (1963), which cataloged research and defined the young field. The book is a source of classic papers such as the 1950 paper by Alan Turing (1912-1954), "Computing Machinery and Intelligence." Feldman also published articles based on his continued AI research in connectionism, a controversial theory regarding cognitive phenomena.

Jay Wright Forrester
1918-

Electrical engineer who invented computer memory storage. Forrester was born in Nebraska and spent his early years on a cattle ranch. His first engineering project was to create a 12-volt electrical system powered by wind, which provided his ranch with electricity. After graduating from the University of Nebraska, he worked as a research assistant at the Massachusetts Institute of Technology's high voltage lab, then went on to its servomechanisms lab. In the mid-1940s Forrester began work on the U.S. Navy sponsored Project Whirlwind, a huge computer developed at MIT as part of the United States's defense against the Soviet Union. While working on this project, Forrester invented the multicoordinate digital information storage device, which became known as magnetic-core memory storage, a precursor to modern random access memory (RAM).

Robert Frankston

American computer scientist who, with Daniel Bricklin, developed the first spreadsheet program for personal computers. This program, called "Visicalc," allowed accountants, scientists, and engineers to perform a huge number of interdependent calculations automatically, and to automatically recalculate the results if any input parameters were changed. This development quickly led to industry standard programs such as Lotus 1-2-3 and Excel, which have since improved significantly, becoming exceptionally powerful tools for business management, finance, engineering design, and science.

Charles P. Ginsburg
1920-1992

Engineer who developed the first practical videotape recorder. Ginsburg was born in San Francisco, California, and graduated from San Jose State in 1948. He worked as a studio and transmitter engineer at a San Francisco radio station before joining the Ampex Corporation in 1952. Ginsburg developed a new method for recording a television signal by using a rapidly rotating recording head to apply high-frequency signals to magnetic tape. Ginsburg led Ampex in the development of a special machine that ran the tape at a lower speed, working in conjunction with the high-speed recording heads. His videotape recorder (VTR) changed the face of television. Networks soon replaced live broadcasts with taped and edited shows.

Andy Grove
1936-

American computer scientist, engineer, and businessman who co-founded the computer company Intel. Unlike many giants in the computer industry, Grove not only completed a college degree, but went on to earn a doctorate. He co-founded Intel in 1968 and became the company's president in 1979. Since that time, Intel has rapidly emerged as one of the world's most influential companies. For his achievements, Grove has received numerous honors from American and international organizations.

John L. Gustafson

American computer scientist who has won three R & D 100 awards for his innovations. Two of these awards were given for making important advances in benchmarking tests used to compare the performance of various configurations of computer equipment. The other was awarded for being the first person to demonstrate parallel computer processing in a problem of practical significance.

William Edward Hanford
1908-1996

American chemist whose exploratory research in organic and polymer chemistry yielded over 120 U.S. patents as well as hundreds of articles on industrial chemistry and research management. Hanford's research determined some of the basic chemistry for the polyurethane industry. Under his direction, the first liquid detergent for home marketing was developed as was a refining process for the manufacture of petroleum products from coal. Hanford also directed vital research into the use of polyols for flame-retardant products.

William R. Hewlett
1913-

Electrical engineer who invented the audio oscillator and cofounded the Hewlett-Packard electronics corporation. William Hewlett was born in Ann Arbor, Michigan. He graduated from Stanford University with a B.A., then completed his master's degree in electrical engineering at the Massachusetts Institute of Technology. While in graduate school, he developed the design for his audio oscillator. The machine, which generated low-frequency audio signals, was used by scientists, researchers, and even for the soundtrack to the Walt Disney film *Fantasia*. Along with David Packard, whom he met while they were undergraduates at Stanford, Hewlett formed the Hewlett-Packard company in 1938 and marketed the audio oscillator as its first product. Hewlett served as the company's president from 1964 to 1977 and its chief executive officer from 1969 to 1978, after which he served on the board of directors.

James Hillier
1915-

Canadian-born American physicist and inventor whose pioneering research on the electron microscope led to the development of the first commercially available electron microscope in North America. Hillier, who holds more than 40 patents, is credited with the invention of the electron lens correction device and the electron microprobe microanalyser as well as being the first to picture bacterial viruses, an achievement that led to the use of the electron microscope as a practical research tool.

Jean Hoerni
1924-1997

Swiss-American physicist who invented the planar process, which led to the first integrated circuit. The Swiss-born Hoerni completed doctorates at Cambridge University and the University of Geneva before immigrating to the United States in 1952. In the early 1950s, while working at the California Institute of Technology, Hoerni was recruited by Nobel laureate William Shockley to join his new Shockley Transistor Laboratories. Hoerni and several of his colleagues soon formed a new company to develop their own integrated circuit. His planar process, which fused a layer of silicon dioxide onto a chip before applying the conducting metal circuitry, helped give his Fairchild Semiconductor company the leading edge in the semiconductor industry.

Marcian Edward Hoff Jr.
1937-

American electrical engineer who designed the first microprocessor. In 1969 Hoff was assigned to work on Intel's Busicom contract to produce a 12-chip hand-held calculator. Employing Intel's silicon-gated metal-oxide semiconductor (MOS) technology, Hoff proposed an alternate single-chip architecture that combined the separate functions. Refinements to his 2000 transistor CPU were later made by Stanely Mazor and Frederico Faggin, and the product was delivered in February, 1971. In 1980 Hoff became the first Intel Fellow.

Donald Fletcher Holmes
1910-1980

American chemist and inventor who with William Hanford (1908-1996), a colleague at E.I. du Pont de Nemours & Co., developed polyurethane in 1942. Holmes spent his entire career with DuPont, working in the company's textile divisions researching synthetic materials that formed the basis for many of the global business segments of an organization once called "the world's largest chemical company."

Nick Holonyak Jr.
1928-

American engineer and inventor who first developed the light-emitting diode (LED), subsequently used in a wide variety of products. LEDs are solid-state devices that emit light when energized. First used in digital watches and calculators, LEDs required far less power and were more compact than their predecessors. Though largely replaced by even more efficient liquid crystal displays (LCDs), LEDs are still used in many devices, such as miniature LED lasers.

Eugene Jules Houdry
1892-1962

French-born, American chemist whose method for catalytically cracking crude petroleum to produce high-octane gasoline revolutionized the refining industry. During World War II approximately 90 percent of all Allied aviation fuel was produced by his process. Also during the war, Houdry developed a catalytic process for producing synthetic rubber. After World War II he focused on reducing carcinogens in automobile and industrial exhausts. His 1962-patented catalytic converter is now standard on all American cars.

Jean Ichbiah

French computer scientist who helped develop the Ada computer language, a programming language that won a U.S. Department of Defense contest in the 1970s. Searching for a single language to replace over 2000 then in use in a variety of Defense Department applications, Ichbiah lead the group that produced Ada, named for Lady Ada Lovelace. Ada has subsequently been released to the public and is considered by many a powerful and easily learned high-level language.

Narinder S. Kapany

Indian physicist whose 1955 doctoral research in optics led to the development of fiber optics. Originally viewed as either a novelty item or a way to transmit images from place to place, fiber optics subsequently found a myriad of uses, including data transmission and communications. During the 1980s, a number of telecommunications companies installed high-capacity fiber optic communications systems that have since been put to use in data transmission. Fiber optics are also used for imaging, including use as visual guides during surgical procedures.

Alan Kay

American computer scientist who helped develop a number of commonly used computer features, notably the graphical user interface (GUI) system. While working at the Xerox Palo Alto Research Center (PARC), Kay helped design GUI, the user-friendly graphics interface that utilized a "mouse" to point to "icons" and "windows" on the screen. This system became the standard graphical interface adopted by Macintosh and the Windows 95, 98, and NT operating systems. Kay also made important contributions toward developing the first laptop computer, and was primarily responsible for the Smalltalk computer language, an early object-oriented programming language.

Donald B. Keck
1941-

American physicist who, with Robert Maurer and Peter Schultz, helped make fiber optics practical and useful. Early optical fibers were fragile and suffered from high signal loss. Keck, Maurer, and Schultz found that doping the glass with titanium and, later, germanium, markedly improved the strength and optical properties of the fibers. This, in turn, made it possible to transmit signals farther and with greater accuracy, paving the way for their use in telecommunications and data transmission.

John George Kemeny
1926-1992

Hungarian computer scientist who, with Thomas Kurtz, developed the computer language BASIC. A far simpler language than FORTRAN, COBOL, or other languages that existed at the time, BASIC was eventually transformed into more flexible and powerful variants, such as Visual Basic, used in many consumer software products. BASIC, an acronym for "Beginners All-purpose Symbolic Instruction Code," was originally designed in 1964 to help teach computer programming to students at Dartmouth University.

Thomas Eugene Kurtz
1928-

American software engineer who, with John Kemeny, developed the computer language BASIC, an easy-to-learn language initially designed as a teaching tool for programming novices at Dartmouth College. Kurtz once commented that if FORTRAN was the lingua franca (common language) of the computer world, BASIC was the "lingua playpen." Available at no charge when developed in 1964, later versions of BASIC and its variants became standard computer languages used around the world.

Raymond Kurzweil

American computer and software engineer who was awarded the Grace M. Hopper Award for developing a device that scans and reads printed pages to the blind. This machine is able to scan and recognize printed characters reliably, as well as implement pronunciation rules to make the synthesized voice understandable. Kurzweil's invention is expected to greatly increase the number of works available to the blind, in addition to the relatively few that are published in Braille.

Stephanie Louise Kwolek
1923-

American chemist who invented Kevlar and other high-strength synthetic fibers. Kwolek studied chemistry in college, working with polymers at DuPont after graduation. Actually, Kwolek started working at DuPont as a way to save money to attend medical school, only later deciding to make chemistry her career. At DuPont she developed or helped develop hundreds of new fibers, including Kevlar, used in bulletproof vests and other high-strength applications. Since its invention, Kevlar has saved the lives of many police, soldiers, and others.

William P. Lear
1902-1978

American engineer best known for his work with corporate jet aircraft. Lear also designed one of the first practical automobile radios, cramming a large volume of equipment into a package small enough to fit into a standard car. He later designed a radio amplifier that became the basis for all RCA radios for many years, and invented the once-popular 8-track tape player. In the aviation industry, Lear developed a lightweight automatic pilot device, aircraft navigational aids, and in 1962 founded Lear Jet Corporation, the renowned manufacturer of small jet aircraft for private and business travel.

Robert Steven Ledley
1926-

American biophysicist who revolutionized the science of radiology imaging with his invention of the whole body computed tomography (CT) scanner. Ledley also pioneered the development of automated chromosome analysis for prenatal diagnosis of birth defects. He authored numerous articles, primarily concerned with the use of computers in biology and medicine and biomedical engineering, as well as several books, and holds at least 5 biomedical patents.

Robert D. Maurer
1924-

American physicist who, with Donald Keck and Peter Schultz, helped develop the first practical glass for use in fiber optics systems. The key problem that Maurer and his collaborators faced was finding a way to reduce signal loss and degradation to the point where at least 1% of the original signal remained after traveling one kilometer. This was accomplished by adding germanium impurities to the glass in a process called "doping." At present, over 90% of long-distance telephone calls in the United States travel over fiber optic lines, attesting to their importance.

Stanely Mazor
1941-

American computer designer who, along with Marcian Hoff and Federico Faggin, developed the first microprocessor. He joined Intel in 1969 and worked with Faggin to improve Hoff's microprocessor architecture. Developed under contract for the Japanese calculator manufacturer Busicom, the resulting 4004-chip was formally announced to the industry in November 1971. Mazor also shares patents on the Symbol, which was developed while he was at Fairchild Semiconductor (1964-1969).

Georges de Mestral
1907-1990

Swiss engineer who made "fastener history" when he developed Velcro—after a walk in the woods, during which he noticed how burrs caught on his clothing could be removed without damaging the fabric. He used this discovery to design a "locking tape" based on the microscopic hooks and loops of the thistle specimens he had collected during his nature walk. De Mestral named his invention, which he patented in 1955, for the French words *velour* (meaning "velvet") and *crochet* ("hook").

Robert Metcalfe
1946-

American computer engineer who invented the Ethernet system of transmitting data from one computer to another. Ethernet helped make Local Area Networks (LANs) possible by providing simple, high-speed data connections to transmit and share data and e-mail. Ethernet was invented in 1973 and, a few years later, Metcalfe founded the 3Com company, an early leader in Ethernet technology. Metcalfe was awarded the IEEE Medal of Honor in 1996 for his contributions in this field.

Marvin Lee Minsky
1927-

American computer scientist who was a pioneer in the field of robotics and artificial intelligence. Minsky co-founded MIT's Artificial Intelligence Laboratory in 1959 and has been a lifelong proponent of machine intelligence. In 1951 he built the SNARC, the first neural network simulator. He has also developed many other robots and robotic devices, such as hands. He continues to work toward developing computers that can think and reason in a human fashion.

Gary R. Montry

American computer engineer who, with Robert Benner and John Gustafson, made significant contributions to the field of parallel computer processing. Parallel processing is the practice of linking a number of small computers together to form a "virtual computer" that performs at very high speeds. Properly written operating systems, programming languages, and computer programs can take advantage of parallel and massively parallel computer architecture to perform at supercomputer speeds. Such systems have the benefit of providing supercomputer performance for the cost of only several desktop computers and software.

Robert Moog
1934-

American musician and inventor who designed the Moog synthesizer, a device that can replicate virtually any sound from any musical instrument. Moog combined an interest in music with skills in electronics and computers to create the Moog synthesizer. His goal, to build a keyboard instrument that could electronically re-create the sound of any musical instrument, was realized in 1964 with the release of the first Moog synthesizer. Using semiconductor technology, the Moog was relatively inexpensive and exceptionally versatile, winning converts throughout the music industry and spawning the proliferation of electronic keyboard music.

Gordon E. Moore
1929-

American chemist who advanced in 1965 what is now known as Moore's law. Moore's law states that the number of circuits that can be printed on computer chips (and therefore their processing power) will double every 18 months. This prediction has proved phenomenally accurate. Moore has recently scaled down the doubling time of Moore's law to every two years. He is Chairman Emeritus of Intel Corporation, the computer chip company he co-founded in 1968.

Allen Newell
1927-1992

American computer scientist whose work with artificial intelligence contributed much toward greater understanding of both computer science and human cognition. Newell's primary work focused on the problem-solving and cognitive architecture that could lead to a computer replication of human thought. Unlike many so-called "AI" programs that merely apply simple rules to a set of alternatives, Newell was working on developing computer systems that would actually think in a manner analogous to humans.

Kirsten Nygaard

Norwegian computer scientist who, with Ole-Johan Dahl, developed the Simula computer programming language, one of the first object-oriented computer languages. Simula became the basis for many follow-on languages, including Beta. Object-oriented computer languages offer many benefits, providing both programmers and users with the ability to assemble and use complex programs composed of "plug and play" modules, or objects. Object-oriented programming became increasingly popular in the 1980s and 1990s.

Kenneth H. Olsen
1926-

American computer scientist who invented magnetic core memory and founded Digital Equipment Corporation (DEC), which invented the minicomputer. Magnetic core memory formed the basis for early computers, although it is not commonly used in more recent machines. However, at the time, it was fundamental to the construction of computers and was a major advance in data storage. In addition, DEC's development of the minicomputer helped bring computing power to smaller businesses and academic researchers, starting a trend that culminated with the personal computer.

John T. Parsons
1913-

American engineer who developed automated controls for machine tools, helping to turn many milling and shaping processes into precision, computer-guided tasks. Prior to Parsons's inventions, most of the processes on assembly lines still took place by hand at metal lathes and milling machines, resulting in a lack of precision and consistency. Parsons was able to transform factories into more automated areas that turned out an endless stream of precision-machined components, enhancing the overall quality of most manufactured goods.

John Robinson Pierce
1910-

American electrical engineer who has been called the "father" of communication satellites for his research into passive and active satellites in synchronous and non-synchronous orbits. His work influenced the launch of the first NASA satellite, *Echo 1,* as well as that of the first active relay satellite, *Telstar.* Pierce has won numerous awards for his satellite work, including the National Medal of Science. He has also researched psychoacoustics and computer-generated music.

Charles Plank
1915-1989

American chemical engineer who, with Edward Rosinski, developed the first zeolite catalytic cracking system for petroleum refining. In catalytic cracking, materials called catalysts are used to speed up chemical reactions, helping to produce gasoline, fuel oil, and other petroleum products made from crude oil. More efficient and less dangerous than traditional methods of fractional distillation, this process marked a major step forward for the petrochemical industry.

Roy J. Plunkett
1910-1994

American chemical engineer who discovered polytetrafluoroetheylene (PTFE) in 1938. PTFE, or Teflon, is chemically inert and used as a non-stick cookware coating. Teflon and its fluropolymer family members, such as Tefzel, are also widely used in the aerospace, electronics, and plastics industries. Plunkett oversaw much of DuPont's research, development, and production of new fluorochemical products and processes. He was elected into the Plastics Hall of Fame in 1973 and retired from DuPont in 1975.

Robert H. Rines
1922-

American engineer who helped invent technologies leading to high-resolution image-scanning radar and sonar systems. Rines's inventions have found widespread use in military technology, including that used in the Persian Gulf War. In addition, devices based on his discoveries are widely used for ultrasound imaging in the body, in deep-sea exploration (as side-scan sonar sleds), and related areas. He holds in excess of 60 patents for his inventions. Rines has also written music for more than 10 Broadway and off-Broadway plays.

Dennis M. Ritchie

American software engineer who won the Alan M. Turing Award for developing, with Kenneth Thompson, the Unix operating system in 1969. Widely regarded as one of the fastest and the most stable operating systems available, Unix forms the basis for a very high percentage of scientific and business computer networks. It also serves as the foundation for the increasingly popular Linux operating system. In addition to his work on Unix, Ritchie played a major role in developing the C programming language.

Larry Roberts

American computer engineer who invented the concept of computer networking and data packet transmission, making possible the Internet, e-mail, and the World Wide Web. Telephone companies initially dismissed Roberts's concept of "packet switching," now the backbone of Internet data transmission, as impractical. However, sending data in small packets by often varying routes and reassembling it at the destination proved much more efficient and reliable than other methods. Roberts's method subsequently became the global standard for data transmission.

Edward J. Rosinski
1921-

American chemical engineer who, with Charles Plank, developed the first zeolite catalytic cracking system for petroleum refining. In catalytic cracking, materials called catalysts are used to speed up chemical reactions, helping to produce gasoline, fuel oil, and other petroleum products made from crude oil. More efficient and less dangerous than traditional methods of fractional distillation, this process marked a major step forward for the petrochemical industry. Rosinski held or shared a total of 76 patents, most dealing with catalytic cracking, leading to new or improved techniques for processing hydrocarbons. A paper that he and Plank co-authored was voted one of the 12 most important papers to be published in the journal *Industrial and Engineering Chemistry*.

Arthur Leonard Schawlow
1921-

American physicist whose early collaborations with Charles Townes on masers resulted in their 1955 *Microwave Spectroscopy*. They extended maser principles to light in 1958 when they published the first detailed proposal for building a laser. Schawlow won a share of the 1981 Nobel Prize in physics for developments in laser spectroscopy, especially for the advanced techniques used by his Stanford research group to reveal details of atomic spectra and give improved values for fundamental constants.

Peter C. Schultz
1947-

American physicist whose work with Robert Maurer and Donald Keck led to the development of the first useable optical fibers able to transmit data over long distances. Scientists had attempted to transmit images over long distances through glass fibers for nearly a century, but were unsuccessful because of the relative opacity of thick glass with high levels of impurities that existed at the time.

Earl D. Shaw
1937-

American physicist who has made important contributions to laser technology. Shaw helped develop the free electron laser, infrared lasers, and tunable lasers. Tunable lasers are important because, unlike conventional lasers, the wavelength of the emitted light can be changed, or tuned, for a particular application. Shaw has applied his studies of tunable lasers to research into the physical properties of DNA and other biologically important molecules.

Michael Shrayer

American software developer who wrote the first word processing program, Electric Pencil, heralded by many as not only the first, but one of the best word processing programs. Electric Pencil had many of the functions found in later word processors, including the first "on-the-fly" spell checkers. Electric Pencil's chief advantages were simplicity, compact code (as compared to the bloated programs now in use), and speed. It formed the basis for all subsequent word processing software.

Alan Shugart

American computer engineer who invented the floppy disk drive and founded Seagate, a disk drive manufacturing company. The floppy disk, which first came as an 8 inch disk and has since shrunk to a standard 3.5 inches in size, helped to make data portable for the first time, allowing files to be transported from one computer to another or to be stored in a separate location.

Shugart also invented the disk drive interface that remains the industry standard.

Herbert Alexander Simon
1916-

American computer scientist and economist who was awarded the 1978 Nobel Prize in Economics for his research into decision-making processes within economic organizations. Simon has investigated the intellectual processes behind decision-making in an effort to help construct computer programs that can replicate human thought processes. Along the way, he helped develop list processing computer languages that are commonly used among artificial intelligence researchers.

Ivan Edward Sutherland
1938-

American electrical and computer engineer who is thought by many to have invented the field of virtual reality. Sutherland also developed the first computer light pen, allowing direct interaction with the computer. This device is often used by television sportscasters during "instant replays," in addition to its other office uses. Sutherland won the Alan M. Turing Award for developing the Sketchpad program, considered the direct ancestor of virtually all computer graphics software today.

Kenneth Thompson
1943-

American computer and software engineer who won recognition for his role in the development of the Unix operating system in 1969. Thompson and Dennis Ritchie developed Unix as an operating system that could be used on any type of computer, giving a stable, flexible, and consistent computing environment. The widespread acceptance of Unix in science and industry settings is testimony to its success, as is the increasing popularity of the Linux operating system, a Unix derivative.

Linus Torvalds
1970-

Finnish computer scientist responsible for the development of the Linux operating system for computers. This system, which is available for free over the Internet, was begun by Torvalds while he was a university student in Finland. The Linux system has become a favorite of a small number of computer programmers seeking an alternative to Microsoft's Windows operating system. Many programmers who favor free software cite the success and reliability of Linux as a model for future products.

Earl Silas Tupper
1907-1983

American chemist and inventor who revolutionized the plastics industry with the development and introduction of Tupperware, which he invented in the 1930s but did not produce until 1947 because of World War II. While a chemist at E.I. duPont de Nemours, Inc., Tupper developed a synthetic polymer that produced a pliable but sturdy plastic, which he called Poly T. In 1942 he founded the Tupperware Corporation, which became famous for its nesting plastic bowls with airtight lids.

Frederic Waller
1886-1954

American photographer who invented Cinerama. Waller gained technical skills at his father's Brooklyn, New York, photography studio. Securing fifty patents, Waller initiated peripheral audiovisual presentation methods that enhanced movie viewers' depth perception. Cinerama consisted of three projectors and a concave screen with plastic strips that absorbed reflected light while a sound system surrounded the audience. Waller won a 1954 Academy Award for this invention. His techniques were applied to flight simulators, computer games, and virtual reality processes.

An Wang
1920-1990

Chinese-American computer engineer who invented the magnetic pulse controlling device, forming the basis for magnetic core memory. This device helped to make possible the first computers with a large (by earlier standards) memory, making them more flexible and versatile. Wang developed many other office automation and information processing devices, and eventually founded Wang Industries to manufacture and market these and other inventions.

Maurice Vincent Wilkes
1913-

English mathematician and computer scientist who developed some of the earliest computers, cache memory, and wide-bandwidth local area networks. He was awarded the prestigious Kyoto Prize for Advanced Technology in 1992. In addition to his work on the EDSAC (Electronic Delay Storage Automatic Computer), Wilkes helped set standards and paradigms for the emerging field of computer programming for stored program computers—that is, computers that store the entire set of instructions (or program) in internal memory.

Nicolaus Wirth
1934-

Swiss computer scientist who developed Pascal, a computer programming language that revolutionized programming on personal computers. Originally designed by Wirth as a "toy" to teach people how to write and compile programs, Pascal was simple and powerful enough to become popular, especially in the revised "Turbo Pascal" version. Wirth later developed the Oberon language, a more serious and powerful successor to Pascal, and Modula-3, an object-oriented programming language that was briefly popular.

Nathaniel Convers Wyeth
1911-1990

American mechanical engineer and inventor who specialized in plastics (and was the brother of the artist Andrew Wyeth). In 1936 he received his B.S. from the University of Pennsylvania. Employed by duPont from 1936 until his retirement in 1976, he developed several basic manufacturing processes and earned 25 patents. His research in the 1960s into the extrusion of nonwoven fibers led to the development of plastic shotgun shells. In 1973 he patented the biaxially oriented polyethylene terephthalate (PET) bottle, which became familiar as a container for carbonated beverages.

Niel Zierler

American computer scientist who helped develop the first computer programming compiler for the MIT "Whirlwind" computer in 1953. A compiler is a program that translates a high-level computer language into machine language. Compilers make it possible to write sophisticated programs without having to write code in either binary (zeros and ones) or hexadecimal (base 16) languages. Compilers also make it possible to write complex programs and to program in high-level languages such as FORTRAN, C++, and BASIC.

Bibliography of Primary Sources

Books

Feldman, Julian, and Edward Feigenbaum. *Computers and Thought* (1963). The first published book on artificial intelligence (AI), consisting of a collection of papers that cataloged research and defined the young field.

Schumacher, Ernst Friedrich. *Small is Beautiful: Economics as if People Mattered* (1973). Contained Schumacher's assertion, immediately adopted by political progressives, that economic production should be a means to an end—that of increased wellbeing—and that this should be achieved with a minimum of resource consumption and resulting environmental degradation. Schumacher argued that production from local resources should be used to fill local needs and that full employment, whether in the home or in an outside setting, is necessary for a healthy society.

Townes, Charles Hard, and Arthur Schawlow. *Microwave Spectroscopy* (1955). Included description of the maser (Microwave Amplification by Stimulated Emission of Radiation), a precursor to the development of lasers.

Periodical Articles

Townes, Charles Hard, and Arthur Schawlow. "Infrared and Optical Masers." *Physical Review* (December 1958). Provided the first detailed theoretical description of a laser, leading to the construction of the first working version in 1960.

Turing, Alan. "Computing Machinery and Intelligence" (1950). A classic early paper on artificial intelligence (AI).

General Bibliography

Books

Agassi, Joseph. *The Continuing Revolution: A History of Physics from the Greeks to Einstein.* New York: McGraw-Hill, 1968.

Allen, Garland E. *Life Science in the Twentieth Century.* New York: Cambridge University Press, 1978.

Anderson, E. W. *Man the Aviator.* London: Priory Press, 1973.

Arnold, Caroline. *Genetics: From Mendel to Gene Splicing.* New York: F. Watts, 1986.

Asimov, Isaac. *Adding a Dimension: Seventeen Essays on the History of Science.* Garden City, NY: Doubleday, 1964.

Bahn, Paul G., editor. *The Cambridge Illustrated History of Archaeology.* New York: Cambridge University Press, 1996.

Basalla, George. *The Evolution of Technology.* New York: Cambridge University Press, 1988.

Benson, Don S. *Man and the Wheel.* London: Priory Press, 1973.

Berridge, Virginia and Philip Strong. *AIDS and Contemporary History.* New York: Cambridge University Press, 1993.

Bowler, Peter J. *The Norton History of the Environmental Sciences.* New York: W. W. Norton, 1993.

Brock, W. H. *The Norton History of Chemistry.* New York: W. W. Norton, 1993.

Bruno, Leonard C. *Science and Technology Firsts.* Edited by Donna Olendorf, guest foreword by Daniel J. Boorstin. Detroit: Gale, 1997.

Bud, Robert and Deborah Jean Warner, editors. *Instruments of Science: An Historical Encyclopedia.* New York: Garland, 1998.

Bunch, Bryan H. *Handbook on Current Science & Technology.* Detroit: Gale, 1996.

Bynum, W. F., et al., editors. *Dictionary of the History of Science.* Princeton, NJ: Princeton University Press, 1981.

Carnegie Library of Pittsburgh. *Science and Technology Desk Reference: 1,500 Frequently Asked or Difficult-to-Answer Questions.* Washington, D.C.: Gale, 1993.

Crone, G. R. *Man the Explorer.* London: Priory Press, 1973.

Elliott, Clark A. *History of Science in the United States: A Chronology and Research Guide.* New York: Garland, 1996.

Erlen, Jonathan. *The History of the Health Care Sciences and Health Care, 1700-1980: A Selective Annotated Bibliography.* New York: Garland, 1984.

Fearing, Franklin. *Reflex Action: A Study in the History of Physiological Psychology.* Introduction by Richard Held. Cambridge: MIT Press, 1970.

Good, Gregory A., editor. *Sciences of the Earth: An Encyclopedia of Events, People, and Phenomena.* New York: Garland, 1998.

Graham, Loren R. *Science in Russia and the Soviet Union: A Short History.* New York: Cambridge University Press, 1993.

Grattan-Guiness, Ivor. *The Norton History of the Mathematical Sciences: The Rainbow of Mathematics.* New York: W. W. Norton, 1998.

Gullberg, Jan. *Mathematics: From the Birth of Numbers.* Technical illustrations by Peter Hilton. New York: W. W. Norton, 1997.

Hellemans, Alexander and Bryan Bunch. *The Timetables of Science: A Chronology of the*

Most Important People and Events in the History of Science. New York: Simon and Schuster, 1988.

Hellyer, Brian. *Man the Timekeeper.* London: Priory Press, 1974.

A History of Science Policy in the United States, 1940-1985: Report Prepared for the Task Force on Science Policy, Committee on Science and Technology, House of Representatives, Ninety-Ninth Congress, Second Session. Washington, D.C. U.S. Government Printing Office, 1986.

Holmes, Edward and Christopher Maynard. *Great Men of Science.* Edited by Jennifer L. Justice. New York: Warwick Press, 1979.

Hoskin, Michael. *The Cambridge Illustrated History of Astronomy.* New York: Cambridge University Press, 1997.

Lankford, John, editor. *History of Astronomy: An Encyclopedia.* New York: Garland, 1997.

Lincoln, Roger J. and G. A. Boxshall. *The Cambridge Illustrated Dictionary of Natural History.* Illustrations by Roberta Smith. New York: Cambridge University Press, 1987.

Porter, Roy. *The Cambridge Illustrated History of Medicine.* New York: Cambridge University Press, 1996.

Rassias, Themistocles M. and George M. Rassias, editors. *Selected Studies, Physics-Astrophysics, Mathematics, History of Science: A Volume Dedicated to the Memory of Albert Einstein.* New York: Elsevier North-Holland Publishing Company, 1982.

Reingold, Nathan, editor. *Science in America Since 1820.* New York: Science History Publications, 1976.

Rothenberg, Marc. *The History of Science in the United States: An Encyclopedia.* New York: Garland, 2000.

Rudwick, M. J. S. *The Meaning of Fossils: Episodes in the History of Paleontology.* New York: American Elsevier, 1972.

Sarton, George. *Introduction to the History of Science.* Huntington, NY: R. E. Krieger Publishing Company, 1975.

Sarton, George. *The History of Science and the New Humanism.* New Brunswick, NJ: Transaction Books, 1987.

Schneiderman, Ron. *Computers: From Babbage to the Fifth Generation.* New York: F. Watts, 1986.

Smith, Roger. *The Norton History of the Human Sciences.* New York: W. W. Norton, 1997.

Spangenburg, Ray and Diane K. Moser. *The History of Science from 1946 to the 1990s.* New York: Facts on File, 1994.

Stiffler, Lee Ann. *Science Rediscovered: A Daily Chronicle of Highlights in the History of Science.* Durham, NC: Carolina Academic Press, 1995.

Stwertka, Albert and Eve Stwertka. *Physics: From Newton to the Big Bang.* New York: F. Watts, 1986.

Travers, Bridget, editor. *The Gale Encyclopedia of Science.* Detroit: Gale, 1996.

Willmore, A. P. and S. R. Willmore, consultant editors. *Aerospace Research Index: A Guide to World Research in Aeronautics, Meteorology, Astronomy, and Space Science.* Harlow, England: F. Hodgson, 1981.

World of Scientific Discovery. Detroit: Gale, 1994.

Young, Robyn V., editor. *Notable Mathematicians: From Ancient Times to the Present.* Detroit: Gale, 1998.

JUDSON KNIGHT

Index

Numbers in bold refer to main biographical entries

D

G

H

I

J

K

M

N

O

P

Q

R

S

T

U

V

W

X

Y

Z